Gmelin Handbook of Inorganic Chemistry

8th Edition

The following Gmelin Formula Index volumes have been published up to now:

Formula Index

Volume 1	Ac–Au
Volume 2	$B-Br_2$
Volume 3	Br_3-C_3
Volume 4	C_4-C_7
Volume 5	C_8-C_{12}
Volume 6	$C_{13}-C_{23}$
Volume 7	$C_{24}-Ca$
Volume 8	Cb–Cl
Volume 9	Cm–Fr
Volume 10	Ga–I
Volume 11	In–Ns
Volume 12	O–Zr
	Elements 104–132

Formula Index 1st Supplement

Volume 1	Ac–Au
Volume 2	$B-B_{1.9}$
Volume 3	B_2-B_{100}
Volume 4	$Ba-C_7$ (present volume)

Gmelin Handbook of Inorganic Chemistry

8th Edition

Gmelin Handbuch der Anorganischen Chemie

Achte, völlig neu bearbeitete Auflage

Prepared and issued by
Gmelin-Institut für Anorganische Chemie
der Max-Planck-Gesellschaft
zur Förderung der Wissenschaften
Director: Ekkehard Fluck

Founded by
Leopold Gmelin

8th Edition
8th Edition begun under the auspices of the
Deutsche Chemische Gesellschaft by R. J. Meyer

Continued by
E.H.E. Pietsch and A. Kotowski, and by
Margot Becke-Goehring

Springer-Verlag Berlin · Heidelberg · New York · Tokyo 1985

Gmelin Handbook of Inorganic Chemistry

8th Edition

INDEX

Formula Index

1st Supplement Volume 4

Ba–C_7

AUTHORS Marie-Louise Gerwien, Helga Hartwig, Uwe Nohl, Hans-Jürgen Richter-Ditten, Paul Velić, Rudolf Warncke

CHIEF EDITOR Rudolf Warncke

Springer-Verlag Berlin · Heidelberg · New York · Tokyo 1985

The volumes of the Gmelin Handbook are evaluated from 1974 up to the end of 1979

Library of Congress Catalog Card Number: Agr 25-1383

ISBN 3-540-93484-7 Springer-Verlag, Berlin · Heidelberg · New York · Tokyo
ISBN 0-387-93484-7 Springer-Verlag, New York · Heidelberg · Berlin · Tokyo

Typesetting, printing and bookbinding: Universitätsdruckerei H. Stürtz AG, Würzburg

Foreword

The Gmelin Formula Index published between 1975 and 1980 covered all volumes of the Eighth Edition of the Gmelin Handbook that had appeared up to the end of 1974 in the case of Main Volumes and up to the end of 1973 in the case of Supplement Volumes. The Gmelin Formula Index, First Supplement, continues from there and covers the handbook volumes published up to the end of 1979.

This First Supplement will consist of eight volumes, which will appear at intervals of four to six months. The basic structure of the Formula Index has been fully retained in the First Supplement: The index lists all elements, compounds, ions, and systems having definite composition that are described in the handbook text. The first column gives the empirical formula, while the second gives the conventional formula. The third column lists the pertinent pages. (The details are available in "Instructions for the Formula Index", on the next pages.)

This First Supplement was prepared and printed with extensive use of computers. In the future this will allow publication of cumulative indexes. The procedures were worked out together with the Technical Section of the *Gesellschaft für Information und Dokumentation mbH (GID)*, Frankfurt, and I take this occasion to thank them for their generous help. I would also like to thank our printers, *Universitätsdruckerei H. Stürtz AG*, Würzburg, for their advice and cooperation.

Frankfurt am Main
September 1983

Rudolf Warncke

Instructions for the Formula Index

The formula index consists of three columns. The first gives the empirical formula, the second gives the conventional formula, as well as any supplementary information or subdivisions, and the third gives the pertinent volume and page numbers.

First Column (Empirical Formula)

In the empirical formula the symbols of the elements are arranged alphabetically; C and H are not placed first. The list of empirical formulas is arranged alphabetically and by the magnitude of the subscripts. Any indefinite subscripts are placed last. Ions are always placed after the neutral species, positive ions preceding the negative.

The unsubscripted symbol is used for the element unless a specific diatomic or polyatomic species is meant (e.g., Br_2, Br_3). Transuranium elements that do not yet have an internationally recognized symbol are listed under their atomic number and placed at the end of the index. Special superscripted symbols are not used for isotopes.

H_2O is included in the empirical formula only if it is an integral part of a complex as written in the second column. Polymers of the type $(AB)_n$ are listed under AB. Multicomponent systems (mixed crystals, melts, etc.) are found under the empirical formulas of their components. However, solutions are found only under the solute.

Second Column (Conventional Formula)

The formula is written as it appears in the handbook text. However, in many cases another form is shown if there is adequate space and if it presents additional structural details. If this is not possible for isomers, then they are numbered consecutively. For elements the name is given in the second column.

Entries having the same empirical formula are arranged as follows: compounds, isotopic species, polymers, hydrates, multicomponent systems. Elements are treated in the same way.

For multicomponent systems the components are arranged in the sequence "inorganic components — organic components — water." The inorganic components are arranged alphabetically; the organic components are arranged by number of carbon atoms. If an element is a component, it is always represented by its unsubscripted symbol. Isotopic species are listed immediately after the normal species.

The concept *system* is used in its restricted sense in this index: it represents equilibrium mixtures described in phase diagrams. Ionic systems are included under their parent compounds.

The location of solubility data for compounds mentioned only briefly in the text is included under the main empirical-formula entry.

Entries for elements and compounds treated extensively in the handbook are subdivided by topics, e.g., geochemistry, preparation, or toxicity.

The concepts *solubility, solutions,* and *systems* partially overlap, and in these cases the user should always look at all three places. That is also true for the concepts *diffusion* and *systems* and for the concepts *sorption* and *system*.

In referrals to another entry in the index both the empirical and conventional formula are given. For example, "see $Al_2Na_2O_4$...$Na_2O \cdot Al_2O_3$." For referrals within the topics of a particular compound, then only the topic is given. For example "see Deposits."

Third Column (Volume and Page Numbers)

The first symbol is that of the element to which the volume belongs. Next is the abbreviated form of the type of volume followed usually by the Part or Section. The page numbers are given after a hyphen. The following abbreviations are used for type of volume:

MVol.	Main Volume (Hauptband)
SVol.	Supplement Volume (Ergänzungsband)
Org.Comp.	Organic Compounds
Org.Verb.	Organische Verbindungen
PerFHalOrg.	Perfluorohalogenoorgano Compounds of Main Group Elements
SVol.GD	Gmelin-Durrer, Metallurgy of Iron
TrU.	Transuranium Elements
Water Desalt.	Water Desalting

For example, the entry "Ag: MVol.B7-237/9" indicates that the information is to be found on pages 237 to 239 of the Silver Main Volume B 7. The entry "Fe: Org.Comp.C3-89" indicates that the desired information is to be found on page 89 of "Organic Compounds C 3" for the element iron (Fe).

Comments on System Numbers and Element Symbols

In the Formula Index itself the volume and page number citation was based on the traditional System Number (Main Volume Series) or the New Supplement Series Volume number. Today the volumes are usually arranged by the symbols of the elements. However, the symbols can be deduced easily from the system numbers. Most citations have the element symbol immediately after the system number. For example, 61 (Ag) refers to the silver volumes. The exceptions are

1 (EG)	is now	He
8 (J)	is now	I
39 (SE)	is now	Sc
69 (Ma)	is now	Tc

The old abbreviation for MVol was Hb (Hauptband), and the old abbreviation for SVol was Eb (Ergänzungsband). The volumes of the New Supplement Series are associated with the symbols of the elements as follows:

Erg.W. 1	He
Erg.W. 2	V
Erg.W. 3	Cr
Erg.W. 4, 7a, 7b, 8	Np
Erg.W. 5, 6	Co
Erg.W. 9, 12	F
Erg.W. 10	Zr
Erg.W. 11	Hf

The New Supplement Series Volumes 2 and 3 and the Volumes 10 and 11 are bound together as double volumes.

Ba — C_7

$Ba_{0.33}O_{6.33}U_2$	$Ba_{0.33}U_2O_{6.33} \cdot 3.67\ H_2O$	U: SVol.C3-127
$Ba_{0.5}O_4Th_{0.5}V$	$Ba_{0.5}Th_{0.5}VO_4$	Th: SVol.C2-125/6
Ba	Barium systems	
	Ba-Am	Np: TrU.B3-267
	Ba-Cm	Np: TrU.B3-272
	Ba-Np	Np: TrU.B2-1
	Ba-Pu	Np: TrU.B2-65/6
$BaBi_2C_{24}H_{20}$	$[(C_6H_5)_2Bi]_2Ba$	Bi: Org.Verb.-114
$BaBr_2$	$BaBr_2$ systems	
	$BaBr_2$-$MnBr_2$	Mn: MVol.C5-306
	$BaBr_2$-HBr-$RaBr_2$-H_2O	Ra: SVol.2-180/5
$BaBr_4Sn$	$BaSnBr_4 \cdot H_2O$	Sn: MVol.C3-199
$BaCF_3O_3P$	CF_3PO_3Ba	F: PerFHalOrg.3-50
$BaCSSe_2$	$BaCSe_2S$	C: MVol.D6-216
–	$BaCSe_2S$ solid solutions	
	$BaCSe_2S$-$BaCSe_3$	C: MVol.D6-216
$BaCS_2Se$	$BaCSeS_2$	C: MVol.D6-219
$BaCS_3$	$BaCS_3$	C: MVol.D4-223
$BaCSe_3$	$BaCSe_3$	C: MVol.D6-215
–	$BaCSe_3$ solid solutions	
	$BaCSe_3$-$BaCSe_2S$	C: MVol.D6-216
$BaC_2H_4N_2S_4$	$Ba[S_2CNH_2]_2$	C: MVol.D6-140
$BaC_4Cl_{12}H_6O_2Sn_2$	$Ba[SnCl_5 \cdot CH_3COCl]_2$	Sn: MVol.C5-126
BaC_4FeO_4	$Ba[Fe(CO)_4]$	Fe: Org.Verb.B2-22, 31
$BaC_4H_6O_2S_4$	$Ba[S_2COCH_3]_2$	C: MVol.D4-253
$BaC_4H_6O_4$	$Ba(CH_3COO)_2$ systems	
	$Ba(CH_3COO)_2$-$Ra(CH_3COO)_2$-H_2O	Ra: SVol.2-371
BaC_4O_8SSn	$BaSnS(C_2O_4)_2$	Sn: MVol.C3-205
$BaC_6H_6O_{12}Sn_2$	$Ba[Sn(HCOO)_3]_2$	Sn: MVol.C3-205
$BaC_6H_{10}O_2S_4$	$Ba[S_2COC_2H_5]_2$	C: MVol.D4-253
$BaC_8Fe_2H_2O_8$	$Ba[HFe(CO)_4]_2$	Fe: Org.Verb.B2-18, 25
		Fe: Org.Verb.B3-115
$BaC_8H_8O_{13}Sn$	$BaSnO(C_4H_4O_6)_2 \cdot 4.5\ H_2O$	Sn: MVol.C3-205
$BaC_8H_{14}O_2S_4$	$Ba[S_2COC_3H_7]_2$	C: MVol.D4-253
$BaC_8H_{20}N_2O_6S_2$	$Ba[(C_2H_5)_2NSO_3]_2 \cdot 2\ H_2O$	S: S-N-Verb.1-97
$BaC_{10}H_{12}N_2O_8Sn$	$Ba[Sn(C_2H_4N_2(CH_2COO)_4)] \cdot 4\ H_2O$	Sn: MVol.C5-37
$BaC_{10}H_{18}O_2S_4$	$Ba[S_2COC_4H_9]_2$	C: MVol.D4-253

$BaC_{12}Cl_6H_{12}O_{12}Sn_2$	$Ba[Sn(CH_2ClCOO)_3]_2$	Sn: MVol.C3-205
$BaC_{12}H_{18}O_{12}Sn_2$	$Ba[Sn(CH_3COO)_3]_2$	Sn: MVol.C3-205
$BaC_{12}H_{22}O_2S_4$	$Ba[S_2COC_5H_{11}]_2$	C: MVol.D4-253
$BaC_{18}H_{12}O_6Sn$	$Ba[Sn(C_6H_4O_2)_3]$	Sn: MVol.C5-77
$BaC_{20}Cl_{10}H_{30}N_{10}Sn_2$	$[Ba(CH_3CN)_8][SnCl_5(CH_3CN)]_2$	Sn: MVol.C6-67/8
$BaCaMgO_6U$	$BaCaMgUO_6$	U: SVol.C3-85, 98, 129
$BaCaO_6SrU$	$BaSrCaUO_6$	U: SVol.C3-129/30
$BaCa_2O_6U$	$BaCa_2UO_6$	U: SVol.C3-98, 129
$BaCe_{0.5}O_3Pa_{0.5}$	$Ba(Ce_{0.5}Pa_{0.5})O_3$	Pa: SVol.2-21/2
$BaCl_2$	$BaCl_2$ systems	
	$BaCl_2-HCl-RaCl_2-H_2O$	Ra: SVol.2-174/9
	$BaCl_2-LiCl-RaCl_2$	Ra: SVol.2-364
	$BaCl_2-MnCl_2$	Mn: MVol.C5-219/20
	$BaCl_2-MnCl_2-H_2O$	Mn: MVol.C5-219/20
	$BaCl_2-PuCl_3$	Np: TrU.D2-444/5
	$BaCl_2-UCl_3$	U: SVol.C9-9
	$BaCl_2-UCl_4$	U: SVol.C9-23/4, 52
$BaCl_2O_6$	$Ba(ClO_3)_2$ systems	
	$Ba(ClO_3)_2-Ra(ClO_3)_2-H_2O$	Ra: SVol.2-197
$BaCl_2O_8$	$Ba(ClO_4)_2$ solid solutions	
	$Ba[(Cl,Mn)O_4]_2 \cdot 3\,H_2O$	Mn: MVol.C2-36/7, 44/6, 276
$BaCl_4Mn$	$BaMnCl_4 = BaCl_2 \cdot MnCl_2$	Mn: MVol.C5-219/20
$BaCl_4Sn$	$BaSnCl_4 \cdot 2\,H_2O$	Sn: MVol.C3-199
–	$BaSnCl_4 \cdot 4\,H_2O$	Sn: MVol.C3-199
$BaCl_6Sn$	$Ba[SnCl_6]$	Sn: MVol.C3-199
$BaCl_6U$	$BaUCl_6$	U: SVol.C9-51
$BaCl_{10}S_8$	$BaCl_2 \cdot 4\,S_2Cl_2$	S: SVol.2-260
$BaCl_{12}U_2$	$Ba(UCl_6)_2$	U: SVol.C9-60
$BaCrO_4$	$BaCrO_4$ solid solutions	
	$BaCrO_4-KMnO_4$	Mn: MVol.C2-174
–	$BaCrO_4$ systems	
	$BaCrO_4-RaCrO_4-H_2O$	Ra: SVol.2-185/92
$BaDy_{0.5}O_3Pa_{0.5}$	$Ba(Dy_{0.5}Pa_{0.5})O_3$	Pa: SVol.2-21
$BaEr_{0.5}O_3Pa_{0.5}$	$Ba(Er_{0.5}Pa_{0.5})O_3$	Pa: SVol.2-21
$BaEu_{0.5}O_3Pa_{0.5}$	$Ba(Eu_{0.5}Pa_{0.5})O_3$	Pa: SVol.2-21
BaF_2	BaF_2 solid solutions	
	BaF_2-YF_3	Sc: MVol.C3-151
–	BaF_2 systems	
	BaF_2-MnF_2	Mn: MVol.C4-230/1
	BaF_2-SnF_2	Sn: MVol.C3-196/7
BaF_3Li	$LiBaF_3$ systems	
	$LiBaF_3-KMnF_3$	Mn: MVol.C4-235/6
BaF_4Mn	$BaMnF_4 = BaF_2 \cdot MnF_2$	Mn: MVol.C4-230/5
BaF_5HMn	$BaMnF_4 \cdot HF = BaF_2 \cdot MnF_2 \cdot HF$	Mn: MVol.C4-235
BaF_5HOSn	$Ba[Sn(OH)F_5] \cdot H_2O$	Sn: MVol.C3-199
BaF_5Mn	$BaMnF_5 = BaF_2 \cdot MnF_3$	Mn: MVol.C4-235/6
BaF_6Ir	$Ba[IrF_6]$	Ir: SVol.2-109

BaF_6Mn	$BaMnF_6 = BaF_2 \cdot MnF_4$	Mn: MVol.C4-95, 238
BaF_6S	$BaSF_6$	S: SVol.2-183
BaF_6Sn	$Ba[SnF_6]$	Sn: MVol.C3-198/9
–	$Ba[SnF_6] \cdot 3\ H_2O$	Sn: MVol.C3-198/9
BaF_6Sn_2	$Ba(SnF_3)_2$	Sn: MVol.C3-196/8
BaF_8Te	$BaTeF_8 = BaF \cdot TeF_7$	Te: SVol.B2-30, 31
$BaF_{10}Sn_4$	$Ba(Sn_2F_5)_2$	Sn: MVol.C3-196/7
$BaF_{12}Ir_2$	$Ba[IrF_6]_2$	Ir: SVol.2-109
$BaF_{14}Te_2$	$Ba(TeF_7)_2$	Te: SVol.B2-30
$BaGd_{0.5}O_3Pa_{0.5}$	$Ba(Gd_{0.5}Pa_{0.5})O_3$	Pa: SVol.2-21
$BaGe_3O_9Sn$	$BaSnGe_3O_9$	Sn: MVol.C4-231
–	$BaSnGe_3O_9$ solid solutions	
	$BaSn(Ge_{3-x}Si_x)O_9$	Sn: MVol.C4-232
	$BaTi_{1-x}Sn_xGe_3O_9$	Sn: MVol.C4-233
$BaGe_3O_9Ti$	$BaTiGe_3O_9$ solid solutions	
	$BaTi_{1-x}Sn_xGe_3O_9$	Sn: MVol.C4-233
$BaH_2N_2O_6S_2$	$Ba[NH_2N(SO_3)_2] \cdot 2\ H_2O$	S: S-N-Verb.1-219
$BaH_4I_2O_{16}U_2$	$Ba(H_2UO_2IO_6)_2 \cdot x\ H_2O$	U: SVol.C9-181
$BaH_4N_4O_6S_3$	$Ba[SO_2(NSO_2NH_2)_2] \cdot 2\ H_2O$	S: S-N-Verb.1-179
$BaH_4O_5Sn_2$	$Ba[(OH)_2SnOSn(OH)_2]$	Sn: MVol.C3-192
BaH_4O_6Te	BaH_4TeO_6	Te: SVol.B1-132
$BaH_4O_{14}U_2V_2$	$Ba[UO_2(OH)_2(VO_3)]_2$	U: SVol.C3-302
BaH_5MnO_5	$Ba[Mn(OH)_5]$	Mn: MVol.C2-227
$BaH_6N_4O_6S_2$	$Ba[SO_3NHNH_2]_2 \cdot H_2O$	S: S-N-Verb.1-218
BaH_6O_6Sn	$Ba[Sn(OH)_6]$	Sn: MVol.C3-195/6
–	$Ba[Sn(OH)_6] \cdot 3\ H_2O$	Sn: MVol.C3-196
–	$Ba[Sn(OH)_6] \cdot 4\ H_2O$	Sn: MVol.C3-196
$BaH_6O_6Sn_2$	$Ba[Sn(OH)_3]_2$	Sn: MVol.C3-191/2
–	$Ba[Sn(OH)_3]_2 \cdot 2\ H_2O$	Sn: MVol.C3-191/2
$BaH_6O_{22}U_6$	$BaU_6O_{16}(OH)_6 \cdot 8\ H_2O$	U: SVol.C3-127
–	$BaU_6O_{16}(OH)_6$ solid solutions	
	$(Pb,Ba)U_6O_{16}(OH)_6 \cdot n\ H_2O$	U: SVol.C3-169
$BaH_8N_6O_8S_4$	$Ba[N(SO_2NH_2)_2]_2$	S: S-N-Verb.1-160
$BaH_{14}O_{26}U_6$	$[UO_2(OH)_2]_6Ba(OH)_2 \cdot 4\ H_2O$	U: SVol.C3-127
$BaHo_{0.5}O_3Pa_{0.5}$	$Ba(Ho_{0.5}Pa_{0.5})O_3$	Pa: SVol.2-21
BaI_2	BaI_2 systems	
	BaI_2-MnI_2	Mn: MVol.C5-326
BaI_2O_6	$Ba(IO_3)_2$ systems	
	$Ba(IO_3)_2$-$Ra(IO_3)_2$-H_2O	Ra: SVol.2-197/8
BaI_6MnO_{18}	$BaMn(IO_3)_6$	Mn: MVol.C5-337
$BaIn_{0.5}O_3Pa_{0.5}$	$Ba(In_{0.5}Pa_{0.5})O_3$	Pa: SVol.2-21
$BaIn_{0.67}Mo_{0.33}O_3$	$Ba(In_{0.67}Mo_{0.33})O_3$	Mo: SVol.B2-94
$BaIr_{0.33}O_3Ti_{0.67}$	$BaIr_{0.33}Ti_{0.67}O_3$	Ir: SVol.2-19
$BaIrO_3$	$BaIrO_3$	Ir: SVol.2-16
–	$BaIrO_3$ solid solutions	
	$BaIr_xTi_{1-x}O_3$	Ir: SVol.2-19
–	$BaIrO_3$ systems	
	$BaIrO_3$-$SrIrO_3$	Ir: SVol.2-16

$BaKMnO_4$	$KBaMnO_4$	Mn: MVol.C2-259
BaK_2O_4Th	K_2BaThO_4	Th: SVol.C2-2
$BaLa_{0.5}O_3Pa_{0.5}$	$Ba(La_{0.5}Pa_{0.5})O_3$	Pa: SVol.2-21
$BaLaLiMoO_6$	$LiBaLaMoO_6$	Mo: SVol.B2-210
$BaLaMnO_6Ti$	$BaLaMnTiO_6$ solid solutions	
	$Ba_{1-x}Sr_xLaMnTiO_6$	Mn: MVol.C3-125
$BaLi_{0.33}Mo_{0.33}Nb_{0.33}O_3$		
	$BaLi_{0.33}Nb_{0.33}Mo_{0.33}O_3$	Mo: SVol.B2-305
$BaLi_{0.33}Mo_{0.33}O_3Ta_{0.33}$		
	$BaLi_{0.33}Ta_{0.33}Mo_{0.33}O_3$	Mo: SVol.B2-309
$BaLu_{0.5}O_3Pa_{0.5}$	$Ba(Lu_{0.5}Pa_{0.5})O_3$	Pa: SVol.2-21
$BaMgO_6SrU$	$BaSrMgUO_6$	U: SVol.C3-85
$BaMg_2O_6U$	$BaMg_2UO_6$	U: SVol.C3-85
$BaMn_{0.5}O_3Sb_{0.5}$	$BaMn_{0.5}Sb_{0.5}O_3$	Mn: MVol.C3-135/6
$BaMnNaO_4$	$NaBaMnO_4$	Mn: MVol.C2-259/60
$BaMnO_2$	$BaMnO_2$	Mn: MVol.C2-216
$BaMnO_3$	$BaMnO_3$	Mn: MVol.C2-241/5
–	$BaMnO_{3-x}$	Mn: MVol.C2-240/1
–	$BaMnO_3 \cdot x\,H_2O$	Mn: MVol.C2-251
–	$BaMnO_3$ solid solutions	
	$BaMnO_3$-$BaNbO_3$	Mn: MVol.C3-175/6
	$BaMnO_3$-$BaTaO_3$	Mn: MVol.C3-185
	$BaMnO_3$-$BaTiO_3$	Mn: MVol.C3-120/1
	$BaMn_xU_{1-x}O_3$	Mn: MVol.C3-229
	$BaMnO_3$-$BiMnO_3$-$CaMnO_3$	Mn: MVol.C3-145
	$La_{1-x}Ba_xMnO_3$	Mn: MVol.C3-26/40
	$Nd_{1-x}Ba_xMnO_3$	Mn: MVol.C3-66/7
–	$BaMnO_3$ systems	
	$BaMnO_{3-x}$-$SrMnO_{3-x}$	Mn: MVol.C2-245/6
$BaMnO_4$	$BaMnO_4$	Mn: MVol.C2-13/4, 262/5
–	$BaMnO_4$ solid solutions	
	$BaMnO_4$-$BaSO_4$	Mn: MVol.C2-13/5, 19, 265
	$BaMnO_4$-$BaSeO_4$	Mn: MVol.C2-265
$BaMnS_2$	$BaMnS_2$	Mn: MVol.C6-44
$BaMn_2O_4$	$BaMn_2O_4$	Mn: MVol.C2-226
$BaMn_2O_5$	$BaO \cdot 2\,MnO_2$	Mn: MVol.C2-228
$BaMn_2O_6$	$BaMn_2O_6$	Mn: MVol.C2-256
$BaMn_2O_8$	$Ba(MnO_4)_2$	Mn: MVol.C2-31/4, 37, 41, 265/6, 273/8
–	$Ba(MnO_4)_2$ solid solutions	
	$Ba[(Mn,Cl)O_4]_2 \cdot 3\,H_2O$	Mn: MVol.C2-36/7, 44/6, 276
$BaMn_4O_8$	$BaMn_4O_8$	Mn: MVol.C2-227
$BaMn_5O_{11}$	$BaO \cdot 5\,MnO_2$	Mn: MVol.C2-227/8
$BaMoNa_2Nb_4O_{15}$	$Na_2BaNb_4MoO_{15}$ solid solutions	
	$Na_{1+x}Ba_{2-x}Nb_{5-x}Mo_xO_{15}$	Mo: SVol.B2-305
$BaMoO_3$	$BaMoO_3$ — $BaO \cdot MoO_2$	Mo: SVol.B2-1/2, 50
–	$BaMoO_3$ solid solutions	
	$Ba_xSr_{1-x}MoO_3$	Mo: SVol.B2-50

$BaMoO_4$	$BaMoO_4$	Mo: SVol.B2-2/15, 50/1, 53/5
–	$BaMoO_4$, doped with rare earths	Mo: SVol.B2-208/9
–	$BaMoO_4$, doped with U	U: SVol.C3-334
–	$BaMoO_4$ solid solutions	
	$BaMoO_4$-$PbMoO_4$	Mo: SVol.B2-12/3, 266
	$(Ba,Sr)MoO_4$	Mo: SVol.B2-12/3
–	$BaMoO_4$ systems	
	$BaMoO_4$-$BaTiO_3$	Mo: SVol.B2-219
	$BaMoO_4$-KCl	Mo: SVol.B2-15/6
	$BaMoO_4$-LiCl	Mo: SVol.B2-15/6
	$BaMoO_4$-Li_2MoO_4	Mo: SVol.B2-56
	$BaMoO_4$-NaCl	Mo: SVol.B2-15/6
	$BaMoO_4$-RbCl	Mo: SVol.B2-15/6
$BaMoO_6Sr_2$	$BaSr_2MoO_6$	Mo: SVol.B2-57
$BaMo_2O_6$	$Ba(MoO_3)_2$	Mo: SVol.B2-2
$BaMo_2O_7$	$BaMo_2O_7$ = BaO · 2 MoO_3	Mo: SVol.B2-8, 51, 55
$BaMo_2O_8Rb_2$	$Rb_2Ba(MoO_4)_2$	Mo: SVol.B2-56
$BaMo_3O_{10}$	BaO · 3 MoO_3	Mo: SVol.B2-3
$BaMo_4O_{13}$	BaO · 4 MoO_3	Mo: SVol.B2-3
BaN_2O_6	$Ba(NO_3)_2$ systems	
	$Ba(NO_3)_2$-$Al(NO_3)_3$-$Mn(NO_3)_2$-H_2O	Mn: MVol.C3-301
	$Ba(NO_3)_2$-$Mn(NO_3)_2$-H_2O	Mn: MVol.C3-300
	$Ba(NO_3)_2$-$Ra(NO_3)_2$-H_2O	Ra: SVol.2-192/6
	$Ba(NO_3)_2$-SnF_2-H_2O	Sn: MVol.C3-199
$BaNbO_3$	$BaNbO_3$ solid solutions	
	$BaNbO_3$-$BaMnO_3$	Mn: MVol.C3-175/6
$BaNd_{0.5}O_3Pa_{0.5}$	$Ba(Nd_{0.5}Pa_{0.5})O_3$	Pa: SVol.2-21
BaO	BaO glasses	
	BaO-Al_2O_3-B_2O_3	B: B-Verb.7-110
	BaO-B_2O_3-Na_2O	B: B-Verb.7-94
	BaO-SiO_2-UO_2	U: SVol.C3-158/60
	BaO-TeO_2	Te: SVol.B1-62
–	BaO solid solutions	
	BaO-$BaUO_3$	U: SVol.C3-121
	BaO-SnO_2-TiO_2	Sn: MVol.C4-200
–	BaO systems	
	BaO-B_2O_3	B: B-Verb.7-84
	BaO-B_2O_3-PbO	B: B-Verb.7-95
	BaO-B_2O_3-SiO_2	B: B-Verb.7-214
	BaO-B_2O_3-ZnO	B: B-Verb.7-95
	BaO-$BaThO_3$	Th: SVol.C2-15
	BaO-MoO_3	Mo: SVol.B2-50/1
	BaO-SnO_2	Sn: MVol.C3-192
	BaO-SnO_2-SrO	Sn: MVol.C3-196
	BaO-SnO_2-TiO_2	Sn: MVol.C4-173/5
	BaO-ThO_2	Th: SVol.C2-14/5
	BaO-UO_2	U: SVol.C3-119/20
	BaO-UO_2-UO_3	U: SVol.C3-110

BaO_2Sn .... $BaSnO_2$ .... Sn: MVol.C3-191
$BaO_3Pa_{0.5}Pr_{0.5}$ ..... $Ba(Pr_{0.5}Pa_{0.5})O_3$ .... Pa: SVol.2-21
$BaO_3Pa_{0.5}Pu_{0.5}$ ..... $Ba(Pu_{0.5}Pa_{0.5})O_3$ .... Pa: SVol.2-21/2
$BaO_3Pa_{0.5}Sc_{0.5}$ ..... $Ba(Sc_{0.5}Pa_{0.5})O_3$ .... Pa: SVol.2-21
$BaO_3Pa_{0.5}Sm_{0.5}$ .... $Ba(Sm_{0.5}Pa_{0.5})O_3$ .... Pa: SVol.2-21
$BaO_3Pa_{0.5}Tb_{0.5}$ ..... $Ba(Tb_{0.5}Pa_{0.5})O_3$ .... Pa: SVol.2-21
$BaO_3Pa_{0.5}Tm_{0.5}$ ..... $Ba(Tm_{0.5}Pa_{0.5})O_3$ .... Pa: SVol.2-21
$BaO_3Pa_{0.5}Y_{0.5}$ ...... $Ba(Y_{0.5}Pa_{0.5})O_3$ .... Pa: SVol.2-21
$BaO_3Pa_{0.5}Yb_{0.5}$ ..... $Ba(Yb_{0.5}Pa_{0.5})O_3$ .... Pa: SVol.2-21
BaO_3Pa .... $BaPaO_3$ .... Pa: SVol.2-20/1
BaO_3Si .... $BaSiO_3$ solid solutions
$Ba(Ti,Sn,Si)O_3$ .... Sn: MVol.C4-200/1

BaO_3Sn .... $BaSnO_3$ .... Sn: MVol.C3-192/5
– .... $BaSnO_3$ solid solutions
$BaSnO_3$–$Al_2O_3 \cdot Bi_2O_3$ .... Sn: MVol.C4-101
$Ba(Ti,Sn,Si)O_3$ .... Sn: MVol.C4-200/1
$Ba(Sn,Ti)O_3$ .... Sn: MVol.C4-175/200, 215/6
$Ba(Sn,Ti)O_3$, doped with metal oxides . Sn: MVol.C4-175/200, 215/6
$Ba(Ti,Zr,Sn)O_3$ .... Sn: MVol.C4-215/6, 220/4
$(Ba,Ca)(Ti,Zr,Sn)O_3$ .... Sn: MVol.C4-224
$(Ba,Sr,Mg)(Ti,Zr,Sn)O_3$ .... Sn: MVol.C4-224
$(Ba,Ca)(Ti,Sn)O_3$ .... Sn: MVol.C4-206
$(Ba,Sr,Ca)(Ti,Sn)O_3$ .... Sn: MVol.C4-210
$BaSnO_3$–$BaTiO_3$–$CaZrO_3$ .... Sn: MVol.C4-215
$(Ba,Cd)(Ti,Sn)O_3$ .... Sn: MVol.C4-212
$(Ba,Mg)(Ti,Sn)O_3$ .... Sn: MVol.C4-203/5
$(Ba,Sr)(Ti,Sn)O_3$ .... Sn: MVol.C4-206/10
$(Ba,Sr)(Sn,Ti)O_3$, doped with metal oxides
Sn: MVol.C4-209/10
$(Ba,Zn)(Ti,Sn)O_3$ .... Sn: MVol.C4-211
$(Ba,Sr)SnO_3$ .... Sn: MVol.C3-196

BaO_3Ta .... $BaTaO_3$ solid solutions
$BaTaO_3$–$BaMnO_3$ .... Mn: MVol.C3-185

BaO_3Th .... $BaThO_3$ .... Th: SVol.C2-14/5
– .... $BaThO_3$ systems
$BaThO_3$–BaO .... Th: SVol.C2-15

BaO_3Ti .... $BaTiO_3$, doped with metal oxides .... Sn: MVol.C4-215/6
– .... $BaTiO_3$ solid solutions
$BaTiO_3$–$BaIrO_3$ .... Ir: SVol.2-19
$BaTiO_3$–$BaMnO_3$ .... Mn: MVol.C3-120/1
$Ba(Ti,Sn,Si)O_3$ .... Sn: MVol.C4-200/1
$Ba(Ti,Sn)O_3$ .... Sn: MVol.C4-175/200, 215/6
$Ba(Ti,Sn)O_3$, doped with metal oxides . Sn: MVol.C4-175/200, 215/6
$Ba(Ti,Zr,Sn)O_3$ .... Sn: MVol.C4-215/6, 220/4
$(Ba,Ca)(Ti,Zr,Sn)O_3$ .... Sn: MVol.C4-224
$(Ba,Sr,Mg)(Ti,Zr,Sn)O_3$ .... Sn: MVol.C4-224
$(Ba,Ca)(Ti,Sn)O_3$ .... Sn: MVol.C4-206
$(Ba,Sr,Ca)(Ti,Sn)O_3$ .... Sn: MVol.C4-210

BaO_3Ti .... $BaTiO_3$ solid solutions
$BaTiO_3$-$BaSnO_3$-$CaZrO_3$ .... Sn: MVol.C4-215
(Ba,Cd)(Ti,Sn)O_3 .... Sn: MVol.C4-212
(Ba,Mg)(Ti,Sn)O_3 .... Sn: MVol.C4-203/5
(Ba,Sr)(Ti,Sn)O_3 .... Sn: MVol.C4-206/10
(Ba,Sr)(Ti,Sn)O_3, doped with metal oxides
Sn: MVol.C4-209/10
(Ba,Zn)(Ti,Sn)O_3 .... Sn: MVol.C4-211
$BaTiO_3$-$Bi_2(SnO_3)_3$ .... Sn: MVol.C4-201/3
$BaTiO_3$-$(Er,Bi)_2(SnO_3)_3$ .... Sn: MVol.C4-215
$BaTiO_3$-$(La,Bi)_2(SnO_3)_3$ .... Sn: MVol.C4-215
$BaTiO_3$-$CaSnO_3$ .... Sn: MVol.C4-215/6
$BaTiO_3$-$Ce_2(SnO_3)_3$ .... Sn: MVol.C4-213/5
$BaTiO_3$-$Dy_2(SnO_3)_3$ .... Sn: MVol.C4-213/5
$BaTiO_3$-$Er_2(SnO_3)_3$ .... Sn: MVol.C4-213/5
$BaTiO_3$-$Gd_2(SnO_3)_3$ .... Sn: MVol.C4-213/5
$BaTiO_3$-$Ho_2(SnO_3)_3$ .... Sn: MVol.C4-213/5
$BaTiO_3$-$In_2(SnO_3)_3$ .... Sn: MVol.C4-212
$Ba_xLa_{1-x}Mn_{1-x}Ti_xO_3$ .... Mn: MVol.C3-123/4
$BaTiO_3$-$La_2(SnO_3)_3$ .... Sn: MVol.C4-213/5
$BaTiO_3$-$MnNb_2O_6$ .... Mn: MVol.C3-176
$BaTiO_3$-2 MnO · Nb_2O_5 .... Mn: MVol.C3-176
$BaTiO_3$-MnO .... Mn: MVol.C3-119/20
$BaTiO_3$-MnO_2 .... Mn: MVol.C3-119/20
$BaTiO_3$-$MnTa_2O_6$ .... Mn: MVol.C3-185/6
$BaTiO_3$-2 MnO · Ta_2O_5 .... Mn: MVol.C3-185/6
$BaTiO_3$-$Nd_2(SnO_3)_3$ .... Sn: MVol.C4-213/5
$BaTiO_3$-$Pr_2(SnO_3)_3$ .... Sn: MVol.C4-213/5
$BaTiO_3$-$Sm_2(SnO_3)_3$ .... Sn: MVol.C4-213/5
$BaTiO_3$-SnO_2 .... Sn: MVol.C4-175/200
(Ba,Sn)TiO_3 .... Sn: MVol.C4-173
$BaTiO_3$-$Sr_{0.3}La_{0.7}MnO_3$ .... Mn: MVol.C3-125/6
$BaTiO_3$-$Tb_2(SnO_3)_3$ .... Sn: MVol.C4-213/5
$BaTiO_3$-UO_2 .... U: SVol.C3-275
$BaTiO_3$-$Y_2(SnO_3)_3$ .... Sn: MVol.C4-213/5
$BaTiO_3$-$Yb_2(SnO_3)_3$ .... Sn: MVol.C4-213/5

– .... $BaTiO_3$ systems
$BaTiO_3$-$BaMoO_4$ .... Mo: SVol.B2-219
$BaTiO_3$-K_2MoO_4 .... Mo: SVol.B2-219
$BaTiO_3$-MnO-Nb_2O_5 .... Mn: MVol.C3-176
$BaTiO_3$-MnO-Ta_2O_5 .... Mn: MVol.C3-185/6
$BaTiO_3$-Na_2MoO_4 .... Mo: SVol.B2-219

BaO_3U .... $BaUO_3$ .... U: SVol.C3-119/21
– .... $BaUO_3$ solid solutions
$BaMn_xU_{1-x}O_3$ .... Mn: MVol.C3-229
$BaUO_3$-BaO .... U: SVol.C3-121
$K_xBa_{1-x}UO_3$ .... U: SVol.C3-120
$Na_xBa_{1-x}UO_3$ .... U: SVol.C3-120
(Ba,Sr)UO_3 .... U: SVol.C3-103, 112, 121

BaO_3Zr	$BaZrO_3$ solid solutions	
	$Ba(Ti,Zr,Sn)O_3$	Sn: MVol.C4-215/6, 220/4
	$(Ba,Ca)(Ti,Zr,Sn)O_3$	Sn: MVol.C4-224
	$(Ba,Sr,Mg)(Ti,Zr,Sn)O_3$	Sn: MVol.C4-224
	$BaZrO_3$-$Ba_3Sc_2UO_9$	U: SVol.C3-137
	$BaZrO_3$-$Ba_3Y_2UO_9$	U: SVol.C3-137
	$BaZrO_3$-$Sr_{1.5}Ba_{1.5}ScYUO_9$	U: SVol.C3-137
–	$BaZrO_3$ systems	
	$BaZrO_3$-Ba_3UO_6	U: SVol.C3-126
BaO_4PbSrU	$BaSrPbUO_4$	U: SVol.C3-171
BaO_4S	$BaSO_4$ solid solutions	
	$BaSO_4$-$BaMnO_4$	Mn: MVol.C2-13/5, 19, 265
	$BaSO_4$-$KMnO_4$	Mn: MVol.C2-33, 173/4
	$BaSO_4$-$LiMnO_4$	Mn: MVol.C2-140, 173
	$BaSO_4$-NH_4MnO_4	Mn: MVol.C2-34, 173, 200/1
	$BaSO_4$-$NaMnO_4$	Mn: MVol.C2-145, 173
	$BaSO_4$-$RbMnO_4$	Mn: MVol.C2-173, 193
–	$BaSO_4$ systems	
	$BaSO_4$-$RaSO_4$-electrolyte-H_2O	Ra: SVol.2-368
BaO_4Se	$BaSeO_4$ solid solutions	
	$BaSeO_4$-$BaMnO_4$	Mn: MVol.C2-265
	$BaSeO_4$-$KMnO_4$	Mn: MVol.C2-174
BaO_4U	$BaUO_4$	U: SVol.C3-122, 124/5
BaO_4W	$BaWO_4$	W: SVol.B2-201, 214/5
–	$BaWO_4$, doped with U	U: SVol.C3-137, 343
BaO_5Te_2	$BaTe_2O_5$	Te: SVol.B1-68
$BaO_{5.5}PaSr_2$	$BaSr_2PaO_{5.5}$	Pa: SVol.2-21
BaO_6Pb_2U	$BaPb_2UO_6$	U: SVol.C3-171
BaO_6Sr_2U	$BaSr_2UO_6$	U: SVol.C3-128/30
BaO_6U_2	$Ba(UO_3)_2$	U: SVol.C3-121/2
BaO_7Pu_2	$BaPu_2O_7 \cdot x\ H_2O$	Np: TrU.D1-113
BaO_7Th_3	$BaTh_3O_7$	Th: SVol.C2-14
BaO_7U_2	BaU_2O_7	U: SVol.C3-122/4, 137
$BaO_{7.6}U_{2.2}$	$BaO \cdot 2.2\ UO_3$	U: SVol.C3-124
BaO_8P_2Sn	$BaSn(PO_4)_2$	Sn: MVol.C3-206
$BaO_{8.5}U_{2.5}$	$BaO \cdot 2.5\ UO_3$	U: SVol.C3-124
$BaO_9Si_3Sn_{0.5}Ti_{0.5}$	$BaSn_{0.5}Ti_{0.5}Si_3O_9$	Sn: MVol.C4-200
BaO_9Si_3Sn	$BaSnSi_3O_9$	Sn: MVol.C3-205/6
–	$BaSnSi_3O_9$ solid solutions	
	$BaSn(Ge_{3-x}Si_x)O_9$	Sn: MVol.C4-232
$BaO_{10}U_3$	BaU_3O_{10}	U: SVol.C3-122
–	$BaO \cdot 3\ UO_3 \cdot 4\ H_2O$	U: SVol.C3-128
–	$BaU_3O_{10} \cdot 4.5\ H_2O$	U: SVol.C3-128
$BaO_{12}S_3Sn$	$BaSn(SO_4)_3 \cdot 3\ H_2O$	Sn: MVol.C3-204
$BaO_{12}U_2V_2$	$Ba(UO_2)_2(VO_4)_2 \cdot 5\ H_2O$	U: SVol.C3-137, 308
–	$Ba(UO_2)_2(VO_4)_2$ solid solutions	
	$(Ba,Pb)(UO_2)_2(VO_4)_2$	U: SVol.C3-305, 307

$BaO_{13}U_4$	BaU_4O_{13}	U:	SVol.C3-122
$BaO_{17}Te_8$	$BaO \cdot 8\ TeO_2$	Te:	SVol.B1-68
$BaO_{19}U_6$	BaU_6O_{19}	U:	SVol.C3-122
–	$BaO \cdot 6\ UO_3 \cdot 10\ H_2O$	U:	SVol.C3-127
–	$BaU_6O_{19} \cdot 11\ H_2O$	U:	SVol.C3-127
–	$BaO \cdot 6\ UO_3$ solid solutions		
	$(Pb,Ba)O \cdot 6\ UO_3 \cdot 10\ H_2O$	U:	SVol.C3-169
BaS	BaS systems		
	BaS-SnS	Sn:	MVol.C3-200
BaS_2Sn	$BaSnS_2$	Sn:	MVol.C3-200/1
BaS_3Sn	$BaSnS_3$	Sn:	MVol.C3-202
–	$BaSnS_3 \cdot 8\ H_2O$	Sn:	MVol.C3-202
BaS_3Sn_2	$BaSn_2S_3$	Sn:	MVol.C3-200/1
BaS_3Te	$BaTeS_3 \cdot 2\ H_2O$	Te:	SVol.B1-153
BaS_4Sn_2	$BaSn_2S_4$	Sn:	MVol.C3-202
$BaSe_3Sn$	$BaSnSe_3$	Sn:	MVol.C3-204
BaTe	BaTe	Te:	SVol.B1-62
$Ba_{1.13}O_{7.13}U_2$	$1.13\ BaO \cdot 2\ UO_3 \cdot 1.88\ H_2O$	U:	SVol.C3-128
$Ba_{1.5}H_6IO_6Sn_2$	$Ba_{1.5}Sn_2(OH)_6I \cdot H_2O$	Sn:	MVol.C3-200
$Ba_{1.5}O_9ScSr_{1.5}UY$	$Sr_{1.5}Ba_{1.5}ScYUO_9$ solid solutions		
	$Sr_{1.5}Ba_{1.5}ScYUO_9$-$BaZrO_3$	U:	SVol.C3-137
	$Sr_{1.5}Ba_{1.5}ScYUO_9$-$CaZrO_3$	U:	SVol.C3-137
	$Sr_{1.5}Ba_{1.5}ScYUO_9$-$SrZrO_3$	U:	SVol.C3-137
$Ba_{1.8}H_7MnNa_{0.4}O_7$	$Ba_{1.8}Na_{0.4}[Mn(OH)_7]$	Mn:	MVol.C2-227
Ba_2BeO_6U	Ba_2BeUO_6	U:	SVol.C3-75, 128/9
$Ba_2C_8O_{16}Sn$	$Ba_2Sn(C_2O_4)_4 \cdot 8\ H_2O$	Sn:	MVol.C3-205
Ba_2CaMoO_5	Ba_2CaMoO_5	Mo:	SVol.B2-57
Ba_2CaMoO_6	Ba_2CaMoO_6	Mo:	SVol.B2-8, 57
Ba_2CaO_5U	Ba_2CaUO_5	U:	SVol.C3-121
Ba_2CaO_6U	Ba_2CaUO_6	U:	SVol.C3-128/30
–	Ba_2CaUO_6, doped with rare earths	U:	SVol.C3-100
Ba_2CdMoO_6	Ba_2CdMoO_6	Mo:	SVol.B1-130
Ba_2CdO_5U	Ba_2CdUO_5	U:	SVol.C3-121, 195
Ba_2CdO_6U	Ba_2CdUO_6	U:	SVol.C3-128/9
$Ba_2Ce_{0.67}O_{6.17}U_{1.33}$	$Ba_2(Ce_{0.67}U_{0.33})UO_{6.17}$	U:	SVol.C3-135/6
Ba_2CeMoO_6	Ba_2CeMoO_6	Mo:	SVol.B2-209/10
Ba_2CeO_6U	Ba_2CeUO_6	U:	SVol.C3-137
Ba_2Cl_6Mn	$Ba_2MnCl_6 = 2\ BaCl_2 \cdot MnCl_2$	Mn:	MVol.C5-219/20
Ba_2Cl_7U	$2\ BaCl_2 \cdot UCl_3$	U:	SVol.C9-9, 50
Ba_2Cl_8U	$2\ BaCl_2 \cdot UCl_4$	U:	SVol.C9-24, 27
Ba_2CoO_6U	Ba_2CoUO_6	U:	SVol.C3-128/9, 355/6
Ba_2CrO_6U	Ba_2CrUO_6	U:	SVol.C3-131/3, 323
Ba_2CuO_6U	Ba_2CuUO_6	U:	SVol.C3-128/9, 187
$Ba_2Dy_{0.67}O_{5.5}U$	$Ba_2Dy_{0.67}UO_{5.5}$	U:	SVol.C3-132/3
$Ba_2Dy_{0.67}O_6U$	$Ba_2Dy_{0.67}UO_6$	U:	SVol.C3-134/5
$Ba_2Dy_{0.67}O_{6.17}U_{1.33}$	$Ba_2(Dy_{0.67}U_{0.33})UO_{6.17}$	U:	SVol.C3-135/6
Ba_2DyMoO_6	Ba_2DyMoO_6	Mo:	SVol.B2-209/10
Ba_2DyO_6U	Ba_2DyUO_6	U:	SVol.C3-131/3

$Ba_2Er_{0.67}O_{5.5}U$	$Ba_2Er_{0.67}UO_{5.5}$	U: SVol.C3-132/3
$Ba_2Er_{0.67}O_6U$	$Ba_2Er_{0.67}UO_6$	U: SVol.C3-134/5
$Ba_2Er_{0.67}O_{6.17}U_{1.33}$	$Ba_2(Er_{0.67}U_{0.33})UO_{6.17}$	U: SVol.C3-135/6
Ba_2ErMoO_6	Ba_2ErMoO_6	Mo: SVol.B2-209/10
Ba_2ErO_6U	Ba_2ErUO_6	U: SVol.C3-131/3
$Ba_2Eu_{0.67}O_{5.5}U$	$Ba_2Eu_{0.67}UO_{5.5}$	U: SVol.C3-132/3
$Ba_2Eu_{0.67}O_6U$	$Ba_2Eu_{0.67}UO_6$	U: SVol.C3-134/5
$Ba_2Eu_{0.67}O_{6.17}U_{1.33}$	$Ba_2(Eu_{0.67}U_{0.33})UO_{6.17}$	U: SVol.C3-135/6
Ba_2EuMoO_6	Ba_2EuMoO_6	Mo: SVol.B2-209/10
Ba_2EuO_6U	Ba_2EuUO_6	U: SVol.C3-131/3
$Ba_2FLiMoO_5$	$LiBa_2MoO_5F$	Mo: SVol.B2-56
$Ba_2FMoNaO_5$	$NaBa_2MoO_5F$	Mo: SVol.B2-56
$Ba_2F_3LiMoO_3$	$LiBa_2MoO_3F_3$	Mo: SVol.B2-56
$Ba_2F_3MoNaO_3$	$NaBa_2MoO_3F_3$	Mo: SVol.B2-56
Ba_2F_6Mn	$Ba_2MnF_6 = 2\ BaF_2 \cdot MnF_2$	Mn: MVol.C4-230/1
Ba_2FeO_6U	Ba_2FeUO_6	U: SVol.C3-131/3, 350
–	Ba_2FeUO_6 solid solutions	
	$Ba_2(Fe_{1-x}Ga_x)UO_6$	U: SVol.C3-146/7, 350
	$Ba_2(Fe_{1-x}In_x)UO_6$	U: SVol.C3-147/9, 350
	$Ba_2(Fe,In,Mg)UO_6$	U: SVol.C3-149
	$Ba_2(Fe_{1-x}Mg_x)UO_6$	U: SVol.C3-130, 350
	$Ba_2(Fe_{1-x}Zn_x)UO_6$	U: SVol.C3-130, 191, 350
Ba_2GaO_6U	Ba_2GaUO_6	U: SVol.C3-146
–	Ba_2GaUO_6 solid solutions	
	$Ba_2(Fe_{1-x}Ga_x)UO_6$	U: SVol.C3-146/7, 350
$Ba_2Gd_{0.67}O_{5.5}U$	$Ba_2Gd_{0.67}UO_{5.5}$	U: SVol.C3-132/3
$Ba_2Gd_{0.67}O_6U$	$Ba_2Gd_{0.67}UO_6$	U: SVol.C3-134/5
–	$Ba_2Gd_{0.67}UO_6$ solid solutions	
	$Ba_2Gd_{0.67}U_{1-x}W_xO_6$	U: SVol.C3-135
$Ba_2Gd_{0.67}O_6W$	$Ba_2Gd_{0.67}WO_6$ solid solutions	
	$Ba_2Gd_{0.67}U_{1-x}W_xO_6$	U: SVol.C3-135
$Ba_2Gd_{0.67}O_{6.17}U_{1.33}$	$Ba_2(Gd_{0.67}U_{0.33})UO_{6.17}$	U: SVol.C3-135/6
Ba_2GdMoO_6	Ba_2GdMoO_6	Mo: SVol.B2-209/10
Ba_2GdO_6U	Ba_2GdUO_6	U: SVol.C3-131/3
Ba_2GeO_6U	Ba_2GeUO_6	U: SVol.C3-137, 165
$Ba_2Ge_2MnO_7$	$Ba_2MnGe_2O_7$	Mn: MVol.C3-84
$Ba_2H_3IrO_{13}P_3$	$Ba_2H[Ir(PO_4)_3H_2O]$	Ir: SVol.2-37
$Ba_2H_6MnO_6$	$Ba_2[Mn(OH)_6]$	Mn: MVol.C2-216/7
$Ba_2H_7I_3MnO_{18}$	$Ba_2H_7Mn(IO_6)_3$	Mn: MVol.C5-338
$Ba_2H_7MnO_7$	$Ba_2[Mn(OH)_7]$	Mn: MVol.C2-227
Ba_2HfO_6U	Ba_2HfO_6U	U: SVol.C3-137, 294
Ba_2HgO_6U	Ba_2HgUO_6	U: SVol.C3-128/9, 196
$Ba_2Ho_{0.67}O_{5.5}U$	$Ba_2Ho_{0.67}UO_{5.5}$	U: SVol.C3-132/3
$Ba_2Ho_{0.67}O_6U$	$Ba_2Ho_{0.67}UO_6$	U: SVol.C3-134/5
$Ba_2Ho_{0.67}O_{6.17}U_{1.33}$	$Ba_2(Ho_{0.67}U_{0.33})UO_{6.17}$	U: SVol.C3-135/6
Ba_2HoMoO_6	Ba_2HoMoO_6	Mo: SVol.B2-209/10
Ba_2HoO_6U	Ba_2HoUO_6	U: SVol.C3-131/3
$Ba_2InO_{5.5}U$	$Ba_2InUO_{5.5}$	U: SVol.C3-131/3, 147
Ba_2InO_6U	Ba_2InUO_6	U: SVol.C3-131/3, 147

Ba_2InO_6U	Ba_2InUO_6 solid solutions	
	$Ba_2(Fe_{1-x}In_x)UO_6$	U: SVol.C3-147/9, 350
	$Ba_2(In,Fe,Mg)UO_6$	U: SVol.C3-149
	$Ba_2Mn_{1-x}In_xUO_6$	Mn: MVol.C3-230 U: SVol.C3-130, 147/9
$Ba_2La_{0.67}O_{5.5}U$	$Ba_2La_{0.67}UO_{5.5}$	U: SVol.C3-132/3
$Ba_2La_{0.67}O_6U$	$Ba_2La_{0.67}UO_6$	U: SVol.C3-134/5
$Ba_2La_{0.67}O_{6.17}U_{1.33}$	$Ba_2(La_{0.67}U_{0.33})UO_{6.17}$	U: SVol.C3-135/6
Ba_2LaMoO_6	Ba_2LaMoO_6	Mo: SVol.B2-209/10
Ba_2LaO_6U	Ba_2LaUO_6	U: SVol.C3-131/3
$Ba_2LiMoO_{5.5}$	$LiBa_2MoO_{5.5}$	Mo: SVol.B2-56
–	$LiBa_2MoO_{5.5}$ solid solutions	
	$Li_xBa_{3-x}MoO_{6-0.5x}$	Mo: SVol.B2-56
$Ba_2Lu_{0.67}O_{5.5}U$	$Ba_2Lu_{0.67}UO_{5.5}$	U: SVol.C3-132/3
$Ba_2Lu_{0.67}O_6U$	$Ba_2Lu_{0.67}UO_6$	U: SVol.C3-134/5
$Ba_2Lu_{0.67}O_{6.17}U_{1.33}$	$Ba_2(Lu_{0.67}U_{0.33})UO_{6.17}$	U: SVol.C3-135/6
Ba_2LuMoO_6	Ba_2LuMoO_6	Mo: SVol.B2-209/10
Ba_2LuO_6U	Ba_2LuUO_6	U: SVol.C3-131/3
Ba_2MgMoO_5	Ba_2MgMoO_5	Mo: SVol.B2-57
Ba_2MgMoO_6	Ba_2MgMoO_6	Mo: SVol.B2-8, 57
Ba_2MgO_6U	Ba_2MgUO_6	U: SVol.C3-85, 128/9
–	Ba_2MgUO_6 solid solutions	
	$Ba_2(Fe_{1-x}Mg_x)UO_6$	U: SVol.C3-130, 350
	$Ba_2(Mg,Fe,In)UO_6$	U: SVol.C3-149
	$Ba_2Mg(W_{1-x}U_x)O_6$	U: SVol.C3-343
	Ba_2MgUO_6-$Ba_3Fe_2UO_9$	U: SVol.C3-137, 351
Ba_2MgO_6W	Ba_2MgWO_6, doped with U	U: SVol.C3-137
–	Ba_2MgWO_6 solid solutions	
	$Ba_2Mg(W_{1-x}U_x)O_6$	U: SVol.C3-343
Ba_2MnO_3	Ba_2MnO_3	Mn: MVol.C2-216
Ba_2MnO_4	Ba_2MnO_4	Mn: MVol.C2-251
Ba_2MnO_6Te	Ba_2MnTeO_6	Mn: MVol.C6-356
Ba_2MnO_6U	Ba_2MnUO_6	Mn: MVol.C3-229/30 U: SVol.C3-128/30, 137
–	Ba_2MnUO_6 solid solutions	
	$Ba_2Mn_{1-x}In_xUO_6$	Mn: MVol.C3-230 U: SVol.C3-130, 147/9
Ba_2MnS_3	Ba_2MnS_3	Mn: MVol.C6-44/5
Ba_2MnSe_3	Ba_2MnSe_3	Mn: MVol.C6-287
$Ba_2Mn_2O_7$	$Ba_2Mn_2O_7$	Mn: MVol.C2-256
$Ba_2MoNaO_{5.5}$	$NaBa_2MoO_{5.5}$	Mo: SVol.B2-56
–	$NaBa_2MoO_{5.5}$ solid solutions	
	$Na_xBa_{3-x}MoO_{6-0.5x}$	Mo: SVol.B2-56
Ba_2MoNdO_6	Ba_2NdMoO_6	Mo: SVol.B2-209/10
Ba_2MoO_4	$Ba_2MoO_4 = 2\ BaO \cdot MoO_2$	Mo: SVol.B2-49/50
Ba_2MoO_5	$Ba_2MoO_5 = 2\ BaO \cdot MoO_3$	Mo: SVol.B2-50/1, 52/3
Ba_2MoO_6Sr	Ba_2SrMoO_5	Mo: SVol.B2-57
Ba_2MoO_6Pb	Ba_2PbMoO_6	Mo: SVol.B2-266

Ba_2MoO_6Pr	Ba_2PrMoO_6	Mo: SVol.B2-209/10
Ba_2MoO_6Sc	Ba_2ScMoO_6	Mo: SVol.B2-209/10
Ba_2MoO_6Sm	Ba_2SmMoO_6	Mo: SVol.B2-209/10
Ba_2MoO_6Sr	Ba_2SrMoO_6	Mo: SVol.B2-57
Ba_2MoO_6Tb	Ba_2TbMoO_6	Mo: SVol.B2-209/10
Ba_2MoO_6Tm	Ba_2TmMoO_6	Mo: SVol.B2-209/10
Ba_2MoO_6Y	Ba_2YMoO_6	Mo: SVol.B2-209/10
Ba_2MoO_6Yb	Ba_2YbMoO_6	Mo: SVol.B2-209/10
Ba_2MoO_6Zn	Ba_2ZnMoO_6	Mo: SVol.B1-130
$Ba_2N_2O_{12}S_4$	$Ba_2[(SO_3)_2NN(SO_3)_2] \cdot 4\ H_2O$	S: S-N-Verb.1-221
$Ba_2N_4O_8S_4$	$Ba_2[NSO_2]_4 \cdot 6\ H_2O$	S: S-N-Verb.1-4
$Ba_2NaNb_5O_{15}$	$NaBa_2Nb_5O_{15}$ solid solutions	
	$Na_{1+x}Ba_{2-x}Nb_{5-x}Mo_xO_{15}$	Mo: SVol.B2-305/6
$Ba_2Nd_{0.67}O_{5.5}U$	$Ba_2Nd_{0.67}UO_{5.5}$	U: SVol.C3-132/3
$Ba_2Nd_{0.67}O_6U$	$Ba_2Nd_{0.67}UO_6$	U: SVol.C3-134/5
$Ba_2Nd_{0.67}O_{6.17}U_{1.33}$	$Ba_2(Nd_{0.67}U_{0.33})UO_{6.17}$	U: SVol.C3-135/6
Ba_2NdO_6U	Ba_2NdUO_6	U: SVol.C3-131/3
Ba_2NiO_6U	Ba_2NiUO_6	U: SVol.C3-128/9, 358/9
Ba_2O_2SSn	$2\ BaO \cdot SnS$	Sn: MVol.C3-204
Ba_2O_4Sn	Ba_2SnO_4	Sn: MVol.C3-195
–	$2\ BaO \cdot SnO_2 \cdot 10\ H_2O$	Sn: MVol.C3-196
–	Ba_2SnO_4 solid solutions	
	$(Ba,Sr)_2SnO_4$	Sn: MVol.C3-196
Ba_2O_4Th	Ba_2ThO_4	Th: SVol.C2-14
Ba_2O_4U	Ba_2UO_4	U: SVol.C3-121
$Ba_2O_{4.5}U$	$Ba_2UO_{4.5}$	U: SVol.C3-122
Ba_2O_5Te	Ba_2TeO_5, doped with U	U: SVol.C3-137
Ba_2O_5U	Ba_2UO_5	U: SVol.C3-122, 125
Ba_2O_5W	Ba_2WO_5	W: SVol.B2-201, 214/5
–	Ba_2WO_5, doped with U	U: SVol.C3-137
$Ba_2O_{5.5}PaSr$	$Ba_2SrPaO_{5.5}$	Pa: SVol.2-21
$Ba_2O_{5.5}Pr_{0.67}U$	$Ba_2Pr_{0.67}UO_{5.5}$	U: SVol.C3-132/3
$Ba_2O_{5.5}Sm_{0.67}U$	$Ba_2Sm_{0.67}UO_{5.5}$	U: SVol.C3-132/3
$Ba_2O_{5.5}Tb_{0.67}U$	$Ba_2Tb_{0.67}UO_{5.5}$	U: SVol.C3-132/3
$Ba_2O_{5.5}Tm_{0.67}U$	$Ba_2Tm_{0.67}UO_{5.5}$	U: SVol.C3-132/3
$Ba_2O_{5.5}UY_{0.67}$	$Ba_2Y_{0.67}UO_{5.5}$	U: SVol.C3-132/3
$Ba_2O_{5.5}UYb_{0.67}$	$Ba_2Yb_{0.67}UO_{5.5}$	U: SVol.C3-132/3
Ba_2O_6PbU	Ba_2PbUO_6	U: SVol.C3-128/9, 171
$Ba_2O_6Pr_{0.67}U$	$Ba_2Pr_{0.67}UO_6$	U: SVol.C3-134/5
Ba_2O_6RhU	Ba_2RhUO_6	U: SVol.C3-131/3, 360
Ba_2O_6ScU	Ba_2ScUO_6	U: SVol.C3-131/3
$Ba_2O_6Sm_{0.67}U$	$Ba_2Sm_{0.67}UO_6$	U: SVol.C3-134/5
–	$Ba_2Sm_{0.67}UO_6$ solid solutions	
	$Ba_2Sm_{0.67}U_{1-x}W_xO_6$	U: SVol.C3-135
$Ba_2O_6Sm_{0.67}W$	$Ba_2Sm_{0.67}WO_6$ solid solutions	
	$Ba_2Sm_{0.67}U_{1-x}W_xO_6$	U: SVol.C3-135
Ba_2O_6SmU	Ba_2SmUO_6	U: SVol.C3-131/3
Ba_2O_6SrU	Ba_2SrUO_6	U: SVol.C3-128/30
$Ba_2O_6Tb_{0.67}U$	$Ba_2Tb_{0.67}UO_6$	U: SVol.C3-134/5

Ba_2O_6TbU	Ba_2TbUO_6	U: SVol.C3-131/3
Ba_2O_6TiU	Ba_2TiUO_6	U: SVol.C3-137, 275
$Ba_2O_6Tm_{0.67}U$	$Ba_2Tm_{0.67}UO_6$	U: SVol.C3-134/5
$Ba_2O_6UY_{0.67}$	$Ba_2Y_{0.67}UO_6$	U: SVol.C3-134/5
Ba_2O_6UY	Ba_2YUO_6	U: SVol.C3-131/3
$Ba_2O_6UYb_{0.67}$	$Ba_2Yb_{0.67}UO_6$	U: SVol.C3-134/5
Ba_2O_6UYb	Ba_2YbUO_6	U: SVol.C3-131/3
Ba_2O_6UZn	Ba_2ZnUO_6	U: SVol.C3-128/9, 191
–	Ba_2ZnUO_6 solid solutions $Ba_2(Fe_{1-x}Zn_x)UO_6$	U: SVol.C3-130, 191, 350
Ba_2O_6UZr	Ba_2ZrUO_6	U: SVol.C3-137, 291
$Ba_2O_{6.17}Pr_{0.67}U_{1.33}$	$Ba_2(Pr_{0.67}U_{0.33})UO_{6.17}$	U: SVol.C3-135/6
$Ba_2O_{6.17}Sm_{0.67}U_{1.33}$	$Ba_2(Sm_{0.67}U_{0.33})UO_{6.17}$	U: SVol.C3-135/6
$Ba_2O_{6.17}Tb_{0.67}U_{1.33}$	$Ba_2(Tb_{0.67}U_{0.33})UO_{6.17}$	U: SVol.C3-135/6
$Ba_2O_{6.17}Tm_{0.67}U_{1.33}$	$Ba_2(Tm_{0.67}U_{0.33})UO_{6.17}$	U: SVol.C3-135/6
$Ba_2O_{6.17}U_{1.33}Y_{0.67}$	$Ba_2(Y_{0.67}U_{0.33})UO_{6.17}$	U: SVol.C3-135/6
$Ba_2O_{6.17}U_{1.33}Yb_{0.67}$	$Ba_2(Yb_{0.67}U_{0.33})UO_{6.17}$	U: SVol.C3-135/6
$Ba_2O_7U_2$	$Ba_2U_2O_7$	U: SVol.C3-121/2
$Ba_2O_9U_2Zn$	$Ba_2ZnU_2O_9$	U: SVol.C3-191
$Ba_2O_{10}U_3$	$Ba_2U_3O_{10}$	U: SVol.C3-121/2
$Ba_2O_{11}U_3$	$Ba_2U_3O_{11}$	U: SVol.C3-122/3
$Ba_2O_{17}U_5$	$2\ BaO \cdot 5\ UO_3 \cdot 8\ H_2O$	U: SVol.C3-128
$Ba_2O_{18}Si_6SnTi$	$Ba_2SnTiSi_6O_{18}$	Sn: MVol.C4-200
$Ba_2O_{23}U_7$	$Ba_2U_7O_{23} \cdot 11\ H_2O$	U: SVol.C3-128
Ba_2S_4Sn	Ba_2SnS_4	Sn: MVol.C3-203
$Ba_{2.18}Dy_{0.73}O_6U_{1.09}$	$Ba_2(Ba_{0.18}Dy_{0.73}U_{1.09})O_6$	U: SVol.C3-132
$Ba_{2.18}Er_{0.73}O_6U_{1.09}$	$Ba_2(Ba_{0.18}Er_{0.73}U_{1.09})O_6$	U: SVol.C3-132
$Ba_{2.18}Eu_{0.73}O_6U_{1.09}$	$Ba_2(Ba_{0.18}Eu_{0.73}U_{1.09})O_6$	U: SVol.C3-132
$Ba_{2.18}Gd_{0.73}O_6U_{1.09}$	$Ba_2(Ba_{0.18}Gd_{0.73}U_{1.09})O_6$	U: SVol.C3-132
$Ba_{2.18}Ho_{0.73}O_6U_{1.09}$	$Ba_2(Ba_{0.18}Ho_{0.73}U_{1.09})O_6$	U: SVol.C3-132
$Ba_{2.18}La_{0.73}O_6U_{1.09}$	$Ba_2(Ba_{0.18}La_{0.73}U_{1.09})O_6$	U: SVol.C3-132
$Ba_{2.18}Lu_{0.73}O_6U_{1.09}$	$Ba_2(Ba_{0.18}Lu_{0.73}U_{1.09})O_6$	U: SVol.C3-132
$Ba_{2.18}Nd_{0.73}O_6U_{1.09}$	$Ba_2(Ba_{0.18}Nd_{0.73}U_{1.09})O_6$	U: SVol.C3-132
$Ba_{2.18}O_6Pr_{0.73}U_{1.09}$	$Ba_2(Ba_{0.18}Pr_{0.73}U_{1.09})O_6$	U: SVol.C3-132
$Ba_{2.18}O_6Sm_{0.73}U_{1.09}$	$Ba_2(Ba_{0.18}Sm_{0.73}U_{1.09})O_6$	U: SVol.C3-132
$Ba_{2.18}O_6Tb_{0.73}U_{1.09}$	$Ba_2(Ba_{0.18}Tb_{0.73}U_{1.09})O_6$	U: SVol.C3-132
$Ba_{2.18}O_6Tm_{0.73}U_{1.09}$	$Ba_2(Ba_{0.18}Tm_{0.73}U_{1.09})O_6$	U: SVol.C3-132
$Ba_{2.18}O_6U_{1.09}Y_{0.73}$	$Ba_2(Ba_{0.18}Y_{0.73}U_{1.09})O_6$	U: SVol.C3-132
$Ba_{2.18}O_6U_{1.09}Yb_{0.73}$	$Ba_2(Ba_{0.18}Yb_{0.73}U_{1.09})O_6$	U: SVol.C3-132
$Ba_{2.67}O_6U_{1.33}$	$Ba_2Ba_{0.67}U_{1.33}O_6$	U: SVol.C3-121/2
$Ba_3Bi_2MoO_9$	$Ba_3Bi_2MoO_9 = 3\ BaO \cdot Bi_2O_3 \cdot MoO_3$	Mo: SVol.B2-55/6
$Ba_3CaO_9U_2$	$Ba_3CaU_2O_9$	U: SVol.C3-130/1
$Ba_3CoO_9U_2$	$Ba_3CoU_2O_9$	U: SVol.C3-130/1, 355/6
$Ba_3Cr_2MoO_9$	$Ba_3Cr_2MoO_9$	Mo: SVol.B2-319
$Ba_3Cr_2O_9U$	$Ba_3Cr_2UO_9$	U: SVol.C3-136/7, 323
$Ba_3Dy_2O_9U$	$Ba_3Dy_2UO_9$	U: SVol.C3-136/7
–	$Ba_3Dy_2UO_9$ solid solutions $Ba_3Fe_{2-x}Dy_xUO_9$	U: SVol.C3-137, 351
$Ba_3Er_{0.1}O_{6.15}U$	$Ba_3UO_6 \cdot 0.05\ Er_2O_3$	U: SVol.C3-126

$Ba_3Er_{1.84}O_{8.76}U$ ..... $Ba_3UO_6 \cdot 0.92\ Er_2O_3$ .... U: SVol.C3-126
$Ba_3Er_2O_9U$ .... $Ba_3Er_2UO_9$ .... U: SVol.C3-126, 136/7
– .... $Ba_3Er_2UO_9$ solid solutions
$Ba_3Fe_{2-x}Er_xUO_9$ .... U: SVol.C3-137, 351
$Ba_3FeO_9U_2$ .... $Ba_3FeU_2O_9$ .... U: SVol.C3-130/1
$Ba_3Fe_2O_9U$ .... $Ba_3Fe_2UO_9$ .... U: SVol.C3-136/7, 351
– .... $Ba_3Fe_2UO_9$ solid solutions
$Ba_3Fe_2UO_9$–Ba_2MgUO_6 .... U: SVol.C3-137, 351
$Ba_3Fe_{2-x}Dy_xUO_9$ .... U: SVol.C3-137, 351
$Ba_3Fe_{2-x}Er_xUO_9$ .... U: SVol.C3-137, 351
$Ba_3Fe_{2-x}Ho_xUO_9$ .... U: SVol.C3-137, 351
$Ba_3Fe_{2-x}Y_xUO_9$ .... U: SVol.C3-137, 351
$Ba_3Gd_2O_9U$ .... $Ba_3Gd_2UO_9$ .... U: SVol.C3-136/7
$Ba_3Ho_2O_9U$ .... $Ba_3Ho_2UO_9$ solid solutions
$Ba_3Fe_{2-x}Ho_xUO_9$ .... U: SVol.C3-137, 351
$Ba_3In_2O_9U$ .... $Ba_3In_2UO_9$ .... U: SVol.C3-136/7, 147
$Ba_3MgO_9U_2$ .... $Ba_3MgU_2O_9$ .... U: SVol.C3-85, 130/1
$Ba_3MnMo_9O_{32}$ ..... $Ba_3MnMo_9O_{32} \cdot 12\ H_2O$ .... Mn: MVol.C3-211
Ba_3MnO_5 .... Ba_3MnO_5 .... Mn: MVol.C2-252
$Ba_3MnO_9U_2$ .... $Ba_3MnU_2O_9$ .... Mn: MVol.C3-229
U: SVol.C3-130/1
$Ba_3Mn_2O_8$ .... $Ba_3(MnO_4)_2$ .... Mn: MVol.C2-4, 9, 258/9
– .... $Ba_3(MnO_4)_2 \cdot H_2O$ .... Mn: MVol.C2-258/9
– .... $Ba_3(MnO_4)_2$ solid solutions
$Ba_3(MnO_4)_2$–$Ba_3(AsO_4)_2$ .... Mn: MVol.C2-9, 258/9
$Ba_3(MnO_4)_2$–$Ba_3(PO_4)_2$ .... Mn: MVol.C2-9, 258/9
$Ba_3(MnO_4)_2$–$Ba_3(VO_4)_2$ .... Mn: MVol.C2-258/9
Mn: MVol.C3-167
$Ba_3Mn_2O_9U$ .... $Ba_3Mn_2UO_9$ .... Mn: MVol.C3-230
U: SVol.C3-136/7
Ba_3MoO_6 .... $Ba_3MoO_6 = 3\ BaO \cdot MoO_3$ .... Mo: SVol.B2-8, 50/2
– .... Ba_3MoO_6 solid solutions
$Li_xBa_{3-x}MoO_{6-0.5\ x}$ .... Mo: SVol.B2-56
$Na_xBa_{3-x}MoO_{6-0.5\ x}$ .... Mo: SVol.B2-56
$Ba_3MoO_9Y_2$ .... $Ba_3Y_2MoO_9$ .... Mo: SVol.B2-208
$Ba_3Mo_6O_{21}$ .... $Ba_3Mo_6O_{21} = 3\ BaO \cdot 6\ MoO_3$ .... Mo: SVol.B2-55
$Ba_3Mo_7O_{24}$ .... $Ba_3Mo_7O_{24} = 3\ BaO \cdot 7\ MoO_3$ .... Mo: SVol.B2-3, 55
$Ba_3Mo_8O_{27}$ .... $3\ BaO \cdot 8\ MoO_3$ .... Mo: SVol.B2-3
$Ba_3NiO_9U_2$ .... $Ba_3NiU_2O_9$ .... U: SVol.C3-130/1, 358/9
Ba_3O_5U .... Ba_3UO_5 .... U: SVol.C3-119/21
– .... Ba_3UO_5 solid solutions
$Ba_{3-x}Sr_xUO_5$ .... U: SVol.C3-103, 121
$Ba_3O_{5.5}Pa$ .... $Ba_3PaO_{5.5}$ .... Pa: SVol.2-21
– .... $Ba_3PaO_{5.5}$ solid solutions
$Ba_3PaO_{5.5}$–$Sr_3PaO_{5.5}$ .... Pa: SVol.2-21
Ba_3O_6Te .... Ba_3TeO_6 .... Te: SVol.B1-62, 68
Ba_3O_6U .... Ba_3UO_6 .... U: SVol.C3-122, 125/7

Ba_3O_6U .... Ba_3UO_6 solid solutions
$(Ba,Sr)_3UO_6$ .... U: SVol.C3-130

– .... Ba_3UO_6 systems
Ba_3UO_6-$BaZrO_3$ .... U: SVol.C3-126
Ba_3UO_6-BeO .... U: SVol.C3-126
Ba_3UO_6-La_2O_3 .... U: SVol.C3-126
Ba_3UO_6-MgO .... U: SVol.C3-126
Ba_3UO_6-Sc_2O_3 .... U: SVol.C3-126/7
Ba_3UO_6-ThO_2 .... U: SVol.C3-126
Ba_3UO_6-UO_2 .... U: SVol.C3-110
Ba_3UO_6-Y_2O_3 .... U: SVol.C3-126/7

Ba_3O_6W .... Ba_3WO_6 .... W: SVol.B2-201, 214/5
$Ba_3O_7Sn_2$ .... $Ba_3Sn_2O_7$ .... Sn: MVol.C3-195
– .... $Ba_3Sn_2O_7$ solid solutions
$Ba_3Sn_2O_7$-$Sr_3Sn_2O_7$ .... Sn: MVol.C3-196

$Ba_3O_8P_2$ .... $Ba_3(PO_4)_2$ solid solutions
$Ba_3(PO_4)_2$-$Ba_3(MnO_4)_2$ .... Mn: MVol.C2-9, 258/9

$Ba_3O_8V_2$ .... $Ba_3(VO_4)_2$ solid solutions
$Ba_3(VO_4)_2$-$Ba_3(MnO_4)_2$ .... Mn: MVol.C2-258/9
Mn: MVol.C3-167

$Ba_3O_9Pr_2U$ .... $Ba_3Pr_2UO_9$ .... U: SVol.C3-136/7
$Ba_3O_9Sc_2U$ .... $Ba_3Sc_2UO_9$ .... U: SVol.C3-136/7
– .... $Ba_3Sc_2UO_9$ solid solutions
$Ba_3Sc_2UO_9$-$BaZrO_3$ .... U: SVol.C3-137
$Ba_3Sc_2UO_9$-$Ba_3Y_2UO_9$ .... U: SVol.C3-137
$Ba_3Sc_2UO_9$-$Sr_3Sc_2UO_9$ .... U: SVol.C3-137

$Ba_3O_9SrU_2$ .... $Ba_3SrU_2O_9$ .... U: SVol.C3-130/1
$Ba_3O_9UY_2$ .... $Ba_3Y_2UO_9$ .... U: SVol.C3-136/7
– .... $Ba_3Y_2UO_9$ solid solutions
$Ba_3Y_2UO_9$-$BaZrO_3$ .... U: SVol.C3-137
$Ba_3Fe_{2-x}Y_xUO_9$ .... U: SVol.C3-137, 351
$Ba_3Y_2UO_9$-$Ba_3Sc_2UO_9$ .... U: SVol.C3-137
$Ba_3Y_2UO_9$-$Sr_3Sc_2UO_9$ .... U: SVol.C3-137
$Ba_3Y_2UO_9$-$Sr_3Y_2UO_9$ .... U: SVol.C3-137

$Ba_3O_9U_2Zn$ .... $Ba_3ZnU_2O_9$ .... U: SVol.C3-130/1, 191
$Ba_3S_7Sn_2$ .... $Ba_3Sn_2S_7$ .... Sn: MVol.C3-202/3
$Ba_4H_{12}Ir_2O_{29}P_6$ .... $Ba_4H_2[Ir_2O(OH)_4(HPO_4)_6]$ · 3 H_2O .... Ir: SVol.2-38
$Ba_4Mn_2O_9$ .... 4 BaO · Mn_2O_5 .... Mn: MVol.C2-256
Ba_4O_6Th .... Ba_4ThO_6 .... Th: SVol.C2-14

$Ba_5ClMn_3O_{12}$ .... $Ba_5(MnO_4)_3Cl$ .... Mn: MVol.C2-261
$Ba_5FMn_3O_{12}$ .... $Ba_5(MnO_4)_3F$ .... Mn: MVol.C2-261
$Ba_5HI_3MnO_{18}$ .... $Ba_5HMn(IO_6)_3$ · 10 H_2O .... Mn: MVol.C5-341
$Ba_5HMn_3O_{13}$ .... $Ba_5(MnO_4)_3OH$ .... Mn: MVol.C2-260/1
– .... $Ba_5(MnO_4)_3OH$ solid solutions
$Ba_5(MnO_4,AsO_4)_3OH$ .... Mn: MVol.C2-9, 261
$Ba_5(MnO_4,PO_4)_3OH$ .... Mn: MVol.C2-9, 261
$Ba_5(MnO_4)_3OH$-$Ba_5(VO_4)_3OH$ .... Mn: MVol.C3-167

$Ba_5HO_{13}P_3$	$Ba_5(PO_4)_3OH$ solid solutions	
	$Ba_5(PO_4,MnO_4)_3OH$	Mn: MVol.C2-9, 261
$Ba_5HO_{13}V_3$	$Ba_5(VO_4)_3OH$ solid solutions	
	$Ba_5(VO_4)_3OH-Ba_5(MnO_4)_3OH$	Mn: MVol.C3-167
$Ba_5Li_2O_{12.96}U_3$	$Ba_5U_3Li_2O_{12.96}$	U: SVol.C3-10
$Ba_5Li_2O_{14.05}U_3$	$Ba_5U_3Li_2O_{14.05}$	U: SVol.C3-10
$Ba_5Mn_2O_{74}W_{22}$	$Ba_5Mn_2W_{22}O_{74} \cdot 20\ H_2O$	Mn: MVol.C3-223/4
$Ba_5Mo_3O_{14}$	$5\ BaO \cdot 3\ MoO_3$	Mo: SVol.B2-3
$Ba_{10}Cl_2Mn_6O_{24}$	$Ba_{10}(MnO_4)_6Cl_2$	Mn: MVol.C2-261
$Ba_{10}F_2Mn_6O_{24}$	$Ba_{10}(MnO_4)_6F_2$	Mn: MVol.C2-261
Be	Beryllium systems	
	Be-Am	Np: TrU.B3-270
	Be-Np	Np: TrU.B2-12
	Be-Pu	Np: TrU.B2-19, 23
		Np: TrU.B3-111/9
$BeBr_6Sn$	$Be[SnBr_6] \cdot 10\ H_2O$	Sn: MVol.C3-158
$BeC_8Cl_6H_{12}N_4Sn$	$[Be(CH_3CN)_4][SnCl_6]$	Sn: MVol.C3-157/8
$BeC_{11}Cl_6H_{16.5}N_{5.5}Sn$	$[Be(CH_3CN)_{5.5}][SnCl_6]$	Sn: MVol.C3-157/8
$BeC_{11}F_5H_5$	$C_6F_5BeC_5H_5$	F: PerFHalOrg.4-60
$BeC_{12}F_{10}$	$(C_6F_5)_2Be$	F: PerFHalOrg.4-60
$BeC_{16}Cl_{10}H_{24}N_8Sn_2$	$[Be(CH_3CN)_6][SnCl_5(CH_3CN)]_2$	Sn: MVol.C6-67/8
$BeC_{16}F_5H_{25}N_2$	$C_6F_5BeC(CH_3)_3 \cdot [(CH_3)_2N]_2C_2H_4$	F: PerFHalOrg.4-60
$BeC_{18}F_{10}H_{16}N_2$	$(C_6F_5)_2Be \cdot (CH_3)_2NC_2H_4N(CH_3)_2$	F: PerFHalOrg.4-60
$BeC_{20}F_{10}H_{16}O_2$	$(C_6F_5)_2Be \cdot 2\ (CH_2)_4O$	F: PerFHalOrg.4-60
$BeC_{22}F_{10}H_{10}N_2$	$(C_6F_5)_2Be \cdot 2\ C_5H_5N$	F: PerFHalOrg.4-60
$BeCl_2$	$BeCl_2$ systems	
	$BeCl_2-UCl_4$	U: SVol.C9-24, 52
$BeCl_6Sn$	$Be[SnCl_6] \cdot 8\ H_2O$	Sn: MVol.C3-157
$BeCl_6U$	$BeUCl_6$	U: SVol.C9-53, 55
BeF_3HMnO	$MnBe(OH)F_3 \cdot H_2O$	Mn: MVol.C4-268
–	$MnBe(OH)F_3 \cdot 4\ H_2O$	Mn: MVol.C4-268
BeF_3H_4N	NH_4BeF_3 systems	
	$NH_4BeF_3-NH_4MnF_3-H_2O$	Mn: MVol.C4-226
$BeF_4H_8MnN_2O_4S$	$(NH_4)_2BeF_4 \cdot MnSO_4 \cdot 6\ H_2O$	Mn: MVol.C6-109, 166
$BeF_4H_8N_2$	$(NH_4)_2BeF_4$ systems	
	$(NH_4)_2BeF_4-NH_4MnF_3-H_2O$	Mn: MVol.C4-226/7
BeF_4Mn	$MnBeF_4$	Mn: MVol.C4-225
BeF_4Ra	$RaBeF_4$	Ra: SVol.2-373/4
BeF_6Sn	$Be[SnF_6]$	Sn: MVol.C3-157
$BeMn_2O_8$	$Be(MnO_4)_2 \cdot 5\ H_2O$	Mn: MVol.C2-266/7
$BeMoO_4$	$BeMoO_4$	Mo: SVol.B2-8/9, 13, 14
BeO	BeO ceramics	
	$BeO-Al_2O_3-MgO-ThO_2$	Th: SVol.C2-5
	$BeO-Al_2O_3-UO_2$	U: SVol.C3-145
	$BeO-MgO-ThO_2$	Th: SVol.C2-5
	$BeO-ThO_2$	Th: SVol.C2-3/4

BeO	BeO glasses	
	$BeO-Al_2O_3-ThO_2$	Th: SVol.C2-5
	$BeO-B_2O_3-K_2O$	B: B-Verb.7-93
	$BeO-B_2O_3-Li_2O$	B: B-Verb.7-93
	$BeO-B_2O_3-Na_2O$	B: B-Verb.7-93
	$BeO-SiO_2-UO_2$	U: SVol.C3-160
–	BeO systems	
	$BeO-Al_2O_3-ThO_2$	Th: SVol.C2-5
	$BeO-Ba_3UO_6$	U: SVol.C3-126
	$BeO-Ca_3UO_6$	U: SVol.C3-99
	$BeO-MgO-ThO_2$	Th: SVol.C2-5
	$BeO-Mn_3O_4$	Mn: MVol.C2-203
	$BeO-MoO_3$	Mo: SVol.B2-17
	$BeO-Sr_3UO_6$	U: SVol.C3-111
	$BeO-ThO_2$	Th: SVol.C2-3/4
	$BeO-UO_2$	U: SVol.C3-70/5
	$BeO-UO_{2+x}$	U: SVol.C3-75
BeO_3Sn	$BeSnO_3$	Sn: MVol.C3-157
BeO_3W	$WO_2 \cdot BeO$	W: SVol.B2-42
BeO_4S	$BeSO_4$ systems	
	$BeSO_4-MnSO_4-H_2O$	Mn: MVol.C6-233
	$BeSO_4-MnSO_4-ZnSO_4-H_2O$	Mn: MVol.C6-242/3
BeO_4U	$BeUO_4$	U: SVol.C3-75
BeO_6U_2	$Be(UO_3)_2$	U: SVol.C3-75
BeS_4Sn_2	$BeSn_2S_4$	Sn: MVol.C3-158
$Be_2Br_6O_2U$	$Be_2UO_2Br_6 \cdot 8\ H_2O$	U: SVol.C9-151, 154
Be_2Cl_8U	Be_2UCl_8	U: SVol.C9-53, 55
$Be_2F_8H_8MnN_2$	$(NH_4)_2Mn(BeF_4)_2 \cdot 2\ H_2O$	Mn: MVol.C4-227
–	$(NH_4)_2Mn(BeF_4)_2 \cdot 6\ H_2O$	Mn: MVol.C4-227
Be_2Ir	$IrBe_2$	Ir: SVol.1-48
Be_2MoO_5	$2\ BeO \cdot MoO_3$	Mo: SVol.B2-8
Be_2O_4Sn	Be_2SnO_4	Sn: MVol.C3-157
Be_2O_4W	$WO_2 \cdot (BeO)_2$	W: SVol.B2-42
$Be_2O_8Si_2Th$	$ThBe_2Si_2O_8$	Th: SVol.C2-4/5, 26
Be_2S_4Sn	Be_2SnS_4	Sn: MVol.C3-158
$Be_3CdCsF_{12}MnRb$	$RbCsCdMn(BeF_4)_3$	Mn: MVol.C4-241/2
$Be_3CdCs_2F_{12}Mn$	$Cs_2CdMn(BeF_4)_3$	Mn: MVol.C4-241/2
$Be_3CdF_{12}KMnRb$	$KRbCdMn(BeF_4)_3$	Mn: MVol.C4-241/2
$Be_3CdF_{12}K_2Mn$	$K_2CdMn(BeF_4)_3$	Mn: MVol.C4-241/2
$Be_3CdF_{12}MnRb_2$	$Rb_2CdMn(BeF_4)_3$	Mn: MVol.C4-241/2
$Be_3Cd_{1.55}F_{12}KMn_{0.45}Rb$	$KRbCd_{1.55}Mn_{0.45}(BeF_4)_3$	Mn: MVol.C4-241/2
$Be_3CsF_{12}Mn_2Rb$	$RbCsMn_2(BeF_4)_3$	
	$= Rb_2BeF_4 \cdot Cs_2BeF_4 \cdot 4\ MnBeF_4$	Mn: MVol.C4-228
$Be_3Cs_2F_{12}Mn_2$	$Cs_2Mn_2(BeF_4)_3$	Mn: MVol.C4-228
Be_3F_9KMn	$KMn(BeF_3)_3$	Mn: MVol.C4-225/6
Be_3F_9MnRb	$RbMn(BeF_3)_3$	Mn: MVol.C4-228
$Be_3F_{12}H_8Mn_2N_2$	$(NH_4)_2Mn_2(BeF_4)_3$	Mn: MVol.C4-226

$Be_3F_{12}KMn_2Rb$	$KRbMn_2(BeF_4)_3$	
	$= K_2BeF_4 \cdot Rb_2BeF_4 \cdot 4\ MnBeF_4$	Mn: MVol.C4-228
$Be_3F_{12}K_2Mn_2$	$K_2Mn_2(BeF_4)_3 = K_2BeF_4 \cdot 2\ MnBeF_4$	Mn: MVol.C4-225
$Be_3F_{12}Mn_2Rb_2$	$Rb_2Mn_2(BeF_4)_3$	Mn: MVol.C4-227
$Be_3F_{12}Mn_2Tl_2$	$Tl_2Mn_2(BeF_4)_3 = Tl_2BeF_4 \cdot 2\ MnBeF_4$	Mn: MVol.C4-249
Be_3MoO_6	$3\ BeO \cdot MoO_3$	Mo: SVol.B2-8
Be_3O_5W	$WO_2 \cdot (BeO)_3$	W: SVol.B2-42
Be_4MoO_7	$4\ BeO \cdot MoO_3$	Mo: SVol.B2-8
Be_4O_6W	$WO_2 \cdot (BeO)_4$	W: SVol.B2-42
Be_5MoO_8	$5\ BeO \cdot MoO_3$	Mo: SVol.B2-8
$Be_6Cs_2F_{24}Mn_4Rb_2$	$Rb_2BeF_4 \cdot Cs_2BeF_4 \cdot 4\ MnBeF_4$	
	$= RbCsMn_2(BeF_4)_3$	Mn: MVol.C4-228
$Be_6F_{24}K_2Mn_4Rb_2$	$K_2BeF_4 \cdot Rb_2BeF_4 \cdot 4\ MnBeF_4$	
	$= KRbMn_2(BeF_4)_3$	Mn: MVol.C4-228
Be_6MoO_9	$6\ BeO \cdot MoO_3$	Mo: SVol.B2-8
$Be_{13}Np$	$NpBe_{13}$	Np: TrU.B2-12
$Be_{13}Pa$	$PaBe_{13}$	Pa: SVol.2-4
$Be_{13}Pu$	$PuBe_{13}$	Np: TrU.B2-40
		Np: TrU.B3-111/9
$Be_{13}Th$	$ThBe_{13}$	Th: SVol.C1-179
$Be_{13}U$	UBe_{13}	U: SVol.C3-70
$Be_{17}Ir_2$	Ir_2Be_{17}	Ir: SVol.1-48
$Bi_{0.015}H_{3.75}Th_{0.985}$	$Th_{0.985}Bi_{0.015}H_{3.75}$	Th: SVol.C1-38
$Bi_{0.5}K_{0.5}O_3Ti$	$K_{0.5}Bi_{0.5}TiO_3$ solid solutions	
	$K_{0.5}Bi_{0.5}TiO_3$-$Sr_{0.3}La_{0.7}MnO_3$	Mn: MVol.C3-148/9
Bi	Bismuth systems	
	Bi-Am	Np: TrU.B3-271
	Bi-Cl-Sn	Sn: MVol.C2-297/8
	Bi-Cm	Np: TrU.B3-273
	Bi-Ir	Ir: SVol.1-45
	Bi-Pu	Np: TrU.B3-257/66
	Bi-Sn-Sn_4As_3	Sn: MVol.C2-300
	Bi-$SnCl_2$	Sn: MVol.C2-298
$BiBrC_2H_6$	$(CH_3)_2BiBr$	Bi: Org.Verb.-100
$BiBrC_4H_{10}$	$(C_2H_5)_2BiBr$	Bi: Org.Verb.-100
$BiBrC_6Cl_2H_4$	$BrC_6H_4BiCl_2$	Bi: Org.Verb.-125
$BiBrC_8H_{18}$	$(i\text{-}C_4H_9)_2BiBr$	Bi: Org.Verb.-100
$BiBrC_{10}H_{22}$	$(i\text{-}C_5H_{11})_2BiBr$	Bi: Org.Verb.-53
$BiBrC_{12}Cl_2H_8$	$(ClC_6H_4)_2BiBr$	Bi: Org.Verb.-110
$BiBrC_{12}H_{10}$	$(C_6H_5)_2BiBr$	Bi: Org.Verb.-109/10
$BiBrC_{14}H_{14}$	$(CH_3C_6H_4)_2BiBr$	Bi: Org.Verb.-109
$BiBrC_{18}H_{16}O$	$(C_6H_5)_3Bi(OH)Br$	Bi: Org.Verb.-43
$BiBrC_{20}H_{14}$	$(C_{10}H_7)_2BiBr$	Bi: Org.Verb.-96
$BiBrC_{22}H_{19}NO_2$	$[(C_6H_5)_3BiN(COCH_2)_2]Br$	Bi: Org.Verb.-146
$BiBrC_{24}H_{20}$	$[(C_6H_5)_4Bi]Br$	Bi: Org.Verb.-8
$BiBrC_{27}H_{21}NO$	$(C_6H_5)_3Bi(OC_9H_6N)Br$	Bi: Org.Verb.-43
$BiBr_2CH_3$	CH_3BiBr_2	Bi: Org.Verb.-119
$BiBr_2C_2H_5$	$C_2H_5BiBr_2$	Bi: Org.Verb.-119

$BiBr_2C_3H_5$	$CH_2C(CH_3)BiBr_2$	Bi: Org.Verb.-120
$BiBr_2C_3H_9$	$(CH_3)_3BiBr_2$	Bi: Org.Verb.-14
$BiBr_2C_4H_9$	$(i\text{-}C_4H_9)BiBr_2$	Bi: Org.Verb.-119
$BiBr_2C_5H_{11}$	$(i\text{-}C_5H_{11})BiBr_2$	Bi: Org.Verb.-120
$BiBr_2C_6ClH_4$	$ClC_6H_4BiBr_2$	Bi: Org.Verb.-126
$BiBr_2C_6H_5$	$C_6H_5BiBr_2$	Bi: Org.Verb.-126
$BiBr_2C_7H_7$	$CH_3C_6H_4BiBr_2$	Bi: Org.Verb.-126
$BiBr_2C_9H_{15}$	$(CH_3CHCH)_3BiBr_2$	Bi: Org.Verb.-14
$BiBr_2C_{10}H_7$	$C_{10}H_7BiBr_2$	Bi: Org.Verb.-126
$BiBr_2C_{12}ClH_8$	$(BrC_6H_4)_2BiCl$	Bi: Org.Verb.-108
$BiBr_2C_{16}H_{13}N_2$	$C_6H_5BiBr_2 \cdot NC_5H_4C_5H_4N$	Bi: Org.Verb.-142
$BiBr_2C_{16}H_{15}N_2$	$C_6H_5BiBr_2 \cdot 2\ C_5H_5N$	Bi: Org.Verb.-142
$BiBr_2C_{16}H_{33}$	$C_{16}H_{33}BiBr_2$	Bi: Org.Verb.-120
$BiBr_2C_{18}Cl_3H_{12}$	$(ClC_6H_4)_3BiBr_2$	Bi: Org.Verb.-30
$BiBr_2C_{18}H_{13}N_2$	$C_6H_5BiBr_2 \cdot C_{12}H_8N_2$	Bi: Org.Verb.-142/3
$BiBr_2C_{18}H_{15}$	$(C_6H_5)_3BiBr_2$	Bi: Org.Verb.-28/9
$BiBr_2C_{20}ClH_{18}$	$(CH_3C_6H_4)_2(ClC_6H_4)BiBr_2$	Bi: Org.Verb.-31
$BiBr_2C_{21}H_{21}$	$(CH_3C_6H_4)_3BiBr_2$	Bi: Org.Verb.-30
–	$(CH_3C_6H_4)_3{}^{210}BiBr_2$	Bi: Org.Verb.-30
$BiBr_2C_{21}H_{21}O_3$	$(CH_3OC_6H_4)_3BiBr_2$	Bi: Org.Verb.-31
$BiBr_2C_{22}Cl_2H_{15}$	$(ClC_6H_4)_2(C_{10}H_7)BiBr_2$	Bi: Org.Verb.-32
$BiBr_2C_{22}H_{17}$	$(C_6H_5)_2(C_{10}H_7)BiBr_2$	Bi: Org.Verb.-30
$BiBr_2C_{22}H_{19}O_4$	$(C_6H_5)_3Bi(OOCCH_2Br)_2$	Bi: Org.Verb.-38/9
$BiBr_2C_{24}H_{21}$	$(CH_3C_6H_4)_2(C_{10}H_7)BiBr_2$	Bi: Org.Verb.-32
$BiBr_2C_{24}H_{27}$	$[(CH_3)_2C_6H_3]_3BiBr_2$	Bi: Org.Verb.-31
–	$[(CH_3)_2C_6H_3]_3{}^{210}BiBr_2$	Bi: Org.Verb.-31
$BiBr_2C_{24}H_{27}O_3$	$(C_2H_5OC_6H_4)_3BiBr_2$	Bi: Org.Verb.-31
$BiBr_2C_{27}H_{33}$	$[(CH_3)_3C_6H_2]_3BiBr_2$	Bi: Org.Verb.-32
–	$[(i\text{-}C_3H_7)C_6H_4]_3BiBr_2$	Bi: Org.Verb.-30
$BiBr_2C_{30}H_{21}$	$(C_{10}H_7)_3BiBr_2$	Bi: Org.Verb.-32/3
$BiBr_2C_{30}H_{39}$	$[(CH_3)(i\text{-}C_3H_7)C_6H_3]_3BiBr_2$	Bi: Org.Verb.-32
$BiBr_2C_{36}H_{27}$	$(C_6H_5C_6H_4)_3BiBr_2$	Bi: Org.Verb.-30
$BiBr_3$	$BiBr_3$ systems	
	$BiBr_3$-$SnBr_4$	Sn: MVol.C2-298
$BiBr_3C_{18}Cl_2H_{12}$	$(BrC_6H_4)_3BiCl_2$	Bi: Org.Verb.-24
$BiBr_3C_{18}H_{12}$	$Bi(C_6H_4Br)_3$	Bi: Org.Verb.-88/9
$BiBr_3C_{21}H_{18}$	$Bi(CH_2C_6H_4Br)_3$	Bi: Org.Verb.-56
$BiBr_3C_{24}H_{20}$	$[(C_6H_5)_4Bi]Br_3$	Bi: Org.Verb.-8
$BiBr_3C_{30}Fe_3H_{24}$	$Bi(C_5H_4FeC_5H_4Br)_3$	Bi: Org.Verb.-61
		Fe: Org.Verb.A6-266/7
$BiBr_5C_{18}H_{12}$	$(BrC_6H_4)_3BiBr_2$	Bi: Org.Verb.-31
$BiCClH_2O$	CH_2ClBiO	Bi: Org.Verb.-131
$BiCCl_2H_3$	CH_3BiCl_2	Bi: Org.Verb.-118
$BiCCl_3H_2$	$ClCH_2BiCl_2$	Bi: Org.Verb.-119
$BiCF_3I_2$	CF_3BiI_2	Bi: Org.Verb.-58
$BiCH_3I_2$	CH_3BiI_2	Bi: Org.Verb.-120
$BiCH_3O$	CH_3BiO	Bi: Org.Verb.-131
$BiCH_3S$	CH_3BiS	Bi: Org.Verb.-131
$BiCH_5$	CH_3BiH_2	Bi: Org.Verb.-118
BiC_2ClH_6	$(CH_3)_2BiCl$	Bi: Org.Verb.-100

$BiC_2Cl_2H_3$	$CH_2CHBiCl_2$	Bi:	Org.Verb.-119
$BiC_2Cl_2H_5$	$C_2H_5BiCl_2$	Bi:	Org.Verb.-119
BiC_2F_6I	$(CF_3)_2BiI$	Bi:	Org.Verb.-58
$BiC_2H_5I_2$	$C_2H_5BiI_2$	Bi:	Org.Verb.-120
$BiC_2H_5N_2O_6$	$C_2H_5Bi(NO_3)_2$	Bi:	Org.Verb.-122
BiC_2H_5O	C_2H_5BiO	Bi:	Org.Verb.-131
$BiC_2H_5O_2$	C_2H_5OBiO	Bi:	Org.Verb.-102
BiC_2H_6	$Bi(CH_3)_2$	Bi:	Org.Verb.-118
$BiC_2H_6N_3$	$(CH_3)_2BiN_3$	Bi:	Org.Verb.-101/2
BiC_2H_7	$(CH_3)_2BiH$	Bi:	Org.Verb.-99/100
BiC_2H_7O	$(CH_3)_2BiOH$	Bi:	Org.Verb.-102
$BiC_3Cl_2H_9$	$(CH_3)_3BiCl_2$	Bi:	Org.Verb.-14
$BiC_3F_3H_6$	$(CH_3)_2BiCF_3$	Bi:	Org.Verb.-57/8
		F:	PerFHalOrg.3-233
$BiC_3F_6H_3$	$(CF_3)_2BiCH_3$	Bi:	Org.Verb.-58
		F:	PerFHalOrg.3-233
BiC_3F_9	$Bi(CF_3)_3$	Bi:	Org.Verb.-49
		F:	PerFHalOrg.3-233
BiC_3H_6NS	$(CH_3)_2BiSCN$	Bi:	Org.Verb.-102
$BiC_3H_6N_3S_6$	$Bi[S_2CNH_2]_3$	C:	MVol.D6-140
$BiC_3H_7O_2$	$CH_3Bi[O_2(CH_2)_2]$	Bi:	Org.Verb.-123
$BiC_3H_7S_2$	$CH_3Bi[S_2(CH_2)_2]$	Bi:	Org.Verb.-124
BiC_3H_9	$Bi(CH_3)_3$	Bi:	Org.Verb.-46/9
–	$^{210}Bi(CH_3)_3$	Bi:	Org.Verb.-48
$BiC_3H_{12}N_6S_3^{3+}$	$[Bi((NH_2)_2CS)_3]^{3+}$	C:	MVol.D6-106
$BiC_4F_5H_6$	$(CH_3)_2BiC_2F_5$	Bi:	Org.Verb.-58
		F:	PerFHalOrg.3-233
$BiC_4F_6H_5$	$(CF_3)_2BiC_2H_5$	Bi:	Org.Verb.-58
		F:	PerFHalOrg.3-233
$BiC_4H_9O_2$	$CH_3Bi[O_2CH_2CH(CH_3)]$	Bi:	Org.Verb.-123
$BiC_4H_{11}O$	$(C_2H_5)_2BiOH$	Bi:	Org.Verb.-102
$BiC_4H_{12}^+$	$[(CH_3)_4Bi]^+$	Bi:	Org.Verb.-13
BiC_5ClH_6	C_5H_6BiCl	Bi:	Org.Verb.-117
–	$C_5H_4D_2BiCl$	Bi:	Org.Verb.-117
$BiC_5Cl_2H_{11}$	$C_5H_{11}BiCl_2$	Bi:	Org.Verb.-119
$BiC_5F_3H_{10}$	$(C_2H_5)_2BiCF_3$	Bi:	Org.Verb.-59
		F:	PerFHalOrg.3-233
$BiC_5F_3H_{13}N$	$(CH_3)_2BiCF_3 \cdot (CH_3)_2NH$	Bi:	Org.Verb.-140
$BiC_5F_6H_{10}N$	$CH_3Bi(CF_3)_2 \cdot (CH_3)_2NH$	Bi:	Org.Verb.-140
$BiC_5F_7H_6$	$(CH_3)_2BiC_3F_7$	Bi:	Org.Verb.-58
		F:	PerFHalOrg.3-233
$BiC_5F_{10}H_3$	$(C_2F_5)_2BiCH_3$	Bi:	Org.Verb.-59
		F:	PerFHalOrg.3-233
BiC_5H_5	C_5H_5Bi	Bi:	Org.Verb.-116/7
$BiC_5H_{13}O_2$	$CH_3Bi(OC_2H_5)_2$	Bi:	Org.Verb.-120/1
$BiC_5H_{13}O_2S_2$	$CH_3Bi(SCH_2CH_2OH)_2$	Bi:	Org.Verb.-121
$BiC_5H_{13}S_2$	$CH_3Bi(SC_2H_5)_2$	Bi:	Org.Verb.-121
$BiC_5H_{15}OSi$	$(CH_3)_2BiOSi(CH_3)_3$	Bi:	Org.Verb.-103
$BiC_6Cl_2H_5$	$C_6H_5BiCl_2$	Bi:	Org.Verb.-125

BiC_6F_9	$Bi(CFCF_2)_3$	Bi: Org.Verb.-56
		F: PerFHalOrg.3-233
$BiC_6F_9O_6$	$(CF_3COO)_3Bi$	F: PerFHalOrg.3-233
$BiC_6F_{18}N_3O_3$	$[(CF_3)_2NO]_3Bi$	F: PerFHalOrg.7-101/2
BiC_6H_5	$[C_6H_5Bi]_n$	Bi: Org.Verb.-132
$BiC_6H_5I_2$	$C_6H_5BiI_2$	Bi: Org.Verb.-127
–	$C_6H_5{}^{210}BiI_2$	Bi: Org.Verb.-127
BiC_6H_5O	C_6H_5BiO	Bi: Org.Verb.-131
$BiC_6H_5O_2S$	$C_6H_5BiSO_2$	Bi: Org.Verb.-132
BiC_6H_7	$C_6H_5BiH_2$	Bi: Org.Verb.-132
$BiC_6H_8IN_2$	$(NCCH_2CH_2)_2BiI$	Bi: Org.Verb.-100
BiC_6H_9	$Bi(CHCH_2)_3$	Bi: Org.Verb.-56
$BiC_6H_9O_3S_6$	$Bi[S_2COCH_3]_3$	C: MVol.D4-254
$BiC_6H_{11}N_2O_2$	$(CH_3)_2BiC(N_2)COOC_2H_5$	Bi: Org.Verb.-58
BiC_6H_{15}	$Bi(C_2H_5)_3$	Bi: Org.Verb.-49/51
$BiC_6H_{15}O$	$(C_2H_5)_2BiOC_2H_5$	Bi: Org.Verb.-102
$BiC_6H_{15}O_3$	$Bi(OC_2H_5)_3$	Bi: Org.Verb.-127
$BiC_6H_{15}S$	$(C_2H_5)_3BiS$	Bi: Org.Verb.-51
$BiC_6H_{18}NSi$	$(CH_3)_2BiN(CH_3)Si(CH_3)_3$	Bi: Org.Verb.-103
$BiC_7Cl_2H_7$	$CH_3C_6H_4BiCl_2$	Bi: Org.Verb.-125
$BiC_7Cl_2H_7O$	$CH_3OC_6H_4BiCl_2$	Bi: Org.Verb.-125
$BiC_7F_7H_{13}N$	$(CH_3)_2BiC_3F_7 \cdot (CH_3)_2NH$	Bi: Org.Verb.-140
$BiC_7F_{14}H_3$	$(C_3F_7)_2BiCH_3$	Bi: Org.Verb.-59
		F: PerFHalOrg.3-233
$BiC_7H_7O_2$	$CH_3Bi(O_2C_6H_4)$	Bi: Org.Verb.-124
BiC_7H_{11}	$(CH_3)_2BiC_5H_5$	Bi: Org.Verb.-98
BiC_7H_{15}	$(CH_2)_5BiC_2H_5$	Bi: Org.Verb.-59
$BiC_7H_{15}O_2$	$CH_3Bi[O_2(C(CH_3)_2)_2]$	Bi: Org.Verb.-123
$BiC_7H_{21}O_2Si_2$	$CH_3Bi[OSi(CH_3)_3]_2$	Bi: Org.Verb.-121
$BiC_8Cl_2H_7O_2$	$CH_3OOCC_6H_4BiCl_2$	Bi: Org.Verb.-125
$BiC_8Cl_2H_{10}N$	$(CH_3)_2NC_6H_4BiCl_2$	Bi: Org.Verb.-125
$BiC_8CrH_9O_5$	$(CH_3)_3BiCr(CO)_5$	Bi: Org.Verb.-147
$BiC_8H_5N_2S_2$	$C_6H_5Bi(SCN)_2$	Bi: Org.Verb.-73, 112
$BiC_8H_9S_2$	$CH_3Bi[S_2C_6H_3(CH_3)]$	Bi: Org.Verb.-124
–	$C_6H_5Bi(SCH_2)_2$	Bi: Org.Verb.-131
$BiC_9F_6H_5$	$BiC_7H_5(CF_3)_2$	Bi: Org.Verb.-59
$BiC_9F_{14}H_{10}N$	$CH_3Bi(C_3F_7)_2 \cdot (CH_3)_2NH$	Bi: Org.Verb.-140
$BiC_9H_{12}N_3$	$Bi(CH_2CH_2CN)_3$	Bi: Org.Verb.-55
BiC_9H_{15}	$Bi[C(CH_3)CH_2]_3$	Bi: Org.Verb.-57
–	$Bi(CHCHCH_3)_3$	Bi: Org.Verb.-57
–	$Bi(CH_2CHCH_2)_3$	Bi: Org.Verb.-55
$BiC_9H_{15}NiO_3$	$(CO)_3NiBi(C_2H_5)_3$	Bi: Org.Verb.-51
		Ni: Org.Verb.1-188
$BiC_9H_{15}O_3S_6$	$Bi[S_2COC_2H_5]_3$	C: MVol.D4-254
BiC_9H_{21}	$Bi(C_3H_7)_3$	Bi: Org.Verb.-52
–	$(C_2H_5)_2BiC_5H_{11}$	Bi: Org.Verb.-59
$BiC_9H_{27}N_2Si_2$	$CH_3Bi[N(CH_3)Si(CH_3)_3]_2$	Bi: Org.Verb.-122
$BiC_{10}ClH_{10}$	$(C_5H_5)_2BiCl$	Bi: Org.Verb.-104
$BiC_{10}Cl_4FeH_{10}$	$[Fe(C_5H_5)_2]BiCl_4$	Fe: Org.Verb.A1-224

$BiC_{10}FeH_{15}O_4$	$(CO)_4FeBi(C_2H_5)_3$	Fe:	Org.Verb.B2-81
$BiC_{10}GeH_{25}$	$(C_2H_5)_2BiGe(C_2H_5)_3$	Bi:	Org.Verb.-103
$BiC_{10}H_{10}NO_4$	$C_6H_5Bi[(OOCCH_2)_2NH]$	Bi:	Org.Verb.-128
$BiC_{10}H_{11}O_6$	$C_6H_5Bi(OOCCH_2OH)_2$	Bi:	Org.Verb.-128
$BiC_{10}H_{15}O_2$	$C_6H_5Bi(OC_2H_5)_2$	Bi:	Org.Verb.-127
$BiC_{10}H_{15}O_2S_2$	$C_6H_5Bi(SCH_2CH_2OH)_2$	Bi:	Org.Verb.-129
$BiC_{10}H_{15}S_2$	$C_6H_5Bi(SC_2H_5)_2$	Bi:	Org.Verb.-129
$BiC_{11}CrH_{15}O_5$	$(C_2H_5)_3BiCr(CO)_5$	Bi:	Org.Verb.-147
$BiC_{11}H_{15}MoO_5$	$(C_2H_5)_3BiMo(CO)_5$	Bi:	Org.Verb.-148
$BiC_{12}ClH_{10}$	$(C_6H_5)_2BiCl$	Bi:	Org.Verb.-104/6
–	$(C_6H_5)_2{}^{210}BiCl$	Bi:	Org.Verb.-104
$BiC_{12}Cl_2H_8I$	$(ClC_6H_4)_2BiI$	Bi:	Org.Verb.-111
$BiC_{12}Cl_2H_9O$	$(ClC_6H_4)_2BiOH$	Bi:	Org.Verb.-113
$BiC_{12}Cl_3H_8$	$(ClC_6H_4)_2BiCl$	Bi:	Org.Verb.-107
$BiC_{12}Ge_3H_{27}NiO_3$	$(CO)_3NiBi(Ge(CH_3)_3)_3$	Ni:	Org.Verb.1-188, 190
$BiC_{12}H_8I$	$C_{12}H_8BiI$	Bi:	Org.Verb.-116
$BiC_{12}H_9S_3$	$Bi(C_4H_3S)_3$	Bi:	Org.Verb.-62
$BiC_{12}H_{10}$	$Bi(C_6H_5)_2$	Bi:	Org.Verb.-118
$BiC_{12}H_{10}^-$	$[(C_6H_5)_2Bi]^-$	Bi:	Org.Verb.-68, 133
$BiC_{12}H_{10}I$	$(C_6H_5)_2BiI$	Bi:	Org.Verb.-111
–	$(C_6H_5)_2{}^{210}BiI$	Bi:	Org.Verb.-111
$BiC_{12}H_{10}K$	$(C_6H_5)_2BiK$	Bi:	Org.Verb.-114
$BiC_{12}H_{10}Li$	$(C_6H_5)_2BiLi$	Bi:	Org.Verb.-114
$BiC_{12}H_{10}N_3$	$(C_6H_5)_2BiN_3$	Bi:	Org.Verb.-112
$BiC_{12}H_{10}Na$	$(C_6H_5)_2BiNa$	Bi:	Org.Verb.-114/5
$BiC_{12}H_{11}$	$(C_6H_5)_2BiH$	Bi:	Org.Verb.-104
$BiC_{12}H_{11}O$	$(C_6H_5)_2BiOH$	Bi:	Org.Verb.-113
$BiC_{12}H_{17}N_2S_4$	$C_6H_5Bi[SSCN(CH_3)_2]_2$	Bi:	Org.Verb.-130
$BiC_{12}H_{21}$	$Bi(CHCHC_2H_5)_3$	Bi:	Org.Verb.-57
$BiC_{12}H_{21}O_3S_6$	$Bi[S_2COC_3H_7]_3$	C:	MVol.D4-254
$BiC_{12}H_{27}$	$Bi(n\text{-}C_4H_9)_3$	Bi:	Org.Verb.-52
–	$Bi(i\text{-}C_4H_9)_3$	Bi:	Org.Verb.-52
–	$Bi(t\text{-}C_4H_9)_3$	Bi:	Org.Verb.-52
$BiC_{12}H_{27}NiO_3Sn_3$	$(CO)_3NiBi(Sn(CH_3)_3)_3$	Ni:	Org.Verb.1-188, 190
$BiC_{12}H_{33}Si_3$	$Bi[CH_2Si(CH_3)_3]_3$	Bi:	Org.Verb.-55/6
$BiC_{13}H_9O_2S$	$C_6H_5Bi(OOCC_6H_4S)$	Bi:	Org.Verb.-129
$BiC_{13}H_{10}N$	$(C_6H_5)_2BiCN$	Bi:	Org.Verb.-112
		C:	MVol.D6-261
$BiC_{13}H_{10}NS$	$(C_6H_5)_2BiNCS$	C:	MVol.D6-34
–	$(C_6H_5)_2BiSCN$	Bi:	Org.Verb.-112
$BiC_{13}H_{10}NSe$	$Bi(C_6H_5)_2SeCN$	C:	MVol.D6-239
$BiC_{13}H_{13}S_2$	$CH_3Bi(SC_6H_5)_2$	Bi:	Org.Verb.-122
$BiC_{14}ClH_{14}$	$(CH_3C_6H_4)_2BiCl$	Bi:	Org.Verb.-107
$BiC_{14}ClH_{14}O_2$	$(CH_3OC_6H_4)_2BiCl$	Bi:	Org.Verb.-108
$BiC_{14}Ge_2H_{35}$	$C_2H_5Bi[Ge(C_2H_5)_3]_2$	Bi:	Org.Verb.-123
$BiC_{14}H_{13}O_2$	$(C_6H_5)_2BiOOCCH_3$	Bi:	Org.Verb.-113
$BiC_{14}H_{14}I$	$(CH_3C_6H_4)_2BiI$	Bi:	Org.Verb.-111
$BiC_{14}H_{15}O$	$(CH_3C_6H_4)_2BiOH$	Bi:	Org.Verb.-113
$BiC_{14}H_{15}O_4$	$C_6H_5Bi[OOCC(CH_3)CH_2]_2$	Bi:	Org.Verb.-128
$BiC_{14}H_{15}S_2$	$C_2H_5Bi(SC_6H_5)_2$	Bi:	Org.Verb.-122

$BiC_{15}H_{11}N_2S_4$	$CH_3Bi(SC_7H_4NS)_2$	Bi:	Org.Verb.-122
$BiC_{15}H_{15}$	$Bi(C_5H_5)_3$	Bi:	Org.Verb.-60/1
$BiC_{15}H_{15}O_2$	$CH_3Bi[O_2(CHC_6H_5)_2]$	Bi:	Org.Verb.-124
$BiC_{15}H_{17}S_2$	$CH_3Bi(SCH_2C_6H_5)_2$	Bi:	Org.Verb.-121
$BiC_{15}H_{19}OSi$	$(C_6H_5)_2BiOSi(CH_3)_3$	Bi:	Org.Verb.-113
$BiC_{15}H_{27}NiO_3$	$(t\text{-}C_4H_9)_3BiNi(CO)_3$	Bi:	Org.Verb.-151/2
		Ni:	Org.Verb.1-188, 190
$BiC_{15}H_{27}O_3S_6$	$Bi[S_2COC_4H_9]_3$	C:	MVol.D4-254
$BiC_{15}H_{33}$	$Bi(C_5H_{11})_3$	Bi:	Org.Verb.-53
–	$Bi(i\text{-}C_5H_{11})_3$	Bi:	Org.Verb.-53
$BiC_{16}ClH_{14}O_4$	$(CH_3OOCC_6H_4)_2BiCl$	Bi:	Org.Verb.-107
$BiC_{16}ClH_{20}N_2$	$[(CH_3)_2NC_6H_4]_2BiCl$	Bi:	Org.Verb.-108
$BiC_{16}Cl_2H_{13}N_2$	$C_6H_5BiCl_2 \cdot NC_5H_4C_5H_4N$	Bi:	Org.Verb.-141/2
$BiC_{16}Cl_2H_{15}N_2$	$C_6H_5BiCl_2 \cdot 2\ C_5H_5N$	Bi:	Org.Verb.-141
$BiC_{16}Cl_2H_{22}N$	$[(CH_3)_4N][(C_6H_5)_2BiCl_2]$	Bi:	Org.Verb.-140
$BiC_{16}H_{13}I_2N_2$	$C_6H_5BiI_2 \cdot NC_5H_4C_5H_4N$	Bi:	Org.Verb.-143
$BiC_{16}H_{13}N_2O_2S_2$	$C_6H_5Bi(SC_5H_4NO)_2$	Bi:	Org.Verb.-144/5
$BiC_{16}H_{15}$	$(CH_3C_6H_4)_2BiCCH$	Bi:	Org.Verb.-98
$BiC_{16}H_{15}O_2$	$(C_6H_5)_2BiOOCC(CH_3)CH_2$	Bi:	Org.Verb.-113
$BiC_{16}H_{25}N_2S_4$	$C_6H_5Bi[SSCN(C_2H_5)_2]_2$	Bi:	Org.Verb.-130
$BiC_{17}CrH_{27}O_5$	$(t\text{-}C_4H_9)_3BiCr(CO)_5$	Bi:	Org.Verb.-148
$BiC_{17}H_{10}O_5Re$	$(C_6H_5)_2BiRe(CO)_5$	Bi:	Org.Verb.-115
$BiC_{17}H_{27}MoO_5$	$(t\text{-}C_4H_9)_3BiMo(CO)_5$	Bi:	Org.Verb.-149
$BiC_{17}H_{27}O_5W$	$(t\text{-}C_4H_9)_3BiW(CO)_5$	Bi:	Org.Verb.-149/50
$BiC_{18}ClH_{14}$	$(C_6H_5)_2BiC_6H_4Cl$	Bi:	Org.Verb.-81
$BiC_{18}ClH_{14}S_2$	$ClC_6H_4Bi(SC_6H_5)_2$	Bi:	Org.Verb.-130
$BiC_{18}ClH_{16}O$	$(C_6H_5)_3Bi(OH)Cl$	Bi:	Org.Verb.-43
$BiC_{18}ClH_{18}O_4$	$(C_2H_5OOCC_6H_4)_2BiCl$	Bi:	Org.Verb.-107
$BiC_{18}Cl_2FH_{12}$	$(ClC_6H_4)_2BiC_6H_4F$	Bi:	Org.Verb.-93
$BiC_{18}Cl_2H_9N_6O_{12}$	$(NO_2)_6C_{18}H_9BiCl_2$	Bi:	Org.Verb.-26
$BiC_{18}Cl_2H_{11}N_4O_8$	$(NO_2)_4C_{18}H_{11}BiCl_2$	Bi:	Org.Verb.-26
$BiC_{18}Cl_2H_{12}N_3O_6$	$(NO_2C_6H_4)_3BiCl_2$	Bi:	Org.Verb.-24, 25
$BiC_{18}Cl_2H_{13}N_2$	$C_6H_5BiCl_2 \cdot C_{12}H_8N_2$	Bi:	Org.Verb.-142
$BiC_{18}Cl_2H_{13}N_2O_4$	$(NO_2)_2C_{18}H_{13}BiCl_2$	Bi:	Org.Verb.-26
$BiC_{18}Cl_2H_{14}NO_2$	$(NO_2C_6H_4)(C_6H_5)_2BiCl_2$	Bi:	Org.Verb.-21
$BiC_{18}Cl_2H_{15}$	$(C_6H_5)_3BiCl_2$	Bi:	Org.Verb.-15/21
$BiC_{18}Cl_2H_{15}O_8$	$(C_6H_5)_3Bi(ClO_4)_2$	Bi:	Org.Verb.-138
$BiC_{18}Cl_3H_{12}$	$Bi(C_6H_4Cl)_3$	Bi:	Org.Verb.-86/8
$BiC_{18}Cl_3H_{12}N_2O_6$	$(ClC_6H_4)_3Bi(NO_3)_2$	Bi:	Org.Verb.-36
$BiC_{18}Cl_5H_{12}$	$(ClC_6H_4)_3BiCl_2$	Bi:	Org.Verb.-24
$BiC_{18}Cl_5H_{15}Nb$	$(C_6H_5)_3Bi \cdot NbCl_5$	Bi:	Org.Verb.-146
$BiC_{18}Cl_{10}H_{15}Nb_2$	$(C_6H_5)_3Bi \cdot 2\ NbCl_5$	Bi:	Org.Verb.-146
$BiC_{18}Cl_{10}H_{15}U_2$	$[(C_6H_5)_3BiU_2Cl_9]Cl$		
	$= 2\ UCl_5 \cdot (C_6H_5)_3Bi$	Bi:	Org.Verb.-152
		U:	SVol.E1-52
$BiC_{18}CrH_{15}O_4$	$(C_6H_5)_3BiCrO_4$	Bi:	Org.Verb.-45
$BiC_{18}FH_{14}$	$(C_6H_5)_2BiC_6H_4F$	Bi:	Org.Verb.-81
$BiC_{18}FH_{14}S$	$(C_6H_5)_2BiSC_6H_4F$	Bi:	Org.Verb.-114
$BiC_{18}F_2H_{15}$	$(C_6H_5)_3BiF_2$	Bi:	Org.Verb.-14/5
$BiC_{18}F_3H_{12}$	$Bi(C_6H_4F)_3$	Bi:	Org.Verb.-86

$BiC_{18}F_{12}H_{15}P_2$	$(C_6H_5)_3Bi(PF_6)_2$	Bi:	Org.Verb.-138
$BiC_{18}F_{15}$	$Bi(C_6F_5)_3$	Bi:	Org.Verb.-94/5
		F:	PerFHalOrg.3-233
$BiC_{18}Ge_3H_{45}$	$[(C_2H_5)_3Ge]_3Bi$	Bi:	Org.Verb.-51, 103
$BiC_{18}H_9N_8O_{18}$	$(NO_2)_6C_{18}H_9Bi(NO_3)_2$	Bi:	Org.Verb.-37
$BiC_{18}H_{11}N_6O_{14}$	$(NO_2)_4C_{18}H_{11}Bi(NO_3)_2$	Bi:	Org.Verb.-37
$BiC_{18}H_{12}I_3$	$Bi(C_6H_4I)_3$	Bi:	Org.Verb.-89
$BiC_{18}H_{12}N_3O_6$	$Bi(C_6H_4NO_2)_3$	Bi:	Org.Verb.-91/2
$BiC_{18}H_{12}N_5O_{12}$	$(NO_2C_6H_4)_3Bi(NO_3)_2$	Bi:	Org.Verb.-36
$BiC_{18}H_{13}$	$C_6H_5BiC_{12}H_8$	Bi:	Org.Verb.-99
$BiC_{18}H_{13}I_2N_2$	$C_6H_5BiI_2 \cdot C_{12}H_8N_2$	Bi:	Org.Verb.-143
$BiC_{18}H_{13}N_4O_{10}$	$(C_6H_5)(NO_2C_6H_4)_2Bi(NO_3)_2$	Bi:	Org.Verb.-35
–	$(NO_2)_2C_{18}H_{13}Bi(NO_3)_2$	Bi:	Org.Verb.-37
$BiC_{18}H_{14}N_3O_8$	$(C_6H_5)_2(NO_2C_6H_4)Bi(NO_3)_2$	Bi:	Org.Verb.-35
$BiC_{18}H_{15}$	$Bi(C_6H_5)_3$		
	Chemical reactions	Bi:	Org.Verb.-72/8
	Physical properties	Bi:	Org.Verb.-68/72
	Preparation	Bi:	Org.Verb.-62/8
	Uses	Bi:	Org.Verb.-78/80
–	$^{210}Bi(C_6H_5)_3$	Bi:	Org.Verb.-66
$BiC_{18}H_{15}I_2$	$(C_6H_5)_3BiI_2$	Bi:	Org.Verb.-33
$BiC_{18}H_{15}N_2O_6$	$(C_6H_5)_3Bi(NO_3)_2$	Bi:	Org.Verb.-34/5
$BiC_{18}H_{15}N_6$	$(C_6H_5)_3Bi(N_3)_2$	Bi:	Org.Verb.-33
$BiC_{18}H_{15}O$	$(C_6H_5)_3BiO$	Bi:	Org.Verb.-44
$BiC_{18}H_{15}O_2S$	$(C_6H_5)_2BiOOSC_6H_5$	Bi:	Org.Verb.-113
$BiC_{18}H_{15}O_3S$	$(C_6H_5)_3Bi \cdot SO_3$	Bi:	Org.Verb.-146
$BiC_{18}H_{15}O_4S$	$(C_6H_5)_3BiSO_4$	Bi:	Org.Verb.-44/5
$BiC_{18}H_{15}O_4S_2$	$C_6H_5Bi(OOSC_6H_5)_2$	Bi:	Org.Verb.-128
–	$(C_6H_5)_3BiS_2O_4$	Bi:	Org.Verb.-75
$BiC_{18}H_{15}O_4Se$	$(C_6H_5)_3BiSeO_4$	Bi:	Org.Verb.-45
$BiC_{18}H_{15}O_6S_3$	$Bi(O_2SC_6H_5)_3$	Bi:	Org.Verb.-73
$BiC_{18}H_{15}S$	$(C_6H_5)_2BiSC_6H_5$	Bi:	Org.Verb.-114
$BiC_{18}H_{15}S_2$	$C_6H_5Bi(SC_6H_5)_2$	Bi:	Org.Verb.-129/30
$BiC_{18}H_{15}Tl^+$	$[(C_6H_5)_3BiTl]^+$	Bi:	Org.Verb.-74
$BiC_{18}H_{17}O_2$	$(C_6H_5)_3Bi(OH)_2$	Bi:	Org.Verb.-37
$BiC_{18}H_{18}N_3O_6S_3$	$Bi(C_6H_4SO_2NH_2)_3$	Bi:	Org.Verb.-91
$BiC_{18}H_{21}$	$Bi(C_5H_4CH_3)_3$	Bi:	Org.Verb.-61
$BiC_{18}H_{25}Sn$	$(C_6H_5)_2BiSn(C_2H_5)_3$	Bi:	Org.Verb.-115
$BiC_{18}H_{33}$	$Bi(C_6H_{11})_3$	Bi:	Org.Verb.-55
$BiC_{18}H_{45}Si_3$	$[(C_2H_5)_3Si]_3Bi$	Bi:	Org.Verb.-51
$BiC_{18}H_{45}Sn_3$	$[(C_2H_5)_3Sn]_3Bi$	Bi:	Org.Verb.-51
$BiC_{19}Cl_2H_{15}$	$(ClC_6H_4)_2BiC_6H_4CH_3$	Bi:	Org.Verb.-93
$BiC_{19}Cl_2H_{17}$	$(CH_3C_6H_4)(C_6H_5)_2BiCl_2$	Bi:	Org.Verb.-21
$BiC_{19}Cl_4H_{15}$	$(ClC_6H_4)_2(CH_3C_6H_4)BiCl_2$	Bi:	Org.Verb.-25
$BiC_{19}H_{15}O_3$	$(C_6H_5)_3BiCO_3$	Bi:	Org.Verb.-44
$BiC_{19}H_{16}NO$	$(C_6H_5)_3Bi(OH)CN$	Bi:	Org.Verb.-33
$BiC_{19}H_{17}S_2$	$CH_3C_6H_4Bi(SC_6H_5)_2$	Bi:	Org.Verb.-130
$BiC_{20}ClH_{14}$	$(C_{10}H_7)_2BiCl$	Bi:	Org.Verb.-108
$BiC_{20}ClH_{18}$	$(CH_3C_6H_4)_2BiC_6H_4Cl$	Bi:	Org.Verb.-92
$BiC_{20}Cl_2FeH_{15}NO_3P$	$(CO)_2Fe(BiCl_2)(NO)P(C_6H_5)_3$	Fe:	Org.Verb.B1-133/5

$BiC_{20}Cl_2H_{13}$	$(ClC_6H_4)_2BiCCC_6H_5$	Bi: Org.Verb.-99
$BiC_{20}FH_{18}$	$(CH_3C_6H_4)_2BiC_6H_4F$	Bi: Org.Verb.-92
$BiC_{20}FH_{18}O_2$	$(CH_3OC_6H_4)_2BiC_6H_4F$	Bi: Org.Verb.-93
$BiC_{20}Fe_2H_{20}I_5$	$[Fe(C_5H_5)_2]_2BiI_5$	Fe: Org.Verb.A1-224
$BiC_{20}H_{15}$	$(C_6H_5)_2BiCCC_6H_5$	Bi: Org.Verb.-98
$BiC_{20}H_{15}N_2$	$(C_6H_5)_3Bi(CN)_2$	Bi: Org.Verb.-33
$BiC_{20}H_{15}N_2O_2$	$(C_6H_5)_3Bi(OCN)_2$	Bi: Org.Verb.-34
$BiC_{20}H_{15}O_4$	$C_6H_5Bi(OOCC_6H_5)_2$	Bi: Org.Verb.-128
–	$(C_6H_5)_3BiC_2O_4$	Bi: Org.Verb.-40
$BiC_{20}H_{15}O_6$	$C_6H_5Bi(OOCC_6H_4OH)_2$	Bi: Org.Verb.-128
$BiC_{20}H_{17}$	$(C_6H_5)_2BiC_6H_4CHCH_2$	Bi: Org.Verb.-81
–	$[(C_6H_5)_2BiC_6H_4CHCH_2]_n$	Bi: Org.Verb.-81
$BiC_{20}H_{19}O_4S_2$	$C_6H_5Bi(OOSC_6H_4CH_3)_2$	Bi: Org.Verb.-128
$BiC_{20}H_{19}S_2$	$C_6H_5Bi(SCH_2C_6H_5)_2$	Bi: Org.Verb.-129
$BiC_{20}H_{20}N$	$(C_6H_5)_2BiC_6H_4N(CH_3)_2$	Bi: Org.Verb.-82
$BiC_{20}H_{40}NO_4S_8$	$[N(C_2H_5)_4][Bi(S_2COC_2H_5)_4]$	C: MVol.D4-254
$BiC_{21}ClH_{20}O_5$	$[(C_6H_5)_3BiCH_2COCH_3]ClO_4$	Bi: Org.Verb.-13
$BiC_{21}Cl_2H_{12}N_3O_{12}$	$[(NO_2)(HOOC)C_6H_3]_3BiCl_2$	Bi: Org.Verb.-26
$BiC_{21}Cl_2H_{15}O_6$	$(HOOCC_6H_4)_3BiCl_2$	Bi: Org.Verb.-23
$BiC_{21}Cl_2H_{18}N_3O_6$	$[(CH_3)(NO_2)C_6H_3]_3BiCl_2$	Bi: Org.Verb.-26
$BiC_{21}Cl_2H_{21}$	$(CH_3C_6H_4)_3BiCl_2$	Bi: Org.Verb.-21/3
–	$(CH_3C_6H_4)_3{}^{210}BiCl_2$	Bi: Org.Verb.-22
$BiC_{21}Cl_2H_{21}O_3$	$(CH_3OC_6H_4)_3BiCl_2$	Bi: Org.Verb.-23
–	$(CH_3OC_6H_4)_3{}^{210}BiCl_2$	Bi: Org.Verb.-23
$BiC_{21}Cl_3H_{18}$	$Bi(CH_2C_6H_4Cl)_3$	Bi: Org.Verb.-56
$BiC_{21}F_2H_{21}$	$(CH_3C_6H_4)_3{}^{210}BiF_2$	Bi: Org.Verb.-15
$BiC_{21}H_{15}NiO_3$	$(C_6H_5)_3BiNi(CO)_3$	Bi: Org.Verb.-151
$BiC_{21}H_{15}O_6$	$Bi(C_6H_4COOH)_3$	Bi: Org.Verb.-85
$BiC_{21}H_{18}N_3O_6$	$Bi[C_6H_3(NO_2)(CH_3)]_3$	Bi: Org.Verb.-94
$BiC_{21}H_{18}N_5O_{12}$	$[(NO_2)(CH_3)C_6H_3]_3Bi(NO_3)_2$	Bi: Org.Verb.-37
$BiC_{21}H_{21}$	$Bi(CH_2C_6H_5)_3$	Bi: Org.Verb.-56
–	$Bi(C_6H_4CH_3)_3$	Bi: Org.Verb.-82/4
$BiC_{21}H_{21}N_2O_6$	$(CH_3C_6H_4)_3Bi(NO_3)_2$	Bi: Org.Verb.-36
$BiC_{21}H_{21}O_3$	$Bi(C_6H_4OCH_3)_3$	Bi: Org.Verb.-90
$BiC_{21}H_{21}O_6S_3$	$Bi(O_2SC_6H_4CH_3)_3$	Bi: Org.Verb.-73
$BiC_{21}H_{21}O_7S_2$	$(CH_3C_6H_4)_2[(CH_3)(HO_3S)C_6H_3]BiSO_4$	Bi: Org.Verb.-45
$BiC_{21}H_{21}O_{10}S_3$	$(CH_3C_6H_4)[(CH_3)(HO_3S)C_6H_3]_2BiSO_4$	Bi: Org.Verb.-45
$BiC_{22}Cl_2H_{15}$	$(ClC_6H_4)_2Bi(C_{10}H_7)$	Bi: Org.Verb.-06
$BiC_{22}Cl_2H_{17}$	$(C_{10}H_7)(C_6H_5)_2BiCl_2$	Bi: Org.Verb.-21
$BiC_{22}Cl_2H_{19}O_4$	$(C_6H_5)_3Bi(OOCCH_2Cl)_2$	Bi: Org.Verb.-38/9
$BiC_{22}Cl_2H_{27}O_{10}S_2$	$[(C_6H_5)_3Bi[(CH_3)_2SO]_2](ClO_4)_2$	Bi: Org.Verb.-138/9
–	$[(C_6H_5)_3Bi[(CD_3)_2SO]_2](ClO_4)_2$	Bi: Org.Verb.-138/9
$BiC_{22}Cl_4H_{15}$	$(ClC_6H_4)_2(C_{10}H_7)BiCl_2$	Bi: Org.Verb.-27
$BiC_{22}Cl_4H_{17}O_4$	$(C_6H_5)_3Bi(OOCCHCl_2)_2$	Bi: Org.Verb.-38/9
$BiC_{22}Cl_6H_{15}O_4$	$(C_6H_5)_3Bi(OOCCCl_3)_2$	Bi: Org.Verb.-38/9
$BiC_{22}FH_{24}N_2$	$[(CH_3)_2NC_6H_4]_2BiC_6H_4F$	Bi: Org.Verb.-93
$BiC_{22}F_2H_{19}O_4$	$(C_6H_5)_3Bi(OOCCH_2F)_2$	Bi: Org.Verb.-38/9
$BiC_{22}F_4H_{17}O_4$	$(C_6H_5)_3Bi(OOCCHF_2)_2$	Bi: Org.Verb.-38/9
$BiC_{22}F_6H_{15}O_4$	$(C_6H_5)_3Bi(OOCCF_3)_2$	Bi: Org.Verb.-38/9
$BiC_{22}FeH_{15}O_4$	$(CO)_4FeBi(C_6H_5)_3$	Fe: Org.Verb.B2-81

$BiC_{22}H_{17}$	$(C_6H_5)_2Bi(C_{10}H_7)$	Bi:	Org.Verb.-80/1
$BiC_{22}H_{17}S_2$	$C_{10}H_7Bi(SC_6H_5)_2$	Bi:	Org.Verb.-130
$BiC_{22}H_{19}$	$(CH_3C_6H_4)_2BiCCC_6H_5$	Bi:	Org.Verb.-99
$BiC_{22}H_{19}O_4S_2$	$C_6H_5Bi(SC_6H_4COOCH_3)_2$	Bi:	Org.Verb.-130
$BiC_{22}H_{21}O_4$	$(C_6H_5)_3Bi(OOCCH_3)_2$	Bi:	Org.Verb.-38
$BiC_{22}H_{27}O_2S_2^{2+}$	$[(C_6H_5)_3Bi((CH_3)_2SO)_2]^{2+}$	Bi:	Org.Verb.-138/9
$BiC_{23}CrH_{15}O_5$	$(C_6H_5)_3BiCr(CO)_5$	Bi:	Org.Verb.-148
$BiC_{23}CrH_{33}O_5$	$(C_6H_{11})_3BiCr(CO)_5$	Bi:	Org.Verb.-147
$BiC_{23}H_{15}MoO_5$	$(C_6H_5)_3BiMo(CO)_5$	Bi:	Org.Verb.-149
$BiC_{23}H_{15}O_5W$	$(C_6H_5)_3BiW(CO)_5$	Bi:	Org.Verb.-150
$BiC_{23}H_{21}N_2Se_2$	$(CH_3C_6H_4)_3Bi(SeCN)_2$	Bi:	Org.Verb.-23
$BiC_{23}H_{23}O_5$	$(C_6H_5)_3Bi(OOCCH_3)_2 \cdot 0.5\ CH_3COOH$	Bi:	Org.Verb.-38
$BiC_{24}ClH_{18}$	$(C_6H_5C_6H_4)_2BiCl$	Bi:	Org.Verb.-107
$BiC_{24}ClH_{20}$	$[(C_6H_5)_4Bi]Cl$	Bi:	Org.Verb.-7/8
$BiC_{24}ClH_{20}O_4$	$[(C_6H_5)_4Bi]ClO_4$	Bi:	Org.Verb.-11
$BiC_{24}Cl_2H_{18}N_3O_{12}$	$[(NO_2)(CH_3OOC)C_6H_3]_3BiCl_2$	Bi:	Org.Verb.-26
$BiC_{24}Cl_2H_{19}$	$[(C_6H_5)_3BiC_6H_4Cl]Cl$	Bi:	Org.Verb.-6
$BiC_{24}Cl_2H_{21}$	$(CH_3C_6H_4)_2(C_{10}H_7)BiCl_2$	Bi:	Org.Verb.-27
$BiC_{24}Cl_2H_{21}O_6$	$[(CH_3)(HOOC)C_6H_3]_3BiCl_2$	Bi:	Org.Verb.-26
–	$(CH_3OOCC_6H_4)_3BiCl_2$	Bi:	Org.Verb.-24
$BiC_{24}Cl_2H_{27}$	$[(CH_3)_2C_6H_3]_3BiCl_2$	Bi:	Org.Verb.-25/6
–	$[(CH_3)_2C_6H_3]_3{}^{210}BiCl_2$	Bi:	Org.Verb.-26
$BiC_{24}Cl_2H_{30}N_3$	$[(CH_3)_2NC_6H_4]_3BiCl_2$	Bi:	Org.Verb.-24
$BiC_{24}Cl_{12}H_{30}N_3Sb_2$	$[(CH_3)_2NC_6H_4]_3Bi[SbCl_6]_2$	Bi:	Org.Verb.-24
$BiC_{24}F_2H_{27}$	$[(CH_3)_2C_6H_3]_3{}^{210}BiF_2$	Bi:	Org.Verb.-15
$BiC_{24}F_6H_{20}P$	$[(C_6H_5)_4Bi]PF_6$	Bi:	Org.Verb.-10/1
$BiC_{24}H_{12}Mn_3O_9$	$Bi[C_5H_4Mn(CO)_3]_3$	Bi:	Org.Verb.-61
$BiC_{24}H_{19}N_2O_4$	$(C_6H_5)_3Bi(OOCCH_2CN)_2$	Bi:	Org.Verb.-38/9
$BiC_{24}H_{19}O_4$	$C_6H_5Bi(OOCCHCHC_6H_5)_2$	Bi:	Org.Verb.-128
–	$C_6H_5Bi(OOCC_6H_4CHCH_2)_2$	Bi:	Org.Verb.-128
$BiC_{24}H_{20}{}^+$	$[(C_6H_5)_4Bi]^+$	Bi:	Org.Verb.-11
$BiC_{24}H_{20}NO_3$	$[(C_6H_5)_4Bi]NO_3$	Bi:	Org.Verb.-12
$BiC_{24}H_{21}$	$Bi(CHCHC_6H_5)_3$	Bi:	Org.Verb.-57
–	$Bi(C_6H_4CHCH_2)_3$	Bi:	Org.Verb.-84
–	$(CH_3C_6H_4)_2Bi(C_{10}H_7)$	Bi:	Org.Verb.-95
$BiC_{24}H_{21}N_2O_{12}$	$(CH_3OOCC_6H_4)_3Bi(NO_3)_2$	Bi:	Org.Verb.-36
$BiC_{24}H_{21}O$	$[(C_6H_5)_4Bi]OH$	Bi:	Org.Verb.-9
$BiC_{24}H_{21}O_2$	$(CH_3OC_6H_4)_2Bi(C_{10}H_7)$	Bi:	Org.Verb.-95
$BiC_{24}H_{25}O_4$	$(C_6H_5)_3Bi(OOCC_2H_5)_2$	Bi:	Org.Verb.-39
$BiC_{24}H_{25}O_4S_2$	$(C_6H_5)_3Bi(OOCCH_2SCH_3)_2$	Bi:	Org.Verb.-40
$BiC_{24}H_{27}$	$Bi[C_6H_3(CH_3)_2]_3$	Bi:	Org.Verb.-93
$BiC_{24}H_{27}O_3$	$Bi(C_6H_4OC_2H_5)_3$	Bi:	Org.Verb.-90/1
$BiC_{24}H_{30}N_3$	$Bi[C_6H_4N(CH_3)_2]_3$	Bi:	Org.Verb.-92
$BiC_{25}CoH_{29}N_5O_4$	$(C_6H_5)_2BiCo(ON(CCH_3)_2NOH)_2(C_5H_5N)$	Bi:	Org.Verb.-115
$BiC_{25}H_{20}MnO_2$	$(C_6H_5)_3BiMn(CO)_2(C_5H_5)$	Bi:	Org.Verb.-150
$BiC_{25}H_{20}NO$	$[(C_6H_5)_4Bi]NCO$	Bi:	Org.Verb.-8/9
$BiC_{25}H_{20}NS$	$[(C_6H_5)_4Bi]NCS$	Bi:	Org.Verb.-9
$BiC_{25}H_{22}NO_2S$	$(C_6H_5)_3BiNSO_2C_6H_4CH_3$	Bi:	Org.Verb.-34
$BiC_{25}H_{27}O_4$	$(CH_3C_6H_4)_3Bi(OOCCH_3)_2$	Bi:	Org.Verb.-38

$BiC_{26}Cl_3H_{20}O_2$	$[(C_6H_5)_4Bi]OOCCCl_3$	Bi: Org.Verb.-12
$BiC_{26}H_{19}O_4$	$(C_6H_5)_3Bi[(OOC)_2C_6H_4]$	Bi: Org.Verb.-42
$BiC_{26}H_{25}O_2$	$(C_2H_5OC_6H_4)_2Bi(C_{10}H_7)$	Bi: Org.Verb.-95
$BiC_{26}H_{25}O_4$	$(C_6H_5)_3Bi[OOCC(CH_3)CH_2]_2$	Bi: Org.Verb.-39
$BiC_{26}H_{29}O_4S_2$	$(C_6H_5)_3Bi(OOCCH_2SC_2H_5)_2$	Bi: Org.Verb.-40
$BiC_{27}ClH_{21}NO$	$(C_6H_5)_3Bi(OC_9H_6N)Cl$	Bi: Org.Verb.-43
$BiC_{27}Cl_2H_{27}O_6$	$(C_2H_5OOCC_6H_4)_3BiCl_2$	Bi: Org.Verb.-24
$BiC_{27}Cl_2H_{33}$	$[(CH_3)_3C_6H_2]_3BiCl_2$	Bi: Org.Verb.-27
–	$[(i\text{-}C_3H_7)C_6H_4]_3BiCl_2$	Bi: Org.Verb.-23
$BiC_{27}H_{18}N_5O_{18}$	$[(NO_2)(CH_3OOCC)C_6H_3]_3Bi(NO_3)_2$	Bi: Org.Verb.-37
$BiC_{27}H_{27}O_6$	$Bi(C_6H_4COOC_2H_5)_3$	Bi: Org.Verb.-85
$BiC_{27}H_{33}$	$Bi[C_6H_2(CH_3)_3]_3$	Bi: Org.Verb.-94
–	$Bi(C_6H_4C_3H_7\text{-}i)_3$	Bi: Org.Verb.-84
$BiC_{28}Cl_2H_{25}N_2O_{10}$	$[(C_6H_5)_3Bi(C_5H_5NO)_2](ClO_4)_2$	Bi: Org.Verb.-138/9
$BiC_{28}CoH_{29}S_4$	$(C_6H_5)_3Bi \cdot Co[CH(CSCH_3)_2]_2$	Bi: Org.Verb.-151
$BiC_{28}H_{25}N_2O_2{}^{2+}$	$[(C_6H_5)_3Bi(C_5H_5NO)_2]^{2+}$	Bi: Org.Verb.-138/9
$BiC_{28}H_{29}$	$[(CH_3)_3C_6H_2]_2Bi(C_{10}H_7)$	Bi: Org.Verb.-96
$BiC_{28}H_{33}O_4S_2$	$(C_6H_5)_3Bi(OOCCH_2SC_3H_7)_2$	Bi: Org.Verb.-40
–	$(C_6H_5)_3Bi(OOCCH_2SC_3H_7\text{-}i)_2$	Bi: Org.Verb.-40
$BiC_{28}H_{46}N$	$(C_6H_5)_2Bi[N(C_4H_9)_4]$	Bi: Org.Verb.-115
$BiC_{30}ClH_{24}$	$[(C_6H_5)_3BiC_{12}H_9]Cl$	Bi: Org.Verb.-8
$BiC_{30}Cl_2H_{21}$	$(C_{10}H_7)_3BiCl_2$	Bi: Org.Verb.-27
–	$(C_{10}H_7)_3{}^{210}BiCl_2$	Bi: Org.Verb.-27
$BiC_{30}Cl_2H_{23}$	$(C_6H_5)_3Bi(C_6H_4Cl)_2$	Bi: Org.Verb.-6
$BiC_{30}Cl_2H_{39}$	$[(CH_3)(i\text{-}C_3H_7)C_6H_3]_3BiCl_2$	Bi: Org.Verb.-26
$BiC_{30}Cl_3Fe_3H_{24}$	$Bi(C_5H_3ClFeC_5H_5)_3$	Bi: Org.Verb.-61 Fe: Org.Verb.A6-266/7
–	$Bi(C_5H_4FeC_5H_4Cl)_3$	Bi: Org.Verb.-61 Fe: Org.Verb.A6-266/7
$BiC_{30}H_{21}$	$Bi(C_{10}H_7)_3$	Bi: Org.Verb.-96/7
$BiC_{30}H_{23}$	$(C_6H_5)_3BiC_{12}H_8$	Bi: Org.Verb.-6/7
$BiC_{30}H_{24}NO_3$	$[(C_6H_5)_3BiC_{12}H_9]NO_3$	Bi: Org.Verb.-12
$BiC_{30}H_{25}$	$Bi(C_6H_5)_5$	Bi: Org.Verb.-5/6
$BiC_{30}H_{25}S_2$	$(C_6H_5)_3Bi(SC_6H_5)_2$	Bi: Org.Verb.-42
$BiC_{30}H_{37}O_4S_2$	$(C_6H_5)_3Bi(OOCCH_2SC_4H_9)_2$	Bi: Org.Verb.-40
$BiC_{30}H_{39}$	$Bi[C_6H_3(CH_3)(i\text{-}C_3H_7)]_3$	Bi: Org.Verb.-94
$BiC_{32}Cl_3H_{22}O_6$	$(ClC_6H_4)_3Bi(OOCC_6H_4OH)_2$	Bi: Org.Verb.-42
$BiC_{32}H_{25}O_4$	$(C_6H_5)_3Bi(OOCC_6H_5)_2$	Bi: Org.Verb.-41
$BiC_{32}H_{25}O_6$	$(C_6H_5)_3Bi(OOCC_6H_4OH)_2$	Bi: Org.Verb.-41
$BiC_{32}H_{27}N_2O_4$	$(C_6H_5)_3Bi(OOCC_6H_4NH_2)_2$	Bi: Org.Verb.-41
$BiC_{33}H_{23}MoN_2O_3$	$(C_6H_5)_3BiMo(CO)_3(C_{12}H_8N_2)$	Bi: Org.Verb.-149
$BiC_{34}H_{25}O_8$	$(C_6H_5)_3Bi(OOCC_6H_4COOH)_2$	Bi: Org.Verb.-41
$BiC_{34}H_{29}O_4S_2$	$(C_6H_5)_3Bi(OOCCH_2SC_6H_5)_2$	Bi: Org.Verb.-40
$BiC_{34}H_{29}O_6$	$(C_6H_5)_3Bi(OOCCH_2OC_6H_5)_2$	Bi: Org.Verb.-39
$BiC_{35}H_{31}O_4$	$(CH_3C_6H_4)_3Bi(OOCC_6H_5)_2$	Bi: Org.Verb.-42
$BiC_{35}H_{31}O_6$	$(CH_3C_6H_4)_3Bi(OOCC_6H_4OH)_2$	Bi: Org.Verb.-42
$BiC_{35}H_{31}O_7$	$(C_6H_5)_3Bi(OOCC_6H_4OH)_2 \cdot (CH_3)_2CO$	Bi: Org.Verb.-45
$BiC_{36}Cl_2H_{27}$	$(C_6H_5C_6H_4)_3BiCl_2$	Bi: Org.Verb.-23
$BiC_{36}H_{27}$	$Bi(C_6H_4C_6H_5)_3$	Bi: Org.Verb.-85

$BiC_{36}H_{27}N_2O_6$	$(C_6H_5C_6H_4)_3Bi(NO_3)_2$	Bi: Org.Verb.-36
$BiC_{36}H_{27}O_4$	$(C_{10}H_7)(C_6H_5)_2Bi(OOCC_6H_5)_2$	Bi: Org.Verb.-42
$BiC_{36}H_{28}Li$	$Li[(C_6H_5)_4BiC_{12}H_8]$	Bi: Org.Verb.-138
$BiC_{36}H_{29}O_4$	$(C_6H_5)_3Bi(OOCCHCHC_6H_5)_2$	Bi: Org.Verb.-39
–	$(C_6H_5)_3Bi(OOCC_6H_4CHCH_2)_2$	Bi: Org.Verb.-41
$BiC_{36}H_{30}Li$	$Li[Bi(C_6H_5)_6]$	Bi: Org.Verb.-138
$BiC_{36}H_{33}O_4S_2$	$(C_6H_5)_3Bi(OOCCH_2SCH_2C_6H_5)_2$	Bi: Org.Verb.-40
$BiC_{36}H_{75}$	$Bi(C_{12}H_{25})_3$	Bi: Org.Verb.-53
$BiC_{38}H_{29}O_4$	$(C_6H_5C_6H_4)(C_6H_5)_2Bi(OOCC_6H_5)_2$	Bi: Org.Verb.-42
$BiC_{38}H_{33}N_2O_4$	$(C_6H_5)_3Bi(OOCC_6H_4NH_2)_2 \cdot C_6H_6$	Bi: Org.Verb.-41
$BiC_{41}H_{37}O_6$	$(CH_3C_6H_4)_3Bi(OOCC_6H_4OH)_2 \cdot C_6H_6$	Bi: Org.Verb.-42
$BiC_{42}H_{35}OSi$	$[(C_6H_5)_4Bi]OSi(C_6H_5)_3$	Bi: Org.Verb.-9
$BiC_{47}H_{35}$	$(C_6H_5)_3BiC_5(C_6H_5)_4$	Bi: Org.Verb.-13
$BiC_{48}H_{99}$	$Bi(C_{16}H_{33})_3$	Bi: Org.Verb.-53
$BiC_{54}Cl_2H_{45}O_{10}P_2$	$[(C_6H_5)_3Bi[(C_6H_5)_3PO]_2](ClO_4)_2$	Bi: Org.Verb.-138/9
$BiC_{54}F_{12}H_{45}O_2P_4$	$[(C_6H_5)_3Bi[(C_6H_5)_3PO]_2](PF_6)_2$	Bi: Org.Verb.-138/9
$BiC_{54}Ge_3H_{45}$	$[(C_6H_5)_3Ge]_3Bi$	Bi: Org.Verb.-51
$BiC_{54}H_{45}O_2P_2{}^{2+}$	$[(C_6H_5)_3Bi((C_6H_5)_3PO)_2]^{2+}$	Bi: Org.Verb.-138/9
$BiC_{54}H_{45}Si_3$	$[(C_6H_5)_3Si]_3Bi$	Bi: Org.Verb.-51
$BiC_{55}ClH_{45}IrO_5P_2$	$[(C_6H_5)_3Bi[(C_6H_5)_3P]_2(CO)Ir]ClO_4$	Bi: Org.Verb.-151
$BiC_{75}H_{60}P_3S_9$	$[(C_6H_5)_4P]_3[Bi(CS_3)_3]$	C: MVol.D4-223
$BiCl_3$	$BiCl_3$ systems	
	$BiCl_3$-Sn	Sn: MVol.C2-297
	$BiCl_3$-$SnCl_2$	Sn: MVol.C2-298
	$BiCl_3$-$TeCl_4$	Te: SVol.B2-95
$BiCl_3OS_2$	$BiOCl \cdot S_2Cl_2$	S: SVol.2-260
$BiCl_7Sn_2$	$2\ SnCl_2 \cdot BiCl_3$	Sn: MVol.C2-298
BiCm	CmBi	Np: TrU.B3-273
$BiCsMo_2O_8$	$CsBi(MoO_4)_2$	Mo: SVol.B1-240
$BiCsO_8W_2$	$CsBi(WO_4)_2$	W: SVol.B3-262
BiF_3	BiF_3 solid solutions	
	BiF_3-EuF_3	Sc: MVol.C3-214
$BiHO_5U$	$Bi(OH)UO_4 = Bi_2O_3 \cdot 2\ UO_3 \cdot 2\ H_2O$	U: SVol.C3-183
BiI_5O_2U	$BiUO_2I_5$	U: SVol.C9-179
BiIr	IrBi	Ir: SVol.1-45/6
BiIrS	IrBiS	Ir: SVol.1-46/7
BiIrSe	IrBiSe	Ir: SVol.1-46/7
BiIrTe	IrBiTe	Ir: SVol.1-46/7
$BiIr_2$	Ir_2Bi	Ir: SVol.1-45/6
$BiKMo_2O_8$	$KBi(MoO_4)_2$	Mo: SVol.B1-223
$BiKO_8W_2$	$KBi(WO_4)_2$	W: SVol.B3-231
$BiK_5Mo_4O_{16}$	$K_5Bi(MoO_4)_4$	Mo: SVol.B1-223
$BiK_5O_{16}W_4$	$K_5Bi(WO_4)_4$	W: SVol.B3-231
$BiLi_{0.2}O_4W_{0.8}$	$Bi(Li_{0.2}W_{0.8})O_4$	W: SVol.B3-185
$BiLiMo_2O_8$	$LiBi(MoO_4)_2$	Mo: SVol.B1-203
$BiLiO_8W_2$	$LiBi(WO_4)_2$	W: SVol.B3-185
BiMn	MnBi solid solutions	
	$MnBi_{1-x}Te_x$	Mn: MVol.C6-349
$BiMnO_3$	$BiMnO_3$	Mn: MVol.C3-137/8

$BiMnO_3$	$BiMnO_3$ solid solutions	
	$BiMnO_3$-$BaMnO_3$-$CaMnO_3$	Mn: MVol.C3-145
	$BiMnO_3$-$CaMnO_3$	Mn: MVol.C3-141/4
	$BiMnO_3$-$CaMnO_3$-$LaMnO_3$	Mn: MVol.C3-145/7
	$BiMnO_3$-$CaMnO_3$-$PbMnO_3$	Mn: MVol.C3-147
	$BiMnO_3$-$CaMnO_3$-$SrMnO_3$	Mn: MVol.C3-145
	$BiMnO_3$-Ca_2MnO_4	Mn: MVol.C3-144
	$BiMnO_3$-$LaMnO_3$-$SrMnO_3$	Mn: MVol.C3-145/7
	$BiMnO_3$-$PbTiO_3$	Mn: MVol.C3-147/8
	$BiMnO_3$-$SrMnO_3$	Mn: MVol.C3-144/5
	$BiMnO_3$-$YMnO_3$	Mn: MVol.C3-145
$BiMn_2Na_2O_{12}V_3$	$Na_2BiMn_2V_3O_{12}$	Mn: MVol.C3-164
$BiMo_2NaO_8$	$NaBi(MoO_4)_2$	Mo: SVol.B1-214
$BiMo_2O_8Rb$	$RbBi(MoO_4)_2$	Mo: SVol.B1-234
$BiMo_4Na_5O_{16}$	$Na_5Bi(MoO_4)_4$	Mo: SVol.B1-214
$BiNaO_8W_2$	$NaBi(WO_4)_2$	W: SVol.B3-210/1
$BiNa_5O_{16}W_4$	$Na_5Bi(WO_4)_4$	W: SVol.B3-211
$BiNbO_4$	$BiNbO_4$, doped with U	U: SVol.C3-183, 314
BiO_4U	$BiUO_4$	U: SVol.C3-181/2
–	$BiUO_4$ solid solutions	
	$BiUO_4$-U_4O_9	U: SVol.C3-181
BiO_4V	$BiVO_4$ solid solutions	
	$BiVO_4$-$Bi_2(MoO_4)_3$	Mo: SVol.B2-297/8
BiO_8RbW_2	$RbBi(WO_4)_2$	W: SVol.B3-251
$BiO_{16}Rb_5W_4$	$Rb_5Bi(WO_4)_4$	W: SVol.B3-251/2
BiPu	PuBi	Np: TrU.B2-41
		Np: TrU.B3-258/65
$BiPu_2$	Pu_2Bi	Np: TrU.B2-41
		Np: TrU.B3-261
$BiTe_2Tl$	$TlBiTe_2$ solid solutions	
	$TlBiTe_2$-SnTe	Sn: MVol.C1-142
$Bi_{1.33}Mo_{0.67}O_6Pb_2$	$Pb_2Bi_{1.33}Mo_{0.67}O_6$	Mo: SVol.B2-261/2
$Bi_2Br_2C_{36}H_{30}O$	$[((C_6H_5)_3Bi)_2O]Br_2$	Bi: Org.Verb.-135
$Bi_2Br_4C_{30}H_{24}$	$[(C_6H_5)_2BiBr_2]_2C_6H_4$	Bi: Org.Verb.-135
$Bi_2C_4H_{12}$	$[Bi(CH_3)_2]_2$	Bi: Org.Verb.-133
$Bi_2C_4H_{12}S$	$[(CH_3)_2Bi]_2S$	Bi: Org.Verb.-137
$Bi_2C_5H_{12}N_2$	$[(CH_3)_2Bi]_2CN_2$	Bi: Org.Verb.-134
$Bi_2C_{10}H_{10}$	$Bi_2C_{10}H_{10}$	Bi: Org.Verb.-133
$Bi_2C_{16}H_{30}MoO_4$	$[(C_2H_5)_3Bi]_2Mo(CO)_4$	Bi: Org.Verb.-148
$Bi_2C_{20}Fe_5O_{20}$	$Bi_2Fe_5(CO)_{20}$	Fe: Org.Verb.B2-27
$Bi_2C_{24}CaH_{20}$	$[(C_6H_5)_2Bi]_2Ca$	Bi: Org.Verb.-114
$Bi_2C_{24}H_{20}$	$[Bi(C_6H_5)_2]_2$	Bi: Org.Verb.-133
$Bi_2C_{24}H_{20}S$	$[(C_6H_5)_2Bi]_2S$	Bi: Org.Verb.-137
$Bi_2C_{26}Cl_4H_{16}$	$[(ClC_6H_4)_2BiC]_2$	Bi: Org.Verb.-134
$Bi_2C_{26}H_{20}$	$[(C_6H_5)_2BiC]_2$	Bi: Org.Verb.-134
$Bi_2C_{30}H_{24}$	$[(C_6H_5)_2Bi]_2C_6H_4$	Bi: Org.Verb.-134/5
$Bi_2C_{30}H_{28}$	$[(CH_3C_6H_4)_2BiC]_2$	Bi: Org.Verb.-134
$Bi_2C_{36}Cl_2H_{30}O$	$[((C_6H_5)_3Bi)_2O]Cl_2$	Bi: Org.Verb.-135

$Bi_2C_{36}Cl_2H_{30}O_9$	$[((C_6H_5)_3Bi)_2O](ClO_4)_2$	Bi:	Org.Verb.-136/7
–	$[((C_6H_5)_3Bi)_2O](ClO_4)_2 \cdot 2\ H_2O$	Bi:	Org.Verb.-145
$Bi_2C_{36}H_{30}N_2O_7$	$[((C_6H_5)_3Bi)_2O](NO_3)_2$	Bi:	Org.Verb.-136
$Bi_2C_{38}H_{30}N_2O_3$	$[((C_6H_5)_3Bi)_2O](OCN)_2$	Bi:	Org.Verb.-135
$Bi_2C_{39}Cl_6H_{30}O_6S_3$	$[(Bi(C_6H_5)_3)_2C_3Cl_4S_3O_6]Cl_2$	Bi:	Org.Verb.-137
$Bi_2C_{39}FeH_{30}O_3$	$(CO)_3Fe(Bi(C_6H_5)_3)_2$	Fe:	Org.Verb.B1-149
$Bi_2C_{40}Cl_2H_{42}O_{11}S_2$	$[[(C_6H_5)_3Bi[(CH_3)_2SO]]_2O](ClO_4)_2$	Bi:	Org.Verb.-145
–	$[[(C_6H_5)_3Bi[(CD_3)_2SO]]_2O](ClO_4)_2$	Bi:	Org.Verb.-145
$Bi_2C_{40}F_6H_{30}O_5$	$[[(C_6H_5)_3Bi]_2O](OOCCF_3)_2$	Bi:	Org.Verb.-135
$Bi_2C_{40}FeH_{30}O_4$	$[(C_6H_5)_3Bi]_2Fe(CO)_4$	Bi:	Org.Verb.-151
$Bi_2C_{46}Cl_2H_{40}N_2O_{11}$	$[[(C_6H_5)_3Bi(C_5H_5NO)]_2O](ClO_4)_2$	Bi:	Org.Verb.-145
$Bi_2C_{52}H_{80}Mn_3N_{12}O_{20}Se_{12}$			
	$Mn_3[Bi(NCSe)_6]_2 \cdot 10\ C_4H_8O_2$	Mn:	MVol.D1-156
$Bi_2C_{72}Cl_2H_{60}O_{11}P_2$	$[[(C_6H_5)_3Bi[(C_6H_5)_3PO]]_2O](ClO_4)_2$	Bi:	Org.Verb.-145
$Bi_2Fe_4O_9$	$Bi_2Fe_4O_9$ solid solutions		
	$Bi_2(Mn_{1-x}Fe_x)_4O_9$	Mn:	MVol.C3-139
Bi_2HMnO_7	$[Bi_2O_2(OH)]MnO_4 \cdot 1.5\ H_2O$		
	$= [Bi_6O_6](OH)_3(MnO_4)_3 \cdot 4.5\ H_2O$	Mn:	MVol.C3-140
$Bi_2H_4O_{11}U_2$	$Bi_2(OH)_4U_2O_7$	U:	SVol.C3-183
Bi_2Ir	$IrBi_2$	Ir:	SVol.1-45/6
$Bi_2Ir_2O_7$	$Bi_2Ir_2O_7$	Ir:	SVol.2-9
Bi_2MnO_4	$Bi_2O_3 \cdot MnO$	Mn:	MVol.C3-136
$Bi_2Mn_2O_5$	$Bi_2O_3 \cdot 2\ MnO$	Mn:	MVol.C3-136
$Bi_2Mn_3N_8O_{26}$	$3\ Mn(NO_3)_2 \cdot 2\ BiONO_3$	Mn:	MVol.C3-305
$Bi_2Mn_3N_{12}O_{36}$	$3\ Mn(NO_3)_2 \cdot 2\ Bi(NO_3)_3 \cdot 24\ H_2O$	Mn:	MVol.C3-305
$Bi_2Mn_3O_6$	$Bi_2O_3 \cdot 3\ MnO$	Mn:	MVol.C3-136
$Bi_2Mn_4O_9$	$Bi_2Mn_4O_9$	Mn:	MVol.C3-138
–	$Bi_2Mn_4O_9$ solid solutions		
	$Bi_2(Mn_{1-x}Fe_x)_4O_9$	Mn:	MVol.C3-139
$Bi_2Mn_4O_{10}$	$Bi_2Mn_4O_{10}$	Mn:	MVol.C3-138/40
Bi_2MoO_6	$Bi_2MoO_6 = MoO_3 \cdot Bi_2O_3$	Mo:	SVol.B1-146/51, 153/5
–	Bi_2MoO_6 solid solutions		
	$Bi_xGd_{2-x}MoO_6$	Mo:	SVol.B2-161
–	Bi_2MoO_6 systems		
	Bi_2MoO_6-$Bi_4Ti_3O_{12}$	Mo:	SVol.B2-217
	Bi_2MoO_6-La_2MoO_6	Mo:	SVol.B2-123
$Bi_2Mo_2O_9$	$Bi_2Mo_2O_9 = 2\ MoO_3 \cdot Bi_2O_3$	Mo:	SVol.B1-146/53
$Bi_2Mo_3O_{12}$	$Bi_2Mo_3O_{12} = 3\ MoO_3 \cdot Bi_2O_3$	Mo:	SVol.B1-146/52
–	$Bi_2(MoO_4)_3$ melts		
	$Bi_2(MoO_4)_3$-$PbMoO_4$	Mo:	SVol.B2-260/1
–	$Bi_2(MoO_4)_3$ solid solutions		
	$Bi_2\ (MoO_4)_3$-$BiVO_4$	Mo:	SVol.B2-297/8
Bi_2O_3	Bi_2O_3 glasses		
	Bi_2O_3-B_2O_3-PbO	B:	B-Verb.7-111
–	Bi_2O_3 solid solutions		
	Bi_2O_3-Bi_2UO_6	U:	SVol.C3-183
	Bi_2O_3-Bi_6WO_{12}	W:	SVol.B3-18
	Bi_2O_3-ThO_2	Th:	SVol.C2-33/4

Bi_2O_3	Bi_2O_3 solid solutions	
	Bi_2O_3-UO_2	U: SVol.C3-181
	Bi_2O_3-WO_3	W: SVol.B3-20
–	Bi_2O_3 systems	
	Bi_2O_3-B_2O_3	B: B-Verb.7-98
	Bi_2O_3-B_2O_3-Nd_2O_3	B: B-Verb.7-117
	Bi_2O_3-MoO_3	Mo: SVol.B1-146/8
	Bi_2O_3-MoO_3-PbO	Mo: SVol.B2-259/61
	Bi_2O_3-SnO_2	Sn: MVol.C2-296
	Bi_2O_3-TeO_2	Te: SVol.B1-60
	Bi_2O_3-WO_3	W: SVol.B3-2/5
Bi_2O_5Te	Bi_2TeO_5 = Bi_2O_3 · TeO_2	Te: SVol.B1-60
Bi_2O_6SnZn	$ZnBi_2SnO_6$	Sn: MVol.C4-7
Bi_2O_6U	Bi_2UO_6	U: SVol.C3-182/3
–	Bi_2UO_6 solid solutions	
	Bi_2UO_6-Bi_2O_3	U: SVol.C3-183
Bi_2O_6W	Bi_2WO_6 = Bi_2O_3 · WO_3	W: SVol.B3-2/4, 7, 11/7
–	Bi_2WO_6 solid solutions	
	Bi_2WO_6-$Bi_2(WO_4)_3$	W: SVol.B3-9
	Bi_2WO_6-Bi_6WO_{12}	W: SVol.B3-18
–	Bi_2WO_6 systems	
	Bi_2WO_6-Na_2WO_4	W: SVol.B3-210
$Bi_2O_7Sn_2$	$Bi_2Sn_2O_7$ = Bi_2O_3 · 2 SnO_2	Sn: MVol.C2-296/7
Bi_2O_7Te	Bi_2TeO_7 = Bi_2O_3 · 2 TeO_2	Te: SVol.B1-60
$Bi_2O_{7.2}W_{1.4}$	Bi_2O_3 · 1.4 WO_3	W: SVol.B3-3, 17
$Bi_2O_9Sn_3$	$Bi_2(SnO_3)_3$	Sn: MVol.C2-296
–	$Bi_2(SnO_3)_3$ · 5 H_2O	Sn: MVol.C2-296
–	$Bi_2(SnO_3)_3$ solid solutions	
	$Bi_2(SnO_3)_3$-$BaTiO_3$	Sn: MVol.C4-201/3
	$(Bi,Er)_2(SnO_3)_3$-$BaTiO_3$	Sn: MVol.C4-215
	$(Bi,La)_2(SnO_3)_3$-$BaTiO_3$	Sn: MVol.C4-215
$Bi_2O_9ThTi_2$	$ThBi_2Ti_2O_9$	Th: SVol.C2-34
$Bi_2O_9U_2$	Bi_2O_3 · 2 UO_3 · 2 H_2O = $Bi(OH)UO_4$	U: SVol.C3-183
$Bi_2O_9W_2$	$Bi_2W_2O_9$ = Bi_2O_3 · 2 WO_3	W: SVol.B3-2/4, 6, 10/1
$Bi_2O_{11}Te_4$	$Bi_2Te_4O_{11}$ = Bi_2O_3 · 4 TeO_2	Te: SVol.B1-60
$Bi_2O_{12}W_3$	$Bi_2(WO_4)_3$ = Bi_2O_3 · 3 WO_3	W: SVol.B3-2/3, 6, 9/10
–	$Bi_2(WO_4)_3$ solid solutions	
	$Bi_2(WO_4)_3$-Bi_2WO_6	W: SVol.B3-9
Bi_2Pu	$PuBi_2$	Np: TrU.B2-41
		Np: TrU.B3-258/65
Bi_2Pu_3	Pu_3Bi_2	Np: TrU.B2-41
		Np: TrU.B3-261
Bi_2S_3	Bi_2S_3 systems	
	Bi_2S_3-SnS	Sn: MVol.C2-298
Bi_2SnTe_4	SnTe · Bi_2Te_3	Sn: MVol.C2-299
Bi_2Te_3	Bi_2Te_3 solid solutions	
	Bi_2Te_3-(Sn,Ge)Te	Sn: MVol.C4-243

Bi_2Te_3	Bi_2Te_3 systems	
	Bi_2Te_3-SnTe	Sn: MVol.C2-299/300
Bi_2Th	$ThBi_2$	Th: SVol.C1-179
$Bi_3C_{54}Cl_2H_{45}O_2Ru$	$[(C_6H_5)_3Bi]_3RuCl_2 \cdot O_2$	Bi: Org.Verb.-151
$Bi_3C_{54}Cl_2H_{45}Ru$	$[(C_6H_5)_3Bi]_3RuCl_2$	Bi: Org.Verb.-151
$Bi_3FeMo_2O_{12}$	$Bi_3FeO_4(MoO_4)_2$	Mo: SVol.B1-135
$Bi_3GaMo_2O_{12}$	$Bi_3(GaO_4)(MoO_4)_2$	Mo: SVol.B1-130
$Bi_3H_2MnO_9$	$[Bi_3O_3(OH)_2]MnO_4$	Mn: MVol.C3-140
Bi_3Ir	$IrBi_3$	Ir: SVol.1-45/6
Bi_3Pu	$PuBi_3$	Np: TrU.B2-25
		Np: TrU.B3-261, 265
Bi_3Pu_2	Pu_2Bi_3	Np: TrU.B3-261/2
$Bi_4Cl_{14}Mn$	$MnBi_4Cl_{14} \cdot 12\ H_2O$	Mn: MVol.C5-234
Bi_4MoO_9	$Bi_4MoO_9 = MoO_3 \cdot 2\ Bi_2O_3$	Mo: SVol.B1-147, 155/6
Bi_4O_9U	$Bi_2UO_6 \cdot Bi_2O_3$	U: SVol.C3-183
Bi_4O_9W	$Bi_4WO_9 = 2\ Bi_2O_3 \cdot WO_3$	W: SVol.B3-2/4, 6, 18
$Bi_4O_{12}Ti_3$	$Bi_4Ti_3O_{12}$ systems	
	$Bi_4Ti_3O_{12}$-Bi_2MoO_6	Mo: SVol.B2-217
Bi_4SnTe_7	$SnTe \cdot 2\ Bi_2Te_3$	Sn: MVol.C2-299
$Bi_{5.9}O_{30}W_{7.05}$	$2.95\ Bi_2O_3 \cdot 7.05\ WO_3$	W: SVol.B3-10
$Bi_6H_2Mn_4O_{24}$	$[Bi_6O_6](OH)_2(MnO_4)_4 \cdot 6\ H_2O$	
	$= [Bi_6(OH)_{12}](OH)_2(MnO_4)_4$	Mn: MVol.C3-140
$Bi_6H_3Mn_3O_{21}$	$[Bi_6O_6](OH)_3(MnO_4)_3 \cdot 4.5\ H_2O$	
	$= [Bi_2O_2(OH)]MnO_4 \cdot 1.5\ H_2O$	Mn: MVol.C3-140
$Bi_6H_4Mn_2O_{18}$	$[Bi_6O_6](OH)_4(MnO_4)_2$	Mn: MVol.C3-140
$Bi_6H_{14}Mn_4O_{30}$	$[Bi_6(OH)_{12}](OH)_2(MnO_4)_4$	
	$= [Bi_6O_6](OH)_2(MnO_4)_4 \cdot 6\ H_2O$	Mn: MVol.C3-140
Bi_6MoO_{12}	$Bi_6MoO_{12} = MoO_3 \cdot 3\ Bi_2O_3$	Mo: SVol.B1-147/50, 156/7
$Bi_6O_{12}W$	$Bi_6WO_{12} = 3\ Bi_2O_3 \cdot WO_3$	W: SVol.B3-2/7, 18/9
–	Bi_6WO_{12} solid solutions	
	Bi_6WO_{12}-Bi_2O_3	W: SVol.B3-18
	Bi_6WO_{12}-Bi_2WO_6	W: SVol.B3-18
$Bi_{10}O_{19}Te_2$	$Bi_{10}Te_2O_{19} = 5\ Bi_2O_3 \cdot 2\ TeO_2$	Te: SVol.B1-60
$Bi_{12}O_{21}W$	$Bi_{12}WO_{21} = 6\ Bi_2O_3 \cdot WO_3$	W: SVol.B3-2/4, 20
$Bi_{14}Mo_3O_{42}Ti_6$	$2\ Bi_4Ti_3O_{12} \cdot 3\ Bi_2MoO_6$	Mo: SVol.B2-217
$Bi_{14}O_{24}U$	$Bi_2UO_6 \cdot 6\ Bi_2O_3$	U: SVol.C3-183
$Bi_{14}O_{24}W$	$Bi_{14}WO_{24} = 7\ Bi_2O_3 \cdot WO_3$	W: SVol.B3-4, 20
$Bi_{14}O_{27}W_2$	$Bi_{14}W_2O_{27} = 3.5\ Bi_2O_3 \cdot WO_3$	W: SVol.B3-4, 6, 18/9
$Bi_{16}O_{27}U$	$Bi_2UO_6 \cdot 7\ Bi_2O_3$	U: SVol.C3-183
$Bi_{16}O_{34}Te_5$	$Bi_{16}Te_5O_{34} = 8\ Bi_2O_3 \cdot 5\ TeO_2$	Te: SVol.B1-60
$Bi_{19}O_{31.5}W$	$9.5\ Bi_2O_3 \cdot WO_3$	W: SVol.B3-20
$Bi_{20}MoO_{33}$	$MoO_3 \cdot 10\ Bi_2O_3$	Mo: SVol.B1-147, 150, 157
$Bi_{20}O_{33}W$	$10\ Bi_2O_3 \cdot WO_3$	W: SVol.B3-20
$Bi_{82}O_{126}U$	$Bi_2UO_6 \cdot 40\ Bi_2O_3$	U: SVol.C3-183
Bi_xO_3W	Bi_xWO_3 (tungsten oxide bronzes)	W: SVol.B3-39/41
Bk	Berkelium	
	Chemical reactions	Np: TrU.B1-76

Bk .... Berkelium

Chemistry in organic solvents
see Extraction

Complex chemistry in aqueous solutions
Np: TrU.D1-129/76

Electrochemical behavior .... Np: TrU.D1-5/18

Extraction

by β-dicarbonyl and related compounds .... Np: TrU.D2-236/48

by neutral organophosphorus compounds .... Np: TrU.D2-177/220

by onium compounds .... Np: TrU.D2-325/44

by organophosphoric acids ..... Np: TrU.D2-249/311

Ion exchange reactions .... Np: TrU.D2-373/437

Oxidation states .... Np: TrU.D1-5/7

Physical properties .... Np: TrU.B1-74/6

Precipitation from aqueous solutions Np: TrU.D1-114, 116, 126

Preparation .... Np: TrU.B1-72/4

Redox reactions .... Np: TrU.D1-80/2

Separation by coprecipitation ...... Np: TrU.D1-106/11

– .... Berkelium ions

Chemistry in organic solvents
see Bk...Berkelium: Extraction

Complex chemistry .... Np: TrU.D1-137/66

Coprecipitation .... Np: TrU.D1-106/11

Hydration .... Np: TrU.D1-132/3

Hydrolysis .... Np: TrU.D1-133/7

Ion exchange reactions

Anion exchange reactions ...... Np: TrU.D2-409/17, 422

Cation exchange reactions ..... Np: TrU.D2-373/95

Exchange reactions with inorganic ion exchangers ..... Np: TrU.D2-402/7

Redox exchange reactions ...... Np: TrU.D2-437

Precipitation from aqueous solutions Np: TrU.D1-114, 116, 126

Redox reactions .... Np: TrU.D1-80/2

– .... Berkelium systems

Bk-Hg .... Np: TrU.B3-274

Bk-Li .... Np: TrU.B3-274

Bk^{3+} .... Bk^{3+}

Complex chemistry .... Np: TrU.D1-145, 151/66

Coprecipitation .... Np: TrU.D1-106/11

Hydration .... Np: TrU.D1-132/3

Hydrolysis .... Np: TrU.D1-133/6

Ion exchange reactions .... Np: TrU.D2-373/95, 402/5, 409/17, 437

Precipitation .... Np: TrU.D1-114, 116, 126

Redox reactions .... Np: TrU.D1-80/2

Bk^{4+}	Bk^{4+}		
	Hydration	Np:	TrU.D1-132/3
	Ion exchange reactions	Np:	TrU.D2-405, 422
	Precipitation	Np:	TrU.D1-114, 116, 126
	Redox reactions	Np:	TrU.D1-80/2
$BkCNS^{2+}$	$Bk(SCN)^{2+}$	Np:	TrU.D1-145
$BkC_2H_3O_2^{2+}$	$Bk(CH_3COO)^{2+}$	Np:	TrU.D1-153
$BkC_9H_{15}O_9$	$Bk(CH_3CH(OH)COO)_3$	Np:	TrU.D1-154
$BkC_{10}H_{12}N_2O_8^-$	$BkC_2H_4N_2(CH_2COO)_4^-$	Np:	TrU.D1-165
$BkC_{12}H_{21}O_9$	$Bk((CH_3)_2C(OH)COO)_3$	Np:	TrU.D1-154
$BkC_{14}H_{18}N_3O_{10}^{2-}$	$Bk(CH_2COO)N(C_2H_4N(CH_2COO)_2)_2^{2-}$	Np:	TrU.D1-166
$BkC_{58}H_{77}N_5O_8P$	$Bk(N(C_6H_5)NC(CH_3)C(COC_6H_5)CO)_2(NO_3)$ · $(C_8H_{17})_3PO$	Np:	TrU.D2-365, 368
$BkC_{75}H_{93}N_6O_{14}P_2$	$Bk(N(C_6H_5)NC(CH_3)C(COC_6H_5)CO)_3$ · 2 $(C_4H_9O)_3PO$	Np:	TrU.D2-365, 368
$BkC_{99}H_{141}N_6O_8P_2$	$Bk(N(C_6H_5)NC(CH_3)C(COC_6H_5)CO)_3$ · 2 $(C_8H_{17})_3PO$	Np:	TrU.D2-365, 368
$BkHO^{2+}$	$Bk(OH)^{2+}$	Np:	TrU.D1-136
$Br_{0.5}F_{8.5}IrK_2$	$K[IrF_6]$ · KF · 0.5 BrF_3	Ir:	SVol.2-107
$Br_{0.8}C_8Cl_{1.2}H_{18}Sn$	$(C_4H_9)_2SnBr_{0.8}Cl_{1.2}$	Sn:	Org.Verb.6-88
$Br_{0.9}K_{0.9}O_3U$	$K_{0.9}UO_3Br_{0.9}$	U:	SVol.C9-151/6
Br	Bromine systems		
	Br-Te	Te:	SVol.B3-1/2
$BrCClF_2N_2O_4S_2$	$ClN(SO_2F)_2$ · BrCN	S:	S-N-Verb.1-52
BrCClO	COBrCl	C:	MVol.D3-120/2
BrCClS	ClBrCS	C:	MVol.D6-164/5
$BrCCl_2H_3Sn$	$CH_3SnBrCl_2$	Sn:	Org.Verb.6-218
$BrCCl_3S$	$BrCl_2CSCl$	C:	MVol.D4-27
$BrCCl_5NSb$	BrCN · $SbCl_5$	C:	MVol.D3-236
BrCFO	COBrF	C:	MVol.D3-118/20
$BrCF_2NO$	$BrCF_2NO$	F:	PerFHalOrg.7-173/4, 182
$BrCF_3Mg$	CF_3MgBr	F:	PerFHalOrg.4-61, 71
$BrCF_5Si$	CF_3SiF_2Br	F:	PerFHalOrg.4-147
$BrCHNO^-$	$[BrNCOH]^-$	C:	MVol.D3-233
$BrCH_2NO$	$NH_2C(O)Br$	C:	MVol.D3-132/3
$BrCH_3N_2O$	$(NHBr)(NH_2)CO$	C:	MVol.D3-129
$BrCH_3N_2S$	$(NHBr)(NH_2)CS$	C:	MVol.D6-131
$BrCH_6N_3O_2$	$[NH_2C(NHOH)_2]Br$	C:	MVol.D3-235
$BrCIN^-$	$[IBrCN]^-$	C:	MVol.D3-237
–	$[ICNBr]^-$	C:	MVol.D3-263
$BrCIrO_3$	$IrBrCO_3$	Ir:	SVol.2-143
BrCN	BrCN		
	Chemical reactions		
	in molecular beams	C:	MVol.D3-230
	with electrons	C:	MVol.D3-230
	with elements	C:	MVol.D3-231/2
	with inorganic compounds	C:	MVol.D3-232/40
	with organic compounds	C:	MVol.D3-234/5, 241

BrCN .... BrCN
- Crystallographic properties .... C: MVol.D3-220
- Electrical properties .... C: MVol.D3-227/8
- Electrochemical behavior .... C: MVol.D3-227/8
- Formation and preparation .... C: MVol.D3-217/20
- Hydrolysis .... C: MVol.D3-232/3
- Isotope exchange .... C: MVol.D3-230/1
- Mechanical properties .... C: MVol.D3-221
- Molecule .... C: MVol.D3-151/80
- Photolysis .... C: MVol.D3-229/30
- Polymerization .... C: MVol.D3-228
- Pyrolysis .... C: MVol.D3-229
- Review .... C: MVol.D3-149/50
- Solubility .... C: MVol.D3-240
- Spectra .... C: MVol.D3-223/7
- Stability .... C: MVol.D3-228
- Thermal properties .... C: MVol.D3-221/3
- Toxicity .... C: MVol.D3-150/1

– .... BrNC .... C: MVol.D3-149
$BrCN^+$ .... $BrCN^+$ .... C: MVol.D3-265
BrCNO .... BrNCO .... C: MVol.D3-142/3
$BrCNO_2S$ .... BrO_2SCN .... C: MVol.D5-189
BrCNS .... BrSCN .... C: MVol.D6-176/7
– .... $(BrSCN)_x$ .... C: MVol.D6-178
$BrCNS^-$ .... $BrSCN^-$ .... C: MVol.D5-177
$BrCSe^+$ .... $BrCSe^+$ .... C: MVol.D6-251
BrC_2ClFH_2NO ...... $CFClBrC(O)NH_2$ .... F: PerFHalOrg.7-17, 19, 45
BrC_2ClF_3NO .... $BrCF_2CFClNO$ .... F: PerFHalOrg.7-175/6, 183, 192
BrC_2ClF_4HN .... $CF_3N(H)CFClBr$ .... F: PerFHalOrg.7-13/4, 41
$BrC_2Cl_3H_{12}N_6O_2U$ ... $[CN_3H_6]_2UO_2BrCl_3$ .... U: SVol.C9-162
BrC_2FN_2S .... NCFSCBrN .... F: PerFHalOrg.5-67/8
$BrC_2F_2H_2NO$ .... $CF_2BrC(O)NH_2$ .... F: PerFHalOrg.7-17, 19, 45
$BrC_2F_2H_6NOSSn$ .... $(CH_3)_2SnBr[NS(O)F_2]$ .... S: S-N-Verb.1-64
BrC_2F_2IMg .... $CF_2CBrMgI$ .... F: PerFHalOrg.4-61, 90
BrC_2F_3HNO .... $CF_2BrC(F)NOH$ .... F: PerFHalOrg.7-88
– .... $CF_3C(Br)NOH$ .... F: PerFHalOrg.7-88, 146
– .... $CF_3C(O)NHBr$ .... F: PerFHalOrg.7-17, 22, 45, 57
– .... $NHOCFBrCF_2$ .... F: PerFHalOrg.5-29/30, 38
BrC_2F_3Mg .... $CF_2CFMgBr$ .... F: PerFHalOrg.4-61, 71/3, 89/90, 110
BrC_2F_3O .... CF_3COBr .... F: PerFHalOrg.6-114
$BrC_2F_4H_6NS$ .... $(CH_3)_2NSF_4Br$ .... S: SVol.2-305
BrC_2F_4NO .... $BrCF_2CF_2NO$ .... F: PerFHalOrg.7-175/6, 183, 192
BrC_2F_5HN .... $CF_2BrN(H)CF_3$ .... F: PerFHalOrg.7-13/4, 41, 55
$BrC_2F_5H_2S$ .... $SF_5CHCHBr$ .... S: SVol.2-305

$BrC_2F_5H_4S$	$SF_5CH_2CH_2Br$	S:	SVol.2-304
BrC_2F_5Mg	C_2F_5MgBr	F:	PerFHalOrg.4-61/2, 90
BrC_2F_6OP	$(CF_3)_2P(O)Br$	F:	PerFHalOrg.3-52, 53, 55, 57
BrC_2F_6P	$(CF_3)_2PBr$	F:	PerFHalOrg.3-74, 82/3, 92, 98/101, 120
BrC_2F_6PS	$(CF_3)_2P(S)Br$	F:	PerFHalOrg.3-140, 143, 148, 152/3
BrC_2F_6Sb	$(CF_3)_2SbBr$	F:	PerFHalOrg.3-226, 229/31
$BrC_2F_7H_2S$	$SF_5CHFCHBrF$	S:	SVol.2-304
$BrC_2F_{10}H_3S_2$	$SF_5CHBrCH_2SF_5$	S:	SVol.2-304
$BrC_2FeN_3O_2$	$(CO)_2Fe(Br)N_3$	Fe:	Org.Verb.B3-119
$BrC_2H_3N_2NiO_2$	$N(OH)CHCHNi(NO)Br$	Ni:	Org.Verb.2-277
$BrC_2H_3NiO_2$	$Ni(O_2CCH_3)Br$	Ni:	Org.Verb.2-278
$BrC_2H_6NO_2S$	$BrSO_2N(CH_3)_2$	S:	S-N-Verb.1-134
$BrC_2H_8N_4S_2Sn^+$	$[SnBr(CS(NH_2)_2)_2]^+$	Sn:	MVol.C5-56
BrC_2NO_2	$BrC(O)NCO$	C:	MVol.D3-143
$BrC_2N_2^-$	$[CNBrCN]^-$	C:	MVol.D3-237
$BrC_3ClFH_5NO_3P$	$ClCFNOPBr(O)OC_2H_5$	F:	PerFHalOrg.7-199/200
BrC_3ClFNS	$CClNCFSCBr$	F:	PerFHalOrg.5-57/8, 80, 110
$BrC_3ClF_3H_3N_2$	$BrCF_2CFClNNCH_3$	F:	PerFHalOrg.7-192
BrC_3ClF_6NO	$(BrCF_2CFCl)(CF_3)NO$	F:	PerFHalOrg.7-85/6
$BrC_3Cl_2N_3$	$C_3N_3Cl_2Br$	C:	MVol.D3-289
$BrC_3F_2H_3O$	CF_2CBrCH_2OH	F:	PerFHalOrg.4-90
$BrC_3F_2N_3$	$C_3N_3F_2Br$	C:	MVol.D3-289/90
$BrC_3F_3H_5NiP$	$C_3H_5Ni(PF_3)Br$	Ni:	Org.Verb.2-12
BrC_3F_3Mg	$CF_3CCMgBr$	F:	PerFHalOrg.4-62/3, 74, 96
$BrC_3F_4H_3N_2$	$BrCF_2CF_2NNCH_3$	F:	PerFHalOrg.7-192
BrC_3F_5HNO	$C_2F_5C(O)NHBr$	F:	PerFHalOrg.7-18, 22, 46
BrC_3F_6IMg	$CF_2ICF_2CF_2MgBr$	F:	PerFHalOrg.4-62
BrC_3F_6NO	$[N(CF_3)OCF_2CFBr]_n$	F:	PerFHalOrg.7-207/8
BrC_3F_7Mg	$(n\text{-}C_3F_7)MgBr$	F:	PerFHalOrg.4-61/2, 71, 91/2
–	$(CF_3)_2CFMgBr$	F:	PerFHalOrg.4-61/2, 71, 90/1
$BrC_3F_9FeH_5P_3$	$C_3H_5Fe(PF_3)_3Br$	Fe:	Org.Verb.B5-2
BrC_3F_9Ge	$(CF_3)_3GeBr$	F:	PerFHalOrg.4-166/7, 176
BrC_3F_9Si	$(CF_3)_3SiBr$	F:	PerFHalOrg.4-147
$BrC_3F_{10}NS$	$SF_3Br[NCF(CF_3)_2]$	S:	SVol.2-305
$BrC_3FeHgNO_4$	$(CO)_3Fe(NO)HgBr$	Fe:	Org.Verb.B1-196/7
BrC_3H_3Ni	$CHCCH_2NiBr$	Ni:	Org.Verb.1-15
$BrC_3H_5N_2NiO_2$	$CH_2CHCHN(OH)Ni(NO)Br$	Ni:	Org.Verb.2-281
$BrC_3H_5OPdS_2$	$[PdBr(S_2COC_2H_5)]_2$	C:	MVol.D4-261
BrC_3H_7NiO	$C_3H_5Ni(H_2O)Br$	Ni:	Org.Verb.2-277
BrC_3H_8NNi	$C_3H_5Ni(NH_3)Br$	Ni:	Org.Verb.2-9
$BrC_3H_{11}N_2Ni$	$C_3H_5Ni(NH_3)_2Br$	Ni:	Org.Verb.2-19
BrC_3IrO_3	$Ir(CO)_3Br$	Ir:	SVol.2-142

BrC_4ClF_4Mg	$(CF_2)_2CClCMgBr$	F:	PerFHalOrg.4-62/3, 69, 71, 107
BrC_4ClH_9LiSn	$(CH_3)_3SnCBrClLi$	Sn:	Org.Verb.2-16, 18
$BrC_4ClH_{10}Sn$	$(CH_3)_3SnCHBrCl$	Sn:	Org.Verb.2-8
–	$(C_2H_5)_2SnBrCl$	Sn:	Org.Verb.6-53
$BrC_4ClH_{16}IIrN_4O_4$	$[Ir(C_2H_4(NH_2)_2)_2IBr]ClO_4 \cdot H_2O$	Ir:	SVol.2-85
$BrC_4ClH_{16}IrN_4^+$	$[Ir(C_2H_4(NH_2)_2)_2BrCl]^+$	Ir:	SVol.2-83, 84
$BrC_4Cl_2H_9Sn$	$(CH_3)_3SnCBrCl_2$	Sn:	Org.Verb.2-11
$BrC_4Cl_2H_{12}NOS$	$SOCl_2 \cdot [(CH_3)_4N]Br$	S:	SVol.1-45/6
$BrC_4Cl_2H_{16}IrN_4O_4$	$[Ir(C_2H_4(NH_2)_2)_2BrCl]ClO_4$	Ir:	SVol.2-84
$BrC_4Cl_3H_{10}O_2Ti$	$TiCl_3Br \cdot CH_3OCH_2CH_2OCH_3$	Ti:	Org.Verb.1-41
$BrC_4F_2FeO_4P$	$(CO)_4FePF_2Br$	Fe:	Org.Verb.B2-85, 93, 114
$BrC_4F_3H_5NO$	$CF_3N(OH)CHCHCH_2Br$	F:	PerFHalOrg.7-166/7
$BrC_4F_3N_2$	NCBrCFNCFCF	F:	PerFHalOrg.6-28/30, 49, 60
BrC_4F_4I	$IC(CF_2CF_2)CBr$	F:	PerFHalOrg.4-107
$BrC_4F_4NO_2$	$BrN(COCF_2)_2$	F:	PerFHalOrg.5-51/2, 96
BrC_4F_5Mg	$FC(CF_2)_2CMgBr$	F:	PerFHalOrg.4-62/3, 71
$BrC_4F_6H_2NO$	$CF_2BrCF_2CF_2C(O)NH_2$	F:	PerFHalOrg.7-18/21, 46
$BrC_4F_6H_3N_2$	$Br(CF_2)_3C(NH)NH_2$	F:	PerFHalOrg.7-3/4, 31
BrC_4F_6NO	$CF_3NOCBrCCF_3$	F:	PerFHalOrg.5-33/4, 41
BrC_4F_7HNO	$C_3F_7C(O)NHBr$	F:	PerFHalOrg.7-18, 22, 46, 57
BrC_4F_7NNaO	$Na[C_3F_7C(O)NBr]$	F:	PerFHalOrg.7-18, 22, 56
$BrC_4FeGeH_6NO_3$	$(CO)_2(NO)FeBrGe(CH_3)_2$	Fe:	Org.Verb.B1-133
BrC_4FeHO_4	$(CO)_4Fe(Br)H$	Fe:	Org.Verb.C2-27
BrC_4FeH_{10}	$(C_2H_5)_2FeBr$	Fe:	Org.Verb.B1-7
BrC_4FeIO_4	$(CO)_4Fe(Br)I$	Fe:	Org.Verb.B2-69
BrC_4FeNaO_4	$Na[(CO)_4FeBr]$	Fe:	Org.Verb.B2-51
$BrC_4H_4MnO_2^+$	$[MnBr(CH_3COCHCO)]^+$	Mn:	MVol.D1-102
BrC_4H_5NiO	$C_3H_5Ni(CO)Br$	Ni:	Org.Verb.2-42
$BrC_4H_7N_2NiO_2$	$CH_2C(CH_3)CHNOHNi(NO)Br$	Ni:	Org.Verb.2-306
$BrC_4H_9N_2NiS$	$C_3H_5Ni(SC(NH_2)_2)Br$	Ni:	Org.Verb.2-10
BrC_4H_9NiO	$CH_3CHCHCH_2Ni(H_2O)Br$	Ni:	Org.Verb.2-293
$BrC_4H_{10}NO_2S$	$BrSO_2N(C_2H_5)_2$	S:	S-N-Verb.1-135
$BrC_4H_{11}Sn$	$(CH_3)_3SnCH_2Br$	Sn:	Org.Verb.2-6/7
$BrC_4H_{16}IIrN_4^+$	$[Ir(C_2H_4(NH_2)_2)_2IBr]^+$	Ir:	SVol.2-84
$BrC_4H_{16}IrN_7^+$	$[Ir(C_2H_4(NH_2)_2)_2(N_3)Br]^+$	Ir:	SVol.2-84
$BrC_4H_{16}N_9O_9U$	$UO_2(NO_3)Br \cdot 4\ (NH_2)_2CO \cdot H_2O$	U:	SVol.E1-85
$BrC_4H_{18}N_9O_{10}U$	$[UO_2((NH_2)_2CO)_4(H_2O)]Br(NO_3)$	U:	SVol.E1-83, 85, 87
$BrC_5ClFHINO$	$HOC_5FBrClIN$	F:	PerFHalOrg.5-182/3
		F:	PerFHalOrg.6-13
BrC_5ClF_2HNO	HOC_5F_2BrClN	F:	PerFHalOrg.5-182/3
		F:	PerFHalOrg.6-13
BrC_5ClF_2IN	ClF_2C_5BrIN	F:	PerFHalOrg.5-118/9
BrC_5ClF_3N	$BrClC_5F_3N$	F:	PerFHalOrg.5-118/9, 127, 129
$BrC_5Cl_2F_2N$	$BrCl_2C_5F_2N$	F:	PerFHalOrg.5-118/9, 127
$BrC_5Cl_2H_5Ti$	$C_5H_5TiCl_2Br$	Ti:	Org.Verb.1-139, 149
$BrC_5CoH_{18}IrN_{11}$	$[Co(NH_3)_6][Ir(CN)_5Br]$	Ir:	SVol.2-184

BrC_5F_2HINO	HOC_5F_2BrIN	F:	PerFHalOrg.5-182/3
		F:	PerFHalOrg.6-13
BrC_5F_3HNO	HOC_5F_3BrN	F:	PerFHalOrg.5-182/3, 199
		F:	PerFHalOrg.6-13
$BrC_5F_3H_2N_2$	$NH_2C_5F_3BrN$	F:	PerFHalOrg.5-185/6, 201, 216
BrC_5F_3IN	$IBrC_5F_3N$	F:	PerFHalOrg.5-118/9
BrC_5F_4MgN	NC_5F_4MgBr	F:	PerFHalOrg.4-65, 97/8, 110
		F:	PerFHalOrg.5-115/6, 143
BrC_5F_4N	BrC_5F_4N	F:	PerFHalOrg.5-118/9, 126, 128, 136
BrC_5F_6Li	$LiC(CF_2)_3CBr$	F:	PerFHalOrg.4-3, 27, 34
$BrC_5F_6NO_2$	$BrN(COCF_2)_2CF_2$	F:	PerFHalOrg.5-180/1
BrC_5F_9HNO	$C_4F_9C(O)NHBr$	F:	PerFHalOrg.7-18, 22, 46
$BrC_5F_{10}N$	$BrN(CF_2)_5$	F:	PerFHalOrg.5-120/1, 131
$BrC_5F_{11}MgO$	$(CF_3)_2CFO(CF_2)_2MgBr$	F:	PerFHalOrg.4-63, 86
BrC_5FeHO_5	$[(CO)_5FeH]Br$	Fe:	Org.Verb.B3-252
$BrC_5FeH_4NO_3$	$BrC_3H_4Fe(CO)_2NO$	Fe:	Org.Verb.B5-14/5, 17, 22/3
BrC_5FeH_5	$[C_5H_5Fe]Br$	Fe:	Org.Verb.A1-183
$BrC_5FeO_5^+$	$[(CO)_5FeBr]^+$	Fe:	Org.Verb.B3-107, 253
$BrC_5H_7MnO_2^+$	$[MnBr(CH(COCH_3)_2)]^+$	Mn:	MVol.D1-102
BrC_5H_7NiO	$CH_3C_3H_4Ni(CO)Br$	Ni:	Org.Verb.2-42
BrC_5H_9Sn	$(CH_3)_3SnCCBr$	Sn:	Org.Verb.2-112
$BrC_5H_{11}N_2NiS$	$CH_3C_3H_4Ni(SC(NH_2)_2)Br$	Ni:	Org.Verb.2-10
$BrC_5H_{12}NiO_2P$	$NiBr(OCOCH_3)P(CH_3)_3$	Ni:	Org.Verb.2-14
$BrC_5H_{13}N_4NiS_2$	$C_3H_5Ni(SC(NH_2)_2)_2Br$	Ni:	Org.Verb.2-22
$BrC_5H_{15}N_2O_2S$	$[(CH_3)_2NSO_2N(CH_3)_3]Br$	S:	S-N-Verb.1-172
$BrC_5H_{20}MnO_5^+$	$[MnBr(CH_3OH)_5]^+$	Mn:	MVol.D1-24/5
$BrC_5IrK_3N_5$	$K_3[Ir(CN)_5Br]$	Ir:	SVol.2-184
$BrC_5IrN_5^{3-}$	$[Ir(CN)_5Br]^{3-}$	Ir:	SVol.2-183
BrC_6ClCuF_5Mg	$C_6F_5Cu(MgBrCl)$	F:	PerFHalOrg.4-111
BrC_6ClF_4	BrC_6F_4Cl	F:	PerFHalOrg.4-16
BrC_6ClF_4Mg	ClC_6F_4MgBr	F:	PerFHalOrg.4-98, 109
BrC_6ClF_5P	C_6F_5PBrCl	F:	PerFHalOrg.3-124, 128
$BrC_6Cl_3H_4Sn$	$BrC_6H_4SnCl_3$	Sn:	Org.Verb.6-284
$BrC_6FFeH_2O_4$	$BrCHCHFFe(CO)_4$	Fe:	Org.Verb.B4-269, 278/80
$BrC_6F_2FeHO_4$	$F_2CCHBrFe(CO)_4$	Fe:	Org.Verb.B4-315/6
$BrC_6F_2H_6N_3$	$NCBrCFNCFC(N(CH_3)_2)$	F:	PerFHalOrg.6-60, 67
BrC_6F_3	BrC_6F_3	F:	PerFHalOrg.4-9/10, 11
$BrC_6F_3FeO_4$	$CF_2CFBrFe(CO)_4$	Fe:	Org.Verb.B4-320, 322/4
$BrC_6F_3H_3NO$	$(Br)(CH_3O)C_5F_3N$	F:	PerFHalOrg.5-136
$BrC_6F_3H_{11}NO_2$	$CF_2BrC(F)NOH \cdot (C_2H_5)_2O$	F:	PerFHalOrg.7-88, 105
–	$CF_3CBrNOH \cdot (C_2H_5)_2O$	F:	PerFHalOrg.7-88, 105, 146
$BrC_6F_4H_2N$	$BrC_6F_4NH_2$	F:	PerFHalOrg.7-9, 37, 62
BrC_6F_4I	IC_6F_4Br	F:	PerFHalOrg.4-16
BrC_6F_4Li	BrC_6F_4Li	F:	PerFHalOrg.4-5/6, 10/1, 13/5, 25, 28/9, 39, 53

BrC_6F_4NO	NOC_6F_4Br	F:	PerFHalOrg.7-180, 186, 198/9, 209
BrC_6F_5	C_6F_5Br	F:	PerFHalOrg.4-10, 107, 136
BrC_6F_5Mg	C_6F_5MgBr	F:	PerFHalOrg.4-63/5, 67/71, 79/85, 99/102, 106/12
$BrC_6F_5N_2$	C_6F_5NNBr	F:	PerFHalOrg.7-61/2
BrC_6F_8Li	$BrC(CF_2)_4CLi$	F:	PerFHalOrg.4-3/4, 15
$BrC_6F_8N_3$	$(CF_3)_2(CF_2Br)C_3N_3$	F:	PerFHalOrg.6-81/2, 101
BrC_6F_9	$BrC(CF_2)_4CF$	F:	PerFHalOrg.4-14
$BrC_6F_{11}HNO$	$O(CF_2CF_2)_2NC_2HBrF_3$	F:	PerFHalOrg.6-12
$BrC_6F_{13}Mg$	n-$C_6F_{13}MgBr$	F:	PerFHalOrg.4-62/3, 68
$BrC_6F_{15}N_2O_2$	$(CF_3)_2NOCFBrCF_2ON(CF_3)_2$	F:	PerFHalOrg.7-95/6, 113
$BrC_6F_{18}GeN_3O_3$	$[(CF_3)_2NO]_3GeBr$	F:	PerFHalOrg.7-102
$BrC_6FeH_2IO_4$	$BrCHCHIFe(CO)_4$	Fe:	Org.Verb.B4-269, 278/80
$BrC_6FeH_3O_4$	$BrCHCH_2Fe(CO)_4$	Fe:	Org.Verb.B4-233/4, 239/40
$BrC_6FeH_5O_3$	$C_3H_5Fe(CO)_3Br$	Fe:	Org.Verb.B5-35/7, 47/50, 54/8
$BrC_6H_4S_{1.5}Sn$	$(BrC_6H_4SnS_{1.5})_x$	Sn:	Org.Verb.6-284
$BrC_6H_5O_2Sn$	$(BrC_6H_4SnOOH)_x$	Sn:	Org.Verb.6-284
$BrC_6H_8NiO_4$	$(CH(COOCH_3)CHCOOCH_3)NiBr$	Ni:	Org.Verb.1-338/9
$BrC_6H_9N_2NiO_4$	$CH_2C(COOC_2H_5)CHNOHNi(NO)Br$	Ni:	Org.Verb.2-307
$BrC_6H_9O_6Sn$	$SnBr(CH_3COO)_3$	Sn:	MVol.C2-222
$BrC_6H_{11}N_2NiS$	$C_3H_5Ni(SC(NHCH_2)_2)Br$	Ni:	Org.Verb.2-279
$BrC_6H_{14}NiP$	$C_3H_5Ni(P(CH_3)_3)Br$	Ni:	Org.Verb.2-12, 14
$BrC_7ClH_5NO_2S$	$BrC_6H_4CHNSO_2Cl$	S:	S-N-Verb.1-126
$BrC_7ClH_{18}SiSn$	$(CH_3)_3SnCBrClSi(CH_3)_3$	Sn:	Org.Verb.2-17/8
$BrC_7F_2H_6NO_2$	$(Br)(CH_3O)_2C_5F_2N$	F:	PerFHalOrg.5-136
$BrC_7F_3H_3N_5O_2$	$NC_5Br(OH)(NNH)N(OH)C(CF_3)N$	F:	PerFHalOrg.6-126/7
$BrC_7F_4HO_2$	BrC_6F_4COOH	F:	PerFHalOrg.4-102
$BrC_7F_4H_2NO$	$BrC_6F_4C(O)NH_2$	F:	PerFHalOrg.7-19, 22, 48
BrC_7F_7Mg	$CF_3C_6F_4MgBr$	F:	PerFHalOrg.4-63, 65, 109
BrC_7F_8N	$(CF_2)_2C_5F_4BrN$	F:	PerFHalOrg.5-156/7, 170
BrC_7F_9	$FC(CF_2CF_2)(CF_2)(CFCF)CBr$	F:	PerFHalOrg.4-69
BrC_7F_9Mg	$FC(CFCF)(CF_2)(CF_2CF_2)CMgBr$	F:	PerFHalOrg.4-66, 71
$BrC_7F_{10}Li$	$BrC(CF_2)(CF_2CF_2)(CFLiCF_2)CF$	F:	PerFHalOrg.4-4, 9, 15
BrC_7F_{11}	$BrC(CF_2)(CF_2CF_2)_2CF$	F:	PerFHalOrg.4-14
$BrC_7F_{11}Mg$	$FC(CF_2)(CF_2CF_2)_2CMgBr$	F:	PerFHalOrg.4-66, 69, 71
$BrC_7F_{15}MgO$	$(CF_3)_2CFO(CF_2)_4MgBr$	F:	PerFHalOrg.4-63, 86
$BrC_7F_{15}N_2O_2$	$CF_3C(ON(CF_3)_2)C(ON(CF_3)_2)Br$	F:	PerFHalOrg.7-97/8, 115
$BrC_7F_{18}N_3O_3$	$[(CF_3)_2NO]_3CBr$	F:	PerFHalOrg.7-94, 109
$BrC_7FeH_7O_3$	$CH_3C_3H_4Fe(CO)_3Br$	Fe:	Org.Verb.B5-41, 45, 48, 52/3, 60/1, 63
$BrC_7FeH_7O_4$	$CH_3OC_3H_4Fe(CO)_3Br$	Fe:	Org.Verb.B5-53, 63
$BrC_7Fe_2GeH_{13}N_4O_4$	$(C_3H_5Fe(NO)_2)_2Ge(Br)CH_3$	Fe:	Org.Verb.C2-165
$BrC_7Fe_2H_{13}N_4O_4Pb$	$(C_3H_5Fe(NO)_2)_2Pb(Br)CH_3$	Fe:	Org.Verb.C2-165
$BrC_7Fe_2H_{13}N_4O_4Sn$	$(C_3H_5Fe(NO)_2)_2Sn(Br)CH_3$	Fe:	Org.Verb.C2-165
$BrC_7H_{15}N_2NiS$	$C_3H_5Ni(SC(NH_2)(NHCH(CH_3)_2))Br$	Ni:	Org.Verb.2-279

$BrC_7H_{16}NNi$	$C_3H_5Ni(NH(C_2H_5)_2)Br$	Ni:	Org.Verb.2-9
$BrC_7H_{17}NiO_2S_2$	$C_3H_5Ni(OS(CH_3)_2)_2Br$	Ni:	Org.Verb.2-19
$BrC_7H_{17}Sn$	$(C_2H_5)_3SnCH_2Br$	Sn:	Org.Verb.2-164
$BrC_7H_{19}N_2O_2S$	$[(C_2H_5)_2NSO_2N(CH_3)_3]Br$	S:	S-N-Verb.1-172
$BrC_7H_{20}NSn$	$[(CH_3)_3SnCH_2N(CH_3)_3]Br$	Sn:	Org.Verb.2-14, 16
$BrC_7H_{21}NiP_2$	$CH_3Ni(P(CH_3)_3)_2Br$	Ni:	Org.Verb.1-12, 19
$BrC_8Cl_2H_{20}NSn$	$[(C_2H_5)_4N]SnBrCl_2$	Sn:	MVol.C3-113
$BrC_8Cl_3F_3HN_2$	$C_6Cl_3BrNC(CF_3)NH$	F:	PerFHalOrg.6-125/6, 139
$BrC_8Cl_3F_3HN_2O_2S$	$C_6Cl_2Br(SO_2Cl)NC(CF_3)NH$	F:	PerFHalOrg.6-125/6, 160, 162
$BrC_8Cl_3H_{16}O_2Ti$	$TiCl_3Br \cdot 2\ C_4H_8O$	Ti:	Org.Verb.1-41
$BrC_8Cl_4H_{20}NTe$	$[(C_2H_5)_4N][TeBrCl_4]$	Te:	SVol.B3-86
$BrC_8FFe_2H_2O_6$	$(CO)_3Fe(CHCHF)BrFe(CO)_3$	Fe:	Org.Verb.C2-79/81
$BrC_8F_4H_3O_2$	$CH_3COOC_6F_4Br$	F:	PerFHalOrg.4-56
$BrC_8F_5H_4O$	$CH_2BrCH(OH)C_6F_5$	F:	PerFHalOrg.4-101
BrC_8F_5Mg	$C_6F_5CCMgBr$	F:	PerFHalOrg.4-62/3
BrC_8F_{13}	$CF_3C(CF_2)(CF_2CF_2)_2CBr$	F:	PerFHalOrg.4-9, 14
BrC_8F_{15}	$C_6F_{13}CFCFBr$	F:	PerFHalOrg.4-68/9
$BrC_8F_{16}^-$	$C_6F_{13}CFCF_2Br^-$	F:	PerFHalOrg.4-69
$BrC_8F_{16}IMg$	$CF_2I(CF_2)_7MgBr$	F:	PerFHalOrg.4-62/3
$BrC_8F_{17}Mg$	$(n\text{-}C_8F_{17})MgBr$	F:	PerFHalOrg.4-62/3, 68/9, 71, 86, 103/4
$BrC_8FeH_5O_4$	$(CH_2CHCBrCH_2)Fe(CO)_4$	Fe:	Org.Verb.C2-31
$BrC_8FeH_7MgO_5$	$[MgBr][CH_3(CH_2)_2C(O)Fe(CO)_4]$	Fe:	Org.Verb.B4-156, 167
$BrC_8FeH_7O_4$	$CH_3COC_3H_4Fe(CO)_3Br$	Fe:	Org.Verb.B5-52
$BrC_8FeH_7O_5$	$CH_3OCOC_3H_4Fe(CO)_3Br$	Fe:	Org.Verb.B5-52, 62
$BrC_8FeH_9O_3$	$(CH_3)_2C_3H_3Fe(CO)_3Br$	Fe:	Org.Verb.B5-46, 53, 64
$BrC_8Fe_2H_2IO_6$	$(CO)_3Fe(CHCHBr)IFe(CO)_3$	Fe:	Org.Verb.C2-79/80, 82, 84
$BrC_8H_7NTe^-$	$C_6H_5CH_2Te(Br)CN^-$	C:	MVol.D6-260
$BrC_8H_7S_3$	$SC(SCH_2C_6H_4Br)(SH)$	C:	MVol.D4-264/5
$BrC_8H_{10}NNi$	$C_3H_5Ni(C_5H_5N)Br$	Ni:	Org.Verb.2-278
$BrC_8H_{12}Ni$	$C_8H_{12}NiBr$	Ni:	Org.Verb.2-74
–	$(C_8H_{12}NiBr)_x$	Ni:	Org.Verb.2-399/400
$BrC_8H_{13}N_2NiO_2$	$C_8H_{12}NOHNi(NO)Br$	Ni:	Org.Verb.2-320
$BrC_8H_{15}Sn$	$(C_2H_5)_3SnCCBr$	Sn:	Org.Verb.2-218
$BrC_8H_{16}NNi$	$C_8H_{13}Ni(NH_3)Br$	Ni:	Org.Verb.2-10
$BrC_8H_{17}N_2NiS$	$C_3H_5Ni(SC(N(CH_3)_2)_2)Br$	Ni:	Org.Verb.2-10
$BrC_8H_{19}Sn$	$(CH_3)_3Sn(CH_2)_5Br$	Sn:	Org.Verb.2-53, 55
$BrC_8H_{20}I_2NSn$	$[(C_2H_5)_4N]SnBrI_2$	Sn:	MVol.C3-120
$BrC_8H_{21}NiOP_2$	$CH_3C(O)Ni(P(CH_3)_3)_2Br$	Ni:	Org.Verb.1-14
$BrC_9ClH_{12}Sn$	$(CH_3)_2(BrC_6H_4CH_2)SnCl$	Sn:	Org.Verb.5-194
$BrC_9Cl_2F_3H_5N_3O_2S$	$(CH_3NHSO_2)(CF_3)C_7BrCl_2HN_2$	F:	PerFHalOrg.6-160
$BrC_9Cl_3H_{16}Sn$	$(C_2H_5)_3SnCBrCHCCl_3$	Sn:	Org.Verb.2-193
–	$(C_2H_5)_3SnC(CCl_3)CHBr$	Sn:	Org.Verb.2-193
$BrC_9Cl_3H_{18}Sn$	$(C_2H_5)_3SnCHBrCH_2CCl_3$	Sn:	Org.Verb.2-165
$BrC_9Cl_4H_8NO_2Sn$	$SnCl_4 \cdot BrC_6H_4CONHCOCH_3$	Sn:	MVol.C6-50/1
$BrC_9F_4H_9Si$	$(CH_3)_3SiC_6F_4Br$	F:	PerFHalOrg.4-25
$BrC_9F_4H_9Sn$	$BrC_6F_4Sn(CH_3)_3$	F:	PerFHalOrg.4-28
		Sn:	Org.Verb.2-135/6

BrC_9F_5HNO	HOC_9F_5BrN	F:	PerFHalOrg.6-130/1, 143, 160
$BrC_9F_6H_6N$	$BrC_6H_4C(CF_3)_2NH_2$	F:	PerFHalOrg.7-69, 72
BrC_9F_6N	C_9F_6BrN	F:	PerFHalOrg.6-130/1, 143, 153
–	i-C_9F_6BrN	F:	PerFHalOrg.6-132/3, 146
BrC_9F_9	$BrC_6F_3(CF_2)_3$	F:	PerFHalOrg.7-62
$BrC_9F_{19}MgO$	$(CF_3)_2CFO(CF_2)_6MgBr$	F:	PerFHalOrg.4-63, 71, 87
$BrC_9FeH_9MgO_5$	$[MgBr][CH_3(CH_2)_3C(O)Fe(CO)_4]$	Fe:	Org.Verb.B4-156, 167
$BrC_9FeH_{11}O_3$	$(CH_3)_3C_3H_2Fe(CO)_3Br$	Fe:	Org.Verb.B5-46, 54, 64
$BrC_9H_{12}N_3Sn$	$BrSn(CH_2CH_2CN)_3$	Sn:	MVol.C6-184
$BrC_9H_{13}MgSn$	$(CH_3)_3SnC_6H_4MgBr$	Sn:	Org.Verb.2-140/1
$BrC_9H_{13}Sn$	$(CH_3)_3SnC_6H_4Br$	Sn:	Org.Verb.2-133/4
$BrC_9H_{15}O_2Ti$	$C_5H_5Ti(OC_2H_5)_2Br$	Ti :	Org.Verb.1-184/7
$BrC_9H_{16}NNi$	$CH_3C_3H_4Ni(CNC_4H_9\text{-t})Br$	Ni:	Org.Verb.2-45
$BrC_9H_{16}Ni$	$C_8H_{13}Ni(CH_3)Br$	Ni:	Org.Verb.2-320
$BrC_9H_{17}N_4NiS_2$	$C_3H_5Ni(SC(NHCH_2)_2)_2Br$	Ni:	Org.Verb.2-21
$BrC_9H_{17}Sn$	$(C_2H_5)_3SnCCCH_2Br$	Sn:	Org.Verb.2-221, 234
$BrC_9H_{20}NNi$	$C_3H_5Ni(N(C_2H_5)_3)Br$	Ni:	Org.Verb.2-9/10
$BrC_9H_{20}NiP$	$C_3H_5Ni(P(C_2H_5)_3)Br$	Ni:	Org.Verb.2-12, 14
$BrC_9H_{23}NiP_2$	$C_3H_5Ni(P(CH_3)_3)_2Br$	Ni:	Org.Verb.2-278
$BrC_{10}ClFeH_4O_6S$	$BrC_6H_4SO_2Fe(CO)_4Cl$	Fe:	Org.Verb.B3-187
$BrC_{10}Cl_2F_3H_7N_3O_2S$	$((CH_3)_2NSO_2)(CF_3)C_7BrCl_2HN_2$	F:	PerFHalOrg.6-162
–	$(C_2H_5NHSO_2)(CF_3)C_7Cl_2BrHN_2$	F:	PerFHalOrg.6-160
$BrC_{10}Cl_3H_{12}N_2O_2U$	$[C_5H_5NH]_2UO_2BrCl_3$	U:	SVol.C9-162
$BrC_{10}Cl_4H_{10}NO_2Sn$	$SnCl_4 \cdot BrC_6H_4CONHCOC_2H_5$	Sn:	MVol.C6-50/1
$BrC_{10}Cl_6H_9Sn$	$(CH_3)_3SnC_7Cl_6Br$	Sn:	Org.Verb.2-102
$BrC_{10}F_5H_3NO$	$CH_3OC_9F_5BrN$	F:	PerFHalOrg.6-153, 160
$BrC_{10}F_6H_5MgNiO$	$C_5H_5Ni(CO)CF(MgBr)CF_2CFCF_2$	Ni:	Org.Verb.2-154
$BrC_{10}F_6Mn_2O_8P$	$Mn_2(CO)_8P(CF_3)_2Br$	F:	PerFHalOrg.3-110/1, 120
$BrC_{10}F_7Mg$	$C_{10}F_7MgBr$	F:	PerFHalOrg.4-64/5
$BrC_{10}F_{21}Mg$	$(n\text{-}C_{10}F_{21})MgBr$	F:	PerFHalOrg.4-62/3, 68, 71, 86
$BrC_{10}FeH_9$	$C_5H_5FeC_5H_4Br$	Fe:	Org.Verb.A1-368/70, 372/80
$BrC_{10}FeH_9^+$	$[C_5H_5FeC_5H_4Br]^+$	Fe:	Org.Verb.A1-373
$BrC_{10}FeH_{10}$	$[Fe(C_5H_5)_2]Br$	Fe:	Org.Verb.A1-215
$BrC_{10}FeH_{11}MgO_5$	$MgBr[C_6H_{11}C(O)Fe(CO)_4]$	Fe:	Org.Verb.B4-150, 160
$BrC_{10}FeH_{11}O_3$	$C_6H_9CH_2Fe(CO)_3Br$	Fe:	Org.Verb.B5-53, 64
$BrC_{10}FeH_{19}NO_5P$	$BrC_3H_4Fe(CO)(NO)P(OC_2H_5)_3$	Fe:	Org.Verb.B5-108/11
$BrC_{10}Fe_2HO_9$	$(CO)_3Fe(CCHCOOCO)BrFe(CO)_3$	Fe:	Org.Verb.C2-79/80, 82, 84
$BrC_{10}Fe_2H_5O_8$	$(CO)_3Fe(CHCHCOOCH_3)BrFe(CO)_3$	Fe:	Org.Verb.C2-79/81, 84
$BrC_{10}H_{11}Ni$	$[C_5H_5NiC_5H_6]Br$	Ni:	Org.Verb.2-205
$BrC_{10}H_{14}MnO_4$	$[MnBr(CH(COCH_3)_2)_2]$	Mn:	MVol.D1-101/2
$BrC_{10}H_{15}MnN_5^+$	$[Mn(CH_3CN)_5Br]^+$	Mn:	MVol.C5-287
$BrC_{10}H_{15}Ni$	$C_3H_5Ni(C_7H_{10})Br$	Ni:	Org.Verb.2-46
$BrC_{10}H_{15}Sn$	$(CH_3)_3SnCH_2C_6H_4Br$	Sn:	Org.Verb.2-23
–	$(CH_3)_3SnC_6H_4CH_2Br$	Sn:	Org.Verb.2-127
$BrC_{10}H_{16}Ni$	$(C_{10}H_{16}NiBr)_x$	Ni:	Org.Verb.2-400

$BrC_{10}H_{17}N_2NiO_2$	$(CH_3)_2CC_3H_5C(CH_3)C_2H_2NOHNi(NO)Br$	Ni: Org.Verb.2-311
$BrC_{10}H_{18}MnO_6$	$[Mn(CH(COCH_3)_2)_2(H_2O)_2]Br \cdot 2\ H_2O$	Mn: MVol.D1-99
$BrC_{10}H_{18}N_2Ni$	$(Ni(CNC_4H_9\text{-}t)_2Br)_x$	Ni: Org.Verb.2-397
$BrC_{10}H_{19}N_4NiS_2$	$CH_3C_3H_4Ni(SC(NHCH_2)_2)_2Br$	Ni: Org.Verb.2-21
$BrC_{10}H_{19}Sn$	$(CH_3)_3SnC_7H_{10}Br$	Sn: Org.Verb.2-62, 67
–	$(C_2H_5)_3SnCCCHBrCH_3$	Sn: Org.Verb.2-223
$BrC_{10}H_{21}N_2NiS$	$C_3H_5Ni(SC(NHCH(CH_3)_2)_2)Br$	Ni: Org.Verb.2-279
$BrC_{11}Cl_2F_2H_3N_4O_4$	$NC_5F_2Cl_2NHC_6H_2Br(NO_2)_2$	F: PerFHalOrg.5-140
$BrC_{11}Cl_2F_2H_4N_3O_2$	$NC_5F_2Cl_2NHC_6H_3Br(NO_2)$	F: PerFHalOrg.5-140
$BrC_{11}Cl_2F_6HN_2$	$NC_5F_2Cl_2NHC_6F_4Br$	F: PerFHalOrg.5-188/9
		F: PerFHalOrg.6-13
$BrC_{11}Cl_4H_{12}NO_2Sn$	$SnCl_4 \cdot BrC_6H_4CONHCOC_3H_7$	Sn: MVol.C6-50/1
$BrC_{11}F_5H_5Ti$	$C_5H_5Ti(C_6F_5)Br$	F: PerFHalOrg.4-112
$BrC_{11}F_{23}MgO$	$(CF_3)_2CFO(CF_2)_8MgBr$	F: PerFHalOrg.4-63, 71, 87
$BrC_{11}FeH_5MgO_5$	$[MgBr][C_6H_5C(O)Fe(CO)_4]$	Fe: Org.Verb.B4-157, 167
$BrC_{11}FeH_6MgNO_5$	$[MgBr][C_6H_5NHC(O)Fe(CO)_4]$	Fe: Org.Verb.B4-156, 165/6
$BrC_{11}FeH_8N$	$BrC_5H_4FeC_5H_4CN$	Fe: Org.Verb.A1-374
$BrC_{11}FeH_9O_2$	$BrC_5H_4FeC_5H_4COOH$	Fe: Org.Verb.A3-58/9
$BrC_{11}FeH_{10}^+$	$[C_5H_5FeC_6H_5Br]^+$	Fe: Org.Verb.A1-118
$BrC_{11}FeH_{11}MgO_5$	$[MgBr][C_6H_{11}C(O)Fe(CO)_4]$	Fe: Org.Verb.B3-134
$BrC_{11}FeH_{11}N_2O_3$	$(CH_3)_2C_3BrN_2CH_2CHCH_2Fe(CO)_3$	Fe: Org.Verb.B4-216, 225
$BrC_{11}FeH_{12}MgNO_5$	$[MgBr][C_6H_{11}NHC(O)Fe(CO)_4]$	Fe: Org.Verb.B4-156, 165/6
$BrC_{11}FeH_{13}MgO_5$	$MgBr[C_6H_{13}C(O)Fe(CO)_4]$	Fe: Org.Verb.B4-151, 160
$BrC_{11}Fe_2GeH_{19}N_4O_4$	$(C_3H_5Fe(NO)_2)_2Ge(Br)C_5H_9$	Fe: Org.Verb.C2-165
$BrC_{11}Fe_2GeH_{21}N_4O_4$	$((CH_3)_2C_3H_3Fe(NO)_2)_2Ge(Br)CH_3$	Fe: Org.Verb.C2-165
$BrC_{11}Fe_2H_{19}N_4O_4Pb$	$(C_3H_5Fe(NO)_2)_2Pb(Br)C_5H_9$	Fe: Org.Verb.C2-165
$BrC_{11}Fe_2H_{19}N_4O_4Sn$	$(C_3H_5Fe(NO)_2)_2Sn(Br)C_5H_9$	Fe: Org.Verb.C2-165
$BrC_{11}Fe_2H_{21}N_4O_4Pb$	$((CH_3)_2C_3H_3Fe(NO)_2)_2Pb(Br)CH_3$	Fe: Org.Verb.C2-165
$BrC_{11}Fe_2H_{21}N_4O_4Sn$	$((CH_3)_2C_3H_3Fe(NO)_2)_2Sn(Br)CH_3$	Fe: Org.Verb.C2-165
$BrC_{11}H_{20}NiP$	$C_5H_5Ni(P(C_2H_5)_3)Br$	Ni: Org.Verb.2-117, 119
$BrC_{11}H_{25}Sn$	$(C_2H_5)_3SnCH_2(CH_2)_4Br$	Sn: Org.Verb.2-168, 177
$BrC_{11}H_{27}N_2Ni$	$C_3H_5Ni(NH(C_2H_5)_2)_2Br$	Ni: Org.Verb.2-19
$BrC_{12}ClH_{10}Sn$	$(C_6H_5)_2SnBrCl$	Sn: Org.Verb.5-173
$BrC_{12}Cl_2F_3H_{11}N_3O_2S$	$((C_2H_5)_2NSO_2)(CF_3)C_7BrCl_2HN_2$	F: PerFHalOrg.6-162
$BrC_{12}Cl_2F_5H_4N_2$	$NC_5F_2Cl_2NHC_6H_3Br(CF_3)$	F: PerFHalOrg.5-140
$BrC_{12}Cl_2FeH_{20}NO_4Sn$	$[N(C_2H_5)_4][(CO)_4FeSnCl_2Br]$	Fe: Org.Verb.B2-143
$BrC_{12}Cl_2H_{28}NSn$	$[(C_3H_7)_4N]SnBrCl_2$	Sn: MVol.C3-113
$BrC_{12}Cl_3H_{22}Sn$	$(C_3H_7)_3SnCBrCHCCl_3$	Sn: Org.Verb.2-260
$BrC_{12}Cl_4H_7IrKN_2$	$K[Ir(C_{12}H_7BrN_2)Cl_4]$	Ir: SVol.2-93
$BrC_{12}Cl_4H_{14}NO_2Sn$	$SnCl_4 \cdot BrC_6H_4CONHCOC_4H_9$	Sn: MVol.C6-50/1
$BrC_{12}F_4H_8NO$	$OC_6H_8NC_6F_4Br$	F: PerFHalOrg.7-198/9
$BrC_{12}F_7$	$BrC_6F_3C_6F_4$	F: PerFHalOrg.4-14
$BrC_{12}F_7N_2$	$C_6F_3BrNNC_6F_4$	F: PerFHalOrg.6-136/7, 150, 158
$BrC_{12}F_8HN_2O_2$	$(NO_2C_6F_3Br)(C_6F_5)NH$	F: PerFHalOrg.7-14/5, 43
$BrC_{12}F_8H_3N_2$	$(NH_2C_6F_3Br)(C_6F_5)NH$	F: PerFHalOrg.7-14/5, 40, 55
$BrC_{12}F_8Li$	$BrC_6F_4C_6F_4Li$	F: PerFHalOrg.4-6/7, 7/8
$BrC_{12}F_9$	$BrC_6F_4C_6F_5$	F: PerFHalOrg.4-11, 14/9
$BrC_{12}F_9HN$	$(BrC_6F_4)(C_6F_5)NH$	F: PerFHalOrg.7-14/5, 42

$BrC_{12}F_9Mg$	$C_6F_5C_6F_4MgBr$	F:	PerFHalOrg.4-63, 65, 70, 71
$BrC_{12}F_9MgO$	$C_6F_5OC_6F_4MgBr$	F:	PerFHalOrg.4-63, 65, 70/1, 104, 112
$BrC_{12}F_9O$	$C_6F_5OC_6F_4Br$	F:	PerFHalOrg.4-16
		F:	PerFHalOrg.7-62
$BrC_{12}F_{10}GeH$	$(C_6F_5)_2GeBrH$	F:	PerFHalOrg.4-165/6, 174, 179
$BrC_{12}F_{10}In$	$(C_6F_5)_2InBr$	F:	PerFHalOrg.4-128/9
$BrC_{12}F_{10}P$	$(C_6F_5)_2PBr$	F:	PerFHalOrg.3-124, 125, 128
$BrC_{12}F_{10}PS$	$(C_6F_5)_2P(S)Br$	F:	PerFHalOrg.3-141, 148
$BrC_{12}F_{10}Tl$	$(C_6F_5)_2TlBr$	F:	PerFHalOrg.4-131/2, 135/8
$BrC_{12}FeH_7MgO_5$	$[MgBr][CH_3C_6H_4C(O)Fe(CO)_4]$	Fe:	Org.Verb.B4-157, 169
$BrC_{12}FeH_7O_4$	$BrC_6H_4CHCH_2Fe(CO)_4$	Fe:	Org.Verb.B4-239, 254
$BrC_{12}FeH_8MgNO_5$	$[MgBr][CH_3C_6H_4NHC(O)Fe(CO)_4]$	Fe:	Org.Verb.B4-156, 165/6
$BrC_{12}FeH_9O_3$	$C_6H_5C_3H_4Fe(CO)_3Br$	Fe:	Org.Verb.B5-43, 48, 52/3, 62, 64
$BrC_{12}FeH_{11}$	$C_5H_5FeC_5H_4CHCHBr$	Fe:	Org.Verb.A1-384, 386
$BrC_{12}FeH_{13}$	$C_5H_5FeC_5H_4CHBrCH_3$	Fe:	Org.Verb.A1-381, 383
–	$C_5H_5FeC_5H_4CH_2CH_2Br$	Fe:	Org.Verb.A1-381
$BrC_{12}FeH_{13}N_2O_3$	$(CH_3)_2C_3BrN_2CH_2C(CH_3)CH_2Fe(CO)_3$	Fe:	Org.Verb.B4-216, 225
$BrC_{12}Fe_2GeH_{15}N_4O_4$	$(C_3H_5Fe(NO)_2)_2Ge(Br)C_6H_5$	Fe:	Org.Verb.C2-165
$BrC_{12}Fe_2H_{15}N_4O_4Pb$	$(C_3H_5Fe(NO)_2)_2Pb(Br)C_6H_5$	Fe:	Org.Verb.C2-165
$BrC_{12}Fe_2H_{15}N_4O_4Sn$	$(C_3H_5Fe(NO)_2)_2Sn(Br)C_6H_5$	Fe:	Org.Verb.C2-165
$BrC_{12}H_{15}Sn$	$(CH_3)_3SnC_9H_6Br$	Sn:	Org.Verb.2-103, 106
$BrC_{12}H_{19}MgSn$	$(C_2H_5)_3SnC_6H_4MgBr$	Sn:	Org.Verb.2-242
$BrC_{12}H_{19}Sn$	$(C_2H_5)_3SnC_6H_4Br$	Sn:	Org.Verb.2-238/40, 243
$BrC_{12}H_{26}NiP$	$C_3H_5Ni(P(C_3H_7\text{-}i)_3)Br$	Ni:	Org.Verb.2-12, 14
$BrC_{12}H_{27}Sn$	$(C_4H_9)_3SnBr$	Sn:	Org.Verb.5-28, 131
$BrC_{12}H_{30}N_3Ti$	$BrTi(N(C_2H_5)_2)_3$	Ti:	Org.Verb.1-190
$BrC_{13}ClFeH_6O_5$	$ClCHCHCOC_6H_4BrFe(CO)_4$	Fe:	Org.Verb.B4-268, 288/9
$BrC_{13}Cl_2H_{25}MnN_2$	$(C_2H_5)_4N[MnBrCl_2(C_5H_5N)]$	Mn:	MVol.C5-309
$BrC_{13}Cl_4H_{10}NTe$	$TeCl_4 \cdot C_6H_4BrCHNC_6H_5$	Te:	SVol.B3-172/3
–	$TeCl_4 \cdot C_6H_5CHNC_6H_4Br$	Te:	SVol.B3-172/3
$BrC_{13}FH_{12}Sn$	$(C_6H_5)_2(CH_2Br)SnF$	Sn:	Org.Verb.5-40
$BrC_{13}F_3H_{30}NiP_2$	$CF_3Ni(P(C_2H_5)_3)_2Br$	Ni:	Org.Verb.1-28
$BrC_{13}F_5H_5NO$	$HOC_6H_3BrCHNC_6F_5$	F:	PerFHalOrg.7-59
$BrC_{13}FeH_{13}O$	$C_5H_5FeC_5H_4CHCBrCH_2OH$	Fe:	Org.Verb.A2-74, 79
–	$C_5H_5FeC_5H_4CHOHCBrCH_2$	Fe:	Org.Verb.A2-45, 49
–	$C_5H_5FeC_5H_4COCH_2CH_2Br$	Fe:	Org.Verb.A2-242/3, 245
$BrC_{13}FeH_{16}N$	$C_5H_5FeC_5H_3(Br)CH_2N(CH_3)_2$	Fe:	Org.Verb.A6-36
$BrC_{13}FeH_{17}MgO_5$	$[MgBr][CH_3(CH_2)_7C(O)Fe(CO)_4]$	Fe:	Org.Verb.B4-156, 167
$BrC_{13}Fe_2GeH_{17}N_4O_4$	$(C_3H_5Fe(NO)_2)_2Ge(Br)CH_2C_6H_5$	Fe:	Org.Verb.C2-165
–	$(C_3H_5Fe(NO)_2)_2Ge(Br)C_6H_4CH_3$	Fe:	Org.Verb.C2-165
$BrC_{13}Fe_2H_{17}N_4O_4Pb$	$(C_3H_5Fe(NO)_2)_2Pb(Br)CH_2C_6H_5$	Fe:	Org.Verb.C2-165
–	$(C_3H_5Fe(NO)_2)_2Pb(Br)C_6H_4CH_3$	Fe:	Org.Verb.C2-165
$BrC_{13}Fe_2H_{17}N_4O_4Sn$	$(C_3H_5Fe(NO)_2)_2Sn(Br)CH_2C_6H_5$	Fe:	Org.Verb.C2-165
–	$(C_3H_5Fe(NO)_2)_2Sn(Br)C_6H_4CH_3$	Fe:	Org.Verb.C2-165

$BrC_{13}H_{13}N_2Ni$	$C_3H_5Ni(NC_5H_4C_5H_4N)Br$	Ni:	Org.Verb.2-10
$BrC_{13}H_{15}N_2Ni$	$C_3H_5Ni(NC_5H_5)_2Br$	Ni:	Org.Verb.2-19
$BrC_{13}H_{21}NiO$	$CH_3(t-C_4H_9)_2C_3Ni(CO)Br$	Ni:	Org.Verb.2-42
$BrC_{13}H_{23}O_2Ti$	$C_5H_5Ti(OC_4H_9)_2Br$	Ti:	Org.Verb.1-185/6
$BrC_{13}H_{25}N_2Ti$	$C_5H_5Ti(N(C_2H_5)_2)_2Br$	Ti:	Org.Verb.1-190
$BrC_{13}H_{28}NiP$	$C_3H_5Ni((C_4H_9-t)_2PC_2H_5)Br$	Ni:	Org.Verb.2-12
$BrC_{13}H_{29}Sn$	$(C_4H_9)_3SnCH_2Br$	Sn:	Org.Verb.2-280
$BrC_{13}H_{39}NiP_4$	$CH_3Ni(P(CH_3)_3)_4Br$	Ni:	Org.Verb.1-12, 20
$BrC_{14}Cl_2H_{27}MnN_2$	$(C_2H_5)_4N[MnBrCl_2(CH_3C_5H_4N)]$	Mn:	MVol.C5-309
$BrC_{14}Cl_3H_{30}NiP_2$	$CCl_2CClNi(P(C_2H_5)_3)_2Br$	Ni:	Org.Verb.1-28, 30
$BrC_{14}F_3H_{30}NiP_2$	$[(C_2H_5)_3P]_2Ni(Br)CFCF_2$	F:	PerFHalOrg.4-110
		Ni:	Org.Verb.1-27, 29
$BrC_{14}F_3H_{30}P_2Pt$	$[(C_2H_5)_3P]_2Pt(Br)CFCF_2$	F:	PerFHalOrg.4-110
$BrC_{14}F_{10}H_5OSi$	$(C_6F_5)_2Si(OC_2H_5)Br$	F:	PerFHalOrg.4-151
$BrC_{14}F_{11}$	$CFBrCFC_6F_4C_6F_5$	F:	PerFHalOrg.4-17
$BrC_{14}FeH_{13}$	$C_5H_5FeC_5H_4C(CH_2)CBrCH_2$	Fe:	Org.Verb.A1-385, 386
–	$C_5H_5FeC_5H_4CHCBrCHCH_2$	Fe:	Org.Verb.A1-385, 386
$BrC_{14}FeH_{15}O$	$C_5H_5FeC_5H_4C(CH_3)CBrCH_2OH$	Fe:	Org.Verb.A2-74, 79/80
–	$C_5H_5FeC_5H_4C(CH_3)(CBrCH_2)OH$	Fe:	Org.Verb.A2-59, 62
$BrC_{14}FeH_{18}N_5O_5$	$COFe(NC_5H_4Br)(C_2HN_2O_2(CH_3)_2)_2$	Fe:	Org.Verb.B1-33
$BrC_{14}Fe_2GeH_{23}N_4O_4$	$(C_3H_5Fe(NO)_2)_2Ge(Br)C_7H_{10}CH_3$	Fe:	Org.Verb.C2-165
$BrC_{14}Fe_2H_7O_7S$	$(BrC_6H_3CH(OCH_3)S)Fe_2(CO)_6$	Fe:	Org.Verb.C2-114
–	$(CO)_6Fe_2(CH_3OC(C_6H_4Br)S)$	Fe:	Org.Verb.C1-238/9, 241
$BrC_{14}Fe_2H_{23}N_4O_4Pb$	$(C_3H_5Fe(NO)_2)_2Pb(Br)C_7H_{10}CH_3$	Fe:	Org.Verb.C2-165
$BrC_{14}Fe_2H_{23}N_4O_4Sn$	$(C_3H_5Fe(NO)_2)_2Sn(Br)C_7H_{10}CH_3$	Fe:	Org.Verb.C2-165
$BrC_{14}H_{15}N_2NiS$	$C_3H_5Ni(SC(NH_2)(NHC_{10}H_7))Br$	Ni:	Org.Verb.2-279
$BrC_{14}H_{22}MnO_6$	$[MnBr(CH(COCH_3)_2)_2(C_4H_8O_2)] \cdot H_2O$	Mn:	MVol.D1-105
$BrC_{14}H_{30}NiP$	$C_3H_5Ni[(C_4H_9-t)_2PC_3H_7-i]Br$	Ni:	Org.Verb.2-12
$BrC_{15}Cl_2H_{27}MnN_2$	$(C_2H_5)_4N[MnBrCl_2(CH_2CHC_5H_4N)]$	Mn:	MVol.C5-309
$BrC_{15}Cl_2H_{29}MnN_2$	$(C_2H_5)_4N[MnBrCl_2((CH_3)_2C_5H_3N)]$	Mn:	MVol.C5-309
$BrC_{15}Cl_3H_{28}Sn$	$(C_4H_9)_3SnCBrCHCCl_3$	Sn:	Org.Verb.2-310
$BrC_{15}F_{10}HN_2$	$NHC(C_6F_5)CBrC(C_6F_5)N$	F:	PerFHalOrg.5-61, 83
$BrC_{15}FeH_{15}$	$C_5H_5FeC_5H_4C(CHCH_3)CBrCH_2$	Fe:	Org.Verb.A1-385, 387
$BrC_{15}FeH_{17}O$	$C_5H_5FeC_5H_4CHCBrCH_2OC_2H_5$	Fe:	Org.Verb.A2-131, 135
–	$C_5H_5FeC_5H_4CH(CBrCH_2)OC_2H_5$	Fe:	Org.Verb.A2-118, 120
–	$C_5H_5FeC_5H_4COCHBrC_3H_7$	Fe:	Org.Verb.A2-244, 246
–	$C_5H_5FeC_5H_4CO(CH_2)_4Br$	Fe:	Org.Verb.A2-244, 246
$BrC_{15}Fe_2GeH_{27}N_4O_4$	$((CH_3)_2C_3H_3Fe(NO)_2)_2Ge(Br)C_5H_9$	Fe:	Org.Verb.C2-165
$BrC_{15}Fe_2H_{27}N_4O_4Pb$	$((CH_3)_2C_3H_3Fe(NO)_2)_2Pb(Br)C_5H_9$	Fe:	Org.Verb.C2-165
$BrC_{15}Fe_2H_{27}N_4O_4Sn$	$((CH_3)_2C_3H_3Fe(NO)_2)_2Sn(Br)C_5H_9$	Fe:	Org.Verb.C2-165
$BrC_{15}H_{11}Ti$	$CH_2CHC_{13}H_8TiBr$	Ti:	Org.Verb.1-133
$BrC_{15}H_{18}NO_6U$	$[UO_2(CH(COCH_3)_2)_2(BrC_5H_4N)]$	U:	SVol.E1-31, 36/7
$BrC_{15}H_{23}N_2Ni$	$[C_5H_5Ni(CNC(CH_3)_3)_2]Br$	Ni:	Org.Verb.2-172
$BrC_{15}H_{25}Sn$	$(C_3H_7)_3SnC_6H_4Br$	Sn:	Org.Verb.2-267
$BrC_{15}H_{32}NiP$	$C_3H_5Ni(P(C_4H_9)_3)Br$	Ni:	Org.Verb.2-12
$BrC_{15}H_{35}NiP_2$	$C_3H_5Ni(P(C_2H_5)_3)_2Br$	Ni:	Org.Verb.2-22
$BrC_{16}Cl_2H_{36}NSn$	$[(C_4H_9)_4N]SnBrCl_2$	Sn:	MVol.C3-113
$BrC_{16}F_{33}Mg$	$[C_6F_{13}CFCF_2C_8F_{17}]MgBr$	F:	PerFHalOrg.4-69
$BrC_{16}FeH_{13}$	$C_5H_5FeC_5H_4C_6H_4Br$	Fe:	Org.Verb.A1-390
$BrC_{16}FeH_{13}^+$	$[C_5H_5FeC_5H_4C_6H_4Br]^+$	Fe:	Org.Verb.A1-390

$BrC_{16}FeH_{13}Mg$	$C_5H_5FeC_5H_4C_6H_4MgBr$	Fe: Org.Verb.A1-390
$BrC_{16}FeH_{13}O$	$C_5H_5FeC_5H_4OC_6H_4Br$	Fe: Org.Verb.A2-108
$BrC_{16}FeH_{15}$	$C_5H_5FeC_5H_4C(CBrCH_2)CHCHCH_2$	Fe: Org.Verb.A1-385, 387/8
$BrC_{16}FeH_{17}$	$C_5H_5FeC_5H_4C(CHC_2H_5)CBrCH_2$	Fe: Org.Verb.A1-385, 387
$BrC_{16}FeH_{19}O$	$C_5H_5FeC_5H_4C(CH_3)CBrCH_2OC_2H_5$	Fe: Org.Verb.A2-131, 135
–	$C_5H_5FeC_5H_4C(CH_3)(CBrCH_2)OC_2H_5$	Fe: Org.Verb.A2-119, 123
–	$C_5H_5FeC_5H_4CO(CH_2)_5Br$	Fe: Org.Verb.A2-244
$BrC_{16}Fe_2GeH_{17}N_4O_4$	$(C_3H_5Fe(NO)_2)_2Ge(Br)C_{10}H_7$	Fe: Org.Verb.C2-165
$BrC_{16}Fe_2GeH_{23}N_4O_4$	$((CH_3)_2C_3H_3Fe(NO)_2)_2Ge(Br)C_6H_5$	Fe: Org.Verb.C2-165
$BrC_{16}Fe_2H_{17}N_4O_4Pb$	$(C_3H_5Fe(NO)_2)_2Pb(Br)C_{10}H_7$	Fe: Org.Verb.C2-165
$BrC_{16}Fe_2H_{17}N_4O_4Sn$	$(C_3H_5Fe(NO)_2)_2Sn(Br)C_{10}H_7$	Fe: Org.Verb.C2-165
$BrC_{16}Fe_2H_{23}N_4O_4Pb$	$((CH_3)_2C_3H_3Fe(NO)_2)_2Pb(Br)C_6H_5$	Fe: Org.Verb.C2-165
$BrC_{16}Fe_2H_{23}N_4O_4Sn$	$((CH_3)_2C_3H_3Fe(NO)_2)_2Sn(Br)C_6H_5$	Fe: Org.Verb.C2-165
$BrC_{16}H_{13}N_2Ni$	$C_6H_5Ni(NC_5H_4C_5H_4N)Br$	Ni: Org.Verb.1-38, 44
$BrC_{16}H_{20}NSn$	$(CH_3)_3SnC_6H_4N(CH_3)C_6H_4Br$	Sn: Org.Verb.2-143
$BrC_{16}H_{21}MnNO_4$	$[MnBr(CH(COCH_3)_2)_2(CH_3C_5H_4N)]$	Mn: MVol.D1-105
$BrC_{16}H_{27}NiO$	$(t\text{-}C_4H_9)_3C_3Ni(CO)Br$	Ni: Org.Verb.2-42
$BrC_{16}H_{32}NiOP$	$C_3H_5Ni(CO)(P(C_4H_9)_3)Br$	Ni: Org.Verb.1-169
$BrC_{17}Cl_4H_{12}NOTe$	$TeCl_4 \cdot HOC_{10}H_6CHNC_6H_4Br$	Te: SVol.B3-172, 174
$BrC_{17}FeH_{13}O$	$C_5H_5FeC_5H_4COC_6H_4Br$	Fe: Org.Verb.A2-285/8
$BrC_{17}FeH_{15}$	$C_5H_5FeC_5H_4CHBrC_6H_5$	Fe: Org.Verb.A1-394
–	$C_5H_5FeC_5H_4CH_2C_6H_4Br$	Fe: Org.Verb.A1-388/9
$BrC_{17}FeH_{25}MgO_5$	$[MgBr][CH_3(CH_2)_{11}C(O)Fe(CO)_4]$	Fe: Org.Verb.B3-134
$BrC_{17}FeH_{26}MgNO_5$	$[MgBr][CH_3(CH_2)_{11}NHC(O)Fe(CO)_4]$	Fe: Org.Verb.B4-156, 166
$BrC_{17}Fe_2GeH_{25}N_4O_4$	$((CH_3)_2C_3H_3Fe(NO)_2)_2Ge(Br)CH_2C_6H_5$	Fe: Org.Verb.C2-165
–	$((CH_3)_2C_3H_3Fe(NO)_2)_2Ge(Br)C_6H_4CH_3$	Fe: Org.Verb.C2-165
$BrC_{17}Fe_2H_{25}N_4O_4Pb$	$((CH_3)_2C_3H_3Fe(NO)_2)_2Pb(Br)CH_2C_6H_5$	Fe: Org.Verb.C2-165
–	$((CH_3)_2C_3H_3Fe(NO)_2)_2Pb(Br)C_6H_4CH_3$	Fe: Org.Verb.C2-165
$BrC_{17}Fe_2H_{25}N_4O_4Sn$	$((CH_3)_2C_3H_3Fe(NO)_2)_2Sn(Br)CH_2C_6H_5$	Fe: Org.Verb.C2-165
–	$((CH_3)_2C_3H_3Fe(NO)_2)_2Sn(Br)C_6H_4CH_3$	Fe: Org.Verb.C2-165
$BrC_{17}H_{15}N_2Ni$	$C_6H_5CH_2Ni(NC_5H_4C_5H_4N)Br$	Ni: Org.Verb.1-18, 24
$BrC_{17}H_{17}N_2O_7U$	$UO_2(C_{13}H_7BrN_2O_4) \cdot C_4H_9OH$	U: SVol.E1-62, 69
$BrC_{17}H_{21}N_2NiS$	$C_3H_5Ni(SC(NHCH(CH_3)_2)(NHC_{10}H_7))Br$	Ni: Org.Verb.2-279
$BrC_{17}H_{25}O_2Ti$	$C_9H_7Ti(OC_4H_9)_2Br$	Ti: Org.Verb.1-187/8
$BrC_{17}H_{27}NiO_2$	$(t\text{-}C_4H_9)_3C_3Ni(CO)_2Br$	Ni: Org.Verb.2-43
$BrC_{17}H_{37}N_4NiS_2$	$C_3H_5Ni(SC(NHC_3H_7)_2)_2Br$	Ni: Org.Verb.2-21
$BrC_{18}ClH_{34}NiP_2$	$BrC_6H_4Ni(P(C_2H_5)_3)_2Cl$	Ni: Org.Verb.1-34, 41
–	$ClC_6H_4Ni(P(C_2H_5)_3)_2Br$	Ni: Org.Verb.1-34, 40
$BrC_{18}Cl_2H_{45}IrP_3$	$Ir(P(C_2H_5)_3)_3Cl_2Br$	Ir: SVol.2-200/2
$BrC_{18}FH_{34}NiP_2$	$FC_6H_4Ni(P(C_2H_5)_3)_2Br$	Ni: Org.Verb.1-38, 45
$BrC_{18}F_5H_{30}NiP_2$	$[(C_2H_5)_3P]_2Ni(C_6F_5)Br$	F: PerFHalOrg.4-112 Ni: Org.Verb.1-52, 54
$BrC_{18}F_5H_{30}P_2Pt$	$[(C_2H_5)_3P]_2Pt(Br)C_6F_5$	F: PerFHalOrg.4-111
$BrC_{18}F_{10}H_{12}O_2Tl$	$(C_6F_5)_2TlBr \cdot 2\ CH_3COCH_3$	F: PerFHalOrg.4-138
$BrC_{18}F_{13}$	$C_6F_5C_6F_4C_6F_4Br$	F: PerFHalOrg.4-16
$BrC_{18}F_{15}Ge$	$(C_6F_5)_3GeBr$	F: PerFHalOrg.4-166/7, 176/7, 180
$BrC_{18}F_{15}Pb$	$(C_6F_5)_3PbBr$	F: PerFHalOrg.4-190/1
$BrC_{18}F_{15}Si$	$(C_6F_5)_3SiBr$	F: PerFHalOrg.4-150/2, 157, 163

$BrC_{18}F_{15}Sn$	$(C_6F_5)_3SnBr$	F:	PerFHalOrg.4-182, 186, 189
$BrC_{18}FeH_{12}NO_3$	$(C_6H_5CHCHCHNC_6H_4Br)Fe(CO)_3$	Fe:	Org.Verb.B5-121/5, 128/9, 138
$BrC_{18}Fe_2GeH_{31}N_4O_4$	$((CH_3)_2C_3H_3Fe(NO)_2)_2Ge(Br)C_7H_{10}CH_3$	Fe:	Org.Verb.C2-165
$BrC_{18}Fe_2H_{31}N_4O_4Pb$	$((CH_3)_2C_3H_3Fe(NO)_2)_2Pb(Br)C_7H_{10}CH_3$	Fe:	Org.Verb.C2-165
$BrC_{18}Fe_2H_{31}N_4O_4Sn$	$((CH_3)_2C_3H_3Fe(NO)_2)_2Sn(Br)C_7H_{10}CH_3$	Fe:	Org.Verb.C2-165
$BrC_{18}H_{12}N_3O_6Sn$	$SnBr(NC_5H_4COO)_3$	Sn:	MVol.C5-205/6
$BrC_{18}H_{15}Sn$	$(C_6H_5)_3SnBr$	Sn:	Org.Verb.5-173
$BrC_{18}H_{18}NiP$	$C_5H_5Ni(P(C_6H_5)_2CH_3)Br$	Ni:	Org.Verb.2-117
$BrC_{18}H_{31}Sn$	$(C_4H_9)_3SnC_6H_4Br$	Sn:	Org.Verb.2-319, 321
$BrC_{18}H_{39}N_4NiS_2$	$CH_3C_3H_4Ni(SC(NHC_3H_7)_2)_2Br$	Ni:	Org.Verb.2-21
$BrC_{19}ClFeH_{18}N_6O_5$	$[COFe(C_{14}H_6N_6(CH_3)_4)Br]ClO_4$	Fe:	Org.Verb.B1-86
$BrC_{19}Cl_2H_{27}MnN_4$	$(CH_3)_4N[MnBrCl_2(C_5H_5N)_3]$	Mn:	MVol.C5-309
$BrC_{19}F_6FeH_{18}N_6OP$	$[COFe(C_{14}H_6N_6(CH_3)_4)Br]PF_6$	Fe:	Org.Verb.B1-86/7
$BrC_{19}FeH_{12}NO_4$	$(CO)_4Fe(BrC_6H_4NCHCHCHC_6H_5)$	Fe:	Org.Verb.B2-71/2, 74
$BrC_{19}FeH_{14}N$	$C_5H_5FeC_5H_4CHC(CN)C_6H_4Br$	Fe:	Org.Verb.A2-161
$BrC_{19}FeH_{15}O$	$C_5H_5FeC_5H_4CHCBrCOC_6H_5$	Fe:	Org.Verb.A3-24/5
–	$C_5H_5FeC_5H_4CHCHCOC_6H_4Br$	Fe:	Org.Verb.A3-18/9
–	$C_5H_5FeC_5H_4COCHCHC_6H_4Br$	Fe:	Org.Verb.A2-261/3
$BrC_{19}FeH_{18}N_6O^+$	$[COFe(C_{14}H_6N_6(CH_3)_4)Br]^+$	Fe:	Org.Verb.B1-86/7
$BrC_{19}FeH_{19}O$	$C_5H_5FeC_5H_4CHOHCH_2CH_2C_6H_4Br$	Fe:	Org.Verb.A2-45, 49
$BrC_{19}H_{15}Ni$	$(C_6H_5)_3CNiBr$	Ni:	Org.Verb.1-17, 23
$BrC_{19}H_{32}NiP$	$(CH_3)_2C_3H_3Ni(P(C_4H_9\text{-t})_2C_6H_5)Br$	Ni:	Org.Verb.2-16
$BrC_{19}H_{36}NNiO_3$	$[N(C_4H_9)_4][(CO)_3NiBr]$	Ni:	Org.Verb.1-191, 192
$BrC_{19}H_{37}NiP_2$	$CH_3C_6H_4Ni(P(C_2H_5)_3)_2Br$	Ni:	Org.Verb.1-35, 41
$BrC_{19}H_{42}NSn$	$[(C_4H_9)_3SnCH_2NCH_3(CH_2)_5]Br$	Sn:	Org.Verb.2-292
$BrC_{19}H_{45}NiP_2$	$CH_3Ni(P(C_3H_7\text{-i})_3)_2Br$	Ni:	Org.Verb.1-13, 20
$BrC_{20}ClH_{45}IrN_4O_4$	$[Ir(C_5H_{10}NH)_4HBr]ClO_4$	Ir:	SVol.2-76
$BrC_{20}Cl_2H_{21}SSn$	$[(C_6H_5)_3S][(CH_3)_2SnBrCl_2]$	Sn:	Org.Verb.6-35
$BrC_{20}Fe_2GeH_{25}N_4O_4$	$((CH_3)_2C_3H_3Fe(NO)_2)_2Ge(Br)C_{10}H_7$	Fe:	Org.Verb.C2-165
$BrC_{20}Fe_2H_{25}N_4O_4Pb$	$((CH_3)_2C_3H_3Fe(NO)_2)_2Pb(Br)C_{10}H_7$	Fe:	Org.Verb.C2-165
$BrC_{20}Fe_2H_{25}N_4O_4Sn$	$((CH_3)_2C_3H_3Fe(NO)_2)_2Sn(Br)C_{10}H_7$	Fe:	Org.Verb.C2-165
$BrC_{20}H_{15}Sn$	$(C_6H_5)_3SnCCBr$	Sn:	Org.Verb.2-403
$BrC_{20}H_{17}N_2O_3Ti$	$Ti(C_9H_6NO)_2(OC_2H_5)Br$	Ti:	Org.Verb.1-196
$BrC_{20}H_{18}NiOPS_2$	$[(C_6H_5)_3P]NiBr[S_2COCH_3]$	C:	MVol.D4-260
$BrC_{20}H_{19}Sn$	$(C_6H_5)_3SnCH_2CH_2Br$	Sn:	Org.Verb.2-362
$BrC_{20}H_{37}NiP_2$	$(CH_2CH)C_6H_4Ni(P(C_2H_5)_3)_2Br$	Ni:	Org.Verb.1-36/7, 42/3
–	$(C_6H_5CHCH)Ni(P(C_2H_5)_3)_2Br$	Ni:	Org.Verb.1-18, 24
$BrC_{20}H_{39}NiP_2$	$(CH_3)_2C_6H_3Ni(P(C_2H_5)_3)_2Br$	Ni:	Org.Verb.1-48, 49
$BrC_{20}H_{40}N_4PaS_8$	$Pa((C_2H_5)_2NCS_2)_4Br$	Pa:	SVol.2-92
$BrC_{20}H_{45}IrN_4^+$	$[Ir(C_5H_{10}NH)_4HBr]^+$	Ir:	SVol.2-76, 123
$BrC_{21}Cl_3H_{18}Sn$	$(C_6H_5)_3SnCHBrCH_2CCl_3$	Sn:	Org.Verb.2-363
$BrC_{21}FeH_{21}O$	$C_5H_5FeC_5H_4C(C_6H_5)CBrCH_2OC_2H_5$	Fe:	Org.Verb.A2-132, 136
$BrC_{21}FeH_{25}N_6$	$[(CH_2CHCH_2NC)_5FeCN]Br \cdot H_2O$	Fe:	Org.Verb.B4-76/8, 82
$BrC_{21}H_{20}NiOPS_2$	$[(C_6H_5)_3P]NiBr[S_2COC_2H_5]$	C:	MVol.D4-260
$BrC_{21}H_{20}NiO_3P$	$C_3H_5Ni(P(OC_6H_5)_3)Br$	Ni:	Org.Verb.2-13, 15
$BrC_{21}H_{20}NiP$	$C_3H_5Ni(P(C_6H_5)_3)Br$	Ni:	Org.Verb.2-12, 14
$BrC_{21}H_{21}Sn$	$(C_6H_5)_3SnCH_2CH_2CH_2Br$	Sn:	Org.Verb.2-363
$BrC_{21}H_{21}Ti$	$(C_6H_5CH_2)_3TiBr$	Ti:	Org.Verb.1-85

$BrC_{21}H_{38}NiP$	$C_3H_5Ni(P(C_6H_{11})_3)Br$	Ni:	Org.Verb.2-13, 15
$BrC_{21}H_{41}NiP_2$	$(CH_3)_3C_6H_2Ni(P(C_2H_5)_3)_2Br$	Ni:	Org.Verb.1-48, 49
$BrC_{21}H_{50}NiP_3$	$CH_2CHCH_2Ni(P(C_2H_5)_3)_3Br$	Ni:	Org.Verb.1-15
$BrC_{22}Cl_2H_{33}MnN_4$	$(CH_3)_4N[MnBrCl_2(CH_3C_5H_4N)_3]$	Mn:	MVol.C5-309
$BrC_{22}Cl_5H_{22}NiP_2$	$C_6Cl_5Ni(P(CH_3)_2C_6H_5)_2Br$	Ni:	Org.Verb.1-53
$BrC_{22}F_5H_{22}NiP_2$	$C_6F_5Ni(P(CH_3)_2C_6H_5)_2Br$	Ni:	Org.Verb.1-52, 54
$BrC_{22}F_{10}H_8N_2Tl$	$(C_6F_5)_2TlBr \cdot C_{10}H_8N_2$	F:	PerFHalOrg.4-137/8, 142
$BrC_{22}F_{10}H_{10}N_2Tl$	$(C_6F_5)_2TlBr \cdot (C_5H_5N)_2$	F:	PerFHalOrg.4-138, 142
$BrC_{22}H_{15}NiO$	$(C_6H_5)_3C_3Ni(CO)Br$	Ni:	Org.Verb.2-42/3
$BrC_{22}H_{20}NiOP$	$C_3H_5Ni(CO)(P(C_6H_5)_3)Br$	Ni:	Org.Verb.2-44
$BrC_{22}H_{22}NiOP$	$(CH_3O)C_3H_4Ni(P(C_6H_5)_3)Br$	Ni:	Org.Verb.2-16
$BrC_{22}H_{22}NiP$	$CH_3C_3H_4Ni(P(C_6H_5)_3)Br$	Ni:	Org.Verb.2-15
$BrC_{22}H_{26}NSn$	$[(C_6H_5)_3SnCH_2CH_2NH(CH_3)_2]Br$	Sn:	Org.Verb.2-365
$BrC_{22}H_{37}NiP_2$	$C_{10}H_7Ni(P(C_2H_5)_3)_2Br$	Ni:	Org.Verb.1-57
$BrC_{22}H_{38}NiOP$	$C_3H_5Ni(CO)(P(C_6H_{11})_3)Br$	Ni:	Org.Verb.2-44/5
$BrC_{22}H_{43}Li_2O_5Ti$	$C_6H_5Ti(OC_3H_7\text{-}i)_3 \cdot LiOC_3H_7 \cdot LiBr \cdot O(C_2H_5)_2$	Ti:	Org.Verb.1-47, 50
$BrC_{23}Cl_2H_{20}NiPSn$	$C_5H_5Ni(P(C_6H_5)_3)SnCl_2Br$	Ni:	Org.Verb.2-131/2
$BrC_{23}Cl_2H_{35}MnN_4$	$(C_2H_5)_4N[MnBrCl_2(C_5H_5N)_3]$	Mn:	MVol.C5-309
$BrC_{23}F_6FeH_{30}O_3P_3$	$[(CO)_3Fe(P(C_2H_5)_2C_6H_5)_2Br]PF_6$	Fe:	Org.Verb.B1-184/5
$BrC_{23}FeH_{18}O_3P$	$(CO)_3Fe(CH(CH_3)C_6H_4P(C_6H_5)_2)Br$	Fe:	Org.Verb.B1-209
$BrC_{23}FeH_{20}O_2P$	$C_3H_5Fe(CO)_2P(C_6H_5)_3Br$	Fe:	Org.Verb.B5-24/5
$BrC_{23}FeH_{21}O_4$	$C_5H_5FeC_5H_4COCH_2CH(CH_3)COOCH_2COC_6H_4Br$	Fe:	Org.Verb.A3-136
$BrC_{23}FeH_{22}NS$	$C_5H_5FeC_5H_4(CH)_4CSC_6H_4NC_2H_5Br$	Fe:	Org.Verb.A2-177
$BrC_{23}H_{17}N_2O_2Ti$	$C_5H_5Ti(C_9H_6NO)_2Br$	Ti:	Org.Verb.1-193, 196
$BrC_{23}H_{20}MgNiP$	$C_5H_5Ni(P(C_6H_5)_3)MgBr$	Ni:	Org.Verb.2-128
$BrC_{23}H_{20}NiP$	$C_5H_5Ni(P(C_6H_5)_3)Br$	Ni:	Org.Verb.2-117, 119
$BrC_{23}H_{22}NiOP$	$CH_3C_3H_4Ni(CO)(P(C_6H_5)_3)Br$	Ni:	Org.Verb.2-44
$BrC_{23}H_{29}NiP_2$	$CH_3C_6H_4Ni(P(CH_3)_2C_6H_5)_2Br$	Ni:	Org.Verb.1-35, 42
$BrC_{24}ClH_{33}NP_3ReS$	$[ReClBr(NS)(C_6H_5P(CH_3)_2)_3]$	S:	SVol.2-261
$BrC_{24}Cl_2H_{33}IrP_3$	$Ir(P(CH_3)_2C_6H_5)_3Cl_2Br$	Ir:	SVol.2-212
$BrC_{24}Cl_3H_{18}N_4O_2U$	$[C_{12}H_8N_2H]_2UO_2BrCl_3$	U:	SVol.C9-162/3
$BrC_{24}Cl_3H_{20}SSn$	$[(C_6H_5)_3S][C_6H_5SnCl_3Br]$	Sn:	Org.Verb.6-279
$BrC_{24}F_{10}H_8N_2Tl$	$(C_6F_5)_2TlBr \cdot C_{12}H_8N_2$	F:	PerFHalOrg.4-137/8, 143
$BrC_{24}F_{17}$	$C_6F_5(C_6F_4)_2C_6F_4Br$	F:	PerFHalOrg.4-16
$BrC_{24}Fe_2H_{28}N$	$[(C_5H_5FeC_5H_4CH_2)_2N(CH_3)_2]Br$	Fe:	Org.Verb.A6-164/5
$BrC_{24}Fe_2H_{28}O$	$[Fe(C_5H_5)_2]Br \cdot Fe(C_5H_5)_2 \cdot CH_3C(O)C_2H_5$	Fe:	Org.Verb.A1-101, 215
$BrC_{24}H_{18}IrN_4O^{2+}$	$[Ir(C_{12}H_8N_2)_2(H_2O)Br]^{2+}$	Ir:	SVol.2-90
$BrC_{24}H_{19}Sn$	$(C_6H_5)_3SnC_6H_4Br$	Sn:	Org.Verb.2-409, 417
$BrC_{24}H_{23}NiO_4P$	$(C_2H_2(COOCH_3)_2)Ni(P(C_6H_5)_3)Br$	Ni:	Org.Verb.1-341
$BrC_{24}H_{33}I_2IrP_3$	$Ir(P(CH_3)_2C_6H_5)_3BrI_2$	Ir:	SVol.2-212
$BrC_{24}H_{37}Sn$	$(C_6H_{11})_3SnC_6H_4Br$	Sn:	Org.Verb.2-339, 343
$BrC_{24}H_{42}NiP$	$C_6H_9Ni(P(C_6H_{11})_3)Br$	Ni:	Org.Verb.2-17
$BrC_{25}Cl_2H_{33}MnN_4$	$(CH_3)_4N[MnBrCl_2(CH_2CHC_5H_4N)_3]$	Mn:	MVol.C5-309
$BrC_{25}Cl_2H_{39}MnN_4$	$(CH_3)_4N[MnBrCl_2((CH_3)_2C_5H_3N)_3]$	Mn:	MVol.C5-309
$BrC_{25}H_{20}MoO_2P$	$C_5H_5Mo(CO)_2[(C_6H_5)_3P]Br$	F:	PerFHalOrg.4-40
$BrC_{25}H_{20}O_2PW$	$C_5H_5W(CO)_2[(C_6H_5)_3P]Br$	F:	PerFHalOrg.4-40
$BrC_{25}H_{33}NiP_2$	$(CH_3)_3C_6H_2Ni(P(CH_3)_2C_6H_5)_2Br$	Ni:	Org.Verb.1-48, 49
$BrC_{26}ClH_{34}NiP_2$	$ClC_6H_4Ni(P(C_2H_5)_2C_6H_5)_2Br$	Ni:	Org.Verb.1-34, 40

$BrC_{26}Cl_2H_{41}MnN_4$	$(C_2H_5)_4N[MnBrCl_2(CH_3C_5H_4N)_3]$	Mn: MVol.C5-309
$BrC_{26}Fe_2GeH_{49}N_4O_4$	$(C_3H_5Fe(NO)_2)_2Ge(Br)C_{20}H_{39}$	Fe: Org.Verb.C2-165
$BrC_{26}Fe_2H_{49}N_4O_4Pb$	$(C_3H_5Fe(NO)_2)_2Pb(Br)C_{20}H_{39}$	Fe: Org.Verb.C2-165
$BrC_{26}Fe_2H_{49}N_4O_4Sn$	$(C_3H_5Fe(NO)_2)_2Sn(Br)C_{20}H_{39}$	Fe: Org.Verb.C2-165
$BrC_{26}H_{19}Sn$	$(C_6H_5)_3SnCCC_6H_4Br$	Sn: Org.Verb.2-405
$BrC_{26}H_{25}N_2P_2S$	$[CH_3S[NP(C_6H_5)_2]_2CH_2]Br$	S: S-N-Verb.1-40
$BrC_{26}H_{26}NSn$	$[(C_6H_5)_3SnCH_2CH_2NH_2C_6H_5]Br$	Sn: Org.Verb.2-370
$BrC_{26}H_{26}N_3P_2S$	$[(CH_3)_2S[NP(C_6H_5)_2]_2N]Br$	S: S-N-Verb.1-37
$BrC_{26}H_{29}NiP_2$	$C_{10}H_7Ni(P(CH_3)_2C_6H_5)_2Br$	Ni: Org.Verb.1-57/8
$BrC_{26}H_{39}NiP_2$	$C_{14}H_9Ni(P(C_2H_5)_3)_2Br$	Ni: Org.Verb.1-57
$BrC_{27}H_{27}N_2P_2S$	$[(CH_3)_2S[NP(C_6H_5)_2]_2CH]Br$	S: S-N-Verb.1-40
$BrC_{27}H_{27}NiP_2$	$CH_3Ni[(C_6H_5)_2P(CH_2)_2P(C_6H_5)_2]Br$	Ni: Org.Verb.1-13
$BrC_{27}H_{28}NSn$	$[(C_6H_5)_3SnCH_2CH_2NH(CH_3)C_6H_5]Br$	Sn: Org.Verb.2-372
$BrC_{27}H_{33}N_6Sn$	$[Sn(C_7H_5(NCH_3)_2)_3]Br$	Sn: MVol.C5-188
$BrC_{27}H_{37}NiOP_2$	$CH_3OC_6H_4Ni(P(C_2H_5)_2C_6H_5)_2Br$	Ni: Org.Verb.1-35, 41
$BrC_{27}H_{37}NiP_2$	$CH_3C_6H_4Ni(P(C_2H_5)_2C_6H_5)_2Br$	Ni: Org.Verb.1-35, 42
$BrC_{28}FeH_{21}O_2$	$C_5H_5FeC_5H_4C_6H_4OCH_2(CC)_2CH_2OC_6H_4Br$	Fe: Org.Verb.A2-126
$BrC_{28}H_{20}O_8Pa$	$Pa(C_7H_5O_2)_4Br$	Pa: SVol.2-87/8
$BrC_{28}H_{27}Ti$	$(BrC_6H_4CH_2)(C_6H_5CH_2)_3Ti$	Ti: Org.Verb.1-95
$BrC_{28}H_{48}NiP$	$C_{10}H_{15}Ni(P(C_6H_{11})_3)Br$	Ni: Org.Verb.2-17
$BrC_{29}ClH_{29}OPSn$	$[(CH_3)_3SnCH(P(C_6H_5)_3)COC_6H_4Br]Cl$	Sn: Org.Verb.2-25
$BrC_{29}Cl_2H_{41}MnN_4$	$(C_2H_5)_4N[MnBrCl_2(CH_2CHC_5H_4N)_3]$	Mn: MVol.C5-309
$BrC_{29}Cl_2H_{47}MnN_4$	$(C_2H_5)_4N[MnBrCl_2((CH_3)_2C_5H_3N)_3]$	Mn: MVol.C5-309
$BrC_{29}Fe_2H_{28}O_2P$	$[(C_{10}H_9Fe)_2P(C_6H_5)CH_2COOCH_3]Br$	Fe: Org.Verb.A6-190/1
$BrC_{29}H_{41}NiP_2$	$(CH_3)_3C_6H_2Ni(P(C_2H_5)_2C_6H_5)_2Br$	Ni: Org.Verb.1-48, 49
$BrC_{29}H_{59}NiP_2$	$[C_5H_5Ni(P(C_4H_9\text{-}n)_3)_2]Br$	Ni: Org.Verb.2-136, 138
$BrC_{30}ClH_{45}NP_3ReS$	$[ReClBr(NS)(C_6H_5P(C_2H_5)_2)_3]$	S: SVol.2-261
$BrC_{30}Cl_2H_{37}NiP_2$	$C_{10}H_7Ni(P(C_2H_5)_2C_6H_5)_2(Br)Cl_2$	Ni: Org.Verb.1-57/8
$BrC_{30}Cl_2H_{45}IrP_3$	$Ir(P(C_2H_5)_2C_6H_5)_3Cl_2Br$	Ir: SVol.2-213
$BrC_{30}F_{10}H_{15}OPTl$	$(C_6F_5)_2TlBr \cdot (C_6H_5)_3PO$	F: PerFHalOrg.4-136/7, 140
$BrC_{30}F_{10}H_{15}PTl$	$(C_6F_5)_2TlBr \cdot (C_6H_5)_3P$	F: PerFHalOrg.4-136/7, 141
$BrC_{30}F_{21}$	$C_6F_5(C_6F_4)_3C_6F_4Br$	F: PerFHalOrg.4-16
$BrC_{30}FeH_{25}N_5S_5^+$	$[(SC_4H_3CH_2NC)_5FeBr]^+$	Fe: Org.Verb.B4-74
$BrC_{30}Fe_2GeH_{57}N_4O_4$	$((CH_3)_2C_3H_3Fe(NO)_2)_2Ge(Br)C_{20}H_{39}$	Fe: Org.Verb.C2-165
$BrC_{30}Fe_2H_{57}N_4O_4Pb$	$((CH_3)_2C_3H_3Fe(NO)_2)_2Pb(Br)C_{20}H_{39}$	Fe: Org.Verb.C2-165
$BrC_{30}Fe_2H_{57}N_4O_4Sn$	$((CH_3)_2C_3H_3Fe(NO)_2)_2Sn(Br)C_{20}H_{39}$	Fe: Org.Verb.C2-165
$BrC_{30}H_{31}NiP_2$	$CH_3C_3H_4Ni[(C_6H_5)_2PC_2H_4P(C_6H_5)_2]Br$	Ni: Org.Verb.2-16
$BrC_{30}H_{37}NiP_2$	$C_{10}H_7Ni(P(C_2H_5)_2C_6H_5)_2Br$	Ni: Org.Verb.1-57/8
$BrC_{30}H_{47}IrP_3$	$Ir(P(C_2H_5)_2(C_6H_5))_3H_2Br$	Ir: SVol.2-207/8
$BrC_{31}FeH_{25}N_6S_5$	$[(SC_4H_3CH_2NC)_5FeCN]Br \cdot 2\ H_2O$	Fe: Org.Verb.B4-76/8
$BrC_{31}FeH_{28}NO_3P_2$	$COFe(NO)[C_2H_4(P(C_6H_5)_2)_2]COCH_2CBrCH_2$	Fe: Org.Verb.B1-40/2
		Fe: Org.Verb.B5-108/11
$BrC_{31}Fe_3H_{27}$	$(C_5H_5FeC_5H_4)_3CBr$	Fe: Org.Verb.A6-244
$BrC_{31}H_{25}N_2Ni$	$C_3(C_6H_5)_3Ni(NC_5H_5)_2Br$	Ni: Org.Verb.2-19, 21
$BrC_{31}H_{29}NiP_2$	$[C_5H_5Ni(C_6H_5)_2P(CH_2)_2P(C_6H_5)_2]Br$	Ni: Org.Verb.2-141/2
$BrC_{32}F_5H_{26}NiP_2$	$C_6F_5Ni(P(C_6H_5)_2CH_3)_2Br$	Ni: Org.Verb.1-52, 54
$BrC_{32}H_{28}N_3Sn$	$(C_6H_5)_3SnC_6H_3(N(CH_3)_2)N_2C_6H_4Br$	Sn: Org.Verb.2-414
$BrC_{32}H_{39}NiP_2$	$C_6H_5C_6H_4Ni(P(C_2H_5)_2C_6H_5)_2Br$	Ni: Org.Verb.1-37, 44
$BrC_{33}H_{25}Ni$	$[C_5H_5Ni(C_4(C_6H_5)_4)]Br$	Ni: Org.Verb.2-188
$BrC_{33}H_{33}NiP_2$	$CH_3C_6H_4Ni(P(C_6H_5)_2CH_3)_2Br$	Ni: Org.Verb.1-35, 42

$BrC_{33}H_{36}NiP$	$(C_6H_5)_2CCHC(C_6H_5)_2Ni(P(C_2H_5)_3)Br$	Ni: Org.Verb.2-17
$BrC_{33}H_{57}O_{12}Sn$	$SnBr(OOC(CH_2)_8COOCH_3)_3$	Sn: MVol.C5-143
$BrC_{34}Cl_2H_{35}IrP_3$	$Ir(P(CH_3)_2C_6H_5)(((C_6H_5)_2P)_2C_2H_4)BrCl_2 \cdot CH_2Cl_2$	Ir: SVol.2-218/9
–	$Ir(P(CH_3)_2C_6H_5)(((C_6H_5)_2P)_2C_2H_4)BrCl_2 \cdot (CH_3)_2CO$	Ir: SVol.2-218/9
$BrC_{34}H_{33}NiP_2$	$C_6H_5Ni(P(C_6H_5)_2(CH_2)_4P(C_6H_5)_2)Br$	Ni: Org.Verb.1-39, 46
$BrC_{34}H_{39}NiP_2$	$C_{14}H_9Ni(P(C_2H_5)_2C_6H_5)_2Br$	Ni: Org.Verb.1-57/8
$BrC_{35}Cl_4H_{37}IrP_3$	$Ir(P(CH_3)_2C_6H_5)(((C_6H_5)_2P)_2C_2H_4)BrCl_2 \cdot CH_2Cl_2$	Ir: SVol.2-218/9
$BrC_{35}FeH_{27}NO_2P$	$(C_6H_5(CH)_3NC_6H_4Br)Fe(CO)_2P(C_6H_5)_3$	Fe: Org.Verb.B5-114, 118
$BrC_{35}H_{33}NiOP_2$	$C_6H_5CONi(P(C_6H_5)_2(CH_2)_4P(C_6H_5)_2)Br$	Ni: Org.Verb.1-46
$BrC_{35}H_{35}NiP_2$	$(CH_3)_3C_6H_2Ni[(C_6H_5)_2PC_2H_4P(C_6H_5)_2]Br$	Ni: Org.Verb.1-48, 49
$BrC_{36}ClH_{30}SSn$	$[(C_6H_5)_3S][(C_6H_5)_3SnClBr]$	Sn: Org.Verb.5-179
$BrC_{36}Cl_2H_{81}IrP_3$	$Ir(P(n\text{-}C_4H_9)_3)_3Cl_2Br$	Ir: SVol.2-200/2
$BrC_{36}F_{25}$	$C_6F_5(C_6F_4)_4C_6F_4Br$	F: PerFHalOrg.4-16
$BrC_{36}H_{30}IrN_2O_2P_2$	$Ir(NO)_2Br(P(C_6H_5)_3)_2$	Ir: SVol.2-53
$BrC_{36}H_{30}IrN_2P_2$	$Ir(N_2)(P(C_6H_5)_3)_2Br$	Ir: SVol.2-40
$BrC_{36}H_{30}N_3Ni$	$C_3(C_6H_5)_3Ni(NC_5H_5)_2Br \cdot NC_5H_5$	Ni: Org.Verb.2-21
$BrC_{36}H_{31}IrNOP_2$	$Ir(NO)HBr(P(C_6H_5)_3)_2$	Ir: SVol.2-52/3
$BrC_{36}H_{31}Sn$	$(CH_3)_2Sn(C_6H_5)(C_4Br(C_6H_5)_4)$	Sn: Org.Verb.3-81, 85
$BrC_{37}Cl_2H_{41}IrOP_3$	$Ir(P(CH_3)_2C_6H_5)(((C_6H_5)_2P)_2C_2H_4)BrCl_2 \cdot (CH_3)_2CO$	Ir: SVol.2-218/9
$BrC_{37}FeH_{30}NO_2P_2$	$COFe(NO)(P(C_6H_5)_3)_2Br$	Fe: Org.Verb.B1-38/40
$BrC_{37}Fe_2H_{34}O_2P$	$[(CH_3COC_5H_4FeC_5H_4)_2P(C_6H_5)CH_2C_6H_5]Br$	Fe: Org.Verb.A6-189
$BrC_{37}H_{33}Ni$	$[C_5H_5Ni(C_4(C_6H_4CH_3)_4)]Br$	Ni: Org.Verb.2-188
$BrC_{38}F_3H_{30}NiP_2$	$CF_2CFNi(P(C_6H_5)_3)_2Br$	Ni: Org.Verb.1-27, 29
$BrC_{38}FeH_{30}NO_3P_2$	$[(CO)_2Fe(P(C_6H_5)_3)_2NO]Br$	Fe: Org.Verb.B1-130/3
$BrC_{38}H_{29}NiP_2$	$((C_6H_5)_2PC_6H_4)_2C_2HNiBr$	Ni: Org.Verb.1-66
$BrC_{38}H_{35}NiP_2$	$C_2H_5Ni(P(C_6H_5)_3)_2Br$	Ni: Org.Verb.1-13
$BrC_{38}H_{37}NiP_2$	$C_{10}H_7Ni(P(C_6H_5)_2C_2H_5)_2Br$	Ni: Org.Verb.1-57
$BrC_{38}H_{38}IrOP_2$	$Ir(P(C_6H_5)_3)_2H_2Br \cdot C_2H_5OH$	Ir: SVol.2-223
$BrC_{38}H_{71}NiP_2$	$C_2H_5Ni(P(C_6H_{11})_3)_2Br$	Ni: Org.Verb.1-14
$BrC_{39}ClH_{39}NP_3ReS$	$[ReClBr(NS)((C_6H_5)_2PCH_3)_3]$	S: SVol.2-261
$BrC_{39}Cl_2H_{39}IrP_3$	$Ir(PCH_3(C_6H_5)_2)_3Cl_2Br$	Ir: SVol.2-213
$BrC_{39}H_{35}NiO_6P_2$	$C_3H_5Ni(P(OC_6H_5)_3)_2Br$	Ni: Org.Verb.2-23
$BrC_{39}H_{35}NiP_2$	$C_3H_5Ni(P(C_6H_5)_3)_2Br$	Ni: Org.Verb.2-23
$BrC_{39}H_{41}IrP_3$	$Ir(PCH_3(C_6H_5)_2)_3H_2Br$	Ir: SVol.2-207/8
$BrC_{39}H_{71}NiP_2$	$C_3H_5Ni(P(C_6H_{11})_3)_2Br$	Ni: Org.Verb.2-278
$BrC_{40}ClFeH_{35}N_5O_4$	$[(CH_3C_6H_4NC)_5FeBr]ClO_4$	Fe: Org.Verb.B4-80, 88
$BrC_{40}F_6Fe_2H_{30}O_4P_3S_2$	$[(CO)_4(P(C_6H_5)_3)_2Fe_2S_2Br]PF_6$	Fe: Org.Verb.C1-12, 15
$BrC_{40}FeH_{31}N_9O_8^+$	$[C_6H_5CH_2NC(NO_2C_6H_4CH_2NC)_4FeBr]^+$	Fe: Org.Verb.B4-91
$BrC_{40}FeH_{35}N_5^+$	$[(CH_3C_6H_4NC)_5FeBr]^+$	Fe: Org.Verb.B4-74, 80, 88
$BrC_{40}Fe_2H_{30}O_4P_2S_2^+$	$[(CO)_4(P(C_6H_5)_3)_2Fe_2S_2Br]^+$	Fe: Org.Verb.C1-12, 15
$BrC_{40}H_{37}NiO_2P_2$	$C_2H_5O_2CCH_2Ni(P(C_6H_5)_3)_2Br$	Ni: Org.Verb.1-15
$BrC_{41}Cl_5FeH_{30}N_6$	$[(ClC_6H_4CH_2NC)_5FeCN]Br$	Fe: Org.Verb.B4-76/7, 79, 87
$BrC_{41}FeH_{33}N_8O_4$	$[(C_6H_5CH_2NC)_3(O_2NC_6H_4CH_2NC)_2FeCN]Br \cdot H_2O$	Fe: Org.Verb.B4-92
$BrC_{41}FeH_{34}N_6^+$	$[(C_7H_7NC)_4(BrC_6H_4CH_2NC)FeCN]^+$	Fe: Org.Verb.B4-92
$BrC_{41}FeH_{35}N_6$	$[(C_6H_5CH_2NC)_5FeCN]Br$	Fe: Org.Verb.B4-76/9, 83/6

$BrC_{41}FeH_{35}N_6$	$[(C_6H_5CH_2NC)_5FeCN]Br \cdot H_2O$	Fe:	Org.Verb.B4-76/9, 83/6
$BrC_{41}Fe_2H_{37}N_2S_2$	$[(C_5H_5FeC_5H_4C_7H_3NS(C_2H_5)CH)_2CH]Br$	Fe:	Org.Verb.A6-177, 180
$BrC_{42}ClH_{34}NiP_2$	$BrC_6H_4Ni(P(C_6H_5)_3)_2Cl$	Ni:	Org.Verb.1-34
–	$ClC_6H_4Ni(P(C_6H_5)_3)_2Br$	Ni:	Org.Verb.1-34
$BrC_{42}ClH_{70}NiP_2$	$BrC_6H_4Ni(P(C_6H_{11})_3)_2Cl$	Ni:	Org.Verb.1-34
–	$ClC_6H_4Ni(P(C_6H_{11})_3)_2Br$	Ni:	Org.Verb.1-34
$BrC_{42}Cl_2H_{39}IrP_3$	$Ir(PC_{14}H_{13})_3Cl_2Br$	Ir:	SVol.2-234
$BrC_{42}Cl_5H_{30}NiP_2$	$C_6Cl_5Ni(P(C_6H_5)_3)_2Br$	Ni:	Org.Verb.1-53, 55
$BrC_{42}F_5H_{30}NiP_2$	$[(C_6H_5)_3P]_2Ni(C_6F_5)Br$	F:	PerFHalOrg.4-111
		Ni:	Org.Verb.1-52, 54
$BrC_{42}F_6Fe_2H_{36}O_4P_3S_2$	$[(CO)_4(P(C_6H_5)_3)_2Fe_2Br(SCH_3)_2]PF_6$	Fe:	Org.Verb.C1-12, 15
$BrC_{42}F_{29}$	$C_6F_5(C_6F_4)_5C_6F_4Br$	F:	PerFHalOrg.4-16
$BrC_{42}Fe_2H_{36}O_4P_2S_2{}^+$	$[(CO)_4(P(C_6H_5)_3)_2Fe_2Br(SCH_3)_2]^+$	Fe:	Org.Verb.C1-12, 15
$BrC_{42}H_{34}INiP_2$	$BrC_6H_4Ni(P(C_6H_5)_3)_2I$	Ni:	Org.Verb.1-34
$BrC_{42}H_{35}NiP_2$	$C_6H_5Ni(P(C_6H_5)_3)_2Br$	Ni:	Org.Verb.1-33, 40
$BrC_{42}H_{36}NNiP_2$	$H_2NC_6H_4Ni(P(C_6H_5)_3)_2Br$	Ni:	Org.Verb.1-32
$BrC_{42}H_{43}NiP_2$	$(C_6H_{13})Ni(P(C_6H_5)_3)_2Br$	Ni:	Org.Verb.1-15
$BrC_{42}H_{70}INiP_2$	$BrC_6H_4Ni(P(C_6H_{11})_3)_2I$	Ni:	Org.Verb.1-34
$BrC_{42}H_{71}NiP_2$	$C_6H_5Ni(P(C_6H_{11})_3)_2Br$	Ni:	Org.Verb.1-33
$BrC_{43}H_{34}NNiP_2$	$NCC_6H_4Ni(P(C_6H_5)_3)_2Br$	Ni:	Org.Verb.1-35
$BrC_{43}H_{37}NiP_2$	$CH_3C_6H_4Ni(P(C_6H_5)_3)_2Br$	Ni:	Org.Verb.1-35, 42
–	$C_6H_5CH_2Ni(P(C_6H_5)_3)_2Br$	Ni:	Org.Verb.1-16
$BrC_{43}H_{70}NNiP_2$	$NCC_6H_4Ni(P(C_6H_{11})_3)_2Br$	Ni:	Org.Verb.1-35
$BrC_{44}ClFeH_{53}N_4O_{14}P_3$	$[(NO_2C_6H_4NC)_2Fe(P(OC_2H_5)_2C_6H_5)_3Br]ClO_4$	Fe:	Org.Verb.B4-21/3
$BrC_{44}ClFeH_{55}N_2O_{10}P_3$	$[(C_6H_5NC)_2Fe(P(OC_2H_5)_2C_6H_5)_3Br]ClO_4$	Fe:	Org.Verb.B4-21/3
$BrC_{44}ClH_{35}OPSn$	$[(C_6H_5)_3SnCH(P(C_6H_5)_3)COC_6H_4Br]Cl$	Sn:	Org.Verb.2-379
$BrC_{44}Cl_3H_{48}O_2P_2U$	$[P(C_6H_5)_3C_4H_9]_2UO_2BrCl_3$	U:	SVol.C9-162
$BrC_{44}F_6Fe_2H_{40}O_4P_3S_2$	$[(CO)_4(P(C_6H_5)_3)_2Fe_2Br(SC_2H_5)_2]PF_6$	Fe:	Org.Verb.C1-12, 16
$BrC_{44}FeH_{37}N_2OP_2$	$COFe(P(C_6H_5)_3)_2(N_2C_6H_4CH_3)Br$	Fe:	Org.Verb.B1-37
$BrC_{44}FeH_{53}N_4O_{10}P_3{}^+$	$[(NO_2C_6H_4NC)_2Fe(P(OC_2H_5)_2C_6H_5)_3Br]^+$	Fe:	Org.Verb.B4-21/3
$BrC_{44}FeH_{55}N_2O_6P_3{}^+$	$[(C_6H_5NC)_2Fe(P(OC_2H_5)_2C_6H_5)_3Br]^+$	Fe:	Org.Verb.B4-21/3
$BrC_{44}Fe_2H_{40}O_4P_2S_2{}^+$	$[(CO)_4(P(C_6H_5)_3)_2Fe_2Br(SC_2H_5)_2]^+$	Fe:	Org.Verb.C1-12, 16
$BrC_{44}H_{35}NiP_2$	$C_6H_5CCNi(P(C_6H_5)_3)_2Br$	Ni:	Org.Verb.1-17
$BrC_{44}H_{37}NiOP_2$	$CH_3C(O)C_6H_4Ni(P(C_6H_5)_3)_2Br$	Ni:	Org.Verb.1-37
$BrC_{44}H_{37}NiP_2$	$(C_6H_5CHCH)Ni(P(C_6H_5)_3)_2Br$	Ni:	Org.Verb.1-18, 25
$BrC_{44}H_{40}NNiP_2$	$(CH_3)_2NC_6H_4Ni(P(C_6H_5)_3)_2Br$	Ni:	Org.Verb.1-32
$BrC_{45}FeH_{35}N_5O_5{}^+$	$[(C_6H_5COCH_2NC)_5FeBr]^+$	Fe:	Org.Verb.B4-76
$BrC_{45}FeH_{43}N_6$	$[C_6H_5CH_2NC(CH_3C_6H_4CH_2NC)_4FeCN]Br$	Fe:	Org.Verb.B4-91
$BrC_{45}H_{36}NiP$	$(C_6H_5)_2CCHC(C_6H_5)_2Ni(P(C_6H_5)_3)Br$	Ni:	Org.Verb.2-17
$BrC_{45}H_{41}NiP_2$	$(CH_3)_3C_6H_2Ni(P(C_6H_5)_3)_2Br$	Ni:	Org.Verb.1-48, 50
$BrC_{46}ClFeH_{59}N_2O_{10}P_3$	$[(CH_3C_6H_4NC)_2Fe(P(OC_2H_5)_2C_6H_5)_3Br]ClO_4$	Fe:	Org.Verb.B4-21/3
$BrC_{46}ClFeH_{59}N_2O_{12}P_3$	$[(CH_3OC_6H_4NC)_2Fe(P(OC_2H_5)_2C_6H_5)_3Br]ClO_4$	Fe:	Org.Verb.B4-21/3
$BrC_{46}FeH_{45}N_6$	$[(CH_3C_6H_4CH_2NC)_5FeCN]Br$	Fe:	Org.Verb.B4-76/7, 81, 89
–	$[(CH_3C_6H_4CH_2NC)_5FeCN]Br \cdot H_2O$	Fe:	Org.Verb.B4-76/7, 81, 89
$BrC_{46}FeH_{59}N_2O_6P_3{}^+$	$[(CH_3C_6H_4NC)_2Fe(P(OC_2H_5)_2C_6H_5)_3Br]^+$	Fe:	Org.Verb.B4-21/3
$BrC_{46}FeH_{59}N_2O_8P_3{}^+$	$[(CH_3OC_6H_4NC)_2Fe(P(OC_2H_5)_2C_6H_5)_3Br]^+$	Fe:	Org.Verb.B4-21/3
$BrC_{46}H_{37}NiP_2$	$C_{10}H_7Ni(P(C_6H_5)_3)_2Br$	Ni:	Org.Verb.1-57/8
$BrC_{47}FeH_{41}N_6O_6$	$[(C_6H_5CH_2NC)_2(CH_3OCOC_6H_4CH_2NC)_3FeCN]Br$	Fe:	Org.Verb.B4-91/2

$BrC_{48}ClFeH_{63}N_2O_{10}P_3$	$[((CH_3)_2C_6H_3NC)_2Fe(P(OC_2H_5)_2C_6H_5)_3Br]ClO_4$	Fe: Org.Verb.B4-21/3
$BrC_{48}FeH_{63}N_2O_6P_3^+$	$[((CH_3)_2C_6H_3NC)_2Fe(P(OC_2H_5)_2C_6H_5)_3Br]^+$	Fe: Org.Verb.B4-21/3
$BrC_{49}Cl_3H_{46}N_2Sn_2$	$[(NC_5H_5CH_2)_2CH_2][(C_6H_5)_6Sn_2Cl_3Br]$	Sn: Org.Verb.5-180
$BrC_{49}FeH_{39}N_6$	$[(C_6H_5CH_2NC)_3(C_{10}H_7CH_2NC)_2FeCN]Br$	Fe: Org.Verb.B4-92
$BrC_{49}FeH_{43}N_6O_8$	$[C_6H_5CH_2NC(CH_3OCOC_6H_4CH_2NC)_4FeCN]Br$	Fe: Org.Verb.B4-91
–	$[C_6H_5CH_2NC(CH_3OCOC_6H_4CH_2NC)_4FeCN]Br \cdot H_2O$	Fe: Org.Verb.B4-91
$BrC_{51}FeH_{45}N_6O_{10}$	$[(CH_3OCOC_6H_4CH_2NC)_5FeCN]Br$	Fe: Org.Verb.B4-76/7, 81/2, 89/90
–	$[(CH_3OCOC_6H_4CH_2NC)_5FeCN]Br \cdot H_2O$	Fe: Org.Verb.B4-76/7, 81/2, 89/90
$BrC_{51}FeH_{55}N_6O_5$	$[(C_6H_5OC_3H_6NC)_5FeCN]Br \cdot H_2O$	Fe: Org.Verb.B4-76/7, 81, 89
$BrC_{52}ClH_{45}IrO_4P_4$	$[Ir(((C_6H_5)_2P)_2C_2H_2)_2HBr]ClO_4$	Ir: SVol.2-242
$BrC_{52}ClH_{49}IrP_4$	$[Ir(((C_6H_5)_2P)_2C_2H_4)_2HBr]Cl$	Ir: SVol.2-240
$BrC_{52}F_6Fe_2H_{40}O_4P_3S_2$	$[(CO)_4(P(C_6H_5)_3)_2Fe_2Br(SC_6H_5)_2]PF_6$	Fe: Org.Verb.C1-12, 16
$BrC_{52}Fe_2H_{40}O_4P_2S_2^+$	$[(CO)_4(P(C_6H_5)_3)_2Fe_2Br(SC_6H_5)_2]^+$	Fe: Org.Verb.C1-12, 16
$BrC_{52}H_{44}IrP_4$	$[Ir(((C_6H_5)_2P)_2C_2H_2)_2]Br$	Ir: SVol.2-241
$BrC_{52}H_{46}IrP_4$	$[Ir(((C_6H_5)_2P)_2C_2H_2)_2H_2]Br$	Ir: SVol.2-242
$BrC_{52}H_{48}IrO_2P_4$	$[Ir(O_2)((C_6H_5)_2PC_2H_4P(C_6H_5)_2)_2]Br$	Ir: SVol.2-23
$BrC_{52}H_{48}IrO_{2x}P_4S_x$	$[Ir(((C_6H_5)_2P)_2C_2H_4)_2]Br \cdot n\ SO_2$	Ir: SVol.2-155
$BrC_{52}H_{48}IrP_4$	$[Ir(((C_6H_5)_2P)_2C_2H_4)_2]Br$	Ir: SVol.2-239
$BrC_{52}H_{50}IrP_4$	$[Ir(((C_6H_5)_2P)_2C_2H_4)_2H_2]Br$	Ir: SVol.2-240
$BrC_{54}Cl_5H_{42}NiP_2$	$C_6Cl_5Ni(P(C_6H_5)_3)_2Br \cdot 2\ C_6H_6$	Ni: Org.Verb.1-55
$BrC_{54}Cl_9H_{36}IrO_9P_3$	$Ir(P(OC_6H_4Cl\text{-}p)_3)_3Br$	Ir: SVol.2-238
$BrC_{54}H_{42}IrP_4$	$Ir(((C_6H_5)_2PC_6H_4)_3P)Br$	Ir: SVol.2-244
$BrC_{54}H_{45}IrO_9P_3$	$Ir(P(OC_6H_5)_3)_3Br$	Ir: SVol.2-238
$BrC_{54}H_{45}IrP_3$	$Ir(P(C_6H_5)_3)_3Br$	Ir: SVol.2-232
$BrC_{54}H_{45}NiP_3$	$Ni(P(C_6H_5)_3)_3Br$	Ni: Org.Verb.1-20
$BrC_{54}H_{47}IrP_3$	$Ir(P(C_6H_5)_3)_3H_2Br$	Ir: SVol.2-222
$BrC_{57}FeH_{44}N_5$	$((C_6H_5)_2CHNC)_4Fe(CN)Br$	Fe: Org.Verb.B4-66
$BrC_{60}ClFeH_{51}N_3O_4P_2$	$[(CH_3C_6H_4NC)_3Fe(P(C_6H_5)_3)_2Br]ClO_4$	Fe: Org.Verb.B4-47
$BrC_{60}FeH_{51}N_3P_2^+$	$[(CH_3C_6H_4NC)_3Fe(P(C_6H_5)_3)_2Br]^+$	Fe: Org.Verb.B4-47
$BrC_{61}FeH_{45}N_6$	$[(C_{10}H_7CH_2NC)_5FeCN]Br \cdot H_2O$	Fe: Org.Verb.B4-77, 82, 90
$BrC_{63}H_{63}IrO_9P_3$	$Ir(P(OCH_3C_6H_4)_3)_3Br$	Ir: SVol.2-238
$BrC_{65}FeH_{115}N_5O_{10}^+$	$[(HOCOC_{11}H_{22}NC)_5FeBr]^+$	Fe: Org.Verb.B4-76
$BrC_{72}H_{62}IrO_{12}P_4$	$[Ir(P(OC_6H_5)_3)_4H_2]Br$	Ir: SVol.2-236
$BrC_{80}Cl_2H_{168}IN_2O_2U$	$[N(C_{10}H_{21})_4]_2[UO_2BrCl_2I]$	U: SVol.C9-185
$BrC_{80}Cl_3H_{168}N_2O_2U$	$[N(C_{10}H_{21})_4]_2UO_2BrCl_3$	U: SVol.C9-162
$BrC_{82}FeH_{60}N_6$	$((C_6H_5)_3CNC)_4Fe(CN)_2Br$	Fe: Org.Verb.B4-71
$BrC_{101}FeH_{75}N_6$	$[((C_6H_5)_3CNC)_5FeCN]Br$	Fe: Org.Verb.B4-82
$BrC_{101}FeH_{155}N_6$	$[(CH_3(CH_2)_{11}C_6H_4CH_2NC)_5FeCN]Br$	Fe: Org.Verb.B4-82
$BrCe^{2+}$	$[CeBr]^{2+}$	Sc: MVol.C6-36/7
$BrCeClO_8S_2^{2-}$	$Ce(SO_4)_2ClBr^{2-}$	Sc: MVol.B7-77
$BrCeO$	$CeOBr$	Sc: MVol.C6-77/82
$BrCeO_8S_2^-$	$Ce(SO_4)_2Br^-$	Sc: MVol.B7-77
$BrClHIU$	$UHBrClI$	U: SVol.C9-161, 184
$BrClH_2U$	UH_2BrCl	U: SVol.C9-161
$BrClI_2U$	$UBrClI_2$	U: SVol.C9-183/4

BrClOS	SOBrCl	S:	SVol.1-71
$BrClS_2$	S_2BrCl	S:	SVol.2-299
$BrClTe^+$	$TeBrCl^+$	Te:	SVol.B3-50
$BrCl_2CsSn$	$CsSnBrCl_2$	Sn:	MVol.C3-153
$BrCl_2HU$	$UHBrCl_2$	U:	SVol.C9-161
$BrCl_2H_4NSn$	$NH_4SnBrCl_2 \cdot H_2O$	Sn:	MVol.C3-113
$BrCl_2H_{15}IrN_5$	$[Ir(NH_3)_5Br]Cl_2$	Ir:	SVol.2-45
$BrCl_2H_{15}IrN_5O_8$	$[Ir(NH_3)_5Br](ClO_4)_2$	Ir:	SVol.2-44/5
$BrCl_2IO_2U^{2-}$	$[UO_2BrCl_2I]^{2-}$	U:	SVol.C9-185
$BrCl_2IU$	$UBrCl_2I$	U:	SVol.C9-183/4
$BrCl_2KSn$	$KSnBrCl_2 \cdot H_2O$	Sn:	MVol.C3-67
$BrCl_2OS^-$	$[SOCl_2Br]^-$	S:	SVol.1-45/6
$BrCl_2S^+$	$SBrCl_2{}^+$	S:	SVol.2-235
$BrCl_2S_2{}^+$	$S_2BrCl_2{}^+$	S:	SVol.2-255
$BrCl_2Te^+$	$TeBrCl_2{}^+$	Te:	SVol.B3-50
$BrCl_2U$	$UBrCl_2$	U:	SVol.C9-158/9
$BrCl_3Cs_2Mn$	$Cs_2MnBrCl_3$	Mn:	MVol.C5-308
$BrCl_3Cs_2O_2U$	$Cs_2UO_2BrCl_3$	U:	SVol.C9-162/3
$BrCl_3H_8N_2O_2U$	$(NH_4)_2UO_2BrCl_3 \cdot 2\ H_2O$	U:	SVol.C9-162/3
$BrCl_3K_2O_2U$	$K_2UO_2BrCl_3 \cdot 2\ H_2O$	U:	SVol.C9-162/3
$BrCl_3MnRb_2$	$Rb_2MnBrCl_3$	Mn:	MVol.C5-308
$BrCl_3O_2Rb_2U$	$Rb_2UO_2BrCl_3 \cdot 2\ H_2O$	U:	SVol.C9-162/3
$BrCl_3O_2U^{2-}$	$[UO_2BrCl_3]^{2-}$	U:	SVol.C9-162/3
$BrCl_3Te^+$	$TeBrCl_3{}^+$	Te:	SVol.B3-50
$BrCl_3U$	$UBrCl_3$	U:	SVol.C9-158/61
$BrCl_4Sn^-$	$SnBrCl_4{}^-$	Sn:	MVol.C2-268
$BrCl_4Te^-$	$TeBrCl_4{}^-$	Te:	SVol.B3-81/3
$BrCl_5H_8N_2Te$	$(NH_4)_2[TeCl_5Br]$	Te:	SVol.B3-86
$BrCl_5Ir^{2-}$	$[IrCl_5Br]^{2-}$	Ir:	SVol.2-150
$BrCl_5IrRb_2$	$Rb_2[IrCl_5Br]$	Ir:	SVol.2-150
$BrCrH_{15}IrN_5O_4$	$[Ir(NH_3)_5Br]CrO_4$	Ir:	SVol.2-45
$BrCr_2H_{15}IrN_5O_7$	$[Ir(NH_3)_5Br]Cr_2O_7$	Ir:	SVol.2-45
BrCs	CsBr glasses		
	$CsBr-B_2O_3-Cs_2O-CuO$	B:	B-Verb.7-208
–	CsBr systems		
	$CsBr-CsBO_2-KBO_2-KBr-NaBO_2-NaBr$ $-RbBO_2-RbBr$	B:	B-Verb.7-206
	$CsBr-CsBO_2-NaBO_2-NaBr$	B:	B-Verb.7-206
	$CsBr-GdBr_3$	Sc:	MVol.C6-114/6, 128/31
	$CsBr-LaBr_3$	Sc:	MVol.C6-114/6, 128/31
	$CsBr-MnBr_2$	Mn:	MVol.C5-296
	$CsBr-NdBr_3$	Sc:	MVol.C6-114/6, 128/31
	$CsBr-PrBr_3$	Sc:	MVol.C6-114/6, 128/31
	$CsBr-ScBr_3$	Sc:	MVol.C6-114/5, 128/31
	$CsBr-SmBr_3$	Sc:	MVol.C6-114/6, 128/31
	$CsBr-SnBr_2$	Sn:	MVol.C3-151
	$CsBr-SnBr_2-H_2O$	Sn:	MVol.C3-151
	$CsBr-TeBr_4$	Te:	SVol.B3-33/4
	$CsBr-TeBr_4-H_2O$	Te:	SVol.B3-33/4

BrCs .... CsBr systems
CsBr-UBr_3 .... U: SVol.C9-120, 132/3
CsBr-UBr_4 .... U: SVol.C9-134
$BrCsSn^+$ .... $CsSnBr^+$ .... Sn: MVol.C3-151
$BrDy^{2+}$ .... $[DyBr]^{2+}$ .... Sc: MVol.C6-36/7
BrDyO .... DyOBr .... Sc: MVol.C6-77/82
$BrEr^{2+}$ .... $[ErBr]^{2+}$ .... Sc: MVol.C6-36/7, 40
BrErO .... ErOBr .... Sc: MVol.C6-77/82
$BrEu^+$ .... $EuBr^+$ .... Sc: MVol.C6-1
$BrEu^{2+}$ .... $[EuBr]^{2+}$ .... Sc: MVol.C6-36/7
BrEuF .... EuBrF .... Sc: MVol.C6-113
$BrEuH_2O_2$ .... $Eu(OH)_2Br$ .... Sc: MVol.C6-87/90
BrEuO .... EuOBr .... Sc: MVol.C6-77/82, 86/7
– .... EuOBr · H_2O
see $BrEuH_2O_2$ ... $Eu(OH)_2Br$
$BrEuO_3^{2+}$ .... $[EuBrO_3]^{2+}$ .... Sc: MVol.C6-105
$BrEu_3O_4$ .... Eu_3O_4Br .... Sc: MVol.C6-83/5, 87
BrFHIU .... UHBrFI .... U: SVol.C9-161, 184
$BrFH_2U$ .... UH_2BrF .... U: SVol.C9-161
BrFOS .... SOBrF .... S: SVol.1-70
BrFSm .... SmBrF .... Sc: MVol.C6-113
BrFTm .... TmBrF .... Sc: MVol.C6-113
BrFYb .... YbBrF .... Sc: MVol.C6-113
BrF_2HU .... $UHBrF_2$ .... U: SVol.C9-161
BrF_2NOS .... $BrNS(O)F_2$ .... S: S-N-Verb.1-67
BrF_2NOS^+ .... $BrNSOF_2^+$ .... S: S-N-Verb.1-67
BrF_2NS .... SF_2BrN .... S: SVol.2-75
$BrF_3H_8N_2Sn$ .... $(NH_4)_2SnBrF_3$ · H_2O .... Sn: MVol.C3-113
BrF_3U .... $UBrF_3$ .... U: SVol.C9-159
BrF_4OTe^+ .... $BrTeF_4O^+$ .... Te: SVol.B2-61
BrF_4S^+ .... $BrSF_4^+$ .... S: SVol.2-304
BrF_5OTe .... $BrTeF_5O$ .... Te: SVol.B2-61
BrF_5OTe^+ .... $BrTeF_5O^+$ .... Te: SVol.B2-61
BrF_5S .... SF_5Br .... S: SVol.2-300/5
BrF_5S^+ .... $BrSF_5^+$ .... S: SVol.2-304
BrF_5Te .... $TeBrF_5$ .... Te: SVol.B3-49
$BrF_6H_{15}IrN_5Si$ .... $[Ir(NH_3)_5Br]SiF_6$ .... Ir: SVol.2-45
BrF_8Ir .... $(BrF_2)^+[IrF_6]^-$ .... Ir: SVol.2-105
$BrF_{15}O_3Te_3$ .... $Br(TeF_5O)_3$ .... Te: SVol.B2-61
BrFe .... FeBr .... Fe: Org.Verb.B3-82
$BrGd^{2+}$ .... $[GdBr]^{2+}$ .... Sc: MVol.C6-36/7
$BrGdH_2O_2$ .... $Gd(OH)_2Br$ .... Sc: MVol.C6-87/90
BrGdO .... GdOBr .... Sc: MVol.C6-77/82
– .... GdOBr · H_2O
see $BrGdH_2O_2$ ... $Gd(OH)_2Br$
$BrHI_2U$ .... $UHBrI_2$ .... U: SVol.C9-161, 184
$BrHMnO_4^-$ .... $[HO_3MnOBr]^-$ .... Mn: MVol.C2-67
BrHNU .... U(NH)Br .... U: SVol.C9-119, 126

$BrHO_3U$ .... $UO_2(OH)Br$ .... U: SVol.C9-146
– .... $UO_2(OH)Br \cdot 2\ H_2O$ .... U: SVol.C9-147
$BrH_{1.2}N_{1.3}U$ .... $UBrN_{1.3}H_{1.2}$ .... U: SVol.C9-157
BrH_2IU .... UH_2BrI .... U: SVol.C9-161, 184
BrH_2LaO_2 .... $La(OH)_2Br$ .... Sc: MVol.C6-87/90
BrH_2MnO^+ .... $[Mn(H_2O)Br]^+$ .... Mn: MVol.C5-286
BrH_2NO_2S .... $BrSO_2NH_2$ .... S: S-N-Verb.1-134
BrH_2NdO_2 .... $Nd(OH)_2Br$ .... Sc: MVol.C6-87/90
BrH_2O_2Pr .... $Pr(OH)_2Br$ .... Sc: MVol.C6-87/90
$BrH_3Mn_2O_3$ .... $Mn_2(OH)_3Br$ .... Mn: MVol.C5-308
BrH_3U .... UH_3Br .... U: SVol.C9-161
BrH_4IrN_2 .... $Ir(NH_2)_2Br$ .... Ir: SVol.2-42
BrH_4N .... NH_4Br solid solutions
$NH_4Br-(NH_4)_2MnBr_4 \cdot H_2O$ .... Mn: MVol.C5-293
– .... NH_4Br systems
$NH_4Br-SnBr_2$ .... Sn: MVol.C3-111
$NH_4Br-SnBr_2-H_2O$ .... Sn: MVol.C3-111
$BrH_8I_5N_2Te$ .... $(NH_4)_2[TeBrI_5]$ .... Te: SVol.B3-86
$BrH_{10}MnO_5^+$ .... $[Mn(H_2O)_5Br]^+$ .... Mn: MVol.C5-285
$BrH_{10}O_5Te^{3+}$ .... $[TeBr(H_2O)_5]^{3+}$ .... Te: SVol.B3-15, 29
$BrH_{15}I_2IrN_5$ .... $[Ir(NH_3)_5Br]I_2$ .... Ir: SVol.2-45
$BrH_{15}IrN_5^{2+}$ .... $[Ir(NH_3)_5Br]^{2+}$ .... Ir: SVol.2-44/5
$BrH_{15}IrN_5O_3S$ .... $[Ir(NH_3)_5SO_3]Br$ .... Ir: SVol.2-46
$BrH_{15}IrN_5O_6S_2$ .... $[Ir(NH_3)_5Br]S_2O_6$ .... Ir: SVol.2-45
$BrH_{15}IrN_7O_4$ .... $[Ir(NH_3)_5Br](NO_2)_2$ .... Ir: SVol.2-45
$BrH_{15}IrN_7O_6$ .... $[Ir(NH_3)_5Br](NO_3)_2$ .... Ir: SVol.2-45
$BrHo^{2+}$ .... $[HoBr]^{2+}$ .... Sc: MVol.C6-36/7, 40
BrHoO .... HoOBr .... Sc: MVol.C6-77/82
BrI_2U .... $UBrI_2$ .... U: SVol.C9-182
BrI_3U .... $UBrI_3$ .... U: SVol.C9-183/4
BrIn .... InBr melts
$InBr-SnBr_2$ .... Sn: MVol.C4-117
BrIr .... IrBr .... Ir: SVol.2-141
BrK .... KBr glasses
$KBr-B_2O_3-CuO-K_2O$ .... B: B-Verb.7-208
– .... KBr solid solutions
$KBr-KMnO_4$ .... Mn: MVol.C2-32/3, 36/7, 45, 171/2
– .... KBr systems
$KBr-AgNO_3-(CH_2)_6N_4-H_2O$ .... Ag: MVol.B6-198
$KBr-CsBO_2-CsBr-KBO_2-NaBO_2-NaBr$
$-RbBO_2-RbBr$ .... B: B-Verb.7-206
$KBr-LaBr_3-H_2O$ .... Sc: MVol.C6-124/5
$KBr-MnBr_2$ .... Mn: MVol.C5-292
$KBr-MnBr_2-H_2O$ .... Mn: MVol.C5-293
$KBr-ScBr_3$ .... Sc: MVol.C6-114/5, 122/3
$KBr-SnBr_2$ .... Sn: MVol.C3-63
$KBr-SnBr_2-H_2O$ .... Sn: MVol.C3-63

BrK	KBr systems		
	$KBr-TeBr_4$	Te:	SVol.B3-31/2
	$KBr-TeBr_4-H_2O$	Te:	SVol.B3-31/2
	$KBr-UBr_3$	U:	SVol.C9-120, 132/3
	$KBr-UBr_4$	U:	SVol.C9-134
$BrKSn^+$	$KSnBr^+$	Sn:	MVol.C3-63
$BrK_3O_{12}S_3Sn_2$	$K_3Sn_2Br(SO_4)_3$	Sn:	MVol.C3-72
BrLa	LaBr	Sc:	MVol.C6-1/2
$BrLa^{2+}$	$[LaBr]^{2+}$	Sc:	MVol.C6-36/7
BrLaO	LaOBr	Sc:	MVol.C6-77/82
–	$LaOBr \cdot H_2O$		
	see BrH_2LaO_2 ... $La(OH)_2Br$		
$BrLaO_4U$	$LaUO_2(O_2)Br$	U:	SVol.C9-155
BrLi	LiBr glasses		
	$LiBr-B_2O_3-CuO-Li_2O$	B:	B-Verb.7-208
–	LiBr systems		
	$LiBr-MnBr_2$	Mn:	MVol.C5-291
	$LiBr-ScBr_3$	Sc:	MVol.C6-114/6
	$LiBr-SnBr_2$	Sn:	MVol.C3-5
	$LiBr-SnBr_2-H_2O$	Sn:	MVol.C3-5
	$LiBr-TeBr_4-H_2O$	Te:	SVol.B3-30/1
BrLuO	LuOBr	Sc:	MVol.C6-77/82
BrMn	MnBr	Mn:	MVol.C5-262/4
$BrMn^+$	$[MnBr]^+$	Mn:	MVol.C5-285/6, 288/90
–	$[Mn(H_2O)Br]^+$	Mn:	MVol.C5-286
–	$[Mn(H_2O)_5Br]^+$	Mn:	MVol.C5-285
$BrMn^{2+}$	$MnBr^{2+} \cdot x H_2O$	Mn:	MVol.C5-290
BrMnS	MnSBr	Mn:	MVol.C6-263
BrNU	UNBr	U:	SVol.C9-119, 157
BrNa	NaBr glasses		
	$NaBr-B_2O_3-CuO-Na_2O$	B:	B-Verb.7-208
–	NaBr systems		
	$NaBr-CsBO_2-CsBr-KBO_2-KBr-NaBO_2$ $-RbBO_2-RbBr$	B:	B-Verb.7-206
	$NaBr-CsBO_2-CsBr-NaBO_2$	B:	B-Verb.7-206
	$NaBr-GdBr_3$	Sc:	MVol.C6-114/6, 121/2
	$NaBr-LaBr_3$	Sc:	MVol.C6-114/6, 118
	$NaBr-MnBr_2$	Mn:	MVol.C5-292
	$NaBr-NdBr_3$	Sc:	MVol.C6-114/6, 119/20
	$NaBr-PrBr_3$	Sc:	MVol.C6-114/6, 119
	$NaBr-ScBr_3$	Sc:	MVol.C6-114/5, 117/8
	$NaBr-SmBr_3$	Sc:	MVol.C6-114/6, 120/1
	$NaBr-SnBr_2$	Sn:	MVol.C3-25
	$NaBr-SnBr_2-H_2O$	Sn:	MVol.C3-25
	$NaBr-TeBr_4$	Te:	SVol.B3-31
	$NaBr-TeBr_4-H_2O$	Te:	SVol.B3-31
	$NaBr-UBr_3$	U:	SVol.C9-120, 132/3
	$NaBr-UBr_4$	U:	SVol.C9-134

$BrNaSn^+$ .... $NaSnBr^+$ .... Sn: MVol.C3-25
$BrNd^{2+}$ .... $[NdBr]^{2+}$ .... Sc: MVol.C6-36/7, 40
BrNdO .... NdOBr .... Sc: MVol.C6-77/82
– .... NdOBr · H_2O
see BrH_2NdO_2 ... $Nd(OH)_2Br$
$BrNd_3O_4$ .... Nd_3O_4Br .... Sc: MVol.C6-83/5
$BrNp^{2+}$ .... $NpBr^{2+}$ .... Np: TrU.D1-144
BrOPm .... PmOBr .... Sc: MVol.C6-79, 85
BrOPr .... PrOBr .... Sc: MVol.C6-77/82
– .... PrOBr · H_2O
see BrH_2O_2Pr ... $Pr(OH)_2Br$
$BrOS^+$ .... $SOBr^+$ .... S: SVol.1-68
BrOSc .... ScOBr .... Sc: MVol.C6-77, 85
– .... ScOBr · H_2O .... Sc: MVol.C6-85
BrOSm .... SmOBr .... Sc: MVol.C6-77/82
BrOTb .... TbOBr .... Sc: MVol.C6-77/82
BrOTm .... TmOBr .... Sc: MVol.C6-77/82
BrOU .... UOBr .... U: SVol.C9-142
BrOY .... YOBr .... Sc: MVol.C6-77/82
BrOYb .... YbOBr .... Sc: MVol.C6-77/82
BrO_2Pa .... PaO_2Br .... Pa: SVol.2-68/9
BrO_2U .... UO_2Br .... U: SVol.C9-144/5
BrO_3Tb^{2+} .... $[TbBrO_3]^{2+}$ .... Sc: MVol.C6-105
BrO_4Sm_3 .... Sm_3O_4Br .... Sc: MVol.C6-83/6
BrO_4Tm_3 .... Tm_3O_4Br .... Sc: MVol.C6-83/5
BrO_4UY .... $YUO_2(O_2)Br$ .... U: SVol.C9-155
BrO_4Yb_3 .... Yb_3O_4Br .... Sc: MVol.C6-83/5
$BrPm^{2+}$ .... $[PmBr]^{2+}$ .... Sc: MVol.C6-36/7
$BrPr^{2+}$ .... $[PrBr]^{2+}$ .... Sc: MVol.C6-36/7, 40
$BrPu^{2+}$ .... $PuBr^{2+}$ .... Np: TrU.D1-144
BrRb .... RbBr solid solutions
$RbBr-KMnO_4$ .... Mn: MVol.C2-32/3, 171/2
– .... RbBr systems
$RbBr-CsBO_2-CsBr-KBO_2-KBr-NaBO_2$
$-NaBr-RbBO_2$ .... B: B-Verb.7-206
$RbBr-MnBr_2$ .... Mn: MVol.C5-293/4
$RbBr-ScBr_3$ .... Sc: MVol.C6-114/5, 127/8
$RbBr-SnBr_2$ .... Sn: MVol.C3-135/6
$RbBr-SnBr_2-H_2O$ .... Sn: MVol.C3-135/6
$RbBr-TeBr_4$ .... Te: SVol.B3-32/3
$RbBr-TeBr_4-H_2O$ .... Te: SVol.B3-32/3
$RbBr-UBr_3$ .... U: SVol.C9-120, 132/3
$RbBr-UBr_4$ .... U: SVol.C9-134
$BrRbSn^+$ .... $RbSnBr^+$ .... Sn: MVol.C3-136
$BrSb_3Te_4$ .... Sb_3Te_4Br = 4 Sb_2Te_3 · $SbBr_3$ .... Te: SVol.B3-12
BrSc .... ScBr .... Sc: MVol.C6-1/2
$BrSc^{2+}$ .... $[ScBr]^{2+}$ .... Sc: MVol.C6-36/7, 51/2
$BrSm^{2+}$ .... $[SmBr]^{2+}$ .... Sc: MVol.C6-36/7, 40

$BrTb^{2+}$	$[TbBr]^{2+}$	Sc: MVol.C6-36/7
BrTe	TeBr	Te: SVol.B3-4
$BrTe^+$	$TeBr^+$	Te: SVol.B3-10
$BrTe^{3+}$	$TeBr^{3+}$	Te: SVol.B3-15/6, 18/22
$BrTe_2$	Te_2Br	Te: SVol.B3-1/3
–	Te_2Br solid solutions	
	$Te_2Br_{1-x}I_x$	Te: SVol.B3-79/80
BrTl	TlBr melts	
	$TlBr-SnBr_2$	Sn: MVol.C4-139
–	TlBr systems	
	$TlBr-MnBr_2$	Mn: MVol.C5-307
$BrTm^{2+}$	$[TmBr]^{2+}$	Sc: MVol.C6-36/7
BrY	YBr	Sc: MVol.C6-1/2
BrY^{2+}	$[YBr]^{2+}$	Sc: MVol.C6-36/7
BrYb	YbBr	Sc: MVol.C6-1
$Br_{1.03}N_{1.03}U$	$UN_{1.03}Br_{1.03}$	U: SVol.C9-157
$Br_{1.06}H_{0.6}N_{1.03}U$	$UBr_{1.06}N_{1.03}H_{0.6}$	U: SVol.C9-157
$Br_{1.5}CNS$	$Br_{1.5}SCN$	C: MVol.D5-229
$Br_{1.5}C_{24}Cl_{4.5}H_{36}NiPt$	$[((CH_3)_6C_6)_2Ni][PtCl_{4.5}Br_{1.5}]$	Ni: Org.Verb.2-249
$Br_{1.5}Sc$	$ScBr_{1.5}$	Sc: MVol.C6-48
$Br_2C_{0.96}H_{2.24}MnO_{0.16}$	$MnBr_2 \cdot 0.16\ (i\text{-}C_3H_7)_2O$	Mn: MVol.D1-140
$Br_2CClFNO_2P$	$CFClNOP(O)Br_2$	F: PerFHalOrg.7-100, 118
Br_2CClFS	Br_2FCSCl	C: MVol.D6-164
Br_2CClH_3Sn	CH_3SnBr_2Cl	Sn: Org.Verb.6-218
Br_2CCl_2S	$Br_2ClCSCl$	C: MVol.D6-164
$Br_2CCl_3H_3Ti^{2-}$	$[CH_3TiCl_3Br_2]^{2-}$	Ti: Org.Verb.1-42/3
Br_2CF_2Se	$F_2BrCSeBr$	C: MVol.D6-250
Br_2CF_3P	CF_3PBr_2	
	Chemical reactions	F: PerFHalOrg.3-99, 111, 120
	Physical properties	F: PerFHalOrg.3-80/2, 92, 98
	Preparation	F: PerFHalOrg.3-74
Br_2CF_3Sb	CF_3SbBr_2	F: PerFHalOrg.3-226, 229/30
$Br_2CH_2MnO_2$	$MnBr_2 \cdot HCOOH$	Mn: MVol.C5-273
$Br_2CH_2N_2O$	$(NHBr)_2CO$	C: MVol.D3-129
$Br_2CH_3IrN_2S$	$[Ir(NCS)Br_2NH_3]_n$	Ir: SVol.2-163
$Br_2CH_4N_2S$	$(NH_2)_2CS \cdot Br_2$	C: MVol.D6-95
$Br_2CH_4N_2SSn$	$SnBr_2 \cdot CS(NH_2)_2$	Sn: MVol.C5-55/6
$Br_2CH_{15}IrN_6S$	$[Ir(NCS)(NH_3)_5]Br_2$	Ir: SVol.2-162
Br_2CN	Br_2CN	C: MVol.D3-230
Br_2CN^-	$[BrBrCN]^-$	C: MVol.D3-237
Br_2CNiO	$Ni(CO)Br_2$	Ni: Org.Verb.1-111
Br_2CO	$COBr_2$	C: MVol.D3-110/8
Br_2CS	Br_2CS	C: MVol.D6-163/4
Br_2CSe^+	Br_2CSe^+	C: MVol.D6-251
$Br_2C_{1.38}H_{3.22}MnO_{0.23}$	$MnBr_2 \cdot 0.23\ (i\text{-}C_3H_7)_2O$	Mn: MVol.D1-140

$Br_2C_2Cl_2H_{12}N_6O_2U$	$[CN_3H_6]_2UO_2Br_2Cl_2$	U: SVol.C9-162
$Br_2C_2Cl_2S_2$	$(ClBrCS)_2$	C: MVol.D6-169
$Br_2C_2Cl_3N_2Sb$	$SbCl_3 \cdot 2\ BrCN$	C: MVol.D3-236
$Br_2C_2Cl_4N_2Ti$	$TiCl_4 \cdot 2\ BrCN$	C: MVol.D3-237
$Br_2C_2Cl_{12}N_2Ti_3$	$3\ TiCl_4 \cdot 2\ BrCN$	C: MVol.D3-237
$Br_2C_2FH_2NO$	$CFBr_2C(O)NH_2$	F: PerFHalOrg.7-17, 19, 45
$Br_2C_2F_6Ge$	$(CF_3)_2GeBr_2$	F: PerFHalOrg.4-166/7, 171/3, 175
$Br_2C_2F_6Si$	$(CF_3)_2SiBr_2$	F: PerFHalOrg.4-147
$Br_2C_2FeH_4$	$CH_2CH_2FeBr_2 \cdot 2\ H_2O$	Fe: Org.Verb.B4-178, 181
$Br_2C_2FeH_5$	$C_2H_5FeBr_2$	Fe: Org.Verb.B1-2/3
$Br_2C_2H_4MnO_2$	$MnBr_2 \cdot CH_3COOH$	Mn: MVol.C5-273
$Br_2C_2H_6MnO$	$MnBr_2 \cdot C_2H_5OH$	Mn: MVol.C5-266
		Mn: MVol.D1-32
$Br_2C_2H_6O_4S_2Ti$	$(CH_3SO_2)_2TiBr_2$	Ti: Org.Verb.1-23
$Br_2C_2H_6Sn$	$(CH_3)_2SnBr_2$	Sn: Org.Verb.5-199
$Br_2C_2H_6Ti$	$(CH_3)_2TiBr_2$	Ti: Org.Verb.1-72
$Br_2C_2H_8MnO_2$	$MnBr_2 \cdot 2\ CH_3OH$	Mn: MVol.D1-26
$Br_2C_2H_8N_4S_2Sn$	$SnBr_2 \cdot 2\ CS(NH_2)_2$	Sn: MVol.C5-55/6
$Br_2C_2H_8N_4S_2Te$	$Te(SC(NH_2)_2)_2Br_2$	Te: SVol.B3-141
$Br_2C_2H_{18}IrN_5O_2$	$[Ir(NH_3)_5(CH_3COO)]Br_2$	Ir: SVol.2-31
$Br_2C_2N_2O_2$	$NBr_2C(O)NCO$	C: MVol.D3-142/3
$Br_2C_2OS_2$	Br_2CS_2CO	C: MVol.D6-172
$Br_2C_{2.16}H_{5.04}MnO_{0.36}$	$MnBr_2 \cdot 0.36\ (i\text{-}C_3H_7)_2O$	Mn: MVol.D1-140
Br_2C_3FNS	CBrNCFSCBr	F: PerFHalOrg.5-57/8, 80, 110
$Br_2C_3FN_3$	$C_3N_3FBr_2$	C: MVol.D3-289/90
$Br_2C_3F_3H_2NO$	$CF_2BrCFBrC(O)NH_2$	F: PerFHalOrg.7-18, 21, 46
$Br_2C_3F_6NO$	$(BrCF_2CFBr)(CF_3)NO$	F: PerFHalOrg.7-85/6
$Br_2C_3F_9P$	$(CF_3)_3PBr_2$	F: PerFHalOrg.3-74, 99
$Br_2C_3F_9Sb$	$(CF_3)_3SbBr_2$	F: PerFHalOrg.3-226/7, 229, 231
$Br_2C_3H_6MnO_{0.6}$	$MnBr_2 \cdot 0.6\ (CH_2)_5O$	Mn: MVol.D1-146/7
$Br_2C_3H_6MnO$	$MnBr_2 \cdot (CH_3)_2CO$	Mn: MVol.C5-273
$Br_2C_3H_8MnO$	$MnBr_2 \cdot C_3H_7OH$	Mn: MVol.C5-266
$Br_2C_3H_{12}N_6Se_3$	$[(NH_2)_2CSe]_3Br_2 \cdot H_2O$	C: MVol.D6-246, 247/8
$Br_2C_{3.5}H_3HgMnN_2O_{0.5}S_2$	$MnHgBr_2(SCN)_2 \cdot 0.5\ (CH_3)_2CO$	Mn: MVol.D1-56
$Br_2C_{3.75}H_{7.5}MnO_{0.75}$	$MnBr_2 \cdot 0.75\ (CH_2)_5O$	Mn: MVol.D1-146/7
$Br_2C_4ClH_9MgSn$	$(CH_3)_3SnCBr_2MgCl$	Sn: Org.Verb.2-16, 18
$Br_2C_4ClH_{16}IrN_4$	$[Ir(C_2H_4(NH_2)_2)_2BrCl]Br$	Ir: SVol.2-84
$Br_2C_4ClH_{16}IrN_4O_4$	$[Ir(C_2H_4(NH_2)_2)_2Br_2]ClO_4$	Ir: SVol.2-84
$Br_2C_4Cl_4H_6O_4Sn$	$SnCl_4 \cdot 2\ CH_2BrCOOH$	Sn: MVol.C5-119
$Br_2C_4F_2N_2$	NCBrCFNCFCBr	F: PerFHalOrg.6-28/30, 49
$Br_2C_4F_4Mg$	$(CF_2)_2CBrCMgBr$	F: PerFHalOrg.4-62/3, 71, 107
$Br_2C_4F_6Fe$	$(CF_2CFBr)_2Fe$	Fe: Org.Verb.B4-356
$Br_2C_4F_8Mg_2$	$BrMg(CF_2)_4MgBr$	F: PerFHalOrg.4-66, 87
$Br_2C_4FeHg_2O_4$	$(CO)_4Fe(HgBr)_2$	Fe: Org.Verb.B2-153/6, 159
$Br_2C_4FeInO_4^-$	$[(CO)_4FeInBr_2]^-$	Fe: Org.Verb.B2-49

$Br_2C_4FeO_4$	$(CO)_4FeBr_2$	Fe:	Org.Verb.B2-54/67
$Br_2C_4Fe_2N_6O_4$	$[(CO)_2Fe(N_3)_2Fe(CO)_2]Br_2$	Fe:	Org.Verb.C1-3
$Br_2C_4H_8MnO_2$	$MnBr_2 \cdot CH_3COOC_2H_5$	Mn:	MVol.C5-273
–	$MnBr_2 \cdot O(CH_2CH_2)_2O$	Mn:	MVol.D1-155
$Br_2C_4H_8OSSn$	$SnBr_2 \cdot S(CH_2CH_2)_2O$	Sn:	MVol.C5-52/3
$Br_2C_4H_8O_2Sn$	$SnBr_2 \cdot C_4H_8O_2$	Sn:	MVol.C5-10
$Br_2C_4H_8S_2Sn$	$SnBr_2 \cdot S(CH_2CH_2)_2S$	Sn:	MVol.C5-52/3
$Br_2C_4H_9LiSn$	$(CH_3)_3SnCBr_2Li$	Sn:	Org.Verb.2-16, 18
$Br_2C_4H_{10}MnO$	$MnBr_2 \cdot (C_2H_5)_2O$	Mn:	MVol.D1-139
$Br_2C_4H_{10}MnO_2$	$MnBr_2 \cdot CH_3OCH_2CH_2OCH_3$	Mn:	MVol.D1-141
$Br_2C_4H_{10}N_2Sn$	$SnBr_2 \cdot NH(CH_2CH_2)_2NH$	Sn:	MVol.C5-23/4
$Br_2C_4H_{10}Sn$	$(CH_3)_2Sn(CH_2Br)_2$	Sn:	Org.Verb.3-6
–	$(CH_3)_3SnCHBr_2$	Sn:	Org.Verb.2-7
$Br_2C_4H_{12}MnO_4$	$Mn(HOC_2H_4OH)_2Br_2$	Mn:	MVol.D1-36
$Br_2C_4H_{12}O_2S_2Sn$	$SnBr_2 \cdot 2\ (CH_3)_2SO$	Sn:	MVol.C5-45
$Br_2C_4H_{16}IrN_4{}^+$	$[Ir(C_2H_4(NH_2)_2)_2Br_2]^+$	Ir:	SVol.2-83, 84
$Br_2C_4H_{16}IrN_5O_3$	$[Ir(C_2H_4(NH_2)_2)_2Br_2]NO_3$	Ir:	SVol.2-84
$Br_2C_4H_{16}N_8S_4Te$	$Te(SC(NH_2)_2)_4Br_2$	Te:	SVol.B3-146
$Br_2C_4H_{18}N_4Sn$	$SnBr_2(CH_3NH)_2 \cdot 2\ CH_3NH_2$	Sn:	MVol.C5-168
$Br_2C_5ClFHNO$	HOC_5FBr_2ClN	F:	PerFHalOrg.5-182/3
		F:	PerFHalOrg.6-13
$Br_2C_5ClF_2N$	$Br_2ClC_5F_2N$	F:	PerFHalOrg.5-118/9
$Br_2C_5ClH_5Ti$	$C_5H_5TiClBr_2$	Ti:	Org.Verb.1-139, 149
$Br_2C_5Cl_2F_9NO$	$BrCF_2CFClN(CF_3)OCF_2CFClBr$	F:	PerFHalOrg.7-95, 112
$Br_2C_5F_2HNO$	$HOC_5F_2Br_2N$	F:	PerFHalOrg.5-182/3, 199
		F:	PerFHalOrg.6-13
$Br_2C_5F_3N$	$Br_2C_5F_3N$	F:	PerFHalOrg.5-118/9, 127
$Br_2C_5F_{11}NO$	$BrCF_2CF_2N(CF_3)OCF_2CF_2Br$	F:	PerFHalOrg.7-95, 112
$Br_2C_5FeHgO_5$	$(CO)_5FeHgBr_2$	Fe:	Org.Verb.B3-250
$Br_2C_5FeO_5$	$(CO)_4Fe(COBr)Br$	Fe:	Org.Verb.B2-176/7
–	$[(CO)_5FeBr]Br$	Fe:	Org.Verb.B3-106/7, 253/4
$Br_2C_5H_4Ti$	$C_5H_4TiBr_2$	Ti:	Org.Verb.1-122
$Br_2C_5H_5NSn$	$SnBr_2 \cdot C_5H_5N$	Sn:	MVol.C5-17
$Br_2C_5H_5Ti$	$C_5H_5TiBr_2 \cdot C_4H_8O$	Ti:	Org.Verb.1-135/6
$Br_2C_5H_{12}N_2SSn$	$SnBr_2 \cdot CS(N(CH_3)_2)_2$	Sn:	MVol.C5-56/7
$Br_2C_5H_{12}N_2STe$	$Te((CH_3)_2NCSN(CH_3)_2)Br_2$	Te:	SVol.B3-148
$Br_2C_6F_2FeO_4$	$CF_2CBr_2Fe(CO)_4$	Fe:	Org.Verb.B4-320, 322/4
$Br_2C_6F_3HN_4$	$CF_3C_5Br_2HN_4$	F:	PerFHalOrg.6-127, 139, 162
$Br_2C_6F_4$	BrC_6F_4Br	F:	PerFHalOrg.4-14/6
$Br_2C_6F_4FeO_4$	$(CO)_4Fe(CF_2CF_2Br)Br$	Fe:	Org.Verb.B2-178, 180, 190
$Br_2C_6F_4Mg$	BrC_6F_4MgBr	F:	PerFHalOrg.4-63, 65, 71, 102
$Br_2C_6F_4Mg_2$	$BrMgC_6F_4MgBr$	F:	PerFHalOrg.4-66/7, 71, 103
$Br_2C_6F_5P$	$C_6F_5PBr_2$	F:	PerFHalOrg.3-124, 125, 127
$Br_2C_6F_5PS$	$C_6F_5P(S)Br_2$	F:	PerFHalOrg.3-141, 148
$Br_2C_6F_6H_8Sn$	$(CH_3)_2Sn(CF_2CHBrF)_2$	Sn:	Org.Verb.3-6, 15

$Br_2C_6F_7N_3$	$(CF_3)(CF_2Br)_2C_3N_3$	F:	PerFHalOrg.6-81/2, 102
$Br_2C_6F_8$	$BrC(CF_2)_4CBr$	F:	PerFHalOrg.4-15
$Br_2C_6F_{10}N_2O_3$	$ON(C_2F_4NO_2)CF_2(CFBr)_2CF_2$	F:	PerFHalOrg.6-4/5, 9
$Br_2C_6F_{11}NO$	$O(CF_2CF_2)_2NCFBrCF_2Br$	F:	PerFHalOrg.6-1/2, 6
$Br_2C_6F_{12}Mg$	$CF_2Br(CF_2)_5MgBr$	F:	PerFHalOrg.4-62/3, 71
$Br_2C_6F_{12}Mg_2$	$BrMg(CF_2)_6MgBr$	F:	PerFHalOrg.4-66/7, 71, 87, 103
$Br_2C_6FeH_2O_4$	$(CO)_4FeCHBrCHBr$	Fe:	Org.Verb.B4-269, 278/80
–	$(CO)_4Fe(CHCHBr)Br$	Fe:	Org.Verb.B2-176
$Br_2C_6FeH_4O_3$	$BrC_3H_4Fe(CO)_3Br$	Fe:	Org.Verb.B5-40, 44, 48, 51, 53, 60/3
$Br_2C_6Fe_2GeH_{10}N_4O_4$	$(CH_2CHCH_2Fe(NO)_2)_2GeBr_2$	Fe:	Org.Verb.C2-156, 165
$Br_2C_6Fe_2H_{10}N_4O_4Pb$	$(CH_2CHCH_2Fe(NO)_2)_2PbBr_2$	Fe:	Org.Verb.C2-156, 165
$Br_2C_6Fe_2H_{10}N_4O_4Sn$	$(CH_2CHCH_2Fe(NO)_2)_2SnBr_2$	Fe:	Org.Verb.C2-156, 165
$Br_2C_6Fe_2O_6$	$(CO)_3FeBr_2Fe(CO)_3$	Fe:	Org.Verb.C1-73
$Br_2C_6H_7NSn$	$SnBr_2 \cdot CH_3C_5H_4N$	Sn:	MVol.C5-18
$Br_2C_6H_8N_2Sn$	$Br_2Sn(CH_2CH_2CN)_2$	Sn:	MVol.C6-184
–	$SnBr_2 \cdot C_6H_4(NH_2)_2$	Sn:	MVol.C5-15
–	$SnBr_2 \cdot C_6H_5NHNH_2$	Sn:	MVol.C5-15
$Br_2C_6H_{10}Ni_2$	$(C_3H_5NiBr)_2$	Ni:	Org.Verb.2-276/80
$Br_2C_6H_{10}O_2Pd_2S_4$	$[PdBr(S_2COC_2H_5)]_2$	C:	MVol.D4-261
$Br_2C_6H_{10}O_3Sn$	$[SnBr_2(CH(COCH_3)_2)OCH_3]$	Sn:	MVol.C5-106
$Br_2C_6H_{12}N_2Se_4Sn$	$[SnBr_2((CH_3)_2NCSeSe)_2]$	Sn:	MVol.C6-124/5
$Br_2C_6H_{12}N_4S_2Te$	$Te(SC(NHCH_2)_2)_2Br_2$	Te:	SVol.B3-151/3
$Br_2C_6H_{18}N_4S_4$	$[(NS(CH_3)_2NS(CH_3))_2]Br_2$	S:	S-N-Verb.1-5
$Br_2C_6H_{18}O_4P_2U$	$UO_2Br_2 \cdot 2\ (CH_3)_3PO$	U:	SVol.E1-148, 156
$Br_2C_6H_{18}O_4S_4$	$3\ (CH_3)_2SO \cdot SOBr_2$	S:	SVol.1-68
$Br_2C_6H_{18}O_5S_3U$	$UO_2Br_2 \cdot 3\ (CH_3)_2SO$	U:	SVol.E1-203, 207
$Br_2C_7ClF_3HN_3O$	$NC_5Br_2ClN(OH)C(CF_3)N$	F:	PerFHalOrg.6-126/7, 162
$Br_2C_7F_3HN_4O_3$	$NC_5Br_2(NO_2)N(OH)C(CF_3)N$	F:	PerFHalOrg.6-126/7, 162
$Br_2C_7F_3H_3N_4O$	$NC_5Br_2(NH_2)N(OH)C(CF_3)N$	F:	PerFHalOrg.6-126/7, 162
$Br_2C_7F_{10}$	$BrC(CF_2)(CF_2CF_2)_2CBr$	F:	PerFHalOrg.4-14
$Br_2C_7F_{10}Mg_2$	$BrMgC(CF_2)(CF_2CF_2)_2CMgBr$	F:	PerFHalOrg.4-66/7, 71
$Br_2C_7F_{13}N$	$CF_2BrCFBrN(CF_2)_5$	F:	PerFHalOrg.5-157/8, 170, 175
$Br_2C_7FeH_6O_3$	$BrCH_2C_3H_4Fe(CO)_3Br$	Fe:	Org.Verb.B5-46, 53, 64
$Br_2C_7H_{10}OTi$	$C_5H_5Ti(OC_2H_5)Br_2$	Ti:	Org.Verb.1-180/2
$Br_2C_7H_{16}Sn$	$(C_2H_5)_3SnCHBr_2$	Sn:	Org.Verb.2-164
$Br_2C_7H_{18}SiSn$	$(CH_3)_3SnCBr_2Si(CH_3)_3$	Sn:	Org.Verb.2-17/8
$Br_2C_8ClH_{20}NSn$	$[(C_2H_5)_4N]SnBr_2Cl$	Sn:	MVol.C3-113
$Br_2C_8Cl_2F_3HN_2O_2S$	$C_6Br_2Cl(SO_2Cl)NC(CF_3)NH$	F:	PerFHalOrg.6-125/6, 160, 162
$Br_2C_8Cl_2H_{24}MnN_2$	$[(CH_3)_4N]_2MnBr_2Cl_2$	Mn:	MVol.C5-309
$Br_2C_8Cl_3H_{11}Ni_2O_2$	$(C_3H_5NiBr)_2 \cdot CCl_3COOH$	Ni:	Org.Verb.2-285
$Br_2C_8Cl_3H_{20}NTe$	$[(C_2H_5)_4N][TeBr_2Cl_3]$	Te:	SVol.B3-86
$Br_2C_8Cl_4H_{24}N_2Sn$	$[(CH_3)_4N]_2[SnBr_2Cl_4]$	Sn:	MVol.C3-119
$Br_2C_8EuH_{16}O_2$	$EuBr_2 \cdot 2\ C_4H_8O$	Sc:	MVol.C6-69
$Br_2C_8F_3H_{15}O_3Si$	$(C_2H_5O)_3SiCFBrCF_2Br$	F:	PerFHalOrg.4-88
$Br_2C_8F_4FeO_4$	$(CBrCF_2)_2Fe(CO)_4$	Fe:	Org.Verb.B4-337, 345
$Br_2C_8F_{12}H_3N_3$	$Br(CF_2)_3C(NH)NC(NH_2)(CF_2)_3Br$	F:	PerFHalOrg.7-7, 34

$Br_2C_8F_{12}MoO_4P_2$	$Mo(CO)_4[(CF_3)_2PBr]_2$	F:	PerFHalOrg.3-110/1, 120
$Br_2C_8F_{18}N_2O_2$	$BrCF_2CF_2[N(CF_3)OCF_2CF_2]_2Br$	F:	PerFHalOrg.7-95, 112
$Br_2C_8FeH_9NO_3$	$(t\text{-}C_4H_9NC)Fe(CO)_3Br_2$	Fe:	Org.Verb.B4-2/3
$Br_2C_8FeH_{12}Hg_2INO_4$	$[N(CH_3)_4][(CO)_4Fe(HgBr)_2I]$	Fe:	Org.Verb.B2-153, 157
$Br_2C_8FeH_{12}N_4$	$(CH_3NC)_4FeBr_2$	Fe:	Org.Verb.B4-48/51, 56
$Br_2C_8FeH_{14}O_2S_2$	$(CO)_2Fe[(C_2H_5)SCH_2CH_2S(C_2H_5)]Br_2$	Fe:	Org.Verb.B1-101/2, 107
$Br_2C_8FeH_{18}O_2P_2$	$(CO)_2Fe(P(CH_3)_3)_2Br_2$	Fe:	Org.Verb.B1-101/3, 109
$Br_2C_8FeH_{18}O_8P_2$	$(CO)_2Fe(P(OCH_3)_3)_2Br_2$	Fe:	Org.Verb.B1-101/2, 105, 110
$Br_2C_8Fe_2H_2O_6$	$(CO)_3Fe(CHCHBr)BrFe(CO)_3$	Fe:	Org.Verb.C2-79/84
–	$(CO)_3Fe(CDCDBr)BrFe(CO)_3$	Fe:	Org.Verb.C2-79/81
$Br_2C_8H_{11}NS_2Ti$	$C_5H_5Ti(S_2CN(CH_3)_2)Br_2$	Ti:	Org.Verb.1-198/9
$Br_2C_8H_{12}O_2Ti$	$C_5H_5Ti(OC_2H_4OCH_3)Br_2$	Ti:	Org.Verb.1-180
$Br_2C_8H_{12}O_8U$	$UO_2Br_2 \cdot 2\ (CH_3CO)_2O$	U:	SVol.E1-123
$Br_2C_8H_{14}Ni_2$	$(CH_2C(CH_3)CH_2NiBr)_2$	Ni:	Org.Verb.2-305
–	$(CH_3CHCHCH_2NiBr)_2$	Ni:	Org.Verb.2-293/4
$Br_2C_8H_{14}Ni_2O_2$	$(CH_2C(OCH_3)CH_2NiBr)_2$	Ni:	Org.Verb.2-306/7
$Br_2C_8H_{14}O_7U$	$UO_2Br_2 \cdot (CH_3CO)_2O \cdot C_4H_8O_2$	U:	SVol.E1-123
$Br_2C_8H_{16}MnO_2$	$MnBr_2 \cdot 2\ C_4H_8O$	Mn:	MVol.D1-143
$Br_2C_8H_{16}MnO_4$	$MnBr_2 \cdot 2\ C_4H_8O_2$	Mn:	MVol.D1-154/5, 157
$Br_2C_8H_{16}N_4S_2Te$	$Te(SC(NHCH_2)_2CH_2)_2Br_2$	Te:	SVol.B3-159
$Br_2C_8H_{18}N_2O_2Sn$	$SnBr_2 \cdot 2\ HN(CH_2CH_2)_2O$	Sn:	MVol.C5-29
$Br_2C_8H_{18}N_2O_3S$	$2\ NH(CH_2CH_2)_2O \cdot SOBr_2$	S:	SVol.1-68
$Br_2C_8H_{18}N_2O_4U$	$UO_2Br_2 \cdot 2\ CH_3CON(CH_3)_2$	U:	SVol.E1-103/4, 106
$Br_2C_8H_{20}INSn$	$[(C_2H_5)_4N]SnBr_2I$	Sn:	MVol.C3-120
$Br_2C_8H_{20}MnO_2$	$MnBr_2 \cdot 2\ C_4H_9OH$	Mn:	MVol.D1-33
$Br_2C_8H_{20}O_2P_2Sn$	$SnBr_2[OP(C_2H_5)_2]_2$	Sn:	MVol.C6-140
$Br_2C_8H_{20}O_6P_2Sn$	$SnBr_2(C_2H_5(C_2H_5O)P(O)O)_2$	Sn:	MVol.C6-162
$Br_2C_8H_{20}P_2S_4Sn$	$SnBr_2((C_2H_5)_2P(S)S)_2$	Sn:	MVol.C6-157
$Br_2C_8H_{22}N_4S_4$	$[(NS(CH_3)_2NS(C_2H_5))_2]Br_2$	S:	S-N-Verb.1-5/6
$Br_2C_8H_{24}O_6S_4U$	$UO_2Br_2 \cdot 4\ (CH_3)_2SO$	U:	SVol.E1-203
$Br_2C_9F_5N$	$C_9F_5Br_2N$	F:	PerFHalOrg.6-130/1, 143
$Br_2C_9FeH_5InNO_4^-$	$[(CO)_4FeInBr_2(NC_5H_5)]^-$	Fe:	Org.Verb.B2-49
$Br_2C_9FeH_{18}HgO_3P_2$	$(CO)_3Fe(P(CH_3)_3)_2 \cdot HgBr_2$	Fe:	Org.Verb.B1-192/4
$Br_2C_9FeH_{18}HgO_9P_2$	$(CO)_3Fe(P(OCH_3)_3)_2 \cdot HgBr_2$	Fe:	Org.Verb.B1-192/4
$Br_2C_9H_{12}Sn$	$(CH_3)_3SnC_6H_3Br_2$	Sn:	Org.Verb.2-136
$Br_2C_9H_{13}OTi$	$C_5H_5TiBr_2 \cdot C_4H_8O$	Ti:	Org.Verb.1-134/6
$Br_2C_9H_{21}N_3O_5U$	$UO_2Br_2 \cdot 3\ HCON(CH_3)_2$	U:	SVol.E1-100/1
$Br_2C_9H_{27}O_{6.5}S_{4.5}U$	$UO_2Br_2 \cdot 4.5\ (CH_3)_2SO$	U:	SVol.E1-203, 207
$Br_2C_{10}ClF_3H_7N_3O_2S$	$(C_2H_5NHSO_2)(CF_3)C_7Br_2ClHN_2$	F:	PerFHalOrg.6-160
$Br_2C_{10}ClH_8N_3OsS$	$[OsClBr_2(NS)(NC_5H_4C_5H_4N)]$	S:	SVol.2-261
$Br_2C_{10}Cl_2F_2H_{12}N_2Te$	$[C_5H_5NH]_2Te[Br_2Cl_2F_2]$	Te:	SVol.B3-87
$Br_2C_{10}Cl_2H_{10}N_2Te$	$TeBr_2Cl_2 \cdot 2\ C_5H_5N$	Te:	SVol.B3-170
$Br_2C_{10}Cl_2H_{10}Ti$	$C_5H_5Ti(C_5H_5Br_2)Cl_2$	Ti:	Org.Verb.1-206
$Br_2C_{10}Cl_2H_{12}I_2N_2Te$	$[C_5H_5NH]_2[TeBr_2Cl_2I_2]$	Te:	SVol.B3-87
$Br_2C_{10}Cl_2H_{12}N_2O_2U$	$[C_5H_5NH]_2UO_2Br_2Cl_2$	U:	SVol.C9-162
$Br_2C_{10}Cl_2H_{12}O_4Sn$	$[SnBr_2(ClC(COCH_3)_2)_2]$	Sn:	MVol.C5-107
$Br_2C_{10}Cl_3H_{15}Ni_2O_2$	$(CH_3CHCHCH_2NiBr)_2 \cdot CCl_3COOH$	Ni:	Org.Verb.2-300
$Br_2C_{10}Cl_4H_4O_2Sn$	$SnCl_4 \cdot C_{10}H_4Br_2O_2$	Sn:	MVol.C5-112
$Br_2C_{10}Cl_4H_{10}Ti$	$C_5H_5Ti(C_5H_5Cl_4)Br_2$	Ti:	Org.Verb.1-206

$Br_2C_{10}Cl_4H_{12}N_2Te$	$[C_5H_5NH]_2[TeCl_4Br_2]$	Te:	SVol.B3-87
$Br_2C_{10}F_6Mg_2$	$C_{10}F_6(MgBr)_2$	F:	PerFHalOrg.4-66/7
$Br_2C_{10}F_{12}Fe_2O_6P_2$	$Br(CO)_3Fe(P(CF_3)_2)_2Fe(CO)_3Br$	Fe:	Org.Verb.C1-6/8, 11
$Br_2C_{10}F_{12}Hg$	$Hg(C_5F_6Br)_2$	F:	PerFHalOrg.4-34
$Br_2C_{10}F_{15}N_3$	$(CF_2BrC_2F_4)_2(CF_3)C_3N_3$	F:	PerFHalOrg.6-86/8, 107
$Br_2C_{10}FeH_{10}$	$C_5H_5FeC_5H_5Br_2$	Fe:	Org.Verb.A1-183
$Br_2C_{10}FeH_{15}InNO_4^-$	$[(CO)_4FeInBr_2(N(C_2H_5)_3)]^-$	Fe:	Org.Verb.B2-49
$Br_2C_{10}FeH_{30}O_5S_5$	$Fe(OS(CH_3)_2)_5Br_2$	Fe:	Org.Verb.B5-58
$Br_2C_{10}Fe_2GeH_{18}N_4O_4$	$((CH_3)_2C_3H_3Fe(NO)_2)_2GeBr_2$	Fe:	Org.Verb.C2-156, 165
$Br_2C_{10}Fe_2H_2O_{10}$	$[(CO)_5FeH]_2Br_2$	Fe:	Org.Verb.B3-252
$Br_2C_{10}Fe_2H_{10}O_6S_2$	$Br(CO)_3Fe(SC_2H_5)_2Fe(CO)_3Br$	Fe:	Org.Verb.C1-6/7, 9, 12
$Br_2C_{10}Fe_2H_{12}O_6P_2$	$Br(CO)_3Fe(P(CH_3)_2)_2Fe(CO)_3Br$	Fe:	Org.Verb.C1-6/7, 10
$Br_2C_{10}Fe_2H_{18}N_4O_4Pb$	$((CH_3)_2C_3H_3Fe(NO)_2)_2PbBr_2$	Fe:	Org.Verb.C2-156, 165
$Br_2C_{10}Fe_2H_{18}N_4O_4Sn$	$((CH_3)_2C_3H_3Fe(NO)_2)_2SnBr_2$	Fe:	Org.Verb.C2-156, 165
$Br_2C_{10}H_4O_2Te$	$C_{10}H_4O_2TeBr_2$	Te:	SVol.B3-14
$Br_2C_{10}H_8N_2O_2S_2Sn$	$SnBr_2(C_5H_4NOS)_2$	Sn:	MVol.C6-106
$Br_2C_{10}H_8N_2Sn$	$SnBr_2 \cdot NC_5H_4C_5H_4N$	Sn:	MVol.C5-21/2
$Br_2C_{10}H_{10}I_2N_2Te$	$TeBr_2I_2 \cdot 2\ C_5H_5N$	Te:	SVol.B3-170
$Br_2C_{10}H_{10}I_4Ti$	$C_5H_5Ti(C_5H_5I_4)Br_2$	Ti:	Org.Verb.1-206/7
$Br_2C_{10}H_{10}N_2OS$	$2\ C_5H_5N \cdot SOBr_2$	S:	SVol.1-68
$Br_2C_{10}H_{10}N_2Sn$	$SnBr_2 \cdot 2\ C_5H_5N$	Sn:	MVol.C5-16/7
$Br_2C_{10}H_{10}N_2Te$	$TeBr_2 \cdot 2\ C_5H_5N$	Te:	SVol.B3-133
$Br_2C_{10}H_{10}Ni$	$C_5H_5NiC_5H_5Br_2$	Ni:	Org.Verb.2-205
$Br_2C_{10}H_{14}O_4Sn$	$[SnBr_2(CH(COCH_3)_2)_2]$	Sn:	MVol.C5-101/6
$Br_2C_{10}H_{15}NS_2Ti$	$C_5H_5Ti(S_2CN(C_2H_5)_2)Br_2$	Ti:	Org.Verb.1-198/9
$Br_2C_{10}H_{16}MnO_4$	$[MnBr_2(CH_2(COCH_3)_2)_2]$	Mn:	MVol.D1-64/5
$Br_2C_{10}H_{18}N_2Ni$	$(t\text{-}C_4H_9NC)_2NiBr_2$	Ni:	Org.Verb.1-316
$Br_2C_{10}H_{18}Ni_2$	$(CH_3CHCHCHCH_3NiBr)_2$	Ni:	Org.Verb.2-314
–	$((CH_3)_2CCHCH_2NiBr)_2$	Ni:	Org.Verb.2-310/1
$Br_2C_{10}H_{20}N_2S_4Sn$	$[SnBr_2((C_2H_5)_2NCSS)_2]$	Sn:	MVol.C6-119/20
$Br_2C_{10}H_{22}N_2O_2Sn$	$SnBr_2 \cdot 2\ CH_3N(CH_2CH_2)_2O$	Sn:	MVol.C5-29
$Br_2C_{10}H_{22}N_2Sn$	$SnBr_2 \cdot 2\ C_5H_{10}NH$	Sn:	MVol.C5-16
$Br_2C_{10}H_{24}N_4S_2Te$	$Te((CH_3)_2NCSN(CH_3)_2)_2Br_2$	Te:	SVol.B3-148
$Br_2C_{11}FeH_7NO_3$	$CH_3C_6H_4NCFe(CO)_3Br_2$	Fe:	Org.Verb.B4-2/3
$Br_2C_{11}H_{11}NOSSn$	$[SnBr_2(SC_6H_4NC(CH_3)CHCOCH_3)]$	Sn:	MVol.C6-90/1
$Br_2C_{12}ClF_3H_{11}N_3O_2S$	$((C_2H_5)_2NSO_2)(CF_3)C_7Br_2ClHN_2$	F:	PerFHalOrg.6-162
$Br_2C_{12}ClFeH_{20}Hg_2NO_4$	$[N(C_2H_5)_4][(CO)_4Fe(HgBr)_2Cl]$	Fe:	Org.Verb.B2-153, 157
$Br_2C_{12}Cl_2H_8Sn$	$(BrC_6H_4)_2SnCl_2$	Sn:	Org.Verb.6-184/6
$Br_2C_{12}Cl_4H_{12}N_2Sn$	$SnCl_4 \cdot 2\ BrC_6H_4NH_2$	Sn:	MVol.C5-179/80
$Br_2C_{12}Cl_6H_{14}N_2Sn$	$[BrC_6H_4NH_3]_2[SnCl_6]$	Sn:	MVol.C3-103/4
$Br_2C_{12}F_5H_4N_3O_2$	$NC_5F_2Br_2NHC_6H_3(CF_3)NO_2$	F:	PerFHalOrg.5-140
$Br_2C_{12}F_6Li_2$	$Li(Br)C_6F_3C_6F_3(Br)Li$	F:	PerFHalOrg.4-7, 14
$Br_2C_{12}F_7Li$	$BrC_6F_4C_6F_3(Br)Li$	F:	PerFHalOrg.4-11
$Br_2C_{12}F_8$	$BrC_6F_4C_6F_4Br$	F:	PerFHalOrg.4-11, 14, 39
–	$Br_2C_6F_3C_6F_5$	F:	PerFHalOrg.4-11
$Br_2C_{12}F_8Hg$	$(BrC_6F_4)_2Hg$	F:	PerFHalOrg.4-39
$Br_2C_{12}F_8HgMg_2$	$(BrMgC_6F_4)_2Hg$	F:	PerFHalOrg.4-66/7, 71
$Br_2C_{12}F_8Mg$	$(BrC_6F_4)_2Mg$	F:	PerFHalOrg.4-67
$Br_2C_{12}F_8Mg_2$	$BrMgC_6F_4C_6F_4MgBr$	F:	PerFHalOrg.4-66/7, 71

$Br_2C_{12}F_8O$	$BrC_6F_4OC_6F_4Br$	F: PerFHalOrg.7-62
$Br_2C_{12}F_{10}Ge$	$(C_6F_5)_2GeBr_2$	F: PerFHalOrg.4-166/7, 176, 180
$Br_2C_{12}F_{10}Si$	$(C_6F_5)_2SiBr_2$	F: PerFHalOrg.4-150/2, 157, 162
$Br_2C_{12}F_{10}Sn$	$(C_6F_5)_2SnBr_2$	F: PerFHalOrg.4-181/2, 186
$Br_2C_{12}F_{10}Tl^-$	$[(C_6F_5)_2TlBr_2]^-$	F: PerFHalOrg.4-137
$Br_2C_{12}F_{12}GeH_6$	$(CH_3)_2Ge(C_5F_6Br)_2$	F: PerFHalOrg.4-27
$Br_2C_{12}F_{14}N_6$	$[(CF_2Br)(CF_3)C_3N_3]_2C_2F_4$	F: PerFHalOrg.6-89/90, 108
$Br_2C_{12}F_{14}N_6O$	$[(CF_3)(CF_2Br)C_3N_3CF_2]_2O$	F: PerFHalOrg.6-90/1, 108
$Br_2C_{12}F_{14}N_6S$	$[(CF_3)(CF_2Br)C_3N_3CF_2]_2S$	F: PerFHalOrg.6-90/1, 108
$Br_2C_{12}F_{19}N_3$	$(CF_2BrC_2F_4)_2(C_3F_7)C_3N_3$	F: PerFHalOrg.6-87/8, 107
$Br_2C_{12}F_{19}N_3O$	$(CF_2BrC_2F_4)_2(CF_3OC_2F_4)C_3N_3$	F: PerFHalOrg.6-87/8, 107
$Br_2C_{12}FeH_{10}$	$C_5H_5FeC_5H_4CHCBr_2$	Fe: Org.Verb.A1-385, 386
$Br_2C_{12}FeH_{10}N_2O_2$	$(CO)_2Fe(NC_5H_5)_2Br_2$	Fe: Org.Verb.B1-101
$Br_2C_{12}FeH_{12}$	$C_5H_5FeC_5H_4CHBrCH_2Br$	Fe: Org.Verb.A1-381/3
$Br_2C_{12}FeH_{18}N_2O_2$	$(t\text{-}C_4H_9NC)_2Fe(CO)_2Br_2$	Fe: Org.Verb.B4-16
$Br_2C_{12}FeH_{18}N_6$	$[(CH_3NC)_6Fe]Br_2$	Fe: Org.Verb.B4-97, 101
–	$[(CH_3NC)_6Fe]Br_2 \cdot 2\ H_2O$	Fe: Org.Verb.B4-101
–	$[(CH_3NC)_6Fe]Br_2 \cdot 3\ H_2O$	Fe: Org.Verb.B4-101
$Br_2C_{12}H_8I_2Sn$	$(BrC_6H_4)_2SnI_2$	Sn: Org.Verb.6-186
$Br_2C_{12}H_8N_2O_4Sn$	$SnBr_2(NC_5H_4COO)_2$	Sn: MVol.C5-205/6
$Br_2C_{12}H_8N_2Sn$	$SnBr_2 \cdot C_{12}H_8N_2$	Sn: MVol.C5-22/3
$Br_2C_{12}H_8OSn$	$[(BrC_6H_4)_2SnO]_x$	Sn: Org.Verb.6-186
$Br_2C_{12}H_{10}N_2Te$	$(C_6H_4NH)_2TeBr_2$	Te: SVol.B3-14
$Br_2C_{12}H_{10}Ni_2O_2Sn$	$(C_5H_5NiCO)_2SnBr_2$	Ni: Org.Verb.2-350
$Br_2C_{12}H_{10}O_8Sn$	$[SnBr_2(C_5H_2O_2(CH_2OH)O)_2]$	Sn: MVol.C5-160/1
$Br_2C_{12}H_{10}Te$	$(C_6H_5)_2TeBr_2$	Te: SVol.B2-99
$Br_2C_{12}H_{10}Ti$	$(C_6H_5)_2TiBr_2$	Ti: Org.Verb.1-73
$Br_2C_{12}H_{12}N_2O_2U$	$[UO_2Br_2((CH_3)_2C_{10}H_6N_2)]$	U: SVol.E1-39/40, 43
$Br_2C_{12}H_{12}N_2Sn$	$SnBr_2 \cdot NH_2C_6H_4C_6H_4NH_2$	Sn: MVol.C5-15
$Br_2C_{12}H_{14}MnN_2$	$MnBr_2 \cdot 2\ C_6H_5NH_2$	Mn: MVol.C5-273
$Br_2C_{12}H_{14}N_2Sn$	$SnBr_2 \cdot 2\ CH_3C_5H_4N$	Sn: MVol.C5-18
–	$SnBr_2 \cdot 2\ C_6H_5NH_2$	Sn: MVol.C5-12/3
$Br_2C_{12}H_{14}N_2Ti$	$(CH_3)_2TiBr_2 \cdot NC_5H_4C_5H_4N$	Ti: Org.Verb.1-75
$Br_2C_{12}H_{18}Ni_2$	$((CH_3)_2C_4H_3NiBr)_2$	Ni: Org.Verb.2-319
–	$(C_6H_9NiBr)_2$	Ni: Org.Verb.2-319
$Br_2C_{12}H_{18}Ni_2O_4$	$(CH(COOC_2H_5)CHCH_2NiBr)_2$	Ni: Org.Verb.2-297
–	$(CH_2C(COOC_2H_5)CH_2NiBr)_2$	Ni: Org.Verb.2-307/8
$Br_2C_{12}H_{18}O_4Sn$	$[SnBr_2(CH_3C(COCH_3)_2)_2]$	Sn: MVol.C5-108
$Br_2C_{12}H_{24}N_8S_4Te$	$Te(SC(NHCH_2)_2)_4Br_2$	Te: SVol.B3-158
–	$Te(SC(NHCH_2)_2)_4Br_2 \cdot 2\ H_2O$	Te: SVol.B3-158
$Br_2C_{12}H_{24}N_8S_4Te_2^{2+}$	$[Te_2(SC(NHCH_2)_2)_4Br_2]^{2+}$	Te: SVol.B3-152/3
$Br_2C_{12}H_{24}N_8Te$	$TeBr_2 \cdot 2\ (CH_2)_6N_4$	Te: SVol.B3-133/4
$Br_2C_{12}H_{27}O_6PU$	$UO_2Br_2 \cdot (C_4H_9O)_3PO$	U: SVol.E1-182
$Br_2C_{12}H_{30}O_4P_2U$	$UO_2Br_2 \cdot 2\ (C_2H_5)_3PO$	U: SVol.E1-149, 156/7
$Br_2C_{12}H_{32}N_8S_4Te$	$Te(SC(NH_2)_2)_2((CH_3)_2NCSN(CH_3)_2)_2Br_2$	Te: SVol.B3-151
$Br_2C_{12}H_{36}IrNO_3P_4$	$[Ir(P(CH_3)_3)_4Br_2]NO_3$	Ir: SVol.2-199
$Br_2C_{12}H_{36}N_6O_4P_2U$	$UO_2Br_2 \cdot 2\ [(CH_3)_2N]_3PO$	U: SVol.E1-163/4
$Br_2C_{13}Cl_2H_9NSn$	$CH_3N(C_6H_3Br)_2SnCl_2$	Sn: Org.Verb.6-208

$Br_2C_{13}FeGeH_{10}NiO_3$	$C_5H_5Ni(CO)GeBr_2Fe(CO)_2C_5H_5$	Ni:	Org.Verb.2-164/6
$Br_2C_{13}FeH_{10}NiO_3Sn$	$C_5H_5Ni(CO)SnBr_2Fe(CO)_2C_5H_5$	Ni:	Org.Verb.2-164/6
$Br_2C_{13}FeH_{12}$	$C_5H_5FeC_5H_4C_3H_3Br_2$	Fe:	Org.Verb.A1-392
$Br_2C_{13}FeH_{12}O$	$C_5H_5FeC_5H_4COCHBrCH_2Br$	Fe:	Org.Verb.A2-243
$Br_2C_{13}H_9NOSSn$	$[SnBr_2(SC_6H_4NCHC_6H_4O)]$	Sn:	MVol.C6-91/2
$Br_2C_{13}H_9NO_2Sn$	$[SnBr_2(OC_6H_4CHNC_6H_4O)]$	Sn:	MVol.C6-21
$Br_2C_{13}H_{28}Sn$	$(C_4H_9)_3SnCHBr_2$	Sn:	Org.Verb.2-280
$Br_2C_{14}Cl_2H_{12}Sn$	$(BrC_6H_4CH_2)_2SnCl_2$	Sn:	Org.Verb.6-134
$Br_2C_{14}Cl_2H_{16}Ni_2O_2$	$(CH_3CHCHCH_2NiCl)_2 \cdot C_6H_2Br_2O_2$	Ni:	Org.Verb.2-300
$Br_2C_{14}Cl_4H_6O_2Sn$	$SnCl_4 \cdot C_{14}H_6Br_2O_2$	Sn:	MVol.C5-114
$Br_2C_{14}Cl_4H_{14}Ni_2O_2$	$(CH_3CHCHCH_2NiBr)_2 \cdot C_6Cl_4O_2$	Ni:	Org.Verb.2-301
$Br_2C_{14}F_8H_6Sn$	$(BrC_6F_4)_2Sn(CH_3)_2$	F:	PerFHalOrg.4-29
		Sn:	Org.Verb.3-10, 18
$Br_2C_{14}FeH_8N_2O_2$	$(CO)_2Fe(N_2C_{12}H_8)Br_2$	Fe:	Org.Verb.B1-101/2, 106
$Br_2C_{14}FeH_{14}$	$C_5H_5FeC_5H_4C_3H_2Br_2CH_3$	Fe:	Org.Verb.A1-392/4
$Br_2C_{14}FeH_{26}Hg_2N_2O_2$	$(CO)_2Fe(NH_2CH(CH_2)_5)_2(HgBr)_2$	Fe:	Org.Verb.B1-116/7
$Br_2C_{14}FeH_{30}O_2P_2$	$(CO)_2Fe(P(C_2H_5)_3)_2Br_2$	Fe:	Org.Verb.B1-101/3
$Br_2C_{14}FeH_{30}O_8P_2$	$(CO)_2Fe(P(OC_2H_5)_3)_2Br_2$	Fe:	Org.Verb.B1-101/2, 105, 110
$Br_2C_{14}H_8MnO_2$	$MnBr_2 \cdot C_{14}H_8O_2$	Mn:	MVol.D1-137
$Br_2C_{14}H_8N_2O_2Te_2$	$BrC_6H_3(CO)NHTeTe(NH)COC_6H_3Br$	Te:	SVol.B3-14
$Br_2C_{14}H_{10}O_4Sn$	$[SnBr_2(C_7H_5O_2)_2]$	Sn:	MVol.C5-97
–	$SnBr_2(OC_6H_4CHO)_2$	Sn:	MVol.C5-81
$Br_2C_{14}H_{12}N_2O_2U$	$[UO_2Br_2((CH_3)_2C_{12}H_6N_2)]$	U:	SVol.E1-44/6
$Br_2C_{14}H_{14}Ti$	$(C_6H_5CH_2)_2TiBr_2$	Ti:	Org.Verb.1-73, 75
$Br_2C_{14}H_{16}MnO_4$	$MnBr_2 \cdot 2\ O(CCH_3CH)_2CO \cdot 8\ H_2O$	Mn:	MVol.D1-149
$Br_2C_{14}H_{16}N_2O_2Sn$	$SnBr_2 \cdot NH_2(C_6H_3OCH_3)_2NH_2$	Sn:	MVol.C5-15
$Br_2C_{14}H_{16}N_2Sn$	$SnBr_2 \cdot NH_2(C_6H_3CH_3)_2NH_2$	Sn:	MVol.C5-15
$Br_2C_{14}H_{16}O_2Te$	$TeBr_2 \cdot 2\ C_6H_5CH_2OH$	Te:	SVol.B3-133
$Br_2C_{14}H_{18}N_2OS$	$2\ C_6H_5CH_2NH_2 \cdot SOBr_2$	S:	SVol.1-68
$Br_2C_{14}H_{18}N_2O_2Sn$	$SnBr_2 \cdot 2\ CH_3OC_6H_4NH_2$	Sn:	MVol.C5-13
$Br_2C_{14}H_{18}N_2Sn$	$SnBr_2 \cdot 2\ CH_3C_6H_4NH_2$	Sn:	MVol.C5-13
–	$SnBr_2 \cdot 2\ C_6H_5CH_2NH_2$	Sn:	MVol.C5-13
–	$SnBr_2 \cdot 2\ C_6H_5NHCH_3$	Sn:	MVol.C5-13
$Br_2C_{14}H_{22}Ni_2$	$(CH_2CHCH_2CH_2CHCHCH_2NiBr)_2$	Ni:	Org.Verb.2-298
$Br_2C_{14}H_{22}Ni_2O_4$	$(C_2H_5O_2CCHC(CH_3)CH_2NiBr)_2$	Ni:	Org.Verb.2-312
$Br_2C_{14}H_{22}O_4Sn$	$[SnBr_2(C_2H_5C(COCH_3)_2)_2]$	Sn:	MVol.C5-108
$Br_2C_{14}H_{34}O_6P_2Sn$	$SnBr_2 \cdot 2\ CH_3PO(OCH(CH_3)_2)_2$	Sn:	MVol.C5-60/1
$Br_2C_{15}FeH_{10}O_3Te$	$(CO)_3Fe(Te(C_6H_5)_2)Br_2$	Fe:	Org.Verb.B1-180/3
$Br_2C_{15}FeH_{16}$	$C_5H_5FeC_5H_4C_3H_2Br_2C_2H_5$	Fe:	Org.Verb.A1-392/3
$Br_2C_{15}H_{10}O_5Sn$	$SnBr_2 \cdot (HO)_2C_9H_4OC_6H_3O(OH)$	Sn:	MVol.C5-11
$Br_2C_{15}H_{10}O_6Sn$	$SnBr_2 \cdot (HO)_2C_9H_4OC_6H_2O(OH)_2$	Sn:	MVol.C5-11
$Br_2C_{15}H_{15}NSn$	$(CH_3)_2Sn(BrC_6H_3N(CH_3)C_6H_3Br)$	Sn:	Org.Verb.3-118, 122
$Br_2C_{15}H_{36}MnO_3$	$MnBr_2 \cdot 3\ C_5H_{11}OH$	Mn:	MVol.D1-33
$Br_2C_{16}ClH_{36}NSn$	$[(C_4H_9)_4N]SnBr_2Cl$	Sn:	MVol.C3-113
$Br_2C_{16}Cl_2F_2H_{40}N_2Te$	$[(C_2H_5)_4N]_2[TeCl_2Br_2F_2]$	Te:	SVol.B3-87
$Br_2C_{16}Cl_2H_{40}I_2N_2Te$	$[(C_2H_5)_4N]_2[TeCl_2Br_2I_2]$	Te:	SVol.B3-87
$Br_2C_{16}Cl_2H_{40}MnN_2$	$[(C_2H_5)_4N]_2MnBr_2Cl_2$	Mn:	MVol.C5-309
$Br_2C_{16}Cl_4H_{18}N_4Sn$	$(C_2H_5)_2SnCl_2 \cdot 2\ BrC_6H_4N_2Cl$	Sn:	Org.Verb.6-61
$Br_2C_{16}Cl_4H_{40}N_2Sn$	$[(C_2H_5)_4N]_2[SnBr_2Cl_4]$	Sn:	MVol.C3-119

$Br_2C_{16}Cl_4H_{40}N_2Te$	$[(C_2H_5)_4N]_2[TeCl_4Br_2]$	Te: SVol.B3-83, 87
$Br_2C_{16}FeH_{16}$	$C_5H_5FeC_5H_4C(CBrCH_2)CHCH_2CH_2Br$	Fe: Org.Verb.A1-387, 388
–	$C_5H_5FeC_5H_4C_3H_2Br_2C_3H_5$	Fe: Org.Verb.A1-392/3
$Br_2C_{16}FeH_{18}$	$C_5H_5FeC_5H_4C_3H_2Br_2C_3H_7$-n	Fe: Org.Verb.A1-392/3
$Br_2C_{16}FeH_{26}Hg_2N_2O_4$	$(CO)_4Fe(HgBr)_2(C_6H_{11}NH_2)_2$	Fe: Org.Verb.B2-153/5, 158
$Br_2C_{16}H_{13}NO_2Sn$	$[SnBr_2(C_6H_5COCHC(CH_3)NC_6H_4O)]$	Sn: MVol.C6-27/8
$Br_2C_{16}H_{14}N_2O_2Sn$	$[SnBr_2(C_2H_4(NCHC_6H_4O)_2)]$	Sn: MVol.C6-22/3
$Br_2C_{16}H_{18}Sn$	$(C_2H_5)_2Sn(C_6H_4Br)_2$	Sn: Org.Verb.3-29
$Br_2C_{16}H_{20}O_4Sn$	$[SnBr_2(CH(COCH_3)_2)_2] \cdot C_6H_6$	Sn: MVol.C5-106
$Br_2C_{16}H_{22}N_2O_2Sn$	$SnBr_2 \cdot 2\ C_2H_5OC_6H_4NH_2$	Sn: MVol.C5-13
$Br_2C_{16}H_{22}N_2Sn$	$SnBr_2 \cdot 2\ C_6H_5NHC_2H_5$	Sn: MVol.C5-13
$Br_2C_{16}H_{26}Ni_2$	$(C_8H_{13}NiBr)_2$	Ni: Org.Verb.2-320
$Br_2C_{16}H_{40}I_2MnN_2$	$[(C_2H_5)_4N]_2MnI_2Br_2$	Mn: MVol.C5-343
$Br_2C_{16}H_{40}I_4N_2Sn$	$[(C_2H_5)_4N]_2[SnBr_2I_4]$	Sn: MVol.C3-121
$Br_2C_{16}H_{42}Si_4Sn$	$[((CH_3)_3Si)_2CBr]_2Sn(CH_3)_2$	Sn: Org.Verb.6-18
$Br_2C_{17}Cl_3H_{43}N_2Ti$	$[(C_2H_5)_4N]_2[CH_3TiCl_3Br_2]$	Ti: Org.Verb.1-42/3
$Br_2C_{17}FeH_{18}$	$C_5H_5FeC_5H_4C(CBrCHCH_3)CHCH_2CH_2Br$	Fe: Org.Verb.A1-387, 388
$Br_2C_{17}FeH_{20}$	$C_5H_5FeC_5H_4C_3H_2Br_2C(CH_3)_3$	Fe: Org.Verb.A1-392/3
$Br_2C_{18}ClH_{15}OTi$	$C_6H_5CH_2C_5H_4Ti(OC_6H_4Cl)Br_2$	Ti: Org.Verb.1-180
$Br_2C_{18}ClH_{45}IrP_3$	$Ir(P(C_2H_5)_3)_3ClBr_2$	Ir: SVol.2-200/2
$Br_2C_{18}Cl_3H_{45}HgIrP_3$	$Ir(P(C_2H_5)_3)_3ClBr_2(HgCl_2)$	Ir: SVol.2-264/5
–	$Ir(P(C_2H_5)_3)_3Cl_3(HgBr_2)$	Ir: SVol.2-264/5
$Br_2C_{18}F_{15}I_2P$	$(C_6F_5)_3PBr_2 \cdot I_2$	F: PerFHalOrg.3-125
$Br_2C_{18}F_{15}P$	$(C_6F_5)_3PBr_2$	F: PerFHalOrg.3-124
$Br_2C_{18}F_{21}N_9$	$Br[CF_2C_3N_3(CF_3)CF_2]_3Br$	F: PerFHalOrg.6-89/90, 108
$Br_2C_{18}F_{26}N_6$	$[(CF_3)(CF_2Br)C_3N_3(CF_2)_3]_2C_2F_4$	F: PerFHalOrg.6-91/2, 108
$Br_2C_{18}FeH_{14}N_2O_2$	$(CH_3C_6H_4NC)_2Fe(CO)_2Br_2$	Fe: Org.Verb.B4-16/7
$Br_2C_{18}H_{10}MnO_2$	$MnBr_2 \cdot C_{18}H_{10}O_2$	Mn: MVol.D1-137
$Br_2C_{18}H_{10}N_2S_2Sn$	$Sn(C_9H_5BrNS)_2$	Sn: MVol.C5-51/2
$Br_2C_{18}H_{12}N_2O_2Sn$	$SnBr_2(C_9H_6NO)_2$	Sn: MVol.C5-214
$Br_2C_{18}H_{14}N_2OS$	$2\ C_9H_7N \cdot SOBr_2$	S: SVol.1-68
$Br_2C_{18}H_{14}N_2O_4U$	$UO_2Br_2 \cdot 2\ C_9H_7NO$	U: SVol.E1-134, 143/4
$Br_2C_{18}H_{15}OPSn$	$SnBr_2 \cdot (C_6H_5)_3PO$	Sn: MVol.C5-59
$Br_2C_{18}H_{15}PSn$	$SnBr_2 \cdot P(C_6H_5)_3$	Sn: MVol.C5-58
$Br_2C_{18}H_{18}N_2O_5U$	$UO_2(C_2H_4(NCHC_6H_3(Br)O)_2) \cdot C_2H_5OH$	U: SVol.E1-62, 68/9
$Br_2C_{18}H_{18}Ni_2$	$(CH_2C(C_6H_5)CH_2NiBr)_2$	Ni: Org.Verb.2-307
$Br_2C_{18}H_{18}Ni_2O_4$	$(C_3H_5NiBr)_2 \cdot 2\ C_6H_4O_2$	Ni: Org.Verb.2-285
$Br_2C_{18}H_{26}N_4S_4$	$[NS(CH_3)_2NS(CH_2C_6H_5)]_2Br_2$	S: S-N-Verb.1-6
$Br_2C_{18}H_{26}Ni_2$	$[(C_2H_5)_2C_5H_3NiBr]_2$	Ni: Org.Verb.2-339, 341
$Br_2C_{18}H_{36}IrNO_2$	$(C_4H_9)_4N[Ir(CO)_2Br_2]$	Ir: SVol.2-146
$Br_2C_{18}H_{46}IrP_3$	$Ir(P(C_2H_5)_3)_3HBr_2$	Ir: SVol.2-196/7
$Br_2C_{19}F_{28}N_6O_2$	$[CF_3(CF_2Br)C_3N_3CF(CF_3)OC_2F_4]_2CF_2$	F: PerFHalOrg.6-91/2, 108
–	$[(CF_3)(CF_2Br)C_3N_3]_2[CF(CF_3)OCF_2CF(CF_3)O(CF_2)_4]$	F: PerFHalOrg.6-91/2, 108
$Br_2C_{19}FeH_{16}$	$C_5H_5FeC_5H_4C_3H_2Br_2C_6H_5$	Fe: Org.Verb.A1-392/3
$Br_2C_{19}FeH_{16}O$	$C_5H_5FeC_5H_4COCHBrCHBrC_6H_5$	Fe: Org.Verb.A2-263, 265
$Br_2C_{19}FeH_{22}$	$C_5H_5FeC_5H_4C_3H_2Br_2C_6H_{11}$	Fe: Org.Verb.A1-392, 394
$Br_2C_{20}ClH_{20}IrN_4$	$[Ir(C_5H_5N)_4ClBr]Br$	Ir: SVol.2-69
$Br_2C_{20}Cl_2H_{32}MnN_2$	$[(CH_3)_3(C_6H_5CH_2)N]_2MnBr_2Cl_2$	Mn: MVol.C5-309
$Br_2C_{20}F_{30}N_6$	$[(CF_2BrC_2F_4)(CF_3)C_3N_3(CF_2)_3]_2$	F: PerFHalOrg.6-91/2, 109

$Br_2C_{20}FeH_{36}N_4$	$(t\text{-}C_4H_9NC)_4FeBr_2$	Fe:	Org.Verb.B4-48/51, 58
$Br_2C_{20}Fe_2H_{16}$	$BrC_5H_4FeC_5H_4C_5H_4FeC_5H_4Br$	Fe:	Org.Verb.A6-18/9, 23/4
$Br_2C_{20}Fe_2H_{16}Hg$	$(BrC_5H_4FeC_5H_4)_2Hg$	Fe:	Org.Verb.A6-208/10
–	$(C_5H_5FeC_5H_3Br)_2Hg$	Fe:	Org.Verb.A6-209/10
$Br_2C_{20}Fe_2H_{18}Hg$	$[(C_5H_5FeC_5H_4)_2Hg]Br_2$	Fe:	Org.Verb.A6-207/8
$Br_2C_{20}H_{14}Sn$	$(C_{10}H_7)_2SnBr_2$	Sn:	Org.Verb.6-184
$Br_2C_{20}H_{16}MnN_4$	$[MnBr_2(NC_5H_4C_5H_4N)_2]$	Mn:	MVol.D1-102
$Br_2C_{20}H_{18}N_2Sn$	$SnBr_2 \cdot 2\ C_{10}H_7NH_2$	Sn:	MVol.C5-13
$Br_2C_{20}H_{18}O_4Sn$	$[SnBr_2(C_6H_5COCHCOCH_3)_2]$	Sn:	MVol.C5-110
$Br_2C_{20}H_{22}N_2O_5U$	$UO_2(C_2H_4(NCHC_6H_3(Br)O)_2) \cdot C_4H_9OH$	U:	SVol.E1-62, 69
$Br_2C_{20}H_{22}Ni_2O_4$	$(CH_2C(CH_3)CH_2NiBr)_2 \cdot 2\ C_6H_4O_2$	Ni:	Org.Verb.2-308/9
–	$(CH_3CHCHCH_2NiBr)_2 \cdot 2\ C_6H_4O_2$	Ni:	Org.Verb.2-300, 301
–	$(C_3H_5NiBr)_2 \cdot 2\ CH_3C_6H_3O_2$	Ni:	Org.Verb.2-285
$Br_2C_{20}H_{26}Sn$	$(t\text{-}C_4H_9)_2Sn(C_6H_4Br)_2$	Sn:	Org.Verb.3-69
$Br_2C_{20}H_{30}Ni_2$	$((CH_3)_2C_8H_9NiBr)_2$	Ni:	Org.Verb.2-318
$Br_2C_{20}H_{32}NiS_2$	$S(CH_2C(CH_3)_2)_2C_4(C(CH_3)_2CH_2)_2SNiBr_2$	Ni:	Org.Verb.2-329
$Br_2C_{20}H_{34}Ni_2$	$((CH_3)_2CCHC_2H_4C(CH_3)CHCH_2NiBr)_2$	Ni:	Org.Verb.2-311
$Br_2C_{20}H_{42}IrP_3$	$Ir(P(C_2H_5)_3)_2(P(CH_3)_2(C_6H_5))HBr_2$	Ir:	SVol.2-197/8
$Br_2C_{20}H_{45}IrN_4$	$[Ir(C_5H_{10}NH)_4HBr]Br \cdot H_2O$	Ir:	SVol.2-76
$Br_2C_{21}FeH_{15}O_3P$	$(CO)_3Fe(P(C_6H_5)_3)Br_2$	Fe:	Org.Verb.B1-180/2
$Br_2C_{21}FeH_{15}O_6P$	$(CO)_3Fe(P(OC_6H_5)_3)Br_2$	Fe:	Org.Verb.B1-180/1
$Br_2C_{21}H_{17}N_3O_4U$	$[UO_2(C_2H_4(NCHC_6H_3(Br)O)_2)(C_5H_5N)]$	U:	SVol.E1-32, 37
$Br_2C_{22}F_{34}N_6$	$[(CF_2BrC_2F_4)(CF_3)C_3N_3(CF_2)_4]_2$	F:	PerFHalOrg.6-91/2, 109
$Br_2C_{22}FeH_{20}$	$C_5H_5FeC_5H_4C_3H_2Br_2C_3H_4C_6H_5$	Fe:	Org.Verb.A1-392/4
$Br_2C_{22}FeH_{30}O_2P_2$	$(CO)_2Fe(P(C_2H_5)_2C_6H_5)_2Br_2$	Fe:	Org.Verb.B1-101/2, 104
$Br_2C_{22}Fe_2H_{12}N_2O_6$	$(CO)_6Fe_2[(NCC_6H_4Br)_2C_2H_4]$	Fe:	Org.Verb.C1-132
$Br_2C_{22}H_{26}Ni_2O_4$	$(CH_2C(CH_3)CH_2NiBr)_2 \cdot 2\ CH_3C_6H_3O_2$	Ni:	Org.Verb.2-308/9
–	$((CH_3)_2CCHCH_2NiBr)_2 \cdot 2\ C_6H_4O_2$	Ni:	Org.Verb.2-312
$Br_2C_{22}H_{34}Ni_2$	$(CH_2CHCH_2(CH_2CHCHCH_2)_2NiBr)_2$	Ni:	Org.Verb.2-298
$Br_2C_{22}H_{38}O_4Sn$	$[SnBr_2(CH(COC(CH_3)_3)_2)_2]$	Sn:	MVol.C5-108
$Br_2C_{23}ClFeH_{18}Hg_2O_4P$	$[P(C_6H_5)_3CH_3][(CO)_4FeHg_2Br_2Cl]$	Fe:	Org.Verb.B2-153, 158, 161
$Br_2C_{23}Cl_3Fe_2H_{30}O_3P_2$	$[(CO)_3Fe(P(C_2H_5)_2C_6H_5)_2Br]FeCl_3Br$	Fe:	Org.Verb.B1-184/5
$Br_2C_{23}FeH_{18}Hg_2IO_4P$	$[P(C_6H_5)_3CH_3][(CO)_4Fe(HgBr)_2I]$	Fe:	Org.Verb.B2-153, 157
$Br_2C_{23}FeH_{22}O$	$C_5H_5FeC_5H_4C_3H_2Br_2C_3H_4C_6H_4OCH_3$	Fe:	Org.Verb.A2-133, 137
$Br_2C_{23}Fe_2H_{10}N_2O_7$	$(CO)_6Fe_2(CO(NC(C_6H_4Br)CH)_2)$	Fe:	Org.Verb.C1-143/4, 146
$Br_2C_{23}Fe_2H_{12}N_2O_7$	$(CO)_6Fe_2[(CH_2C(C_6H_4Br)N)_2CO]$	Fe:	Org.Verb.C1-145
$Br_2C_{23}Fe_2H_{20}O$	$C_5H_5FeC_5H_4(CHBr)_2COC_5H_4FeC_5H_5$	Fe:	Org.Verb.A6-108
$Br_2C_{24}ClH_{16}IrN_4O_4$	$[Ir(C_{12}H_8N_2)_2Br_2]ClO_4$	Ir:	SVol.2-90
$Br_2C_{24}ClH_{33}IrP_3$	$Ir(P(CH_3)_2C_6H_5)_3Br_2Cl$	Ir:	SVol.2-212
$Br_2C_{24}F_6H_{20}Sn$	$[(C_6H_5)_2Br]_2[SnF_6]$	Sn:	MVol.C3-91
$Br_2C_{24}F_{20}Ge_2$	$(C_6F_5)_2BrGeGeBr(C_6F_5)_2$	F:	PerFHalOrg.4-166/7, 178
$Br_2C_{24}F_{28}N_{12}$	$Br[CF_2C_3N_3(CF_3)CF_2]_4Br$	F:	PerFHalOrg.6-89/90, 108
$Br_2C_{24}F_{38}N_6$	$[(CF_2BrC_2F_4)(C_3F_7)C_3N_3(CF_2)_3]_2$	F:	PerFHalOrg.6-91/2, 109
$Br_2C_{24}F_{38}N_6O_2$	$[(CF_3OC_2F_4)(CF_2BrC_2F_4)C_3N_3(CF_2)_3]_2$	F:	PerFHalOrg.6-91/2, 109
$Br_2C_{24}Fe_2H_{16}$	$BrC_5H_4FeC_5H_4(CC)_2C_5H_4FeC_5H_4Br$	Fe:	Org.Verb.A6-132, 135
$Br_2C_{24}H_{16}IrN_4^+$	$[Ir(C_{12}H_8N_2)_2Br_2]^+$	Ir:	SVol.2-92
$Br_2C_{24}H_{16}MnN_4$	$[MnBr_2(C_{12}H_8N_2)_2]$	Mn:	MVol.D1-102
$Br_2C_{24}H_{16}N_4Sn$	$SnBr_2 \cdot 2\ C_{12}H_8N_2$	Sn:	MVol.C5-22/3

$Br_2C_{24}H_{20}N_2Te$	$(C_6H_5)_2NTeBr_2N(C_6H_5)_2$	Te:	SVol.B3-14
$Br_2C_{24}H_{20}O_4P_2Sn$	$[SnBr_2((C_6H_5)_2PO_2)_2]$	Sn:	MVol.C6-151
$Br_2C_{24}H_{20}O_4S_2U$	$UO_2Br_2 \cdot 2\ (C_6H_5)_2SO$	U:	SVol.E1-213, 215
$Br_2C_{24}H_{20}P_2S_4Sn$	$SnBr_2((C_6H_5)_2P(S)S)_2$	Sn:	MVol.C6-160
$Br_2C_{24}H_{30}Ni_2O_4$	$(CH_2C(CH_3)CH_2NiBr)_2 \cdot 2\ (CH_3)_2C_6H_2O_2$	Ni:	Org.Verb.2-308/9
$Br_2C_{24}H_{30}Sn$	$(C_6H_{11})_2Sn(C_6H_4Br)_2$	Sn:	Org.Verb.3-70
$Br_2C_{24}H_{34}IrP_3$	$Ir(P(CH_3)_2C_6H_5)_3HBr_2$	Ir:	SVol.2-209/10
$Br_2C_{24}H_{38}Ni_2$	$(CH_3(CHCH(CH_2)_2)_2CHCHCH_2NiBr)_2$	Ni:	Org.Verb.2-298
$Br_2C_{24}H_{40}N_6S_4Sn$	$[(C_2H_5)_4N]_2[SnBr_2(NC(S)CC(S)CN)_2]$	Sn:	MVol.C6-88/9
$Br_2C_{24}H_{54}O_4P_2U$	$UO_2Br_2 \cdot 2\ (C_4H_9)_3PO$	U:	SVol.E1-150, 158
$Br_2C_{24}H_{54}O_{10}P_2U$	$UO_2Br_2 \cdot 2\ (C_4H_9O)_3PO$	U:	SVol.E1-182, 184
$Br_2C_{25}FeH_{21}N_3O$	$(CH_3C_6H_4NC)_3Fe(CO)Br_2$	Fe:	Org.Verb.B4-45
$Br_2C_{25}H_{19}NSn$	$(C_6H_5)_2Sn(BrC_6H_3N(CH_3)C_6H_3Br)$	Sn:	Org.Verb.3-118, 122
$Br_2C_{26}ClFeH_{22}Hg_2O_4P$	$[P(C_6H_5)_3C_3H_4(CH_3)][(CO)_4Fe(HgBr)_2Cl]$	Fe:	Org.Verb.B2-153, 157
$Br_2C_{26}Cl_4H_{14}O_2Sn$	$SnCl_4 \cdot 2\ C_{13}H_7BrO$	Sn:	MVol.C5-100
$Br_2C_{26}Cl_4H_{18}N_4O_6Sn$	$SnCl_4 \cdot 2\ BrC_6H_4CHNC_6H_3(OH)NO_2$	Sn:	MVol.C6-25
$Br_2C_{26}Cl_4H_{20}N_2O_2Sn$	$SnCl_4 \cdot 2\ BrC_6H_4CHNC_6H_4OH$	Sn:	MVol.C6-25
$Br_2C_{26}Cl_4H_{20}N_2Te$	$TeCl_4 \cdot 2\ C_6H_4BrCHNC_6H_5$	Te:	SVol.B3-172/3
–	$TeCl_4 \cdot 2\ C_6H_5CHNC_6H_4Br$	Te:	SVol.B3-172/3
$Br_2C_{26}F_{42}N_6$	$[(CF_2BrC_2F_4)(C_3F_7)C_3N_3(CF_2)_4]_2$	F:	PerFHalOrg.6-91/2, 109
$Br_2C_{26}F_{42}N_6O_2$	$[(CF_2BrC_2F_4)(CF_3OC_2F_4)C_3N_3(CF_2)_4]_2$	F:	PerFHalOrg.6-91/2, 109
$Br_2C_{26}Fe_2H_{21}OP$	$(C_5H_5FeC_5H_3Br)_2(C_6H_5)PO$	Fe:	Org.Verb.A6-193/4, 196/7
$Br_2C_{26}H_{14}N_8Sn$	$[SnBr_2(C_{26}H_{14}N_8)]$	Sn:	MVol.C5-237
$Br_2C_{26}H_{20}N_2O_4Sn$	$[SnBr_2(C_6H_5CON(C_6H_5)O)_2]$	Sn:	MVol.C6-37
$Br_2C_{26}H_{21}NSn$	$(C_6H_5)_2Sn(BrC_6H_3N(C_2H_5)C_6H_3Br)$	Sn:	Org.Verb.3-119
$Br_2C_{26}H_{24}O_4P_2U$	$UO_2Br_2 \cdot C_2H_4(PO(C_6H_5)_2)_2$	U:	SVol.E1-179/80
$Br_2C_{26}H_{34}NiP_2$	$BrC_6H_4Ni(P(C_2H_5)_2C_6H_5)_2Br$	Ni:	Org.Verb.1-34, 41
$Br_2C_{26}H_{34}Ni_2O_4$	$((CH_3)_2CCHCH_2NiBr)_2 \cdot 2\ (CH_3)_2C_6H_2O_2$	Ni:	Org.Verb.2-312
$Br_2C_{26}H_{34}Ni_2O_8$	$(CH_2C(CH_3)CH_2NiBr)_2 \cdot 2\ (CH_3O)_2C_6HO_2CH_3$	Ni:	Org.Verb.2-308/9
$Br_2C_{26}H_{54}Ni_2P_2$	$[((CH_3)_2CH)_3PNiBr]_2C_8H_{12}$	Ni:	Org.Verb.2-366/7
$Br_2C_{27}H_{28}N_2P_2S$	$[(CH_3)_2S[NP(C_6H_5)_2]_2CH_2]Br_2$	S:	S-N-Verb.1-39/40
$Br_2C_{28}Cl_4FeH_{16}N_4$	$(ClC_6H_4NC)_4FeBr_2$	Fe:	Org.Verb.B4-48/51
$Br_2C_{28}Cl_4H_{14}O_4Sn$	$SnCl_4 \cdot 2\ C_{14}H_7BrO_2$	Sn:	MVol.C5-115
$Br_2C_{28}Cl_4H_{24}N_2Sn$	$SnCl_4 \cdot 2\ BrC_6H_4CHNC_6H_4CH_3$	Sn:	MVol.C6-24
$Br_2C_{28}FeH_{16}K_4N_4O_{12}S_4$	$(KO_3SC_6H_4NC)_4FeBr_2$	Fe:	Org.Verb.B4-48/9, 52/3, 59
$Br_2C_{28}FeH_{20}O_2$	$C_5H_5FeC_5H_4C_6H_4OCH_2(CC)_2CH_2OC_6H_3Br_2$	Fe:	Org.Verb.A2-126
$Br_2C_{28}FeH_{24}O_2P_2$	$(CO)_2Fe[(C_6H_5)_2PCH_2CH_2P(C_6H_5)_2]Br_2$	Fe:	Org.Verb.B1-101/2, 106
$Br_2C_{28}FeH_{26}O_2P_2$	$(CO)_2Fe(P(C_6H_5)_2CH_3)_2Br_2$	Fe:	Org.Verb.B1-101/2, 104
$Br_2C_{28}H_{20}Ni$	$C_4(C_6H_5)_4NiBr_2 \cdot P(C_6H_5)_3$	Ni:	Org.Verb.2-75
$Br_2C_{28}H_{26}Ni_2O_4$	$(CH_2C(CH_3)CH_2NiBr)_2 \cdot 2\ C_{10}H_6O_2$	Ni:	Org.Verb.2-308/9
$Br_2C_{28}H_{28}O_4P_2U$	$UO_2Br_2 \cdot C_4H_8(PO(C_6H_5)_2)_2$	U:	SVol.E1-179/80
$Br_2C_{28}H_{32}N_4P_2S_2$	$[(NS(CH_3)_2NP(C_6H_5)_2)_2]Br_2$	S:	S-N-Verb.1-7
$Br_2C_{28}H_{56}N_4P_2S_2$	$[NS(CH_3)_2NP(C_6H_{11})_2]_2Br_2$	S:	S-N-Verb.1-7
$Br_2C_{30}ClH_{45}IrP_3$	$Ir(P(C_2H_5)_2C_6H_5)_3ClBr_2$	Ir:	SVol.2-213
$Br_2C_{30}Cl_2H_{54}N_6O_{14}U$	$[UBr_2(HN(CH_2)_4CO)_6](ClO_4)_2$	U:	SVol.E1-118/20
$Br_2C_{30}Cl_4H_{24}N_2O_4Sn$	$SnCl_4 \cdot 2\ BrC_6H_4CONHC_6H_4OCHCH_2$	Sn:	MVol.C6-49

$Br_2C_{30}FeH_{25}N_5S_5$	$[(C_4H_3SCH_2NC)_5FeBr]Br \cdot H_2O$	Fe:	Org.Verb.B4-74
$Br_2C_{30}FeH_{30}O_2P_2$	$(CO)_2Fe(P(C_6H_5)_2C_2H_5)_2Br_2$	Fe:	Org.Verb.B1-101/2, 104
$Br_2C_{30}Fe_3H_{24}$	$Fe(C_5H_4C_5H_4FeC_5H_4Br)_2$	Fe:	Org.Verb.A6-233
$Br_2C_{30}H_{22}O_4Sn$	$[SnBr_2(CH(COC_6H_5)_2)_2]$	Sn:	MVol.C5-110/1
$Br_2C_{30}H_{26}Ni_2O_8$	$(C_3H_5NiBr)_2 \cdot 4\ C_6H_4O_2$	Ni:	Org.Verb.2-285
$Br_2C_{30}H_{30}MnN_6$	$MnBr_2 \cdot 6\ C_5H_5N$	Mn:	MVol.C5-273
$Br_2C_{30}H_{30}Ni_2P_2$	$C_3H_5Ni(Br)P(C_6H_5)_2P(C_6H_5)_2(Br)NiC_3H_5$	Ni:	Org.Verb.2-285/6
$Br_2C_{30}H_{46}IrP_3$	$Ir(P(C_2H_5)_2C_6H_5)_3HBr_2$	Ir:	SVol.2-209/10
$Br_2C_{30}H_{54}Ni_2$	$((t\text{-}C_4H_9)_3C_3NiBr)_2$	Ni:	Org.Verb.2-319
$Br_2C_{32}Cl_2F_2H_{72}N_2Te$	$[(C_4H_9)_4N]_2[TeCl_2Br_2F_2]$	Te:	SVol.B3-87
$Br_2C_{32}Cl_2FeH_{28}HgN_4$	$(CH_3C_6H_4NC)_4FeCl_2 \cdot HgBr_2$	Fe:	Org.Verb.B4-69
$Br_2C_{32}Cl_2H_{72}I_2N_2Te$	$[(C_4H_9)_4N]_2[TeCl_2Br_2I_2]$	Te:	SVol.B3-85, 87
$Br_2C_{32}Cl_2H_{72}MnN_2$	$[(C_4H_9)_4N]_2MnBr_2Cl_2$	Mn:	MVol.C5-309
$Br_2C_{32}Cl_4H_{72}N_2Te$	$[(C_4H_9)_4N]_2[TeCl_4Br_2]$	Te:	SVol.B3-87
$Br_2C_{32}F_6H_{44}IrO_8P_5$	$[Ir(P(OCH_3)_2C_6H_5)_4Br_2]PF_6$	Ir:	SVol.2-238
$Br_2C_{32}FeH_{28}HgI_2N_4$	$(CH_3C_6H_4NC)_4FeBr_2 \cdot HgI_2$	Fe:	Org.Verb.B4-69
$Br_2C_{32}FeH_{28}N_4$	$(CH_3C_6H_4NC)_4FeBr_2$	Fe:	Org.Verb.B4-48/9, 52/5, 60/2
$Br_2C_{32}H_{16}N_8Sn$	$[SnBr_2(C_{32}H_{16}N_8)]$	Sn:	MVol.C5-240/1
$Br_2C_{32}H_{26}N_2Ni$	$C_6H_5CCC_6H_5Ni(CNC_6H_2(CH_3)_2Br)_2$	Ni:	Org.Verb.1-385
$Br_2C_{32}H_{36}N_4Sn$	$[SnBr_2(C_{20}H_4N_4(C_2H_5)_4(CH_3)_4)] \cdot 2\ H_2O$	Sn:	MVol.C5-231/2
$Br_2C_{32}H_{38}N_4Sn$	$[SnBr_2(C_{20}H_6N_4(C_2H_5)_4(CH_3)_4)]$	Sn:	MVol.C5-232
$Br_2C_{32}H_{40}N_4P_2S_2$	$[NS(CH_3)_2NP(CH_2C_6H_5)_2]_2Br_2$	S:	S-N-Verb.1-7
$Br_2C_{34}H_{28}N_4Ni$	$(BrC_6H_4CH_2)_2Ni(NC_5H_4C_5H_4N)_2$	Ni:	Org.Verb.1-71
$Br_2C_{34}H_{34}Ni_2O_8$	$(C_3H_5NiBr)_2 \cdot 4\ CH_3C_6H_3O_2$	Ni:	Org.Verb.2-285
$Br_2C_{36}Cl_6H_{24}IrNOP_2$	$Ir(NO)Br_2(P(p\text{-}C_6H_4Cl)_3)_2$	Ir:	SVol.2-55
$Br_2C_{36}F_{10}H_{20}PTl$	$[(C_6H_5)_4P][Tl(C_6F_5)_2Br_2]$	F:	PerFHalOrg.4-137, 141
$Br_2C_{36}F_{30}P_2Pd$	$[(C_6F_5)_3P]_2PdBr_2$	F:	PerFHalOrg.3-174
$Br_2C_{36}F_{30}P_2Pt$	$[(C_6F_5)_3P]_2PtBr_2$	F:	PerFHalOrg.3-174
$Br_2C_{36}H_{30}IrNOP_2$	$Ir(NO)Br_2(P(C_6H_5)_3)_2$	Ir:	SVol.2-55
$Br_2C_{36}H_{30}Ir_2N_2O_3P_2$	$[Ir(NO)Br(P(C_6H_5)_3)]_2O$	Ir:	SVol.2-57
$Br_2C_{36}H_{30}O_2P_2Sn$	$SnBr_2 \cdot 2\ (C_6H_5)_3PO$	Sn:	MVol.C5-59/60
$Br_2C_{36}H_{30}O_2P_2U$	$[UO_2Br_2(P(C_6H_5)_3)_2]$	U:	SVol.E1-51
$Br_2C_{36}H_{30}O_4P_2U$	$UO_2Br_2 \cdot 2\ (C_6H_5)_3PO$	U:	SVol.E1-170, 176
$Br_2C_{36}H_{30}O_5S_3U$	$UO_2Br_2 \cdot 3\ (C_6H_5)_2SO$	U:	SVol.E1-213, 215
$Br_2C_{36}H_{31}IrP_2$	$Ir(P(C_6H_5)_3)_2HBr_2$	Ir:	SVol.2-225
$Br_2C_{36}H_{34}IrNP_2$	$Ir(P(C_6H_5)_3)_2HBr_2(NH_3)$	Ir:	SVol.2-229
$Br_2C_{38}FeH_{30}O_2P_2$	$(CO)_2Fe(P(C_6H_5)_3)_2Br_2$	Fe:	Org.Verb.B1-101/2, 105, 109
		Fe:	Org.Verb.B3-205
$Br_2C_{38}FeH_{30}O_2Sb_2$	$(CO)_2Fe(Sb(C_6H_5)_3)_2Br_2$	Fe:	Org.Verb.B1-101/2, 106, 110
$Br_2C_{38}FeH_{30}O_8P_2$	$(CO)_2Fe(P(OC_6H_5)_3)_2Br_2$	Fe:	Org.Verb.B1-101/2, 105
$Br_2C_{39}Cl_3Fe_2H_{30}O_3P_2$	$[(CO)_3Fe(P(C_6H_5)_3)_2Br]FeCl_3Br$	Fe:	Org.Verb.B1-184/5
$Br_2C_{39}FeH_{30}HgO_3P_2$	$(CO)_3Fe(P(C_6H_5)_3)_2 \cdot HgBr_2$	Fe:	Org.Verb.B1-192/3
$Br_2C_{39}FeH_{30}HgO_3Sb_2$	$(CO)_3Fe(Sb(C_6H_5)_3)_2 \cdot HgBr_2$	Fe:	Org.Verb.B1-192/3
$Br_2C_{39}H_{40}IrP_3$	$Ir(P(CH_3)(C_6H_5)_2)_3HBr_2$	Ir:	SVol.2-209/10
$Br_2C_{40}FeH_{30}InNO_4P_2$	$[(C_6H_5)_3PNP(C_6H_5)_3][(CO)_4FeInBr_2]$	Fe:	Org.Verb.B2-49
$Br_2C_{40}FeH_{31}N_9O_8$	$[C_6H_5CH_2NC(NO_2C_6H_4CH_2NC)_4FeBr]Br \cdot 2\ H_2O$	Fe:	Org.Verb.B4-91

$Br_2C_{40}FeH_{35}N_5$	$[(CH_3C_6H_4NC)_5FeBr]Br$	Fe: Org.Verb.B4-74
$Br_2C_{40}H_{78}Ni_2$	$(CH_3(CH(CH_3)C_3H_6)_3C(CH_3)CHCH_2NiBr)_2$	Ni: Org.Verb.2-311
$Br_2C_{41}FeH_{34}N_6$	$[(C_6H_5CH_2NC)_4(BrC_6H_4CH_2NC)FeCN]Br$	Fe: Org.Verb.B4-92
$Br_2C_{41}H_{36}IrNP_2$	$Ir(P(C_6H_5)_3)_2HBr_2(C_5H_5N)$	Ir: SVol.2-229
$Br_2C_{42}ClFe_2H_{47}N_2$	$[(C_5H_5FeC_5H_4CH_2N(CH_3)_2CH_2C_6H_4)_2C_2HCl]Br_2$	Fe: Org.Verb.A6-175
$Br_2C_{42}Fe_2H_{46}N_2$	$[(C_5H_5FeC_5H_4CH_2N(CH_3)_2CH_2C_6H_4)_2C_2]Br_2$	Fe: Org.Verb.A6-175
$Br_2C_{42}Fe_2H_{48}N_2$	$[(C_5H_5FeC_5H_4CH_2N(CH_3)_2CH_2C_6H_4)_2C_2H_2]Br_2$	Fe: Org.Verb.A6-175
$Br_2C_{42}H_{30}Ni_2$	$((C_6H_5)_3C_3NiBr)_2$	Ni: Org.Verb.2-319
$Br_2C_{42}H_{34}NiP_2$	$BrC_6H_4Ni(P(C_6H_5)_3)_2Br$	Ni: Org.Verb.1-34
$Br_2C_{42}H_{40}IrP_3$	$Ir(PC_{14}H_{13})_3HBr_2$	Ir: SVol.2-234
$Br_2C_{42}H_{46}IrP_3$	$Ir(PC_2H_5(C_6H_5)_2)_3HBr_2$	Ir: SVol.2-209/11
$Br_2C_{42}H_{70}NiP_2$	$BrC_6H_4Ni(P(C_6H_{11})_3)_2Br$	Ni: Org.Verb.1-34
$Br_2C_{43}FeH_{39}I_2O_2P_3$	$(CO)_2Fe[(P(C_6H_5)_2CH_2)_3CCH_3]I_2Br_2$	Fe: Org.Verb.B1-110
$Br_2C_{44}Cl_2H_{48}O_2P_2U$	$[P(C_6H_5)_3C_4H_9]_2UO_2Br_2Cl_2$	U: SVol.C9-162
$Br_2C_{44}Cl_4H_{48}P_2U$	$[P(C_6H_5)_3C_4H_9]_2UBr_2Cl_4$	U: SVol.C9-161/2
$Br_2C_{44}H_{42}Ni_2P_2$	$C_8H_{12}Ni_2(P(C_6H_5)_3)_2Br_2$	Ni: Org.Verb.2-400
$Br_2C_{44}H_{48}I_2O_2P_2U$	$[P(C_6H_5)_3C_4H_9]_2[UO_2Br_2I_2]$	U: SVol.C9-185
$Br_2C_{45}FeH_{35}InN_2O_4P_2$	$[((C_6H_5)_3P)_2N][(CO)_4FeInBr_2(NC_5H_5)]$	Fe: Org.Verb.B2-49/50
$Br_2C_{45}FeH_{35}N_5O_5$	$[(C_6H_5COCH_2NC)_5FeBr]Br$	Fe: Org.Verb.B4-76
$Br_2C_{46}Cl_2H_{56}N_2Sn_2$	$[(CH_3)_2(C_2H_5)NCH_2]_2[(C_6H_5)_3SnBrCl]_2$	Sn: Org.Verb.5-180
$Br_2C_{46}FeH_{45}InN_2O_4P_2$	$[N(P(C_6H_5)_3)_2][(CO)_4FeInBr_2N(C_2H_5)_3]$	Fe: Org.Verb.B2-49/50
$Br_2C_{46}H_{35}NiP$	$C_4(C_6H_5)_4NiBr_2 \cdot P(C_6H_5)_3$	Ni: Org.Verb.2-75
$Br_2C_{46}H_{48}N_2Ni_2P_2$	$(Ni(CNC(CH_3)_3)(P(C_6H_5)_3)Br)_2$	Ni: Org.Verb.2-267
$Br_2C_{46}H_{58}Ni_2O_{16}$	$((CH_3)_2CCHCH_2NiBr)_2 \cdot 4\ (CH_3O)_2C_6HO_2CH_3$	Ni: Org.Verb.2-312
$Br_2C_{48}ClH_{41}N_2O_8P_4Sn$	$[SnBr_2H(N(PO(C_6H_5)_2)_2)_2]ClO_4$	Sn: MVol.C6-156
$Br_2C_{48}Cl_2H_{42}N_2O_{12}P_4Sn$	$[SnBr_2(NH(PO(C_6H_5)_2)_2)_2](ClO_4)_2$	Sn: MVol.C6-156
$Br_2C_{48}Cl_2H_{44}N_2Sn_2$	$[C_5H_5NCH_2]_2[(C_6H_5)_3SnBrCl]_2$	Sn: Org.Verb.5-180
$Br_2C_{48}Cl_2H_{60}N_2Sn_2$	$[(CH_3)_2(C_3H_7)NCH_2]_2[(C_6H_5)_3SnBrCl]_2$	Sn: Org.Verb.5-180
$Br_2C_{48}Cl_{10}H_{30}Ni_2P_2$	$(C_6Cl_5Ni(P(C_6H_5)_3)Br)_2$	Ni: Org.Verb.2-255
$Br_2C_{48}FeH_{42}N_6$	$[(C_6H_5CH_2NC)_6Fe]Br_2$	Fe: Org.Verb.B4-98, 104
–	$[(C_6H_5CH_2NC)_6Fe]Br_2 \cdot H_2O$	Fe: Org.Verb.B4-104
$Br_2C_{48}H_{40}N_2O_4P_4Sn$	$[SnBr_2((C_6H_5)_2P(O)NP(O)(C_6H_5)_2)_2]$	Sn: MVol.C6-155
$Br_2C_{50}Cl_2H_{48}N_2Sn_2$	$[CH_3C_5H_4NCH_2]_2[(C_6H_5)_3SnBrCl]_2$	Sn: Org.Verb.5-180
–	$[C_5H_5N(CH_2)_2]_2[(C_6H_5)_3SnBrCl]_2$	Sn: Org.Verb.5-180
$Br_2C_{50}Cl_2H_{50}N_4Sn_2$	$[NH_2C_5H_4N(CH_2)_2]_2[(C_6H_5)_3SnBrCl]_2$	Sn: Org.Verb.5-180
$Br_2C_{50}Cl_2H_{60}N_2O_2Sn_2$	$[O(CH_2CH_2)_2NCH_3(CH_2)_2]_2[(C_6H_5)_3SnBrCl]_2$	Sn: Org.Verb.5-180
$Br_2C_{50}Cl_2H_{64}N_2Sn_2$	$[(CH_3)_2(C_2H_5)N(CH_2)_3]_2[(C_6H_5)_3SnClBr]_2$	Sn: Org.Verb.5-179
$Br_2C_{52}Cl_2H_{52}N_2Sn_2$	$[CH_3C_5H_4N(CH_2)_2]_2[(C_6H_5)_3SnBrCl]_2$	Sn: Org.Verb.5-180
$Br_2C_{52}Cl_2H_{68}N_2Sn_2$	$[(CH_3)_2(C_3H_7)N(CH_2)_3]_2[(C_6H_5)_3SnClBr]_2$	Sn: Org.Verb.5-179
$Br_2C_{52}H_{45}IrP_4$	$[Ir(((C_6H_5)_2P)_2C_2H_2)_2HBr]Br$	Ir: SVol.2-242
$Br_2C_{54}Cl_9H_{37}IrO_9P_3$	$Ir(P(OC_6H_4Cl\text{-}p)_3)_3HBr_2$	Ir: SVol.2-237/8
$Br_2C_{54}H_{42}Ni_2$	$((C_6H_5)_2CCHC(C_6H_5)_2NiBr)_2$	Ni: Org.Verb.2-318
$Br_2C_{54}H_{45}IrP_3$	$Ir(P(C_6H_5)_3)_3Br_2$	Ir: SVol.2-232
$Br_2C_{54}H_{46}IrO_9P_3$	$Ir(P(OC_6H_5)_3)_3HBr_2$	Ir: SVol.2-237/8
$Br_2C_{54}H_{46}IrP_2Sb$	$Ir(P(C_6H_5)_3)_2HBr_2(Sb(C_6H_5)_3)$	Ir: SVol.2-229
$Br_2C_{54}H_{46}IrP_3$	$Ir(P(C_6H_5)_3)_3HBr_2$	Ir: SVol.2-225

$Br_2C_{54}H_{46}IrSb_3$	$IrHBr_2(Sb(C_6H_5)_3)_3$	Ir:	SVol.2-263
$Br_2C_{56}Cl_2H_{60}N_2Sn_2$	$[C_5H_5N(CH_2)_{10}NC_5H_5][(C_6H_5)_3SnBrCl]_2$	Sn:	Org.Verb.5-180
$Br_2C_{58}Cl_2H_{52}N_2Sn_2$	$[C_9H_7N(CH_2)_2]_2[(C_6H_5)_3SnBrCl]_2$	Sn:	Org.Verb.5-180
$Br_2C_{58}Cl_2H_{80}N_2Sn_2$	$[(CH_3)_2(C_6H_{13})N(CH_2)_3]_2[(C_6H_5)_3SnClBr]_2$	Sn:	Org.Verb.5-179
$Br_2C_{60}Cl_4H_{66}N_2Sn_2$	$[(CH_3)_2(ClC_6H_4CH_2)N(CH_2)_3]_2[(C_6H_5)_3SnClBr]_2$	Sn:	Org.Verb.5-179
$Br_2C_{60}H_{52}IrP_3$	$Ir(P(C_6H_5)_3)_3HBr_2 \cdot C_6H_6$	Ir:	SVol.2-225
$Br_2C_{60}H_{60}Ni_2P_4$	$[((C_6H_5)_2P)_2C_2H_4NiBr]_2C_8H_{12}$	Ni:	Org.Verb.2-367
$Br_2C_{62}Cl_2H_{88}N_2Sn_2$	$[(CH_3)_2(C_8H_{17})N(CH_2)_3]_2[(C_6H_5)_3SnClBr]_2$	Sn:	Org.Verb.5-179
$Br_2C_{62}Cl_6H_{62}Ni_2P_4$	$[((C_6H_5)_2P)_2C_2H_4NiBr]_2C_8H_{12} \cdot 2\ CHCl_3$	Ni:	Org.Verb.2-368
$Br_2C_{63}H_{64}IrO_9P_3$	$Ir(P(O(CH_3C_6H_4))_3)_3HBr_2$	Ir:	SVol.2-237/8
$Br_2C_{65}FeH_{115}N_5O_{10}$	$[(HO_2C(CH_2)_{10}CH_2NC)_5FeBr]Br \cdot 3\ H_2O$	Fe:	Org.Verb.B4-76
$Br_2C_{66}H_{66}Ni_2P_4$	$[((C_6H_5)_2P)_2C_2H_4NiBr]_2C_8H_{12} \cdot C_6H_6$	Ni:	Org.Verb.2-368
$Br_2C_{70}Cl_2H_{104}N_2Sn_2$	$[(CH_3)_2(C_{12}H_{25})N(CH_2)_3]_2[(C_6H_5)_3SnClBr]_2$	Sn:	Org.Verb.5-179
$Br_2C_{72}F_{60}P_4Rh_2$	$[(C_6F_5)_3P]_4Rh_2Br_2$	F:	PerFHalOrg.3-174
$Br_2C_{72}H_{60}Ni_2P_4$	$[Ni(P(C_6H_5)_3)_2Br]_2$	Ni:	Org.Verb.1-351
$Br_2C_{80}Cl_2H_{168}N_2O_2U$	$[N(C_{10}H_{21})_4]_2UO_2Br_2Cl_2$	U:	SVol.C9-162
$Br_2C_{84}Cl_2H_{70}Ir_2P_4S_2$	$[Ir(SC_6H_4Br)HCl(P(C_6H_5)_3)_2]_2$	Ir:	SVol.2-163
$Br_2C_{90}H_{146}Ni_2$	$(CH_3C(CH_3)CHCH_2[CH_2C(CH_3)CHCH_2]_7$ $CH_2C(CH_3)CHCH_2NiBr)_2$	Ni:	Org.Verb.2-311
$Br_2C_{210}F_{245}N_{105}$	$Br[CF_2C_3N_3(CF_3)CF_2]_{35}Br$	F:	PerFHalOrg.6-89/90, 108
$Br_2C_{240}F_{280}N_{120}$	$Br[CF_2C_3N_3(CF_3)CF_2]_{40}Br$	F:	PerFHalOrg.6-89/90, 108
$Br_2C_{2100}F_{2450}N_{1050}$	$Br[CF_2C_3N_3(CF_3)CF_2]_{350}Br$	F:	PerFHalOrg.6-89/90, 108
Br_2Cd	$CdBr_2$ melts		
	$CdBr_2$-Sn	Sn:	MVol.C4-55/6
Br_2Ce^+	$[CeBr_2]^+$	Sc:	MVol.C6-36/7
Br_2CeO_4U	$CeUO_2(O_2)Br_2$	U:	SVol.C9-155
$Br_2ClCsSn$	$CsSnBr_2Cl$	Sn:	MVol.C3-153
Br_2ClHU	$UHBr_2Cl$	U:	SVol.C9-161
Br_2ClH_4NSn	$NH_4SnBr_2Cl \cdot H_2O$	Sn:	MVol.C3-113
$Br_2ClH_{15}IrN_5$	$[Ir(NH_3)_5Cl]Br_2$	Ir:	SVol.2-43
Br_2ClIU	UBr_2ClI	U:	SVol.C9-183/4
Br_2ClKSn	$KSnBr_2Cl \cdot H_2O$	Sn:	MVol.C3-67
Br_2ClTe^+	$TeBr_2Cl^+$	Te:	SVol.B3-81
Br_2ClU	UBr_2Cl	U:	SVol.C9-158/9
$Br_2Cl_2Cs_2F_2Te$	$Cs_2TeCl_2Br_2F_2$	Te:	SVol.B3-87
$Br_2Cl_2Cs_2I_2Te$	$Cs_2TeCl_2Br_2I_2$	Te:	SVol.B3-85, 87
$Br_2Cl_2Cs_2Mn$	$Cs_2MnBr_2Cl_2$	Mn:	MVol.C5-308
$Br_2Cl_2Cs_2O_2U$	$Cs_2UO_2Br_2Cl_2$	U:	SVol.C9-162/3
$Br_2Cl_2F_2Te^{2-}$	$TeBr_2Cl_2F_2^{2-}$	Te:	SVol.B3-85/6
$Br_2Cl_2H_2O_2U$	$H_2UO_2Br_2Cl_2$	U:	SVol.C9-162
$Br_2Cl_2H_8MnN_2$	$(NH_4)_2MnBr_2Cl_2$	Mn:	MVol.C5-308
$Br_2Cl_2H_8N_2O_2U$	$[NH_4]_2UO_2Br_2Cl_2$	U:	SVol.C9-162/3
–	$(NH_4)_2UO_2Br_2Cl_2 \cdot 2\ H_2O$	U:	SVol.C9-162/3
$Br_2Cl_2H_8N_2Sn$	$(NH_4)_2SnBr_2Cl_2$	Sn:	MVol.C3-113
$Br_2Cl_2I_2Te^{2-}$	$TeBr_2Cl_2I_2^{2-}$	Te:	SVol.B3-85/6
$Br_2Cl_2K_2O_2U$	$K_2UO_2Br_2Cl_2$	U:	SVol.C9-162/3
–	$K_2UO_2Br_2Cl_2 \cdot 2\ H_2O$	U:	SVol.C9-162/3
$Br_2Cl_2K_2O_5U_2$	$K_2U_2O_5Br_2Cl_2$	U:	SVol.C9-162/3

$Br_2Cl_2K_2Sn$	$K_2SnBr_2Cl_2 \cdot 2\ H_2O$	Sn: MVol.C3-67
$Br_2Cl_2Mn^{2-}$	$[MnBr_2Cl_2]^{2-}$	Mn: MVol.C5-309
$Br_2Cl_2MnRb_2$	$Rb_2MnBr_2Cl_2$	Mn: MVol.C5-308
$Br_2Cl_2Na_2Sn$	$Na_2SnBr_2Cl_2$	Sn: MVol.C3-25
$Br_2Cl_2O_2Rb_2U$	$Rb_2UO_2Br_2Cl_2 \cdot 2\ H_2O$	U: SVol.C9-162
$Br_2Cl_2O_2U^{2-}$	$[UO_2Br_2Cl_2]^{2-}$	U: SVol.C9-162/3
$Br_2Cl_2O_5U_2{}^{2-}$	$[U_2O_5Br_2Cl_2]^{2-}$	U: SVol.C9-162
Br_2Cl_2Te	$TeBr_2Cl_2$	Te: SVol.B3-49/50
$Br_2Cl_2Te^+$	$TeBr_2Cl_2{}^+$	Te: SVol.B3-50
Br_2Cl_2U	UBr_2Cl_2	U: SVol.C9-158/61
$Br_2Cl_3Te^-$	$TeBr_2Cl_3{}^-$	Te: SVol.B3-81/3
$Br_2Cl_4Cs_2Te$	$Cs_2TeCl_4Br_2$	Te: SVol.B3-83, 87
$Br_2Cl_4H_8N_2Sn$	$(NH_4)_2[SnBr_2Cl_4]$	Sn: MVol.C3-119
$Br_2Cl_4H_8N_2Te$	$(NH_4)_2[TeCl_4Br_2]$	Te: SVol.B3-86
$Br_2Cl_4Ir^{2-}$	$[IrCl_4Br_2]^{2-}$	Ir: SVol.2-124, 150
$Br_2Cl_4IrRb_2$	$Rb_2[IrCl_4Br_2]$	Ir: SVol.2-150
$Br_2Cl_4Na_2U$	$Na_2UBr_2Cl_4$	U: SVol.C9-161/2
$Br_2Cl_4Te^{2-}$	$TeCl_4Br_2{}^{2-}$	Te: SVol.B3-83/4
Br_2CsSn^+	$CsSnBr_2{}^+$	Sn: MVol.C3-151
Br_2Dy	$DyBr_2$	Sc: MVol.C6-7/11, 74
Br_2Dy^+	$[DyBr_2]^+$	Sc: MVol.C6-36/7
Br_2Er^+	$[ErBr_2]^+$	Sc: MVol.C6-36/7
Br_2Eu	$EuBr_2$	Sc: MVol.C6-2/3, 7/12, 66/9
–	$EuBr_2 \cdot H_2O$	Sc: MVol.C6-69/70
Br_2Eu^+	$[EuBr_2]^+$	Sc: MVol.C6-2, 36/7
$Br_2Eu_3H_5O_6$	$Eu_3O(OH)_5Br_2$	Sc: MVol.C6-90/2
Br_2FHU	$UHBr_2F$	U: SVol.C9-161
$Br_2F_2O_6S_2Sn$	$SnBr_2(SO_3F)_2$	Sn: MVol.C2-79
$Br_2F_{10}Sn$	$(BrF_2)_2[SnF_6] = SnF_4 \cdot 2\ BrF_3$	Sn: MVol.C3-90/1
$Br_2F_{14}Sn$	$(BrF_4)_2[SnF_6] = SnF_4 \cdot 2\ BrF_5$	Sn: MVol.C3-90/1
Br_2Gd^+	$[GdBr_2]^+$	Sc: MVol.C6-36/7
$Br_2Gd_3H_5O_6$	$Gd_3O(OH)_5Br_2$	Sc: MVol.C6-90/2
Br_2HIU	$UHBr_2I$	U: SVol.C9-161, 184
Br_2HNU	$UBr_2(NH)$	U: SVol.C9-126
Br_2HOSc	$Sc(OH)Br_2 \cdot 4\ H_2O$	Sc: MVol.C6-53, 89/90
–	$Sc(OH)Br_2 \cdot 5\ H_2O$	Sc: MVol.C6-52/3, 87/90
Br_2H_2IrN	$Ir(NH_2)Br_2$	Ir: SVol.2-42
$Br_2H_2O_6U_2$	$[UO_2(OH)_2UO_2]Br_2 \cdot 4\ H_2O$ $= [(UO_2)_2(OH)_2Br_2(H_2O)_4]$	U: SVol.C9-146/7
Br_2H_2U	UH_2Br_2	U: SVol.C9-161
Br_2H_3MnN	$[Mn(NH_3)]Br_2$	Mn: MVol.C5-272
Br_2H_3NSn	$SnBr_2 \cdot NH_3$	Sn: MVol.C5-6
$Br_2H_4O_4P_2Sn_2$	$SnBr_2 \cdot Sn(H_2PO_2)_2$	Sn: MVol.C2-268
$Br_2H_5O_6Sm_3$	$Sm_3O(OH)_5Br_2$	Sc: MVol.C6-90/2
$Br_2H_6MnN_2$	$[Mn(NH_3)_2]Br_2$	Mn: MVol.C5-272
$Br_2H_6N_2Sn$	$SnBr_2 \cdot 2\ NH_3$	Sn: MVol.C5-6
$Br_2H_6N_2Te$	$TeBr_2 \cdot 2\ NH_3$	Te: SVol.B3-133
$Br_2H_8I_4N_2Te$	$(NH_4)_2[TeBr_2I_4]$	Te: SVol.B3-86
$Br_2H_8IrO_4{}^+$	$[IrBr_2(H_2O)_4]^+$	Ir: SVol.2-146, 150
$Br_2H_8MnO_4$	$[Mn(H_2O)_4]Br_2$	Mn: MVol.C5-276, 285

$Br_2H_8N_2O_3U$ .... $[NH_4]_2UO_3Br_2$ .... U: SVol.C9-151/6
$Br_2H_8O_4Te^{2+}$ .... $[TeBr_2(H_2O)_4]^{2+}$ .... Te: SVol.B3-15, 29
$Br_2H_9N_3O_2U$ .... $UO_2NH(NH_4Br)_2$ .... U: SVol.C9-156
$Br_2H_9N_3Sn$ .... $SnBr_2 \cdot 3\ NH_3$ .... Sn: MVol.C5-6
$Br_2H_{10}O_{10}U_2$ .... $[(UO_2)_2(OH)_2Br_2(H_2O)_4]$
= $[UO_2(OH)_2UO_2]Br_2 \cdot 4\ H_2O$ .... U: SVol.C9-146/7
$Br_2H_{12}N_4Sn$ .... $SnBr_2 \cdot 4\ NH_3$ .... Sn: MVol.C5-6
$Br_2H_{12}O_{12}P_6Sn_4$ .... $SnBr_2 \cdot 3\ Sn(H_2PO_2)_2$ .... Sn: MVol.C2-268
$Br_2H_{15}N_5OU$ .... $[UO(NH_2)_2(NH_3)] \cdot 2\ NH_4Br$ .... U: SVol.E1-12, 15
$Br_2H_{15}N_5O_2U$ .... $[UO_2NH(NH_3)_2](NH_4Br)_2$ .... U: SVol.C9-156
$Br_2H_{15}N_5Sn$ .... $SnBr_2 \cdot 5\ NH_3$ .... Sn: MVol.C5-6
$Br_2H_{18}MnN_6$ .... $[Mn(NH_3)_6]Br_2$ .... Mn: MVol.C5-272
$Br_2H_{27}N_9Sn$ .... $SnBr_2 \cdot 9\ NH_3$ .... Sn: MVol.C5-6
$Br_2H_{30}MnN_{10}$ .... $[Mn(NH_3)_{10}]Br_2$ .... Mn: MVol.C5-272
Br_2Ho .... $HoBr_2$ .... Sc: MVol.C6-7
Br_2Ho^+ .... $[HoBr_2]^+$ .... Sc: MVol.C6-36/7
Br_2ITe^+ .... $TeBr_2I^+$ .... Te: SVol.B3-81
Br_2IU .... UBr_2I .... U: SVol.C9-182
$Br_2I_2K_2Sn$ .... $K_2SnBr_2I_2$ .... Sn: MVol.C3-68
$Br_2I_2O_2U^{2-}$ .... $[UO_2Br_2I_2]^{2-}$ .... U: SVol.C9-185
Br_2I_2Te .... $TeBr_2I_2$ .... Te: SVol.B3-80
Br_2I_2U .... UBr_2I_2 .... U: SVol.C9-183/4
Br_2Ir .... $IrBr_2$ .... Ir: SVol.2-141
Br_2KSn^+ .... $KSnBr_2^+$ .... Sn: MVol.C3-63
$Br_2K_2O_3U$ .... $K_2UO_3Br_2$ .... U: SVol.C9-151/6
$Br_2K_2O_6U_2$ .... $K_2U_2O_6Br_2$ .... U: SVol.C9-151/6
$Br_2K_2O_9U_3$ .... $K_2U_3O_9Br_2$ .... U: SVol.C9-151/4
Br_2La .... $LaBr_2$ .... Sc: MVol.C6-2/3
Br_2La^+ .... $[LaBr_2]^+$ .... Sc: MVol.C6-36/7
Br_2Mn .... $MnBr_2$ .... Mn: MVol.C5-264/73
– .... $MnBr_2 \cdot H_2O$ .... Mn: MVol.C5-285
– .... $MnBr_2 \cdot 2\ H_2O$ .... Mn: MVol.C5-274, 284
– .... $MnBr_2 \cdot 4\ H_2O$ .... Mn: MVol.C5-46, 274/84
– .... $MnBr_2 \cdot 4\ (H,D)_2O$ .... Mn: MVol.C5-275
– .... $MnBr_2 \cdot 4\ D_2O$ .... Mn: MVol.C5-275, 277, 281
– .... $MnBr_2 \cdot 6\ H_2O$ .... Mn: MVol.C5-274
– .... $MnBr_2$ solutions
$MnBr_2$-H_2O .... Mn: MVol.C5-285/7
$MnBr_2$-organic solvents .... Mn: MVol.C5-287/8
– .... $MnBr_2$ systems
$MnBr_2$-AgBr .... Mn: MVol.C5-305
$MnBr_2$-$AlBr_3$ .... Mn: MVol.C5-306
$MnBr_2$-$BaBr_2$ .... Mn: MVol.C5-306
$MnBr_2$-CsBr .... Mn: MVol.C5-296
$MnBr_2$-H_2O .... Mn: MVol.C5-274
$MnBr_2$-$InBr_3$ .... Mn: MVol.C5-306/7
$MnBr_2$-KBr .... Mn: MVol.C5-292
$MnBr_2$-KBr-H_2O .... Mn: MVol.C5-293
$MnBr_2$-LiBr .... Mn: MVol.C5-291

Br_2Mn	$MnBr_2$ systems	
	$MnBr_2$-$MnSO_4$-H_2O	Mn: MVol.C6-268
	$MnBr_2$-NaBr	Mn: MVol.C5-292
	$MnBr_2$-RbBr	Mn: MVol.C5-293/4
	$MnBr_2$-TlBr	Mn: MVol.C5-307
Br_2MnO_6	$Mn(BrO_3)_2$	Mn: MVol.C5-308
Br_2MnSe	$MnSeBr_2$	Mn: MVol.C6-307
Br_2Mn_2Se	Mn_2SeBr_2	Mn: MVol.C6-307/8
Br_2N_2S	BrNSNBr	S: SVol.2-75
Br_2NaSn^+	$NaSnBr_2^+$	Sn: MVol.C3-25
Br_2Nd	$NdBr_2$	Sc: MVol.C6-7/11, 60
Br_2Nd^+	$[NdBr_2]^+$	Sc: MVol.C6-36/7
Br_2Np^+	$NpBr_2^+$	Np: TrU.D1-138, 144
Br_2OPa	$PaOBr_2$	Pa: SVol.2-67/8
Br_2OS	$SOBr_2$	S: SVol.1-57/9, 63/9
Br_2OTe	$TeBr_2O$	Te: SVol.B3-48
Br_2OU	$UOBr_2$	U: SVol.C9-142/3
Br_2O_2U	UO_2Br_2	U: SVol.C9-145/6
–	$UO_2Br_2 \cdot H_2O$	U: SVol.C9-146/7
–	$UO_2Br_2 \cdot 3\ H_2O$	U: SVol.C9-146/7
–	$UO_2Br_2 \cdot 7\ H_2O$	U: SVol.C9-146/7
$Br_2O_3U^{2-}$	$[UO_3Br_2]^{2-}$	U: SVol.C9-151/6
$Br_2O_7S_2Te$	$TeBr_2O \cdot 2\ SO_3$	Te: SVol.B3-108
$Br_2O_{11}Te_6$	$Te_6Br_2O_{11}$	Te: SVol.B3-47/9
Br_2Pm	$PmBr_2$	Sc: MVol.C6-7
Br_2Pm^+	$[PmBr_2]^+$	Sc: MVol.C6-36/7
Br_2Pr^+	$[PrBr_2]^+$	Sc: MVol.C6-36/7
Br_2Pu^+	$PuBr_2^+$	Np: TrU.D1-144
Br_2Ra	$RaBr_2$	Ra: SVol.2-364/5, 377
–	$RaBr_2 \cdot 0.5\ H_2O$	Ra: SVol.2-365
–	$RaBr_2 \cdot 2\ H_2O$	Ra: SVol.2-365/6
–	$RaBr_2$ systems	
	$RaBr_2$-$BaBr_2$-HBr-H_2O	Ra: SVol.2-180/5
Br_2RbSn^+	$RbSnBr_2^+$	Sn: MVol.C3-136
Br_2S	SBr_2	S: SVol.2-293
Br_2STe	$TeBr_2S$	Te: SVol.B3-108
Br_2S_2	S_2Br_2	S: SVol.2-293/9
–	S_2Br_2 systems	
	S_2Br_2-$SnBr_4$	Sn: MVol.C2-78
Br_2S_3	S_3Br_2	S: SVol.2-299/300
Br_2S_4	S_4Br_2	S: SVol.2-299/300
Br_2S_5	S_5Br_2	S: SVol.2-299/300
Br_2S_6	S_6Br_2	S: SVol.2-299/300
Br_2S_7	S_7Br_2	S: SVol.2-299/300
Br_2S_8	S_8Br_2	S: SVol.2-299/300
Br_2Sc	$ScBr_2$	Sc: MVol.C6-2/3, 49
Br_2Sc^+	$[ScBr_2]^+$	Sc: MVol.C6-36/7, 51/2
Br_2SeSn	$SnSeBr_2$	Sn: MVol.C2-124

Br_2Sm	$SmBr_2$	Sc:	MVol.C6-2/3, 7/12, 64/5
–	$SmBr_2 \cdot H_2O$	Sc:	MVol.C6-65
Br_2Sm^+	$[SmBr_2]^+$	Sc:	MVol.C6-36/7
Br_2Sn	$SnBr_2$ melts		
	$SnBr_2$-InBr	Sn:	MVol.C4-117
	$SnBr_2$-TlBr	Sn:	MVol.C4-139
–	$SnBr_2$ systems		
	$SnBr_2$-$AlBr_3$	Sn:	MVol.C4-103
	$SnBr_2$-CsBr	Sn:	MVol.C3-151
	$SnBr_2$-CsBr-H_2O	Sn:	MVol.C3-151
	$SnBr_2$-KBr	Sn:	MVol.C3-63
	$SnBr_2$-KBr-H_2O	Sn:	MVol.C3-63
	$SnBr_2$-LiBr	Sn:	MVol.C3-5
	$SnBr_2$-LiBr-H_2O	Sn:	MVol.C3-5
	$SnBr_2$-NH_3	Sn:	MVol.C5-6
	$SnBr_2$-NH_4Br	Sn:	MVol.C3-111
	$SnBr_2$-NH_4Br-H_2O	Sn:	MVol.C3-111
	$SnBr_2$-NaBr	Sn:	MVol.C3-25
	$SnBr_2$-NaBr-H_2O	Sn:	MVol.C3-25
	$SnBr_2$-RbBr	Sn:	MVol.C3-135/6
	$SnBr_2$-RbBr-H_2O	Sn:	MVol.C3-135/6
	$SnBr_2$-SnS	Sn:	MVol.C2-74, 78
	$SnBr_2$-SnSe	Sn:	MVol.C2-123
	$SnBr_2$-SnTe	Sn:	MVol.C2-209/10
Br_2Sr	$SrBr_2$ systems		
	$SrBr_2$-UBr_3	U:	SVol.C9-120, 133
Br_2Tb^+	$[TbBr_2]^+$	Sc:	MVol.C6-36/7
Br_2Te	$TeBr_2$	Te:	SVol.B3-1/2, 4/6
Br_2Te^+	$TeBr_2^+$	Te:	SVol.B3-10
Br_2Te^{2+}	$TeBr_2^{2+}$	Te:	SVol.B3-15/6, 18/21
Br_2Tm	$TmBr_2$	Sc:	MVol.C6-7/11, 75
Br_2Tm^+	$[TmBr_2]^+$	Sc:	MVol.C6-36/7
Br_2U	UBr_2	U:	SVol.C9-119
Br_2Y	YBr_2	Sc:	MVol.C6-2/3
Br_2Y^+	$[YBr_2]^+$	Sc:	MVol.C6-36/7
Br_2Yb	$YbBr_2$	Sc:	MVol.C6-2/3, 7/12, 75/6
$Br_{2.4}Pr$	$PrBr_{2.4}$	Sc:	MVol.C6-59/60
$Br_3CCl_2H_3Ti^{2-}$	$[CH_3TiCl_2Br_3]^{2-}$	Ti:	Org.Verb.1-42/3
Br_3CF_3Ge	CF_3GeBr_3	F:	PerFHalOrg.4-166, 171/2, 175
Br_3CF_3Si	CF_3SiBr_3	F:	PerFHalOrg.4-147
Br_3CH_3Sn	CH_3SnBr_3	Sn:	Org.Verb.6-218
Br_3CH_3Ti	CH_3TiBr_3	Ti:	Org.Verb.1-14, 23
Br_3CH_6MnN	$CH_3NH_3MnBr_3$	Mn:	MVol.C5-301
Br_3CNS	BrSC(Br)NBr	C:	MVol.D6-180
Br_3CNSSn_2	Sn_2Br_3SCN	Sn:	MVol.C2-229
$Br_3C_2ClH_{12}N_6O_2U$	$[CN_3H_6]_2UO_2Br_3Cl$	U:	SVol.C9-162
$Br_3C_2F_6P$	$(CF_3)_2PBr_3$	F:	PerFHalOrg.3-74, 76, 92, 99, 100

$Br_3C_2H_4O_{0.5}Sm$	$SmBr_3 \cdot 0.5\ C_4H_8O$	Sc:	MVol.C6-47
$Br_3C_2H_5O_3STe$	$CH_3O(SO)OCH_2TeBr_3$	Te:	SVol.B3-14
$Br_3C_2H_5Ti$	$C_2H_5TiBr_3$	Ti:	Org.Verb.1-15
$Br_3C_2H_6OPS_2Sn$	$SnBr_3(CH_3(CH_3O)P(S)S)$	Sn:	MVol.C6-168
$Br_3C_2H_6O_2PS_2Sn$	$SnBr_3((CH_3O)_2P(S)S)$	Sn:	MVol.C6-175/6
$Br_3C_2H_8IrN_2$	$Ir(C_2H_4(NH_2)_2)Br_3$	Ir:	SVol.2-85
$Br_3C_2H_8MnN$	$(CH_3)_2NH_2MnBr_3$	Mn:	MVol.C5-301
$Br_3C_3F_2N_3$	NCFBrNFCBrNCBr	F:	PerFHalOrg.6-74/6
$Br_3C_3F_4N_3$	NCFBrNFCFBrNFCBr	F:	PerFHalOrg.6-75/6, 97
$Br_3C_3F_6NO$	$(CF_3)_2NOCBr_3$	F:	PerFHalOrg.7-93/4, 109, 124
$Br_3C_3F_6N_3$	$FN(CFBrNF)_2CFBr$	F:	PerFHalOrg.6-75/6
$Br_3C_3F_9Sb^-$	$[(CF_3)_3SbBr_3]^-$	F:	PerFHalOrg.3-230
$Br_3C_3FeHg_2O_3^-$	$[(CO)_3Fe(HgBr)_2Br]^-$	Fe:	Org.Verb.B2-157
$Br_3C_3FeNO_4Sn$	$(CO)_3Fe(NO)SnBr_3$	Fe:	Org.Verb.B1-197/8
$Br_3C_3H_4NSn$	$Br_3Sn(CH_2CH_2CN)$	Sn:	MVol.C6-184
$Br_3C_3H_{10}MnN$	$(CH_3)_3NHMnBr_3$	Mn:	MVol.C5-301/2
–	$(CH_3)_3NHMnBr_3 \cdot 2\ H_2O$	Mn:	MVol.C5-301/2
$Br_3C_3N_3$	$C_3N_3Br_3$	C:	MVol.D3-287/8
$Br_3C_3N_3O_3$	$(BrNCO)_3$	C:	MVol.D3-146, 147
$Br_3C_3N_3S_3$	$(BrSCN)_3$	C:	MVol.D6-178
$Br_3C_4CeH_8O$	$CeBr_3 \cdot C_4H_8O$	Sc:	MVol.C6-47
$Br_3C_4FeGaO_4^{2-}$	$[(CO)_4FeGaBr_3]^{2-}$	Fe:	Org.Verb.B2-49
$Br_3C_4FeInO_4^{2-}$	$[(CO)_4FeInBr_3]^{2-}$	Fe:	Org.Verb.B2-49
$Br_3C_4H_4NTe$	$(CH)_4NTeBr_3$	Te:	SVol.B3-14
$Br_3C_4H_8LaO$	$LaBr_3 \cdot C_4H_8O$	Sc:	MVol.C6-47
$Br_3C_4H_8OPr$	$PrBr_3 \cdot C_4H_8O$	Sc:	MVol.C6-47
$Br_3C_4H_8OSn^+$	$[SnBr_3(CH_3COC_2H_5)]^+$	Sn:	MVol.C5-87
$Br_3C_4H_9Sn$	$(CH_3)_3SnCBr_3$	Sn:	Org.Verb.2-11
$Br_3C_4H_{10}O_2PS_2Sn$	$SnBr_3((C_2H_5O)_2P(S)S)$	Sn:	MVol.C6-175/6
$Br_3C_4H_{11}MnO$	$HMnBr_3 \cdot (C_2H_5)_2O$	Mn:	MVol.C5-288
		Mn:	MVol.D1-139
$Br_3C_4H_{11}O_2Sn$	$SnBr_3OH \cdot (C_2H_5)_2O \cdot H_2O$	Sn:	MVol.C5-150/1
$Br_3C_4H_{12}MnN$	$(CH_3)_4NMnBr_3$	Mn:	MVol.C5-303
$Br_3C_4H_{16}IrN_4$	$[Ir(C_2H_4(NH_2)_2)_2Br_2]Br$	Ir:	SVol.2-84
–	$[Ir(C_2H_4(NH_2)_2)_2Br_2]Br \cdot H_2O$	Ir:	SVol.2-84
$Br_3C_4H_{16}O_4Sc$	$ScBr_3 \cdot 4\ CH_3OH$	Sc:	MVol.C6-51
$Br_3C_5CeH_5N$	$CeBr_3 \cdot C_5H_5N$	Sc:	MVol.C6-58/9
Br_3C_5FHNO	HOC_5FBr_3N	F:	PerFHalOrg.5-182/3
		F:	PerFHalOrg.6-13
$Br_3C_5F_2FeO_4$	$(CO)_4Fe(CF_2Br_2)Br$	Fe:	Org.Verb.B3-160
$Br_3C_5F_2N$	$Br_3C_5F_2N$	F:	PerFHalOrg.5-118/9, 128/9
$Br_3C_5FeHO_4$	$(CO)_4Fe(CHBr_2)Br$	Fe:	Org.Verb.B2-176
		Fe:	Org.Verb.B3-90, 171
$Br_3C_5H_5Ti$	$C_5H_5TiBr_3$	Ti:	Org.Verb.1-139, 148/9
$Br_3C_5H_6NSn$	$[C_5H_6N]SnBr_3$	Sn:	MVol.C3-112
$Br_3C_5H_7O_2Sn$	$SnBr_3(CH(COCH_3)_2) \cdot 2\ H_2O$ = $[Sn(CH(COCH_3)_2)_2(H_2O)_4][SnBr_6]$	Sn:	MVol.C5-107
$Br_3C_5H_8N_3Sn$	$SnBr_3N_2H_3 \cdot C_5H_5N$	Sn:	MVol.C5-199

$Br_3C_5H_{20}O_5Sc$	$ScBr_3 \cdot 5\ CH_3OH$	Sc:	MVol.C6-51
$Br_3C_6F_4FeO_4$	$(CO)_4Fe(C_2F_4Br_2)Br$	Fe:	Org.Verb.B3-160
$Br_3C_6F_5Ge$	$C_6F_5GeBr_3$	F:	PerFHalOrg.4-166/7, 176, 180
$Br_3C_6F_5Si$	$C_6F_5SiBr_3$	F:	PerFHalOrg.4-150/2, 157, 162
$Br_3C_6F_6N_3$	$(CF_2Br)_3C_3N_3$	F:	PerFHalOrg.6-81/2, 102
$Br_3C_6FeH_5O_3Sn$	$C_3H_5Fe(CO)_3SnBr_3$	Fe:	Org.Verb.B5-67/8
$Br_3C_6GdH_9N_3$	$GdBr_3 \cdot 3\ CH_3CN$	Sc:	MVol.C6-46
$Br_3C_6GeH_5NiO$	$C_5H_5Ni(CO)GeBr_3$	Ni:	Org.Verb.2-164/6
$Br_3C_6H_3O_2Te$	$C_6H_3O_2TeBr_3$	Te:	SVol.B3-14
$Br_3C_6H_5NiOSn$	$C_5H_5Ni(CO)SnBr_3$	Ni:	Org.Verb.2-164/6
$Br_3C_6H_5Sn$	$C_6H_5SnBr_3$	Sn:	Org.Verb.6-275/6
$Br_3C_6H_5Ti$	$C_6H_5TiBr_3$	Ti:	Org.Verb.1-17
$Br_3C_6H_7N_2Sn$	$[SnBr_3(C_6H_5NHNH)]$	Sn:	MVol.C5-189
$Br_3C_6H_7Ti$	$CH_3C_5H_4TiBr_3$	Ti:	Org.Verb.1-140, 150
$Br_3C_6H_9LaN_3$	$LaBr_3 \cdot 3\ CH_3CN$	Sc:	MVol.C6-46
$Br_3C_6H_9N_3Nd$	$NdBr_3 \cdot 3\ CH_3CN$	Sc:	MVol.C6-46
$Br_3C_6H_{12}N_3$	$(CH_3)_4NBr \cdot 2\ BrCN$	C:	MVol.D3-235
$Br_3C_6H_{15}NSc$	$ScBr_3 \cdot N(C_2H_5)_3$	Sc:	MVol.C6-50
–	$ScBr_3 \cdot NH(CH(CH_3)_2)_2$	Sc:	MVol.C6-50
$Br_3C_6H_{18}IrS_3$	$Ir(S(CH_3)_2)_3Br_3$	Ir:	SVol.2-164/5
$Br_3C_6H_{18}O_3Sc$	$ScBr_3 \cdot 3\ C_2H_5OH$	Sc:	MVol.C6-51
$Br_3C_6H_{24}IN_{12}O_6U$	$UIBr_3 \cdot 6\ (NH_2)_2CO \cdot 3\ H_2O$	U:	SVol.E1-82/3
$Br_3C_7H_7OTi$	$CH_3C(O)C_5H_4TiBr_3$	Ti:	Org.Verb.1-140, 150
$Br_3C_7H_{11}O_4Sn$	$SnBr_3OOC(CH_2)_4COOCH_3$	Sn:	MVol.C5-143
$Br_3C_7H_{15}Sn$	$(C_2H_5)_3SnCBr_3$	Sn:	Org.Verb.2-163, 164
$Br_3C_8ClF_3HN_2O_2S$	$C_6Br_3(SO_2Cl)NC(CF_3)NH$	F:	PerFHalOrg.6-125/6, 160, 162
$Br_3C_8Cl_2H_{20}NTe$	$[(C_2H_5)_4N][TeBr_3Cl_2]$	Te:	SVol.B3-86
$Br_3C_8H_{12}N_4Sm$	$SmBr_3 \cdot 4\ CH_3CN$	Sc:	MVol.C6-46
$Br_3C_8H_{18}O_2PS_2Sn$	$SnBr_3((C_4H_9O)_2P(S)S)$	Sn:	MVol.C6-175/6
$Br_3C_8H_{20}MnN$	$(C_2H_5)_4NMnBr_3$	Mn:	MVol.C5-303
$Br_3C_8H_{20}NSn$	$[(C_2H_5)_4N]SnBr_3$	Sn:	MVol.C3-113
$Br_3C_8H_{24}O_4Sc$	$ScBr_3 \cdot 4\ C_2H_5OH$	Sc:	MVol.C6-51
$Br_3C_8H_{26}IrN_6$	$[Ir((NH_2CH_2CH_2)_2NH)_2]Br_3$	Ir:	SVol.2-95
$Br_3C_9F_{12}N_3$	$(CF_3CFBr)_3C_3N_3$	F:	PerFHalOrg.6-82/3, 103
$Br_3C_9H_{27}IrP_3$	$Ir(P(CH_3)_3)_3Br_3$	Ir:	SVol.2-199/200
$Br_3C_{10}CeH_{10}N_2$	$CeBr_3 \cdot 2\ C_5H_5N$	Sc:	MVol.C6-58/9
$Br_3C_{10}ClH_{12}N_2O_2U$	$[C_5H_5NH]_2UO_2Br_3Cl$	U:	SVol.C9-162
$Br_3C_{10}F_3H_7N_3O_2S$	$(C_2H_5NHSO_2)(CF_3)C_7Br_3HN_2$	F:	PerFHalOrg.6-160
$Br_3C_{10}F_4H_2N_3$	$NC_5F_4NHC_5Br_3HN$	F:	PerFHalOrg.5-139
$Br_3C_{10}FeH_{10}$	$[Fe(C_5H_5)_2]Br_3$	Fe:	Org.Verb.A1-215
$Br_3C_{10}Fe_2H_{10}$	$[Fe(C_5H_5)_2]FeBr_3$	Fe:	Org.Verb.A1-222
$Br_3C_{10}H_{10}IrN_2$	$[Ir(C_5H_5N)_2Br_3]_n$	Ir:	SVol.2-72
$Br_3C_{10}H_{10}LaN_2$	$LaBr_3 \cdot 2\ C_5H_5N$	Sc:	MVol.C6-46
$Br_3C_{10}H_{15}N_5Pr$	$PrBr_3 \cdot 5\ CH_3CN$	Sc:	MVol.C6-46
$Br_3C_{10}H_{15}Ti$	$(CH_3)_5C_5TiBr_3$	Ti:	Org.Verb.1-141, 151
$Br_3C_{10}H_{22.5}N_{2.5}O_{3.5}U$	$UOBr_3 \cdot 2.5\ CH_3CON(CH_3)_2$	U:	SVol.E1-106
$Br_3C_{11}FeH_{20}Hg_2NO_3$	$[N(C_2H_5)_4][(CO)_3Fe(HgBr)_2Br]$	Fe:	Org.Verb.B1-192

$Br_3C_{11}H_{11}N_2Ti$	$CH_3TiBr_3 \cdot NC_5H_4C_5H_4N$	Ti:	Org.Verb.1-34/5
$Br_3C_{11}H_{12}N_3Sn$	$SnBr_3N_2H_2C_6H_5 \cdot C_5H_5N$	Sn:	MVol.C5-199
$Br_3C_{12}F_3H_{11}N_3O_2S$	$((C_2H_5)_2NSO_2)(CF_3)C_7Br_3HN_2$	F:	PerFHalOrg.6-162
$Br_3C_{12}F_4H_4N_3O_2$	$NC_5FBr_3NHC_6H_3(CF_3)NO_2$	F:	PerFHalOrg.5-140
$Br_3C_{12}F_7$	$Br_2C_6F_3C_6F_4Br$	F:	PerFHalOrg.4-11
$Br_3C_{12}FeH_{11}O$	$C_5H_5FeC_5H_4CHOHCBr_3$	Fe:	Org.Verb.A2-45, 48
$Br_3C_{12}FeH_{20}Hg_2NO_4$	$[N(C_2H_5)_4][(CO)_4Fe(HgBr)_2Br]$	Fe:	Org.Verb.B2-153, 157
$Br_3C_{12}FeH_{36}O_6S_6$	$Fe(OS(CH_3)_2)_6Br_3$	Fe:	Org.Verb.B5-58
$Br_3C_{12}H_{10}IrN_2O$	$Ir(C_{12}H_8N_2)Br_3(H_2O)$	Ir:	SVol.2-92
$Br_3C_{12}H_{11}Ti$	$C_6H_5CH_2C_5H_4TiBr_3$	Ti:	Org.Verb.1-141, 150
$Br_3C_{12}H_{12}IrN_6$	$Ir(C_4H_4N_2)_3Br_3$	Ir:	SVol.2-77
$Br_3C_{12}H_{24}IrS_3$	$IrBr_3(C_4H_8S)_3$	Ir:	SVol.2-166
$Br_3C_{12}H_{24}LuO_3$	$LuBr_3 \cdot 3\ C_4H_8O$	Sc:	MVol.C6-46/7
$Br_3C_{12}H_{24}O_3Sn^+$	$[SnBr_3(CH_3COC_2H_5)_3]^+$	Sn:	MVol.C5-87
$Br_3C_{12}H_{24}O_3Tm$	$TmBr_3 \cdot 3\ C_4H_8O$	Sc:	MVol.C6-46/7
$Br_3C_{12}H_{24}O_3Yb$	$YbBr_3 \cdot 3\ C_4H_8O$	Sc:	MVol.C6-46/7
$Br_3C_{12}H_{25}Sn$	$C_{12}H_{25}SnBr_3$	Sn:	Org.Verb.6-253
$Br_3C_{12}H_{30}IrS_3$	$Ir(S(C_2H_5)_2)_3Br_3$	Ir:	SVol.2-166
$Br_3C_{12}H_{30}MnO_3$	$MnBr_3 \cdot 3\ (C_2H_5)_2O$	Mn:	MVol.C5-290
$Br_3C_{12}H_{31}MnO_3$	$HMnBr_3 \cdot 3\ (C_2H_5)_2O$	Mn:	MVol.D1-139
$Br_3C_{13}H_{27}Sn$	$(C_4H_9)_3SnCBr_3$	Sn:	Org.Verb.2-278, 280
$Br_3C_{14}Cl_5H_{14}N_2Sn$	$[C_6H_5C(Cl)NH_2]_2[SnBr_3Cl_3]$	Sn:	MVol.C3-119
$Br_3C_{14}DyH_{28}O_{3.5}$	$DyBr_3 \cdot 3.5\ C_4H_8O$	Sc:	MVol.C6-46/7
$Br_3C_{14}ErH_{28}O_{3.5}$	$ErBr_3 \cdot 3.5\ C_4H_8O$	Sc:	MVol.C6-46/7
$Br_3C_{14}EuH_{28}O_{3.5}$	$EuBr_3 \cdot 3.5\ C_4H_8O$	Sc:	MVol.C6-46/7
$Br_3C_{14}GdH_{28}O_{3.5}$	$GdBr_3 \cdot 3.5\ C_4H_8O$	Sc:	MVol.C6-46/7
$Br_3C_{14}H_7O_2Te$	$C_{14}H_7O_2TeBr_3$	Te:	SVol.B3-14
$Br_3C_{14}H_{28}HoO_{3.5}$	$HoBr_3 \cdot 3.5\ C_4H_8O$	Sc:	MVol.C6-46/7
$Br_3C_{14}H_{28}NdO_{3.5}$	$NdBr_3 \cdot 3.5\ C_4H_8O$	Sc:	MVol.C6-46/7
$Br_3C_{14}H_{28}O_{3.5}Sm$	$SmBr_3 \cdot 3.5\ C_4H_8O$	Sc:	MVol.C6-46/7
$Br_3C_{14}H_{28}O_{3.5}Tb$	$TbBr_3 \cdot 3.5\ C_4H_8O$	Sc:	MVol.C6-46/7
$Br_3C_{14}H_{28}O_{3.5}Y$	$YBr_3 \cdot 3.5\ C_4H_8O$	Sc:	MVol.C6-46/7
$Br_3C_{15}CeH_{15}N_3$	$CeBr_3 \cdot 3\ C_5H_5N$	Sc:	MVol.C6-58/9
$Br_3C_{15}FeH_{13}$	$C_5H_5FeC_5H_4C_3H_2Br_2CBrCH_2$	Fe:	Org.Verb.A1-392/3
$Br_3C_{15}Fe_3H_3O_{15}$	$[(CO)_5FeH]_3Br_3$	Fe:	Org.Verb.B3-252
$Br_3C_{15}H_{15}IrN_3$	$Ir(C_5H_5N)_3Br_3$	Ir:	SVol.2-70
$Br_3C_{15}H_{15}N_3Sc$	$ScBr_3 \cdot 3\ C_5H_5N$	Sc:	MVol.C6-50
$Br_3C_{15}H_{15}N_3Sm$	$SmBr_3 \cdot 3\ C_5H_5N$	Sc:	MVol.C6-46
$Br_3C_{15}H_{30}IrS_3$	$IrBr_3(C_5H_{10}S)_3$	Ir:	SVol.2-166
$Br_3C_{16}CeH_{32}O_4$	$CeBr_3 \cdot 4\ C_4H_8O$	Sc:	MVol.C6-46/7
$Br_3C_{16}H_{32}LaO_4$	$LaBr_3 \cdot 4\ C_4H_8O$	Sc:	MVol.C6-46/7
$Br_3C_{16}H_{32}O_4Pr$	$PrBr_3 \cdot 4\ C_4H_8O$	Sc:	MVol.C6-46/7
$Br_3C_{16}H_{36}NSn$	$[(n\text{-}C_4H_9)_4N]SnBr_3$	Sn:	MVol.C3-113
$Br_3C_{16}H_{44}N_4Sc$	$ScBr_3 \cdot 4\ NH(C_2H_5)_2$	Sc:	MVol.C6-50
$Br_3C_{17}Cl_2H_{43}N_2Ti$	$[(C_2H_5)_4N]_2[CH_3TiCl_2Br_3]$	Ti:	Org.Verb.1-42/3
$Br_3C_{17}FeH_{15}Hg$	$[C_5H_5FeC_5H_4CHC_6H_5]HgBr_3$	Fe:	Org.Verb.A1-351, 354
$Br_3C_{17.5}GdH_{17.5}N_{3.5}$	$GdBr_3 \cdot 3.5\ C_5H_5N$	Sc:	MVol.C6-46
$Br_3C_{17.5}H_{17.5}N_{3.5}Nd$	$NdBr_3 \cdot 3.5\ C_5H_5N$	Sc:	MVol.C6-46
$Br_3C_{17.5}H_{17.5}N_{3.5}Pr$	$PrBr_3 \cdot 3.5\ C_5H_5N$	Sc:	MVol.C6-46
$Br_3C_{18}FH_{12}Sn$	$(BrC_6H_4)_3SnF$	Sn:	Org.Verb.5-37

$Br_3C_{18}F_{15}IP$	$(C_6F_5)_3PBr_2 \cdot BrI$	F:	PerFHalOrg.3-124/5
$Br_3C_{18}H_9IrN_9$	$Ir(C_6H_3BrN_3)_3 \cdot 3\ H_2O$	Ir:	SVol.2-96
$Br_3C_{18}H_{12}IrO_6Se_3$	$Ir(O_2SeC_6H_4Br)_3 \cdot 2\ H_2O$	Ir:	SVol.2-38
$Br_3C_{18}H_{45}IrP_3$	$Ir(P(C_2H_5)_3)_3Br_3$	Ir:	SVol.2-200, 202
$Br_3C_{19}H_{15}Sn$	$(C_6H_5)_3SnCBr_3$	Sn:	Org.Verb.2-362
$Br_3C_{20}FeH_{15}NO_3PSn$	$(CO)_2Fe(SnBr_3)(NO)P(C_6H_5)_3$	Fe:	Org.Verb.B1-133/5
$Br_3C_{20}H_{16}IrN_4$	$[Ir(C_5H_4NC_5H_4N)_2Br_2]Br \cdot 3\ H_2O$	Ir:	SVol.2-88
$Br_3C_{20}H_{20}IrN_4$	$[Ir(C_5H_5N)_4Br_2]Br$	Ir:	SVol.2-69
$Br_3C_{20}H_{20}N_4Sc$	$ScBr_3 \cdot 4\ C_5H_5N$	Sc:	MVol.C6-50
$Br_3C_{20}H_{44}N_4Sc$	$ScBr_3 \cdot 4\ (CH_2)_5NH$	Sc:	MVol.C6-50
$Br_3C_{21}FeH_{30}O_{12}$	$Fe(BrC(COOC_2H_5)_2)_3$	Fe:	Org.Verb.B3-177
$Br_3C_{22}H_{25}O_3PPa$	$Pa(OC_2H_5)_2Br_3 \cdot (C_6H_5)_3PO$	Pa:	SVol.2-82
$Br_3C_{22}H_{26}HgPSn$	$[(CH_3)_3SnCH_2P(C_6H_5)_3]HgBr_3$	Sn:	Org.Verb.2-16
$Br_3C_{23}FeH_{18}Hg_2O_4P$	$[P(C_6H_5)_3CH_3][(CO)_4Fe(HgBr)_2Br]$	Fe:	Org.Verb.B2-153, 157
$Br_3C_{23}GeH_{20}NiP$	$C_5H_5Ni(P(C_6H_5)_3)GeBr_3$	Ni:	Org.Verb.2-130
$Br_3C_{23}H_{20}NiPSn$	$C_5H_5Ni(P(C_6H_5)_3)SnBr_3$	Ni:	Org.Verb.2-131/2
$Br_3C_{24}H_{16}IrN_4$	$[Ir(C_{12}H_8N_2)_2Br_2]Br$	Ir:	SVol.2-90
–	$[Ir(C_{12}H_8N_2)_2Br_2]Br \cdot 2\ H_2O$	Ir:	SVol.2-90
–	$[Ir(C_{12}H_8N_2)_2Br_2]Br \cdot 3\ H_2O$	Ir:	SVol.2-90
$Br_3C_{24}H_{20}IrN_4O_2$	$[Ir(C_{12}H_8N_2)_2(H_2O)_2]Br_3 \cdot 3\ H_2O$	Ir:	SVol.2-91
$Br_3C_{24}H_{33}IrP_3$	$Ir(P(CH_3)_2C_6H_5)_3Br_3$	Ir:	SVol.2-212
$Br_3C_{24}H_{56}N_8NdO_8$	$NdBr_3 \cdot 8\ HCON(CH_3)_2$	Sc:	MVol.C6-46
$Br_3C_{25}FeH_{20}Hg_2O_4P$	$[P(C_6H_5)_3C_3H_5][(CO)_4Fe(HgBr)_2Br]$	Fe:	Org.Verb.B2-153, 157
$Br_3C_{25}H_{26}MnN_5$	$HMnBr_3 \cdot 5\ C_5H_5N$	Mn:	MVol.C5-288
$Br_3C_{26}Fe_2H_{21}$	$C_5H_5FeC_5H_4(CBrCH)_3C_5H_4FeC_5H_5$	Fe:	Org.Verb.A6-140
$Br_3C_{27}H_{15}IrN_3S_3$	$Ir(NSC_9H_5Br)_3$	Ir:	SVol.2-172
$Br_3C_{28}FeH_{19}O_2$	$C_5H_5FeC_5H_4C_6H_4OCH_2(CC)_2CH_2OC_6H_2Br_3$	Fe:	Org.Verb.A2-126
$Br_3C_{30}H_{27}IrN_9$	$[Ir(C_{10}H_9N_3)_3]Br_3 \cdot 3\ H_2O$	Ir:	SVol.2-95
$Br_3C_{30}H_{37}NiP_2$	$C_{10}H_7Ni(P(C_2H_5)_2C_6H_5)_2Br_3$	Ni:	Org.Verb.1-57/8
$Br_3C_{30}H_{45}IrP_3$	$Ir(P(C_2H_5)_2C_6H_5)_3Br_3$	Ir:	SVol.2-213/4
$Br_3C_{36}H_{24}IrN_6$	$[Ir(C_{12}H_8N_2)_3]Br_3$	Ir:	SVol.2-89
$Br_3C_{36}H_{30}IrNOP_2$	$Ir(NO)Br_3(P(C_6H_5)_3)_2$	Ir:	SVol.2-56
$Br_3C_{36}H_{30}IrSb_2$	$IrBr_3(Sb(C_6H_5)_3)_2$	Ir:	SVol.2-263
$Br_3C_{36}H_{30}O_3P_2Pa$	$PaOBr_3 \cdot 2\ (C_6H_5)_3PO$	Pa:	SVol.2-93
$Br_3C_{36}H_{33}IrNOP_2$	$Ir(NH_2OH)Br_3(P(C_6H_5)_3)_2$	Ir:	SVol.2-63
$Br_3C_{36}H_{33}IrNSb_2$	$IrBr_3(NH_3)(Sb(C_6H_5)_3)_2$	Ir:	SVol.2-263
$Br_3C_{36}H_{81}IrP_3$	$Ir(P(n\text{-}C_4H_9)_3)_3Br_3$	Ir:	SVol.2-200/2
$Br_3C_{37}H_{30}IrOSb_2$	$IrBr_3(CO)(Sb(C_6H_5)_3)_2$	Ir:	SVol.2-263
$Br_3C_{37.3}H_{35.2}IrO_{1.3}Sb_2$	$IrBr_3(Sb(C_6H_5)_3)_2 \cdot 1.3\ CH_3OH$	Ir:	SVol.2-263
$Br_3C_{39}H_{39}IrP_3$	$Ir(PCH_3(C_6H_5)_2)_3Br_3$	Ir:	SVol.2-213
$Br_3C_{40}H_{84}NO_2U$	$[N(C_{10}H_{21})_4]UO_2Br_3$	U:	SVol.C9-151, 154
$Br_3C_{41}H_{35}IrNSb_2$	$IrBr_3(C_5H_5N)(Sb(C_6H_5)_3)_2$	Ir:	SVol.2-263
$Br_3C_{42}H_{39}IrP_3$	$Ir(PC_{14}H_{13})_3Br_3$	Ir:	SVol.2-234
$Br_3C_{44}ClH_{48}O_2P_2U$	$[P(C_6H_5)_3C_4H_9]_2UO_2Br_3Cl$	U:	SVol.C9-162
$Br_3C_{54}H_{45}IrSb_3$	$IrBr_3(Sb(C_6H_5)_3)_3$	Ir:	SVol.2-263
$Br_3C_{66}H_{57}IrSb_3$	$IrBr_3(Sb(C_6H_5)_3)_3 \cdot 2\ C_6H_6$	Ir:	SVol.2-263
$Br_3C_{72}H_{60}N_3O_6P_6Sn$	$[Sn((C_6H_5)_2P(O)NP(O)(C_6H_5)_2)_3]Br_3$	Sn:	MVol.C6-155
$Br_3C_{76}FeGaH_{60}N_2O_4P_4$	$[(C_6H_5)_3PNP(C_6H_5)_3]_2[(CO)_4FeGaBr_3]$	Fe:	Org.Verb.B2-50

$Br_3C_{76}FeH_{60}InN_2O_4P_4$ $[(C_6H_5)_3PNP(C_6H_5)_3]_2[(CO)_4FeInBr_3]$ ...... Fe: Org.Verb.B2-49
$Br_3C_{80}ClH_{168}N_2O_2U$ .. $[N(C_{10}H_{21})_4]_2UO_2Br_3Cl$ .... U: SVol.C9-162
$Br_3C_{80}H_{168}IN_2O_2U$ ... $[N(C_{10}H_{21})_4]_2[UO_2Br_3I]$ .... U: SVol.C9-185
Br_3Ce .... $CeBr_3$
Chemical reactions .... Sc: MVol.C6-33/6, 59
Formation and preparation .... Sc: MVol.C6-13/7, 26/8, 58
Molecule .... Sc: MVol.C6-3/6
Physical properties .... Sc: MVol.C6-17/33, 58
– .... $CeBr_3 \cdot 6\ H_2O$ .... Sc: MVol.C6-43/6
– .... $CeBr_3 \cdot 7\ H_2O$ .... Sc: MVol.C6-44
– .... $CeBr_3$ solutions
$CeBr_3$-H_2O .... Sc: MVol.C6-36/40
$CeBr_3$-KBr-H_2O .... Sc: MVol.C6-124/5
$CeBr_3$-NaCl-H_2O .... Sc: MVol.C6-119
– .... $CeBr_3$ systems
$CeBr_3$-Ce .... Sc: MVol.C6-7/8, 58
Br_3CeO_9 .... $Ce(BrO_3)_3 \cdot 9\ H_2O$ .... Sc: MVol.C6-94/104, 107
– .... $Ce(BrO_3)_3 \cdot n\ H_2O$ .... Sc: MVol.C6-94/7, 104
$Br_3ClCs_2O_2U$ ...... $Cs_2UO_2Br_3Cl$ .... U: SVol.C9-162/3
$Br_3ClH_8N_2O_2U$ ..... $(NH_4)_2UO_2Br_3Cl \cdot 2\ H_2O$ .... U: SVol.C9-162/3
$Br_3ClK_2O_2U$ .... $K_2UO_2Br_3Cl \cdot 2\ H_2O$ .... U: SVol.C9-162/3
$Br_3ClO_2Rb_2U$ ...... $Rb_2UO_2Br_3Cl \cdot 2\ H_2O$ .... U: SVol.C9-162
$Br_3ClO_2U^{2-}$ .... $[UO_2Br_3Cl]^{2-}$ .... U: SVol.C9-162/3
$Br_3ClO_6U^{2-}$ .... $[UO_2Br_3(ClO_4)]^{2-}$ .... U: SVol.C9-108
Br_3ClU .... UBr_3Cl .... U: SVol.C9-158/61
$Br_3Cl_2Te^-$ .... $TeBr_3Cl_2^-$ .... Te: SVol.B3-81/3
$Br_3Cl_3H_8N_2Te$ ...... $(NH_4)_2[TeCl_3Br_3]$ .... Te: SVol.B3-86
$Br_3Cl_3Ir^{2-}$ .... $[IrCl_3Br_3]^{2-}$ .... Ir: SVol.2-150
$Br_3Cl_3IrRb_2$ .... $Rb_2[IrCl_3Br_3]$ .... Ir: SVol.2-150
Br_3Cl_4OPSn .... $SnCl_4 \cdot POBr_3$ .... Sn: MVol.C2-268
Br_3CsMg .... $CsMgBr_3$ solid solutions
$CsMg_xMn_{1-x}Br_3$ .... Mn: MVol.C5-306
Br_3CsMn .... $CsMnBr_3 = CsBr \cdot MnBr_2$ .... Mn: MVol.C5-296/9
– .... $CsMnBr_3 \cdot 2\ H_2O$ .... Mn: MVol.C5-300
– .... $CsMnBr_3$ solid solutions
$CsMg_xMn_{1-x}Br_3$ .... Mn: MVol.C5-306
Br_3CsSn .... $CsSnBr_3$ .... Sn: MVol.C3-151/2, 153
Br_3Dy .... $DyBr_3$
Chemical reactions .... Sc: MVol.C6-33/6
Formation and preparation .... Sc: MVol.C6-13/7, 27
Molecule .... Sc: MVol.C6-4/6
Physical properties .... Sc: MVol.C6-17/33
– .... $DyBr_3 \cdot 6\ H_2O$ .... Sc: MVol.C6-43/6
– .... $DyBr_3$ melts
$DyBr_3$-CsBr .... Sc: MVol.C6-115/6
$DyBr_3$-KBr .... Sc: MVol.C6-115/6
$DyBr_3$-NaBr .... Sc: MVol.C6-115/6

Br_3Dy .... $DyBr_3$ solutions
$DyBr_3$-H_2O .... Sc: MVol.C6-36/40
$DyBr_3$-KBr-H_2O .... Sc: MVol.C6-124/5
$DyBr_3$-NaBr-H_2O .... Sc: MVol.C6-122

Br_3DyO_9 .... $Dy(BrO_3)_3$ .... Sc: MVol.C6-93/4
– .... $Dy(BrO_3)_3 \cdot 9\ H_2O$ .... Sc: MVol.C6-94/105
– .... $Dy(BrO_3)_3 \cdot n\ H_2O$ .... Sc: MVol.C6-94/7, 104
– .... $Dy(BrO_3)_3$ solutions
$Dy(BrO_3)_3$-H_2O .... Sc: MVol.C6-105/6

Br_3Er .... $ErBr_3$
Chemical reactions .... Sc: MVol.C6-33/6
Formation and preparation .... Sc: MVol.C6-13/7, 27, 74
Molecule .... Sc: MVol.C6-4/6
Physical properties .... Sc: MVol.C6-17/33, 74

– .... $ErBr_3 \cdot 6\ H_2O$ .... Sc: MVol.C6-43/6, 74
– .... $ErBr_3$ melts
$ErBr_3$-CsBr .... Sc: MVol.C6-115/6
$ErBr_3$-KBr .... Sc: MVol.C6-115/6
$ErBr_3$-NaBr .... Sc: MVol.C6-115/6

– .... $ErBr_3$ solutions
$ErBr_3$-H_2O .... Sc: MVol.C6-36/40
$ErBr_3$-KBr-H_2O .... Sc: MVol.C6-124/5
$ErBr_3$-NaBr-H_2O .... Sc: MVol.C6-122
$ErBr_3$-organic solvents .... Sc: MVol.C6-40/1

Br_3ErO_9 .... $Er(BrO_3)_3$ .... Sc: MVol.C6-93/4
– .... $Er(BrO_3)_3 \cdot 9\ H_2O$ .... Sc: MVol.C6-94/105
– .... $Er(BrO_3)_3 \cdot n\ H_2O$ .... Sc: MVol.C6-94/7, 104
– .... $Er(BrO_3)_3$ solutions
$Er(BrO_3)_3$-H_2O .... Sc: MVol.C6-105/6

Br_3Eu .... $EuBr_3$
Chemical reactions .... Sc: MVol.C6-33/6, 71
Formation and preparation .... Sc: MVol.C6-13/6, 27, 70
Molecule .... Sc: MVol.C6-3/6
Physical properties .... Sc: MVol.C6-17/33, 70

– .... $EuBr_3 \cdot H_2O$ .... Sc: MVol.C6-45
– .... $EuBr_3 \cdot 6\ H_2O$ .... Sc: MVol.C6-43/6
– .... $EuBr_3$ solutions
$EuBr_3$-H_2O .... Sc: MVol.C6-36/7
$EuBr_3$-KBr-H_2O .... Sc: MVol.C6-124/5
$EuBr_3$-NaBr-H_2O .... Sc: MVol.C6-121
$EuBr_3$-organic solvents .... Sc: MVol.C6-42

Br_3EuO_9 .... $Eu(BrO_3)_3$ .... Sc: MVol.C6-93/4
– .... $Eu(BrO_3)_3 \cdot 4\ H_2O$ .... Sc: MVol.C6-95, 104
– .... $Eu(BrO_3)_3 \cdot 9\ H_2O$ .... Sc: MVol.C6-94/105, 111/2
– .... $Eu(BrO_3)_3 \cdot n\ H_2O$ .... Sc: MVol.C6-94/7, 104
– .... $Eu(BrO_3)_3$ solutions
$Eu(BrO_3)_3$-H_2O .... Sc: MVol.C6-105/6

$Br_3Eu_7H_{18}O_{18}$ ...... $Eu_7(OH)_{18}Br_3$ ...... Sc: MVol.C6-87/90
Br_3FNO_2PS ...... FSO_2NPBr_3 ...... S: S-N-Verb.1-122
Br_3F_6SSb ...... $[SBr_3][SbF_6]$ ...... S: SVol.2-305
Br_3Gd ...... $GdBr_3$
Chemical reactions ...... Sc: MVol.C6-33/6
Formation and preparation ...... Sc: MVol.C6-13/7, 27, 71
Molecule ...... Sc: MVol.C6-3/6
Physical properties ...... Sc: MVol.C6-17/33, 71/3
– ...... $GdBr_3 \cdot 6\ H_2O$ ...... Sc: MVol.C6-43/6
– ...... $GdBr_3$ melts
$GdBr_3$-Gd ...... Sc: MVol.C6-7/8
$GdBr_3$-KBr ...... Sc: MVol.C6-115/6
– ...... $GdBr_3$ solutions
$GdBr_3$-H_2O ...... Sc: MVol.C6-36/40
$GdBr_3$-KBr-H_2O ...... Sc: MVol.C6-124/5
$GdBr_3$-NaBr-H_2O ...... Sc: MVol.C6-122
$GdBr_3$-organic solvents ...... Sc: MVol.C6-40/3
– ...... $GdBr_3$ systems
$GdBr_3$-CsBr ...... Sc: MVol.C6-114/6, 128/31
$GdBr_3$-NaBr ...... Sc: MVol.C6-114/6, 121/2
Br_3GdO_9 ...... $Gd(BrO_3)_3$ ...... Sc: MVol.C6-93/4
– ...... $Gd(BrO_3)_3 \cdot 9\ H_2O$ ...... Sc: MVol.C6-94/105
– ...... $Gd(BrO_3)_3 \cdot n\ H_2O$ ...... Sc: MVol.C6-94/7, 104
– ...... $Gd(BrO_3)_3$ solutions
$Gd(BrO_3)_3$-H_2O ...... Sc: MVol.C6-105/6
$Br_3Gd_7H_{18}O_{18}$ ...... $Gd_7(OH)_{18}Br_3$ ...... Sc: MVol.C6-87/90
Br_3HU ...... $UHBr_3$ ...... U: SVol.C9-161
Br_3H_2NU ...... $UBr_3(NH_2)$ ...... U: SVol.C9-126
$Br_3H_2O_2U$ ...... $UO_2Br \cdot 2\ HBr$ ...... U: SVol.C9-144
$Br_3H_4La_7O_{11}$ ...... $La_7O_7(OH)_4Br_3$ ...... Sc: MVol.C6-90
$Br_3H_4NO_5U_2$ ...... $[NH_4]U_2O_5Br_3$ ...... U: SVol.C9-151, 154
Br_3H_4NSn ...... NH_4SnBr_3 ...... Sn: MVol.C3-111
– ...... $NH_4SnBr_3 \cdot H_2O$ ...... Sn: MVol.C3-111
$Br_3H_5N_2Sn$ ...... $N_2H_5SnBr_3$ ...... Sn: MVol.C3-112
$Br_3H_6IrO_3$ ...... $[IrBr_3(H_2O)_3]$ ...... Ir: SVol.2-146, 150
$Br_3H_6MnO_3^-$ ...... $[Mn(H_2O)_3Br_3]$ ...... Mn: MVol.C5-285
$Br_3H_6O_3Te^+$ ...... $[TeBr_3(H_2O)_3]^+$ ...... Te: SVol.B3-15, 29
$Br_3H_8I_3N_2Te$ ...... $(NH_4)_2[TeBr_3I_3]$ ...... Te: SVol.B3-86
$Br_3H_9N_3U$ ...... $UBr_3 \cdot 3\ NH_3$ ...... U: SVol.E1-12, 13
$Br_3H_{12}IrN_6$ ...... $Ir(N_2H_4)_3Br_3 \cdot 0.5\ H_2O$ ...... Ir: SVol.2-47
$Br_3H_{12}N_4U$ ...... $UBr_3 \cdot 4\ NH_3$ ...... U: SVol.E1-12, 13
$Br_3H_{15}IrN_5$ ...... $[Ir(NH_3)_5Br]Br_2$ ...... Ir: SVol.2-43/5
$Br_3H_{15}N_5Sc$ ...... $ScBr_3 \cdot 5\ NH_3$ ...... Sc: MVol.C6-34
$Br_3H_{17}IrN_5O$ ...... $[Ir(NH_3)_5H_2O]Br_3$ ...... Ir: SVol.2-45
$Br_3H_{18}La_7O_{18}$ ...... $La_7(OH)_{18}Br_3$ ...... Sc: MVol.C6-87/90
$Br_3H_{18}N_6U$ ...... $UBr_3 \cdot 6\ NH_3$ ...... U: SVol.E1-12, 13
$Br_3H_{18}Nd_7O_{18}$ ...... $Nd_7(OH)_{18}Br_3$ ...... Sc: MVol.C6-87/90
$Br_3H_{18}O_{18}Pr_7$ ...... $Pr_7(OH)_{18}Br_3$ ...... Sc: MVol.C6-87/90

$Br_3H_{18}O_{18}Sm_7$ ...... $Sm_7(OH)_{18}Br_3$ .... Sc: MVol.C6-87/90
$Br_3H_{22.5}N_{7.5}Sc$ ...... $ScBr_3 \cdot 7.5\ NH_3$ .... Sc: MVol.C6-34, 50
$Br_3H_{30}N_{10}Sm$ ...... $SmBr_3 \cdot 10\ NH_3$ .... Sc: MVol.C6-34
$Br_3H_{33}N_{11}Sm$ ...... $SmBr_3 \cdot 11\ NH_3$ .... Sc: MVol.C6-34
Br_3Ho .... $HoBr_3$
Chemical reactions .... Sc: MVol.C6-33/6
Formation and preparation .... Sc: MVol.C6-13/7, 27
Molecule .... Sc: MVol.C6-4/6
Physical properties .... Sc: MVol.C6-17/33
– .... $HoBr_3 \cdot 6\ H_2O$ .... Sc: MVol.C6-43/6
– .... $HoBr_3$ solutions
$HoBr_3$-H_2O .... Sc: MVol.C6-36/40
$HoBr_3$-KBr-H_2O .... Sc: MVol.C6-124/5
$HoBr_3$-NaBr-H_2O .... Sc: MVol.C6-122
$HoBr_3$-organic solvents .... Sc: MVol.C6-40/1
Br_3HoO_9 .... $Ho(BrO_3)_3$ .... Sc: MVol.C6-93/4
– .... $Ho(BrO_3)_3 \cdot 9\ H_2O$ .... Sc: MVol.C6-94/105
– .... $Ho(BrO_3)_3 \cdot n\ H_2O$ .... Sc: MVol.C6-94/7, 104
– .... $Ho(BrO_3)_3$ solutions
$Ho(BrO_3)_3$-H_2O .... Sc: MVol.C6-105/6
$Br_3IO_2U^{2-}$ .... $[UO_2Br_3I]^{2-}$ .... U: SVol.C9-185
Br_3IU .... UBr_3I .... U: SVol.C9-183/4
Br_3I_2Pa .... $PaBr_3I_2$ .... Pa: SVol.2-66, 71
Br_3In .... $InBr_3$ systems
$InBr_3$-$MnBr_2$ .... Mn: MVol.C5-306/7
$InBr_3$-$TeBr_4$ .... Te: SVol.B3-13
Br_3InSn .... $SnBr_2 \cdot InBr$ .... Sn: MVol.C4-117
Br_3Ir .... $IrBr_3$ .... Ir: SVol.2-141/2
Br_3KMn .... $KMnBr_3 = KBr \cdot MnBr_2$ .... Mn: MVol.C5-292/3
Br_3KSn .... $KSnBr_3$ .... Sn: MVol.C3-63/4
– .... $KSnBr_3 \cdot H_2O$ .... Sn: MVol.C3-64
– .... $KSnBr_3 \cdot 2\ H_2O$ .... Sn: MVol.C3-63/4
Br_3La .... $LaBr_3$
Chemical reactions .... Sc: MVol.C6-33/6, 57
Formation and preparation .... Sc: MVol.C6-13/7, 26/7, 54
Molecule .... Sc: MVol.C6-3/6
Physical properties .... Sc: MVol.C6-17/33, 54/6
– .... $LaBr_3$, doped with rare earth ions .... Sc: MVol.C6-55/6
– .... $LaBr_3 \cdot 7\ H_2O$ .... Sc: MVol.C6-43/6
– .... $LaBr_3$ melts
$LaBr_3$-KBr .... Sc: MVol.C6-115/6
– .... $LaBr_3$ solid solutions
$LaBr_3$-$PrBr_3$ .... Sc: MVol.C6-60
– .... $LaBr_3$ solutions
$LaBr_3$-H_2O .... Sc: MVol.C6-36/40
$LaBr_3$-NaBr-H_2O .... Sc: MVol.C6-118
$LaBr_3$-organic solvents .... Sc: MVol.C6-40/3, 52

Br_3La .... $LaBr_3$ systems
  $LaBr_3$-CsBr .... Sc: MVol.C6-114/6, 128/31
  $LaBr_3$-KBr-H_2O .... Sc: MVol.C6-124/5
  $LaBr_3$-La .... Sc: MVol.C6-7/8, 54
  $LaBr_3$-NaBr .... Sc: MVol.C6-114/6, 118
Br_3LaO_9 .... $La(BrO_3)_3$ .... Sc: MVol.C6-93/4
– .... $La(BrO_3)_3 \cdot 4\ H_2O$ .... Sc: MVol.C6-94/7, 103/4
– .... $La(BrO_3)_3 \cdot 9\ H_2O$ .... Sc: MVol.C6-94/105
– .... $La(BrO_3)_3 \cdot n\ H_2O$ .... Sc: MVol.C6-94/7, 104
– .... $La(BrO_3)_3$ solutions
  $La(BrO_3)_3$-H_2O .... Sc: MVol.C6-105/7
Br_3Lu .... $LuBr_3$
    Chemical reactions .... Sc: MVol.C6-33/6
    Formation and preparation .... Sc: MVol.C6-13/7, 27
    Molecule .... Sc: MVol.C6-3/6
    Physical properties .... Sc: MVol.C6-17/33
– .... $LuBr_3 \cdot 6\ H_2O$ .... Sc: MVol.C6-43/6
– .... $LuBr_3 \cdot 8\ H_2O$ .... Sc: MVol.C6-44
Br_3LuO_9 .... $Lu(BrO_3)_3$ .... Sc: MVol.C6-93/4
– .... $Lu(BrO_3)_3 \cdot 9\ H_2O$ .... Sc: MVol.C6-94/105
– .... $Lu(BrO_3)_3$ solutions
  $Lu(BrO_3)_3$-H_2O .... Sc: MVol.C6-105/6
Br_3Mn .... $MnBr_3 \cdot 3\ (C_2H_5)_2O$ .... Mn: MVol.C5-290
Br_3Mn^- .... $[MnBr_3]^-$ .... Mn: MVol.C5-287/90
– .... $[Mn(H_2O)_3Br_3]^-$ .... Mn: MVol.C5-285
Br_3MnRb .... $RbMnBr_3 = RbBr \cdot MnBr_2$ .... Mn: MVol.C5-293/5
Br_3MnTl .... $TlMnBr_3 = TlBr \cdot MnBr_2$ .... Mn: MVol.C5-307
Br_3NaSn .... $NaSnBr_3$ .... Sn: MVol.C3-25
Br_3Nd .... $NdBr_3$
    Chemical reactions .... Sc: MVol.C6-33/6, 61
    Formation and preparation .... Sc: MVol.C6-13/7, 26/8
    Molecule .... Sc: MVol.C6-3/6
    Physical properties .... Sc: MVol.C6-17/33, 61
– .... $NdBr_3 \cdot H_2O$ .... Sc: MVol.C6-45
– .... $NdBr_3 \cdot 6\ H_2O$ .... Sc: MVol.C6-43/6
– .... $NdBr_3$ melts
  $NdBr_3$-KBr .... Sc: MVol.C6-115/6
  $NdBr_3$-Nd .... Sc: MVol.C6-7/8
– .... $NdBr_3$ solutions
  $NdBr_3$-H_2O .... Sc: MVol.C6-36/40
  $NdBr_3$-KBr-H_2O .... Sc: MVol.C6-124/5
  $NdBr_3$-NaBr-H_2O .... Sc: MVol.C6-120
  $NdBr_3$-organic solvents .... Sc: MVol.C6-40/3, 61
– .... $NdBr_3$ systems
  $NdBr_3$-CsBr .... Sc: MVol.C6-114/6, 128/31
  $NdBr_3$-NaBr .... Sc: MVol.C6-114/6, 119/20
Br_3NdO_9 .... $Nd(BrO_3)_3$ .... Sc: MVol.C6-93/4

Br_3NdO_9 .... $Nd(BrO_3)_3 \cdot 4\ H_2O$ .... Sc: MVol.C6-94/7, 103/4
– .... $Nd(BrO_3)_3 \cdot 9\ H_2O$ .... Sc: MVol.C6-94/105, 108/9
– .... $Nd(BrO_3)_3 \cdot n\ H_2O$ .... Sc: MVol.C6-94/7, 104
– .... $Nd(BrO_3)_3$ solid solutions
$(Nd,Pr)(BrO_3)_3 \cdot 9\ H_2O$ .... Sc: MVol.C6-109/10
– .... $Nd(BrO_3)_3$ solutions
$Nd(BrO_3)_3$-H_2O .... Sc: MVol.C6-105/6
Br_3OP .... $POBr_3$ systems
$POBr_3$-$SnBr_4$ .... Sn: MVol.C2-267
$POBr_3$-$SnCl_4$ .... Sn: MVol.C2-268
Br_3OPa .... $PaOBr_3$ .... Pa: SVol.2-68/9
Br_3OU .... $UOBr_3$ .... U: SVol.C9-143/4
$Br_3O_2U^-$ .... $[UO_2Br_3]^-$ .... U: SVol.C9-151/4
Br_3O_9Pr .... $Pr(BrO_3)_3$ .... Sc: MVol.C6-93/4
– .... $Pr(BrO_3)_3 \cdot 4\ H_2O$ .... Sc: MVol.C6-94/7, 103/4
– .... $Pr(BrO_3)_3 \cdot 9\ H_2O$ .... Sc: MVol.C6-94/105, 107/8
– .... $Pr(BrO_3)_3 \cdot n\ H_2O$ .... Sc: MVol.C6-94/7, 104
– .... $Pr(BrO_3)_3$ solid solutions
$(Pr,Nd)(BrO_3)_3 \cdot 9\ H_2O$ .... Sc: MVol.C6-109/10
– .... $Pr(BrO_3)_3$ solutions
$Pr(BrO_3)_3$-H_2O .... Sc: MVol.C6-105/6
Br_3O_9Sc .... $Sc(BrO_3)_3 \cdot 3\ H_2O$ .... Sc: MVol.C6-94/7, 105
Br_3O_9Sm .... $Sm(BrO_3)_3$ .... Sc: MVol.C6-93/4
– .... $Sm(BrO_3)_3 \cdot 9\ H_2O$ .... Sc: MVol.C6-94/105, 110/1
– .... $Sm(BrO_3)_3 \cdot n\ H_2O$ .... Sc: MVol.C6-94/7, 104
– .... $Sm(BrO_3)_3$ solutions
$Sm(BrO_3)_3$-H_2O .... Sc: MVol.C6-105/6
Br_3O_9Tb .... $Tb(BrO_3)_3$ .... Sc: MVol.C6-93/4
– .... $Tb(BrO_3)_3 \cdot 9\ H_2O$ .... Sc: MVol.C6-94/105, 112
– .... $Tb(BrO_3)_3 \cdot n\ H_2O$ .... Sc: MVol.C6-94/7, 104
– .... $Tb(BrO_3)_3$ solutions
$Tb(BrO_3)_3$-H_2O .... Sc: MVol.C6-105/6
Br_3O_9Tm .... $Tm(BrO_3)_3 \cdot 9\ H_2O$ .... Sc: MVol.C6-94/104
– .... $Tm(BrO_3)_3 \cdot n\ H_2O$ .... Sc: MVol.C6-94/7, 104
Br_3O_9Y .... $Y(BrO_3)_3$ .... Sc: MVol.C6-93/4
– .... $Y(BrO_3)_3 \cdot 9\ H_2O$ .... Sc: MVol.C6-94/104
– .... $Y(BrO_3)_3 \cdot n\ H_2O$ .... Sc: MVol.C6-94/7, 104
Br_3O_9Yb .... $Yb(BrO_3)_3$ .... Sc: MVol.C6-93/4
– .... $Yb(BrO_3)_3 \cdot 9\ H_2O$ .... Sc: MVol.C6-94/105
– .... $Yb(BrO_3)_3$ solutions
$Yb(BrO_3)_3$-H_2O .... Sc: MVol.C6-105/6
Br_3P .... PBr_3 systems
PBr_3-$SnBr_4$ .... Sn: MVol.C2-267
Br_3Pm .... $PmBr_3$ .... Sc: MVol.C6-3/6, 17/33, 61/2
– .... $PmBr_3$ solutions
$PmBr_3$-H_2O .... Sc: MVol.C6-36/7

Br_3Pr .... $PrBr_3$
Chemical reactions .... Sc: MVol.C6-33/6, 60
Formation and preparation .... Sc: MVol.C6-13/7, 26/8
Molecule .... Sc: MVol.C6-3/6
Physical properties .... Sc: MVol.C6-17/33, 60
– .... $PrBr_3 \cdot H_2O$ .... Sc: MVol.C6-45
– .... $PrBr_3 \cdot 6\,H_2O$ .... Sc: MVol.C6-43/6
– .... $PrBr_3 \cdot 7\,H_2O$ .... Sc: MVol.C6-43
– .... $PrBr_3$ melts
$PrBr_3$-KBr .... Sc: MVol.C6-115/6
– .... $PrBr_3$ solid solutions
$PrBr_3$-$LaBr_3$ .... Sc: MVol.C6-60
– .... $PrBr_3$ solutions
$PrBr_3$-H_2O .... Sc: MVol.C6-36/40
$PrBr_3$-KBr-H_2O .... Sc: MVol.C6-124/5
$PrBr_3$-NaBr-H_2O .... Sc: MVol.C6-119
$PrBr_3$-organic solvents .... Sc: MVol.C6-40/3
– .... $PrBr_3$ systems
$PrBr_3$-CsBr .... Sc: MVol.C6-114/6, 128/31
$PrBr_3$-NaBr .... Sc: MVol.C6-114/6, 119
$PrBr_3$-Pr .... Sc: MVol.C6-7/8, 59
Br_3Pu .... $PuBr_3$
Vapor pressure .... Np: TrU.D2-451
Br_3RbSn .... $RbSnBr_3$ .... Sn: MVol.C3-136/7
Br_3S^+ .... SBr_3^+ .... S: SVol.2-305
Br_3Sb .... $SbBr_3$ systems
$SbBr_3$-$SnBr_4$ .... Sn: MVol.C2-287
$Br_3Sb_9Te_{12}$ .... $4\,Sb_2Te_3 \cdot SbBr_3 = Sb_3Te_4Br$ .... Te: SVol.B3-12
Br_3Sc .... $ScBr_3$
Chemical reactions .... Sc: MVol.C6-33/4, 50/1
Formation and preparation .... Sc: MVol.C6-26/7, 49
Molecule .... Sc: MVol.C6-3/6
Physical properties .... Sc: MVol.C6-17/33, 50
– .... $ScBr_3 \cdot 3\,H_2O$ .... Sc: MVol.C6-52/3
– .... $ScBr_3 \cdot 5\,H_2O$ .... Sc: MVol.C6-52/3
– .... $ScBr_3 \cdot 6\,H_2O$ .... Sc: MVol.C6-43/6, 52/3
– .... $ScBr_3$ solutions
$ScBr_3$-H_2O .... Sc: MVol.C6-36/7, 51/2
– .... $ScBr_3$ systems
$ScBr_3$-CsBr .... Sc: MVol.C6-114/5, 128/31
$ScBr_3$-KBr .... Sc: MVol.C6-114/5, 122/3
$ScBr_3$-LiBr .... Sc: MVol.C6-114/6
$ScBr_3$-NaBr .... Sc: MVol.C6-114/5, 117/8
$ScBr_3$-RbBr .... Sc: MVol.C6-114/5, 127/8
$ScBr_3$-Sc .... Sc: MVol.C6-7/8, 48
Br_3Sc_2 .... Sc_2Br_3 .... Sc: MVol.C6-48

Br_3Sm .... $SmBr_3$
Chemical reactions .... Sc: MVol.C6-33/6, 66
Formation and preparation .... Sc: MVol.C6-13/6, 26/7, 65/6
Molecule .... Sc: MVol.C6-3/6
Physical properties .... Sc: MVol.C6-17/33, 66
– .... $SmBr_3 \cdot H_2O$ .... Sc: MVol.C6-45
– .... $SmBr_3 \cdot 6\ H_2O$ .... Sc: MVol.C6-43/6
– .... $SmBr_3$ melts
$SmBr_3$-KBr .... Sc: MVol.C6-115/6
– .... $SmBr_3$ solutions
$SmBr_3$-H_2O .... Sc: MVol.C6-36/40
$SmBr_3$-KBr-H_2O .... Sc: MVol.C6-124/5
$SmBr_3$-NaBr-H_2O .... Sc: MVol.C6-120/1
$SmBr_3$-organic solvents .... Sc: MVol.C6-40/3
$SmBr_3$-RbBr-H_2O .... Sc: MVol.C6-128
– .... $SmBr_3$ systems
$SmBr_3$-CsBr .... Sc: MVol.C6-114/6, 128/31
$SmBr_3$-NaBr .... Sc: MVol.C6-114/6, 120/1
$SmBr_3$-Sm .... Sc: MVol.C6-7/8, 62/4
Br_3Tb .... $TbBr_3$
Chemical reactions .... Sc: MVol.C6-33/6
Formation and preparation .... Sc: MVol.C6-13/7, 27
Molecule .... Sc: MVol.C6-4/6
Physical properties .... Sc: MVol.C6-17/33
– .... $TbBr_3 \cdot 6\ H_2O$ .... Sc: MVol.C6-43/6
– .... $TbBr_3$ solutions
$TbBr_3$-H_2O .... Sc: MVol.C6-36/8, 73
Br_3Te^+ .... $TeBr_3^+$ .... Te: SVol.B3-15/6, 18/21
Br_3Tm .... $TmBr_3$ .... Sc: MVol.C6-4/6, 17/36
– .... $TmBr_3 \cdot 6\ H_2O$ .... Sc: MVol.C6-43/6
– .... $TmBr_3$ solutions
$TmBr_3$-H_2O .... Sc: MVol.C6-36/7
Br_3U .... UBr_3 .... U: SVol.C9-113/20
– .... $UBr_3 \cdot 6\ H_2O$ .... U: SVol.C9-118, 119
– .... UBr_3 systems
UBr_3-CsBr .... U: SVol.C9-120, 132/3
UBr_3-KBr .... U: SVol.C9-120, 132/3
UBr_3-NaBr .... U: SVol.C9-120, 132/3
UBr_3-RbBr .... U: SVol.C9-120, 132/3
UBr_3-$SrBr_2$ .... U: SVol.C9-120, 133
UBr_3-SrI_2 .... U: SVol.C9-120
UBr_3-UBr_4 .... U: SVol.C9-119/20
Br_3Y .... YBr_3 .... Sc: MVol.C6-3/6, 13/4, 17/33, 53/4
– .... $YBr_3 \cdot 6\ H_2O$ .... Sc: MVol.C6-43/6
– .... YBr_3 solutions
YBr_3-H_2O .... Sc: MVol.C6-36/8

Br_3Yb	$YbBr_3$		
	Chemical reactions	Sc:	MVol.C6-33/6, 76
	Formation and preparation	Sc:	MVol.C6-13/7, 27, 76
	Molecule	Sc:	MVol.C6-4/6
	Physical properties	Sc:	MVol.C6-17/33
–	$YbBr_3 \cdot 6\ H_2O$	Sc:	MVol.C6-43/6
–	$YbBr_3$ solutions		
	$YbBr_3$-H_2O	Sc:	MVol.C6-39
	$YbBr_3$-KBr-H_2O	Sc:	MVol.C6-124/5
	$YbBr_3$-NaBr-H_2O	Sc:	MVol.C6-122
	$YbBr_3$-organic solvents	Sc:	MVol.C6-40/1
$Br_{3.5}Ir$	$IrBr_{3.5}$	Ir:	SVol.2-142
Br_4CF_3P	CF_3PBr_4	F:	PerFHalOrg.3-74, 76, 92, 99
Br_4CS	Br_3CSBr	C:	MVol.D4-27
Br_4CS_2	CBr_3SSBr	C:	MVol.D4-183
$Br_4C_2CsH_8IrN_2$	$Cs[Ir(C_2H_4(NH_2)_2)Br_4]$	Ir:	SVol.2-85
$Br_4C_2F_5P$	$CF_3P(CF_2Br)Br_3$	F:	PerFHalOrg.3-13
$Br_4C_2H_4IrKN_2$	$K[Ir(HNCHCHNH)Br_4]$	Ir:	SVol.2-86
$Br_4C_2H_7O_4PSn$	$SnBr_4 \cdot (CH_3O)_2P(O)OH$	Sn:	MVol.C6-168/9
$Br_4C_2H_8IrKN_2$	$K[Ir(C_2H_4(NH_2)_2)Br_4] \cdot 2\ H_2O$	Ir:	SVol.2-85
$Br_4C_2H_8N_4O_2Sn$	$SnBr_4 \cdot 2\ CO(NH_2)_2$	Sn:	MVol.C6-55
$Br_4C_2H_8N_4O_2U$	$UBr_4 \cdot 2\ (NH_2)_2CO$	U:	SVol.E1-82/3
–	$UBr_4 \cdot 2\ (NH_2)_2CO \cdot 9\ H_2O$	U:	SVol.E1-82/3
$Br_4C_2H_8N_4S_2Sn$	$SnBr_4 \cdot 2\ CS(NH_2)_2$	Sn:	MVol.C6-120
$Br_4C_2H_8N_4S_2Te$	$Te(SC(NH_2)_2)_2Br_4$	Te:	SVol.B3-179
$Br_4C_2H_{12}MnN_2$	$(CH_3NH_3)_2MnBr_4$	Mn:	MVol.C5-300/1
$Br_4C_2S_2$	$(Br_2CS)_2$	C:	MVol.D6-169
$Br_4C_2Se_2$	$(Br_2CSe)_2$	C:	MVol.D6-251
$Br_4C_3F_7P$	$CF_3P(CF_2Br)_2Br_2$	F:	PerFHalOrg.3-13
$Br_4C_3H_6N_2O_2Sn$	$SnBr_4 \cdot CH_3CONHCONH_2$	Sn:	MVol.C6-58/9
$Br_4C_3H_9NSn$	$SnBr_4 \cdot (CH_3)_3N$	Sn:	MVol.C5-172/3
$Br_4C_3H_9PSn$	$SnBr_4 \cdot P(CH_3)_3$	Sn:	MVol.C6-132
–	$SnBr_4 \cdot P(CD_3)_3$	Sn:	MVol.C6-132
$Br_4C_3H_{12}N_6O_3U$	$UBr_4 \cdot 3\ (NH_2)_2CO$	U:	SVol.E1-82/3
–	$UBr_4 \cdot 3\ (NH_2)_2CO \cdot 8\ H_2O$	U:	SVol.E1-82/3
$Br_4C_4Cl_2H_{12}O_2P_2Sn$	$SnBr_4 \cdot 2\ (CH_3)_2P(O)Cl$	Sn:	MVol.C6-156
$Br_4C_4FeO_4Sn$	$(CO)_4Fe(SnBr_3)Br$	Fe:	Org.Verb.B2-139, 146
$Br_4C_4H_4N_2Sn$	$SnBr_4 \cdot N(CHCH)_2N$	Sn:	MVol.C5-225/6
$Br_4C_4H_4N_2Te$	$TeBr_4 \cdot C_4H_4N_2$	Te:	SVol.B3-172
$Br_4C_4H_4O_4Sn^{2-}$	$[SnBr_4(CH_2COO)_2]^{2-}$	Sn:	MVol.C6-11
$Br_4C_4H_5NO_2Sn$	$SnBr_4 \cdot C_2H_4(CO)_2NH$	Sn:	MVol.C6-53
$Br_4C_4H_6N_2Sn$	$SnBr_4 \cdot 2\ CH_3CN$	Sn:	MVol.C6-68
$Br_4C_4H_6N_2Te$	$[(CH_3CN)_2TeBr_3]Br$	Te:	SVol.B3-17
$Br_4C_4H_6N_4O_2Sn$	$SnBr_4 \cdot C_4H_6N_4O_2$	Sn:	MVol.C5-225
$Br_4C_4H_7NO_2Sn$	$SnBr_4 \cdot CH_3CONHCOCH_3$	Sn:	MVol.C6-50/1
$Br_4C_4H_8N_2O_4Sn^{2-}$	$[SnBr_4(NH_2CH_2COO)_2]^{2-}$	Sn:	MVol.C6-10
$Br_4C_4H_8OSn$	$[SnBr_4(CH_3COC_2H_5)]$	Sn:	MVol.C5-87
$Br_4C_4H_8OTe$	$TeBr_4 \cdot C_4H_8O$	Te:	SVol.B3-166

$Br_4C_4H_8O_2Sn$	$SnBr_4 \cdot CH_3COOC_2H_5$	Sn: MVol.C5-130
–	$SnBr_4 \cdot O(CH_2CH_2)_2O$	Sn: MVol.C5-159
$Br_4C_4H_8O_2Te$	$TeBr_4 \cdot O(CH_2CH_2)_2O$	Te: SVol.B3-166
$Br_4C_4H_8SSn$	$SnBr_4 \cdot (CH_2)_4S$	Sn: MVol.C6-109/10
$Br_4C_4H_8Sn$	$Sn(CH_2Br)_4$	Sn: Org.Verb.1-122
$Br_4C_4H_{10}N_2O_2Sn$	$SnBr_4 \cdot 2\ CH_3CONH_2$	Sn: MVol.C6-43
$Br_4C_4H_{10}N_2O_4Sn$	$SnBr_4 \cdot 2\ NH_2CH_2COOH$	Sn: MVol.C6-11
$Br_4C_4H_{10}OU$	$UBr_4 \cdot (C_2H_5)_2O$	U: SVol.E1-70, 74/5
$Br_4C_4H_{10}O_2Pa^-$	$[Pa(OC_2H_5)_2Br_4]^-$	Pa: SVol.2-82
$Br_4C_4H_{10}O_2Sn$	$SnBr_4 \cdot C_2H_4(OCH_3)_2$	Sn: MVol.C5-154
$Br_4C_4H_{10}S_2Sn$	$SnBr_4 \cdot C_2H_4(SCH_3)_2$	Sn: MVol.C6-96/7
$Br_4C_4H_{11}O_4PSn$	$SnBr_4 \cdot (C_2H_5O)_2P(O)OH$	Sn: MVol.C6-168/9
$Br_4C_4H_{12}MnN^-$	$[(CH_3)_4NMnBr_4]^-$	Mn: MVol.C5-288
$Br_4C_4H_{12}N_2O_2Sn$	$SnBr_4 \cdot (CH_3)_3N \cdot CH_3NO_2$	Sn: MVol.C5-172/3
$Br_4C_4H_{12}N_4O_2Sn$	$SnBr_4 \cdot 2\ CH_3CONHNH_2$	Sn: MVol.C6-59/60
$Br_4C_4H_{12}N_4Sn$	$SnBr_4 \cdot 2\ CH_3NNCH_3$	Sn: MVol.C6-31
$Br_4C_4H_{12}O_2S_2Sn$	$SnBr_4 \cdot 2\ (CH_3)_2SO$	Sn: MVol.C6-100/1
$Br_4C_4H_{12}O_2S_2Te$	$TeBr_4 \cdot 2\ (CH_3)_2SO$	Te: SVol.B3-176
$Br_4C_4H_{12}O_2Se_2Sn$	$SnBr_4 \cdot 2\ (CH_3)_2SeO$	Sn: MVol.C6-123
$Br_4C_4H_{12}O_2Te$	$TeBr_4 \cdot 2\ C_2H_5OH$	Te: SVol.B3-165
$Br_4C_4H_{12}O_4Sn$	$SnBr_4 \cdot 2\ C_2H_4(OH)_2$	Sn: MVol.C5-76
$Br_4C_4H_{12}P_2S_2Sn$	$SnBr_4 \cdot (CH_3)_2PSPS(CH_3)_2$	Sn: MVol.C6-148
$Br_4C_4H_{12}S_2Sn$	$SnBr_4 \cdot 2\ CH_3SCH_3$	Sn: MVol.C6-92
$Br_4C_4H_{14}N_2Sn$	$SnBr_4 \cdot 2\ (CH_3)_2NH$	Sn: MVol.C5-170
$Br_4C_4H_{14}O_4P_2Sn$	$SnBr_4 \cdot 2\ (CH_3)_2P(O)OH$	Sn: MVol.C6-150/1
$Br_4C_4H_{14}O_6P_2Sn$	$SnBr_4 \cdot 2\ (CH_3)(CH_3O)P(O)OH$	Sn: MVol.C6-161
–	$SnBr_4 \cdot 2\ HP(O)(OCH_3)_2$	Sn: MVol.C6-160/1
$Br_4C_4H_{14}O_8P_2Sn$	$SnBr_4 \cdot 2\ (CH_3O)_2P(O)OH$	Sn: MVol.C6-168/9
$Br_4C_4H_{16}MnN_2$	$[(CH_3)_2NH_2]_2MnBr_4$	Mn: MVol.C5-301
–	$(C_2H_5NH_3)_2MnBr_4$	Mn: MVol.C5-300/1
$Br_4C_4H_{16}N_4Sn$	$SnBr_4 \cdot 2\ C_2H_8N_2$	Sn: MVol.C5-184
$Br_4C_4H_{16}O_4Sn$	$SnBr_4 \cdot 4\ CH_3OH$	Sn: MVol.C5-69
$Br_4C_4H_{20}N_4Sn$	$SnBr_4 \cdot 4\ CH_3NH_2$	Sn: MVol.C5-168
$Br_4C_4Hg_2N_4S_4Sn$	$SnBr_4 \cdot 2\ Hg(SCN)_2$	Sn: MVol.C4-97/8
$Br_4C_4Hg_2N_4Sn$	$SnBr_4 \cdot 2\ Hg(CN)_2$	Sn: MVol.C4-97/8
$Br_4C_4N_2$	$N(CBrCBr)_2N$	F: PerFHalOrg.6-57
Br_4C_5FN	FC_5Br_4N	F: PerFHalOrg.5-118/9, 128/9
$Br_4C_5H_5NTe$	$TeBr_4 \cdot C_5H_5N$	Te: SVol.B3-169
$Br_4C_5H_8O_4Sn$	$SnBr_4 \cdot CH_2(COOCH_3)_2$	Sn: MVol.C5-140/1
$Br_4C_5H_9NO_2Sn$	$SnBr_4 \cdot CH_3CONHCOC_2H_5$	Sn: MVol.C6-50/1
$Br_4C_5H_{10}O_3Sn$	$SnBr_4 \cdot CO(OC_2H_5)_2$	Sn: MVol.C5-139
$Br_4C_5H_{12}Se_2Te$	$TeBr_4 \cdot CH_3Se(CH_2)_3SeCH_3$	Te: SVol.B3-183
$Br_4C_5H_{20}N_{10}S_5Sn_2$	$2\ SnBr_2 \cdot 5\ CS(NH_2)_2$	Sn: MVol.C5-55
$Br_4C_{5.4}H_{18.9}N_{2.7}Sn$	$SnBr_4 \cdot 2.7\ (CH_3)_2NH$	Sn: MVol.C5-170
$Br_4C_6ClH_5N_2O_2Sn$	$SnBr_4 \cdot NO_2C_6H_3(Cl)NH_2$	Sn: MVol.C5-177
$Br_4C_6F_6MoO_4P_2$	$Mo(CO)_4(CF_3PBr_2)_2$	F: PerFHalOrg.3-110/1, 120
$Br_4C_6FeO_4$	$Br_2CCBr_2Fe(CO)_4$	Fe: Org.Verb.B4-321, 325
$Br_4C_6Fe_2O_6$	$Br(CO)_3FeBr_2Fe(CO)_3Br$	Fe: Org.Verb.C1-5/6
$Br_4C_6H_6N_2O_2Sn$	$SnBr_4 \cdot NO_2C_6H_4NH_2$	Sn: MVol.C5-177

$Br_4C_6H_8N_2Sn$	$SnBr_4 \cdot C_6H_4(NH_2)_2$	Sn: MVol.C5-186/7
$Br_4C_6H_8N_2Te$	$TeBr_4 \cdot C_6H_4(NH_2)_2$	Te: SVol.B3-169
$Br_4C_6H_8N_4O_2Sn$	$SnBr_4 \cdot 2\ NCCH_2CONH_2$	Sn: MVol.C6-75
$Br_4C_6H_8O_4Sn$	$SnBr_4 \cdot C_2H_2(COOCH_3)_2$	Sn: MVol.C5-143/4
$Br_4C_6H_{10}Ni_3$	$2\ C_3H_5NiBr \cdot NiBr_2$	Ni: Org.Verb.2-277
$Br_4C_6H_{10}O_4Sn$	$SnBr_4 \cdot C_2H_4(COOCH_3)_2$	Sn: MVol.C5-141/3
$Br_4C_6H_{11}NO_2Sn$	$SnBr_4 \cdot CH_3CONHCOC_3H_7$	Sn: MVol.C6-50/1
$Br_4C_6H_{12}N_2Na_2O_2Sn$.	$SnBr_4 \cdot 2\ NaCON(CH_3)_2$	Sn: MVol.C6-42
$Br_4C_6H_{12}N_4O_2Sn$	$SnBr_4 \cdot 2\ CO(NHCH_2)_2$	Sn: MVol.C5-224/5
$Br_4C_6H_{12}N_4S_2Sn$	$SnBr_4 \cdot 2\ CH_2NHC(SH)NCH_2$	Sn: MVol.C6-104/5
$Br_4C_6H_{12}N_4Sn$	$SnBr_4 \cdot (CH_2)_6N_4 \cdot CHCl_3$	Sn: MVol.C5-229
$Br_4C_6H_{12}O_2Sn$	$SnBr_4 \cdot 2\ CH_3COCH_3$	Sn: MVol.C5-86/7
$Br_4C_6H_{12}O_2Te$	$[TeBr_3((CH_3)_2CO)_2]Br$	Te: SVol.B3-165/6
$Br_4C_6H_{12}O_4Sn$	$SnBr_4 \cdot 2\ HCOOC_2H_5$	Sn: MVol.C5-127
$Br_4C_6H_{13}O_3U$	$U(OC_2H_5)Br_4 \cdot CH_3COOC_2H_5$	U: SVol.E1-120/2
$Br_4C_6H_{14}N_2O_2Sn$	$SnBr_4 \cdot 2\ OCHN(CH_3)_2$	Sn: MVol.C6-41/2
$Br_4C_6H_{14}N_2O_2Te$	$TeBr_4 \cdot 2\ HCON(CH_3)_2$	Te: SVol.B3-174
$Br_4C_6H_{14}N_2O_4Sn$	$SnBr_4 \cdot 2\ CH_3CH(NH_2)COOH$	Sn: MVol.C6-12
–	$SnBr_4 \cdot 2\ NH_2CH_2CH_2COOH$	Sn: MVol.C6-12
$Br_4C_6H_{16}N_4O_2Sn$	$SnBr_4 \cdot 2\ (CH_3)_2NCONH_2$	Sn: MVol.C6-56
–	$SnBr_4 \cdot 2\ (CH_3NH)_2CO$	Sn: MVol.C6-56
$Br_4C_6H_{18}N_2Sn$	$SnBr_4 \cdot 2\ (CH_3)_3N$	Sn: MVol.C5-172/3
$Br_4C_6H_{18}O_2P_2S_2Sn$ ..	$SnBr_4 \cdot 2\ (CH_3)_2P(O)SCH_3$	Sn: MVol.C6-158/60
–	$SnBr_4 \cdot 2\ (CH_3)_2P(S)OCH_3$	Sn: MVol.C6-158/60
$Br_4C_6H_{18}O_2P_2Sn$	$SnBr_4 \cdot 2\ (CH_3)_3PO$	Sn: MVol.C6-143/4
$Br_4C_6H_{18}O_2P_2U$	$UBr_4 \cdot 2\ (CH_3)_3PO$	U: SVol.E1-148, 156
$Br_4C_6H_{18}O_3S_3Te$	$TeBr_4 \cdot 3\ (CH_3)_2SO$	Te: SVol.B3-176
$Br_4C_6H_{18}O_4P_2S_4Sn$ ..	$SnBr_4 \cdot 2\ (CH_3O)(CH_3S)_2PO$	Sn: MVol.C6-176
$Br_4C_6H_{18}O_4P_2Sn$	$SnBr_4 \cdot 2\ (CH_3)_2P(O)OCH_3$	Sn: MVol.C6-153/4
–	$SnBr_4 \cdot 2\ C_2H_5(H)P(O)OCH_3$	Sn: MVol.C6-151/2
$Br_4C_6H_{18}O_6P_2S_2Sn$	$SnBr_4 \cdot 2\ (CH_3O)_2(CH_3S)PO$	Sn: MVol.C6-174
$Br_4C_6H_{18}O_6P_2Sn$	$SnBr_4 \cdot 2\ CH_3P(O)(OCH_3)_2$	Sn: MVol.C6-162/3
–	$SnBr_4 \cdot 2\ P(OCH_3)_3$	Sn: MVol.C6-141/2
$Br_4C_6H_{18}O_8P_2Sn$	$SnBr_4 \cdot 2\ (CH_3O)_3PO$	Sn: MVol.C6-169
$Br_4C_6H_{18}P_2S_2Sn$	$SnBr_4 \cdot 2\ (CH_3)_3PS$	Sn: MVol.C6-148
$Br_4C_6H_{20}MnN_2$	$(C_3H_7NH_3)_2MnBr_4$	Mn: MVol.C5-300/1
$Br_4C_6H_{24}N_6Sn$	$SnBr_4 \cdot 3\ C_2H_8N_2$	Sn: MVol.C5-183
$Br_4C_6H_{24}N_{12}O_6U$	$UBr_4 \cdot 6\ (NH_2)_2CO$	U: SVol.E1-82/3
–	$UBr_4 \cdot 6\ (NH_2)_2CO \cdot 2\ H_2O$	U: SVol.E1-82/3
$Br_4C_6MnO_2$	$[Mn(BrC_6Br_3O_2)]$	Mn: MVol.D1-45
$Br_4C_7Cl_3H_{13}N_4Sn$ ...	$SnBr_4 \cdot (CH_2)_6N_4 \cdot CHCl_3$	Sn: MVol.C5-229
$Br_4C_7H_5NSn$	$SnBr_4 \cdot C_6H_5CN$	Sn: MVol.C6-70
$Br_4C_7H_8N_2O_2Sn$	$SnBr_4 \cdot NO_2C_6H_3(CH_3)NH_2$	Sn: MVol.C5-178
$Br_4C_7H_{10}N_2Te$	$TeBr_4 \cdot CH_3C_6H_3(NH_2)_2$	Te: SVol.B3-169
$Br_4C_7H_{12}O_4Sn$	$SnBr_4 \cdot CH_2(COOC_2H_5)_2$	Sn: MVol.C5-140/1
$Br_4C_7H_{13}NO_2Sn$	$SnBr_4 \cdot CH_3CONHCOC_4H_9$	Sn: MVol.C6-50/1
$Br_4C_7H_{14}O_2Sn$	$SnBr_4 \cdot (CH_3)_3CCOOC_2H_5$	Sn: MVol.C5-136
–	$SnBr_4 \cdot C_4H_9COOC_2H_5$	Sn: MVol.C5-136/7
$Br_4C_{7.5}H_{17.5}N_{2.5}O_{2.5}Te$	$TeBr_4 \cdot 2.5\ HCON(CH_3)_2$	Te: SVol.B3-174
$Br_4C_8ClH_{20}NTe$	$[(C_2H_5)_4N][TeBr_4Cl]$	Te: SVol.B3-86

$Br_4C_8Cl_2H_{24}N_2Sn$	$[(CH_3)_4N]_2[SnBr_4Cl_2]$	Sn:	MVol.C3-119
$Br_4C_8F_3HN_2$	$C_6Br_4NC(CF_3)NH$	F:	PerFHalOrg.6-125/6
$Br_4C_8F_3HN_2O$	$C_6Br_4NC(CF_3)N(OH)$	F:	PerFHalOrg.6-125/6, 161/2
$Br_4C_8Fe_2Ge_2O_8$	$[(CO)_4FeGeBr_2]_2$	Fe:	Org.Verb.C1-42, 45
$Br_4C_8Fe_2O_8Sn_2$	$[(CO)_4FeSnBr_2]_2$	Fe:	Org.Verb.C1-42, 46
$Br_4C_8H_5NO_2Sn$	$SnBr_4 \cdot C_6H_4(CO,CO)NH$	Sn:	MVol.C6-54
–	$SnBr_4 \cdot C_6H_4(CO,NH)CO$	Sn:	MVol.C5-192
$Br_4C_8H_6O_2Sn$	$SnBr_4 \cdot C_6H_4(CHO)_2$	Sn:	MVol.C5-83/4
$Br_4C_8H_7NSn$	$SnBr_4 \cdot CH_3C_6H_4CN$	Sn:	MVol.C6-71
$Br_4C_8H_8IrNaN_4$	$Na[Ir(C_4H_4N_2)_2Br_4] \cdot 0.5\ H_2O$	Ir:	SVol.2-78
$Br_4C_8H_8N_2O_2Sn$	$SnBr_4 \cdot C_6H_5CONHCONH_2$	Sn:	MVol.C6-58/9
$Br_4C_8H_8N_4Sn$	$SnBr_4 \cdot 2\ N(CHCH)_2N$	Sn:	MVol.C5-225/6
$Br_4C_8H_{11}NTe$	$TeBr_4 \cdot C_6H_5N(CH_3)_2$	Te:	SVol.B3-168
$Br_4C_8H_{12}N_4Pa$	$PaBr_4 \cdot 4\ CH_3CN$	Pa:	SVol.2-88/9
$Br_4C_8H_{12}N_4U$	$[UBr_4(CH_3CN)_4]$	U:	SVol.E1-48/50
$Br_4C_8H_{14}O_4Sn$	$SnBr_4 \cdot C_2H_4(COOC_2H_5)_2$	Sn:	MVol.C5-141/3
–	$SnBr_4 \cdot C_4H_8(COOCH_3)_2$	Sn:	MVol.C5-141/3
$Br_4C_8H_{15}NO_2Sn$	$SnBr_4 \cdot CH_3CONHCOC_5H_{11}$	Sn:	MVol.C6-50/1
$Br_4C_8H_{16}N_4O_2Sn$	$SnBr_4 \cdot 2\ CH_2CHCH_2NHCONH_2$	Sn:	MVol.C6-57
$Br_4C_8H_{16}N_4S_2Sn$	$SnBr_4 \cdot 2\ NH_2CSNHCH_2CHCH_2$	Sn:	MVol.C6-121
$Br_4C_8H_{16}O_2S_2Sn$	$SnBr_4 \cdot 2\ S(CH_2CH_2)_2O$	Sn:	MVol.C6-113/4
$Br_4C_8H_{16}O_2Se_2Sn$	$SnBr_4 \cdot 2\ Se(CH_2CH_2)_2O$	Sn:	MVol.C6-124
$Br_4C_8H_{16}O_2Sn$	$SnBr_4 \cdot 2\ (CH_2)_4O$	Sn:	MVol.C5-157
–	$[SnBr_4(CH_3COC_2H_5)_2]$	Sn:	MVol.C5-87
$Br_4C_8H_{16}O_2U$	$UBr_4 \cdot 2\ (CH_2)_4O$	U:	SVol.E1-70, 74/5
$Br_4C_8H_{16}O_4Sn$	$SnBr_4 \cdot 2\ CH_3COOC_2H_5$	Sn:	MVol.C5-130
–	$SnBr_4 \cdot 2\ O(CH_2CH_2)_2O$	Sn:	MVol.C5-159
$Br_4C_8H_{16}O_4U$	$UBr_4 \cdot 2\ CH_3COOC_2H_5$	U:	SVol.E1-120/2
–	$UBr_4 \cdot 2\ O(CH_2CH_2)_2O$	U:	SVol.E1-70, 74/5
$Br_4C_8H_{16}S_2Sn$	$SnBr_4 \cdot 2\ (CH_2)_4S$	Sn:	MVol.C6-109/10
$Br_4C_8H_{18}N_2O_2Te$	$[TeBr_3(CH_3CON(CH_3)_2)_2]Br$	Te:	SVol.B3-174
$Br_4C_8H_{18}N_2O_8Sn$	$SnBr_4 \cdot 2\ NH_2CH_2COOH \cdot 2\ CH_3COOH$	Sn:	MVol.C6-11
$Br_4C_8H_{20}MnN^-$	$[(C_2H_5)_4NMnBr_4]^-$	Mn:	MVol.C5-288
$Br_4C_8H_{20}N_4O_8Sn$	$SnBr_4 \cdot 4\ NH_2CH_2COOH$	Sn:	MVol.C6-10
$Br_4C_8H_{20}N_4Te$	$TeBr_4 \cdot 2\ NH(CH_2CH_2)_2NH$	Te:	SVol.B3-172
$Br_4C_8H_{20}O_2Sn$	$SnBr_4 \cdot 2\ C_2H_5OC_2H_5$	Sn:	MVol.C5-150/1
$Br_4C_8H_{20}P_2S_2Sn$	$SnBr_4 \cdot (C_2H_5)_2PSPS(C_2H_5)_2$	Sn:	MVol.C6-148
$Br_4C_8H_{20}S_2Sn$	$SnBr_4 \cdot 2\ C_2H_5SC_2H_5$	Sn:	MVol.C6-92/3
$Br_4C_8H_{22}N_2U$	$UBr_4 \cdot 2\ (C_2H_5)_2NH$	U:	SVol.E1-19, 22
$Br_4C_8H_{22}O_2P_2S_2Sn$	$SnBr_4 \cdot 2\ (C_2H_5)_2POSH$	Sn:	MVol.C6-156/7
$Br_4C_8H_{22}O_4P_2Sn$	$SnBr_4 \cdot 2\ C_2H_5(H)P(O)OC_2H_5$	Sn:	MVol.C6-151/2
$Br_4C_8H_{22}O_6P_2Sn$	$SnBr_4 \cdot 2\ HP(O)(OC_2H_5)_2$	Sn:	MVol.C6-160/1
$Br_4C_8H_{22}O_8P_2Sn$	$SnBr_4 \cdot 2\ (C_2H_5O)_2P(O)OH$	Sn:	MVol.C6-168/9
$Br_4C_8H_{24}MnN_2$	$[(CH_3)_4N]_2MnBr_4$	Mn:	MVol.C5-288/9, 302
$Br_4C_8H_{24}N_2O_2U$	$[N(CH_3)_4]_2UO_2Br_4$	U:	SVol.C9-151/4
$Br_4C_8H_{24}N_4Sn$	$SnBr_4 \cdot 2\ NH_2(CH_2)_4NH_2$	Sn:	MVol.C5-185
$Br_4C_8H_{28}N_4Sn$	$SnBr_4 \cdot 4\ (CH_3)_2NH$	Sn:	MVol.C5-170
$Br_4C_8H_{32}N_8Sn$	$SnBr_4 \cdot 4\ C_2H_8N_2$	Sn:	MVol.C5-183
$Br_4C_8H_{32}N_{16}O_8U$	$UBr_4 \cdot 8\ (NH_2)_2CO$	U:	SVol.E1-82/3

$Br_4C_9FeH_{18}Hg_2O_9P_2$	$(CO)_3Fe(P(OCH_3)_3)_2 \cdot (HgBr_2)_2$	Fe: Org.Verb.B1-192/4
$Br_4C_9H_9NO_2Sn$	$SnBr_4 \cdot C_6H_5CONHCOCH_3$	Sn: MVol.C6-50/1
$Br_4C_9H_{17}NO_2Sn$	$SnBr_4 \cdot CH_3CONHCOC_6H_{13}$	Sn: MVol.C6-50/1
$Br_4C_9H_{18}O_6Sn$	$SnBr_4 \cdot 3\ HCOOC_2H_5$	Sn: MVol.C5-127
$Br_4C_9H_{19}O_4PSn$	$SnBr_4 \cdot CH_3COCH_2P(O)(OC_3H_7)_2$	Sn: MVol.C6-165/6
$Br_4C_{10}Cl_2H_{10}Ti$	$C_5H_5Ti(C_5H_5Br_4)Cl_2$	Ti: Org.Verb.1-206
$Br_4C_{10}F_2H_{10}Ti$	$C_5H_5Ti(C_5H_5Br_4)F_2$	Ti: Org.Verb.1-206
$Br_4C_{10}FeH_{10}$	$C_5H_5FeC_5H_5Br_4$	Fe: Org.Verb.A1-183
$Br_4C_{10}Fe_2H_{10}$	$[Fe(C_5H_5)_2]FeBr_4$	Fe: Org.Verb.A1-222/3
$Br_4C_{10}H_8N_2Sn$	$SnBr_4 \cdot NC_5H_4C_5H_4N$	Sn: MVol.C5-217/9
$Br_4C_{10}H_8N_2Te$	$TeBr_4 \cdot NC_5H_4C_5H_4N$	Te: SVol.B3-171
$Br_4C_{10}H_8O_4Sn$	$SnBr_4 \cdot 2\ C_4H_3OCHO$	Sn: MVol.C5-84
$Br_4C_{10}H_{10}IrN_2^-$	$[Ir(C_5H_5N)_2Br_4]^-$	Ir: SVol.2-71
$Br_4C_{10}H_{10}N_2O_2Sn$	$SnBr_4 \cdot 2\ C_5H_5NO$	Sn: MVol.C5-206/7
$Br_4C_{10}H_{10}N_2Sn$	$SnBr_4 \cdot 2\ C_5H_5N$	Sn: MVol.C5-198/9
$Br_4C_{10}H_{10}N_2Te$	$TeBr_4 \cdot 2\ C_5H_5N$	Te: SVol.B3-170
$Br_4C_{10}H_{10}N_2U$	$UBr_4 \cdot 2\ C_5H_5N$	U: SVol.E1-28, 33
$Br_4C_{10}H_{10}Ni$	$C_5H_5NiC_5H_5Br_4$	Ni: Org.Verb.2-205
$Br_4C_{10}H_{11}NO_2Sn$	$SnBr_4 \cdot C_6H_5CONHCOC_2H_5$	Sn: MVol.C6-50/1
$Br_4C_{10}H_{12}I_2N_2Te$	$[C_5H_5NH]_2[TeBr_4I_2]$	Te: SVol.B3-87
$Br_4C_{10}H_{12}MnN_2$	$(C_5H_5NH)_2MnBr_4$	Mn: MVol.C5-289, 303/4
$Br_4C_{10}H_{12}N_2O_2U$	$[C_5H_5NH]_2UO_2Br_4$	U: SVol.C9-151, 154
$Br_4C_{10}H_{12}N_4Sn$	$SnBr_4 \cdot 2\ NC_5H_4NH_2$	Sn: MVol.C5-203
$Br_4C_{10}H_{13}NO_2Sn$	$C_5H_5NH[SnBr_4(CH(COCH_3)_2)]$	Sn: MVol.C5-107
$Br_4C_{10}H_{13}N_3O_4Sn$	$SnBr_4 \cdot (NO_2)_2C_6H_3N(C_2H_5)_2$	Sn: MVol.C5-180
$Br_4C_{10}H_{14}Ti$	$(CH_3)_4(CH_2Br)C_5TiBr_3$	Ti: Org.Verb.1-141, 151/2
$Br_4C_{10}H_{20}O_2Sn$	$SnBr_4 \cdot 2\ C_5H_{10}O$	Sn: MVol.C5-158
$Br_4C_{10}H_{20}S_2Sn$	$SnBr_4 \cdot 2\ (CH_2)_5S$	Sn: MVol.C6-113
$Br_4C_{10}H_{22}N_2O_8Sn$	$SnBr_4 \cdot 2\ CH_3CH(NH_2)COOH \cdot 2\ CH_3COOH$	Sn: MVol.C6-12
–	$SnBr_4 \cdot 2\ NH_2C_2H_4COOH \cdot 2\ CH_3COOH$	Sn: MVol.C6-12
$Br_4C_{10}H_{22}S_2Sn$	$SnBr_4 \cdot C_4H_9S(CH_2)_2SC_4H_9$	Sn: MVol.C6-97/8
$Br_4C_{10}H_{22.5}N_{2.5}O_{2.5}Pa$	$PaBr_4 \cdot 2.5\ CH_3CON(CH_3)_2$	Pa: SVol.2-89/90
$Br_4C_{10}H_{22.5}N_{2.5}O_{2.5}Te$	$TeBr_4 \cdot 2.5\ CH_3CON(CH_3)_2$	Te: SVol.B3-174
$Br_4C_{10}H_{22.5}N_{2.5}O_{2.5}U$	$UBr_4 \cdot 2.5\ CH_3CON(CH_3)_2$	U: SVol.E1-103/5
$Br_4C_{10}H_{24}N_4O_2Sn$	$SnBr_4 \cdot 2\ (CH_3)_2NCON(CH_3)_2$	Sn: MVol.C6-57
$Br_4C_{10}H_{24}N_4O_2U$	$UBr_4 \cdot 2\ (CH_3)_2NCON(CH_3)_2$	U: SVol.E1-91/2, 94
$Br_4C_{10}H_{24}N_4S_2Sn$	$SnBr_4 \cdot 2\ (CH_3)_2NCSN(CH_3)_2$	Sn: MVol.C6-121
$Br_4C_{10}H_{24}N_4S_2Te$	$Te(SC(N(CH_3)_2)_2)_2Br_4$	Te: SVol.B3-181/2
$Br_4C_{10}H_{26}O_2P_2S_2Sn$	$SnBr_4 \cdot 2\ (C_2H_5)_2P(O)SCH_3$	Sn: MVol.C6-158/60
$Br_4C_{10}H_{26}O_4P_2Sn$	$SnBr_4 \cdot 2\ C_2H_5(H)P(O)OCH(CH_3)_2$	Sn: MVol.C6-151/2
–	$SnBr_4 \cdot 2\ C_2H_5(H)P(O)OC_3H_7\text{-}n$	Sn: MVol.C6-151/2
$Br_4C_{10}H_{26}P_2S_4Sn$	$SnBr_4 \cdot 2\ (C_2H_5)_2P(S)SCH_3$	Sn: MVol.C6-158
$Br_4C_{10}H_{28}N_2O_4U$	$[HOC_2H_4N(CH_3)_3]_2UO_2Br_4$	U: SVol.C9-154/5
$Br_4C_{11}H_{13}NO_2Sn$	$SnBr_4 \cdot C_6H_5CONHCOC_3H_7$	Sn: MVol.C6-50/1
$Br_4C_{11}H_{23}NO_2Sn$	$(C_2H_5)_3NH[SnBr_4(CH(COCH_3)_2)]$	Sn: MVol.C5-107
$Br_4C_{11}H_{24}S_2Sn$	$SnBr_4 \cdot C_4H_9S(CH_2)_3SC_4H_9$	Sn: MVol.C6-97/8
$Br_4C_{11}H_{25}N_3O_2Sn$	$SnBr_4 \cdot 2\ OCHN(CH_3)_2 \cdot (CH_2)_5NH$	Sn: MVol.C6-42
$Br_4C_{12}Cl_2H_{12}N_2Te$	$TeBr_4 \cdot 2\ ClC_6H_4NH_2$	Te: SVol.B3-167
$Br_4C_{12}Fe_2H_{14}$	$[C_5H_5FeC_5H_4C_2H_5]FeBr_4$	Fe: Org.Verb.A1-278

$Br_4C_{12}H_8N_2Sn$	$SnBr_4 \cdot C_{12}H_8N_2$	Sn: MVol.C5-220/1
$Br_4C_{12}H_{11}NSn$	$SnBr_4 \cdot (C_6H_5)_2NH$	Sn: MVol.C5-180
$Br_4C_{12}H_{12}N_2Te$	$TeBr_4 \cdot NH_2C_6H_4C_6H_4NH_2$	Te: SVol.B3-169
$Br_4C_{12}H_{12}N_4O_2Sn$	$SnBr_4 \cdot 2\ NC_5H_4CONH_2$	Sn: MVol.C5-206
$Br_4C_{12}H_{14}N_2O_2Sn$	$SnBr_4 \cdot 2\ CH_3N(CH)_4C(O)$	Sn: MVol.C5-204
–	$SnBr_4 \cdot 2\ HN(CH)_3C(COCH_3)$	Sn: MVol.C5-190
$Br_4C_{12}H_{14}N_2Sn$	$SnBr_4 \cdot 2\ CH_3C_5H_4N$	Sn: MVol.C5-201
–	$SnBr_4 \cdot 2\ C_6H_5NH_2$	Sn: MVol.C5-175/6
$Br_4C_{12}H_{14}N_2Te$	$TeBr_4 \cdot 2\ C_6H_5NH_2$	Te: SVol.B3-167
$Br_4C_{12}H_{15}NO_2Sn$	$SnBr_4 \cdot C_6H_5CONHCOC_4H_9$	Sn: MVol.C6-50/1
$Br_4C_{12}H_{16}N_4Sn$	$SnBr_4 \cdot 2\ C_6H_4(NH_2)_2$	Sn: MVol.C5-186/7
$Br_4C_{12}H_{16}N_4Sn_2$	$SnBr_4 \cdot Sn(CH_2CH_2CN)_4$	Sn: MVol.C6-184
$Br_4C_{12}H_{16}N_4Te$	$TeBr_4 \cdot 2\ C_6H_4(NH_2)_2$	Te: SVol.B3-169
$Br_4C_{12}H_{18}N_2O_2Sn$	$SnBr_4 \cdot 2\ CH_2CHCONHCH_2CHCH_2$	Sn: MVol.C6-45/6
–	$SnBr_4 \cdot (C_2H_5)_2C(OH)CONHNHC_6H_5$	Sn: MVol.C6-62
$Br_4C_{12}H_{20}N_4U$	$[UBr_4(C_2H_5CN)_4]$	U: SVol.E1-48/50
$Br_4C_{12}H_{20}O_8Sn$	$SnBr_4 \cdot 2\ C_2H_5OCOCOOC_2H_5$	Sn: MVol.C5-139/40
$Br_4C_{12}H_{22}O_4Sn$	$SnBr_4 \cdot C_8H_{16}(COOCH_3)_2$	Sn: MVol.C5-143
$Br_4C_{12}H_{24}Mn_2O_6$	$2\ MnBr_2 \cdot (C_2H_4O)_6 \cdot 8\ H_2O$	Mn: MVol.D1-159
$Br_4C_{12}H_{24}N_3PSSn$	$SnBr_4 \cdot (C_4H_8N)_3PS$	Sn: MVol.C6-149
$Br_4C_{12}H_{24}O_3S_3Te$	$TeBr_4 \cdot 3\ (CH_2)_4SO$	Te: SVol.B3-176
$Br_4C_{12}H_{24}O_6Sn$	$SnBr_4 \cdot 3\ CH_3COOC_2H_5$	Sn: MVol.C5-130
$Br_4C_{12}H_{24}O_6U$	$UBr_4 \cdot 3\ CH_3COOC_2H_5$	U: SVol.E1-120/2
–	$UBr_4 \cdot 3\ O(CH_2CH_2)_2O$	U: SVol.E1-70, 74/5
$Br_4C_{12}H_{24}O_8Sn$	$SnBr_4 \cdot 4\ C_2H_5COOH$	Sn: MVol.C5-120/1
$Br_4C_{12}H_{25}O_3U$	$U(OC_5H_{11})Br_4 \cdot CH_3COOC_5H_{11}$	U: SVol.E1-120/2
$Br_4C_{12}H_{26}N_2O_2U$	$UBr_4 \cdot 2\ C_3H_7CON(CH_3)_2$	U: SVol.E1-109/11
$Br_4C_{12}H_{26}N_2O_4Sn$	$SnBr_4 \cdot 2\ (CH_3)_2CHCH_2CH(NH_2)COOH$	Sn: MVol.C6-12
$Br_4C_{12}H_{27}O_4PSn$	$SnBr_4 \cdot (C_4H_9O)_3PO$	Sn: MVol.C6-170/1
$Br_4C_{12}H_{28}MnN_2$	$(CH_3C_5H_9NH_2)_2MnBr_4$	Mn: MVol.C5-301
$Br_4C_{12}H_{28}N_4O_8Sn$	$SnBr_4 \cdot 4\ CH_3CH(NH_2)COOH$	Sn: MVol.C6-11/2
–	$SnBr_4 \cdot 4\ NH_2CH_2CH_2COOH$	Sn: MVol.C6-11/2
$Br_4C_{12}H_{30}NO_2Pa$	$[N(C_2H_5)_4][Pa(OC_2H_5)_2Br_4]$	Pa: SVol.2-69/70
$Br_4C_{12}H_{30}NO_2U$	$[N(C_2H_5)_4]U(OC_2H_5)_2Br_4$	U: SVol.C9-141
$Br_4C_{12}H_{30}N_2Sn$	$SnBr_4 \cdot 2\ (C_2H_5)_3N$	Sn: MVol.C5-174
$Br_4C_{12}H_{30}O_2P_2S_2Sn$	$SnBr_4 \cdot 2\ (CH(CH_3)_2)_2POSH$	Sn: MVol.C6-156/7
–	$SnBr_4 \cdot 2\ (C_3H_7)_2POSH$	Sn: MVol.C6-156/7
$Br_4C_{12}H_{30}O_2P_2Sn$	$SnBr_4 \cdot 2\ (C_2H_5)_2POC_2H_5$	Sn: MVol.C6-139/40
–	$SnBr_4 \cdot 2\ (C_2H_5)_3PO$	Sn: MVol.C6-143/4
$Br_4C_{12}H_{30}O_2P_2U$	$UBr_4 \cdot 2\ (C_2H_5)_3PO$	U: SVol.E1-149, 156/7
$Br_4C_{12}H_{30}O_3S_3Te$	$TeBr_4 \cdot 3\ (C_2H_5)_2SO$	Te: SVol.B3-176
$Br_4C_{12}H_{30}O_4P_2Sn$	$SnBr_4 \cdot 2\ C_2H_5(H)P(O)(OC_4H_9\text{-}n)$	Sn: MVol.C6-151/2
$Br_4C_{12}H_{30}O_6P_2Sn$	$SnBr_4 \cdot 2\ C_2H_5P(O)(OC_2H_5)_2$	Sn: MVol.C6-162/3
–	$SnBr_4 \cdot 2\ HP(O)(OC_3H_7)_2$	Sn: MVol.C6-160/1
–	$SnBr_4 \cdot 2\ P(OC_2H_5)_3$	Sn: MVol.C6-141/2
$Br_4C_{12}H_{30}P_2S_2Sn$	$SnBr_4 \cdot 2\ (C_2H_5)_3PS$	Sn: MVol.C6-148
$Br_4C_{12}H_{30}P_2Sn$	$SnBr_4 \cdot 2\ P(C_2H_5)_3$	Sn: MVol.C6-133
$Br_4C_{12}H_{30}P_2U$	$UBr_4[(C_2H_5)_3P]_2$	U: SVol.E1-51
$Br_4C_{12}H_{32}N_4Sn$	$SnBr_4 \cdot 2\ NH_2(CH_2)_6NH_2$	Sn: MVol.C5-185/6

$Br_4C_{12}H_{32}O_2P_2U$	$[P(C_2H_5)_3H]_2UO_2Br_4$	U:	SVol.C9-151, 154
		U:	SVol.E1-51
$Br_4C_{12}H_{32}O_3Sn$	$H_2SnBr_4 \cdot 3\ (C_2H_5)_2O$	Sn:	MVol.C5-7/8
$Br_4C_{12}H_{36}N_6O_2P_2Pa$	$PaBr_4 \cdot 2\ [(CH_3)_2N]_3PO$	Pa:	SVol.2-93/4
$Br_4C_{12}H_{36}N_6O_2P_2Sn$	$SnBr_4 \cdot 2\ [(CH_3)_2N]_3PO$	Sn:	MVol.C6-171/2
$Br_4C_{12}H_{36}N_6O_2P_2U$	$UBr_4 \cdot 2\ [(CH_3)_2N]_3PO$	U:	SVol.E1-161/2, 164
$Br_4C_{12}H_{36}N_6Sn$	$SnBr_4 \cdot 3\ NH_2(CH_2)_4NH_2$	Sn:	MVol.C5-185
$Br_4C_{12}H_{36}O_4P_4U$	$UBr_4 \cdot 4\ (CH_3)_3PO$	U:	SVol.E1-147
$Br_4C_{12}H_{36}O_6S_6U$	$UBr_4 \cdot 6\ (CH_3)_2SO$	U:	SVol.E1-203, 207
$Br_4C_{13}H_{17}NO_2Sn$	$SnBr_4 \cdot C_6H_5CONHCOC_5H_{11}$	Sn:	MVol.C6-50/1
$Br_4C_{14}Cl_2H_{14}Ni_2O_2$	$(CH_3CHCHCH_2NiCl)_2 \cdot C_6Br_4O_2$	Ni:	Org.Verb.2-301
$Br_4C_{14}H_8O_2Sn$	$SnBr_4 \cdot C_{14}H_8O_2$	Sn:	MVol.C5-115
$Br_4C_{14}H_{10}N_2Sn$	$SnBr_4 \cdot 2\ C_6H_5CN$	Sn:	MVol.C6-70
$Br_4C_{14}H_{11}NO_2Sn$	$SnBr_4 \cdot C_6H_5CONHCOC_6H_5$	Sn:	MVol.C6-50/1
$Br_4C_{14}H_{12}N_2O_2Sn$	$SnBr_4 \cdot C_6H_5CONHNHCOC_6H_5$	Sn:	MVol.C6-59/60
$Br_4C_{14}H_{12}O_2Sn$	$SnBr_4 \cdot 2\ C_6H_5CHO$	Sn:	MVol.C5-79/80
$Br_4C_{14}H_{12}O_4Sn$	$SnBr_4 \cdot 2\ HOC_6H_4CHO$	Sn:	MVol.C5-80, 82
–	$SnBr_4 \cdot 2\ HOC_6H_4CHO \cdot 2\ H_2O$	Sn:	MVol.C5-80
$Br_4C_{14}H_{14}N_2O_2Sn$	$SnBr_4 \cdot 2\ CH_3COC_5H_4N$	Sn:	MVol.C5-205
–	$SnBr_4 \cdot 2\ C_6H_5CONH_2$	Sn:	MVol.C6-46/7
$Br_4C_{14}H_{16}N_4O_4Sn$	$SnBr_4 \cdot 2\ NO_2C_6H_3(CH_3)NH_2$	Sn:	MVol.C5-179
$Br_4C_{14}H_{18}N_2Te$	$TeBr_4 \cdot 2\ CH_3C_6H_4NH_2$	Te:	SVol.B3-167
–	$TeBr_4 \cdot 2\ (CH_3)_2C_5H_3N$	Te:	SVol.B3-171
$Br_4C_{14}H_{19}NO_2Sn$	$SnBr_4 \cdot C_6H_5CONHCOC_6H_{13}$	Sn:	MVol.C6-50/1
$Br_4C_{14}H_{20}Mn_2O_5$	$2\ MnBr_2 \cdot C_6H_4(OC_2H_4OC_2H_4)_2O \cdot 1.5\ (CH_3)_2CO \cdot H_2O$	Mn:	MVol.D1-158
$Br_4C_{14}H_{30}N_2O_2U$	$UBr_4 \cdot 2\ (CH_3)_2CHCH_2CON(CH_3)_2$	U:	SVol.E1-109/11
–	$UBr_4 \cdot 2\ (CH_3)_3CCON(CH_3)_2$	U:	SVol.E1-109/11
$Br_4C_{14}H_{30}O_8P_2Sn$	$SnBr_4 \cdot 2\ (CH_3)_3CCOP(O)(OCH_3)_2$	Sn:	MVol.C6-164
$Br_4C_{14}H_{32}MnN_2$	$(C_2H_5C_5H_9NH_2)_2MnBr_4$	Mn:	MVol.C5-301
$Br_4C_{14}H_{32}N_2O_6U$	$[CH_3COOC_2H_4N(CH_3)_3]_2UO_2Br_4$	U:	SVol.C9-151/5
$Br_4C_{14}H_{32}N_4O_2U$	$UBr_4 \cdot 2\ (CH_3)_2NCON(C_2H_5)_2$	U:	SVol.E1-91/2, 94
$Br_4C_{14}H_{34}O_2P_2Sn$	$SnBr_4 \cdot 2\ (C_2H_5)_2(C_3H_7)PO$	Sn:	MVol.C6-143/4
–	$SnBr_4 \cdot 2\ (C_2H_5)_2POC_3H_7$	Sn:	MVol.C6-139/40
$Br_4C_{14}H_{34}O_6P_2Sn$	$SnBr_4 \cdot 2\ CH_3P(O)(OCH(CH_3)_2)_2$	Sn:	MVol.C6-162/3
$Br_4C_{15}H_{11}N_3Te$	$TeBr_4 \cdot NC_5H_3(C_5H_4N)_2$	Te:	SVol.B3-171
$Br_4C_{15}H_{15}N_3O_2Sn_2$	$2\ SnOBr_2 \cdot 3\ C_5H_5N$	Sn:	MVol.C5-199
$Br_4C_{15}H_{21}N_3O_2Sn$	$SnBr_4 \cdot 2\ OCHN(CH_3)_2 \cdot C_9H_7N$	Sn:	MVol.C6-42
$Br_4C_{15}H_{27}N_3U$	$[UBr_4(C_4H_9CN)_3]$	U:	SVol.E1-48/50
$Br_4C_{16}Cl_2H_{40}N_2Sn$	$[(C_2H_5)_4N]_2[SnBr_4Cl_2]$	Sn:	MVol.C3-119
$Br_4C_{16}Cl_2H_{40}N_2Te$	$[(C_2H_5)_4N]_2[TeBr_4Cl_2]$	Te:	SVol.B3-87
$Br_4C_{16}F_2H_{40}N_2Sn$	$[(C_2H_5)_4N]_2[SnBr_4F_2]$	Sn:	MVol.C3-118/9
$Br_4C_{16}F_2H_{40}N_2Te$	$[(C_2H_5)_4N]_2[TeBr_4F_2]$	Te:	SVol.B3-87
$Br_4C_{16}H_{12}O_6Sn$	$SnBr_4 \cdot 2\ C_7H_5O_2CHO$	Sn:	MVol.C5-82
$Br_4C_{16}H_{16}O_2Sn$	$SnBr_4 \cdot 2\ CH_3COC_6H_5$	Sn:	MVol.C5-88
$Br_4C_{16}H_{16}O_6Sn$	$SnBr_4 \cdot 2\ HOC_6H_4COOCH_3 \cdot 2\ H_2O$	Sn:	MVol.C5-146/7
$Br_4C_{16}H_{18}N_2O_2Sn$	$SnBr_4 \cdot 2\ CH_3CONHC_6H_5$	Sn:	MVol.C6-44/5
$Br_4C_{16}H_{18}N_2O_4Sn$	$SnBr_4 \cdot 2\ (CH_3CO)_2C_4H_2NH$	Sn:	MVol.C5-190
$Br_4C_{16}H_{22}N_2O_2Sn$	$SnBr_4 \cdot 2\ CH_3NC_5H_2O(CH_3)_2$	Sn:	MVol.C5-204

$Br_4C_{16}H_{22}N_2Te$	$TeBr_4 \cdot 2\ C_6H_5N(CH_3)_2$	Te: SVol.B3-168
$Br_4C_{16}H_{22}P_2S_2Sn$	$SnBr_4 \cdot 2\ (C_6H_5)(CH_3)_2PS$	Sn: MVol.C6-148/9
$Br_4C_{16}H_{22}P_2Sn$	$SnBr_4 \cdot 2\ C_6H_5P(CH_3)_2$	Sn: MVol.C6-136/7
$Br_4C_{16}H_{24}MnN_2$	$(CH_3(C_2H_5)C_5H_3NH)_2MnBr_4$	Mn: MVol.C5-289, 304
$Br_4C_{16}H_{24}Ni_2$	$((CH_3)_4C_4NiBr_2)_2$	Ni: Org.Verb.2-328/9
$Br_4C_{16}H_{26}N_2O_2Sn$	$SnBr_4 \cdot (C_4H_9)_2C(OH)CONHNHC_6H_5$	Sn: MVol.C6-62
$Br_4C_{16}H_{28}Mg_4NiO_4$	$(BrMgO(C_3H_7)C)_4Ni$	Ni: Org.Verb.1-336
$Br_4C_{16}H_{28}N_4U$	$[UBr_4(C_3H_7CN)_4]$	U: SVol.E1-48/50
$Br_4C_{16}H_{32}O_8Sn$	$SnBr_4 \cdot 4\ C_3H_7COOH$	Sn: MVol.C5-120/1
$Br_4C_{16}H_{34}N_2O_2U$	$UBr_4 \cdot 2\ CH_3CON(C_3H_7)_2$	U: SVol.E1-106/8
$Br_4C_{16}H_{34}N_2O_8Sn$	$SnBr_4 \cdot 2\ (CH_3)_2CHCH_2CH(NH_2)COOH \cdot 2\ CH_3COOH$	Sn: MVol.C6-12
$Br_4C_{16}H_{36}IrN$	$((C_4H_9)_4N)[IrBr_4] \cdot 2\ C_2H_5OH$	Ir: SVol.2-147
$Br_4C_{16}H_{36}MnN^-$	$[(C_4H_9)_4NMnBr_4]^-$	Mn: MVol.C5-288
$Br_4C_{16}H_{36}N_4O_4U$	$UBr_4 \cdot 4\ CH_3CON(CH_3)_2$	U: SVol.E1-103/5
–	$UBr_4 \cdot 4\ CH_3CONHC_2H_5$	U: SVol.E1-96/8
$Br_4C_{16}H_{36}N_4U$	$UBr_4 \cdot 4\ (CH_2)_4NH$	U: SVol.E1-38/9
$Br_4C_{16}H_{36}O_2S_2U$	$UBr_4 \cdot 2\ (C_4H_9)_2SO$	U: SVol.E1-210/1
$Br_4C_{16}H_{38}O_2P_2S_2Sn$	$SnBr_4 \cdot 2\ (C_4H_9)_2POSH$	Sn: MVol.C6-156/7
$Br_4C_{16}H_{38}O_2P_2Sn$	$SnBr_4 \cdot 2\ (C_4H_9)_2HPO$	Sn: MVol.C6-146/7
$Br_4C_{16}H_{38}O_4P_2Sn$	$SnBr_4 \cdot 2\ (C_4H_9)_2P(O)OH$	Sn: MVol.C6-150/1
$Br_4C_{16}H_{38}O_6P_2Sn$	$SnBr_4 \cdot 2\ HP(O)(OC_4H_9)_2$	Sn: MVol.C6-160/1
$Br_4C_{16}H_{40}I_2N_2Sn$	$[(C_2H_5)_4N]_2[SnBr_4I_2]$	Sn: MVol.C3-121
$Br_4C_{16}H_{40}I_2N_2Te$	$[(C_2H_5)_4N]_2[TeBr_4I_2]$	Te: SVol.B3-87
$Br_4C_{16}H_{40}I_2N_2U$	$[N(C_2H_5)_4]_2UBr_4I_2$	U: SVol.C9-184
$Br_4C_{16}H_{40}MnN_2$	$[(C_2H_5)_4N]_2MnBr_4$	Mn: MVol.C5-288/9, 302/3
$Br_4C_{16}H_{40}N_2O_2U$	$[N(C_2H_5)_4]_2UO_2Br_4$	U: SVol.C9-151/6
$Br_4C_{16}H_{48}N_8Sn$	$SnBr_4 \cdot 4\ NH_2(CH_2)_4NH_2$	Sn: MVol.C5-185
$Br_4C_{16}H_{48}O_8S_8U$	$UBr_4 \cdot 8\ (CH_3)_2SO$	U: SVol.E1-203, 207
$Br_4C_{17}H_{10}OSn$	$SnBr_4 \cdot C_{17}H_{10}O$	Sn: MVol.C5-99
$Br_4C_{17}H_{10}OTe$	$TeBr_4 \cdot C_{17}H_{10}O$	Te: SVol.B3-166
$Br_4C_{17}H_{22}N_2Te$	$TeBr_4 \cdot (CH_3)_2NC_6H_4CH_2C_6H_4N(CH_3)_2$	Te: SVol.B3-169
$Br_4C_{17}H_{28}N_2O_2Sn$	$SnBr_4 \cdot (C_4H_9)_2C(OH)CON_2H_2C_6H_4CH_3$	Sn: MVol.C6-62
$Br_4C_{18}CmH_9N_2O_3$	$Cm(OH)(NC_9H_4Br_2O)_2$	Np: TrU.D2-317
$Br_4C_{18}F_{15}P$	$[(C_6F_5)_3PBr]Br_3$	F: PerFHalOrg.3-124
$Br_4C_{18}H_8N_2O_2Sn$	$Sn(C_9H_4Br_2NO)_2$	Sn: MVol.C5-21
$Br_4C_{18}H_{10}O_2Sn$	$SnBr_4 \cdot C_{18}H_{10}O_2$	Sn: MVol.C5-115
$Br_4C_{18}H_{14}N_2O_2Sn$	$SnBr_4 \cdot 2\ HOC_9H_6N$	Sn: MVol.C5-213/4
$Br_4C_{18}H_{14}N_2Sn$	$SnBr_4 \cdot 2\ C_9H_7N$	Sn: MVol.C5-210
$Br_4C_{18}H_{15}SbSn$	$SnBr_4 \cdot (C_6H_5)_3Sb$	Sn: MVol.C6-182
$Br_4C_{18}H_{16}MnN_2$	$(C_9H_7NH)_2MnBr_4$	Mn: MVol.C5-304
–	$(C_9H_7NH)_2MnBr_4 \cdot 2\ H_2O$	Mn: MVol.C5-304
$Br_4C_{18}H_{16}O_2Sn$	$SnBr_4 \cdot 2\ C_6H_5CHCHCHO$	Sn: MVol.C5-83
$Br_4C_{18}H_{21}N_3Sn$	$SnBr_4 \cdot 3\ C_6H_5NH_2$	Sn: MVol.C5-175/6
$Br_4C_{18}H_{22}N_2O_2Sn$	$SnBr_4 \cdot 2\ (CH_3)_2NC_6H_4CHO$	Sn: MVol.C6-4/5
$Br_4C_{18}H_{22}N_2O_4Sn$	$SnBr_4 \cdot 2\ NH_2C_6H_4COOC_2H_5$	Sn: MVol.C6-14
$Br_4C_{18}H_{22}O_8P_2Sn$	$SnBr_4 \cdot 2\ C_6H_5COP(O)(OCH_3)_2$	Sn: MVol.C6-164
$Br_4C_{18}H_{26}MnN_4$	$((CH_3)_2C_7H_6N_2H)_2MnBr_4$	Mn: MVol.C5-304
$Br_4C_{18}H_{26}N_2O_2Sn$	$SnBr_4 \cdot 2\ C_2H_5NC_5H_2O(CH_3)_2$	Sn: MVol.C5-204
$Br_4C_{18}H_{30}N_2O_2Sn$	$SnBr_4 \cdot (C_5H_{11})_2C(OH)CONHNHC_6H_5$	Sn: MVol.C6-62

$Br_4C_{18}H_{36}IrNO_2$	$(n-C_4H_9)_4N[IrBr_4(CO)_2]$	Ir:	SVol.2-147
$Br_4C_{18}H_{38}O_8P_2Sn$	$SnBr_4 \cdot 2\ CH_3COCH_2P(O)(OC_3H_7)_2$	Sn:	MVol.C6-165/6
$Br_4C_{18}H_{40}N_4O_2U$	$UBr_4 \cdot 2\ (C_2H_5)_2NCON(C_2H_5)_2$	U:	SVol.E1-91/2, 94
$Br_4C_{18}H_{42}O_6P_2Sn$	$SnBr_4 \cdot 2\ (CH_3)_2CHP(O)(OCH(CH_3)_2)_2$	Sn:	MVol.C6-162/3
–	$SnBr_4 \cdot 2\ C_3H_7P(O)(OC_3H_7)_2$	Sn:	MVol.C6-162/3
–	$SnBr_4 \cdot 2\ P(OCH(CH_3)_2)_3$	Sn:	MVol.C6-141/2
–	$SnBr_4 \cdot 2\ P(OC_3H_7-n)_3$	Sn:	MVol.C6-141/2
$Br_4C_{18}H_{42}P_2S_2Sn$	$SnBr_4 \cdot 2\ (C_3H_7)_3PS$	Sn:	MVol.C6-148
$Br_4C_{18}H_{42}P_2S_4Sn$	$SnBr_4 \cdot 2\ (C_4H_9)_2P(S)SCH_3$	Sn:	MVol.C6-158/60
$Br_4C_{18}H_{42}P_2Sn$	$SnBr_4 \cdot 2\ P(C_3H_7)_3$	Sn:	MVol.C6-133
$Br_4C_{18}H_{42}P_2U$	$UBr_4[(n-C_3H_7)_3P]_2$	U:	SVol.E1-51
$Br_4C_{18}H_{44}O_2P_2U$	$[P(C_3H_7)_3H]_2UO_2Br_4$	U:	SVol.C9-151, 154
		U:	SVol.E1-51
$Br_4C_{18}H_{46}IrNP_2$	$((n-C_3H_7)_4N)[Ir(P(CH_3)_3)_2Br_4]$	Ir:	SVol.2-202/3
$Br_4C_{18}H_{48}N_6Sn$	$SnBr_4 \cdot 3\ NH_2(CH_2)_6NH_2$	Sn:	MVol.C5-185/6
$Br_4C_{18}H_{54}O_6P_6U$	$UBr_4 \cdot 6\ (CH_3)_3PO$	U:	SVol.E1-148, 156
$Br_4C_{18}H_{54}O_9S_9U$	$UBr_4 \cdot 9\ (CH_3)_2SO$	U:	SVol.E1-207
$Br_4C_{18.5}H_{29}Mn_2O_{6.5}$	$2\ MnBr_2 \cdot C_6H_4(C_2H_4OC_2H_4O)_2O \cdot 1.5\ (CH_3)_2CO \cdot H_2O$	Mn:	MVol.D1-158
$Br_4C_{19}H_{42}N_4O_5U$	$UBr_4 \cdot 4\ CH_3CON(CH_3)_2 \cdot (CH_3)_2CO$	U:	SVol.E1-104/5
$Br_4C_{20}Cl_2H_{16}N_2O_2Sn$	$SnBr_4 \cdot (ClC_6H_4)_2C(OH)CONHNHC_6H_5$	Sn:	MVol.C6-62
$Br_4C_{20}F_2H_{16}N_2O_2Sn$	$SnBr_4 \cdot (FC_6H_4)_2C(OH)CONHNHC_6H_5$	Sn:	MVol.C6-62
$Br_4C_{20}Fe_2H_{20}Hg$	$[Fe(C_5H_5)_2]_2HgBr_4$	Fe:	Org.Verb.A1-225
$Br_4C_{20}Fe_4H_4O_{20}$	$[(CO)_5FeH]_4Br_4$	Fe:	Org.Verb.B3-252
$Br_4C_{20}H_{16}N_4U$	$[UBr_4(NC_5H_4C_5H_4N)_2]$	U:	SVol.E1-39/40
$Br_4C_{20}H_{17}O_4PSn$	$SnBr_4 \cdot C_6H_5COCH_2P(O)(OC_6H_5)_2$	Sn:	MVol.C6-166
$Br_4C_{20}H_{18}N_2O_2Sn$	$SnBr_4 \cdot (C_6H_5)_2C(OH)CONHNHC_6H_5$	Sn:	MVol.C6-62
$Br_4C_{20}H_{18}N_2Te$	$TeBr_4 \cdot 2\ C_{10}H_7NH_2$	Te:	SVol.B3-167
$Br_4C_{20}H_{20}MnN_2$	$(CH_3C_9H_6NH)_2MnBr_4$	Mn:	MVol.C5-304
$Br_4C_{20}H_{20}NPSn$	$SnBr_4 \cdot P(C_6H_5)_2C_6H_4N(CH_3)_2$	Sn:	MVol.C6-136/7
$Br_4C_{20}H_{32}MnN_2$	$[(CH_3)_3(C_6H_5CH_2)N]_2MnBr_4$	Mn:	MVol.C5-303
$Br_4C_{20}H_{34}N_4O_2P_2U$	$UBr_4 \cdot 2\ [(CH_3)_2N]_2(C_6H_5)PO$	U:	SVol.E1-166/7
$Br_4C_{20}H_{36}Mg_4NiO_4$	$(BrMgO(C_4H_9)C)_4Ni$	Ni:	Org.Verb.1-336
$Br_4C_{20}H_{36}N_4U$	$[UBr_4(C_4H_9CN)_4]$	U:	SVol.E1-48/50
$Br_4C_{20}H_{36}O_2Sn$	$SnBr_4 \cdot 2\ CH_3C_6H_8C(CH_3)_2OH$	Sn:	MVol.C5-75
–	$SnBr_4 \cdot 2\ C_6H_8O(CH_3)CH(CH_3)_2$	Sn:	MVol.C5-161/2
$Br_4C_{20}H_{42}O_{10}P_2Sn$	$SnBr_4 \cdot 2\ C_2H_5OCOCH_2P(O)(OC_3H_7)_2$	Sn:	MVol.C6-165/6
$Br_4C_{20}H_{44}N_4O_4U$	$UBr_4 \cdot 4\ CH_3CONHC_3H_7$	U:	SVol.E1-96/8
$Br_4C_{20}H_{44}N_4U$	$UBr_4 \cdot 4\ (CH_2)_5NH$	U:	SVol.E1-38/9
$Br_4C_{20}H_{44}S_2Sn$	$SnBr_4 \cdot 2\ (i-C_5H_{11})_2S$	Sn:	MVol.C6-94
$Br_4C_{20}H_{45}N_5O_5Pa$	$PaBr_4 \cdot 5\ CH_3CON(CH_3)_2$	Pa:	SVol.2-89/90
$Br_4C_{20}H_{45}N_5O_5U$	$UBr_4 \cdot 5\ CH_3CON(CH_3)_2$	U:	SVol.E1-103/5
$Br_4C_{20}H_{46}O_2P_2Sn$	$SnBr_4 \cdot 2\ (C_2H_5)_2(C_6H_{13})PO$	Sn:	MVol.C6-143/4
–	$SnBr_4 \cdot 2\ (C_2H_5)_2POC_6H_{13}$	Sn:	MVol.C6-139/40
$Br_4C_{20}H_{48}IrNO_2$	$((C_4H_9)_4N)[IrBr_4] \cdot 2\ C_2H_5OH$	Ir:	SVol.2-147
$Br_4C_{20}H_{50}O_5S_5U$	$UBr_4 \cdot 5\ (C_2H_5)_2SO$	U:	SVol.E1-210/1
$Br_4C_{21}FeGeH_{15}O_3P$	$(CO)_3Fe(P(C_6H_5)_3)Br(GeBr_3)$	Fe:	Org.Verb.B1-160
$Br_4C_{21}FeH_{15}O_3PSn$	$(CO)_3Fe(P(C_6H_5)_3)(Br)SnBr_3$	Fe:	Org.Verb.B1-195/6
$Br_4C_{21}FeH_{15}O_3SbSn$	$(CO)_3Fe(Sb(C_6H_5)_3)(Br)SnBr_3$	Fe:	Org.Verb.B1-195/6
$Br_4C_{21}H_{15}N_3Sn$	$SnBr_4 \cdot 3\ C_6H_5CN$	Sn:	MVol.C6-70

$Br_4C_{21}H_{20}N_2O_2Sn$	$SnBr_4 \cdot (C_6H_5)_2C(OH)CON_2H_2C_6H_4CH_3$	Sn: MVol.C6-62
$Br_4C_{21}H_{36}N_2O_2Sn$	$SnBr_4 \cdot (C_6H_{13})_2C(OH)CON_2H_2C_6H_4CH_3$	Sn: MVol.C6-62
$Br_4C_{22}H_{16}O_2Sn$	$SnBr_4 \cdot 2\ C_{10}H_7CHO$	Sn: MVol.C5-83/4
$Br_4C_{22}H_{22}N_2O_2Sn$	$SnBr_4 \cdot (CH_3C_6H_4)_2C(OH)CON_2H_2C_6H_5$	Sn: MVol.C6-62
$Br_4C_{22}H_{24}MnN_2$	$((CH_3)_2C_9H_5NH)_2MnBr_4$	Mn: MVol.C5-304
$Br_4C_{22}H_{24}O_4Sn$	$SnBr_4 \cdot 2\ C_6H_5C_2H_2COOC_2H_5$	Sn: MVol.C5-145/6
$Br_4C_{22}H_{25}N_2PSn$	$SnBr_4 \cdot C_6H_5P(C_6H_4N(CH_3)_2)_2$	Sn: MVol.C6-136/7
$Br_4C_{23}H_{24}N_2O_2Sn$	$SnBr_4 \cdot (CH_3C_6H_4)_2C(OH)CON_2H_2C_6H_4CH_3$	Sn: MVol.C6-62
–	$SnBr_4 \cdot (C_6H_5CH_2)_2C(OH)CON_2H_2C_6H_4CH_3$	Sn: MVol.C6-62
$Br_4C_{23}H_{24}N_4O_2U$	$UBr_4 \cdot CH_2(C_3N_2O(CH_3)_2C_6H_5)_2$	U: SVol.E1-116/8
$Br_4C_{24}ClH_{20}OPU$	$[P(C_6H_5)_4]UOBr_4Cl$	U: SVol.C9-162
$Br_4C_{24}F_{16}Ge$	$(C_6F_4Br)_4Ge$	F: PerFHalOrg.4-168/9
$Br_4C_{24}Fe_2H_{24}O_2Sn$	$2\ C_5H_5FeC_5H_4COCH_3 \cdot SnBr_4$	Fe: Org.Verb.A2-214
$Br_4C_{24}Fe_2H_{24}O_2Ti$	$2\ C_5H_5FeC_5H_4COCH_3 \cdot TiBr_4$	Fe: Org.Verb.A2-214
$Br_4C_{24}H_{16}N_4U$	$[UBr_4(C_{12}H_8N_2)_2]$	U: SVol.E1-44, 45
$Br_4C_{24}H_{17}IrN_4$	$(C_{12}H_8N_2H)[Ir(C_{12}H_8N_2)Br_4]$	Ir: SVol.2-92
$Br_4C_{24}H_{20}N_4O_2Sn$	$SnBr_4 \cdot 2\ C_6H_5NNC_6H_4OH$	Sn: MVol.C6-31/2
$Br_4C_{24}H_{21}NO_2P_2Sn$	$SnBr_4 \cdot NH(PO(C_6H_5)_2)_2$	Sn: MVol.C6-154
$Br_4C_{24}H_{22}N_2Te$	$TeBr_4 \cdot 2\ (C_6H_5)_2NH$	Te: SVol.B3-167
$Br_4C_{24}H_{22}O_2P_2Sn$	$SnBr_4 \cdot 2\ (C_6H_5)_2HPO$	Sn: MVol.C6-146/7
$Br_4C_{24}H_{22}O_4P_2Sn$	$SnBr_4 \cdot 2\ (C_6H_5)_2P(O)OH$	Sn: MVol.C6-151
$Br_4C_{24}H_{30}N_6Sn$	$SnBr_4 \cdot 2\ (C_{12}H_{15}N_3)$	Sn: MVol.C5-190
$Br_4C_{24}H_{32}N_8Sn$	$SnBr_4 \cdot 4\ C_6H_5NHNH_2$	Sn: MVol.C5-189
$Br_4C_{24}H_{32}N_8U$	$UBr_4 \cdot 4\ C_6H_4(NH_2)_2$	U: SVol.E1-23/5
$Br_4C_{24}H_{46}O_2P_2Sn$	$SnBr_4 \cdot 2\ (C_6H_{11})_2HPO$	Sn: MVol.C6-146/7
$Br_4C_{24}H_{46}O_4P_2Sn$	$SnBr_4 \cdot 2\ (C_6H_{11})_2P(O)OH$	Sn: MVol.C6-150/1
$Br_4C_{24}H_{48}N_{16}Sn$	$SnBr_4 \cdot 4\ (CH_2)_6N_4 \cdot C_2H_2Cl_2$	Sn: MVol.C5-229
–	$SnBr_4 \cdot 4\ (CH_2)_6N_4 \cdot C_2H_2Cl_4$	Sn: MVol.C5-229
$Br_4C_{24}H_{48}O_8Sn$	$SnBr_4 \cdot 4\ C_5H_{11}COOH$	Sn: MVol.C5-120/1
$Br_4C_{24}H_{52}N_4O_8Sn$	$SnBr_4 \cdot 4\ (CH_3)_2CHCH_2CH(NH_2)COOH$	Sn: MVol.C6-11/2
$Br_4C_{24}H_{54}O_2P_2Sn$	$SnBr_4 \cdot 2\ (C_4H_9)_3PO$	Sn: MVol.C6-143/4
–	$SnBr_4 \cdot 2\ (C_6H_{13})_2HPO$	Sn: MVol.C6-146/7
$Br_4C_{24}H_{54}O_2P_2U$	$UBr_4 \cdot 2\ (C_4H_9)_3PO$	U: SVol.E1-149, 157
$Br_4C_{24}H_{54}O_3S_3Te$	$TeBr_4 \cdot 3\ (C_4H_9)_2SO$	Te: SVol.B3-176
$Br_4C_{24}H_{54}O_6P_2Sn$	$SnBr_4 \cdot 2\ C_4H_9P(O)(OC_4H_9)_2$	Sn: MVol.C6-162/3
–	$SnBr_4 \cdot 2\ P(OCH(CH_3)C_2H_5)_3$	Sn: MVol.C6-141/2
–	$SnBr_4 \cdot 2\ P(OC_4H_9\text{-}n)_3$	Sn: MVol.C6-141/2
$Br_4C_{24}H_{54}O_8P_2Sn$	$SnBr_4 \cdot 2\ (C_4H_9O)_3PO$	Sn: MVol.C6-170/1
$Br_4C_{24}H_{54}P_2S_2Sn$	$SnBr_4 \cdot 2\ (C_4H_9)_3PS$	Sn: MVol.C6-148
$Br_4C_{24}H_{54}P_2Sn$	$SnBr_4 \cdot 2\ P(C_4H_9)_3$	Sn: MVol.C6-134
$Br_4C_{24}H_{56}MnN_2$	$[(n\text{-}C_3H_7)_4N]_2MnBr_4$	Mn: MVol.C5-302/3
$Br_4C_{24}H_{56}N_2O_2U$	$[N(C_3H_7)_4]_2UO_2Br_4$	U: SVol.C9-151/4
$Br_4C_{24}H_{60}N_4Sn$	$SnBr_4 \cdot 4\ (C_2H_5)_3N$	Sn: MVol.C5-174
$Br_4C_{24}H_{60}O_6S_6U$	$UBr_4 \cdot 6\ (C_2H_5)_2SO$	U: SVol.E1-210/1
$Br_4C_{24}H_{64}N_8Sn$	$SnBr_4 \cdot 4\ NH_2(CH_2)_6NH_2$	Sn: MVol.C5-185/6
$Br_4C_{26}Cl_2H_{50}N_{16}Sn$	$SnBr_4 \cdot 4\ (CH_2)_6N_4 \cdot C_2H_2Cl_2$	Sn: MVol.C5-229
$Br_4C_{26}Cl_4H_{50}N_{10}Sn$	$SnBr_4 \cdot 4\ (CH_2)_6N_4 \cdot C_2H_2Cl_4$	Sn: MVol.C5-229
$Br_4C_{26}H_{16}O_2Sn$	$SnBr_4 \cdot 2\ C_{13}H_8O$	Sn: MVol.C5-99/100
$Br_4C_{26}H_{16}O_4Sn$	$SnBr_4 \cdot 2\ C_{13}H_8O_2$	Sn: MVol.C5-163
$Br_4C_{26}H_{18}N_2Sn$	$Sn(BrC_6H_3N(CH_3)C_6H_3Br)_2$	Sn: Org.Verb.3-129

$Br_4C_{26}H_{20}O_2Sn$	$SnBr_4 \cdot 2\ C_6H_5COC_6H_5$	Sn: MVol.C5-90
$Br_4C_{26}H_{24}O_2P_2Sn$	$SnBr_4 \cdot (C_6H_5)_2POC_2H_4PO(C_6H_5)_2$	Sn: MVol.C6-147
$Br_4C_{26}H_{24}P_2Sn$	$SnBr_4 \cdot (C_6H_5)_2PCH_2CH_2P(C_6H_5)_2$	Sn: MVol.C6-138/9
$Br_4C_{27}H_{33}N_3O_3U$	$UBr_4 \cdot 3\ C_6H_5CON(CH_3)_2$	U: SVol.E1-109/11
$Br_4C_{28}Cl_2H_{10}O_8Sn$	$SnCl_2(C_{14}H_5Br_2O_4)_2$	Sn: MVol.C5-114
$Br_4C_{28}Cl_4H_{12}O_8Sn$	$SnCl_4 \cdot 2\ C_{14}H_6Br_2O_4$	Sn: MVol.C5-114
$Br_4C_{28}H_{18}N_2O_6Sn$	$SnBr_4 \cdot 2\ NO_2C_{14}H_9O$	Sn: MVol.C5-98/9
$Br_4C_{28}H_{20}Mg_4NiO_4$	$(BrMgO(C_6H_5)C)_4Ni$	Ni: Org.Verb.1-336
$Br_4C_{28}H_{20}N_4Sn$	$SnBr_4 \cdot 4\ C_6H_5CN$	Sn: MVol.C6-70
$Br_4C_{28}H_{20}N_4U$	$[UBr_4(C_6H_5CN)_4]$	U: SVol.E1-48/50
$Br_4C_{28}H_{20}O_2Sn$	$SnBr_4 \cdot 2\ C_{14}H_{10}O$	Sn: MVol.C5-98/9
$Br_4C_{28}H_{22}N_2O_4Sn$	$SnBr_4 \cdot 2\ C_6H_5COC(NOH)C_6H_5$	Sn: MVol.C6-36
$Br_4C_{28}H_{22}N_2Sn$	$Sn(BrC_6H_3N(C_2H_5)C_6H_3Br)_2$	Sn: Org.Verb.3-130
$Br_4C_{28}H_{24}Ti$	$(BrC_6H_4CH_2)_4Ti$	Ti: Org.Verb.1-95
$Br_4C_{28}H_{28}O_2P_2Sn$	$SnBr_4 \cdot (C_6H_5)_2POC_4H_8PO(C_6H_5)_2$	Sn: MVol.C6-147
$Br_4C_{28}H_{52}N_4O_4U$	$UBr_4 \cdot 4\ CH_3N(CH_2)_5CO$	U: SVol.E1-118
$Br_4C_{30}H_{20}O_2Sn$	$SnBr_4 \cdot 2\ C_{15}H_{10}O$	Sn: MVol.C5-98/9
$Br_4C_{30}H_{24}P_2Sn$	$SnBr_4 \cdot (C_6H_5)_2PC_6H_4P(C_6H_5)_2$	Sn: MVol.C6-139
$Br_4C_{30}H_{32}N_4O_2U$	$UBr_4 \cdot 2\ CO(N(C_6H_5)CH_3)_2$	U: SVol.E1-91/2, 94
$Br_4C_{30}H_{46}HgIrP_3$	$Ir(P(C_2H_5)_2(C_6H_5))_3HBr_2(HgBr_2)$	Ir: SVol.2-264/5
$Br_4C_{30}H_{54}N_6O_6U$	$UBr_4 \cdot 6\ NH(CH_2)_4CO$	U: SVol.E1-118/20
$Br_4C_{30}H_{54}O_3Sn$	$SnBr_4 \cdot 3\ CH_3C_6H_8C(CH_3)_2OH$	Sn: MVol.C5-75
$Br_4C_{32}Cl_2H_{72}N_2Te$	$[(C_4H_9)_4N]_2[TeBr_4Cl_2]$	Te: SVol.B3-87
$Br_4C_{32}F_2H_{72}N_2Te$	$[(C_4H_9)_4N]_2[TeBr_4F_2]$	Te: SVol.B3-87
$Br_4C_{32}H_{16}Sn$	$Sn(CCC_6H_4Br)_4$	Sn: Org.Verb.1-140
$Br_4C_{32}H_{48}Ni_2$	$(C_8H_{12}C_8H_{12}NiBr_2)_2$	Ni: Org.Verb.2-329/30
$Br_4C_{32}H_{70}O_2P_2Sn$	$SnBr_4 \cdot 2\ (C_8H_{17})_2HPO$	Sn: MVol.C6-146/7
$Br_4C_{32}H_{72}I_2N_2Te$	$[(C_4H_9)_4N]_2[TeBr_4I_2]$	Te: SVol.B3-87
$Br_4C_{32}H_{72}MnN_2$	$[(n\text{-}C_4H_9)_4N]_2MnBr_4$	Mn: MVol.C5-288/9, 302/3
$Br_4C_{32}H_{72}O_4S_4U$	$UBr_4 \cdot 4\ (C_4H_9)_2SO$	U: SVol.E1-210/1
$Br_4C_{33}FeH_{25}Ni$	$[C_5H_5Ni(C_4(C_6H_5)_4)][FeBr_4]$	Ni: Org.Verb.2-188
$Br_4C_{34}H_{20}O_2Sn$	$SnBr_4 \cdot 2\ C_{17}H_{10}O$	Sn: MVol.C5-98/9
$Br_4C_{34}H_{20}O_2Te$	$TeBr_4 \cdot 2\ C_{17}H_{10}O$	Te: SVol.B3-166
$Br_4C_{34}H_{38}N_2O_6Sn$	$SnBr_4 \cdot 2\ (CH_2)_5NCO(CH)_4C_6H_3O_2CH_2$	Sn: MVol.C5-194
$Br_4C_{36}H_{20}N_4O_4Sn$	$Sn(C_9H_5BrNO)_4$	Sn: MVol.C5-215
$Br_4C_{36}H_{30}HgIr_2N_2O_3P_2$	$[Ir(NO)Br(P(C_6H_5)_3)]_2O \cdot HgBr_2$	Ir: SVol.2-57
$Br_4C_{36}H_{30}IrKSb_2$	$K[IrBr_4(Sb(C_6H_5)_3)_2]$	Ir: SVol.2-263
$Br_4C_{36}H_{30}O_2P_2Pa$	$PaBr_4 \cdot 2\ (C_6H_5)_3PO$	Pa: SVol.2-93
$Br_4C_{36}H_{30}O_2P_2Sn$	$SnBr_4 \cdot 2\ (C_6H_5)_3PO$	Sn: MVol.C6-144/5
$Br_4C_{36}H_{30}O_2P_2U$	$UBr_4 \cdot 2\ (C_6H_5)_3PO$	U: SVol.E1-169/70
$Br_4C_{36}H_{30}P_2Sn$	$SnBr_4 \cdot 2\ (C_6H_5)_3P$	Sn: MVol.C6-135
$Br_4C_{36}H_{31}IrSb_2$	$H[IrBr_4(Sb(C_6H_5)_3)_2]$	Ir: SVol.2-263
$Br_4C_{36}H_{32}MnP_2$	$[(C_6H_5)_3PH]_2MnBr_4$	Mn: MVol.C5-305
$Br_4C_{36}H_{32}N_4Ni$	$(BrC_6H_2(CH_3)_2NC)_4Ni$	Ni: Org.Verb.1-330, 332
$Br_4C_{36}H_{32}O_2P_2U$	$[P(C_6H_5)_3H]_2UO_2Br_4$	U: SVol.C9-151, 154
$Br_4C_{36}H_{66}N_6O_6U$	$UBr_4 \cdot 6\ NH(CH_2)_5CO$	U: SVol.E1-118/20
$Br_4C_{36}H_{72}N_{24}Sn$	$SnBr_4 \cdot 6\ (CH_2)_6N_4 \cdot CHCl_3$	Sn: MVol.C5-229
$Br_4C_{37}Cl_3H_{73}N_{24}Sn$	$SnBr_4 \cdot 6\ (CH_2)_6N_4 \cdot CHCl_3$	Sn: MVol.C5-229
$Br_4C_{37}FeH_{33}Ni$	$[C_5H_5Ni(C_4(C_6H_4CH_3)_4)][FeBr_4]$	Ni: Org.Verb.2-188

$Br_4C_{39}FeH_{30}Hg_2O_9P_2$	$(CO)_3Fe(P(OC_6H_5)_3)_2 \cdot (HgBr_2)_2$	Fe: Org.Verb.B1-192/4
$Br_4C_{40}Cl_2H_{40}Hg_2P_2Sn$	$[(CH_3)_2Sn[CH_2P(C_6H_5)_3]_2][HgBr_2Cl]_2$	Sn: Org.Verb.3-14
$Br_4C_{40}H_{42}IrNSb_2$	$((CH_3)_4N)[IrBr_4(Sb(C_6H_5)_3)_2]$	Ir: SVol.2-263
$Br_4C_{40}H_{42.5}N_{2.5}O_{2.5}U$	$UBr_4 \cdot 2.5\ (C_6H_5)_2CHCON(CH_3)_2$	U: SVol.E1-109/11
$Br_4C_{40}H_{64}Ni_2S_4$	$[C_4[(C(CH_3)_2CH_2)_2S]_2NiBr_2]_2$	Ni: Org.Verb.2-329
$Br_4C_{40}H_{72}O_4Sn$	$SnBr_4 \cdot 4\ CH_3C_6H_8C(CH_3)_2OH$	Sn: MVol.C5-75
$Br_4C_{40}H_{100}Li_2MnO_{10}$	$Li_2MnBr_4 \cdot 10\ (C_2H_5)_2O$	Mn: MVol.C5-291/2
		Mn: MVol.D1-139/40
$Br_4C_{41}H_{36}IrNSb_2$	$(C_5H_5NH)[IrBr_4(Sb(C_6H_5)_3)_2]$	Ir: SVol.2-263
$Br_4C_{42}H_{28}O_2Sn$	$SnBr_4 \cdot 2\ C_6H_5CHC_{14}H_8O$	Sn: MVol.C5-98/9
$Br_4C_{42}H_{98}O_7S_7U$	$UBr_4 \cdot 7\ (C_3H_7)_2SO$	U: SVol.E1-210/1
$Br_4C_{43}FeH_{39}O_2P_3$	$(CO)_2Fe[(P(C_6H_5)_2CH_2)_3CCH_3]Br_4$	Fe: Org.Verb.B1-101/2, 107, 110
$Br_4C_{44}H_{48}O_2P_2U$	$[P(C_6H_5)_3C_4H_9]_2UO_2Br_4$	U: SVol.C9-151, 154
$Br_4C_{48}H_{40}O_4S_4U$	$UBr_4 \cdot 4\ (C_6H_5)_2SO$	U: SVol.E1-210, 213
$Br_4C_{48}H_{42}N_6Sn$	$SnBr_4 \cdot 2\ (HNC_8H_7)_2C_8H_4NH$	Sn: MVol.C5-191
$Br_4C_{48}H_{102}P_2Sn$	$SnBr_4 \cdot 2\ P(C_8H_{17})_3$	Sn: MVol.C6-134
$Br_4C_{50}H_{44}O_2P_2U$	$[P(C_6H_5)_3CH_2C_6H_5]_2UO_2Br_4$	U: SVol.C9-151, 154
$Br_4C_{52}H_{44}N_4O_4U$	$UBr_4 \cdot 4\ HCON(C_6H_5)_2$	U: SVol.E1-100/2
$Br_4C_{52}H_{48}P_4Sn$	$SnBr_4 \cdot 2\ (C_6H_5)_2PCH_2CH_2P(C_6H_5)_2$	Sn: MVol.C6-138/9
$Br_4C_{56}Cl_8H_{32}Ni_2$	$[(ClC_6H_4)_4C_4NiBr_2]_2$	Ni: Org.Verb.2-331
$Br_4C_{56}H_{40}Ni_2$	$[(C_6H_5)_4C_4NiBr_2]_2$	Ni: Org.Verb.2-330/1
$Br_4C_{56}H_{44}O_2Sn$	$SnBr_4 \cdot 2\ (C_6H_5CH_2)_2C_{14}H_8O$	Sn: MVol.C5-98/9
$Br_4C_{62}H_{58}Ni_3P_4$	$[C_5H_5Ni(C_6H_5)_2P(CH_2)_2P(C_6H_5)_2]_2[NiBr_4]$	Ni: Org.Verb.2-141/2
$Br_4C_{64}H_{56}Ni_2$	$[(CH_3C_6H_4)_4C_4NiBr_2]_2$	Ni: Org.Verb.2-331
$Br_4C_{64}H_{56}Ni_2O_8$	$[(CH_3OC_6H_4)_4C_4NiBr_2]_2$	Ni: Org.Verb.2-331
$Br_4C_{64}H_{144}O_8S_8U$	$UBr_4 \cdot 8\ (C_4H_9)_2SO$	U: SVol.E1-210/1
$Br_4C_{80}H_{168}N_2O_2U$	$[N(C_{10}H_{21})_4]_2UO_2Br_4$	U: SVol.C9-151, 154
Br_4CaMn	$CaMnBr_4 \cdot 4\ H_2O$	Mn: MVol.C5-306
Br_4CaO_2U	$CaUO_2Br_4 \cdot 8\ H_2O$	U: SVol.C9-151, 154/5
Br_4CaSn	$CaSnBr_4 \cdot 6\ H_2O$	Sn: MVol.C3-177
Br_4CeK	$K[CeBr_4]$	Sc: MVol.C6-124
Br_4CeNa	$Na[CeBr_4]$	Sc: MVol.C6-119
Br_4CeO_{12}	$Ce(BrO_3)_4$	Sc: MVol.C6-107
Br_4ClOU^-	$[UOBr_4Cl]^-$	U: SVol.C9-162
Br_4ClTe^-	$TeBr_4Cl^-$	Te: SVol.B3-81/3
$Br_4Cl_2Cs_2Sn$	$Cs_2[SnBr_4Cl_2]$	Sn: MVol.C3-153
$Br_4Cl_2Cs_2Te$	$Cs_2TeBr_4Cl_2$	Te: SVol.B3-84, 87
$Br_4Cl_2H_8N_2Te$	$(NH_4)_2[TeCl_2Br_4]$	Te: SVol.B3-86
$Br_4Cl_2Ir^{2-}$	$[IrCl_2Br_4]^{2-}$	Ir: SVol.2-124, 150
$Br_4Cl_2IrRb_2$	$Rb_2[IrCl_2Br_4]$	Ir: SVol.2-150
$Br_4Cl_2Na_2U$	$Na_2UBr_4Cl_2$	U: SVol.C9-161/2
$Br_4Cl_2Rb_2Sn$	$Rb_2[SnBr_4Cl_2]$	Sn: MVol.C3-138
$Br_4Cl_2Te^{2-}$	$TeBr_4Cl_2{}^{2-}$	Te: SVol.B3-83/4
$Br_4Cs_2F_2Te$	$Cs_2TeBr_4F_2$	Te: SVol.B3-84, 87
$Br_4Cs_2I_2Te$	$Cs_2TeBr_4I_2$	Te: SVol.B3-84, 87
$Br_4Cs_2IrNO_2$	$Cs_2[Ir(NO_2)Br_4] \cdot H_2O$	Ir: SVol.2-63
Br_4Cs_2Mn	$Cs_2MnBr_4 = 2\ CsBr \cdot MnBr_2$	Mn: MVol.C5-296/7
–	$Cs_2MnBr_4 \cdot 2\ H_2O$	Mn: MVol.C5-299
$Br_4Cs_2O_2U$	$Cs_2UO_2Br_4$	U: SVol.C9-151/4

Br_4Cs_2Sn .... Cs_2SnBr_4 .... Sn: MVol.C3-152
$Br_4Cs_3IrN_2O_4$ ...... $Cs_3[Ir(NO_2)_2Br_4]$ .... Ir: SVol.2-61
Br_4DyK .... $K[DyBr_4]$ .... Sc: MVol.C6-124/5
Br_4DyNa .... $NaDyBr_4$ .... Sc: MVol.C6-122
Br_4EuNa .... $Na[EuBr_4]$ .... Sc: MVol.C6-121
Br_4Eu_3O .... Eu_3OBr_4 .... Sc: MVol.C6-86
$Br_4F_2Te^{2-}$ .... $TeBr_4F_2^{2-}$ .... Te: SVol.B3-83/4
Br_4GdK .... $K[GdBr_4]$ .... Sc: MVol.C6-124/5
Br_4GdNa .... $Na[GdBr_4]$ .... Sc: MVol.C6-122
Br_4Ge .... $GeBr_4$ systems
$GeBr_4$-$SiBr_4$-$SnBr_4$ .... Sn: MVol.C4-234
$GeBr_4$-$SiBr_4$-$SnBr_4$-$TiBr_4$ .... Sn: MVol.C4-235
$GeBr_4$-$SnBr_4$ .... Sn: MVol.C4-234
$GeBr_4$-$SnBr_4$-$TiBr_4$ .... Sn: MVol.C4-234
Br_4H_3PSn .... $SnBr_4 \cdot PH_3$ .... Sn: MVol.C2-267
$Br_4H_4IrO_2^-$ .... $[IrBr_4(H_2O)_2]^-$ .... Ir: SVol.2-146, 150
$Br_4H_4MnO_2^{2-}$ .... $[Mn(H_2O)_2Br_4]^{2-}$ .... Mn: MVol.C5-285
$Br_4H_4N_4S_4Te$ .... $TeBr_4 \cdot S_4N_4H_4$ .... Te: SVol.B3-182
$Br_4H_4O_2Te$ .... $[TeBr_4(H_2O)_2]$ .... Te: SVol.B3-15, 29
$Br_4H_{4.2}P_{1.4}Sn$ .... $SnBr_4 \cdot 1.4\ PH_3$ .... Sn: MVol.C2-267
$Br_4H_6N_2Sn$ .... $SnBr_4 \cdot 2\ NH_3$ .... Sn: MVol.C5-64
$Br_4H_8I_2N_2Te$ .... $(NH_4)_2[TeBr_4I_2]$ .... Te: SVol.B3-86
$Br_4H_8MnN_2$ .... $(NH_4)_2MnBr_4$ solid solutions
$(NH_4)_2MnBr_4 \cdot H_2O$-$NH_4Br$ .... Mn: MVol.C5-293
$Br_4H_8N_2O_2U$ .... $(NH_4)_2UO_2Br_4 \cdot 2\ H_2O$ .... U: SVol.C9-151, 154/6
$Br_4H_8N_2Sn$ .... $NH_4SnBr_3 \cdot NH_4Br = (NH_4)_2SnBr_4$ .... Sn: MVol.C3-111/2
– .... $NH_4SnBr_3 \cdot NH_4Br \cdot H_2O$ .... Sn: MVol.C3-112
– .... $NH_4SnBr_3 \cdot NH_4Br \cdot 2\ H_2O$ .... Sn: MVol.C3-112
$Br_4H_8N_4Sn$ .... $SnBr_4 \cdot 2\ N_2H_4$ .... Sn: MVol.C5-65
$Br_4H_8O_8P_4Sn_2$ .... $SnBr_4 \cdot Sn(H_2PO_2)_4$ .... Sn: MVol.C2-268
$Br_4H_{10}K_2O_{16}U_4$ .... $K_2[(UO_2)_4O_2(OH)_2Br_4(H_2O)_4] \cdot 2\ H_2O$ .... U: SVol.C9-151/3
$Br_4H_{10}N_4Sn$ .... $(N_2H_5)_2SnBr_4 \cdot H_2O$ .... Sn: MVol.C3-112
$Br_4H_{10}O_{16}Rb_2U_4$ .... $Rb_2[(UO_2)_4O_2(OH)_2Br_4(H_2O)_4] \cdot 2\ H_2O$ .... U: SVol.C9-151/3
$Br_4H_{12}N_4U$ .... $[NH_4]_2[UBr_4(NH_2)_2]$ .... U: SVol.C9-139
– .... $UBr_4 \cdot 4\ NH_3$ .... U: SVol.E1-12, 14
$Br_4H_{15}N_5U$ .... $UBr_4 \cdot 5\ NH_3$ .... U: SVol.E1-12, 14
$Br_4H_{18}N_6U$ .... $UBr_4 \cdot 6\ NH_3$ .... U: SVol.E1-12, 14
$Br_4H_{24}N_{12}Sn$ .... $SnBr_4 \cdot 6\ N_2H_4$ .... Sn: MVol.C5-65
$Br_4H_{30}N_{10}U$ .... $UBr_4 \cdot 10\ NH_3$ .... U: SVol.E1-12, 14
$Br_4HgO_8U_3$ .... $HgU_3O_8Br_4 \cdot 4\ H_2O$ .... U: SVol.C9-151, 154/5
Br_4HoK .... $K[HoBr_4]$ .... Sc: MVol.C6-124/5
Br_4HoNa .... $NaHoBr_4$ .... Sc: MVol.C6-122
$Br_4I_2K_2U$ .... $K_2UBr_4I_2$ .... U: SVol.C9-184
$Br_4I_2Te^{2-}$ .... $TeBr_4I_2^{2-}$ .... Te: SVol.B3-83/4
$Br_4IrK_2NO_2$ .... $K_2[Ir(NO_2)Br_4] \cdot H_2O$ .... Ir: SVol.2-63
Br_4KLa .... $K[LaBr_4]$ .... Sc: MVol.C6-124
Br_4KNd .... $K[NdBr_4]$ .... Sc: MVol.C6-124
Br_4KPr .... $K[PrBr_4]$ .... Sc: MVol.C6-124
Br_4KSm .... $K[SmBr_4]$ .... Sc: MVol.C6-124/5

Br_4K_2OS	$SOBr_2 \cdot 2\ KBr$	S:	SVol.1-56
$Br_4K_2O_2U$	$K_2UO_2Br_4 \cdot 2\ H_2O$	U:	SVol.C9-151/2, 154/5
$Br_4K_2O_5U_2$	$K_2U_2O_5Br_4$	U:	SVol.C9-151/6
$Br_4K_2O_{11}U_4$	$K_2U_4O_{11}Br_4$	U:	SVol.C9-151/3
Br_4K_2Sn	$KSnBr_3 \cdot KBr \cdot H_2O = K_2SnBr_4 \cdot H_2O$	Sn:	MVol.C3-64/5
–	$KSnBr_3 \cdot KBr \cdot 2\ H_2O$	Sn:	MVol.C3-63, 64/5
Br_4La^-	$[LaBr_4]^-$	Sc:	MVol.C6-7
Br_4LaNa	$Na[LaBr_4]$	Sc:	MVol.C6-118
Br_4LiMn^-	$[LiMnBr_4]^-$	Mn:	MVol.C5-288
Br_4Li_2Mn	$Li_2MnBr_4 = 2\ LiBr \cdot MnBr_2$	Mn:	MVol.C5-291/2
Br_4MgO_2U	$MgUO_2Br_4 \cdot 4\ H_2O$	U:	SVol.C9-151, 154
Br_4MgSn	$MgSnBr_4 \cdot 6\ H_2O$	Sn:	MVol.C3-166
Br_4Mn^{2-}	$[MnBr_4]^{2-}$	Mn:	MVol.C5-287/90
–	$[Mn(H_2O)_2Br_4]^{2-}$	Mn:	MVol.C5-285
Br_4MnRb_2	$Rb_2MnBr_4 = 2\ RbBr \cdot MnBr_2$	Mn:	MVol.C5-293/4
–	$Rb_2MnBr_4 \cdot 2\ H_2O$	Mn:	MVol.C5-295/6
Br_4Mn_2	Mn_2Br_4	Mn:	MVol.C5-267
$Br_4N_2O_8TeV$	$VO_2(NO_3)_2 \cdot TeBr_4 \cdot 6\ H_2O$	Te:	SVol.B3-13
$Br_4N_4S_4Te$	$TeBr_4 \cdot S_4N_4$	Te:	SVol.B3-182
$Br_4N_8S_8Sn$	$SnBr_4 \cdot 2\ S_4N_4$	Sn:	MVol.C5-66
Br_4NaNd	$Na[NdBr_4]$	Sc:	MVol.C6-120
Br_4NaPr	$Na[PrBr_4]$	Sc:	MVol.C6-119
Br_4NaSm	$Na[SmBr_4]$	Sc:	MVol.C6-120/1
$Br_4Na_2O_2U$	$Na_2UO_2Br_4 \cdot 2\ H_2O$	U:	SVol.C9-151, 154/5
$Br_4OS_{0.5}Sn$	$SnBr_4 \cdot 0.5\ SO_2$	Sn:	MVol.C2-78
$Br_4O_2Rb_2U$	$Rb_2UO_2Br_4 \cdot 2\ H_2O$	U:	SVol.C9-151, 154
Br_4O_2SrU	$SrUO_2Br_4 \cdot 3\ H_2O$	U:	SVol.C9-151, 154
$Br_4O_2U^{2-}$	$[UO_2Br_4]^{2-}$	U:	SVol.C9-151/6
Br_4O_2UZn	$ZnUO_2Br_4 \cdot 4\ H_2O$	U:	SVol.C9-151, 154/5
$Br_4O_{11}Rb_2U_4$	$Rb_2U_4O_{11}Br_4$	U:	SVol.C9-151/3
Br_4Pa	$PaBr_4$	Pa:	SVol.2-63/4
Br_4RbSm	$Rb[SmBr_4]$	Sc:	MVol.C6-128
Br_4Rb_2Sn	$RbSnBr_3 \cdot RbBr = Rb_2SnBr_4$	Sn:	MVol.C3-135/6, 137
$Br_4S_2Si_2$	$Br_2SiS_2SiBr_2$	S:	SVol.2-83, 186
Br_4Si	$SiBr_4$ systems		
	$SiBr_4$-BBr_3	B:	B-Verb.19-191
	$SiBr_4$-$GeBr_4$-$SnBr_4$	Sn:	MVol.C4-234
	$SiBr_4$-$GeBr_4$-$SnBr_4$-$TiBr_4$	Sn:	MVol.C4-235
Br_4Sn	$SnBr_4$ melts		
	$SnBr_4$-$AlBr_3 \cdot SbBr_3$	Sn:	MVol.C4-104/5
–	$SnBr_4$ systems		
	$SnBr_4$-$AlBr_3$	Sn:	MVol.C4-103/4
	$SnBr_4$-$AsBr_3$	Sn:	MVol.C2-281
	$SnBr_4$-BBr_3	Sn:	MVol.C2-216
	$SnBr_4$-$BiBr_3$	Sn:	MVol.C2-298
	$SnBr_4$-$GeBr_4$	Sn:	MVol.C4-234
	$SnBr_4$-$GeBr_4$-$SiBr_4$	Sn:	MVol.C4-234
	$SnBr_4$-$GeBr_4$-$SiBr_4$-$TiBr_4$	Sn:	MVol.C4-235
	$SnBr_4$-$GeBr_4$-$TiBr_4$	Sn:	MVol.C4-234

Br_4Sn .... $SnBr_4$ systems
$SnBr_4$-PBr_3 .... Sn: MVol.C2-267
$SnBr_4$-PBr_5 .... Sn: MVol.C2-267
$SnBr_4$-$POBr_3$ .... Sn: MVol.C2-267
$SnBr_4$-$POCl_3$ .... Sn: MVol.C2-268
$SnBr_4$-SO_2 .... Sn: MVol.C2-78/9
$SnBr_4$-S_2Br_2 .... Sn: MVol.C2-78
$SnBr_4$-$SbBr_3$ .... Sn: MVol.C2-287
$SnBr_4$-$TiBr_4$ .... Sn: MVol.C4-217

Br_4SnSr .... $SrSnBr_4 \cdot 6\ H_2O$ .... Sn: MVol.C3-188
Br_4Te .... $TeBr_4$ .... Te: SVol.B3-6/15
– .... $TeBr_4$ solutions
$TeBr_4$-HBr-H_2O .... Te: SVol.B3-15/6
$TeBr_4$-H_2O .... Te: SVol.B3-15/6
$TeBr_4$-organic solvents .... Te: SVol.B3-16/8

– .... $TeBr_4$ systems
$TeBr_4$-CsBr .... Te: SVol.B3-33/4
$TeBr_4$-CsBr-H_2O .... Te: SVol.B3-33/4
$TeBr_4$-$InBr_3$ .... Te: SVol.B3-13
$TeBr_4$-KBr .... Te: SVol.B3-31/2
$TeBr_4$-KBr-H_2O .... Te: SVol.B3-31/2
$TeBr_4$-LiBr-H_2O .... Te: SVol.B3-30/1
$TeBr_4$-NaBr .... Te: SVol.B3-31
$TeBr_4$-NaBr-H_2O .... Te: SVol.B3-31
$TeBr_4$-RbBr .... Te: SVol.B3-32/3
$TeBr_4$-RbBr-H_2O .... Te: SVol.B3-32/3
$TeBr_4$-Te .... Te: SVol.B3-1/2
$TeBr_4$-$TeCl_4$ .... Te: SVol.B3-49
$TeBr_4$-$TeCl_4$-TeI_4 .... Te: SVol.B3-80/1
$TeBr_4$-$TeCl_4$-TeO_2 .... Te: SVol.B3-51
$TeBr_4$-TeI_4 .... Te: SVol.B3-79
$TeBr_4$-TeO_2 .... Te: SVol.B3-47/8

Br_4Ti .... $TiBr_4$ systems
$TiBr_4$-$GeBr_4$-$SiBr_4$-$SnBr_4$ .... Sn: MVol.C4-235
$TiBr_4$-$GeBr_4$-$SnBr_4$ .... Sn: MVol.C4-234
$TiBr_4$-$SnBr_4$ .... Sn: MVol.C4-217

Br_4U .... UBr_4 .... U: SVol.C9-120/7
– .... UBr_4 systems
UBr_4-CsBr .... U: SVol.C9-134
UBr_4-KBr .... U: SVol.C9-134
UBr_4-NaBr .... U: SVol.C9-134
UBr_4-RbBr .... U: SVol.C9-134
UBr_4-UBr_3 .... U: SVol.C9-119/20

Br_4U^- .... $[UBr_4]^-$ .... U: SVol.C9-119
$Br_{4.81}I_{1.19}Rb_2Sn$ ..... $Rb_2[SnBr_{4.81}I_{1.19}]$ .... Sn: MVol.C3-139

Br_5CCs_2IrO .... $Cs_2[Ir(CO)Br_5]$ .... Ir: SVol.2-149
Br_5CFe_2N .... $2\ FeBr_2 \cdot BrCN$ .... C: MVol.D3-237
$Br_5CH_3Ti^{2-}$ .... $[CH_3TiBr_5]^{2-}$ .... Ti: Org.Verb.1-23, 42/3

$Br_5CH_5N_2OTe$	$[CON_2H_5]TeBr_5$	Te:	SVol.B3-35
Br_5ClrK_2O	$K_2[Ir(CO)Br_5]$	Ir:	SVol.2-149
$Br_5C_4H_5N_2Te$	$[C_4H_4N_2H]TeBr_5$	Te:	SVol.B3-35
$Br_5C_4H_6N_2U$	$UBr_5 \cdot 2\ CH_3CN$	U:	SVol.E1-48/50
$Br_5C_4H_8OSn^-$	$[SnBr_5(CH_3COC_2H_5)]^-$	Sn:	MVol.C5-87
$Br_5C_4H_{12}IrN_2O$	$N(CH_3)_4[Ir(NO)Br_5]$	Ir:	SVol.2-49
$Br_5C_5H_6NSn$	$[C_5H_6N]SnBr_5$	Sn:	MVol.C3-118
$Br_5C_5H_6NTe$	$[C_5H_5NH]TeBr_5$	Te:	SVol.B3-17
$Br_5C_6Fe_3O_6S_2$	$Fe_3(CO)_6S_2Br_5$	Fe:	Org.Verb.C1-18/20
$Br_5C_6H_9N_3Pa$	$PaBr_5 \cdot 3\ CH_3CN$	Pa:	SVol.2-88/9
$Br_5C_6H_9N_3U$	$UBr_5 \cdot 3\ CH_3CN$	U:	SVol.E1-48/50
$Br_5C_6H_{18}N_3OPU$	$UBr_5 \cdot [(CH_3)_2N]_3PO$	U:	SVol.E1-162, 164
$Br_5C_8Fe_3H_4O_6S_2$	$Fe_3(CO)_6(SCH_2CH_2S)Br_5$	Fe:	Org.Verb.C1-18/20
$Br_5C_8H_{20}NSn$	$[(C_2H_5)_4N]SnBr_5$	Sn:	MVol.C3-116/7
$Br_5C_8H_{20}NTe$	$[(C_2H_5)_4N]TeBr_5$	Te:	SVol.B3-19, 22, 34/5
$Br_5C_9H_{23}NTi$	$[(C_2H_5)_4N][CH_3TiBr_5]$	Ti:	Org.Verb.1-42/3
$Br_5C_{10}Fe_3H_{10}O_6S_2$	$Fe_3(CO)_6(SC_2H_5)_2Br_5$	Fe:	Org.Verb.C1-18/20
$Br_5C_{16}H_{36}NSn$	$[(C_4H_9)_4N]SnBr_5$	Sn:	MVol.C3-116/7
$Br_5C_{16}H_{40}N_2OPa$	$[N(C_2H_5)_4]_2PaOBr_5$	Pa:	SVol.2-69/70
$Br_5C_{16}H_{40}N_2OU$	$[N(C_2H_5)_4]_2UOBr_5$	U:	SVol.C9-149/50
$Br_5C_{18}Fe_2H_{30}N_2O_6S_2Sn_2$	$(C_2H_5)_6N_2Br_5Sn_2S_2Fe_2(CO)_6$	Fe:	Org.Verb.C1-76
$Br_5C_{18}Fe_3H_{10}O_6S_2$	$Fe_3(CO)_6(SC_6H_5)_2Br_5$	Fe:	Org.Verb.C1-18/20
$Br_5C_{18}H_{15}OPPa$	$PaBr_5 \cdot (C_6H_5)_3PO$	Pa:	SVol.2-93
$Br_5C_{18}H_{15}OPU$	$UBr_5 \cdot (C_6H_5)_3PO$	U:	SVol.E1-170, 176
$Br_5C_{18}H_{15}PPaS$	$PaBr_5 \cdot (C_6H_5)_3PS$	Pa:	SVol.2-94
$Br_5C_{18}H_{15}PPaSe$	$PaBr_5 \cdot (C_6H_5)_3PSe$	Pa:	SVol.2-94
$Br_5C_{18}H_{16}NSn$	$[(C_6H_5)_3NH]SnBr_5$	Sn:	MVol.C3-116
$Br_5C_{19}H_{15}Sn$	$[(C_6H_5)_3C]SnBr_5$	Sn:	MVol.C3-118
$Br_5C_{20}H_{17}Sn$	$[CH_3C_6H_4(C_6H_5)_2C][SnBr_5]$	Sn:	MVol.C3-118
$Br_5C_{21}H_{19}Sn$	$[(CH_3C_6H_4)_2(C_6H_5)C][SnBr_5]$	Sn:	MVol.C3-118
$Br_5C_{24}H_{20}OPU$	$[P(C_6H_5)_4]UOBr_5$	U:	SVol.C9-151
$Br_5C_{25}H_{22}O_2P_2Pa$	$PaBr_5 \cdot (C_6H_5)_2POCH_2PO(C_6H_5)_2$	Pa:	SVol.2-93/4
$Br_5C_{25}H_{22}P_2PaS_2$	$PaBr_5 \cdot (C_6H_5)_2PSCH_2PS(C_6H_5)_2$	Pa:	SVol.2-94
$Br_5C_{25}H_{22}P_2PaSe_2$	$PaBr_5 \cdot (C_6H_5)_2PSeCH_2PSe(C_6H_5)_2$	Pa:	SVol.2-94
$Br_5C_{32}H_{72}IrN_2$	$((C_4H_9)_4N)_2[IrBr_5]$	Ir:	SVol.2-147
$Br_5C_{33}H_{72}IrN_2O$	$((C_4H_9)_4N)_2[Ir(CO)Br_5]$	Ir:	SVol.2-147
$Br_5C_{36}H_{30}O_2P_2Pa$	$PaBr_5 \cdot 2\ (C_6H_5)_3PO$	Pa:	SVol.2-93
$Br_5C_{40}Fe_3H_{30}O_4P_2S_2$	$Fe_3(CO)_4(P(C_6H_5)_3)_2S_2Br_5$	Fe:	Org.Verb.C1-18/20
$Br_5C_{44}Fe_3H_{40}O_4P_2S_2$	$Fe_3(CO)_4(P(C_6H_5)_3)_2(SC_2H_5)_2Br_5$	Fe:	Org.Verb.C1-18/20
$Br_5C_{52}Fe_3H_{40}O_4P_2S_2$	$Fe_3(CO)_4(P(C_6H_5)_3)_2(SC_6H_5)_2Br_5$	Fe:	Org.Verb.C1-18/20
$Br_5ClH_8N_2Te$	$(NH_4)_2[TeClBr_5]$	Te:	SVol.B3-86
Br_5ClIr^{2-}	$[IrClBr_5]^{2-}$	Ir:	SVol.2-124, 150
$Br_5ClIrRb_2$	$Rb_2[IrClBr_5]$	Ir:	SVol.2-150
$Br_5CoO_8U_3$	$CoU_3O_8Br_5 \cdot 12\ H_2O$	U:	SVol.C9-151, 154/5
$Br_5CsIrNO$	$Cs[Ir(NO)Br_5]$	Ir:	SVol.2-49
Br_5CsSn_2	$CsSn_2Br_5$	Sn:	MVol.C3-151
Br_5Cs_2Ir	$Cs_2[IrBr_5]$	Ir:	SVol.2-149
Br_5Cs_2U	Cs_2UBr_5	U:	SVol.C9-120, 132/3

Br_5Cs_3Mn	$Cs_3MnBr_5 = 3\ CsBr \cdot MnBr_2$	Mn: MVol.C5-296/7
Br_5HOPa^-	$Pa(OH)Br_5^-$	Pa: SVol.2-148
$Br_5H_2IrO^{2-}$	$[IrBr_5(H_2O)]^{2-}$	Ir: SVol.2-121, 146, 150
$Br_5H_2OTe^-$	$[TeBr_5(H_2O)]^-$	Te: SVol.B3-15, 29
$Br_5H_4NSn_2$	$NH_4Sn_2Br_5$	Sn: MVol.C3-111
Br_5H_4NTe	NH_4TeBr_5	Te: SVol.B3-17
$Br_5H_8IN_2Te$	$(NH_4)_2[TeBr_5I]$	Te: SVol.B3-86
Br_5HoK_2	$K_2[HoBr_5]$	Sc: MVol.C6-124
Br_5IU^-	$[UBr_5I]^-$	U: SVol.C9-184
Br_5IrKNO	$K[Ir(NO)Br_5] \cdot H_2O$	Ir: SVol.2-48
Br_5IrK_2	$K_2[IrBr_5]$	Ir: SVol.2-149
$Br_5IrK_3NO_2$	$K_3[Ir(NO_2)Br_5]$	Ir: SVol.2-63
Br_5IrNO^-	$[Ir(NO)Br_5]^-$	Ir: SVol.2-48
$Br_5IrNO_2^{3-}$	$[Ir(NO_2)Br_5]^{3-}$	Ir: SVol.2-48
Br_5KSn_2	KSn_2Br_5	Sn: MVol.C3-63/4
Br_5K_2Sm	$K_2[SmBr_5]$	Sc: MVol.C6-124
Br_5K_2U	K_2UBr_5	U: SVol.C9-120, 132/3
Br_5K_3Sn	K_3SnBr_5	Sn: MVol.C3-63
Br_5LaO_2U	$LaUO_2Br_5 \cdot 3\ H_2O$	U: SVol.C9-151, 154/5
Br_5Na_2Sm	$Na_2[SmBr_5]$	Sc: MVol.C6-120/1
Br_5OU^{2-}	$[UOBr_5]^{2-}$	U: SVol.C9-149/50
$Br_5O_2U^{3-}$	$[UO_2Br_5]^{3-}$	U: SVol.C9-154
Br_5O_2UY	$YUO_2Br_5 \cdot 5\ H_2O$	U: SVol.C9-151, 154/5
$Br_5O_8ScU_3$	$ScU_3O_8Br_5 \cdot 8\ H_2O$	U: SVol.C9-151, 154
$Br_5O_8U_3^{3-}$	$[U_3O_8Br_5]^{3-}$	U: SVol.C9-154
Br_5P	PBr_5 systems	
	PBr_5-SnBr_4	Sn: MVol.C2-267
Br_5Pa	$PaBr_5$	Pa: SVol.2-63/6, 148
Br_5RbSn_2	$RbSn_2Br_5$	Sn: MVol.C3-135/6
Br_5Rb_2Sm	$Rb_2[SmBr_5]$	Sc: MVol.C6-128
Br_5Rb_2U	Rb_2UBr_5	U: SVol.C9-120, 132/3
Br_5SbSn	$Sn[SbBr_5]$	Sn: MVol.C2-287
$Br_5Sn_2^-$	$[Sn_2Br_5]^-$	Sn: MVol.C4-139
Br_5Sn_2Tl	$2\ SnBr_2 \cdot TlBr$	Sn: MVol.C4-139
Br_5Te^-	$TeBr_5^-$	Te: SVol.B3-15/6, 18/22
Br_5U	UBr_5	U: SVol.C9-127/9
Br_5U^{2-}	$[UBr_5]^{2-}$	U: SVol.C9-132
$Br_{5.35}H_8I_{0.65}N_2Sn$	$(NH_4)_2[SnBr_{5.35}I_{0.65}]$	Sn: MVol.C3-121
$Br_{5.75}I_{0.25}K_2Sn$	$K_2[SnBr_{5.75}I_{0.25}]$	Sn: MVol.C3-68
$Br_6C_2H_{12}IrN_2$	$(CH_3NH_3)_2[IrBr_6]$	Ir: SVol.2-146
$Br_6C_2H_{12}N_2Sn$	$[CH_3NH_3]_2[SnBr_6]$	Sn: MVol.C3-116
$Br_6C_2N_2Sn$	$SnBr_4 \cdot 2\ BrCN$	C: MVol.D3-237
$Br_6C_2N_2Ti$	$TiBr_4 \cdot 2\ BrCN$	C: MVol.D3-237
$Br_6C_2S_3$	$CBr_3SSSCBr_3$	C: MVol.D4-183
$Br_6C_3S_2$	$C_3S_2Br_6$	C: MVol.D4-6
$Br_6C_3S_3$	$(Br_2CS)_3$	C: MVol.D6-172
$Br_6C_4FeGe_2O_4$	$(CO)_4Fe(GeBr_3)_2$	Fe: Org.Verb.B2-140
$Br_6C_4FeO_4Sn_2$	$(CO)_4Fe(SnBr_3)_2$	Fe: Org.Verb.B2-141
$Br_6C_4H_{12}NPa$	$[N(CH_3)_4]PaBr_6$	Pa: SVol.2-67, 148

$Br_6C_4H_{12}NU$	$N(CH_3)_4UBr_6$	U:	SVol.C9-139/41
$Br_6C_4H_{16}IrN_2$	$(C_2H_5NH_3)_2[IrBr_6]$	Ir:	SVol.2-146
$Br_6C_8H_8Ir_2N_4S_4$	$[Ir(S_2N_2C_4H_4)Br_3]_2$	Ir:	SVol.2-172
$Br_6C_8H_{20}NPa$	$[N(C_2H_5)_4]PaBr_6$	Pa:	SVol.2-67, 148
$Br_6C_8H_{20}NU$	$N(C_2H_5)_4UBr_6$	U:	SVol.C9-139/41
$Br_6C_8H_{24}N_2Pa$	$[N(CH_3)_4]_2PaBr_6$	Pa:	SVol.2-66/7
$Br_6C_8H_{24}N_2Sn$	$[(CH_3)_4N]_2[SnBr_6]$	Sn:	MVol.C3-117
$Br_6C_8H_{24}N_2Te$	$[(CH_3)_4N]_2TeBr_6$	Te:	SVol.B3-23/9, 41/2
$Br_6C_8H_{24}N_2U$	$[N(CH_3)_4]_2UBr_6$	U:	SVol.C9-133/9
$Br_6C_{10}H_{10}Ti$	$C_5H_5Ti(C_5H_5Br_4)Br_2$	Ti:	Org.Verb.1-206
$Br_6C_{10}H_{12}N_2Sn$	$[C_5H_6N]_2[SnBr_6]$	Sn:	MVol.C3-118
$Br_6C_{10}H_{12}N_2Te$	$[C_5H_5NH]_2TeBr_6$	Te:	SVol.B3-25/6, 42
$Br_6C_{10}H_{22}O_8Sn_2$	$[Sn(CH(COCH_3)_2)_2(H_2O)_4][SnBr_6]$ $= SnBr_3(CH(COCH_3)_2) \cdot 2\,H_2O$	Sn:	MVol.C5-107
$Br_6C_{10}H_{28}N_2Sn$	$[(CH_3)_3(C_2H_5)N]_2[SnBr_6]$	Sn:	MVol.C3-117/8
$Br_6C_{12}Cl_2H_{14}N_2Sn$	$[ClC_6H_4NH_3]_2[SnBr_6]$	Sn:	MVol.C3-116
$Br_6C_{12}H_{12}N_2Te$	$TeBr_4 \cdot 2\,BrC_6H_4NH_2$	Te:	SVol.B3-167
$Br_6C_{12}H_{16}N_2Sn$	$[C_6H_5NH_3]_2[SnBr_6]$	Sn:	MVol.C3-116
$Br_6C_{12}H_{32}N_2Sn$	$[(CH_3)_2(C_2H_5)_2N]_2[SnBr_6]$	Sn:	MVol.C3-117/8
$Br_6C_{12}H_{32}P_2Sn$	$[(C_2H_5)_3PH]_2[SnBr_6]$	Sn:	MVol.C6-133
$Br_6C_{14}H_{20}N_2Sn$	$[CH_3C_6H_4NH_3]_2[SnBr_6]$	Sn:	MVol.C3-116
$Br_6C_{14}H_{36}N_2Sn$	$[(C_2H_5)_3(CH_3)N]_2[SnBr_6]$	Sn:	MVol.C3-117/8
$Br_6C_{15}Ce_2H_{15}N_3$	$2\,CeBr_3 \cdot 3\,C_5H_5N$	Sc:	MVol.C6-58/9
$Br_6C_{15}DyH_{18}N_3$	$(C_5H_5NH)_3DyBr_6$	Sc:	MVol.C6-125
$Br_6C_{15}ErH_{18}N_3$	$(C_5H_5NH)_3ErBr_6$	Sc:	MVol.C6-125
$Br_6C_{15}EuH_{18}N_3$	$(C_5H_5NH)_3EuBr_6$	Sc:	MVol.C6-125
$Br_6C_{15}GdH_{18}N_3$	$(C_5H_5NH)_3GdBr_6$	Sc:	MVol.C6-125
$Br_6C_{15}H_{10}O_5Sn_3$	$3\,SnBr_2 \cdot (HO)_2C_9H_4OC_6H_3O(OH)$	Sn:	MVol.C5-11
$Br_6C_{15}H_{10}O_6Sn_3$	$3\,SnBr_2 \cdot (HO)_2C_9H_4OC_6H_2O(OH)_2$	Sn:	MVol.C5-11
$Br_6C_{15}H_{18}N_3Nd$	$(C_5H_5NH)_3NdBr_6$	Sc:	MVol.C6-125
$Br_6C_{15}H_{18}N_3Yb$	$(C_5H_5NH)_3YbBr_6$	Sc:	MVol.C6-125
$Br_6C_{16}CeH_{40}N_2$	$[(C_2H_5)_4N]_2CeBr_6$	Sc:	MVol.C6-125/6
$Br_6C_{16}H_{36}NU$	$N(C_4H_9)_4UBr_6$	U:	SVol.C9-139
$Br_6C_{16}H_{40}N_2Pa$	$[N(C_2H_5)_4]_2PaBr_6$	Pa:	SVol.2-42, 44, 66/7
$Br_6C_{16}H_{40}N_2Sn$	$[(C_2H_5)_4N]_2[SnBr_6]$	Sn:	MVol.C3-117
–	$[(n\text{-}C_4H_9)_2NH_2]_2[SnBr_6]$	Sn:	MVol.C3-116
$Br_6C_{16}H_{40}N_2Te$	$[(C_2H_5)_4N]_2TeBr_6$	Te:	SVol.B3-25/9, 41/2
$Br_6C_{16}H_{40}N_2U$	$[N(C_2H_5)_4]_2UBr_6$	U:	SVol.C9-133/9
$Br_6C_{18}H_{14}Ni_2O_2$	$(C_3H_5NiOC_6H_2Br_3)_2$	Ni:	Org.Verb.2-283
$Br_6C_{18}H_{16}N_2Sn$	$[C_9H_8N]_2[SnBr_6]$	Sn:	MVol.C3-118
$Br_6C_{20}H_{32}N_2Sn$	$[(CH_3)_2(C_2H_5)(C_6H_5)N]_2[SnBr_6]$	Sn:	MVol.C3-117/8
$Br_6C_{20}H_{45}N_5O_7U_2$	$[UO_2(CH_3CON(CH_3)_2)_5][UBr_6]$	U:	SVol.E1-103/4, 106
$Br_6C_{24}H_{16}Ir_2N_4$	$[Ir(C_{12}H_8N_2)Br_3]_2$	Ir:	SVol.2-92
$Br_6C_{24}H_{20}PU$	$P(C_6H_5)_4UBr_6$	U:	SVol.C9-139/41
$Br_6C_{24}H_{24}N_2Sn$	$[(C_6H_5)_2NH_2]_2[SnBr_6]$	Sn:	MVol.C3-116
$Br_6C_{24}H_{36}NiPt$	$[((CH_3)_6C_6)_2Ni][PtBr_6]$	Ni:	Org.Verb.2-248
$Br_6C_{24}H_{56}N_2Sn$	$[(C_3H_7)_4N]_2[SnBr_6]$	Sn:	MVol.C3-117
$Br_6C_{26}H_{44}N_2Sn$	$[(C_2H_5)_3(C_6H_5CH_2)N]_2[SnBr_6]$	Sn:	MVol.C3-117/8
$Br_6C_{26}H_{60}N_2Sn$	$[(i\text{-}C_4H_9)_3(CH_3)N]_2[SnBr_6]$	Sn:	MVol.C3-117/8
$Br_6C_{28}H_{64}IrN_2$	$((n\text{-}C_7H_{15})_2NH_2)_2[IrBr_6]$	Ir:	SVol.2-147

$Br_6C_{28}H_{64}N_2Sn$	$[(i\text{-}C_4H_9)_3(C_2H_5)N]_2[SnBr_6]$	Sn: MVol.C3-117/8
$Br_6C_{30}H_{30}MnN_6Sn$	$[Mn(C_5H_5N)_6]SnBr_6$	Mn: MVol.C5-307/8
$Br_6C_{30}H_{36}N_2Sn$	$[(CH_3)_2(C_6H_5)(C_6H_5CH_2)N]_2[SnBr_6]$	Sn: MVol.C3-117/8
$Br_6C_{32}H_{72}IrN_2$	$[(n\text{-}C_4H_9)_4N]_2[IrBr_6]$	Ir: SVol.2-146
$Br_6C_{32}H_{72}N_2Sn$	$[(C_4H_9)_4N]_2[SnBr_6]$	Sn: MVol.C3-117
$Br_6C_{32}H_{72}N_2Te$	$[(C_4H_9)_4N]_2TeBr_6$	Te: SVol.B3-25/30
$Br_6C_{32}H_{72}N_2U$	$[N(C_4H_9)_4]_2UBr_6$	U: SVol.C9-138
$Br_6C_{36}H_{24}S_3Sn_3$	$[(BrC_6H_4)_2SnS]_3$	Sn: Org.Verb.6-186
$Br_6C_{36}H_{32}P_2Sn$	$[(C_6H_5)_3PH]_2[SnBr_6]$	Sn: MVol.C3-118
$Br_6C_{36}H_{32}P_2U$	$[P(C_6H_5)_3H]_2UBr_6$	U: SVol.C9-134, 136
$Br_6C_{36}H_{72}MnN_{24}Sn$	$[Mn((CH_2)_6N_4)_6]SnBr_6$	Mn: MVol.C5-307/8
–	$[Mn((CH_2)_6N_4)_6]SnBr_6 \cdot 2\ H_2O$	Mn: MVol.C5-307/8
$Br_6C_{38}H_{30}Sn$	$[(C_6H_5)_3C]_2[SnBr_6]$	Sn: MVol.C3-118
$Br_6C_{40}H_{34}Sn$	$[(CH_3C_6H_4)(C_6H_5)_2C]_2[SnBr_6]$	Sn: MVol.C3-118
$Br_6C_{40}H_{40}P_2U$	$[P(C_6H_5)_3C_2H_5]_2UBr_6$	U: SVol.C9-134
$Br_6C_{42}H_{38}Sn$	$[(CH_3C_6H_4)_2(C_6H_5)C]_2[SnBr_6]$	Sn: MVol.C3-118
$Br_6C_{42}H_{92}IrN_2$	$((n\text{-}C_7H_{15})_3NH)_2[IrBr_6]$	Ir: SVol.2-147
$Br_6C_{44}H_{42}Sn$	$[(CH_3C_6H_4)_3C]_2[SnBr_6]$	Sn: MVol.C3-118
$Br_6C_{44}H_{48}P_2U$	$[P(C_6H_5)_3(C_4H_9)]_2UBr_6$	U: SVol.C9-134/6
$Br_6C_{50}H_{44}P_2U$	$[P(C_6H_5)_3(C_6H_5CH_2)]_2UBr_6$	U: SVol.C9-134
$Br_6C_{54}DyH_{48}P_3$	$[(C_6H_5)_3PH]_3DyBr_6$	Sc: MVol.C6-126
$Br_6C_{54}ErH_{48}P_3$	$[(C_6H_5)_3PH]_3ErBr_6$	Sc: MVol.C6-126
$Br_6C_{54}EuH_{48}P_3$	$[(C_6H_5)_3PH]_3EuBr_6 \cdot x\ (C_6H_5)_3PHBr$	Sc: MVol.C6-126
$Br_6C_{54}H_{48}LaP_3$	$[(C_6H_5)_3PH]_3LaBr_6 \cdot x\ (C_6H_5)_3PHBr$	Sc: MVol.C6-126
$Br_6C_{54}H_{48}NdP_3$	$[(C_6H_5)_3PH]_3NdBr_6$	Sc: MVol.C6-126
$Br_6C_{54}H_{48}P_3Pr$	$[(C_6H_5)_3PH]_3PrBr_6 \cdot x\ (C_6H_5)_3PHBr$	Sc: MVol.C6-126
$Br_6C_{54}H_{48}P_3Sm$	$[(C_6H_5)_3PH]_3SmBr_6 \cdot x\ (C_6H_5)_3PHBr$	Sc: MVol.C6-126
$Br_6C_{54}H_{48}P_3Yb$	$[(C_6H_5)_3PH]_3YbBr_6$	Sc: MVol.C6-126
$Br_6C_{56}H_{120}IrN_2$	$[(n\text{-}C_7H_{15})_4N]_2[IrBr_6]$	Ir: SVol.2-147
$Br_6CaH_{18}N_6Sn$	$Ca[SnBr_6] \cdot 6\ NH_3$	Sn: MVol.C3-177
Br_6CaSn	$Ca[SnBr_6] \cdot 6\ H_2O$	Sn: MVol.C3-177
Br_6CdSn	$Cd[SnBr_6] \cdot 6\ H_2O$	Sn: MVol.C4-56
Br_6Ce^{2-}	$CeBr_6^{2-}$	Sc: MVol.B1-44
Br_6Ce^{3-}	$CeBr_6^{3-}$	Sc: MVol.B1-41
Br_6CeLi_3	Li_3CeBr_6	Sc: MVol.C6-116/7
Br_6CeO_2U	$CeUO_2Br_6 \cdot 4\ H_2O$	U: SVol.C9-151, 154/5
$Br_6Cl_4O_2P_2Sn$	$SnCl_4 \cdot 2\ POBr_3 = [POBr_2]_2SnBr_2Cl_4$	Sn: MVol.C2-268
Br_6CsU	$CsUBr_6$	U: SVol.C9-139/41
Br_6Cs_2Ir	$Cs_2[IrBr_6]$	Ir: SVol.2-149
Br_6Cs_2Sn	$Cs_2[SnBr_6]$	Sn: MVol.C3-152/3
Br_6Cs_2Te	Cs_2TeBr_6	Te: SVol.B3-23/6, 33/4, 44/6
–	Cs_2TeBr_6 systems	
	$Cs_2TeBr_6\text{-}HBr\text{-}Rb_2TeBr_6\text{-}H_2O$	Te: SVol.B3-46/7
	$Cs_2TeBr_6\text{-}K_2TeBr_6$	Te: SVol.B3-46
	$Cs_2TeBr_6\text{-}K_2TeBr_6\text{-}Rb_2TeBr_6$	Te: SVol.B3-47
	$Cs_2TeBr_6\text{-}Rb_2TeBr_6$	Te: SVol.B3-46
Br_6Cs_2U	Cs_2UBr_6	U: SVol.C9-133/9
Br_6Cs_3La	Cs_3LaBr_6	Sc: MVol.C6-128/9
Br_6Cs_3Nd	Cs_3NdBr_6	Sc: MVol.C6-114/5, 128/9

Br_6Cs_3Pr .... Cs_3PrBr_6 .... Sc: MVol.C6-114/5, 128/9
Br_6Cs_3Sc .... Cs_3ScBr_6 .... Sc: MVol.C6-114/5, 128
Br_6Cs_3Sm .... Cs_3SmBr_6 .... Sc: MVol.C6-114/5, 128/9
Br_6Cs_3U .... Cs_3UBr_6 .... U: SVol.C9-120, 132/3
Br_6Eu^{3-} .... $EuBr_6^{3-}$ .... Sc: MVol.B1-43
Br_6Eu_4O .... Eu_4OBr_6 .... Sc: MVol.C6-86
Br_6GdNa_3 .... Na_3GdBr_6 .... Sc: MVol.C6-114, 121/2
Br_6HU .... $HUBr_6$ .... U: SVol.C9-139
Br_6H_2Ir .... $H_2[IrBr_6] \cdot n\ H_2O$ .... Ir: SVol.2-146
Br_6H_2Te .... H_2TeBr_6 .... Te: SVol.B3-25/6, 30
– .... $H_2TeBr_6 \cdot 2\ H_2O$ .... Te: SVol.B3-30
Br_6H_4NU .... NH_4UBr_6 .... U: SVol.C9-139/41
$Br_6H_8IrN_2$ .... $(NH_4)_2[IrBr_6]$ .... Ir: SVol.2-149
$Br_6H_8N_2Sb$ .... $(NH_4)_2[SbBr_6]$ solid solutions
$(NH_4)_2[(Sb,Sn)Br_6]$ .... Sn: MVol.C3-115
$Br_6H_8N_2Sc$ .... $(NH_4)_2[ScBr_6]$ solid solutions
$(NH_4)_2[(Sc,Sn)Br_6]$ .... Sn: MVol.C3-115
$Br_6H_8N_2Sn$ .... $(NH_4)_2[SnBr_6]$ .... Sn: MVol.C3-113/6
– .... $(NH_4)_2[SnBr_6]$ solid solutions
$(NH_4)_2[(Sn,Sb)Br_6]$ .... Sn: MVol.C3-115
$(NH_4)_2[(Sn,Sc)Br_6]$ .... Sn: MVol.C3-115
$Br_6H_8N_2Te$ .... $(NH_4)_2TeBr_6$ .... Te: SVol.B3-23/9, 40/1
$Br_6H_8Na_2O_4Te$ .... $[Na(H_2O)_2]_2TeBr_6$ .... Te: SVol.B3-36
– .... $[Na(H_2O)_2]_2TeBr_6 \cdot 2\ H_2O$
$= Na_2TeBr_6 \cdot 6\ H_2O$ .... Te: SVol.B3-31, 36/7
$Br_6H_{10}N_4Sn$ .... $(N_2H_5)_2[SnBr_6]$ .... Sn: MVol.C3-116
$Br_6H_{16}N_4Sn$ .... $(NH_4)_4SnBr_6 \cdot H_2O$ .... Sn: MVol.C3-112
$Br_6H_{18}N_6SnSr$ .... $Sr[SnBr_6] \cdot 6\ NH_3$ .... Sn: MVol.C3-188
Br_6HgSn .... $Hg[SnBr_6] \cdot 6\ H_2O$ .... Sn: MVol.C4-97
Br_6Ir^{2-} .... $[IrBr_6]^{2-}$ .... Ir: SVol.2-120, 143/6
Br_6Ir^{3-} .... $[IrBr_6]^{3-}$ .... Ir: SVol.2-143/6
Br_6Ir^{4-} .... $[IrBr_6]^{4-}$ .... Ir: SVol.2-143
Br_6IrK_2 .... $K_2[IrBr_6]$ .... Ir: SVol.2-148/9
Br_6IrK_3 .... $K_3[IrBr_6] \cdot n\ H_2O$ .... Ir: SVol.2-147/8
Br_6IrNa_2 .... $Na_2[IrBr_6] \cdot 6\ H_2O$ .... Ir: SVol.2-147
Br_6IrRb_2 .... $Rb_2[IrBr_6]$ .... Ir: SVol.2-149
Br_6Ir_2O .... Ir_2OBr_6 .... Ir: SVol.2-143
Br_6KU .... $KUBr_6$ .... U: SVol.C9-139/41
Br_6K_2Sn .... $K_2[SnBr_6]$ .... Sn: MVol.C3-65/6
Br_6K_2Te .... K_2TeBr_6 .... Te: SVol.B3-23/9, 31/2, 37/40
– .... K_2TeBr_6 systems
K_2TeBr_6-Cs_2TeBr_6 .... Te: SVol.B3-46
K_2TeBr_6-Cs_2TeBr_6-Rb_2TeBr_6 .... Te: SVol.B3-47
K_2TeBr_6-Rb_2TeBr_6 .... Te: SVol.B3-46
Br_6K_2U .... K_2UBr_6 .... U: SVol.C9-133/9
Br_6K_3Sc .... K_3ScBr_6 .... Sc: MVol.C6-115, 122/3
Br_6K_3Sm .... $K_3[SmBr_6]$ .... Sc: MVol.C6-124
Br_6K_3U .... K_3UBr_6 .... U: SVol.C9-120, 132

Br_6K_4Mn	K_4MnBr_6 = 4 KBr · $MnBr_2$	Mn: MVol.C5-292
Br_6K_4Sn	K_4SnBr_6 · H_2O	Sn: MVol.C3-65
Br_6Li_2Sn	$Li_2[SnBr_6]$ · n H_2O	Sn: MVol.C3-5/6
Br_6Li_2Te	Li_2TeBr_6 · 6 H_2O	Te: SVol.B3-30/1, 35
Br_6Li_3Sc	Li_3ScBr_6	Sc: MVol.C6-116
Br_6MgMn_2	$Mg(MnBr_3)_2$ · 12 H_2O	Mn: MVol.C5-305
Br_6MgSn	$Mg[SnBr_6]$ · 6 H_2O	Sn: MVol.C3-166
–	$Mg[SnBr_6]$ · 10 H_2O	Sn: MVol.C3-166
Br_6Mn^{4-}	$[MnBr_6]^{4-}$	Mn: MVol.C5-285
Br_6MnSn	$MnSnBr_6$ · 6 H_2O	Mn: MVol.C5-307/8
Br_6MnTe	$MnTeBr_6$ · 6 H_2O	Mn: MVol.C6-357
Br_6NaU	$NaUBr_6$	U: SVol.C9-139/41
Br_6Na_2Sn	$Na_2[SnBr_6]$ · 6 H_2O	Sn: MVol.C3-25
Br_6Na_2Te	Na_2TeBr_6	Te: SVol.B3-36
–	Na_2TeBr_6 · 6 H_2O = $[Na(H_2O)_2]_2TeBr_6$ · 2 H_2O	Te: SVol.B3-31, 36/7
Br_6Na_2U	Na_2UBr_6	U: SVol.C9-133/9
Br_6Na_3Sc	Na_3ScBr_6	Sc: MVol.C6-115, 117
Br_6O_2ThU	$ThUO_2Br_6$ · 4 H_2O	U: SVol.C9-151, 154
Br_6Pa^{2-}	$PaBr_6^{2-}$	Pa: SVol.2-43
Br_6RbU	$RbUBr_6$	U: SVol.C9-139/41
Br_6Rb_2Sn	$Rb_2[SnBr_6]$	Sn: MVol.C3-137
Br_6Rb_2Te	Rb_2TeBr_6	Te: SVol.B3-24/6, 32/4, 42/4
–	Rb_2TeBr_6 systems	
	Rb_2TeBr_6-Cs_2TeBr_6	Te: SVol.B3-46
	Rb_2TeBr_6-Cs_2TeBr_6-HBr-H_2O	Te: SVol.B3-46/7
	Rb_2TeBr_6-Cs_2TeBr_6-K_2TeBr_6	Te: SVol.B3-47
	Rb_2TeBr_6-K_2TeBr_6	Te: SVol.B3-46
Br_6Rb_2U	Rb_2UBr_6	U: SVol.C9-133/9
Br_6Rb_3Sc	Rb_3ScBr_6	Sc: MVol.C6-127
Br_6Rb_3U	Rb_3UBr_6	U: SVol.C9-120, 132/3
Br_6SSi_2	$Br_3SiSSiBr_3$	S: SVol.2-83, 186
Br_6Sc^{3-}	$[ScBr_6]^{3-}$	Sc: MVol.C6-130/1
Br_6Sm^{3-}	$SmBr_6^{3-}$	Sc: MVol.B1-43
Br_6SnSr	$Sr[SnBr_6]$ · 6 H_2O	Sn: MVol.C3-188
Br_6SnZn	$Zn[SnBr_6]$ · 6 H_2O	Sn: MVol.C4-12
Br_6Te^{2-}	$TeBr_6^{2-}$	Te: SVol.B2-114/5
		Te: SVol.B3-15/6, 18/30
Br_6Tm^{3-}	$TmBr_6^{3-}$	Sc: MVol.B1-43
Br_6U	UBr_6	U: SVol.C9-113
Br_6U^-	$[UBr_6]^-$	U: SVol.C9-139/41
Br_6U^{2-}	UBr_6^{2-}	U: SVol.C9-133/9
Br_6U^{3-}	$[UBr_6]^{3-}$	U: SVol.C9-119, 132
Br_6Yb^{3-}	$YbBr_6^{3-}$	Sc: MVol.B1-43
$Br_{6.1}CsU_{1.7}$	CsBr · 1.7 UBr_3	U: SVol.C9-133
$Br_7C_2H_6Ti_2^-$	$[(CH_3)_2Ti_2Br_7]^-$	Ti: Org.Verb.1-23
$Br_7C_{14}ClH_{14}N_2Sn$	$[C_6H_5C(Br)NH_2]_2[SnBr_5Cl]$	Sn: MVol.C3-119
Br_7Ce_2K	$K[Ce_2Br_7]$	Sc: MVol.C6-124
Br_7Ce_2Na	$Na[Ce_2Br_7]$	Sc: MVol.C6-119

Br_7CsGd_2	$CsGd_2Br_7$	Sc: MVol.C6-114/5, 128/9
Br_7CsNd_2	$CsNd_2Br_7$	Sc: MVol.C6-114/5, 128/9
Br_7CsPr_2	$CsPr_2Br_7$	Sc: MVol.C6-114/5, 128/9
Br_7CsSm_2	$CsSm_2Br_7$	Sc: MVol.C6-114/5, 128/9
Br_7Eu_2K	$K[Eu_2Br_7]$	Sc: MVol.C6-124
Br_7Eu_2Na	$Na[EuBr_4] \cdot EuBr_3$	Sc: MVol.C6-121
Br_7InTe	$TeBr_4 \cdot InBr_3$	Te: SVol.B3-12
Br_7Ir_2	Ir_2Br_7	Ir: SVol.2-142
Br_7KPr_2	$K[Pr_2Br_7]$	Sc: MVol.C6-124
Br_7La_2Na	$Na[La_2Br_7]$	Sc: MVol.C6-118
Br_7NaPr_2	$Na[Pr_2Br_7]$	Sc: MVol.C6-119
Br_7OPSn	$SnBr_4 \cdot POBr_3$	Sn: MVol.C2-267/8
$Br_{7.5}C_{36}H_{90}N_{4.5}OU$	$[N(C_2H_5)_4]_2UOBr_5 \cdot 2.5\ N(C_2H_5)_4Br$	U: SVol.C9-149/50
$Br_8C_2H_6Ti_2^{2-}$	$[(CH_3)_2Ti_2Br_8]^{2-}$	Ti: Org.Verb.1-23
$Br_8C_4H_{12}NSbTe$	$[(CH_3)_4N][SbBr_2]TeBr_6$	Te: SVol.B3-13
$Br_8C_7H_9NTe_2$	$2\ TeBr_4 \cdot (CH_3)_2C_5H_3N$	Te: SVol.B3-171
$Br_8C_8H_{11}NTe_2$	$2\ TeBr_4 \cdot C_6H_5N(CH_3)_2$	Te: SVol.B3-168
$Br_8C_{10}FeH_{10}Sn_2$	$[Fe(C_5H_5)_2][SnBr_4]_2$	Fe: Org.Verb.A1-225
$Br_8C_{12}H_{14}N_2Sn$	$[BrC_6H_4NH_3]_2[SnBr_6]$	Sn: MVol.C3-116
$Br_8C_{12}MnO_4^{2-}$	$[Mn(BrC_6Br_3O_2)_2]^{2-}$	Mn: MVol.D1-45
$Br_8C_{14}H_{14}N_2Sn$	$[C_6H_5C(Br)NH_2]_2[SnBr_6]$	Sn: MVol.C3-118
$Br_8C_{24}H_{30}N_{12}O_6Sn_2$	$2\ SnBr_4 \cdot 3\ (CH_3)_3C_5HN_4O_2 \cdot 3\ CHCl_3$	Sn: MVol.C5-228
$Br_8C_{27}Cl_9H_{33}N_{12}O_6Sn_2$	$2\ SnBr_4 \cdot 3\ (CH_3)_3C_5HN_4O_2 \cdot 3\ CHCl_3$	Sn: MVol.C5-228
$Br_8C_{34}FeH_{18}O_4$	$[Fe(C_5H_5)_2][OC_6H_4C_6Br_4O]_2$	Fe: Org.Verb.A1-229
$Br_8C_{36}H_{16}N_4NpO_4$	$Np(NC_9H_4Br_2O)_4$	Np: TrU.D1-167
$Br_8Cs_2Sn_3$	$Cs_2Sn_3Br_8$	Sn: MVol.C3-151
$Br_8H_9P_3Sn_2$	$2\ SnBr_4 \cdot 3\ PH_3$	Sn: MVol.C2-267
$Br_8Hg_4I_2Sn$	$SnHg_4I_2Br_8$	Sn: MVol.C4-97
$Br_8Mn_3O_2U$	$Mn_3UO_2Br_8 \cdot 11\ H_2O$	U: SVol.C9-151, 154/5
Br_8SnTe	$SnBr_4 \cdot TeBr_4$	Sn: MVol.C2-211
$Br_9C_3N_3Sb_2$	$2\ SbBr_3 \cdot 3\ BrCN$	C: MVol.D3-236
$Br_9C_{10}Ce_3H_{10}N_2$	$3\ CeBr_3 \cdot 2\ C_5H_5N$	Sc: MVol.C6-58/9
$Br_9C_{36}CeH_{66}N_6O_6$	$[Ce(C_6H_{11}ON)_6][Br(Br_2)]_3$	Sc: MVol.B7-32
$Br_9Cs_3Sc_2$	$Cs_3Sc_2Br_9$	Sc: MVol.C6-128/31
$Br_9K_3Sc_2$	$K_3Sc_2Br_9$	Sc: MVol.C6-115, 122/3
Br_9PSn	$SnBr_4 \cdot PBr_5$	Sn: MVol.C2-267
$Br_9Rb_3Sc_2$	$Rb_3Sc_2Br_9$	Sc: MVol.C6-127
$Br_9Sc_2^{3-}$	$[Sc_2Br_9]^{3-}$	Sc: MVol.C6-130/1
$Br_{10}C_{18}F_{15}P$	$[(C_6F_5)_3PBr]^+Br_9^-$	F: PerFHalOrg.3-124
$Br_{10}C_{60}H_{80}N_6Sn_2$	$2\ HSnBr_5 \cdot 3\ (CH_2)_5C(C_4H_2NH)_2C(CH_2)_5$	Sn: MVol.C5-229
$Br_{10}Na_7Sm$	Na_7SmBr_{10}	Sc: MVol.C6-120/1
$Br_{10}O_2P_2Sn$	$SnBr_4 \cdot 2\ POBr_3$	Sn: MVol.C2-267/8
$Br_{10}S_2Sn_2$	$2\ SnBr_4 \cdot S_2Br_2$	Sn: MVol.C2-78
$Br_{10}Sb_2Te$	$[SbBr_2]_2TeBr_6$	Te: SVol.B3-12
$Br_{11}C_5Fe_3N_5$	$Fe_3Br_8 \cdot 3\ BrCN \cdot (CN)_2$	C: MVol.D3-237/8
$Br_{11}Sm_5$	Sm_5Br_{11}	Sc: MVol.C6-62/4
$Br_{12}C_4Fe_3N_4$	$Fe_3Br_8 \cdot 4\ BrCN$	C: MVol.D3-238
$Br_{12}C_8H_{11}NTe_3$	$3\ TeBr_4 \cdot C_6H_5N(CH_3)_2$	Te: SVol.B3-168

$Br_{12}C_{10}FeH_{10}Ti_3$	$[Fe(C_5H_5)_2][TiBr_4]_3$	Fe: Org.Verb.A1-225
$Br_{12}C_{30}H_{22}N_6Sn_3$	$3\ SnBr_4 \cdot 2\ NC_5H_3(C_5H_4N)_2$	Sn: MVol.C5-219
$Br_{12}C_{40}Fe_7H_{40}$	$3\ [(Fe(C_5H_5)_2)FeBr_4 \cdot H_2O] \cdot Fe(C_5H_5)_2$	Fe: Org.Verb.A1-223
$Br_{12}Sn_2Te$	$Sn_2TeBr_{12} \cdot 8\ H_2O$	Sn: MVol.C2-211
$Br_{13}C_4Fe_3N_4$	$3\ FeBr_3 \cdot 4\ BrCN$	C: MVol.D3-328
$Br_{13}CsLa_4$	$CsLa_4Br_{13}$	Sc: MVol.C6-114/5, 128/30
$Br_{13}Sm_6$	Sm_6Br_{13}	Sc: MVol.C6-62/4
$Br_{13}Sn_4Tl_5$	$4\ SnBr_2 \cdot 5\ TlBr$	Sn: MVol.C4-139
$Br_{14}C_{10}FeH_{10}Hg_7$	$[Fe(C_5H_5)_2][HgBr_2]_7$	Fe: Org.Verb.A1-225
$Br_{14}P_2Sn$	$SnBr_4 \cdot 2\ PBr_5$	Sn: MVol.C2-267
$Br_{14}S_2Sn_9$	$Sn_9S_2Br_{14} = 2\ SnS \cdot 7\ SnBr_2$	Sn: MVol.C2-78
$Br_{15}C_{29}H_{53}I_4N_{16}Sn$	$SnI_4 \cdot 4\ (CH_2)_6N_4 \cdot 5\ CHBr_3$	Sn: MVol.C5-230
$Br_{16}C_{36}H_{72}N_{24}S_{12}Te_7$	$Te_7(SC(NHCH_2)_2)_{12}Br_{16}$	Te: SVol.B3-151
$Br_{16}Cs_7Gd_3$	$Cs_7Gd_3Br_{16} = 7\ CsBr \cdot 3\ GdBr_3$	Sc: MVol.C6-128/9
$Br_{16}Hg_8I_4Sn$	$Hg_8SnI_4Br_{16}$	Sn: MVol.C4-97
$Br_{16}Te_4$	$(TeBr_4)_4$	Te: SVol.B3-9, 19
$Br_{18}Ce_2H_{24}O_{12}Sn_3$	$[Ce(H_2O)_6]_2[SnBr_6]_3 \cdot 12\ H_2O$	Sn: MVol.C4-154
$Br_{18}Er_2H_{24}O_{12}Sn_3$	$[Er(H_2O)_6]_2[SnBr_6]_3 \cdot 12\ H_2O$	Sn: MVol.C4-154
$Br_{18}H_{24}La_2O_{12}Sn_3$	$[La(H_2O)_6]_2[SnBr_6]_3$	Sn: MVol.C4-154
$Br_{18}H_{24}O_{12}Pr_2Sn_3$	$[Pr(H_2O)_6]_2[SnBr_6]_3 \cdot 12\ H_2O$	Sn: MVol.C4-154
$Br_{18}H_{24}O_{12}Sn_3Y_2$	$[Y(H_2O)_6]_2[SnBr_6]_3 \cdot 12\ H_2O$	Sn: MVol.C4-154
$Br_{22}Sm_{10}$	$Sm_{10}Br_{22}$	Sc: MVol.C6-62/4
$Br_{24}C_{38}H_{68}I_4N_{20}Sn$	$SnI_4 \cdot 5\ (CH_2)_6N_4 \cdot 8\ CHBr_3$	Sn: MVol.C5-230
$Br_{24}Sm_{11}$	$Sm_{11}Br_{24}$	Sc: MVol.C6-62/4
$Br_{26}Sm_{12}$	$Sm_{12}Br_{26}$	Sc: MVol.C6-62/4
$C_{0.32}Pr_3Sn$	$Pr_3SnC_{0.32}$	Sn: MVol.C4-159
$C_{0.36}Pr_3Sn$	$Pr_3SnC_{0.36}$	Sn: MVol.C4-160
$C_{0.44}Sm_3Sn$	$Sm_3SnC_{0.44}$	Sn: MVol.C4-159
$C_{0.45}SnYb_3$	$Yb_3SnC_{0.45}$	Sn: MVol.C4-159
$C_{0.46}Ce_3Sn$	$Ce_3SnC_{0.46}$	Sn: MVol.C4-159/60
$C_{0.48}Ho_3Sn$	$Ho_3SnC_{0.48}$	Sn: MVol.C4-159
$C_{0.5}Cl_2O_{2.5}Sb$	$SbO_2Cl \cdot 0.5\ COCl_2$	C: MVol.D3-98
$C_{0.55}La_3Sn$	$La_3SnC_{0.55}$	Sn: MVol.C4-159
$C_{0.8}H_{0.4}Th$	$ThC_{0.8}H_{0.4}$	Th: SVol.C1-38
C	Carbon systems	
	C-Ce-Ir	Ir: SVol.1-58
	C-Fe	Fe: SVol.GD5a-1
		Fe: SVol.GD5b-1
	C-Hf-Sn	Sn: MVol.C4-227/8
	C-Ir	Ir: SVol.2-180
	C-O-Sn	Sn: MVol.C2-217/8
	C-Sn	Sn: MVol.C2-217
	C-Sn-Ti	Sn: MVol.C4-217
	C-Sn-Zr	Sn: MVol.C4-226
C^+	C^+	
	Formation	C: MVol.D4-159
$CCaF_2N_2O_5S_2$	$Ca[FSO_2NCONSO_2F] \cdot 2\ H_2O$	S: S-N-Verb.1-103
$CCaS_3$	$CaCS_3$	C: MVol.D4-223

$CCa_3H_4O_4S_4$	$CaCS_4 \cdot 2\ Ca(OH)_2 \cdot 8\ H_2O$	C:	MVol.D4-231
$CCdS_3$	$CdCS_3$	C:	MVol.D4-224
$CCfNS^{2+}$	$Cf(SCN)^{2+}$	Np:	TrU.D1-145
$CClCuF_3HN_2O_2$	$CF_3N(NO)OH \cdot CuCl$	F:	PerFHalOrg.7-87
CClFHN	ClFCNH	F:	PerFHalOrg.7-25/6
CClFHNO	FClCNOH	F:	PerFHalOrg.7-88, 123
–	$[CFClNOH]_n$	F:	PerFHalOrg.7-88
CClFN	FClCN	C:	MVol.D3-206
$CClFNO_2S^+$	FSO_2NCCl^+	S:	S-N-Verb.1-117
$CClFN_2$	$(F)(Cl)CN_2$	F:	PerFHalOrg.5-4/5, 10/1, 20
CClFO	COClF	C:	MVol.D3-103/10
CClFS	FClCS	C:	MVol.D6-161/3
$CClF_2H_3O_6S_2Sn$	$CH_3SnCl(SO_3F)_2$	Sn:	Org.Verb.6-219/20
$CClF_2H_4N_4O_2PS_2$	$[NS(O)F]_2[NPClNHCH_3]$	S:	S-N-Verb.1-27
$CClF_2N$	CFClNF	F:	PerFHalOrg.5-20/1
$CClF_2NO$	$ClCF_2NO$	F:	PerFHalOrg.7-173/4, 182, 188, 191/2, 195, 199/200
–	$NF_2C(O)Cl$	C:	MVol.D3-135/6
$CClF_2NOS$	ClF_2CNSO	S:	SVol.1-17
$CClF_2NO_5S_2$	$ClC(O)N(SO_2F)_2$	S:	S-N-Verb.1-53
$CClF_2N_3$	$(Cl)(NF_2)CN_2$	F:	PerFHalOrg.5-4/5, 11, 21
$CClF_3FeO_2S$	CF_3SO_2FeCl	Fe:	Org.Verb.B3-188
$CClF_3HNO_4S_2$	$CF_3SO_2NHSO_2Cl$	S:	S-N-Verb.1-55
$CClF_3Mg$	CF_3MgCl	F:	PerFHalOrg.4-61, 71
$CClF_3NiO_2S$	CF_3SO_2NiCl	Ni:	Org.Verb.1-233
$CClF_3S$	CF_3SCl	S:	SVol.2-289
$CClF_4N$	$CClF_2NF_2$	C:	MVol.D3-206
–	CF_3NClF	C:	MVol.D3-206
$CClF_4NO_2S$	FSO_2NClCF_3	S:	S-N-Verb.1-110/1
$CClF_4NS$	$CClF_2NSF_2$	S:	SVol.2-82
$CClF_4P$	CF_3PFCl	F:	PerFHalOrg.3-72, 89
$CClF_8NS$	$SF_5NCl(CF_3)$	S:	S-N-Verb.1-141
$CClHN_2$	HClNCN	C:	MVol.D3-294
CClHOS	$OCS \cdot HCl$	C:	MVol.D5-133
$CClH_2NO$	$NH_2C(O)Cl$	C:	MVol.D3-131/2
$CClH_2NO_4S$	$ClSO_2NHCOOH$	S:	S-N-Verb.1-129
$CClH_2NS$	$HNCS \cdot HCl$	C:	MVol.D6-35
$CClH_3I_2Sn$	CH_3SnClI_2	Sn:	Org.Verb.6-218
$CClH_3N_2O$	$(NHCl)(NH_2)CO$	C:	MVol.D3-126/7
$CClH_3S$	CH_3SCl	S:	SVol.2-237, 262
$CClH_3S_2$	CH_3SSCl	S:	SVol.2-237, 262
$CClH_3Sn^+$	CH_3SnCl^+	Sn:	Org.Verb.5-76
		Sn:	Org.Verb.6-13
$CClH_3Sn^{2+}$	CH_3SnCl^{2+}	Sn:	Org.Verb.6-219/20
$CClH_3Ti$	CH_3TiCl	Ti:	Org.Verb.1-7
$CClH_4NO_2S$	$ClSO_2NHCH_3$	S:	S-N-Verb.1-123
$CClH_5N_2O_4S$	$(NH_2)_2CS \cdot HClO_4$	C:	MVol.D6-110, 122, 125/6
$CClH_5N_2S$	$(NH_2)_2CS \cdot HCl$	C:	MVol.D6-121, 125
$CClH_8N_5$	$HNC(NHNH_2)_2 \cdot HCl$	C:	MVol.D3-211

$CClH_{12}MnO_5^+$	$MnCl(H_2O)_4(CH_3OH)^+$	Mn:	MVol.C5-78
$CClH_{15}IrN_5O_7$	$[Ir(NH_3)_5CO_3]ClO_4 \cdot H_2O$	Ir:	SVol.2-46
$CClIN^-$	$[ICNCl]^-$	C:	MVol.D3-263
CClN	ClCN		
	Chemical reactions	C:	MVol.D3-202/17
	with electrons	C:	MVol.D3-204
	with elements	C:	MVol.D3-206/7
	with ions	C:	MVol.D3-205
	with inorganic compounds	C:	MVol.D3-207/16
	with organic compounds	C:	MVol.D3-217
	Crystallographic properties	C:	MVol.D3-193
	Diffusion in air	C:	MVol.D3-197
	Electrical properties	C:	MVol.D3-202
	Electrochemical behavior	C:	MVol.D3-202
	Formation	C:	MVol.D3-185/93
	Formation of radioactive ClCN	C:	MVol.D3-192
	Hydrolysis	C:	MVol.D3-207/9
	Isotope exchange	C:	MVol.D3-205/6
	Magnetic properties	C:	MVol.D3-197
	Mechanical properties	C:	MVol.D3-193/4, 197
	Molecule	C:	MVol.D3-151/80
	Photolysis	C:	MVol.D3-203/4
	Polymerization	C:	MVol.D3-203
	Preparation	C:	MVol.D3-185/92
	Purification	C:	MVol.D3-191/2
	Pyrolysis	C:	MVol.D3-203/4
	Review	C:	MVol.D3-149/50
	Solubility	C:	MVol.D3-216/7
	Spectra	C:	MVol.D3-197/202
	Stability	C:	MVol.D3-203
	Thermal properties	C:	MVol.D3-194/7
	Toxicity	C:	MVol.D3-150/1
–	$ClC^{13}N$	C:	MVol.D3-192
–	$Cl^{11}CN$	C:	MVol.D3-192
–	ClNC	C:	MVol.D3-149
–	ClCN systems		
	ClCN-HCN	C:	MVol.D3-216
	$ClCN-H_2O$	C:	MVol.D3-216
$CClN^+$	$ClCN^+$	C:	MVol.D3-265
$CClN^-$	$ClCN^-$	C:	MVol.D3-265/6
CClNO	ClNCO	C:	MVol.D3-137/42
$CClNO_2S$	$ClCN \cdot SO_2$	C:	MVol.D3-210
$CClNO_3S$	$ClSO_2NCO$	S:	S-N-Verb.1-129
$CClNO_6S_2$	$ClSO_2OSO_2NCO$	S:	S-N-Verb.1-158
CClNS	ClSCN	C:	MVol.D6-174/6
–	$(ClSCN)_x$	C:	MVol.D6-177/8
$CClNS^-$	$ClSCN^-$	C:	MVol.D5-177
CClO	ClCO	C:	MVol.D3-6/10
$CClSe^+$	$ClCSe^+$	C:	MVol.D6-251

$CCl_2FH_3O_3SSn$	$CH_3SnCl_2(SO_3F)$	Sn:	Org.Verb.6-219
CCl_2FH_3Sn	CH_3SnFCl_2	Sn:	Org.Verb.6-295
CCl_2FNO	Cl_2CFNO	F:	PerFHalOrg.7-173/4, 182, 188, 199/202
CCl_2FNO_2S	FSO_2NCCl_2	S:	S-N-Verb.1-116/7
$CCl_2F_2NO_2P$	$CF_2NOP(O)Cl_2$	F:	PerFHalOrg.7-100, 118
$CCl_2F_2N_2O_4S_2$	$ClCN \cdot ClN(SO_2F)_2$	C:	MVol.D3-216
		S:	S-N-Verb.1-52
CCl_2F_2Se	$F_2ClCSeCl$	C:	MVol.D6-250
$CCl_2F_3HNO_3PS$	$CF_3SO_2NHPOCl_2$	S:	S-N-Verb.1-121
CCl_2F_3N	CF_3NCl_2	C:	MVol.D3-214
–	F_2NCFCl_2	C:	MVol.D6-160
$CCl_2F_3NO_2S_2$	$CF_3SO_2NSCl_2$	S:	S-N-Verb.1-111
CCl_2F_3OP	$CF_3P(O)Cl_2$	F:	PerFHalOrg.3-51/4, 57
CCl_2F_3P	CF_3PCl_2		
	Chemical reactions	F:	PerFHalOrg.3-100/2, 108, 111, 120
	Physical properties	F:	PerFHalOrg.3-80/2, 89, 98
	Preparation	F:	PerFHalOrg.3-72
CCl_2F_3PS	$CF_3P(S)Cl_2$	F:	PerFHalOrg.3-140, 147, 151, 153
CCl_2F_5NS	SF_5NCCl_2	S:	S-N-Verb.1-147/8
CCl_2F_5P	$CF_3PF_2Cl_2$	F:	PerFHalOrg.3-72, 88
CCl_2FeH_3	CH_3FeCl_2	Fe:	Org.Verb.B1-2/3
$CCl_2H_2MnO_{0.25}$	$MnCl_2 \cdot 0.25\ C_4H_8O$	Mn:	MVol.D1-142
$CCl_2H_2N_2O$	$(NCl_2)(NH_2)CO$	C:	MVol.D3-129
–	$(NHCl)_2CO$	C:	MVol.D3-127/9
$CCl_2H_2N_2O_5S_2$	$CO(NHSO_2Cl)_2$	S:	S-N-Verb.1-123
$CCl_2H_{2.5}MnO_{0.5}$	$MnCl_2 \cdot 0.25\ CH_3O(CH_2)_2OCH_3$	Mn:	MVol.D1-141
CCl_2H_3ISn	CH_3SnCl_2I	Sn:	Org.Verb.6-218
$CCl_2H_3NO_4S_2$	$CH_3N(SO_2Cl)_2$	S:	S-N-Verb.1-55
$CCl_2H_3N_3O_3S_3$	$[NS(O)Cl]_2[NS(O)CH_3]$	S:	S-N-Verb.1-19
$CCl_2H_3OTi^+$	$[CH_3OTiCl_2]^+$	Ti:	Org.Verb.1-181
$CCl_2H_3Sn^+$	$CH_3SnCl_2{}^+$	Sn:	Org.Verb.6-219/20
CCl_2H_3Ti	CH_3TiCl_2	Ti:	Org.Verb.1-8
$CCl_2H_3Ti^+$	$[CH_3TiCl_2]^+$	Ti:	Org.Verb.1-43
CCl_2H_4MnO	$MnCl_2 \cdot CH_3OH$	Mn:	MVol.D1-26
$CCl_2H_4N_2OSn$	$SnCl_2 \cdot CO(NH_2)_2$	Sn:	MVol.C5-44
$CCl_2H_4N_2SSn$	$SnCl_2 \cdot CS(NH_2)_2$	Sn:	MVol.C5-55/6
CCl_2H_5NSn	$SnCl_2 \cdot CH_3NH_2$	Sn:	MVol.C5-11/2
$CCl_2H_{10}MnO_4$	$MnCl_2(H_2O)_3(CH_3OH)$	Mn:	MVol.C5-78
$CCl_2H_{12}N_4Pt_2S_3$	$Pt_2(NH_3)_4(CS_3)Cl_2$	C:	MVol.D4-226
$CCl_2H_{15}IrN_6O_8S$	$[Ir(NCS)(NH_3)_5](ClO_4)_2$	Ir:	SVol.2-162
–	$[Ir(SCN)(NH_3)_5](ClO_4)_2$	Ir:	SVol.2-161
$CCl_2H_{16}IrN_5O_{10}$	$[Ir(NH_3)_5(OOCH)](ClO_4)_2$	Ir:	SVol.2-31
CCl_2KNOS_2	$[SOCl_2SCN]K$	S:	SVol.1-46
CCl_2N	Cl_2CN	C:	MVol.D3-203/4
CCl_2N^+	Cl_2CN^+	C:	MVol.D3-265
$CCl_2NOS_2{}^-$	$[SOCl_2SCN]^-$	S:	SVol.1-46

$CCl_2NO_2S^+$ .... $SO_2NCCl_2^+$ .... S: S-N-Verb.1-130

CCl_2O .... $COCl_2$

Chemical reactions .... C: MVol.D3-83/103

Crystallographic properties .... C: MVol.D3-75/6

Decomposition .... C: MVol.D3-84/90

Electrical properties .... C: MVol.D3-80/1

Formation .... C: MVol.D3-47/66

Handling .... C: MVol.D3-47

Hydrolysis .... C: MVol.D3-94/5

Isotope exchange .... C: MVol.D3-90, 92

Magnetic properties .... C: MVol.D3-80/1

Mechanical properties .... C: MVol.D3-76/7, 80

Molecule .... C: MVol.D3-66/75

Preparation .... C: MVol.D3-47/65

Purification .... C: MVol.D3-65

Solubility .... C: MVol.D3-94/5, 98/9, 102/3

Spectra .... C: MVol.D3-81/3

Thermal properties .... C: MVol.D3-76/80

Toxicity .... C: MVol.D3-47

– .... $COCl_2$ systems

$COCl_2-BCl_3$ .... B: B-Verb.19-141

CCl_2S .... Cl_2CS

Chemical reactions .... C: MVol.D6-158/61

Formation .... C: MVol.D6-144, 147

Handling .... C: MVol.D6-147

Molecule .... C: MVol.D6-147/53

Physical properties .... C: MVol.D6-153/8

Physiological behavior .... C: MVol.D6-147

Preparation .... C: MVol.D6-145

Preparation, industrial .... C: MVol.D6-145/6

Preparation of isotope-enriched Cl_2CS C: MVol.D6-145

Purification .... C: MVol.D6-146/7

CCl_2Se .... Cl_2CSe .... C: MVol.D6-250

CCl_2Se^+ .... Cl_2CSe^+ .... C: MVol.D6-251

CCl_3FNO_2P .... $CFClNOP(O)Cl_2$ .... F: PerFHalOrg.7-100, 118

CCl_3F_2N .... F_2NCCl_3 .... C: MVol.D6-160

CCl_3F_2NS .... F_2NCl_2CSCl .... C: MVol.D6-160

CCl_3F_3Ge .... CF_3GeCl_3 .... F: PerFHalOrg.4-166, 171/2, 175, 179

CCl_3F_3HNOSi ...... $CF_3N(H)OSiCl_3$ .... F: PerFHalOrg.7-16/7, 43

$CCl_3F_3NO_2PS$ ...... $CF_3SO_2NPCl_3$ .... S: S-N-Verb.1-121

CCl_3F_3Si .... CF_3SiCl_3 .... F: PerFHalOrg.4-147

CCl_3F_4P .... CF_3PFCl_3 .... F: PerFHalOrg.3-72, 89

$CCl_3F_6NiOP_3$ .... $CONi(PF_2Cl)_3$ .... Ni: Org.Verb.1-112

$CCl_3FeN_2O_3P$ .... $COFe(NO)_2PCl_3$ .... Fe: Org.Verb.B1-42

CCl_3HO_2Sn .... $SnCl_3(HCOO)$ .... Sn: MVol.C2-219

CCl_3HS .... HCl_2CSCl .... C: MVol.D4-28

$CCl_3H_2NO_2Sn$	$NO_2CH_2SnCl_3$	Sn:	Org.Verb.6-255
$CCl_3H_3I_2Ti$	$CH_3TiCl_3 \cdot I_2$	Ti:	Org.Verb.1-34/5, 41
$CCl_3H_3NO_4U^-$	$[UO_2Cl_3(CH_3NO_2)]^-$	U:	SVol.E1-144/5
$CCl_3H_3N_3O_2P_2S^+$	$[NSO_2][NPCl_2][CH_3NPCl]^+$	S:	S-N-Verb.1-35
CCl_3H_3Si	CH_3SiCl_3 systems		
	CH_3SiCl_3-BCl_3	B:	B-Verb.19-141
CCl_3H_3Sn	CH_3SnCl_3	Sn:	Org.Verb.6-210/24
CCl_3H_3Ti	CH_3TiCl_3	Ti:	Org.Verb.1-12/4, 18/22
–	CD_3TiCl_3	Ti:	Org.Verb.1-14, 22/3
$CCl_3H_4N_4O_2PS_2$	$[NS(O)Cl]_2[NPClNHCH_3]$	S:	S-N-Verb.1-29
CCl_3H_5SiSn	$H_3SiCH_2SnCl_3$	Sn:	Org.Verb.6-255
CCl_3H_6MnN	$CH_3NH_3MnCl_3$	Mn:	MVol.C5-188
–	$CH_3NH_3MnCl_3 \cdot H_2O$	Mn:	MVol.C5-188
–	$CH_3NH_3MnCl_3 \cdot 2\ H_2O$	Mn:	MVol.C5-188
$CCl_3H_{12}N_5U$	$UCl_3(CN) \cdot 4\ NH_3$	U:	SVol.E1-12, 14
CCl_3N	$ClNCCl_2$	C:	MVol.D3-206
CCl_3NOSi	$SiCl_3NCO$	C:	MVol.D3-138
CCl_3NO_2S	$ClSO_2NCCl_2$	S:	S-N-Verb.1-130
CCl_3NO_2U	$[UO_2Cl_2(CNCl)]$	U:	SVol.E1-48/50
CCl_3NS	ClSC(Cl)NCl	C:	MVol.D6-179/80
CCl_3NSSn_2	Sn_2Cl_3SCN	Sn:	MVol.C2-229
CCl_3NS_2	ClS_2NCCl_2	S:	SVol.2-260
CCl_3NSi	Cl_3SiCN	C:	MVol.D3-213
$CCl_3N_9O_2U$	$C(N_3)_3UO_2Cl_3$	U:	SVol.C9-93
CCl_4	CCl_4 systems		
	CCl_4-BCl_3	B:	B-Verb.19-141
	CCl_4-$SnCl_4$	Sn:	MVol.C2-227
CCl_4F_2Se	$[F_2ClCSeCl_3]$	C:	MVol.D6-250
CCl_4F_3P	CF_3PCl_4	F:	PerFHalOrg.3-73, 78, 79, 90, 99
CCl_4FeN	$FeCl_3 \cdot ClCN$	C:	MVol.D3-215
CCl_4H_2Sn	$CH_2ClSnCl_3$	Sn:	Org.Verb.6-253/5
$CCl_4H_3NO_2U$	$UCl_4 \cdot CH_3NO_2$	U:	SVol.E1-144/5
$CCl_4H_3N_3O_2P_2S$	$[NSO_2][CH_3NPCl_2][NPCl_2]$	S:	S-N-Verb.1-35
–	$[O_2S[NPCl_2]_2NCH_3]$	S:	S-N-Verb.1-33
CCl_4H_4OSn	$SnCl_4 \cdot CH_3OH$	Sn:	MVol.C5-68/9
CCl_4N_2O	$(NCl_2)_2CO$	C:	MVol.D3-129
CCl_4O_2S	$SOCl_2 \cdot COCl_2$	S:	SVol.1-45
CCl_4S	Cl_3CSCl	C:	MVol.D4-27, 183
		C:	MVol.D6-145/6, 158/9
CCl_4S_2	$CCl_2(SCl)_2$	C:	MVol.D4-183
–	CCl_3SSCl	C:	MVol.D4-183
		C:	MVol.D6-159
CCl_4Se	Cl_3CSeCl	C:	MVol.D6-199, 250
$CCl_5Cs_2HIrO_2$	$Cs_2[Ir(OOCH)Cl_5]$	Ir:	SVol.2-31
CCl_5Cs_2IrO	$Cs_2[Ir(CO)Cl_5]$	Ir:	SVol.2-31
$CCl_5H_3Ti^{2-}$	$[CH_3TiCl_5]^{2-}$	Ti:	Org.Verb.1-21, 42/3
$CCl_5H_4MnO^{3-}$	$[Mn(CH_3OH)Cl_5]^{3-}$	Mn:	MVol.C5-77
		Mn:	MVol.D1-24

$CCl_5H_5N_2OTe$	$(CON_2H_5)TeCl_5$	Te:	SVol.B2-132
CCl_5H_5Sn	$H_2[CH_3SnCl_5]$	Sn:	Org.Verb.6-222
$CCl_5H_8IrN_2O$	$(NH_4)_2[Ir(CO)Cl_5]$	Ir:	SVol.2-31
$CCl_5H_9IrN_2O_2$	$(NH_4)_2[Ir(OOCH)Cl_5]$	Ir:	SVol.2-31
CCl_5INSb	$ICN \cdot SbCl_5$	C:	MVol.D3-261
CCl_5NU	$[UCl_4(CNCl)]$	U:	SVol.E1-48/50
CCl_6H_2SiSn	$Cl_3SiCH_2SnCl_3$	Sn:	Org.Verb.6-254/5
$CCl_6H_2Te_2$	$CH_2(TeCl_3)_2$	Te:	SVol.B2-98
$CCl_6H_3N_3O_4P_2S_2$	$CH_3N(SO_2NPCl_3)_2$	S:	S-N-Verb.1-162
CCl_6NNb	$NbCl_5 \cdot ClCN$	C:	MVol.D3-214/5
CCl_6NSb	$ClCN \cdot SbCl_5$	C:	MVol.D3-214
CCl_6NTa	$TaCl_5 \cdot ClCN$	C:	MVol.D3-214/5
CCl_6N_9U	$[C(N_3)_3]UCl_6$	U:	SVol.C9-60/3
$CCl_7N_3O_2P_2S$	$ClSO_2NPCl_2NCClNPCl_3$	S:	S-N-Verb.1-134
$CCl_9N_4O_2P_3S$	$ClSO_2NPCl_2NC(NPCl_3)_2$	S:	S-N-Verb.1-134
CCl_9NiOP_3	$CONi(PCl_3)_3$	Ni:	Org.Verb.1-112/3
$CCl_{12}Pt_6S_2$	$Pt_6Cl_{12} \cdot CS_2$	C:	MVol.D4-190
$CCl_{14}H_3N_7OP_6S$	$[Cl(O)S[NP(NPCl_3)_2]_2NCH_3]Cl$	S:	S-N-Verb.1-36
$CCmNS^{2+}$	$CmSCN^{2+}$	Np:	TrU.D1-145
$CCoHK_4O_{10}S_5$	$Co(S_2CO)(SO_3H) \cdot 2\ K_2SO_3 \cdot 4\ H_2O$	C:	MVol.D4-208
$CCoHNa_4O_{10}S_5$	$Co(S_2CO)(SO_3H) \cdot 2\ Na_2SO_3 \cdot 4\ H_2O$	C:	MVol.D4-208
$CCoH_9N_3S_3$	$Co(NH_3)_3CS_3$	C:	MVol.D4-225
$CCoH_{10}N_3S_4$	$Co(NH_3)_3(CS_3)(HS)$	C:	MVol.D4-225
$CCoS_3$	$CoCS_3$	C:	MVol.D4-225
$CCsCuS_3$	$CsCuCS_3$	C:	MVol.D4-224
$CCsCuS_4$	$CsCuCS_4$	C:	MVol.D4-231
$CCsF_3O$	$CsOCF_3$	C:	MVol.D3-45
$CCsF_4NO$	$CsOCF_2NF_2$	C:	MVol.D3-134
$CCsF_7HP$	$Cs[CF_3PF_4H]$	F:	PerFHalOrg.3-137
$CCsF_8P$	$Cs[CF_3PF_5]$	F:	PerFHalOrg.3-134, 135, 137
$CCsH_2NS_2$	$Cs[S_2CNH_2]$	C:	MVol.D6-140
$CCs_2H_3NS_3$	$Cs_2CS_3 \cdot NH_3$	C:	MVol.D4-223
CCs_2OS	$CsC(O)SCs$	C:	MVol.D5-126
CCs_2S_3	Cs_2CS_3	C:	MVol.D4-223
–	$Cs_2CS_3 \cdot H_2O$	C:	MVol.D4-218, 223
$CCuH_2NS_2$	$Cu[S_2CNH_2]$	C:	MVol.D6-140
$CCuH_4NS_3$	NH_4CuCS_3	C:	MVol.D4-224
$CCuKS_3$	$KCuCS_3$	C:	MVol.D4-224
$CCuNSe$	$CuSeCN$	C:	MVol.D6-229
$CCuRbS_3$	$RbCuCS_3$	C:	MVol.D4-224
$CCuRbS_4$	$RbCuCS_4$	C:	MVol.D4-231
$CCuS_3$	$CuCS_3$	C:	MVol.D4-224
CCu_2S_4	Cu_2CS_4	C:	MVol.D4-231
$CCu_3H_{18}N_6O_5S_3$	$Cu_3(NH_3)_6(S_2CO)(SO_4)$	C:	MVol.D4-208
$CEuHO_4$	$Eu(OH)CO_3$	Sc:	MVol.B7-26
$CFHNO_3S^+$	FSO_2NHCO^+	S:	S-N-Verb.1-102
$CFHO$	$COHF$	C:	MVol.D3-1/3
$CFHOS$	$OCS \cdot HF$	C:	MVol.D5-133
$CFHSi^+$	$SiHCF^+$	F:	PerFHalOrg.4-152

CFH_2NO	$NH_2C(O)F$	C:	MVol.D3-130/1
		F:	PerFHalOrg.7-22/3
$CFH_2NO_2S^+$	$CH_2NSO_2F^+$	S:	S-N-Verb.1-53
CFH_2NO_4S	$FSO_2NHCOOH$	S:	S-N-Verb.1-115
CFH_3NOS^+	CH_3NSOF^+	S:	S-N-Verb.1-157, 193
$CFH_3NO_2S^+$	$CH_3NSO_2F^+$	S:	S-N-Verb.1-53
CFH_3N_2O	$(NHF)(NH_2)CO$	C:	MVol.D3-123
$CFH_3N_2OS^+$	$NSOFNCH_3^+$	S:	S-N-Verb.1-193
CFH_3OS	CH_3SOF	F:	PerFHalOrg.6-114
CFH_3O_2S	$CH_3OS(O)F$	S:	SVol.2-79
CFH_3Sn^{2+}	CH_3SnF^{2+}	Sn:	Org.Verb.5-51
CFH_4NO_2S	FSO_2NHCH_3	S:	S-N-Verb.1-111
CFIO	COIF	C:	MVol.D3-122
CFN	FCN		
	Chemical reactions	C:	MVol.D3-184/5
	Formation	C:	MVol.D3-180/2
	Formation of radioactive FCN	C:	MVol.D3-181
	Molecule	C:	MVol.D3-151/80
	Physical properties	C:	MVol.D3-182/4
	Preparation	C:	MVol.D3-180/2
	Review	C:	MVol.D3-149/50
	Toxicity	C:	MVol.D3-150/1
–	$FC^{13}N$	C:	MVol.D3-181
–	$FC^{15}N$	C:	MVol.D3-181
–	FNC	C:	MVol.D3-149
CFN^-	FCN^-	C:	MVol.D3-265/6
CFNO	FNCO	C:	MVol.D3-137
$CFNO_3S$	FSO_2NCO	S:	S-N-Verb.1-115/6
CFN_2OS^+	$NCNSOF^+$	S:	S-N-Verb.1-75
CFO	FCO	C:	MVol.D3-1/6
CFO^+	FCO^+	C:	MVol.D3-6
CFO^-	FCO^-	C:	MVol.D3-6
CFOS	FSCO	C:	MVol.D5-125
$CFSe^+$	$FCSe^+$	C:	MVol.D6-250
$CFSe_2^+$	$FCSe_2^+$	C:	MVol.D6-250
CF_2HN	CF_2NH	F:	PerFHalOrg.7-25/6, 51, 56, 69
–	$[CHF_2N]_x$	F:	PerFHalOrg.7-56
CF_2HNO_2	$CF_2N(O)OH$	F:	PerFHalOrg.7-158
$CF_2HN_2O_5S_2^+$	$FSO_2NCONHSO_2F^+$	S:	S-N-Verb.1-102
CF_2HSi^+	$SiHCF_2^+$	F:	PerFHalOrg.4-152
$CF_2H_2NOS^+$	$CH_2NSOF_2^+$	S:	S-N-Verb.1-157
$CF_2H_2NO_4S_2^+$	$CH_2N(SO_2F)_2^+$	S:	S-N-Verb.1-53
$CF_2H_2N_2O$	$(NF_2)(NH_2)CO$	C:	MVol.D3-123/5
$CF_2H_2N_2O_2$	$CF_2N(O)NHOH$	F:	PerFHalOrg.7-158
$CF_2H_2N_2O_5S_2$	$FSO_2NHC(O)NHSO_2F$	S:	S-N-Verb.1-102/3
$CF_2H_2N_2O_5S_2^+$	$FSO_2NHCONHSO_2F^+$	S:	S-N-Verb.1-102
$CF_2H_2N_2S$	$(NF_2)(NH_2)CS$	C:	MVol.D6-131
$CF_2H_2N_3O_4S_3^+$	$(NS(O)F)_2(CH_2NSO_2)^+$	S:	S-N-Verb.1-10
CF_2H_3NOS	$CH_3NS(O)F_2$	S:	S-N-Verb.1-62

$CF_2H_3NOSSi^+$	$CH_3SiNSOF_2^+$	S:	S-N-Verb.1-82
$CF_2H_3NO_3S_2$	$CH_3OS(O)NS(O)F_2$	S:	S-N-Verb.1-69
–	$CH_3SO_2NS(O)F_2$	S:	S-N-Verb.1-69/70
$CF_2H_3NO_4S_2$	$CH_3N(SO_2F)_2$	S:	S-N-Verb.1-52/3
$CF_2H_3NO_4S_2^+$	$CH_3N(SO_2F)_2^+$	S:	S-N-Verb.1-53
CF_2H_3NS	CH_3NSF_2	S:	SVol.2-288
$CF_2H_3N_3O_4S_3$	$NS(O)FNSO_2N(CH_3)S(O)F$	S:	S-N-Verb.1-10
$CF_2H_3N_3O_4S_3^+$	$(NS(O)F)_2(CH_3NSO_2)^+$	S:	S-N-Verb.1-10
$CF_2H_3PS_2$	$CH_3SP(S)F_2$	S:	SVol.2-84
$CF_2H_3Sn^+$	$CH_3SnF_2^+$	Sn:	Org.Verb.5-51
$CF_2H_4N_2O_3S_2$	$CH_3NHS(O)FNSO_2F$	S:	S-N-Verb.1-196
$CF_2H_4N_4O_3S_3$	$[NS(O)F]_2[NS(O)NHCH_3]$	S:	S-N-Verb.1-13
$CF_2K_2N_2O_5S_2$	$K_2[FSO_2NCONSO_2F]$	S:	S-N-Verb.1-103
CF_2N	F_2CN	C:	MVol.D3-291/2
$CF_2NO_2S^+$	OSF_2NCO^+	S:	S-N-Verb.1-73/5, 81
CF_2N_2	F_2CN_2	F:	PerFHalOrg.5-4/5, 7, 10, 14/9, 22/3
–	NF_2CN	C:	MVol.D3-293/4
$CF_2N_2Na_2O_5S_2$	$Na_2[FSO_2NCONSO_2F] \cdot H_2O$	S:	S-N-Verb.1-103
CF_2N_2OS	$NCNS(O)F_2$	S:	S-N-Verb.1-75/6
$CF_2N_2OS^+$	$NCNSOF_2^+$	S:	S-N-Verb.1-75
$CF_2N_2O_4S_2$	$OCNSO_2NS(O)F_2$	S:	S-N-Verb.1-71/2
$CF_2N_2O_4S_2^+$	$OCNSO_2NSOF_2^+$	S:	S-N-Verb.1-72
CF_2N_2S	$NCNSF_2$	S:	SVol.2-82
CF_2N_4O	$NF_2C(O)N_3$	C:	MVol.D3-136
CF_2O	COF_2		
	Chemical reactions	C:	MVol.D3-40/7
	Formation	C:	MVol.D3-11/21
	Handling	C:	MVol.D3-20
	Hydrolysis	C:	MVol.D3-42
	Mechanical properties	C:	MVol.D3-34
	Molecule	C:	MVol.D3-21/34
	Preparation	C:	MVol.D3-11/20
	Spectra	C:	MVol.D3-36/40
	Thermal properties	C:	MVol.D3-34/6
	Thermolysis	C:	MVol.D3-40/2
CF_2O^+	COF_2^+	C:	MVol.D3-22
CF_2O_2	$FC(O)OF$	C:	MVol.D3-14
CF_2S	F_2CS	C:	MVol.D6-141/3
CF_2Se	F_2CSe	C:	MVol.D6-249/50
–	$(F_2CSe)_n$	C:	MVol.D6-251
CF_2Se^+	F_2CSe^+	C:	MVol.D6-250
$CF_2Se_2^+$	$F_2CSe_2^+$	C:	MVol.D6-250
$CF_3FeHN_2O_6S$	$CF_3N(NO)OH \cdot FeSO_4$	F:	PerFHalOrg.7-87
CF_3FeO	$COFeF_3$	Fe:	Org.Verb.B3-119
CF_3GeI_3	CF_3GeI_3	F:	PerFHalOrg.4-166, 171/2, 175, 179
CF_3HKO_2P	$CF_3PH(O)OK$	F:	PerFHalOrg.3-49/51
CF_3HKO_3P	$CF_3P(O)(OH)OK$	F:	PerFHalOrg.3-48, 50
CF_3HNO_2	$CF_3N(O)OH$	F:	PerFHalOrg.7-86/7

$CF_3HNO_2^-$	$[CF_3N(OH)O]^-$	F:	PerFHalOrg.7-158
CF_3HN_2O	CF_3NNOH	F:	PerFHalOrg.7-158
$CF_3HN_2O_2$	$CF_3N(NO)OH$	F:	PerFHalOrg.7-145
CF_3HNaO_2P	$CF_3P(O)(H)ONa$	F:	PerFHalOrg.3-44, 50/1
CF_3HNaO_3P	$CF_3P(O)(OH)ONa$	F:	PerFHalOrg.3-50/1
$CF_3HOPS_2^-$	$CF_3P(S)(OH)S^-$	F:	PerFHalOrg.3-139, 146
$CF_3HO_2P^-$	$[CF_3P(H)(O)O]^-$	F:	PerFHalOrg.3-44/5, 49, 51
$CF_3HO_2PS^-$	$CF_3P(S)(OH)O^-$	F:	PerFHalOrg.3-139, 146
$CF_3HO_3P^-$	$[CF_3P(O)(OH)O]^-$	F:	PerFHalOrg.3-43, 45, 48
CF_3HSi^+	$SiHCF_3^+$	F:	PerFHalOrg.4-152
CF_3H_2N	CF_3NH_2	F:	PerFHalOrg.7-1, 59
CF_3H_2NO	$CF_3N(H)OH$	F:	PerFHalOrg.7-16/7, 43, 55, 58
$CF_3H_2NO_2S$	$CF_3SO_2NH_2$	S:	S-N-Verb.1-107
$CF_3H_2N_3$	CF_3NNNH_2	F:	PerFHalOrg.7-158
$CF_3H_2OPS_2$	$CF_3P(S)(SH)OH$	F:	PerFHalOrg.3-159
$CF_3H_2O_2P$	$CF_3P(OH)_2$	F:	PerFHalOrg.3-44/5, 48/50
–	$CF_3P(OH)_2 \cdot H_2O$	F:	PerFHalOrg.3-99
$CF_3H_2O_3P$	$CF_3P(O)(OH)_2$	F:	PerFHalOrg.3-41/3, 45, 48/50
CF_3H_2P	CF_3PH_2	F:	PerFHalOrg.3-14/27
–	CF_3PHD	F:	PerFHalOrg.3-14/7
–	CF_3PD_2	F:	PerFHalOrg.3-14/7, 21
$CF_3H_3NOSSi^+$	$F(CH_3)SiNSOF_2^+$	S:	S-N-Verb.1-90
$CF_3H_3NOSSn^+$	$CH_3SnFNSOF_2^+$	S:	S-N-Verb.1-83
$CF_3H_3NO_2SSi^+$	$CH_3OSi(F)NSOF_2^+$	S:	S-N-Verb.1-82
$CF_3H_3N_2O_2$	$CF_3N(OH)NHOH$	F:	PerFHalOrg.7-158
CF_3H_3Sn	CH_3SnF_3	Sn:	Org.Verb.5-51
$CF_3H_4N_2P$	$CF_3P(NH_2)_2$	F:	PerFHalOrg.3-57, 64, 68/9
$CF_3H_4N_3O_2$	$CF_3N(NO)ONH_4$	F:	PerFHalOrg.7-87, 104
$CF_3H_4P_3$	$H_2PP(CF_3)PH_2$	F:	PerFHalOrg.3-16, 20, 23
$CF_3H_5NO_2P$	$CF_3P(H)(O)ONH_4$	F:	PerFHalOrg.3-68
$CF_3H_8N_2O_3P$	$CF_3P(O)(ONH_4)_2$	F:	PerFHalOrg.3-50/1
CF_3IMg	CF_3MgI	F:	PerFHalOrg.4-61, 68, 71
CF_3IS	CF_3SI	C:	MVol.D4-189
CF_3I_2P	CF_3PI_2		
	Chemical reactions	F:	PerFHalOrg.3-98/102, 108/9, 111
	Physical properties	F:	PerFHalOrg.3-76, 80/2, 93, 98
	Preparation	F:	PerFHalOrg.3-74/5
CF_3I_2PS	$CF_3P(S)I_2$	F:	PerFHalOrg.3-140
CF_3I_2Sb	CF_3SbI_2	F:	PerFHalOrg.3-226, 229/31
CF_3I_3Si	CF_3SiI_3	F:	PerFHalOrg.4-147
CF_3KNO_4S	$CF_3N(O)SO_3K$	F:	PerFHalOrg.7-92/3, 107, 146

CF_3KO	$KOCF_3$	C:	MVol.D3-45
$CF_3K_2O_3P$	$CF_3P(O)(OK)_2$	F:	PerFHalOrg.3-48, 50
CF_3Li	CF_3Li	F:	PerFHalOrg.4-2
$CF_3LiN_2O_2$	$CF_3N(NO)OLi$	F:	PerFHalOrg.7-87, 104
CF_3N	CF_2NF	F:	PerFHalOrg.5-20
CF_3NNaO_4S	$CF_3N(O)SO_3Na$	F:	PerFHalOrg.7-92/3, 107
CF_3NO	CF_3NO		
	Chemical reactions	F:	PerFHalOrg.7-155/70, 194, 206/14
	Decomposition	F:	PerFHalOrg.7-156/7
	Electrochemical behavior	F:	PerFHalOrg.7-155
	Formation and preparation	F:	PerFHalOrg.7-151/2
	Molecule	F:	PerFHalOrg.7-153/5
	Physical properties	F:	PerFHalOrg.7-153/5
	Solvolysis	F:	PerFHalOrg.7-157/8
–	$NF_2C(O)F$	C:	MVol.D3-133/5
		F:	PerFHalOrg.7-170
–	F_2CNOF	F:	PerFHalOrg.7-170
CF_3NOS	$F_2SNC(O)F$	S:	SVol.2-78, 83, 84
CF_3NOSi	SiF_3NCO	C:	MVol.D3-138
CF_3NO_2S	$FC(O)NS(O)F_2$	S:	S-N-Verb.1-73
$CF_3NO_2S^+$	$FCONSOF_2^+$	S:	S-N-Verb.1-73
–	$FSO_2NCF_2^+$	S:	S-N-Verb.1-111
$CF_3NO_3S_2$	CF_3SO_2NSO	S:	S-N-Verb.1-111/2
$CF_3NO_6S_2$	$FSO_2N(COF)OSO_2F$	S:	S-N-Verb.1-106
CF_3NSSn_2	Sn_2F_3SCN	Sn:	MVol.C2-229
$CF_3N_2NaO_2$	$CF_3N(NO)ONa$	F:	PerFHalOrg.7-87, 104, 145
CF_3N_3	$(F)(F_2N)CN_2$	F:	PerFHalOrg.5-4/5, 11, 20
$CF_3N_3O_2S$	$CF_3SO_2N_3$	S:	S-N-Verb.1-216
$CF_3Na_2O_3P$	$CF_3P(O)(ONa)_2$	F:	PerFHalOrg.3-50
CF_3O^-	$[OCF_3]^-$	C:	MVol.D3-45
CF_3OPS	$OCS \cdot PF_3$	C:	MVol.D5-137
$CF_3OPS_2^{2-}$	$CF_3P(S)SO^{2-}$	F:	PerFHalOrg.3-139, 146, 151
$CF_3OPSe_2^{2-}$	$CF_3POSe_2^{2-}$	F:	PerFHalOrg.3-160
CF_3ORb	$RbOCF_3$	C:	MVol.D3-45
CF_3O_2P	$[CF_3PO_2]_x$	F:	PerFHalOrg.3-32, 37
$CF_3O_2PS^{2-}$	$CF_3P(S)O_2^{2-}$	F:	PerFHalOrg.3-139, 146/7, 151
$CF_3O_2PSe^{2-}$	$CF_3PO_2Se^{2-}$	F:	PerFHalOrg.3-160
$CF_3O_3P^{2-}$	$[CF_3P(O)O_2]^{2-}$	F:	PerFHalOrg.3-43, 46, 48
CF_3P	$(CF_3P)_x$	F:	PerFHalOrg.3-1, 6
CF_3PS_2	$CS_2 \cdot PF_3$	C:	MVol.D4-189
$CF_3P_2S^+$	$CF_3P_2S^+$	F:	PerFHalOrg.3-12
CF_4HNO_2S	$CF_3N(H)SO_2F$	F:	PerFHalOrg.7-17, 44, 57
		S:	S-N-Verb.1-100
$CF_4HNO_4S_2$	$CF_3SO_2NHSO_2F$	S:	S-N-Verb.1-50
$CF_4H_2NTe^+$	$TeF_4NCH_2^+$	Te:	SVol.B2-66
$CF_4H_3NOSSi^+$	$CH_3F_2SiNSOF_2^+$	S:	S-N-Verb.1-90

$CF_4H_3N_2O_2S_2Si^+$	$CH_3Si(NSOF_2)_2^+$	S:	S-N-Verb.1-82, 90
$CF_4H_3N_2O_2S_2Sn^+$	$CH_3Sn(NSOF_2)_2^+$	S:	S-N-Verb.1-83
$CF_4H_3N_3O_2S_2Si^+$	$CH_3Si(N)(NSOF_2)_2^+$	S:	S-N-Verb.1-90
$CF_4H_3Sn^-$	$CH_3SnF_4^-$	Sn:	Org.Verb.5-51
$CF_4H_6MnN_3$	$CNH(NH_2)_2HMnF_4 \cdot 3\ H_2O$	Mn:	MVol.C4-212/3
CF_4KNO	NF_2CF_2OK	C:	MVol.D3-134
CF_4NOS^+	$CF_2NSOF_2^+$	S:	S-N-Verb.1-82
–	CF_3NSOF^+	S:	S-N-Verb.1-62
$CF_4NOS_2^+$	CF_3SNSOF^+	S:	S-N-Verb.1-69
$CF_4NO_2S^+$	$FSO_2NCF_3^+$	S:	S-N-Verb.1-111
CF_4N_2	$F_2C(NF)_2$	F:	PerFHalOrg.5-6/7, 12, 21/2
CF_4N_2O	$(NF_2)_2CO$	C:	MVol.D3-125/6
$CF_4N_2OS_2^+$	$SF_2NCF_2NSO^+$	S:	S-N-Verb.1-76
$CF_4N_2O_3S_2$	$CO[NS(O)F_2]_2$	S:	S-N-Verb.1-81
$CF_4N_2O_3S_2^+$	$CO(NSOF_2)_2^+$	S:	S-N-Verb.1-81
CF_4OP^+	CF_4PO^+	F:	PerFHalOrg.3-33
CF_4O_3S	CF_3OSO_2F	F:	PerFHalOrg.7-170
CF_4P^+	CF_3PF^+	F:	PerFHalOrg.3-12, 33, 179
CF_5Ge^+	$GeF_3CF_2^+$	F:	PerFHalOrg.4-174
CF_5H_2P	$CF_3PF_2H_2$	F:	PerFHalOrg.3-72, 89
$CF_5H_3NTe^+$	$TeF_5NCH_3^+$	Te:	SVol.B2-66
CF_5H_3OTe	$TeF_5(OCH_3)$	Te:	SVol.B2-29, 54
$CF_5H_3Sn^{2-}$	$CH_3SnF_5^{2-}$	Sn:	Org.Verb.5-51
$CF_5H_4OSn^-$	$[SnF_5 \cdot CH_3OH]^-$	Sn:	MVol.C5-68
CF_5ISi	CF_3SiF_2I	F:	PerFHalOrg.4-149, 156
CF_5KN_2O	$KOCN_2F_5$	C:	MVol.D3-134
CF_5NOS	$CF_3NS(O)F_2$	S:	S-N-Verb.1-62
–	SF_5NCO	S:	S-N-Verb.1-148/9
CF_5NOS^+	$CF_3NSOF_2^+$	S:	S-N-Verb.1-62
CF_5NOS_2	$CF_3SNS(O)F_2$	S:	S-N-Verb.1-68/9
$CF_5NOS_2^+$	$CF_3SNSOF_2^+$	S:	S-N-Verb.1-69
$CF_5NO_2S_2$	$CF_3SO_2NSF_2$	S:	S-N-Verb.1-113
$CF_5NO_2S_2^+$	$CF_3SO_2NSF_2^+$	S:	S-N-Verb.1-113
$CF_5NO_3S_2$	$CF_3SO_2NS(O)F_2$	S:	S-N-Verb.1-70
CF_5NS	CF_3NSF_2	S:	SVol.2-78, 82, 91
		S:	S-N-Verb.1-76
CF_5NS^-	SF_5CN^-	S:	SVol.2-216
CF_5NS_2	CF_3SNSF_2	S:	SVol.2-78
–	SF_5NCS	S:	S-N-Verb.1-149
$CF_5N_2OS_2^+$	$SF_2NCF_2NS(O)F^+$	S:	S-N-Verb.1-76
$CF_5N_2O_2S_2^+$	$CF(NSOF_2)_2^+$	S:	S-N-Verb.1-82
CF_5N_3	$(F)(F_2N)C(NF)_2$	F:	PerFHalOrg.5-6/7, 12, 22
CF_5OP	$CF_3P(O)F_2$	F:	PerFHalOrg.3-51/4
CF_5P	CF_3PF_2		
	Chemical reactions	F:	PerFHalOrg.3-98/100, 102, 111, 120/1
	Physical properties	F:	PerFHalOrg.3-78, 80/2, 84, 98
	Preparation	F:	PerFHalOrg.3-70/1

CF_5Sb	CF_3SbF_2	F:	PerFHalOrg.3-231
CF_6Ge	CF_3GeF_3	F:	PerFHalOrg.4-166, 171/2, 175, 179
$CF_6GeH_2O_2^{2-}$	$CF_3GeF_3(OH)_2^{2-}$	F:	PerFHalOrg.4-179
CF_6HNOSi	$CF_3N(H)OSiF_3$	F:	PerFHalOrg.7-16/7, 44, 56
CF_6HP	CF_3PF_3H	F:	PerFHalOrg.3-72, 88
$CF_6H_3N_3O_3S_3Si$	$CH_3Si[NS(O)F_2]_3$	S:	S-N-Verb.1-90
$CF_6H_3N_3O_3S_3Si^+$	$CH_3Si(NSOF_2)_3^+$	S:	S-N-Verb.1-90
CF_6KN_3O	$KOCN_3F_6$	C:	MVol.D3-134
$CF_6N_2OS_2$	$SF_2NCF_2NS(O)F_2$	S:	S-N-Verb.1-76
$CF_6N_2OS_2^+$	$SF_2NCF_2NS(O)F_2^+$	S:	S-N-Verb.1-76
$CF_6N_2O_2S_2$	$CF_2[NS(O)F_2]_2$	S:	S-N-Verb.1-81/2
$CF_6N_2O_2S_2^+$	$CF_2(NSOF_2)_2^+$	S:	S-N-Verb.1-82
$CF_6N_2S_2$	$CF_2(NSF_2)_2$	S:	SVol.2-78, 82
CF_6OS	$SF_5C(O)F$	S:	SVol.2-200
CF_6OS_2	$OCS \cdot SF_6$	C:	MVol.D5-136
CF_6O_2S	$SF_5OC(O)F$	S:	SVol.2-81, 200
CF_6S_3	$SF_6 \cdot CS_2$	C:	MVol.D4-189
CF_6Si	CF_3SiF_3	F:	PerFHalOrg.4-147/8, 152/3, 155, 161/2
CF_7HKP	$K[CF_3PF_4H]$	F:	PerFHalOrg.3-135
CF_7HP^-	$[CF_3PF_4H]^-$	F:	PerFHalOrg.3-102, 129, 134/5, 137
CF_7ISi_2	$CF_3SiF_2SiF_2I$	F:	PerFHalOrg.4-149, 156
CF_7NOS	$SF_4NO(CF_3)$	S:	SVol.2-43
CF_7NS	SF_4NCF_3	S:	S-N-Verb.1-149/50
–	SF_5NCF_2	S:	S-N-Verb.1-147
CF_7P	CF_3PF_4	F:	PerFHalOrg.3-71, 76/9, 85, 98/9
CF_8Ge^{2-}	$[CF_3GeF_5]^{2-}$	F:	PerFHalOrg.4-179
CF_8GeK_2	$K_2[CF_3GeF_5]$	F:	PerFHalOrg.4-166/7, 177
CF_8HNS	$CF_3N(H)SF_5$	F:	PerFHalOrg.7-17, 44, 57
		S:	S-N-Verb.1-136
CF_8OS	CF_3OSF_5	S:	SVol.2-81
CF_8O_2S	CF_3OOSF_5	S:	SVol.2-81
CF_8P^-	$[CF_3PF_5]^-$	F:	PerFHalOrg.3-129, 134/7
CF_8S_2	$(SF_3)_2CF_2$	S:	SVol.2-38
CF_9NOS	$CF_3OSF_4NF_2$	S:	S-N-Verb.1-139/40
CF_9NS	$CF_3SF_4NF_2$	S:	S-N-Verb.1-139
–	$SF_5NF(CF_3)$	S:	S-N-Verb.1-138/9
CF_9NiOP_3	$CONi(PF_3)_3$	Ni:	Org.Verb.1-112
$CF_{12}FeOP_4$	$COFe(PF_3)_4$	Fe:	Org.Verb.B1-27/9
$CF_{12}O_2S_2$	$CF_3OSF_4OSF_5$	S:	SVol.2-81
$CF_{12}O_3S_2$	$CF_3OSF_4OOSF_5$	S:	SVol.2-81
$CF_{13}NOS_2$	$CF_3N(SF_5)OSF_5$	F:	PerFHalOrg.7-91/3, 107
$CFeHO_2^+$	$[FeCOOH]^+$	Fe:	Org.Verb.A3-57
$CFeH_4N_2S^{3+}$	$[Fe((NH_2)_2CS)]^{3+}$	C:	MVol.D6-107
$CFeNO_2^+$	$[Fe(CO)(NO)]^+$	Fe:	Org.Verb.B1-140
$CFeN_2O_3^+$	$[Fe(CO)(NO)_2]^+$	Fe:	Org.Verb.B1-140

$CFeN_3O_4$	$Fe(CO)(NO)_3$	Fe:	Org.Verb.B3-118
$CFeN_4OS_4$	$[Fe(CO)N_4S_4]_n$	Fe:	Org.Verb.B3-120
$CFeO^+$	$Fe(CO)^+$	Fe:	Org.Verb.B3-49, 78
$CFeO^{2+}$	$[Fe(CO)]^{2+}$	Fe:	Org.Verb.B3-49, 78
$CFeO^-$	$[FeCO]^-$	Fe:	Org.Verb.B3-78
$CFeO_{13}P_8{}^+$	$[(CO)Fe(P_4O_6)_2]^+$	Fe:	Org.Verb.B1-164
$CFeS_3$	$FeCS_3$	C:	MVol.D4-225
CFe_2O_3	$Fe_2O_2(CO)$	Fe:	Org.Verb.B3-97
$CHINOS^-$	$ISCNOH^-$	C:	MVol.D5-185
CHN	HCN systems		
	HCN-ClCN	C:	MVol.D3-216
CHNOS	HOSCN	C:	MVol.D5-185/6, 200, 228
$CHNOS^-$	$HOSCN^-$	C:	MVol.D5-145, 176
$CHNO_2S$	HOOSCN	C:	MVol.D5-200
$CHNO_4S$	$OCNSO_3H$	S:	S-N-Verb.1-115
$CHNO_5S_2{}^{2-}$	$HSO_5 \cdot SCN^{2-}$	C:	MVol.D5-192
CHNS	HNCS		
	Chemical reactions	C:	MVol.D6-15/7, 25/7, 33/6
	Crystallographic properties	C:	MVol.D6-10/1
	Formation and preparation	C:	MVol.D6-1/3
	General	C:	MVol.D6-1
	Molecule	C:	MVol.D6-3/10
	Spectra	C:	MVol.D6-12/5
	Thermal properties	C:	MVol.D6-11/2
–	DNCS	C:	MVol.D6-2, 6/14
–	$(HNCS)_3$	C:	MVol.D6-37
–	$(CHNS)_n$	C:	MVol.D6-37
–	HNCS solutions		
	$HNCS-H_2O$	C:	MVol.D6-17/27
	HNCS-organic solvents	C:	MVol.D6-28/34
$CHNS^+$	$HNCS^+$	C:	MVol.D6-36/7
–	$HSCN^+$	C:	MVol.D6-36/7
CHNSe	HNCSe	C:	MVol.D6-228, 239/40
CHNTe	HNCTe	C:	MVol.D6-261
$CHNdO_4$	$Nd(OH)CO_3$	Sc:	MVol.B7-26
CHO_2S^-	HCO_2S^-	C:	MVol.D4-202/4
CHO_2Sn^+	$Sn(HCOO)^+$	Sn:	MVol.C2-218
CHO_4Yb	$Yb(OH)CO_3$	Sc:	MVol.B7-26
CHS^+	CSH^+	C:	MVol.D4-30, 160
CHS_2	SCHS	C:	MVol.D4-163
$CHS_2{}^+$	CS_2H^+	C:	MVol.D4-160, 164
CHS_2Se^-	$HCSeS_2{}^-$	C:	MVol.D6-217
$CHS_3{}^-$	$(HCS_3)^-$	C:	MVol.D4-216/7, 221
$CHSe_3{}^-$	$HCSe_3{}^-$	C:	MVol.D6-212
CH_2KNS_2	$K[S_2CNH_2]$	C:	MVol.D6-140
CH_2NOS^-	$[SOCNH_2]^-$	C:	MVol.D6-132
CH_2NOSe^-	$[OSeCNH_2]^-$	C:	MVol.D6-248
CH_2NRbS_2	$Rb[S_2CNH_2]$	C:	MVol.D6-140

CH_2NS^+	H_2NCS^+	C:	MVol.D6-20, 21, 27, 36/7
CH_2NSSe^-	$[SSeCNH_2]^-$	C:	MVol.D6-248
$CH_2NS_2^-$	$[S_2CNH_2]^-$	C:	MVol.D6-135/9
CH_2NS_2Tl	$Tl[S_2CNH_2]$	C:	MVol.D6-140
CH_2NSe^+	H_2NCSe^+	C:	MVol.D6-228, 240
$CH_2NSe_2^-$	$[Se_2CNH_2]^-$	C:	MVol.D6-248
$CH_2N_4O_4Sn$	$Sn(CH_2(NN(O)O)_2) \cdot H_2O$	Sn:	MVol.C5-43
CH_2OSSe	H_2CSeSO	C:	MVol.D6-211, 219
CH_2OS_2	$OC(SH)_2$	C:	MVol.D4-187, 205
–	$SC(OH)(SH)$	C:	MVol.D4-205
CH_2OSe_2	H_2CSe_2O	C:	MVol.D6-211, 215
CH_2O_2S	$OC(OH)(SH)$	C:	MVol.D4-202
–	$SC(OH)_2$	C:	MVol.D4-202
CH_2O_2Se	H_2CSeO_2	C:	MVol.D6-211, 215
CH_2SSe_2	H_2CSe_2S	C:	MVol.D6-211, 215
CH_2S_2Se	H_2CSeS_2	C:	MVol.D6-211, 216/7
CH_2S_3	$CS_2 \cdot H_2S$	C:	MVol.D4-195, 197
–	$SC(SH)_2$	C:	MVol.D4-208/16
–	$SC(SH)_2$ systems		
	$SC(SH)_2$-S	C:	MVol.D4-216
CH_2S_4	$SC(SH)(SSH)$	C:	MVol.D4-227/8
CH_2Se_3	$SeC(SeH)_2$	C:	MVol.D6-211/2
CH_2Th_2	$ThC \cdot ThH_2$	Th:	SVol.C1-8, 35/8
CH_3I_3Sn	CH_3SnI_3	Sn:	Org.Verb.6-218
$CH_3K_2NS_3$	$K_2CS_3 \cdot NH_3$	C:	MVol.D4-222
CH_3MnO^{2+}	$[MnOCH_3]^{2+}$	Mn:	MVol.D1-24/5
$CH_3NNa_2S_3$	$Na_2CS_3 \cdot NH_3$	C:	MVol.D4-222
$CH_3NNa_2S_4$	$Na_2CS_4 \cdot NH_3$	C:	MVol.D4-231
CH_3NOS	$NH_2C(O)SH$	C:	MVol.D6-132
–	$NH_2C(S)OH$	C:	MVol.D6-132
–	$OCS \cdot NH_3$	C:	MVol.D5-132
CH_3NOSe	$H[OSeCNH_2]$	C:	MVol.D6-248
CH_3NOSi	SiH_3NCO	C:	MVol.D3-138
CH_3NO_6SU	$UO_2SO_3 \cdot HCONH_2 \cdot 2\ H_2O$	U:	SVol.E1-96/7
CH_3NPtS_3	$Pt(NH_3)CS_3$	C:	MVol.D4-225
$CH_3NRb_2S_3$	$Rb_2CS_3 \cdot NH_3$	C:	MVol.D4-222
CH_3NS^{2+}	H_3NCS^{2+}	C:	MVol.D6-20, 27
CH_3NSSe	$H[SSeCNH_2]$	C:	MVol.D6-248
CH_3NSSi	SiH_3NCS	C:	MVol.D3-216
CH_3NS_2	$NH_2C(S)SH$	C:	MVol.D6-133/5
CH_3NSe_2	$H[Se_2CNH_2]$	C:	MVol.D6-248
CH_3N_2NaOS	$NH_2NHOCSNa$	C:	MVol.D5-133
$CH_3N_2S^-$	$NH_2C(S)NH^-$	C:	MVol.D6-107
$CH_3N_3O_2S$	$(NHNO_2)(NH_2)CS$	C:	MVol.D6-125
$CH_3N_3O_9Sn$	$CH_3Sn(NO_3)_3$	Sn:	Org.Verb.6-219/20
$CH_3N_3O_{10}U$	$UO_2(NO_3)_2 \cdot CH_3NO_2$	U:	SVol.E1-144/5
$CH_3O_{1.5}Sn$	$(CH_3SnO_{1.5})_x$	Sn:	Org.Verb.6-219/20
$CH_3O_6Sn^-$	$Sn(OH)_3CO_3^-$	Sn:	MVol.C2-218
CH_3S^+	CH_3S^+	C:	MVol.D4-160

CH_3Sn^+	$SnCH_3^+$	Sn:	Org.Verb.1-47, 70/1
		Sn:	Org.Verb.4-10
CH_3Th_2	Th_2CH_3	Th:	SVol.C1-36
$CH_4I_2N_2SSn$	$SnI_2 \cdot CS(NH_2)_2$	Sn:	MVol.C5-55
CH_4MnO^{3+}	$[Mn(CH_3OH)]^{3+}$	Mn:	MVol.D1-24/5
CH_4NOS^+	$(NH_3^+)C(O)SH$	C:	MVol.D6-27
CH_4N_2O	$(NH_2)_2CO$ systems		
	$(NH_2)_2CO-AgNO_3-H_2O$	Ag:	MVol.B6-333/4
	$(NH_2)_2CO-CeI_3-H_2O$	Sc:	MVol.C6-186
	$(NH_2)_2CO-LaI_3-H_2O$	Sc:	MVol.C6-184
	$(NH_2)_2CO-Mn(NO_3)_2$	Mn:	MVol.C3-294/5
	$(NH_2)_2CO-(NH_2)_2CS$	C:	MVol.D6-123
	$(NH_2)_2CO-(NH_2)_2CS-H_2O$	C:	MVol.D6-123
	$(NH_2)_2CO-NdI_3-H_2O$	Sc:	MVol.C6-191
	$(NH_2)_2CO-PrI_3-H_2O$	Sc:	MVol.C6-188
	$(NH_2)_2CO-SmI_3-H_2O$	Sc:	MVol.C6-193
$CH_4N_2O_2S$	$(NH_2)(NH)CSO_2H$	C:	MVol.D6-95, 98, 112
$CH_4N_2O_3S$	$(NH)(NH_2)CSO_3H$	C:	MVol.D6-112
$CH_4N_2O_6S_2U$	$UO_2SO_4 \cdot (NH_2)_2CS$	U:	SVol.E1-219/21
$CH_4N_2O_8S_2U$	$UO_2SO_4 \cdot (NH_2)_2CSO_2 \cdot H_2O$	U:	SVol.E1-216/7
CH_4N_2S	$(NH_2)_2CS$		
	Chemical reactions		
	on heating	C:	MVol.D6-91/3
	on irradiation	C:	MVol.D6-93/4
	with acids	C:	MVol.D6-109/12
	with alkaline solutions	C:	MVol.D6-112
	with elements	C:	MVol.D6-94/8
	with metal compounds	C:	MVol.D6-102/9
	with nonmetal compounds	C:	MVol.D6-98/102, 119
	with organic compounds	C:	MVol.D6-113
	Crystallographic properties	C:	MVol.D6-66/77
	Electrical properties	C:	MVol.D6-81/8
	Electrochemical behavior	C:	MVol.D6-88/91
	Formation	C:	MVol.D6-52/8
	– of thiourea clathrates	C:	MVol.D6-126/31
	General	C:	MVol.D6-51/2
	Magnetic properties	C:	MVol.D6-81
	Mechanical properties	C:	MVol.D6-77/9
	Molecule	C:	MVol.D6-58/65
	Optical properties	C:	MVol.D6-88
	Preparation	C:	MVol.D6-52/8
	Spectra	C:	MVol.D6-59, 62/4
	Thermal properties	C:	MVol.D6-77/81
	Uses	C:	MVol.D6-51/2
–	$(NH_2)_2C^{35}S$	C:	MVol.D6-56/7
–	$(NH_2)_2{}^{14}CS$	C:	MVol.D6-57
–	$(ND_2)_2CS$	C:	MVol.D6-51, 56
–	$(^{15}NH_2)_2CS$	C:	MVol.D6-56

CH_4N_2S	NH_4SCN	
	Chemical reactions	
	on heating	C: MVol.D6-52/3
	with ammonia	C: MVol.D4-187/8
	Formation from thiourea	C: MVol.D6-91/2
–	$(NH_2)_2CS$ solutions	
	$(NH_2)_2CS$-HF	C: MVol.D6-123
	$(NH_2)_2CS$-H_2O	C: MVol.D6-114/8
	$(NH_2)_2CS$-organic solvents	C: MVol.D6-123/4
–	$(NH_2)_2CS$ systems	
	$(NH_2)_2CS$-CH_3COOH	C: MVol.D6-111
	$(NH_2)_2CS$-$HClO_4$-H_2O	C: MVol.D6-122
	$(NH_2)_2CS$-HI-H_2O	C: MVol.D6-122
	$(NH_2)_2CS$-HNO_3-H_2O	C: MVol.D6-109, 121
	$(NH_2)_2CS$-H_2O	C: MVol.D6-113/4
	$(NH_2)_2CS$-H_2SO_4-H_2O	C: MVol.D6-122/3
	$(NH_2)_2CS$-H_3BO_3-H_2O	C: MVol.D6-111
	$(NH_2)_2CS$-H_3PO_4-H_2O	C: MVol.D6-111
	$(NH_2)_2CS$-$(NH_2)_2CO$	C: MVol.D6-123
	$(NH_2)_2CS$-$(NH_2)_2CO$-H_2O	C: MVol.D6-123
	$(NH_2)_2CS$-NH_3	C: MVol.D6-118/20
	$(NH_2)_2CS$-NH_3-H_2O	C: MVol.D6-120/1
$CH_4N_2S^+$	$(NH_2)_2CS^+$	C: MVol.D6-93
CH_4N_2Se	$(NH_2)_2CSe$	C: MVol.D6-241/7, 248
CH_4N_3OS	$(NH_2)_2CS \cdot NO$	C: MVol.D6-99
$CH_4N_4O_9U$	$UO_2(NO_3)_2 \cdot (NH_2)_2CO \cdot H_2O$	U: SVol.E1-83, 88/9
CH_4O_5Sn	$(HO)_3SnOCOH$	Sn: Org.Verb.1-158
CH_4S	CH_3SH	C: MVol.D4-163/4
CH_4S_2	$CH_2(SH)_2$	C: MVol.D4-163/4
CH_4S_4	$CS_2 \cdot 2\ H_2S \cdot 17\ H_2O$	C: MVol.D4-197
CH_4Sn^+	CH_3SnH^+	Sn: Org.Verb.4-10
CH_4Th_3	$ThC \cdot 2\ ThH_2$	Th: SVol.C1-8, 35/8
CH_5IN_2S	$(NH_2)_2CS \cdot HI$	C: MVol.D6-121, 125
CH_5I_2NSn	$SnI_2 \cdot CH_3NH_2$	Sn: MVol.C5-11/2
$CH_5I_5N_2OTe$	$[CON_2H_5]TeI_5$	Te: SVol.B3-65, 69/70
$CH_5MnO_2^{2+}$	$[MnOH(CH_3OH)]^{2+}$	Mn: MVol.D1-24/5
CH_5NO_3S	$NH_2SO_2OCH_3$	S: S-N-Verb.1-97
$CH_5NO_{3.2}W$	$CH_3NH_2 \cdot WO_{3.2}$	W: SVol.B3-241
CH_5NO_5U	$UO_2(OH)(HCOO) \cdot NH_3$	U: SVol.E1-12, 15
$CH_5N_2NaO_3S$	$Na[SO_3NCH_3NH_2]$	S: S-N-Verb.1-219
$CH_5N_2S^+$	$[(NH_2)_2C(SH)]^+$	C: MVol.D6-125
$CH_5N_3O_3S$	$NH_2C(O)NHSO_2NH_2$	S: S-N-Verb.1-167
–	$(NH_2)_2CS \cdot HNO_3$	C: MVol.D6-121, 125
CH_5N_3Se	$NH_2NH_2 \cdot HNCSe$	C: MVol.D6-231
$CH_6N_2Na_2S_3$	$Na_2CS_3 \cdot 2\ NH_3$	C: MVol.D4-222
CH_6N_2OS	$(NH_2)(NH_4S)CO = NH_4[SOCNH_2]$	C: MVol.D5-132/3
		C: MVol.D6-132
$CH_6N_2O_2S$	$CH_3NHSO_2NH_2$	S: S-N-Verb.1-167

$CH_6N_2O_3S$	$CH_3ONHSO_2NH_2$	S:	S-N-Verb.1-168
$CH_6N_2O_4SU$	$UO_2(OH)_2 \cdot (NH_2)_2CS$	U:	SVol.E1-220/1
$CH_6N_2O_4S_2$	$(NH_2)_2CS \cdot H_2SO_4$	C:	MVol.D6-123, 126
$CH_6N_2O_{12}S_2U^{2-}$	$[UO_2(SO_4)_2(H_2O)((NH_2)_2CO)]^{2-}$	U:	SVol.E1-90
$CH_6N_2PtS_3$	$Pt(NH_3)_2CS_3 \cdot H_2O$	C:	MVol.D4-226
$CH_6N_2S^{2+}$	$[(NH_3)(NH_2)C(SH)]^{2+}$	C:	MVol.D6-111
CH_6N_2SSe	$NH_4[SSeCNH_2]$	C:	MVol.D6-248
$CH_6N_2S_2$	$NH_4CS_2NH_2$	C:	MVol.D4-187/8
		C:	MVol.D6-140
$CH_6N_2S_3Zn$	$Zn(NH_3)_2CS_3$	C:	MVol.D4-224
$CH_6N_2Se_2$	$NH_4[Se_2CNH_2]$	C:	MVol.D6-248
CH_6Sn	CH_3SnH_3	Sn:	Org.Verb.4-121/4
–	CH_3SnD_3	Sn:	Org.Verb.4-122/3
–	CD_3SnH_3	Sn:	Org.Verb.4-122, 124
$CH_7N_3O_4S_2$	$CH_3N(SO_2NH_2)_2$	S:	S-N-Verb.1-162
CH_7N_3S	$(NH_2)_2CS \cdot NH_3$	C:	MVol.D6-118/21
$CH_8K_4O_{17}U_2$	$K_4[(UO_2)_2(O_2)_3(CO_3)(H_2O)_4] \cdot H_2O$	U:	SVol.C3-44
$CH_8N_2S_3$	$(NH_4)_2CS_3$	C:	MVol.D4-187, 223
$CH_8N_2S_4$	$(NH_4)_2CS_4$	C:	MVol.D4-230, 231
–	$(NH_4)_2CS_4 \cdot H_2O$	C:	MVol.D4-231
$CH_8N_4O_3S$	$NH_2CON(NH_4)SO_2NH_2$	S:	S-N-Verb.1-167
$CH_9N_3NiS_3$	$Ni(NH_3)_3CS_3$	C:	MVol.D4-225
$CH_9N_3PtS_3$	$Pt(NH_3)_3CS_3$	C:	MVol.D4-226
$CH_{11}N_7O_2P_2S$	$[SO_2[NP(NH_2)_2]_2NCH_3]$	S:	S-N-Verb.1-32
$CH_{12}IrN_5O_3S_2$	$Ir(NCS)(SO_3)(NH_3)_4$	Ir:	SVol.2-163
$CH_{12}S_8$	$CS_2 \cdot 6\ H_2S$	C:	MVol.D4-195, 197
$CH_{13}N_5S$	$(NH_2)_2CS \cdot 3\ NH_3$	C:	MVol.D6-118/21
CIN	ICN		
	Chemical reactions		
	in molecular beams	C:	MVol.D3-256
	with electrons	C:	MVol.D3-255
	with elements	C:	MVol.D3-256/7
	with ions	C:	MVol.D3-263/4
	with metal compounds	C:	MVol.D3-261/3
	with nonmetal compounds	C:	MVol.D3-257/61
	with organic compounds	C:	MVol.D3-264
	Crystallographic properties	C:	MVol.D3-245/6
	Electrical properties	C:	MVol.D3-253/4
	Electrochemical behavior	C:	MVol.D3-253/4
	Formation	C:	MVol.D3-241/5
	– of radioactive ICN	C:	MVol.D3-244
	Hydrolysis	C:	MVol.D3-257/8
	Isotope exchange	C:	MVol.D3-256
	Mechanical properties	C:	MVol.D3-246
	Molecule	C:	MVol.D3-151/80
	Photolysis	C:	MVol.D3-255
	Preparation	C:	MVol.D3-241/4
	Pyrolysis	C:	MVol.D3-254
	Review	C:	MVol.D3-149/50

CIN	ICN		
	Solubility	C:	MVol.D3-263/4
	Spectra	C:	MVol.D3-248/53
	Thermal properties	C:	MVol.D3-246/8
	Toxicity	C:	MVol.D3-150/1
–	$I^{14}CN$	C:	MVol.D3-244
–	^{131}ICN	C:	MVol.D3-244
CIN^+	ICN^+	C:	MVol.D3-265
CINO	INCO	C:	MVol.D3-143/4
CINS	ISCN	C:	MVol.D5-183/6
		C:	MVol.D6-177
–	$(ISCN)_x$	C:	MVol.D6-178
$CINS^-$	$ISCN^-$	C:	MVol.D5-177, 189
$CIN_2O_3^-$	$[ICNNO_3]^-$	C:	MVol.D3-263
CIS^+	CSI^+	C:	MVol.D4-160
CIS_2^+	CS_2I^+	C:	MVol.D4-160
CI_2N^-	$[I_2CN]^-$	C:	MVol.D3-237, 259, 263/4
CI_2NS^-	I_2SCN^-	C:	MVol.D5-177, 183/6
CI_2NiO	$Ni(CO)I_2$	Ni:	Org.Verb.1-111
CI_2O	COI_2	C:	MVol.D3-122
CI_2S	I_2CS	C:	MVol.D6-165
CI_3NS	ISC(I)NI	C:	MVol.D6-181
CI_3NSSn_2	Sn_2I_3SCN	Sn:	MVol.C2-229
CI_4IrKO	$K[IrI_4(CO)]$	Ir:	SVol.2-151
CI_4KNTe	$TeI_4 \cdot KCN$	Te:	SVol.B3-62
CI_4S	CI_3SI	C:	MVol.D4-189
CI_5IrK_2O	$K_2[Ir(CO)I_5]$	Ir:	SVol.2-151, 152
CIr	IrC	Ir:	SVol.2-180
CIr^+	IrC^+	Ir:	SVol.2-180
$CKNO_4S_2Sn$	$KSn(SO_4)SCN$	Sn:	MVol.C3-81
CKNSe	KSeCN	C:	MVol.D6-227
$CKNpO_5$	$KNpO_2CO_3$	Np:	TrU.D1-122
CK_2S_2Se	K_2CSeS_2	C:	MVol.D6-219
CK_2S_3	K_2CS_3	C:	MVol.D4-222
–	$K_2CS_3 \cdot H_2O$	C:	MVol.D4-218, 222
CK_2S_4	K_2CS_4	C:	MVol.D4-231
–	$K_2CS_4 \cdot 0.5\ H_2O$	C:	MVol.D4-231
CK_2Se_3	K_2CSe_3	C:	MVol.D6-215
CLi_2S_3	Li_2CS_3	C:	MVol.D4-214, 222
CMn_4	Mn_4C solid solutions		
	$Mn_4N_{1-x}C_x$	Mn:	MVol.C3-244
$CNOS^-$	$OSCN^-$	C:	MVol.D5-229
$CNOS^{2-}$	$OSCN^{2-}$	C:	MVol.D5-176
CNO_2S^-	O_2SCN^-	C:	MVol.D5-182
$CNO_3S_3^{2-}$	$S_2O_3SCN^{2-}$	C:	MVol.D5-177
$CNPuS^{2+}$	$PuSCN^{2+}$	Np:	TrU.D1-144
CNS	NCS	C:	MVol.D5-145/8
–	$S_x(CN)_x$	C:	MVol.D5-165, 223
CNS^+	NCS^+	C:	MVol.D5-148

CNS^- .... CNS^- .... C: MVol.D5-210
– .... NCS^-
Catalytic behavior .... C: MVol.D5-206
Chemical reactions
with acids or anions .... C: MVol.D5-187/93
with elements .... C: MVol.D5-182/7
with ion exchangers .... C: MVol.D5-204/5
with metal compounds .... C: MVol.D5-193/9
with nonmetal compounds .... C: MVol.D5-200/2
with organic compounds .... C: MVol.D5-206/10
Complex chemical behavior .... C: MVol.D5-202/4
Decomposition .... C: MVol.D5-172/82
Electrochemical behavior .... C: MVol.D5-161/70
Formation .... C: MVol.D5-148/51
Inhibitor effect .... C: MVol.D5-206
Isotopic exchange .... C: MVol.D5-182
Molecule .... C: MVol.D5-151/8
Physical properties .... C: MVol.D5-159/61
Solvation .... C: MVol.D5-170/2
Spectra .... C: MVol.D5-153/6
CNS^{2-} .... NCS^{2-} .... C: MVol.D5-210
$CNSSn^+$ .... $SnSCN^+$ .... Sn: MVol.C2-228
CNS_2 .... $N \cdot CS_2$ .... C: MVol.D4-182
$CNSe^-$ .... $NCSe^-$ .... C: MVol.D6-222/32
CNSeTl .... TlSeCN .... C: MVol.D6-229
$CNTe^-$ .... $NCTe^-$ .... C: MVol.D6-258/60
CN_2NiO_3 .... $CONi(NO)_2$ .... Ni: Org.Verb.1-111
CN_2OS .... NOSCN .... C: MVol.D6-173
CN_2S_2 .... $2 N \cdot CS_2$ .... C: MVol.D4-182
CN_4S_5 .... S_4N_3SCN .... C: MVol.D5-201
CN_6NiO .... $CONi(N_2)_3$ .... Ni: Org.Verb.1-112
$CNaS_2$ .... CS_2Na .... C: MVol.D4-184
CNa_2OS_2 .... $Na_2(S_2CO) \cdot H_2O$ .... C: MVol.D4-208
CNa_2O_3 .... Na_2CO_3 systems
Na_2CO_3-IrO_2 .... Ir: SVol.2-10/1
Na_2CO_3-Na_2TeO_3-H_2O .... Te: SVol.B1-103
CNa_2S_2Se .... Na_2CSeS_2 .... C: MVol.D6-219
CNa_2S_3 .... Na_2CS_3 .... C: MVol.D4-222
– .... $Na_2CS_3 \cdot 2 H_2O$ .... C: MVol.D4-222
CNa_2S_4 .... Na_2CS_4 .... C: MVol.D4-230, 231
– .... $Na_2CS_4 \cdot n H_2O$ .... C: MVol.D4-231
CNb .... NbC
Formation .... C: MVol.D4-185
CNb_2 .... Nb_2C .... C: MVol.D4-185
CNb_2S .... Nb_2CS .... C: MVol.D4-185
CNiO .... NiCO .... Ni: Org.Verb.1-111
$CNiO^+$ .... $Ni(CO)^+$ .... Ni: Org.Verb.1-215
$CNiO^-$ .... $Ni(CO)^-$ .... Ni: Org.Verb.1-215

$CNiO_7P_4$ .... $(P_4O_6Ni(CO))_x$ .... Ni: Org.Verb.2-394/5
$CNiO_{19}P_{12}$ .... $CONi(P_4O_6)_3$ .... Ni: Org.Verb.1-120
CNiS .... NiCS .... C: MVol.D4-28
$CNiS_3$ .... $NiCS_3$ .... C: MVol.D4-225
COS .... OCS
Adsorption on solids .... C: MVol.D5-139/40
Chemical reactions
with ions .... C: MVol.D5-102/4
with metal compounds .... C: MVol.D5-138
with metals .... C: MVol.D5-126
with nonmetal compounds .... C: MVol.D5-127/38
with nonmetals .... C: MVol.D5-105/26
with OH radicals .... C: MVol.D5-126
Crystallographic properties .... C: MVol.D5-43/4
Decomposition .... C: MVol.D5-69/104
Dipole moment .... C: MVol.D5-20/1
Electrical properties .... C: MVol.D5-53
Electrochemical behavior .... C: MVol.D5-69
Force constants .... C: MVol.D5-39/43
Formation .... C: MVol.D5-4/10, 15/6
Handling .... C: MVol.D5-3/4
Hydrolysis .... C: MVol.D5-128/32
Ionization by electrons .... C: MVol.D5-99/102
Ionization energies .... C: MVol.D5-17/9
Magnetic properties .... C: MVol.D5-24, 53
Mechanical properties .... C: MVol.D5-44/7, 51
Molecular vibrations .... C: MVol.D5-33/8
Molecule .... C: MVol.D5-16/43
Occurrence .... C: MVol.D5-2
Optical properties .... C: MVol.D5-53/4
Photolysis .... C: MVol.D5-75/99
Physiological action .... C: MVol.D5-3
Polarizability .... C: MVol.D5-23
Preparation .... C: MVol.D5-10/4
Purification .... C: MVol.D5-10/1, 12
Removal from industrial products .. C: MVol.D5-127
Solubility .... C: MVol.D5-107, 127/8, 140/1
Spectra
Emission in the visible region ... C: MVol.D5-68
IR spectra .... C: MVol.D5-60/5
MW spectra .... C: MVol.D5-54/60
Raman spectrum .... C: MVol.D5-65
UV spectra .... C: MVol.D5-65/8
Storage .... C: MVol.D5-3/4
Structural properties .... C: MVol.D5-43/4
Susceptibility .... C: MVol.D5-24/5
Thermal properties .... C: MVol.D5-44/52
Thermolysis .... C: MVol.D5-69/75

COS	$OC^{33}S$	C:	MVol.D5-14
–	$OC^{35}S$	C:	MVol.D5-14
–	$O^{14}CS$	C:	MVol.D5-14
–	^{18}OCS	C:	MVol.D5-14
–	$^{18}OC^{34}S$	C:	MVol.D5-14
–	OCS mixtures		
	OCS-O	C:	MVol.D5-108/13
COS^+	OCS^+	C:	MVol.D5-141/3
COS^{2+}	OCS^{2+}	C:	MVol.D5-143
COS^{3+}	OCS^{3+}	C:	MVol.D5-143
COS^-	OCS^-	C:	MVol.D5-144
$COSSe^{2-}$	$CSeSO^{2-}$	C:	MVol.D6-211, 219
COS_2	OCS_2	C:	MVol.D4-176
COS_2^+	OCS_2^+	C:	MVol.D5-102
COS_2^{2-}	S_2CO^{2-}	C:	MVol.D4-205/7
COS_2Tl_2	$Tl_2(S_2CO)$	C:	MVol.D4-207, 208
COS_3^{2-}	$[CS_3O]^{2-}$	C:	MVol.D4-220
COSe	OCSe	C:	MVol.D6-202/11
$COSe^+$	$OCSe^+$	C:	MVol.D6-211
$COSe_2^{2-}$	CSe_2O^{2-}	C:	MVol.D6-211, 215
COTe	OCTe	C:	MVol.D6-258
COTi	TiCO	Ti:	Org.Verb.1-122
$COTi^+$	$TiCO^+$	Ti:	Org.Verb.1-122
CO_2S	CO_2S	C:	MVol.D5-89/90
CO_2S^{2-}	CO_2S^{2-}	C:	MVol.D4-202/4
CO_2Se^{2-}	$CSeO_2^{2-}$	C:	MVol.D6-211, 215
CO_3Ra	$RaCO_3$	Ra:	SVol.2-370/1, 377
CO_3S	$OCS \cdot O_2$	C:	MVol.D5-112/3
CO_3Sn	$SnCO_3$	Sn:	MVol.C2-218
CO_5U	$UO_2(CO_3)$	U:	SVol.C2-249
CPa	PaC	Pa:	SVol.2-80
$CPbS_3$	$PbCS_3$	C:	MVol.D4-223
CPu	PuC		
	Formation	Np:	TrU.B2-41
CRb_2S_3	Rb_2CS_3	C:	MVol.D4-222
–	$Rb_2CS_3 \cdot H_2O$	C:	MVol.D4-222
CS	CS		
	Chemical reactions	C:	MVol.D4-25/9
	Formation	C:	MVol.D4-7/11
	Molecule	C:	MVol.D4-11/9
	Spectra	C:	MVol.D4-19/24
	Thermodynamic functions	C:	MVol.D4-24/5
	Transport properties	C:	MVol.D4-25
–	$(CS)_n$	C:	MVol.D4-25/6, 153/4, 182
CS^+	CS^+	C:	MVol.D4-6, 29/31, 159
CS^-	CS^-	C:	MVol.D4-31
CSSe	SCSe	C:	MVol.D6-252/5
$CSSe^+$	$SCSe^+$	C:	MVol.D6-255/6
$CSSe_2^{2-}$	CSe_2S^{2-}	C:	MVol.D6-215/6

CSTe	SCTe	C:	MVol.D6-262/3
CS_2	CS_2		
	Behavior as solvent	C:	MVol.D4-191/3
	Chemical reactions		
	with elementary particles	C:	MVol.D4-157/61
	with elements	C:	MVol.D4-163/86
	with inorganic compounds	C:	MVol.D4-186/91
	Cotton-Mouton effect	C:	MVol.D4-107/8
	Crystallographic properties	C:	MVol.D4-69/71
	Decomposition	C:	MVol.D4-151/6, 161
	Diffusion	C:	MVol.D4-89/90
	Economic importance	C:	MVol.D4-31
	Electrical properties	C:	MVol.D4-92/4
	Electrochemical behavior	C:	MVol.D4-150
	Faraday effect	C:	MVol.D4-108/10
	Formation	C:	MVol.D4-33/9, 41/2
	Handling	C:	MVol.D4-31/2
	Ionization	C:	MVol.D4-45/7, 157/61
	Isotopic exchange	C:	MVol.D4-161/2
	Kerr effect	C:	MVol.D4-102/7
	Magnetic properties	C:	MVol.D4-91/2
	Mechanical properties	C:	MVol.D4-72/4, 76/86
	Miscibility	C:	MVol.D4-193/4
	Molar refraction	C:	MVol.D4-102
	Molecule	C:	MVol.D4-42/69
	Optical properties	C:	MVol.D4-94/150
	Polymerization	C:	MVol.D4-69, 151
	Preparation	C:	MVol.D4-33/41
	Purification	C:	MVol.D4-41
	Rayleigh scattering	C:	MVol.D4-111/30
	Refractive index	C:	MVol.D4-94/101
	Selffocusing	C:	MVol.D4-110/1
	Solubility	C:	MVol.D4-193/4
	Spectra	C:	MVol.D4-120/50
	Structural properties	C:	MVol.D4-69/71
	Thermal properties	C:	MVol.D4-71, 74/5, 86/91
	Toxicity	C:	MVol.D4-32/3
	Uses	C:	MVol.D4-31
–	$CS^{35}S$	C:	MVol.D4-41
–	$2\ CS_2 \cdot H_2O$	C:	MVol.D4-197
–	CS_2 systems		
	CS_2-BCl_3	B:	B-Verb.19-141
	CS_2-H_2S	C:	MVol.D4-195/7
	CS_2-S	C:	MVol.D4-194/5
CS_2^+	CS_2^+	C:	MVol.D4-158/60, 197/200
CS_2^{2+}	CS_2^{2+}	C:	MVol.D4-159, 200
CS_2^{3+}	CS_2^{3+}	C:	MVol.D4-159, 200
CS_2^-	CS_2^-	C:	MVol.D4-163, 200/1
CS_2Se^{2-}	$CSeS_2^{2-}$	C:	MVol.D6-217/8

CS_3	$(CS_3)_n$	C:	MVol.D4-214, 221
CS_3^+	CS_3^+	C:	MVol.D4-159/60
CS_3^{2-}	CS_3^{2-}	C:	MVol.D4-216/22
CS_3Sr	$SrCS_3$	C:	MVol.D4-223
CS_3Tl_2	Tl_2CS_3	C:	MVol.D4-223
CS_3Zn	$ZnCS_3$	C:	MVol.D4-224
CS_4	$(CS_4)_n$	C:	MVol.D4-214
CS_4^+	CS_4^+	C:	MVol.D4-159/60
CS_4^{2-}	CS_4^{2-}	C:	MVol.D4-228/31
CS_4Sr	$SrCS_4$	C:	MVol.D4-230, 231
–	$SrCS_4 \cdot 2\ H_2O$	C:	MVol.D4-231
–	$SrCS_4 \cdot 3\ H_2O$	C:	MVol.D4-231
–	$SrCS_4 \cdot 4\ H_2O$	C:	MVol.D4-231
CS_5	$(CS_5)_n$	C:	MVol.D4-214
CS_6	$(CS_6)_n$	C:	MVol.D4-214
CS_7	$(CS_7)_n$	C:	MVol.D4-214
CS_8	$(CS_8)_n$	C:	MVol.D4-214
CS_9	$(CS_9)_n$	C:	MVol.D4-214
CSe	CSe	C:	MVol.D6-182/6
–	$(CSe)_n$	C:	MVol.D6-182
CSe^+	CSe^+	C:	MVol.D6-197, 251
CSeTe	SeCTe	C:	MVol.D6-264
$CSeTe^+$	$SeCTe^+$	C:	MVol.D6-264
CSe_2	CSe_2		
	Chemical reactions	C:	MVol.D6-197/201
	Electrochemical behavior	C:	MVol.D6-197
	Formation and preparation	C:	MVol.D6-186/9
	Molecule	C:	MVol.D6-189/91
	Physical properties	C:	MVol.D6-191/6
–	$(CSe_2)_n$	C:	MVol.D6-197
CSe_2^+	CSe_2^+	C:	MVol.D6-201/2, 250/1
CSe_3^{2-}	CSe_3^{2-}	C:	MVol.D6-213/5
CSi	SiC systems		
	SiC-Sn	Sn:	MVol.C2-235
CSn	SnC	Sn:	MVol.C2-217
$CSnTi_2$	Ti_2SnC	Sn:	MVol.C4-217
$CSnZr_2$	Zr_2SnC	Sn:	MVol.C4-226
CSn_2Ti_3	Ti_3Sn_2C	Sn:	MVol.C4-217
CSn_2Zr_3	Zr_3Sn_2C	Sn:	MVol.C4-226
CTa	TaC		
	Formation	C:	MVol.D4-185
CTa_2	Ta_2C	C:	MVol.D4-185
CTe	CTe	C:	MVol.D6-258
CTe_2	CTe_2	C:	MVol.D6-258
CTh	ThC systems		
	$ThC-Th-ThH_2$	Th:	SVol.C1-36
$C_{1.05}Ce_3Sn$	$Ce_3SnC_{1.05}$	Sn:	MVol.C4-160
$C_{1.2}Cl_2H_{2.4}MnO_{0.4}$	$MnCl_2 \cdot 0.4\ (CH_3)_2CO$	Mn:	MVol.D1-55
$C_{1.2}Cl_2H_{2.8}MnO_{0.2}$	$MnCl_2 \cdot 0.2\ (i-C_3H_7)_2O$	Mn:	MVol.D1-140

$C_{1.25}Cl_2H_{2.5}MnO_{0.25}$	$MnCl_2 \cdot 0.25\,(CH_2)_5O$	Mn: MVol.D1-146/7
$C_{1.3}ErH_{0.4}O_{4.3}$	$Er(OH)_{0.4}(CO_3)_{1.3} \cdot 2.5\,H_2O$	Sc: MVol.B7-26
$C_{1.8}F_2H_{7.4}O_{3.8}Sn$	$SnF_2(OH)_2 \cdot 0.9\,C_2H_4(OH)_2$	Sn: MVol.C5-76
$C_2CdH_4N_2S_4$	$Cd[S_2CNH_2]_2$	C: MVol.D6-140
$C_2CdN_2Se_2$	$Cd(SeCN)_2$	C: MVol.D6-229
$C_2Ce_2Cl_8H_{12}N_2$	$2\,CeCl_3 \cdot [C_2H_6(NH_3)_2]Cl_2$	Sc: MVol.C5-222
$C_2CfH_3O_2^{2+}$	$Cf(CH_3COO)^{2+}$	Np: TrU.D1-153
C_2ClFH_2NO	$CFClC(O)NH_2$	F: PerFHalOrg.7-56
$C_2ClFH_3NO_2S$	$FSO_2NCClCH_3$	S: S-N-Verb.1-105
$C_2ClFH_3NO_3S$	$FSO_2NCCl(OCH_3)$	S: S-N-Verb.1-118
–	$FSO_2NHC(O)CH_2Cl$	S: S-N-Verb.1-101
$C_2ClFH_4N_4O_4$	$(NO_2)_2CFC(NH)NH_2 \cdot HCl$	F: PerFHalOrg.7-3, 30
C_2ClFH_6Sn	$(CH_3)_2SnFCl$	Sn: Org.Verb.6-293
$C_2ClF_2H_2NO$	$CF_2ClC(O)NH_2$	F: PerFHalOrg.7-17, 19, 45
$C_2ClF_2H_2NO_2S$	$FSO_2NCClCH_2F$	S: S-N-Verb.1-117
$C_2ClF_2H_3NO_3P$	$CF_2NOPCl(O)OCH_3$	F: PerFHalOrg.7-199/200
–	$ClCFNOPF(O)OCH_3$	F: PerFHalOrg.7-199/200
$C_2ClF_2H_3N_2$	$CF_2ClNNCH_3$	F: PerFHalOrg.7-192
$C_2ClF_2H_6NOSSn$	$(CH_3)_2SnCl[NS(O)F_2]$	S: S-N-Verb.1-64
$C_2ClF_2H_6N_4O_2PS_2$	$[NS(O)F]_2[NPClN(CH_3)_2]$	S: S-N-Verb.1-27
–	$[NS(O)F]_2[NPClNHC_2H_5]$	S: S-N-Verb.1-27
C_2ClF_2IMg	$CF_2CClMgI$	F: PerFHalOrg.4-61, 90
C_2ClF_2K	CF_2CClK	F: PerFHalOrg.4-1
C_2ClF_2Li	CF_2CClLi	F: PerFHalOrg.4-2, 19, 45/6
C_2ClF_2N	$CFClCFN$	F: PerFHalOrg.5-2
$C_2ClF_2NO_2$	$ONCF_2C(O)Cl$	F: PerFHalOrg.7-176, 183, 190
$C_2ClF_2NS_2$	$SCClNCF_2S$	F: PerFHalOrg.5-57/8
C_2ClF_2Na	CF_2CClNa	F: PerFHalOrg.4-1
$C_2ClF_2O_{1.5}Si$	$(CFClCFSiO_{1.5})_n$	F: PerFHalOrg.4-147/8
C_2ClF_3	CF_2CFCl	F: PerFHalOrg.4-161
C_2ClF_3HNO	$CF_2ClC(F)NOH$	F: PerFHalOrg.7-88
–	$CF_3C(Cl)NOH$	F: PerFHalOrg.7-88, 105, 145
–	$NHOCFClCF_2$	F: PerFHalOrg.5-29/30, 38
$C_2ClF_3H_4NP$	$CF_3P(Cl)NHCH_3$	F: PerFHalOrg.3-102, 112
$C_2ClF_3H_4P_2$	$H_2PCFClCF_2PH_2$	F: PerFHalOrg.3-15, 20, 23
$C_2ClF_3N_2O_2$	$ONCF_2CFClNO$	F: PerFHalOrg.7-180, 186
$C_2ClF_3N_2O_3$	$O_2NCF_2CFClNO$	F: PerFHalOrg.7-180/1, 186, 191/2, 194/5, 199/200
$C_2ClF_3N_4$	$NNNClC(CF_3)N$	F: PerFHalOrg.5-73/4, 93
C_2ClF_3O	$CF_3C(O)Cl$	F: PerFHalOrg.4-179
		F: PerFHalOrg.6-114
$C_2ClF_3S_2$	$FClCS_2CF_2$	C: MVol.D6-169
$C_2ClF_4H_6NS$	$(CH_3)_2NSF_4Cl$	S: SVol.2-289
C_2ClF_4N	CF_3NCFCl	F: PerFHalOrg.5-45
C_2ClF_4NO	$CF_3CFClNO$	F: PerFHalOrg.7-174/5, 182, 194/5, 199
–	$ClCF_2CF_2NO$	F: PerFHalOrg.7-174/5, 182, 189/92, 208/9

C_2ClF_4NO	$NClOCF_2CF_2$	F:	PerFHalOrg.5-29/30, 38/9, 44/5
–	$NFOCFClCF_2$	F:	PerFHalOrg.5-29/30, 39, 44/5
$C_2ClF_4NO_2$	$CF_2ClCF_2NO_2$	F:	PerFHalOrg.3-109
$C_2ClF_4NO_2S$	$FSO_2NCClCF_3$	S:	S-N-Verb.1-117
C_2ClF_5HN	$CF_2ClN(H)CF_3$	F:	PerFHalOrg.7-13/4, 40
$C_2ClF_5H_2OS$	SF_5CH_2COCl	S:	SVol.2-288
$C_2ClF_5H_2S$	$SF_5CHCHCl$	S:	SVol.2-288
$C_2ClF_5H_4S$	$SF_5CH_2CH_2Cl$	S:	SVol.2-288
$C_2ClF_5H_5OSn^-$	$[SnF_5 \cdot ClC_2H_4OH]^-$	Sn:	MVol.C5-69
$C_2ClF_5N_2$	$CClF_2NNCF_3$	C:	MVol.D3-206, 214
C_2ClF_6N	$(CF_3)_2NCl$	C:	MVol.D3-214
C_2ClF_6NOS	$CClF_2CF_2NS(O)F_2$	S:	S-N-Verb.1-62
C_2ClF_6NS	$(CF_3)_2NSCl$	S:	SVol.2-75, 237, 262
C_2ClF_6OP	$(CF_3)_2P(O)Cl$	F:	PerFHalOrg.3-51, 53/4, 57
C_2ClF_6P	$(CF_3)_2PCl$		
	Chemical reactions	F:	PerFHalOrg.3-98, 100/2, 108/9, 120
	Physical properties	F:	PerFHalOrg.3-82/3, 89, 98
	Preparation	F:	PerFHalOrg.3-72
C_2ClF_6PS	$(CF_3)_2P(S)Cl$	F:	PerFHalOrg.3-140, 143, 147, 151/3, 159
C_2ClF_6Sb	$(CF_3)_2SbCl$	F:	PerFHalOrg.3-226, 229/30
C_2ClF_8HOP	$(CF_3)_2P(O)F_2 \cdot HCl$	F:	PerFHalOrg.3-53
C_2ClF_8NS	$SF_5NCClCF_3$	S:	S-N-Verb.1-148
$C_2ClFeN_3O_2$	$(CO)_2Fe(Cl)N_3$	Fe:	Org.Verb.B3-119
$C_2ClH_2O_4Pu^+$	$PuO_2(ClCH_2COO)^+$	Np:	TrU.D1-153
$C_2ClH_3IrN_2S_2$	$Ir(S_2N_2C_2H_3)Cl$	Ir:	SVol.2-172
C_2ClH_4NOSSn	$SnCl(CH(S)N(CH_3)O)$	Sn:	MVol.C5-57
$C_2ClH_4NO_4S$	$ClSO_2NHCOOCH_3$	S:	S-N-Verb.1-124/5
$C_2ClH_5N_2O_5S_2$	$ClSO_2NHCONHSO_2CH_3$	S:	S-N-Verb.1-125
C_2ClH_5NiO	CH_3OCH_2NiCl	Ni:	Org.Verb.1-14
$C_2ClH_5Sn^+$	$C_2H_5SnCl^+$	Sn:	Org.Verb.5-98
C_2ClH_5Ti	C_2H_5TiCl	Ti:	Org.Verb.1-7
$C_2ClH_6NO_2S$	$ClSO_2N(CH_3)_2$	S:	S-N-Verb.1-124
–	$ClSO_2NHC_2H_5$	S:	S-N-Verb.1-123
$C_2ClH_6NO_2S_2$	$ClSO_2NS(CH_3)_2$	S:	S-N-Verb.1-127
C_2ClH_6NS	$(CH_3)_2NSCl$	S:	SVol.2-237
$C_2ClH_6Sn^+$	$(CH_3)_2SnCl^+$	Sn:	Org.Verb.5-76
		Sn:	Org.Verb.6-13
–	$C_2H_5SnClH^+$	Sn:	Org.Verb.5-98
C_2ClH_6Ti	$(CH_3)_2TiCl$	Ti:	Org.Verb.1-5, 70
$C_2ClH_7O_2Sn$	$C_2H_5Sn(OH)_2Cl \cdot H_2O$	Sn:	Org.Verb.6-230/1
C_2ClH_7Sn	$(CH_3)_2SnHCl$	Sn:	Org.Verb.6-289
$C_2ClH_8NO_5Sn$	$SnCl(OH)_3 \cdot NH_2CH_2COOH$	Sn:	MVol.C6-11
$C_2ClH_8N_2O_4PS$	$(CH_3)_2NSO_2NHPClOOH$	S:	S-N-Verb.1-186

$C_2ClH_9N_4S_2$	$2\ (NH_2)_2CS \cdot HCl$	C: MVol.D6-110, 121, 125
$C_2ClH_{14}MnO_5{}^+$	$MnCl(H_2O)_3(CH_3OH)_2{}^+$	Mn: MVol.C5-78
C_2ClKN_2	$KNCCl(CN)$	C: MVol.D3-211
C_2ClNO_2	$ClCO(NCO)$	C: MVol.D3-100, 141
$C_2ClN_2{}^+$	$Cl(CN)_2{}^+$	C: MVol.D3-265, 282
$C_2ClN_2{}^-$	$[CNClCN]^-$	C: MVol.D3-215
$C_2ClOS_2{}^+$	$ClCS_2CO^+$	C: MVol.D6-172
$C_2Cl_2FH_2NO$	$CFCl_2C(O)NH_2$	F: PerFHalOrg.7-17, 19, 45, 56
$C_2Cl_2FH_2NO_2S$	$FSO_2NCClCH_2Cl$	S: S-N-Verb.1-105
$C_2Cl_2FH_2NO_3S$	$FSO_2NHC(O)CHCl_2$	S: S-N-Verb.1-101
$C_2Cl_2FH_3NO_3P$	$ClCFNOPCl(O)OCH_3$	F: PerFHalOrg.7-199/200
C_2Cl_2FLi	CCl_2CFLi	F: PerFHalOrg.4-2
–	$CFClCClLi$	F: PerFHalOrg.4-2, 19, 46
$C_2Cl_2F_2$	CCl_2CF_2	F: PerFHalOrg.4-161
–	$CFClCFCl$	F: PerFHalOrg.4-161
$C_2Cl_2F_2H_4P_2$	$H_2PCCl_2CF_2PH_2$	F: PerFHalOrg.3-15/6, 20
$C_2Cl_2F_2S_2$	$(FClCS)_2$	C: MVol.D6-168
$C_2Cl_2F_3HNO_2P$	$CF_3C(O)NHP(O)Cl_2$	F: PerFHalOrg.7-24/5, 51
$C_2Cl_2F_3H_{15}IrN_5O_{10}$	$[Ir(NH_3)_5(CF_3COO)](ClO_4)_2$	Ir: SVol.2-32
$C_2Cl_2F_3NO$	CF_3CCl_2NO	F: PerFHalOrg.7-174/5, 182
–	$ClCF_2CFClNO$	F: PerFHalOrg.7-174/5, 182, 189, 191/4, 199/200
–	$NClOCFClCF_2$	F: PerFHalOrg.5-29/30, 39, 44/5
$C_2Cl_2F_3NO_2S$	$ClSO_2NCCl(CF_3)$	S: S-N-Verb.1-130
$C_2Cl_2F_3N_2O_4P$	$NO_2CF_2CFNOP(O)Cl_2$	F: PerFHalOrg.7-100, 118
$C_2Cl_2F_3P$	CF_2CFPCl_2	F: PerFHalOrg.3-73, 91, 99, 100
$C_2Cl_2F_4NiO_2P_2$	$(CO)_2Ni(PF_2Cl)_2$	Ni: Org.Verb.1-124
$C_2Cl_2F_4O$	$F_3COCFCl_2$	C: MVol.D6 160
$C_2Cl_2F_4OS$	$F_3COSCFCl_2$	C: MVol.D6-160
$C_2Cl_2F_4Se$	$(F_2ClC)_2Se$	C: MVol.D6-250
$C_2Cl_2F_5P$	$C_2F_5PCl_2$	F: PerFHalOrg.3-72, 78, 89
$C_2Cl_2F_6Ge$	$(CF_3)_2GeCl_2$	F: PerFHalOrg.4-166/7, 171/3, 175
$C_2Cl_2F_6H_2NP$	$(CF_3)_2PCl_2NH_2$	F: PerFHalOrg.3-61, 68/9
$C_2Cl_2F_6NOP$	$(CF_3)_2NOPCl_2$	F: PerFHalOrg.7-100/1, 118, 126
$C_2Cl_2F_6NO_2P$	$(CF_3)_2NOP(O)Cl_2$	F: PerFHalOrg.7-145
$C_2Cl_2F_6Si$	$CFCl_2CF_2SiF_3$	F: PerFHalOrg.4-148, 161
–	$(CF_3)_2SiCl_2$	F: PerFHalOrg.4-147
$C_2Cl_2F_7P$	$(CF_3)_2PFCl_2$	F: PerFHalOrg.3-72, 89
$C_2Cl_2F_{10}N_2S_2$	$SF_5NCClC(Cl)NSF_5$	S: SVol.2-288
		S: S-N-Verb.1-148
$C_2Cl_2FeGeH_3{}^+$	$[FeClGe(CHCH_2)Cl]^+$	Fe: Org.Verb.B1-133
$C_2Cl_2FeGeH_3NO^+$	$[(NO)FeClGe(CHCH_2)Cl]^+$	Fe: Org.Verb.B1-133
$C_2Cl_2FeH_4$	$CH_2CH_2FeCl_2 \cdot 2\ H_2O$	Fe: Org.Verb.B4-178, 181
$C_2Cl_2FeO_2$	$(CO)_2FeCl_2$	Fe: Org.Verb.B3-121
$C_2Cl_2H_2N_2S_2$	$S_2(CN)_2 \cdot 2\ HCl$	C: MVol.D5-231

$C_2Cl_2H_2O_4Sn$ .... $SnCl_2(HCOO)_2$ .... Sn: MVol.C2-219
$C_2Cl_2H_3IrO_4$ .... $H[Ir(OOCH)_2Cl_2]$ .... Ir: SVol.2-31
$C_2Cl_2H_3NO_2U$ ...... $[UO_2Cl_2(CH_3CN)]$ .... U: SVol.E1-48/50
$C_2Cl_2H_3NO_3S$ ...... $ClSO_2NCCl(OCH_3)$ .... S: S-N-Verb.1-127
$C_2Cl_2H_4MnO_{0.5}$ ..... $MnCl_2 \cdot 0.5\ C_4H_8O$ .... Mn: MVol.D1-142
$C_2Cl_2H_4N_2O_4Sn$ ..... $(NO_2CH_2)_2SnCl_2$ .... Sn: MVol.6-128
$C_2Cl_2H_5MnO$ .... $MnCl_2 \cdot 0.5\ CH_3OC_2H_4OCH_3$ .... Mn: MVol.D1-140/1
$C_2Cl_2H_5NO_2Sn$ ..... $SnCl_2 \cdot NH_2CH_2COOH$ .... Sn: MVol.C5-30/1
$C_2Cl_2H_5NO_4S_2$ ...... $C_2H_5N(SO_2Cl)_2$ .... S: S-N-Verb.1-56
$C_2Cl_2H_5N_2O_2PS^+$ ... $CH_3CH_2NSONPOCl_2^+$ .... S: S-N-Verb.1-197
$C_2Cl_2H_5N_3O_4U$ ...... $UO_2Cl_2 \cdot NH(CONH_2)_2$ .... U: SVol.E1-92/3
$C_2Cl_2H_5Sn^+$ .... $[C_2H_5SnCl_2]^+$ .... Sn: Org.Verb.6-230
$C_2Cl_2H_5Ti$ .... $C_2H_5TiCl_2$ .... Ti: Org.Verb.1-8
$C_2Cl_2H_5Ti^+$ .... $[C_2H_5TiCl_2]^+$ .... Ti: Org.Verb.1-43
$C_2Cl_2H_6MnO$ .... $MnCl_2 \cdot C_2H_5OH$ .... Mn: MVol.D1-32
$C_2Cl_2H_6NO_2PS$ ..... $ClSO_2NP(Cl)(CH_3)_2$ .... S: S-N-Verb.1-127
$C_2Cl_2H_6N_2O_2S$ ...... $(CH_3)_2NSO_2NCl_2$ .... S: S-N-Verb.1-180
$C_2Cl_2H_6N_2O_{14}U$ ..... $UO_2(ClO_4)_2 \cdot 2\ CH_3NO_2$ .... U: SVol.E1-144/5
$C_2Cl_2H_6OSn$ .... $[C_2H_5SnCl_2OH]$ .... Sn: Org.Verb.6-230
– .... $C_2H_5Sn(OH)Cl_2 \cdot H_2O$ .... Sn: Org.Verb.6-230/1
$C_2Cl_2H_6O_2Te$ .... $TeCl_2(OCH_3)_2$ .... Te: SVol.B2-103
$C_2Cl_2H_6O_8S_2Sn$ ..... $SnCl_2(SO_3OCH_3)_2$ .... Sn: MVol.C2-69
$C_2Cl_2H_6SSn$ .... $CH_3SnCl_2(SCH_3)$ .... Sn: Org.Verb.6-218
$C_2Cl_2H_6Sn$ .... $(CH_3)_2SnCl_2$
Analysis .... Sn: Org.Verb.6-3
Chemical reactions .... Sn: Org.Verb.6-15/39
Decomposition .... Sn: Org.Verb.6-15
Formation and preparation .... Sn: Org.Verb.6-1/6
Molecule .... Sn: Org.Verb.6-7
Physical properties .... Sn: Org.Verb.6-13/4
Physiology .... Sn: Org.Verb.6-39
Polarography .... Sn: Org.Verb.6-14/5
Spectra .... Sn: Org.Verb.6-8/13
Structure .... Sn: Org.Verb.6-6/7
Uses .... Sn: Org.Verb.6-39/40
$C_2Cl_2H_6Ti$ .... $(CH_3)_2TiCl_2$ .... Ti: Org.Verb.1-71/2, 74
– .... $(CD_3)_2TiCl_2$ .... Ti: Org.Verb.1-72, 75
$C_2Cl_2H_7N_2O_3PS$ ..... $(CH_3)_2NSO_2NHPOCl_2$ .... S: S-N-Verb.1-186
$C_2Cl_2H_8MnO_2$ ...... $MnCl_2 \cdot 2\ CH_3OH$ .... Mn: MVol.D1-26
$C_2Cl_2H_8N_2Sn$ .... $SnCl_2 \cdot C_2H_8N_2$ .... Sn: MVol.C5-14
$C_2Cl_2H_8N_4O_2S_2U$ .... $UO_2Cl_2 \cdot 2\ (NH_2)_2CS$ .... U: SVol.E1-220/1
$C_2Cl_2H_8N_4O_2Sn$ ..... $SnCl_2 \cdot 2\ CO(NH_2)_2$ .... Sn: MVol.C5-44
$C_2Cl_2H_8N_4O_4U$ ...... $UO_2Cl_2 \cdot 2\ (NH_2)_2CO \cdot H_2O$ .... U: SVol.E1-83/6
$C_2Cl_2H_8N_4S_2Sn$ ..... $SnCl_2 \cdot 2\ CS(NH_2)_2$ .... Sn: MVol.C5-55/6
$C_2Cl_2H_8N_4S_2Te$ ..... $Te(SC(NH_2)_2)_2Cl_2$ .... Te: SVol.B3-140/1
$C_2Cl_2H_8N_4S_3$ .... $[(NH_2)(NH)CSSSC(NH)(NH_2)] \cdot 2\ HCl$ ..... C: MVol.D6-100
$C_2Cl_2H_8N_4Se_2$ ...... $(NH_2NHCSe)_2 \cdot 2\ HCl$ .... C: MVol.D6-246
$C_2Cl_2H_8N_4Se_2Sn$ .... $SnCl_2 \cdot 2\ CSe(NH_2)_2$ .... Sn: MVol.C5-58
$C_2Cl_2H_8N_4Se_2Te$ .... $Te(SeC(NH_2)_2)_2Cl_2$ .... Te: SVol.B3-162
$C_2Cl_2H_8O_2Sn^{2+}$ ..... $[SnCl_2(CH_3OH)_2]^{2+}$ .... Sn: MVol.C5-69

$C_2Cl_2H_{10}N_2O_2U$	$UO_2Cl_2 \cdot 2\ CH_3NH_2$	U:	SVol.E1-18, 21
$C_2Cl_2H_{10}N_2Sn$	$SnCl_2 \cdot 2\ CH_3NH_2$	Sn:	MVol.C5-11/2
$C_2Cl_2H_{12}MnO_4$	$MnCl_2(H_2O)_2(CH_3OH)_2$	Mn:	MVol.C5-78
$C_2Cl_2H_{18}IrN_5O_{10}$	$[Ir(NH_3)_5(CH_3COO)](ClO_4)_2$	Ir:	SVol.2-31
$C_2Cl_2H_{18}N_4Sn$	$(CH_3)_2SnCl_2 \cdot 4\ NH_3$	Sn:	Org.Verb.6-36
$C_2Cl_2IrK_3N_2O_8$	$K_3[Ir(NO_2)_2Cl_2(C_2O_4)]$	Ir:	SVol.2-61
$C_2Cl_2N_2^+$	$[Cl(CN)_2Cl]^+$	C:	MVol.D3-282
$C_2Cl_2N_2O_2$	$(ClCNO)_2$	C:	MVol.D3-144/5
–	$NCl_2C(O)NCO$	C:	MVol.D3-141, 144/5
$C_2Cl_2N_2O_3S$	$[NSO_2][NCCl][OCCl]$	S:	S-N-Verb.1-39
$C_2Cl_2N_2S_2$	$ClNC(Cl)SSCN$	C:	MVol.D6-180
–	$NCSNC(Cl)SCl$	C:	MVol.D6-175
$C_2Cl_2OS_2$	Cl_2CS_2CO	C:	MVol.D6-171/2
$C_2Cl_2O_2$	$(ClCO)_2$	C:	MVol.D3-6/7
$C_2Cl_2S_3$	Cl_2CS_2CS	C:	MVol.D6-169/71
$C_2Cl_3FHNO_2S$	$FSO_2NCClCHCl_2$	S:	S-N-Verb.1-105
$C_2Cl_3FHNO_3S$	$FSO_2NHC(O)CCl_3$	S:	S-N-Verb.1-101
$C_2Cl_3FH_2Sn$	$CH_2CFSnCl_3$	Sn:	Org.Verb.6-265
$C_2Cl_3F_2NO$	$ClCF_2CCl_2NO$	F:	PerFHalOrg.7-174/5, 182
–	$Cl_2CFCFClNO$	F:	PerFHalOrg.7-174/5, 182
$C_2Cl_3F_3HP$	$CF_2ClCFHPCl_2$	F:	PerFHalOrg.3-100
$C_2Cl_3F_3Si$	$CF_2CFSiCl_3$	F:	PerFHalOrg.4-149/50, 156, 161
$C_2Cl_3F_3Sn$	$CF_2CFSnCl_3$	F:	PerFHalOrg.4-181/2, 186
$C_2Cl_3F_4NS$	$CCl_3CF_2NSF_2$	S:	SVol.2-85
$C_2Cl_3F_5Si$	$C_2F_5SiCl_3$	F:	PerFHalOrg.4-147
$C_2Cl_3F_6NOSi$	$(CF_3)_2NOSiCl_3$	F:	PerFHalOrg.7-100, 117
$C_2Cl_3F_6P$	$(CF_3)_2PCl_3$		
	Chemical reactions	F:	PerFHalOrg.3-99, 109, 111
	Physical properties	F:	PerFHalOrg.3-79/80, 91
	Preparation	F:	PerFHalOrg.3-73
$C_2Cl_3F_6Sb$	$(CF_3)_2SbCl_3$	F:	PerFHalOrg.3-226/7, 229, 231
C_2Cl_3HO	$CHCl_2COCl$	C:	MVol.D3-62
C_2Cl_3HSn	$HCCSnCl_3$	Sn:	Org.Verb.6-266
$C_2Cl_3H_3NO_2U^-$	$[UO_2Cl_3(CH_3CN)]^-$	U:	SVol.E1-50
$C_2Cl_3H_3NU$	$[UCl_3(CH_3CN)]$	U:	SVol.E1-48/50
$C_2Cl_3H_3O_2Te$	$(HOCOCH_2)TeCl_3$	Te:	SVol.B2-98
$C_2Cl_3H_3Sn$	$CH_2CHSnCl_3$	Sn:	Org.Verb.6-263/4
$C_2Cl_3H_3Ti$	$CH_2CHTiCl_3$	Ti:	Org.Verb.1-16, 24
$C_2Cl_3H_5OTe$	$TeCl_3OC_2H_5$	Te:	SVol.B2-105
$C_2Cl_3H_5OTi$	$C_2H_5OTiCl_3$	Ti:	Org.Verb.1-23
$C_2Cl_3H_5Sn$	$C_2H_5SnCl_3$	Sn:	Org.Verb.6-224/34
$C_2Cl_3H_5Ti$	$C_2H_5TiCl_3$	Ti:	Org.Verb.1-15, 23
$C_2Cl_3H_6N_2O_2PS$	$(CH_3)_2NS(O)ClNP(O)Cl_2$	S:	S-N-Verb.1-197
	$(CH_3)_2NSO_2NPCl_3$	S:	S-N-Verb.1-187
$C_2Cl_3H_6N_2O_2PS^+$	$(CH_3)_2NSOClNPOCl_2^+$	S:	S-N-Verb.1-197
$C_2Cl_3H_6N_4O_2PS_2$	$[NS(O)Cl]_2[NPClN(CH_3)_2]$	S:	S-N-Verb.1-29
–	$[NS(O)Cl]_2[NPClNHC_2H_5]$	S:	S-N-Verb.1-29

$C_2Cl_3H_6OPS_2Sn$	$SnCl_3(CH_3(CH_3O)P(S)S)$	Sn:	MVol.C6-168
$C_2Cl_3H_6O_2PS_2Sn$	$SnCl_3((CH_3O)_2P(S)S)$	Sn:	MVol.C6-175/6
$C_2Cl_3H_6O_2U$	$UOCl_3 \cdot C_2H_5OH$	U:	SVol.E1-58/9
$C_2Cl_3H_6Sn^-$	$[(CH_3)_2SnCl_3]^-$	Sn:	Org.Verb.6-35, 38
$C_2Cl_3H_7O_2Sn$	$SnCl_3OCH_3 \cdot CH_3OH$	Sn:	MVol.C5-69
$C_2Cl_3H_8IrN_2$	$Ir(C_2H_4(NH_2)_2)Cl_3$	Ir:	SVol.2-85
$C_2Cl_3H_8MnN$	$(CH_3)_2NH_2MnCl_3$	Mn:	MVol.C5-189/90
–	$(CH_3)_2NH_2MnCl_3 \cdot 2\ H_2O$	Mn:	MVol.C5-189
–	$C_2H_5NH_3MnCl_3$	Mn:	MVol.C5-188
–	$C_2H_5NH_3MnCl_3 \cdot H_2O$	Mn:	MVol.C5-188
$C_2Cl_3H_{13}N_4Sn$	$C_2H_5SnCl_3 \cdot 2\ N_2H_4$	Sn:	Org.Verb.6-232
$C_2Cl_3Se^+$	$Cl_2CSeCCl^+$	C:	MVol.D6-251
$C_2Cl_3Se_2{}^+$	$Cl_2CSe_2CCl^+$	C:	MVol.D6-251
$C_2Cl_4Cs_2H_6Sn$	$Cs_2[(CH_3)_2SnCl_4]$	Sn:	Org.Verb.6-35
$C_2Cl_4Cs_2IrO_4$	$Cs_2[Ir(C_2O_4)Cl_4]$	Ir:	SVol.2-36
$C_2Cl_4FH_2NO_2S$	$FSO_2N(CHCl_2)_2$	S:	S-N-Verb.1-104
$C_2Cl_4FNO_2S$	$FSO_2NCClCCl_3$	S:	S-N-Verb.1-105
$C_2Cl_4F_2Si$	$CFClCFSiCl_3$	F:	PerFHalOrg.4-149/50, 156, 161
$C_2Cl_4F_3P$	$CF_2ClCFClPCl_2$	F:	PerFHalOrg.3-73, 92, 100
$C_2Cl_4F_4P_2$	$Cl_2PCF_2CF_2PCl_2$	F:	PerFHalOrg.3-73, 90, 100
$C_2Cl_4F_4Si$	$CF_2ClCF_2SiCl_3$	F:	PerFHalOrg.4-149/50, 156, 161
$C_2Cl_4F_6NOP$	$(CF_3)_2NOPCl_4$	F:	PerFHalOrg.7-100/1, 119, 126, 145
$C_2Cl_4H_2S$	HCl_2CSCCl_2H	C:	MVol.D4-28
$C_2Cl_4H_2Sn$	$CHClCHSnCl_3$	Sn:	Org.Verb.6-265
$C_2Cl_4H_3NSn$	$SnCl_4 \cdot CH_3CN$	Sn:	MVol.C6-65, 67
$C_2Cl_4H_3NTe$	$TeCl_4 \cdot CH_3CN$	Te:	SVol.B3-175
$C_2Cl_4H_3NU$	$[UCl_4(CH_3CN)]$	U:	SVol.E1-49
$C_2Cl_4H_4IrKN_2$	$K[Ir(HNCHCHNH)Cl_4]$	Ir:	SVol.2-86
$C_2Cl_4H_4N_2O_2Sn$	$SnCl_4 \cdot NH_2COCONH_2$	Sn:	MVol.C6-52
$C_2Cl_4H_4Sn$	$(CH_2Cl)_2SnCl_2$	Sn:	Org.Verb.6-128, 137/8
–	$CH_3CHClSnCl_3$	Sn:	Org.Verb.6-255
$C_2Cl_4H_5NO_2Sn$	$SnCl_4 \cdot NH_2CH_2COOH \cdot CH_3NO_2$	Sn:	MVol.C6-11
$C_2Cl_4H_5N_3O_2P_2S$	$[O_2S[NPCl_2]_2NC_2H_5]$	S:	S-N-Verb.1-34/5
$C_2Cl_4H_6N_2O_2Sn$	$SnCl_4 \cdot 2\ OCHNH_2$	Sn:	MVol.C6-40
$C_2Cl_4H_6N_2O_2Te$	$TeCl_4 \cdot 2\ HCONH_2$	Te:	SVol.B3-174
$C_2Cl_4H_6OSTe$	$TeCl_4 \cdot (CH_3)_2SO$	Te:	SVol.B3-176
$C_2Cl_4H_6OSn$	$SnCl_4 \cdot C_2H_5OH$	Sn:	MVol.C5-70/1
$C_2Cl_4H_6OU$	$UCl_4 \cdot (CH_3)_2O$	U:	SVol.E1-69/70, 74
$C_2Cl_4H_6O_2SSn$	$SnCl_4 \cdot (CH_3)_2SO_2$	Sn:	MVol.C6-103
$C_2Cl_4H_6S_2Sn$	$SnCl_4 \cdot HSCH_2CH_2SH$	Sn:	MVol.C6-81
$C_2Cl_4H_7NU$	$UCl_4 \cdot C_2H_5NH_2$	U:	SVol.E1-18, 20
$C_2Cl_4H_7O_4PSn$	$SnCl_4 \cdot (CH_3O)_2P(O)OH$	Sn:	MVol.C6-168/9
$C_2Cl_4H_8IrKN_2$	$K[Ir(C_2H_4(NH_2)_2)Cl_4]$	Ir:	SVol.2-85
–	$K[Ir(C_2H_4(NH_2)_2)Cl_4] \cdot 2\ H_2O$	Ir:	SVol.2-85
$C_2Cl_4H_8IrN_4S_2{}^-$	$[Ir(SC(NH_2)_2)_2Cl_4]^-$	Ir:	SVol.2-124

$C_2Cl_4H_8LaN$	$LaCl_3 \cdot [(CH_3)_2NH_2]Cl$	Sc:	MVol.C5-213/5, 219/20
–	$LaCl_3 \cdot [(CH_3)_2NH_2]Cl \cdot 2\ H_2O$	Sc:	MVol.C5-213/5
$C_2Cl_4H_8MnO_2^{2-}$	$[Mn(CH_3OH)_2Cl_4]^{2-}$	Mn:	MVol.C5-77
		Mn:	MVol.D1-24
$C_2Cl_4H_8N_4O_2Sn$	$SnCl_4 \cdot 2\ CO(NH_2)_2$	Sn:	MVol.C6-55
$C_2Cl_4H_8N_4S_2Sn$	$SnCl_4 \cdot 2\ CS(NH_2)_2$	Sn:	MVol.C6-120
$C_2Cl_4H_8N_4S_2Te$	$Te(SC(NH_2)_2)_2Cl_4$	Te:	SVol.B3-179
$C_2Cl_4H_8O_2Sn$	$SnCl_4 \cdot 2\ CH_3OH$	Sn:	MVol.C5-68/9
$C_2Cl_4H_{10}MnN_2$	$NH_3(CH_2)_2NH_3MnCl_4$	Mn:	MVol.C5-185/8
–	$NH_3(CH_2)_2NH_3MnCl_4 \cdot H_2O$	Mn:	MVol.C5-185, 187/8
$C_2Cl_4H_{10}N_2Sn$	$SnCl_4 \cdot 2\ CH_3NH_2$	Sn:	MVol.C5-168
$C_2Cl_4H_{10}N_2U$	$UCl_4 \cdot 2\ CH_3NH_2$	U:	SVol.E1-18, 20
$C_2Cl_4H_{12}MnN_2$	$(CH_3NH_3)_2MnCl_4$	Mn:	MVol.C5-175/82
–	$(CH_3ND_3)_2MnCl_4$	Mn:	MVol.C5-175/8, 181
–	$(CD_3ND_3)_2MnCl_4$	Mn:	MVol.C5-175, 180/1
$C_2Cl_4H_{12}MnN_6$	$[NHC(NH_2)_2H]_2MnCl_4$	Mn:	MVol.C5-204
$C_2Cl_4HgN_2S_2Sn$	$SnCl_4 \cdot Hg(SCN)_2$	Sn:	MVol.C4-97/8
$C_2Cl_4IrK_2O_4$	$K_2[Ir(C_2O_4)Cl_4]$	Ir:	SVol.2-36
$C_2Cl_4IrK_3O_4$	$K_3[Ir(C_2O_4)Cl_4]$	Ir:	SVol.2-36
$C_2Cl_4IrO_4^{2-}$	$[Ir(C_2O_4)Cl_4]^{2-}$	Ir:	SVol.2-36
$C_2Cl_4MoN_2O_2$	$MoO_2Cl_2 \cdot 2\ ClCN$	C:	MVol.D3-215
$C_2Cl_4N_2S_2$	ClNC(Cl)SSC(Cl)NCl	C:	MVol.D6-180
–	$Cl_2CNSSNCCl_2$	S:	SVol.2-260
$C_2Cl_4N_2S_3$	$ClNC(Cl)S_3C(Cl)NCl$	C:	MVol.D6-180
$C_2Cl_4NiO_2P_2$	$(CO)_2Ni(P_2Cl_4)$	Ni:	Org.Verb.1-149
C_2Cl_4O	CCl_3COCl	C:	MVol.D3-62, 64
$C_2Cl_4O_2$	$ClCOOCCl_3$	C:	MVol.D3-63/4
$C_2Cl_4S_2$	$(Cl_2CS)_2$	C:	MVol.D6-165/8
$C_2Cl_4Se_2$	$(Cl_2CSe)_2$	C:	MVol.D6-251
$C_2Cl_4Se_2^+$	$(Cl_2CSe)_2^+$	C:	MVol.D6-251
$C_2Cl_5F_3Si$	$CFCl_2CF_2SiCl_3$	F:	PerFHalOrg.4-149/50, 156, 161
$C_2Cl_5H_3IrK_2N$	$K_2[Ir(NCCH_3)Cl_5]$	Ir:	SVol.2-65
–	$K_2[Ir(NCCH_3)Cl_5] \cdot H_2O$	Ir:	SVol.2-65
–	$K_2[Ir(NCCH_3)Cl_5] \cdot 2\ H_2O$	Ir:	SVol.2-65
$C_2Cl_5H_3IrN^{2-}$	$[IrCl_5(NCCH_3)]^{2-}$	Ir:	SVol.2-65, 138
$C_2Cl_5H_3NSn^-$	$[SnCl_5(CH_3CN)]^-$	Sn:	MVol.C6-67
$C_2Cl_5H_3Sn$	$(CH_3)(CCl_3)SnCl_2$	Sn:	Org.Verb.6-196
$C_2Cl_5H_5SiSn$	$CH_3SiCl_2CH_2SnCl_3$	Sn:	Org.Verb.6-255
$C_2Cl_5H_6OU$	$UCl_5 \cdot (CH_3)_2O$	U:	SVol.E1-70, 74/5
$C_2Cl_5H_{11}IrN_3$	$(NH_4)_2[Ir(NCCH_3)Cl_5]$	Ir:	SVol.2-65
$C_2Cl_5H_{12}MnN_2$	$(CH_3NH_3)_2MnCl_5$	Mn:	MVol.C5-210
$C_2Cl_5H_{15}IrN_5Pt$	$[Pt(NH_3)_4][Ir(NCCH_3)Cl_5]$	Ir:	SVol.2-65
$C_2Cl_6F_3P$	$CF_2ClCFClPCl_4$	F:	PerFHalOrg.3-73, 100
$C_2Cl_6H_6N_4O_6P_2S_3$	$SO_2[N(CH_3)SO_2NPCl_3]_2$	S:	S-N-Verb.1-180
$C_2Cl_6H_8NU$	$[N(CH_3)_2H_2]UCl_6$	U:	SVol.C9-60/1, 63
$C_2Cl_6H_{10}N_4S_2Sn$	$2\ (NH_2)_2CS \cdot H_2SnCl_6$	C:	MVol.D6-126
$C_2Cl_6H_{12}IrN_2$	$(CH_3NH_3)_2[IrCl_6]$	Ir:	SVol.2-126/7
$C_2Cl_6H_{12}N_2Sb$	$[CH_3NH_3]_2[SbCl_6]$ solid solutions $[CH_3NH_3]_2[(Sb,Sn)Cl_6]$	Sn:	MVol.C3-104

$C_2Cl_6H_{12}N_2Sn$	$[CH_3NH_3]_2[SnCl_6]$	Sn:	MVol.C3-103/4
–	$[CH_3NH_3]_2[SnCl_6]$ solid solutions $[CH_3NH_3]_2[(Sn,Sb)Cl_6]$	Sn:	MVol.C3-104
$C_2Cl_6H_{12}N_2Te$	$[CH_3NH_3]_2TeCl_6$	Te:	SVol.B2-116, 137/8
$C_2Cl_6N_2Th$	$ThCl_4 \cdot 2\ ClCN$	C:	MVol.D3-215
$C_2Cl_6N_2Ti$	$2\ ClCN \cdot TiCl_4$	C:	MVol.D3-214
$C_2Cl_6N_2U$	$[UCl_4(CNCl)_2]$	U:	SVol.E1-48/50
$C_2Cl_6NiO_2P_2$	$(CO)_2Ni(PCl_3)_2$	Ni:	Org.Verb.1-124
$C_2Cl_6S_2$	$Cl_3CSSCCl_3$	C:	MVol.D6-233
$C_2Cl_6S_3$	$Cl_3CS_3CCl_3$	C:	MVol.D6-160
C_2Cl_6Se	$(CCl_3)_2Se$	C:	MVol.D6-199
$C_2Cl_6Se_2$	$(CCl_3)_2Se_2$	C:	MVol.D6-199
$C_2Cl_7H_6Ti_2^-$	$[(CH_3)_2Ti_2Cl_7]^-$	Ti:	Org.Verb.1-21
$C_2Cl_8H_4Si_2Sn$	$(Cl_3SiCH_2)_2SnCl_2$	Sn:	Org.Verb.6-128
$C_2Cl_8H_6Ti_2^{2-}$	$[(CH_3)_2Ti_2Cl_8]^{2-}$	Ti:	Org.Verb.1-21
$C_2Cl_8H_8N_2O_2P_2Sn$	$SnCl_4 \cdot 2\ Cl_2P(O)NHCH_3$	Sn:	MVol.C6-173
$C_2Cl_8O_2Sn$	$SnCl_4 \cdot 2\ COCl_2$	Sn:	MVol.C5-67
$C_2Cl_{11}H_5SnTi_2$	$C_2H_5SnCl_3 \cdot 2\ TiCl_4$	Sn:	Org.Verb.6-233
$C_2CmH_3O_2^{2+}$	$Cm(CH_3COO)^{2+}$	Np:	TrU.D1-153
$C_2CmH_3O_3^{2+}$	$Cm(HOCH_2COO)^{2+}$	Np:	TrU.D1-153
$C_2CmH_4NO_2^{2+}$	$Cm(NH_2CH_2COO)^{2+}$	Np:	TrU.D1-155
$C_2CmN_2S_2^+$	$Cm(SCN)_2^+$	Np:	TrU.D1-145
$C_2CmO_4^+$	$Cm(C_2O_4)^+$	Np:	TrU.D1-158
$C_2CoS_6^{2-}$	$[Co(CS_3)_2]^{2-}$	C:	MVol.D4-221
$C_2Co_2H_6N_2S_7$	$Co_2(NH_3)_2C_2S_7 \cdot CH_3COOH \cdot 4\ H_2O$	C:	MVol.D4-225
$C_2Co_2H_{18}N_6O_3S_8$	$Co_2(NH_3)_6(CS_3)_2(S_2O_3)$	C:	MVol.D4-225
$C_2Co_2H_{18}N_6S_7$	$Co_2(NH_3)_6C_2S_7$	C:	MVol.D4-225
$C_2Co_2H_{19}N_5O_2S_7$	$Co_2(NH_3)_5(H_2O)_2C_2S_7$	C:	MVol.D4-225
$C_2CrH_6O_4Sn$	$[(CH_3)_2SnCrO_4]_x$	Sn:	Org.Verb.6-29
$C_2CrH_8N_4O_8U$	$UO_2CrO_4 \cdot 2\ (NH_2)_2CO$	U:	SVol.E1-85, 89
C_2CsErO_6	$Cs[Er(CO_3)_2] \cdot 3\ H_2O$	Sc:	MVol.B7-26
$C_2CsF_{10}P$	$Cs[(CF_3)_2PF_4]$	F:	PerFHalOrg.3-134, 136, 137
$C_2CsI_3N_2$	$CsI \cdot 2\ ICN$	C:	MVol.D3-260
$C_2Cs_2NiS_6$	$Cs_2[Ni(CS_3)_2]$	C:	MVol.D4-225
$C_2CuH_3OS_2$	$Cu[S_2COCH_3]$	C:	MVol.D4-256
$C_2CuH_3O_2S$	$Cu[SOCOCH_3]$	C:	MVol.D4-237
$C_2CuH_4N_2S_4$	$Cu[S_2CNH_2]_2$	C:	MVol.D6-140
$C_2Cu_4S_7$	$Cu_4C_2S_7$	C:	MVol.D4-224
$C_2FH_2N_3O_5$	$(NO_2)_2CFC(O)NH_2$	F:	PerFHalOrg.7-17, 21/2, 45
$C_2FH_3N_4O_4$	$(NO_2)_2CFC(NH)NH_2 \cdot HCl$	F:	PerFHalOrg.7-3, 30
C_2FH_4NO	$FC(O)NHCH_3$	C:	MVol.D3-46
$C_2FH_4NO_2S$	FSO_2NCHCH_3	S:	S-N-Verb.1-105
$C_2FH_4NO_3S$	$FSO_2NHC(O)CH_3$	S:	S-N-Verb.1-101
$C_2FH_4NO_4S$	$FSO_2NHCOOCH_3$	S:	S-N-Verb.1-101
$C_2FH_5NOS^+$	$SNC_2H_5FO^+$	S:	S-N-Verb.1-194, 195
$C_2FH_5N_2O_2S$	$FSO_2NC(NH_2)CH_3$	S:	S-N-Verb.1-105
$C_2FH_5O_2S$	$C_2H_5OS(O)F$	S:	SVol.2-79
$C_2FH_6NOS^+$	$SNC_2H_6FO^+$	S:	S-N-Verb.1-194, 195

$C_2FH_6NO_2S$	$FSO_2N(CH_3)_2$	S:	S-N-Verb.1-103
$C_2FH_6NO_2S_2$	$FSO_2NS(CH_3)_2$	S:	S-N-Verb.1-107
C_2FH_6PS	$(CH_3)_2PSF$	S:	SVol.2-84
$C_2FH_6Sn^+$	$(CH_3)_2SnF^+$	Sn:	Org.Verb.5-18
C_2FLi	FCCLi	F:	PerFHalOrg.4-2, 9
C_2FNO	COF(CN)	C:	MVol.D3-15
C_2FNOS	FCO(NCS)	C:	MVol.D3-46
C_2FNO_2	FCO(NCO)	C:	MVol.D3-46
C_2FN_3	$(F)(NC)CN_2$	F:	PerFHalOrg.5-4/5, 11
C_2FN_3O	$(F)(OCN)CN_2$	F:	PerFHalOrg.5-4/5
C_2FNa	FCCNa	F:	PerFHalOrg.4-1
$C_2F_2HNO_2$	$[FC(O)]_2NH$	F:	PerFHalOrg.7-23, 50
$C_2F_2H_2INO$	$CF_2IC(O)NH_2$	F:	PerFHalOrg.7-17, 19, 45
$C_2F_2H_2NO$	$CF_2C(O)NH_2$	F:	PerFHalOrg.7-56
$C_2F_2H_2N_2O_3$	$NO_2CF_2C(O)NH_2$	F:	PerFHalOrg.7-17, 19/20, 45
$C_2F_2H_3NOS^+$	$C_2H_3NSOF_2{}^+$	S:	S-N-Verb.1-157
$C_2F_2H_3NO_3S$	$FSO_2N(CH_3)C(O)F$	S:	S-N-Verb.1-104
–	$FSO_2NHC(O)CH_2F$	S:	S-N-Verb.1-101
$C_2F_2H_3N_3O_2$	$NO_2CF_2C(NH)NH_2$	F:	PerFHalOrg.7-3/4, 30
$C_2F_2H_4O$	CH_3OCF_2H	F:	PerFHalOrg.5-19
$C_2F_2H_5NOS^+$	$C_2H_5NSOF_2{}^+$	S:	S-N-Verb.1-157
$C_2F_2H_5NO_3U$	$[UO_2F_2 \cdot (CH_3CONH_2)]_n$	U:	SVol.E1-96/8
$C_2F_2H_5NO_4S_2$	$C_2H_5N(SO_2F)_2$	S:	S-N-Verb.1-53
$C_2F_2H_5NS$	$C_2H_5NSF_2$	S:	SVol.2-85
$C_2F_2H_5N_3O_4S_3$	$NS(O)FNSO_2N(C_2H_5)S(O)F$	S:	S-N-Verb.1-10
$C_2F_2H_6NOSSi^+$	$(CH_3)_2SiNS(O)F_2{}^+$	S:	S-N-Verb.1-64
$C_2F_2H_6N_2O_2S_2$	$(CH_3)_2NSO_2NSF_2$	S:	S-N-Verb.1-182
$C_2F_2H_6N_2O_3S_2$	$(CH_3)_2NS(O)FNSO_2F$	S:	S-N-Verb.1-196
$C_2F_2H_6N_2S$	$NSF_2[N(CH_3)_2]$	S:	S-N-Verb.1-203/4
$C_2F_2H_6N_2S^+$	$NSF_2N(CH_3)_2{}^+$	S:	S-N-Verb.1-204
$C_2F_2H_6N_4O_3S_3$	$[NS(O)F]_2[NS(O)N(CH_3)_2]$	S:	S-N-Verb.1-14
–	$[NS(O)F]_2[NS(O)NHC_2H_5]$	S:	S-N-Verb.1-14
$C_2F_2H_6OSSn$	$SnF_2 \cdot (CH_3)_2SO$	Sn:	MVol.C5-45
$C_2F_2H_6O_3SU$	$UO_2F_2 \cdot (CH_3)_2SO$	U:	SVol.E1-203, 206/7
–	$[UO_2F_2((CH_3)_2SO)]_n$	U:	SVol.E1-207
$C_2F_2H_6O_4Sn$	$SnF_2(OH)_2 \cdot CH_3COOH$	Sn:	MVol.C5-117
$C_2F_2H_6Sn$	$(CH_3)_2SnF_2$	Sn:	Org.Verb.5-41/5
$C_2F_2H_8N_4O_4U$	$UO_2F_2 \cdot 2\ (NH_2)_2CO \cdot H_2O$	U:	SVol.E1-83/5
$C_2F_2H_9N_5O_3S_3$	$[NS(O)F]_2[NS(O)NHCH_3] \cdot CH_3NH_2$	S:	S-N-Verb.1-14
$C_2F_2N_2O$	$NF_2C(O)CN$	C:	MVol.D3-136
$C_2F_2N_2OS$	$NF_2C(O)NCS$	C:	MVol.D3-136
$C_2F_2N_2O_2$	NF_2CONCO	C:	MVol.D3-136
$C_2F_2N_2O_2S$	$C(O)(NCO)NSF_2$	S:	SVol.2-83
$C_2F_2N_2O_3S$	$OCNC(O)NS(O)F_2$	S:	S-N-Verb.1-74/5
$C_2F_2N_2O_3S^+$	$OCNCONSOF_2{}^+$	S:	S-N-Verb.1-75
$C_2F_2OS_2$	F_2CS_2CO	C:	MVol.D6-169
$C_2F_2O_2$	$(FCO)_2$	C:	MVol.D3-3
$C_2F_3HN_2O_3$	$CF_3C(NO_2)NOH$	F:	PerFHalOrg.7-88/9
–	NO_2CF_2CFNOH	F:	PerFHalOrg.7-88

$C_2F_3HN_4$	$NHN_3C(CF_3)$	F:	PerFHalOrg.5-73/4, 93
$C_2F_3H_2NO$	$CF_3C(O)NH_2$	F:	PerFHalOrg.7-17, 22, 27, 45, 55/7, 62/3
–	$CHF_2N(H)C(O)F$	C:	MVol.D3-45
–	$[N(CF_3)OCH_2]_n$	F:	PerFHalOrg.7-209
$C_2F_3H_3NO$	$CF_3N(O)CH_3$	F:	PerFHalOrg.7-169
$C_2F_3H_3NO_3P$	$CF_2NOPF(O)OCH_3$	F:	PerFHalOrg.7-199/200
$C_2F_3H_3N_2$	$CF_3C(NH)NH_2$	F:	PerFHalOrg.7-3/4, 30, 57
–	CF_3NNCH_3	F:	PerFHalOrg.7-158/9, 170
$C_2F_3H_3N_2O$	$CF_3C(NH_2)NOH$	F:	PerFHalOrg.7-88
–	$CF_3C(NOH)NH_2$	F:	PerFHalOrg.7-5, 32
–	CF_3NNOCH_3	F:	PerFHalOrg.7-158/9
–	$CH_3NN(O)CF_3$	F:	PerFHalOrg.7-160/1
$C_2F_3H_3N_2O_2S_2$	$FSO_2NC(NSF_2)CH_3$	S:	S-N-Verb.1-119
$C_2F_3H_3Sn$	CH_2CHSnF_3	Sn:	Org.Verb.5-52
$C_2F_3H_4NO_2SSn^+$	$HSnNCH_3SO_2CF_3{}^+$	S:	S-N-Verb.1-108
$C_2F_3H_4N_3$	$CF_3C(NH_2)NNH_2$	F:	PerFHalOrg.7-6/7, 33
$C_2F_3H_4OP$	CF_3PHOCH_3	F:	PerFHalOrg.3-6
$C_2F_3H_5NOS^+$	$C_2H_5NSOF_3{}^+$	S:	S-N-Verb.1-157
$C_2F_3H_5Sn$	$C_2H_5SnF_3$	Sn:	Org.Verb.5-51
$C_2F_3H_6NOS$	$(CH_3)_2NS(O)F_3$	S:	S-N-Verb.1-156/7
$C_2F_3H_6NOS^+$	$(CH_3)_2NSOF_3{}^+$	S:	S-N-Verb.1-157
$C_2F_3H_6N_2O_2PS$	$[SO_2(NCH_3)_2PF_3]$	S:	S-N-Verb.1-41/2
$C_2F_3H_6N_3O_4S_3$	$(CH_3)_2NSF(NSO_2F)_2$	S:	S-N-Verb.1-208
$C_2F_3H_{10}NSn$	$NH_4[(CH_3)_2SnF_3]$	Sn:	Org.Verb.5-45
C_2F_3IMg	CF_2CFMgI	F:	PerFHalOrg.4-61, 71, 90, 106, 110
C_2F_3K	CF_2CFK	F:	PerFHalOrg.4-1
C_2F_3Li	CF_2CFLi	F:	PerFHalOrg.4-2, 9, 13, 19, 28, 33, 44/5
C_2F_3NO	CF_2CFNO	F:	PerFHalOrg.7-174, 182, 189/91, 208/9
–	CF_3CNO	F:	PerFHalOrg.7-145
–	CF_3NCO	F:	PerFHalOrg.5-46/7
–	$[NOCF_2CF]_n$	F:	PerFHalOrg.5-33/4
$C_2F_3NO_2$	$ONCF_2C(O)F$	F:	PerFHalOrg.7-176, 183, 190
$C_2F_3NO_3S$	CF_3SO_2NCO	S:	S-N-Verb.1-116
$C_2F_3N_4Na$	$NaNNNNC(CF_3)$	F:	PerFHalOrg.5-73, 109
C_2F_3Na	CF_2CFNa	F:	PerFHalOrg.4-1
$C_2F_3O_{1.5}Si$	$(CF_2CFSiO_{1.5})_n$	F:	PerFHalOrg.4-147/8, 161
C_2F_4	CF_2CF_2	F:	PerFHalOrg.4-68
C_2F_4HNO	$CF_3C(F)NOH$	F:	PerFHalOrg.7-88
–	$CF_3NHC(O)F$	F:	PerFHalOrg.7-22/3, 49
–	$NHOCF_2CF_2$	F:	PerFHalOrg.5-29/30, 37
$C_2F_4HNO_3S$	$CF_3C(O)NHSO_2F$	F:	PerFHalOrg.7-24/5, 50
		S:	S-N-Verb.1-101
$C_2F_4H_2N_2$	$NHCF_2CF_2NH$	F:	PerFHalOrg.5-35/6, 42
$C_2F_4H_2N_2O_2S$	$CF_3C(NH_2)NSO_2F$	F:	PerFHalOrg.7-2, 29
		S:	S-N-Verb.1-118

$C_2F_4H_3NO_4S_2$	$CF_3SO_2N(CH_3)SO_2F$	S:	S-N-Verb.1-50
$C_2F_4H_3NU$	$UF_4 \cdot CH_3CN$	U:	SVol.E1-48/50
$C_2F_4H_3P$	$HCF_2CF_2PH_2$	F:	PerFHalOrg.3-16
$C_2F_4H_4N_2S_2$	$F_2SN(CH_2)_2NSF_2$	S:	SVol.2-78, 85
$C_2F_4H_4P_2$	$H_2PCF_2CF_2PH_2$	F:	PerFHalOrg.3-15, 20, 23, 25
$C_2F_4H_6K_2Sn$	$K_2[(CH_3)_2SnF_4]$	Sn:	Org.Verb.5-45
–	$K_2[(CH_3)_2SnF_4] \cdot 2\ H_2O$	Sn:	Org.Verb.5-45
$C_2F_4H_6N_2O_2S_2Si$	$(CH_3)_2Si[NS(O)F_2]_2$	S:	S-N-Verb.1-82
$C_2F_4H_6N_2O_2S_2Si^+$	$(CH_3)_2Si(NSOF_2)_2{}^+$	S:	S-N-Verb.1-82
$C_2F_4H_6N_2O_2S_2Sn$	$(CH_3)_2Sn[NS(O)F_2]_2$	S:	S-N-Verb.1-83
$C_2F_4H_6N_2O_2S_2Sn^+$	$(CH_3)_2Sn(NSOF_2)_2{}^+$	S:	S-N-Verb.1-83
$C_2F_4H_6O_2Te$	$TeF_4(OCH_3)_2$	Te:	SVol.B2-29
$C_2F_4H_6O_4P_2Sn$	$(CH_3)_2Sn(OPOF_2)_2$	Sn:	Org.Verb.6-25
$C_2F_4H_8MnN$	$NH_2(CH_3)_2MnF_4 \cdot 2\ H_2O$	Mn:	MVol.C4-212/3
–	$NH_3(C_2H_5)MnF_4$	Mn:	MVol.C4-212/3
$C_2F_4H_{14}N_2Sn$	$(NH_4)_2[(CH_3)_2SnF_4]$	Sn:	Org.Verb.5-45
$C_2F_4N_2$	F_2CNNCF_2	C:	MVol.D3-213, 291
–	NCF_2CF_2N	F:	PerFHalOrg.5-35/7, 42, 46, 48/9
$C_2F_4N_2O_2$	$ONCF_2CF_2NO$	F:	PerFHalOrg.7-180, 186, 208/9
$C_2F_4N_2O_3$	$NF_2C(O)OC(O)NF_2$	C:	MVol.D3-125
–	$O_2NCF_2CF_2NO$	F:	PerFHalOrg.7-180/1, 186, 189/95, 208/9
$C_2F_4N_2O_6S_2$	$FSO_2(FOC)NN(COF)SO_2F$	S:	S-N-Verb.1-223
$C_2F_4N_2S$	$CF_2(CN)NSF_2$	S:	SVol.2-82
C_2F_4O	CF_3CFO	C:	MVol.D3-17, 19, 45
		F:	PerFHalOrg.6-114
$C_2F_4P^+$	CF_3PCF^+	F:	PerFHalOrg.3-179
$C_2F_4Se_2$	$(F_2CSe)_2$	C:	MVol.D6-250/1
$C_2F_4Se_2{}^+$	$(F_2CSe)_2{}^+$	C:	MVol.D6-250
$C_2F_5HN_2O_3$	$CF_3N(H)CF_2ONO_2$	F:	PerFHalOrg.7-13/4, 41
$C_2F_5HO_3S$	$CF_3SO_3CF_2H$	F:	PerFHalOrg.5-19
$C_2F_5H_2NO$	$C_2F_5N(H)OH$	F:	PerFHalOrg.7-16/7
$C_2F_5H_2NO_4S$	$CF_3N(H)CF_2OSO_3H$	F:	PerFHalOrg.7-13/4, 41
$C_2F_5H_2N_5$	$NCF(NF_2)NFCF(NH_2)N$	F:	PerFHalOrg.5-72/3
$C_2F_5H_3NP$	$CF_3P(NH_2)CF_2H$	F:	PerFHalOrg.3-24
$C_2F_5H_3N_4$	$(NF_2)_2CFC(NH)NH_2$	F:	PerFHalOrg.7-3/4, 31
$C_2F_5H_4S$	$SF_5CH_2CH_2$	S:	SVol.2-288
$C_2F_5H_4Ti^{2-}$	$[C_2H_4TiF_5]^{2-}$	Ti:	Org.Verb.1-125
$C_2F_5H_5NTe^+$	$TeF_5NCH_3CH_2{}^+$	Te:	SVol.B2-65
–	$TeF_5NC_2H_5{}^+$	Te:	SVol.B2-66
$C_2F_5H_5OTe$	$TeF_5OC_2H_5$	Te:	SVol.B2-53
$C_2F_5H_6NS$	$SF_5N(CH_3)_2$	S:	S-N-Verb.1-136
$C_2F_5H_6NTe$	$(CH_3)_2NTeF_5$	Te:	SVol.B2-65
$C_2F_5H_6NTe^+$	$TeF_5N(CH_3)_2{}^+$	Te:	SVol.B2-65
$C_2F_5H_6N_2S_2{}^+$	$SF_5NSN(CH_3)_2{}^+$	S:	S-N-Verb.1-141, 144
$C_2F_5H_6OSn^-$	$[SnF_5 \cdot C_2H_5OH]^-$	Sn:	MVol.C5-69
$C_2F_5H_6O_2Sn^-$	$[SnF_5 \cdot C_2H_4(OH)_2]^-$	Sn:	MVol.C5-75

$C_2F_5H_8N_4S_2U$	$UF_5 \cdot 2\ (NH_2)_2CS$	U:	SVol.E1-219/20
$C_2F_5H_{10}MnN_2$	$(NH_3)_2C_2H_4MnF_5$	Mn:	MVol.C4-212/3
$C_2F_5H_{12}MnN_6$	$(CNH(NH_2)_2H)_2MnF_5$	Mn:	MVol.C4-212/3
$C_2F_5I_2P$	$C_2F_5PI_2$	F:	PerFHalOrg.3-74/5, 93
C_2F_5N	CF_3NCF_2	C:	MVol.D3-213, 214, 291
		F:	PerFHalOrg.5-45/8
		F:	PerFHalOrg.6-114
C_2F_5NO	C_2F_5NO	F:	PerFHalOrg.7-174, 182, 189, 191, 208/9
–	$NFOCF_2CF_2$	F:	PerFHalOrg.5-29/30, 38, 44/5
C_2F_5NOS	$NF_2C(O)SCF_3$	C:	MVol.D3-136
$C_2F_5NO_2S$	$CF_3C(O)NS(O)F_2$	S:	S-N-Verb.1-74
$C_2F_5NO_2S^+$	$CF_3CONSOF_2^+$	S:	S-N-Verb.1-74
$C_2F_5NO_3S$	$CF_3CF(SO_2F)NO$	F:	PerFHalOrg.7-175/6, 183, 189/91, 193
$C_2F_5NO_4S$	$FSO_2N(OCF_3)C(O)F$	S:	S-N-Verb.1-107
C_2F_5P	CF_2CFPF_2	F:	PerFHalOrg.3-71, 85
–	CF_3PCF_2	F:	PerFHalOrg.3-23/4
$C_2F_5P^+$	$CF_3PCF_2^+$	F:	PerFHalOrg.3-179
$C_2F_5Se_2^+$	$[(F_2CSe)_2F]^+$	C:	MVol.D6-251
$C_2F_6Fe_2N_4O_4P$	$Fe_2(NO)_4[P(CF_3)_2]$	F:	PerFHalOrg.3-181
$C_2F_6Fe_2N_4O_4S_2$	$(Fe(NO)_2SCF_3)_2$	Fe:	Org.Verb.C1-86
$C_2F_6Ge^+$	$Ge(CF_3)_2^+$	F:	PerFHalOrg.4-174
$C_2F_6GeI_2$	$(CF_3)_2GeI_2$	F:	PerFHalOrg.4-166/7, 171/3, 176
C_2F_6HN	$(CF_3)_2NH$	F:	PerFHalOrg.7-13/4, 40, 55/6, 58
C_2F_6HNO	$(CF_3)_2NOH$	F:	PerFHalOrg.7-86/7, 102, 104, 121, 125, 138/41
–	$(CF_3)_2NOH \cdot H_2O$	F:	PerFHalOrg.7-86/7, 104
$C_2F_6HNO_2S$	$C_2F_5N(H)SO_2F$	F:	PerFHalOrg.7-17, 44
		S:	S-N-Verb.1-100
$C_2F_6HNP^-$	$[(CF_3)_2PNH]^-$	F:	PerFHalOrg.3-69
C_2F_6HOP	$(CF_3)_2POH$	F:	PerFHalOrg.3-44/5, 49/50
–	$(CF_3)_2POD$	F:	PerFHalOrg.3-44, 46
C_2F_6HOPS	$(CF_3)_2P(S)OH$	F:	PerFHalOrg.3-139/40, 143, 147, 151/3, 159
–	$(CF_3)_2P(S)OD$	F:	PerFHalOrg.3-139/40, 147
$C_2F_6HO_2P$	$(CF_3)_2P(O)OH$	F:	PerFHalOrg.3-43, 45, 48/50
C_2F_6HP	$(CF_3)_2PH$	F:	PerFHalOrg.3-14/29
–	$(CF_3)_2PD$	F:	PerFHalOrg.3-14/7, 22, 27
C_2F_6HPS	$(CF_3)_2PSH$	F:	PerFHalOrg.3-138/9, 145, 151/3, 159
–	$(CF_3)_2PSD$	F:	PerFHalOrg.3-145

$C_2F_6HPS_2$	$(CF_3)_2P(S)SH$	F:	PerFHalOrg.3-138/9, 142, 145, 151/4, 159
$C_2F_6H_2NP$	$(CF_3)_2PNH_2$	F:	PerFHalOrg.3-57, 63/4, 68/9
$C_2F_6H_2NPS$	$(CF_3)_2P(S)NH_2$	F:	PerFHalOrg.3-138/9, 142, 146, 151, 153
$C_2F_6H_2O_5P_2$	$CF_3P(O)(OH)OP(O)(OH)CF_3$	F:	PerFHalOrg.3-37
$C_2F_6H_2P_2$	$CF_3P(H)P(H)CF_3$	F:	PerFHalOrg.3-16, 19, 23
–	$(CF_3)_2PPH_2$	F:	PerFHalOrg.3-16, 19, 23/6
$C_2F_6H_2Si$	$(CF_3)_2SiH_2$	F:	PerFHalOrg.4-147
$C_2F_6H_3OPSi$	$(CF_3)_2POSiH_3$	F:	PerFHalOrg.3-40, 42
$C_2F_6H_3PSi$	$(CF_3)_2PSiH_3$	F:	PerFHalOrg.3-176
$C_2F_6H_4N_2O$	$(CF_3)_2NOH \cdot NH_3$	F:	PerFHalOrg.7-86/7, 104, 138/40
$C_2F_6H_6N_2S_2$	$SF_2NSF_4N(CH_3)_2$	S:	S-N-Verb.1-136
–	$SF_5NSF[N(CH_3)_2]$	S:	S-N-Verb.1-143/4
C_2F_6HgNO	$(CF_3)_2NOHg$	F:	PerFHalOrg.7-102, 144
C_2F_6IP	$(CF_3)_2PI$		
	Chemical reactions	F:	PerFHalOrg.3-98/102, 108/9, 120/1
	Physical properties	F:	PerFHalOrg.3-82/3, 93, 98
	Preparation	F:	PerFHalOrg.3-74/5
C_2F_6IPS	$(CF_3)_2P(S)I$	F:	PerFHalOrg.3-140, 148, 151/3
C_2F_6ISb	$(CF_3)_2SbI$	F:	PerFHalOrg.3-226/7, 229/32
$C_2F_6I_2Si$	$(CF_3)_2SiI_2$	F:	PerFHalOrg.4-147
$C_2F_6I_2Sn$	$(CF_3)_2SnI_2$	F:	PerFHalOrg.4-181/2, 186
$C_2F_6K_2N_2S_2Te$	$TeF_6 \cdot 2\ KSCN = K_2[TeF_6(SCN)_2]$	Te:	SVol.B2-29
C_2F_6LiN	$(CF_3)_2NLi$	F:	PerFHalOrg.7-64
C_2F_6LiP	$Li[P(CF_3)_2]$	F:	PerFHalOrg.3-25
C_2F_6NNaO	$(CF_3)_2NONa$	F:	PerFHalOrg.7-102, 144
C_2F_6NO	$(CF_3)_2NO$	F:	PerFHalOrg.7-85/6, 102/3, 121/38
$C_2F_6NO_3P$	$NO[(CF_3)_2PO_2]$	F:	PerFHalOrg.3-172
C_2F_6NP	$[(CF_3)_2PN]_x$	F:	PerFHalOrg.3-61/2, 66, 68
$C_2F_6N_2$	CF_3NNCF_3	C:	MVol.D3-206, 213, 214, 291
–	$FN(CF_2)_2NF$	F:	PerFHalOrg.5-35/6, 48
–	$NF_2CFCFNF_2$	F:	PerFHalOrg.5-20
–	NF_2CF_2CFNF	F:	PerFHalOrg.5-20
$C_2F_6N_2OS$	$(CF_3)_2NOSN$	F:	PerFHalOrg.7-92/3, 108, 126
$C_2F_6N_2O_2$	$(CF_3)_2NONO$	F:	PerFHalOrg.7-89/90, 106, 124/5, 145
$C_2F_6N_2O_2S_2$	$FSO_2NC(NSF_2)CF_3$	S:	S-N-Verb.1-119

$C_2F_6N_2O_4S_3$	$S(NSO_2CF_3)_2$	S:	S-N-Verb.1-114
$C_2F_6N_2S_3$	$CF_3SNSNSCF_3$	S:	SVol.2-78
$C_2F_6N_3P$	$(CF_3)_2PN_3$	F:	PerFHalOrg.3-57, 64, 68
$C_2F_6N_3P_2$	$[(CF_3P)_2N_3]_n$	F:	PerFHalOrg.3-68
$C_2F_6N_4$	$NCF_2NFCF(NF_2)N$	F:	PerFHalOrg.5-72/3, 92
–	$NF_2CFNNCFNF_2$	F:	PerFHalOrg.5-20
–	$NF_2CF_2NNCFNF$	F:	PerFHalOrg.5-20
C_2F_6NaOPS	$(CF_3)_2P(S)ONa \cdot 2\ H_2O$	F:	PerFHalOrg.3-139/40
$C_2F_6NiO_2P_2$	$(CO)_2Ni(PF_3)_2$	Ni:	Org.Verb.1-124
$C_2F_6OPS^-$	$(CF_3)_2P(S)O^-$	F:	PerFHalOrg.3-140, 147, 151
$C_2F_6OPSe^-$	$(CF_3)_2POSe^-$	F:	PerFHalOrg.3-160
$C_2F_6O_2P^-$	$(CF_3)_2P(O)O^-$	F:	PerFHalOrg.3-43, 45
$C_2F_6P^+$	$(CF_3)_2P^+$	F:	PerFHalOrg.3-33, 179
$C_2F_6PS_2^-$	$(CF_3)_2P(S)S^-$	F:	PerFHalOrg.3-138, 146, 151, 153, 159
$C_2F_6PSe_2^-$	$(CF_3)_2PSe_2^-$	F:	PerFHalOrg.3-160
$C_2F_6P_2S^+$	$(CF_3)_2P_2S^+$	F:	PerFHalOrg.3-12
$C_2F_6P_3^+$	$C_2F_6P_3^+$	F:	PerFHalOrg.3-12
$C_2F_6S_2$	$F_3CS_2CF_3$	C:	MVol.D6-160
C_2F_6Si	CF_2CFSiF_3	F:	PerFHalOrg.4-148, 161
–	$(C_2F_4SiF_2)_n$	F:	PerFHalOrg.4-148/9
$C_2F_7Ge^+$	$GeF_2(CF_3)CF_2^+$	F:	PerFHalOrg.4-174
C_2F_7ISi	$C_2F_5SiF_2I$	F:	PerFHalOrg.4-149, 156
$C_2F_7NO_2S$	$(CF_3)_2NOS(O)F$	F:	PerFHalOrg.7-91/3, 107, 126
$C_2F_7NO_3S$	$(CF_3)_2NOSO_2F$	F:	PerFHalOrg.7-91/3, 107, 126
C_2F_7NS	$CF_3CF_2NSF_2$	S:	SVol.2-78, 82
$C_2F_7N_2S^+$	$SF(NCF_3)_2^+$	S:	S-N-Verb.1-200
$C_2F_7N_3$	$F_2NCFN(CF_3)NF$	F:	PerFHalOrg.5-6/7, 13
C_2F_7OP	$(CF_3)_2P(O)F$	F:	PerFHalOrg.3-51, 53/4
$C_2F_7OP_2^+$	$C_2F_7P_2O^+$	F:	PerFHalOrg.3-33
$C_2F_7O_2S$	$(CF_3O)_2SF$	C:	MVol.D6-141
C_2F_7P	CF_2CFPF_4	F:	PerFHalOrg.3-71, 87
–	$(CF_3)_2PF$		
	Chemical reactions	F:	PerFHalOrg.3-98/100, 102, 111, 120/1
	Physical properties	F:	PerFHalOrg.3-78, 82/4, 98
	Preparation	F:	PerFHalOrg.3-70/1
–	$C_2F_5PF_2$	F:	PerFHalOrg.3-70/1
C_2F_7PS	$(CF_3)_2P(S)F$	F:	PerFHalOrg.3-140, 143, 147
$C_2F_7P_4S^+$	$C_2F_7P_4S^+$	F:	PerFHalOrg.3-12
C_2F_7Sb	$(CF_3)_2SbF$	F:	PerFHalOrg.3-231
C_2F_8Ge	$(CF_3)_2GeF_2$	F:	PerFHalOrg.4-166/7, 171/2, 175
C_2F_8HP	$(CF_3)_2PF_2H$	F:	PerFHalOrg.3-72, 88
$C_2F_8H_8N_2Si$	$(NH_4)_2[CF_2CFSiF_5]$	F:	PerFHalOrg.4-149, 161

$C_2F_8K_2Si$	$K_2[CF_2CFSiF_5]$	F:	PerFHalOrg.4-149, 161
C_2F_8NOP	$(CF_3)_2NOPF_2$	F:	PerFHalOrg.7-100/1, 117/8, 124, 145
$C_2F_8NO_2P$	$(CF_3)_2NOP(O)F_2$	F:	PerFHalOrg.7-100/1, 118
$C_2F_8N_2O$	$(CF_3)_2NONF_2$	F:	PerFHalOrg.7-91, 106
$C_2F_8N_2O_4S_3$	$SF_2(NSO_2CF_3)_2$	S:	S-N-Verb.1-202/3
$C_2F_8N_2S$	$SF_2(NCF_3)_2$	S:	S-N-Verb.1-200
$C_2F_8N_2S^+$	$SF_2(NCF_3)_2^+$	S:	S-N-Verb.1-200
$C_2F_8N_2S_2$	$F_2SNCF_2CF_2NSF_2$	S:	SVol.2-82
$C_2F_8N_4$	$(NF_2CF_2NCF(NF_2)NF)$	F:	PerFHalOrg.5-6/7, 13
$C_2F_8N_4S$	$SF_5NC(N_3)CF_3$	S:	S-N-Verb.1-148
$C_2F_8Na_2Si$	$Na[CF_2CFSiF_5]$	F:	PerFHalOrg.4-149, 161
$C_2F_8OP_2$	$(CF_3)_2POPF_2$	F:	PerFHalOrg.3-32/6
$C_2F_8P_2$	$(CF_3)_2PPF_2$	F:	PerFHalOrg.3-176, 179/80
C_2F_8Si	$C_2F_5SiF_3$	F:	PerFHalOrg.4-147/8, 153, 155, 161/2
$C_2F_8Si_2$	$[SiF_2SiF_2CF_2CF_2]_n$	F:	PerFHalOrg.4-149
$C_2F_9FeO_2P_3$	$(CO)_2Fe(PF_3)_3$	Fe:	Org.Verb.B1-95, 97
C_2F_9HKP	$K[(CF_3)_2PF_3H]$	F:	PerFHalOrg.3-135
C_2F_9HOP	$(CF_3)_2P(O)F_2 \cdot HF$	F:	PerFHalOrg.3-53
$C_2F_9HP^-$	$[(CF_3)_2PF_3H]^-$	F:	PerFHalOrg.3-129, 134/5
C_2F_9NS	SF_5NCFCF_3	S:	S-N-Verb.1-147
C_2F_9P	$(CF_3)_2PF_3$	F:	PerFHalOrg.3-71, 76/7, 86, 98/9
–	$C_2F_5PF_4$	F:	PerFHalOrg.3-71, 76/7, 86
$C_2F_{10}HNS$	$SF_5NH(C_2F_5)$	S:	S-N-Verb.1-136
$C_2F_{10}O_2S$	$(CF_3O)_2SF_4$	S:	SVol.2-81
$C_2F_{10}O_3S$	$CF_3OOSF_4OCF_3$	S:	SVol.2-81
$C_2F_{10}P^-$	$[(CF_3)_2PF_4]^-$	F:	PerFHalOrg.3-129, 134, 136/7
$C_2F_{10}Si_3$	$C_2F_4(SiF_2)_3$	F:	PerFHalOrg.4-149
$C_2F_{11}FeH_3O_2P_4$	$COFeP_4F_{11}(OCH_3)$	Fe:	Org.Verb.B1-28/9
$C_2F_{11}NOS$	$(CF_3)_2NOSF_5$	F:	PerFHalOrg.7-91/3, 107, 124, 126
$C_2F_{11}NS$	$SF_5N(CF_3)_2$	S:	S-N-Verb.1-136/7
–	$SF_5NF(C_2F_5)$	S:	S-N-Verb.1-139
$C_2F_{12}H_6Sb_2Sn$	$(CH_3)_2Sn(SbF_6)_2$	Sn:	Org.Verb.6-29
$C_2F_{14}FeOP_4$	$COFeP_4F_{11}(CF_3)$	Fe:	Org.Verb.B1-28/9
$C_2F_{14}N_2S_2$	$SF_5NCFN(CF_3)SF_5$	S:	S-N-Verb.1-147
$C_2F_{14}O_4S_2$	$CF_3OSF_4OOSF_4OCF_3$	S:	SVol.2-81
$C_2F_{16}N_2S_2$	$SF_5N(CF_3)N(CF_3)SF_5$	S:	S-N-Verb.1-224
$C_2F_{2x+4}Si_x$	$C_2F_4(SiF_2)_m$	F:	PerFHalOrg.4-148/9
C_2FeH^+	$[FeC_2H]^+$	Fe:	Org.Verb.A1-93/4
$C_2FeH_2^+$	$[FeC_2H_2]^+$	Fe:	Org.Verb.A1-93/4
$C_2FeH_4O_3$	$(FeCO \cdot OCH_3 \cdot OH)_x$	Fe:	Org.Verb.B3-174
$C_2FeH_5N_2O_3P^+$	$[Fe(NO)_2POC_2H_5]^+$	Fe:	Org.Verb.B1-48
$C_2FeH_6I_2O_2P_2$	$(CO)_2Fe(PH_3)_2I_2$	Fe:	Org.Verb.B1-101/3, 109
$C_2FeH_6N_4^+$	$[Fe(CH_3NNNNCH_3)]^+$	Fe:	Org.Verb.B1-169

$C_2FeH_8NO_3$	$Fe(NO) \cdot 2\ CH_3OH$	Fe:	Org.Verb.B3-118
$C_2FeIN_3O_2$	$(CO)_2Fe(I)N_3$	Fe:	Org.Verb.B3-119
$C_2FeI_2O_2$	$Fe(CO)_2I_2$	Fe:	Org.Verb.B3-106
$C_2FeK_2N_4O_2$	$K_2[Fe(NO)_2(CN)_2]$	Fe:	Org.Verb.B1-141
$C_2FeNO_3^+$	$[Fe(CO)_2NO]^+$	Fe:	Org.Verb.B1-140
$C_2FeNO_3^-$	$[Fe(CO)_2NO]^-$	Fe:	Org.Verb.B5-19/20
$C_2FeN_2O_4$	$Fe(CO)_2(NO)_2$	Fe:	Org.Verb.B1-137/46
$C_2FeN_2O_4^+$	$[Fe(CO)_2(NO)_2]^+$	Fe:	Org.Verb.B1-140
$C_2FeN_2Se_2$	$Fe(SeCN)_2$	C:	MVol.D6-229
C_2FeO^+	$[Fe(CO)C]^+$	Fe:	Org.Verb.B3-78
C_2FeO_2	$Fe(CO)_2$	Fe:	Org.Verb.B1-93
$C_2FeO_2^+$	$Fe(CO)_2^+$	Fe:	Org.Verb.B3-49, 78
$C_2FeO_2^{2+}$	$[Fe(CO)_2]^{2+}$	Fe:	Org.Verb.B3-78
$C_2FeO_2^-$	$[Fe(CO)_2]^-$	Fe:	Org.Verb.B3-78/9
$C_2FeO_{14}P_8^+$	$[(CO)_2Fe(P_4O_6)_2]^+$	Fe:	Org.Verb.B1-164
$C_2FeS_6^{2-}$	$[Fe(CS_3)_2]^{2-}$	C:	MVol.D4-221
$C_2Fe_2H_{18}N_6S_7$	$Fe_2(NH_3)_6C_2S_7 \cdot 2\ H_2O$	C:	MVol.D4-225
$C_2HN_2OS_2^-$	$HO(SCN)_2^-$	C:	MVol.D5-185
$C_2HN_2S_2^-$	$HN_2C_2S_2^-$	C:	MVol.D6-22
C_2HNpO_6	$NpO_2(HC_2O_4)$	Np:	TrU.D1-158
$C_2HO_4Pa^{4+}$	$Pa(HC_2O_4)^{4+}$	Pa:	SVol.2-170
C_2HO_6Pa	$PaO(OH)(C_2O_4) \cdot x\ H_2O$	Pa:	SVol.2-82, 169/71
C_2HO_6Pu	$PuO_2(HC_2O_4)$	Np:	TrU.D1-157/8
$C_2HS_5^-$	$HC_2S_5^-$	C:	MVol.D4-217
$C_2H_{1.65}Th_3$	$Th_3C_2H_{1.65}$	Th:	SVol.C1-38
$C_2H_2K_4O_{11}U$	$K_4[UO_2(O_2)(CO_3)_2(H_2O)]$	U:	SVol.C3-44
$C_2H_2N_2OS$	$NH_2C(O)NCS$	C:	MVol.D6-17
$C_2H_2N_2O_4U$	$UO_2(CONHNHCO)$	U:	SVol.E1-115/6
$C_2H_2N_2O_5S$	$[O_2S[NHCO]_2O]$	S:	S-N-Verb.1-39
$C_2H_2N_2S_2$	$C_2H_2N_2S_2$	C:	MVol.D6-37/8
–	HSC(S)NHCN	C:	MVol.D6-34
$C_2H_2N_2S_3$	$C_2H_2N_2S_3$	C:	MVol.D6-38/42
$C_2H_2O_4Ra$	$Ra(HCOO)_2$	Ra:	SVol.2-377
$C_2H_2O_4Sn$	$Sn(HCOO)_2$	Sn:	MVol.C2-218/9
$C_2H_2O_6Ra$	$Ra(HCO_3)_2$	Ra:	SVol.2-377
$C_2H_2O_6Sn$	$Sn(HCO_3)_2$	Sn:	MVol.C2-218
$C_2H_2O_9U_2$	$(U_2O_5)(HCOO)_2 \cdot 3\ H_2O$	U:	SVol.C2-291
$C_2H_2S_3$	SCS_2CH_2	C:	MVol.D4-215
C_2H_3HgNO	CH_3HgNCO	C:	MVol.D3-142
C_2H_3HgNSe	$CH_3HgSeCN$	C:	MVol.D6-239
$C_2H_3KOS_2$	$K[S_2COCH_3]$	C:	MVol.D4-246, 249, 252/3
–	$K[S_2COCD_3]$	C:	MVol.D4-246
$C_2H_3KO_2S$	$K[SOCOCH_3]$	C:	MVol.D4-234, 237
–	$K[SOCOCD_3]$	C:	MVol.D4-234
$C_2H_3K_3O_{11}U$	$K_3[UO_2(OOH)(CO_3)_2(H_2O)]$	U:	SVol.C3-44
$C_2H_3NPtS_5$	$Pt(NH_3)C_2S_5$	C:	MVol.D4-226
C_2H_3NSe	CH_3SeCN	C:	MVol.D6-227, 239
$C_2H_3N_2OSSi^+$	$SN_2COSi(CH_3)^+$	S:	S-N-Verb.1-25
$C_2H_3N_2O_2SSi^+$	$SON_2COSi(CH_3)^+$	S:	S-N-Verb.1-25
$C_2H_3N_3O_4S$	$[HNSO_2][HNCO]_2$	S:	S-N-Verb.1-30

$C_2H_3NaOS_2$	$Na[S_2COCH_3]$	C: MVol.D4-252
$C_2H_3NaO_2S$	$Na[SOCOCH_3]$	C: MVol.D4-237
$C_2H_3NpO_4$	$NpO_2(CH_3COO)$	Np: TrU.D1-153
$C_2H_3NpO_4{}^+$	$NpO_2(CH_3COO)^+$	Np: TrU.D1-153
$C_2H_3NpO_5$	$NpO_2(HOCH_2COO)$	Np: TrU.D1-153
$C_2H_3NpO_7{}^+$	$NpO_2(OH)(H_2C_2O_4)^+$	Np: TrU.D1-42
$C_2H_3OS_2{}^-$	$[SOCSCH_3]^-$	C: MVol.D4-264
–	$[S_2COCH_3]^-$	C: MVol.D4-244/52
–	$[S_2COCD_3]^-$	C: MVol.D4-246
$C_2H_3OS_2Tl$	$Tl[S_2COCH_3]$	C: MVol.D4-253
$C_2H_3O_2Pu^{2+}$	$Pu(CH_3COO)^{2+}$	Np: TrU.D1-152
$C_2H_3O_2Pu^{3+}$	$Pu(CH_3COO)^{3+}$	Np: TrU.D1-153
$C_2H_3O_2S^-$	$[SOCOCH_3]^-$	C: MVol.D4-234/7
$C_2H_3O_2STe^+$	$[Te(SCH_2COOH)]^+$	Te: SVol.B3-134
$C_2H_3O_2Sn^+$	$Sn(CH_3COO)^+$	Sn: MVol.C2-219/20
$C_2H_3O_4Pu^+$	$PuO_2(CH_3COO)^+$	Np: TrU.D1-153
$C_2H_3O_5Pu^+$	$PuO_2(HOCH_2COO)^+$	Np: TrU.D1-153
$C_2H_3S^+$	CH_3CS^+	C: MVol.D4-160
$C_2H_4K_2OS_4$	$K_2CS_4 \cdot CH_3OH$	C: MVol.D4-229, 231
$C_2H_4MnO_2$	$Mn(OCH_2CH_2O)$	Mn: MVol.D1-34
$C_2H_4NNpO_4$	$NpO_2(NH_2CH_2COO)$	Np: TrU.D1-155
$C_2H_4NO_4Pu$	$PuO_2(NH_2CH_2COO)$	Np: TrU.D1-155
$C_2H_4NO_6Pu$	$NH_4PuO_2C_2O_4 \cdot 6\ H_2O$	Np: TrU.D1-120
$C_2H_4N_2NiS_4$	$Ni[S_2CNH_2]_2$	C: MVol.D6-140
$C_2H_4N_2O_3S_2Se$	$2\ HNCS \cdot H_2SeO_3$	C: MVol.D6-35
$C_2H_4N_2O_6S$	$SO_2(NHCOOH)_2$	S: S-N-Verb.1-39
$C_2H_4N_2O_6U$	$UO_2C_2O_4 \cdot N_2H_4 \cdot 0.75\ H_2O$	U: SVol.E1-26/7
$C_2H_4N_2PbS_4$	$Pb[S_2CNH_2]_2$	C: MVol.D6-140
$C_2H_4N_2PdS_4$	$Pd[S_2CNH_2]_2$	C: MVol.D6-140
$C_2H_4N_2PtS_4$	$Pt[S_2CNH_2]_2$	C: MVol.D6-140
$C_2H_4N_2S_4$	$H_2NC(S)SSC(S)NH_2$	C: MVol.D6-139
$C_2H_4N_2S_4Zn$	$Zn[S_2CNH_2]_2$	C: MVol.D6-140
$C_2H_4OS_2$	$SC(OCH_3)(SH)$	C: MVol.D4-238/43
$C_2H_4O_2$	CH_3COOH systems	
	$CH_3COOH-(NH_2)_2CS$	C: MVol.D6-111
$C_2H_4O_2Pu^{3+}$	$Pu(CH_3COOH)^{3+}$	Np: TrU.D1-152
$C_2H_4S^+$	$C_2H_4S^+$	C: MVol.D4-160
$C_2H_4S_3$	$SC(SCH_3)(SH)$	C: MVol.D4-264
$C_2H_5NNa_2S_4$	$Na_2CS_4 \cdot CH_3NH_2$	C: MVol.D4-231
$C_2H_5NO_2S$	$C_2H_5NSO_2$	S: S-N-Verb.1-59
$C_2H_5NO_5U$	$UO(HCOO)_2 \cdot NH_3$	U: SVol.E1-12, 15
$C_2H_5NO_6SU$	$UO_2SO_3 \cdot CH_3CONH_2 \cdot 1.5\ H_2O$	U: SVol.E1-96/8
$C_2H_5NO_7SU$	$UO_2SO_4 \cdot CH_3CONH_2 \cdot 2\ H_2O$	U: SVol.E1-96/8
$C_2H_5N_3O_9Sn$	$C_2H_5Sn(NO_3)_3$	Sn: Org.Verb.6-230/1
C_2H_5Ni	NiC_2H_5	Ni: Org.Verb.1-108
$C_2H_5S_{1.5}Sn$	$(C_2H_5SnS_{1.5})_x$	Sn: Org.Verb.6-230/1
$C_2H_5S_3Sn^{3-}$	$(C_2H_5SnS_3)^{3-}$	Sn: Org.Verb.6-230/1
C_2H_5Sn	$[C_2H_5Sn]_x$	Sn: Org.Verb.6-229
$C_2H_5Sn^+$	$(CH_2CH)SnH_2{}^+$	Sn: Org.Verb.3-41
–	$SnC_2H_5{}^+$	Sn: Org.Verb.1-70/1

$C_2H_6I_2O_6Sn$	$(CH_3)_2Sn(IO_3)_2$	Sn:	Org.Verb.6-28
$C_2H_6I_2Sn$	$(CH_3)_2SnI_2$	Sn:	Org.Verb.6-21
$C_2H_6MnO_6S$	$Mn(CH_2OHCH_2OH)SO_4$	Mn:	MVol.D1-36
$C_2H_6MoO_4Sn$	$[(CH_3)_2SnMoO_4]_x$	Sn:	Org.Verb.6-29
$C_2H_6NO_2SSi^+$	$(CH_3)_2SiNSO_2^+$	S:	S-N-Verb.1-25
$C_2H_6N_2NiO_2$	$(CO)_2Ni(NH_3)_2$	Ni:	Org.Verb.1-123
$C_2H_6N_2O_3S_2$	$(CH_3)_2NSO_2NSO$	S:	S-N-Verb.1-181
$C_2H_6N_2O_4S$	$CH_3OC(O)NHSO_2NH_2$	S:	S-N-Verb.1-167
$C_2H_6N_2O_6U$	$UO_2(C_2O_4) \cdot 2\ NH_3$	U:	SVol.E1-12, 15
$C_2H_6N_2O_9S_3$	$[CH_3ONSO_2]_2[OSO_2]$	S:	S-N-Verb.1-23
$C_2H_6N_4O_{12}U$	$UO_2(NO_3)_2 \cdot 2\ CH_3NO_2$	U:	SVol.E1-144/5
$C_2H_6N_4S$	$HNC(NH_2)_2 \cdot HNCS$	C:	MVol.D6-92
–	$(NH_2)(NH)CSC(NH)(NH_2)$	C:	MVol.D6-102
$C_2H_6N_4S_2$	$[(NH_2)(NH)CS]_2$	C:	MVol.D6-99
$C_2H_6N_4S_2Se$	$[(NH_2)(NH)CS]_2Se$	C:	MVol.D6-101
$C_2H_6N_4Se_2$	$(NH_2NHCSe)_2$	C:	MVol.D6-246
C_2H_6Ni	$(CH_3)_2Ni \cdot x\ C_4H_8O$	Ni:	Org.Verb.1-69, 72
$C_2H_6NiO_2$	$(Ni(OCH_3)_2)_n$	Ni:	Org.Verb.2-282
C_2H_6OSn	$[(CH_3)_2SnO]_x$	Sn:	Org.Verb.6-23, 28
–	$[(CD_3)_2SnO]_x$	Sn:	Org.Verb.6-21
$C_2H_6O_2Sn$	$(C_2H_5SnO_2H)_x$	Sn:	Org.Verb.6-230/1
$C_2H_6O_4SSn$	$[(CH_3)_2SnSO_4]_x$	Sn:	Org.Verb.6-28
$C_2H_6O_4SnW$	$[(CH_3)_2SnWO_4]_x$	Sn:	Org.Verb.6-29
$C_2H_6O_5Sn$	$(HO)_3SnOCOCH_3$	Sn:	Org.Verb.1-158
$C_2H_6O_5Th$	$Th(OH)_3(CH_3COO) \cdot H_2O$	Th:	SVol.C1-249
C_2H_6S	$(CH_3)_2S$	C:	MVol.D4-163/4
–	C_2H_5SH	C:	MVol.D4-232
C_2H_6Sn	CH_2CHSnH_3	Sn:	Org.Verb.4-127
–	$[(CH_3)_2Sn]_n$	Sn:	Org.Verb.5-104, 156
$C_2H_6Sn^+$	$Sn(CH_3)_2^+$	Sn:	Org.Verb.1-47
		Sn:	Org.Verb.4-10
$C_2H_6Sn^{2+}$	$[(CH_3)_2Sn]^{2+}$	Sn:	Org.Verb.5-84
C_2H_6Ti	$(CH_3)_2Ti$	Ti:	Org.Verb.1-67
$C_2H_7Li_3Ni$	$C_2H_4Ni(LiH)_3$	Ni:	Org.Verb.1-340
$C_2H_7NO_3S$	$NH_2SO_2OC_2H_5$	S:	S-N-Verb.1-97
$C_2H_7NO_3S_2$	$(CH_3)_2SNSO_3H$	S:	S-N-Verb.1-97
$C_2H_7N_2NaO_3S$	$Na[SO_3NC_2H_5NH_2]$	S:	S-N-Verb.1-219
$C_2H_7N_3O_3S$	$NH_2SO_2NHNHCOCH_3$	S:	S-N-Verb.1-222
$C_2H_7N_5O_6S_2$	$HN(SO_2NHCONH_2)_2$	S:	S-N-Verb.1-162
$C_2H_7Sn^+$	$(CH_3)_2SnH^+$	Sn:	Org.Verb.4-10
–	$H_2SnC_2H_5^+$	Sn:	Org.Verb.1-70/1
$C_2H_8HgN_4S_2$	$Hg[(NH_2)_2CS]_2$	C:	MVol.D6-90
$C_2H_8Hg_2N_4S_2$	$Hg_2[(NH_2)_2CS]_2$	C:	MVol.D6-90
$C_2H_8I_2N_4S_2$	$[(NH_2)_2CS]_2 \cdot I_2$	C:	MVol.D6-95/6
$C_2H_8I_2N_4S_2Te$	$Te(SC(NH_2)_2)_2I_2$	Te:	SVol.B3-141
$C_2H_8IrN_2NaO_6S_2$	$Na[Ir(SO_3)_2(C_2H_4(NH_2)_2)] \cdot 3\ H_2O$	Ir:	SVol.2-159/60
$C_2H_8IrN_2Na_5O_{12}S_4$	$Na_5[Ir(SO_3)_4(C_2H_4(NH_2)_2)] \cdot 2\ H_2O$	Ir:	SVol.2-159/60
$C_2H_8MnN_6O_8$	$Mn(NO_3)_2 \cdot 2\ CO(NH_2)_2 \cdot 4\ H_2O$	Mn:	MVol.C3-294/5
$C_2H_8N_2O_2S$	$CH_3NHSO_2NHCH_3$	S:	S-N-Verb.1-168/9
–	$(CH_3)_2NSO_2NH_2$	S:	S-N-Verb.1-168

$C_2H_8N_2O_4S$	$CH_3ONHSO_2NHOCH_3$	S:	S-N-Verb.1-169
$C_2H_8N_2O_5U$	$UO(HCOO)_2 \cdot 2\ NH_3$	U:	SVol.E1-12, 15
–	$UO_2(OH)_2 \cdot NH_2CONHCH_3$	U:	SVol.E1-93, 95
$C_2H_8N_2O_6U$	$UO_2(HCOO)_2 \cdot 2\ NH_3$	U:	SVol.E1-12, 15
$C_2H_8N_2O_{10}U$	$UO_2(NO_3)_2 \cdot 2\ CH_3OH$	U:	SVol.E1-58/9, 63
$C_2H_8N_2O_{10}U_2$	$[UO_2(OH)]_2C_2O_4 \cdot 2\ NH_3$	U:	SVol.E1-12, 15
$C_2H_8N_4O_4S_3Sn$	$SnSO_4 \cdot 2\ CS(NH_2)_2$	Sn:	MVol.C5-55
$C_2H_8N_4O_6S_3U$	$UO_2SO_4 \cdot 2\ (NH_2)_2CS$	U:	SVol.E1-220/1
$C_2H_8N_4O_6U$	$UO_2C_2O_4 \cdot 2\ N_2H_4 \cdot H_2O$	U:	SVol.E1-26/7
$C_2H_8N_4O_7SU$	$UO_2(SO_3) \cdot 2\ (NH_2)_2CO$	U:	SVol.E1-83, 85, 89
$C_2H_8N_4O_8SU$	$UO_2SO_4 \cdot 2\ (NH_2)_2CO$	U:	SVol.E1-84/5, 89
–	$UO_2SO_4 \cdot 2\ (NH_2)_2CO \cdot H_2O$	U:	SVol.E1-84, 89
$C_2H_8N_4O_8U$	$UO_2(NO_3)_2 \cdot C_2H_8N_2$	U:	SVol.E1-23/5
$C_2H_8N_4S_2Te^{2+}$	$[Te(SC(NH_2)_2)_2]^{2+}$	Te:	SVol.B3-140
$C_2H_8N_4Se_2^{2+}$	$[(NH_2)_2CSe]_2^{2+}$	C:	MVol.D6-247
$C_2H_8N_6O_6S_2$	$[(NH)(NH_2)CS]_2 \cdot 2\ HNO_3$	C:	MVol.D6-109, 112
$C_2H_8N_6O_8S_2U$	$UO_2(NO_3)_2 \cdot 2\ (NH_2)_2CS$	U:	SVol.E1-220/1
$C_2H_8N_6O_{10}U$	$UO_2(NO_3)_2 \cdot 2\ (NH_2)_2CO$	U:	SVol.E1-83, 85/6, 88/9
$C_2H_8O_3Sn$	$C_2H_5Sn(OH)_3$	Sn:	Org.Verb.6-230
C_2H_8Sn	$(CH_3)_2SnH_2$	Sn:	Org.Verb.4-94/8
–	$C_2H_5SnH_3$	Sn:	Org.Verb.4-124/5
$C_2H_9N_2O_5PS$	$(CH_3)_2NSO_2NHPO(OH)_2$	S:	S-N-Verb.1-184
$C_2H_9N_2O_6PU$	$UO_2HPO_4 \cdot C_2H_8N_2$	U:	SVol.E1-23/4
$C_2H_9N_3O_2S$	$(CH_3)_2NSO_2NHNH_2$	S:	S-N-Verb.1-222
$C_2H_9N_5O_4S$	$(NH_4)_2[HNSO_2][NCO]_2 \cdot 0.5\ H_2O$	S:	S-N-Verb.1-30
$C_2H_9N_5O_8U$	$UO_2(OH)NO_3 \cdot 2\ (NH_2)_2CO$	U:	SVol.E1-83, 89
$C_2H_9O_4Sn^-$	$[C_2H_5Sn(OH)_4]^-$	Sn:	Org.Verb.6-230/1
$C_2H_{10}I_2N_2Sn$	$SnI_2 \cdot 2\ CH_3NH_2$	Sn:	MVol.C5-11/2
$C_2H_{10}N_4O_6S_3$	$SO_2[N(CH_3)SO_2NH_2]_2$	S:	S-N-Verb.1-179/80
$C_2H_{10}N_4O_6U$	$UO_2(OH)_2 \cdot 2\ (NH_2)_2CO$	U:	SVol.E1-84, 86/7, 90
$C_2H_{10}N_4O_8U$	$UO_2(NO_3)_2 \cdot 2\ CH_3NH_2$	U:	SVol.E1-18, 21, 25
$C_2H_{10}N_4O_9SU$	$[UO_2SO_4((NH_2)_2CO)_2(H_2O)]$	U:	SVol.E1-89
$C_2H_{12}N_2O_{13}U_4$	$(CH_3NH_3)_2O \cdot 4\ UO_3 \cdot n\ H_2O$	U:	SVol.C3-65/6
$C_2H_{12}N_2O_{25}U_8$	$(CH_3NH_3)_2O \cdot 8\ UO_3 \cdot n\ H_2O$	U:	SVol.C3-65/6
$C_2H_{12}N_2O_{37}U_{12}$	$(CH_3NH_3)_2O \cdot 12\ UO_3 \cdot n\ H_2O$	U:	SVol.C3-65/6
$C_2H_{12}N_6O_4W$	$(CN_3H_6)_2WO_4$	W:	SVol.B3-244
$C_2H_{16}N_6O_6Te$	$[(NH_2)_2CNH_3]_2H_4TeO_6$	Te:	SVol.B1-133
$C_2H_{18}N_6Ni_2S_8$	$[Ni(NH_3)_6][Ni(CS_4)_2]$	C:	MVol.D4-231
$C_2Hf_4Sn_2$	$Hf_4Sn_2C_2$	Sn:	MVol.C4-227/8
$C_2HgN_2Se_2$	$Hg(SeCN)_2$	C:	MVol.D6-229
$C_2Hg_2N_2Se_2$	$Hg_2(SeCN)_2$	C:	MVol.D6-229
$C_2IN_2^-$	$[I(CN)_2]^-$	C:	MVol.D3-259, 263/4
$C_2IN_2S_2^-$	$I(SCN)_2^-$	C:	MVol.D5-183/6, 189/90
$C_2I_2IrO_2$	$IrI_2(CO)_2$	Ir:	SVol.2-151
$C_2I_4IrKO_2$	$K[IrI_4(CO)_2]$	Ir:	SVol.2-151
C_2Ir	IrC_2	Ir:	SVol.2-180
$C_2IrO_4^+$	$[Ir(C_2O_4)]^+$	Ir:	SVol.2-36
C_2IrU_2	IrU_2C_2	Ir:	SVol.1-105
$C_2Ir_3O_4^{7+}$	$[Ir_3(C_2O_4)]^{7+}$	Ir:	SVol.2-36
$C_2Ir_6O_4^{16+}$	$[Ir_6(C_2O_4)]^{16+}$	Ir:	SVol.2-36

$C_2K_2NiS_6$	$K_2[Ni(CS_3)_2]$	C:	MVol.D4-225
$C_2K_2O_4$	$K_2C_2O_4$ systems		
	$K_2C_2O_4$-$Sn(C_2O_4)_2$-H_2O	Sn:	MVol.C3-78
$C_2Li_2O_8U$	$Li_2[UO_2(CO_3)_2]$	U:	SVol.C2-263
C_2MnO_4	MnC_2O_4	Mn:	MVol.C2-82
$C_2MnO_4^+$	$[Mn(C_2O_4)]^+$	Mn:	MVol.C2-82/3
C_2NNiO_3	$(CO)_2NiNO$	Ni:	Org.Verb.1-123
$C_2N_2NiO_4$	$(CO)_2Ni(N_2O_2)$	Ni:	Org.Verb.1-146
$C_2N_2OS_3$	$SO(CNS)_2$	S:	SVol.1-44
–	$SO(SCN)_2$	S:	SVol.1-44, 56
$C_2N_2O_4S$	$SO_2(NCO)_2$	C:	MVol.D3-238/9
		S:	S-N-Verb.1-183/4
$C_2N_2O_7S_2$	$O(SO_2NCO)_2$	C:	MVol.D3-238/9
		S:	S-N-Verb.1-158
$C_2N_2PbSe_2$	$Pb(SeCN)_2$	C:	MVol.D6-229
$C_2N_2PuS_2^+$	$Pu(SCN)_2^+$	Np:	TrU.D1-144
C_2N_2Ra	$Ra(CN)_2$	Ra:	SVol.2-377
$C_2N_2RaS_2$	$Ra(CNS)_2$	Ra:	SVol.2-377
C_2N_2S	NCNCS	C:	MVol.D5-237
–	$S(CN)_2$	C:	MVol.D5-211/22
$C_2N_2S_2$	$S_2(CN)_2$	C:	MVol.D5-222/32
$C_2N_2S_2^-$	$(SCN)_2^-$	C:	MVol.D5-210
$C_2N_2S_2^{2-}$	$(S_2C_2N_2)^{2-}$	C:	MVol.D5-211
$C_2N_2S_2Se$	$Se(SCN)_2$	C:	MVol.D6-256/7
$C_2N_2S_2Se_2$	$Se_2(SCN)_2$	C:	MVol.D6-257
$C_2N_2S_2Sn$	$Sn(SCN)_2$	Sn:	MVol.C2-228/9
$C_2N_2S_2Te$	$Te(SCN)_2$	C:	MVol.D6-263
$C_2N_2S_3$	$S_3(CN)_2$	C:	MVol.D5-233/5
$C_2N_2S_3Se$	$SeS(SCN)_2$	C:	MVol.D6-257
$C_2N_2S_4$	$S_4(CN)_2$	C:	MVol.D5-235/6
$C_2N_2S_5$	$S_5(CN)_2$	C:	MVol.D5-236/7
$C_2N_2S_6$	$S_6(CN)_2$	C:	MVol.D5-236/7
$C_2N_2S_7$	$S_7(CN)_2$	C:	MVol.D5-236/7
$C_2N_2S_8$	$S_8(CN)_2$	C:	MVol.D5-236/7
C_2N_2Se	$Se(CN)_2$	C:	MVol.D6-233/5
$C_2N_2Se^{2-}$	$[NCSeCN]^{2-}$	C:	MVol.D6-227
$C_2N_2Se_2$	$Se_2(CN)_2$	C:	MVol.D6-235/7
$C_2N_2Se_3$	$Se_3(CN)_2$	C:	MVol.D6-237/9
C_2N_2Sn	$Sn(CN)_2$	Sn:	MVol.C2-227
C_2N_2Te	$Te(CN)_2$	C:	MVol.D6-260/1
$C_2N_4NiO_2$	$(CO)_2Ni(N_2)_2$	Ni:	Org.Verb.1-123
$C_2N_6O_4S_5$	$SO_2(NCO)_2 \cdot S_4N_4$	S:	S-N-Verb.1-183
$C_2Na_2O_8U$	$Na_2[UO_2(CO_3)_2]$	U:	SVol.C2-247, 263
C_2NiO_2	$Ni(CO)_2$	Ni:	Org.Verb.1-123
$C_2NiO_2^+$	$Ni(CO)_2^+$	Ni:	Org.Verb.1-215
$C_2NiO_2^{2+}$	$Ni(CO)_2^{2+}$	Ni:	Org.Verb.1-215
$C_2NiO_2^-$	$Ni(CO)_2^-$	Ni:	Org.Verb.1-215
$C_2NiO_8P_4$	$(P_4O_6Ni(CO)_2)_x$	Ni:	Org.Verb.2-394/5
$C_2NiO_{14}P_8$	$(CO)_2Ni(P_4O_6)_2$	Ni:	Org.Verb.1-144
$C_2NiS_6^{2-}$	$[Ni(CS_3)_2]^{2-}$	C:	MVol.D4-221

$C_2NpO_4^{2+}$	$Np(C_2O_4)^{2+}$	Np: TrU.D1-158
C_2NpO_6	$NpO_2C_2O_4$	Np: TrU.D1-157/8
$C_2NpO_6^-$	$NpO_2C_2O_4^-$	Np: TrU.D1-28
$C_2OS_4^{2-}$	$[S_2CSC(O)S]^{2-}$	C: MVol.D4-207
$C_2O_2S_2^+$	$(OCS)_2^+$	C: MVol.D5-100/1
$C_2O_2S_3^{2-}$	$[S_2CSCO_2]^{2-}$	C: MVol.D4-204
C_2O_2Sn	$Sn(CO)_2$	Sn: MVol.C2-218
$C_2O_4Pa^{3+}$	$Pa(C_2O_4)^{3+}$	Pa: SVol.2-170
$C_2O_4Pu^+$	$Pu(C_2O_4)^+$	Np: TrU.D1-158
$C_2O_4Pu^{2+}$	$Pu(C_2O_4)^{2+}$	Np: TrU.D1-158
C_2O_4Ra	$Ra(COO)_2$	Ra: SVol.2-377
C_2O_4Sn	SnC_2O_4	Sn: MVol.C2-224
$C_2O_5Pa^+$	$PaO(C_2O_4)^+$	Pa: SVol.2-169/70
$C_2O_6Np^-$	$NpO_2(C_2O_4)^-$	Np: TrU.D1-157, 158
C_2O_6Pu	$PuO_2C_2O_4$	Np: TrU.D1-157/8
$C_2O_6Pu^-$	$PuO_2(C_2O_4)^-$	Np: TrU.D1-157/8
C_2O_6Sn	$Sn(CO_3)_2$	Sn: MVol.C2-218
$C_2O_8U^{2-}$	$[UO_2(CO_3)_2]^{2-}$	U: SVol.C2-283
C_2Pa	PaC_2	Pa: SVol.2-80
C_2Pu	PuC_2	Np: TrU.B2-41
C_2Pu_3	Pu_3C_2	Np: TrU.B2-41
C_2S^+	CCS^+	C: MVol.D4-6, 160
$C_2S_2^+$	$C_2S_2^+$	C: MVol.D4-6, 30, 159/60
$C_2S_3^+$	$C_2S_3^+$	C: MVol.D4-159/60
$C_2S_4^+$	$C_2S_4^+$	C: MVol.D4-160
$C_2S_4Ta_3$	$Ta_3C_2S_4$	C: MVol.D4-185
$C_2S_5^{2-}$	$C_2S_5^{2-} = [CS_3 \cdot CS_2]^{2-}$	C: MVol.D4-222
$C_2S_6^{2-}$	$C_2S_6^{2-}$	C: MVol.D4-229
$C_{2.04}F_{1.7}GeHO_{1.8}$	$(C_6F_5)_{0.34}GeO_{1.8}H$	F: PerFHalOrg.4-166
$C_{2.4}Cl_2H_{5.6}MnO_{0.4}$	$MnCl_2 \cdot 0.4\ (i\text{-}C_3H_7)_2O$	Mn: MVol.D1-140
$C_{2.5}Cl_2H_5MnO_{0.5}$	$MnCl_2 \cdot 0.5\ (CH_2)_5O$	Mn: MVol.D1-146/7
$C_{2.5}Cl_2H_{5.5}N_{0.5}Sn$	$SnCl_2 \cdot 0.5\ C_5H_{10}NH$	Sn: MVol.C5-15/6
$C_{2.5}Cl_4H_{7.5}N_{2.5}O_5U$	$UCl_4 \cdot 2.5\ CH_3NO_2$	U: SVol.E1-144/5
$C_{2.7}H_{11.8}IN_{5.4}O_{5.7}U$	$UO_2(OH)I \cdot 2.7\ (NH_2)_2CO$	U: SVol.E1-83, 87
$C_{2.7}H_{11.8}I_{1.96}N_{5.4}O_{5.7}U$	$UO_2(OH)I \cdot 2.7\ (NH_2)_2CO \cdot 0.96\ I$	U: SVol.E1-83, 87
$C_3Cd_2FeI_3NO_4$	$(CO)_3Fe(NO)Cd_2I_3$	Fe: Org.Verb.B1-196/7
$C_3ClFH_5NO_3S$	$FSO_2NCCl(OC_2H_5)$	S: S-N-Verb.1-118
$C_3ClFH_6NO_4P$	$ClCFNOPO(OCH_3)_2$	F: PerFHalOrg.7-199/200
$C_3ClFH_6N_2O_2S$	$FSO_2NCCl[N(CH_3)_2]$	S: S-N-Verb.1-117
$C_3ClF_2GeH_9Sn$	$(CH_3)_3SnGeClF_2$	Sn: Org.Verb.5-85
$C_3ClF_2H_3O$	CF_2CClCH_2OH	F: PerFHalOrg.4-90
$C_3ClF_2H_5NO_3P$	$CF_2NOPCl(O)OC_2H_5$	F: PerFHalOrg.7-199/200
–	$ClCFNOPF(O)OC_2H_5$	F: PerFHalOrg.7-199/200
$C_3ClF_2H_{10}N_4O_2PS_2Si$	$[NS(O)F]_2[NPClNHSi(CH_3)_3]$	S: S-N-Verb.1-27
$C_3ClF_2N_3$	$NCFNCFNCCl$	C: MVol.D3-288/90
		F: PerFHalOrg.6-77/8
$C_3ClF_2NiO_3P$	$(CO)_3NiPF_2Cl$	Ni: Org.Verb.1-161/2
$C_3ClF_3H_2N_2O_3$	$CF_3CCl(NO_2)C(O)NH_2$	F: PerFHalOrg.7-18/9
$C_3ClF_3H_3N_2O_5P$	$NO_2CF_2CFNOPCl(O)OCH_3$	F: PerFHalOrg.7-199/200
$C_3ClF_3H_3N_3O_2$	$O_2NCF_2CFClNNCH_3$	F: PerFHalOrg.7-192

$C_3ClF_3H_3O_2P$	$CF_3P(Cl)OC(O)CH_3$	F:	PerFHalOrg.3-101
$C_3ClF_3H_4NO_3P$	$FClCNOPF_2(OCH_2)_2$	F:	PerFHalOrg.7-200, 202
$C_3ClF_3H_5P$	$CF_3P(C_2H_5)Cl$	F:	PerFHalOrg.3-102
$C_3ClF_3H_6NP$	$CF_3P(Cl)N(CH_3)_2$	F:	PerFHalOrg.3-102, 112
$C_3ClF_3H_6Sn$	$(CH_3)_2(CF_3)SnCl$	Sn:	Org.Verb.5-191/2
$C_3ClF_3N_2S$	$CClSNC(CF_3)N$	F:	PerFHalOrg.5-67/8
$C_3ClF_4H_3N_2$	$ClCF_2CF_2NNCH_3$	F:	PerFHalOrg.7-192
$C_3ClF_4NO_2$	$ONCF_2CF_2C(O)Cl$	F:	PerFHalOrg.7-177, 184
C_3ClF_5	$(CF_2)_2CClF$	F:	PerFHalOrg.5-17
C_3ClF_5HN	$CF_3CFNHC(Cl)(F)$	F:	PerFHalOrg.5-2, 9
–	$(CF_3)(CF_2Cl)CNH$	F:	PerFHalOrg.7-25/6, 51/2, 67, 75
$C_3ClF_5HNO_2$	$CF_3CCl(OH)CF_2NO$	F:	PerFHalOrg.7-177, 184, 189
$C_3ClF_5H_3NO$	$(CF_3)(CF_2Cl)C(OH)NH_2$	F:	PerFHalOrg.7-2, 29
$C_3ClF_5H_4N_2$	$C_2F_5C(NH)NH_2 \cdot HCl$	F:	PerFHalOrg.7-3/4, 30
$C_3ClF_5H_4S$	$SF_5CHCClCH_3$	S:	SVol.2-288
$C_3ClF_5N_2$	$(CF_3)(CF_2Cl)CN_2$	F:	PerFHalOrg.5-5
$C_3ClF_6HN_2O_2$	$CF_3N(NO_2)CF_2CFClH$	F:	PerFHalOrg.7-164, 166
–	$CF_3N(ONO)CF_2CFClH$	F:	PerFHalOrg.7-164, 166
C_3ClF_6N	$CF_2ClCF_2NCF_2$	F:	PerFHalOrg.3-109
–	CF_3NCFCF_2Cl	F:	PerFHalOrg.7-145
C_3ClF_6NO	$CF_3CF(NO)CF_2Cl$	F:	PerFHalOrg.7-177, 183
–	$CF_3NOCFClCF_2$	F:	PerFHalOrg.5-31/2, 41, 43
–	$(CF_3)_2CClNO$	F:	PerFHalOrg.7-177, 183, 191, 199
–	$[N(CF_3)OCF_2CFCl]_n$	F:	PerFHalOrg.7-207/8, 213/4
$C_3ClF_6NO_2$	$(CF_3)_2NOC(O)Cl$	F:	PerFHalOrg.7-94/5, 111
$C_3ClF_6NO_2S$	$FSO_2NCClC_2F_5$	S:	S-N-Verb.1-117
C_3ClF_7HNO	$(ClCF_2CF_2)(CF_3)NOH$	F:	PerFHalOrg.7-86/7, 104, 145
C_3ClF_7NO	$(ClCF_2CF_2)(CF_3)NO$	F:	PerFHalOrg.7-85/6, 103
$C_3ClF_7N_2O_2$	$(CF_3)(C_2F_4Cl)NONO$	F:	PerFHalOrg.7-89/90, 106
–	$CF_3N(NO_2)CF_2CF_2Cl$	F:	PerFHalOrg.7-166
–	$CF_3N(ONO)CF_2CF_2Cl$	F:	PerFHalOrg.7-166
$C_3ClF_8H_2NS$	$SF_5N(CH_2)CClCF_3$	S:	S-N-Verb.1-148
$C_3ClF_8NO_4S_2$	$ClCF_2CF(CF_3)N(SO_2F)_2$	S:	S-N-Verb.1-52
–	$(FSO_2)_2NCF_2CFClCF_3$	S:	S-N-Verb.1-52
$C_3ClF_9P^+$	$[(CF_3)_3PCl]^+$	F:	PerFHalOrg.3-111
C_3ClF_9Si	$(CF_3)_3SiCl$	F:	PerFHalOrg.4-147
$C_3ClF_{10}NS$	$SF_3ClNCF(CF_3)_2$	S:	SVol.2-289
$C_3ClF_{11}NP$	$CF_3PF_2ClN(CF_3)_2$	F:	PerFHalOrg.3-61, 66
$C_3ClFeHgNO_4$	$(CO)_3Fe(NO)HgCl$	Fe:	Org.Verb.B1-196/7
$C_3ClH_2N_3O_2$	$C_3N_3Cl(OH)_2$	C:	MVol.D3-283
$C_3ClH_4N_5$	$C_3N_3Cl(NH_2)_2$	C:	MVol.D3-284
C_3ClH_5NiO	$CH_3C(O)CH_2NiCl$	Ni:	Org.Verb.1-15
$C_3ClH_5OPdS_2$	$[PdCl(S_2COC_2H_5)]_2$	C:	MVol.D4-261
$C_3ClH_6NO_4S$	$ClSO_2NHCOOC_2H_5$	S:	S-N-Verb.1-124/5

$C_3ClH_6N_3O_6Sn$	$(O_2NCH_2)_3SnCl$	Sn:	Org.Verb.5-151
$C_3ClH_6N_5S$	$C_3H_2N_3S(NH_2)_2Cl$	C:	MVol.D6-43
$C_3ClH_7N_2O_5S_2$	$ClSO_2NHCONCH_3SO_2CH_3$	S:	S-N-Verb.1-125
$C_3ClH_7O_2Sn$	$(CH_3)_2ClSnOC(O)H$	Sn:	Org.Verb.6-27
$C_3ClH_7Sn^+$	$C_3H_7SnCl^+$	Sn:	Org.Verb.5-112
$C_3ClH_8NO_2S$	$ClSO_2NHC_3H_7$	S:	S-N-Verb.1-123
$C_3ClH_8N_3O_5S_2$	$ClSO_2NHCONHSO_2N(CH_3)_2$	S:	S-N-Verb.1-125
$C_3ClH_8Sn^+$	$C_3H_7SnHCl^+$	Sn:	Org.Verb.5-112, 116
$C_3ClH_9N_2O_4S_2$	$(CH_3)_3NSO_2NSO_2Cl$	S:	S-N-Verb.1-55
C_3ClH_9NaSn	$[(CH_3)_3Sn(Cl)Na]$	Sn:	Org.Verb.5-78
C_3ClH_9OSn	$(CH_3)_2SnCl(OCH_3)$	Sn:	Org.Verb.6-88
$C_3ClH_9O_3SeSn$	$SeO_3 \cdot (CH_3)_3SnCl$	Sn:	Org.Verb.5-84
$C_3ClH_9S_2Sn$	$CH_3SnCl(SCH_3)_2$	Sn:	Org.Verb.6-218
C_3ClH_9Sn	$(CH_3)_3SnCl$		
	Analysis	Sn:	Org.Verb.5-66
	Chemical reactions	Sn:	Org.Verb.5-77/88
	Formation and preparation	Sn:	Org.Verb.5-63/8
	Molecule	Sn:	Org.Verb.5-69
	Physical properties	Sn:	Org.Verb.5-76/7
	Physiological action	Sn:	Org.Verb.5-88
	Polarography	Sn:	Org.Verb.5-77
	Spectra	Sn:	Org.Verb.5-69/76
	Structure	Sn:	Org.Verb.5-68
	Uses	Sn:	Org.Verb.5-89
$C_3ClH_9Sn^+$	$(CH_3)_3SnCl^+$	Sn:	Org.Verb.5-76
C_3ClH_9Ti	$(CH_3)_3TiCl$	Ti:	Org.Verb.1-84
$C_3ClH_{12}MnO_3^+$	$[MnCl(CH_3OH)_3]^+$	Mn:	MVol.D1-24/5
$C_3ClH_{12}NSn$	$(CH_3)_3SnCl \cdot NH_3$	Sn:	Org.Verb.5-87
$C_3ClH_{15}N_2Sn$	$(CH_3)_3SnCl \cdot 2\ NH_3$	Sn:	Org.Verb.5-87
$C_3ClH_{16}MnO_5^+$	$[MnCl(H_2O)_2(CH_3OH)_3]^+$	Mn:	MVol.C5-78
$C_3ClI_2N_3$	$C_3N_3ClI_2$	C:	MVol.D3-284
C_3ClIrO_3	$Ir(CO)_3Cl$	Ir:	SVol.2-113
$C_3Cl_2FH_5NOPS$	$C_2H_5SP(Cl)ONCFCl$	F:	PerFHalOrg.7-200
$C_3Cl_2FH_5NO_3P$	$ClCFNOPCl(O)OC_2H_5$	F:	PerFHalOrg.7-199/200
C_3Cl_2FNS	CClNCFSCCl	F:	PerFHalOrg.5-57/8, 80, 110
$C_3Cl_2FN_3$	NCFNCClNCCl	C:	MVol.D3-288/90
		F:	PerFHalOrg.6-77/8
$C_3Cl_2F_2H_4NO_3P$	$OPF(ONCFCl)OCH_2CH_2Cl$	F:	PerFHalOrg.7-201/2
$C_3Cl_2F_2N_2S$	$CFNC(CFCl_2)NS$	F:	PerFHalOrg.5-67/8
$C_3Cl_2F_3H_2NO$	$CF_2ClCFClC(O)NH_2$	F:	PerFHalOrg.7-18/21, 45
$C_3Cl_2F_3H_3N_2$	$CF_2ClCFClC(NH)NH_2$	F:	PerFHalOrg.7-3/4, 31
–	$CF_2ClCFClNNCH_3$	F:	PerFHalOrg.7-192
$C_3Cl_2F_4$	$(CClF)_2CF_2$	F:	PerFHalOrg.5-17
–	CF_2CClCF_2Cl	F:	PerFHalOrg.5-46/7
–	$(CF_2)_2CCl_2$	F:	PerFHalOrg.5-17
$C_3Cl_2F_4HN$	$(CF_2Cl)_2CNH$	F:	PerFHalOrg.7-25/6, 52, 66/9
$C_3Cl_2F_4H_2N_2$	$(CF_2Cl)_2C(NH)_2$	F:	PerFHalOrg.5-6
$C_3Cl_2F_4H_3NO$	$(CF_2Cl)_2C(OH)NH_2$	F:	PerFHalOrg.7-2, 29

$C_3Cl_2F_4H_4N_2$	$(CF_2Cl)_2C(NH_2)_2$	F:	PerFHalOrg.7-1/2, 28
$C_3Cl_2F_4N_2$	$NCF_2CCl_2CF_2N$	F:	PerFHalOrg.5-60, 82, 97
$C_3Cl_2F_5N$	$CF_3NCClCF_2Cl$	F:	PerFHalOrg.5-46/8
$C_3Cl_2F_5NO$	$CF_3NOCCl_2CF_2$	F:	PerFHalOrg.5-31/2, 41, 47/8
–	$ClCF_2CFClCF_2NO$	F:	PerFHalOrg.7-177, 183
–	$ClCF_2CF_2CFClNO$	F:	PerFHalOrg.7-177, 183
–	$[N(CF_3)OCF_2CCl_2]_n$	F:	PerFHalOrg.7-207/8, 214
$C_3Cl_2F_6HNO$	$(Cl_2CFCF_2)(CF_3)NOH$	F:	PerFHalOrg.7-86/7, 104
$C_3Cl_2F_6H_3NOSi$	$CH_3SiCl_2ON(CF_3)_2$	F:	PerFHalOrg.7-127
$C_3Cl_2F_6H_3NP_2$	$[CF_3P(Cl)]_2NCH_3$	F:	PerFHalOrg.3-102, 108, 113
$C_3Cl_2F_6NO$	$(ClCF_2CFCl)(CF_3)NO$	F:	PerFHalOrg.7-85/6
–	$(Cl_2CFCF_2)(CF_3)NO$	F:	PerFHalOrg.7-85/6, 103
$C_3Cl_2F_6N_2O_2$	$(CF_3)(Cl_2CFCF_2)NONO$	F:	PerFHalOrg.7-89/90, 106
–	$CF_3N(NO_2)CF_2CFCl_2$	F:	PerFHalOrg.7-166
–	$CF_3N(ONO)CF_2CFCl_2$	F:	PerFHalOrg.7-166
$C_3Cl_2F_7P$	$(n\text{-}C_3F_7)PCl_2$	F:	PerFHalOrg.3-72/3, 78, 90, 99/101
$C_3Cl_2F_9H_4O_2Sb$	$[H_3O][(CF_3)_3SbCl_2(OH)]$	F:	PerFHalOrg.3-226
$C_3Cl_2F_9P$	$(CF_3)_3PCl_2$	F:	PerFHalOrg.3-73, 91, 109, 111
–	$(n\text{-}C_3F_7)PF_2Cl_2$	F:	PerFHalOrg.3-72, 76, 88
$C_3Cl_2F_9Sb$	$(CF_3)_3SbCl_2$	F:	PerFHalOrg.3-226/7, 229, 231
–	$(CF_3)_3SbCl_2 \cdot H_2O$	F:	PerFHalOrg.3-226, 229/30
$C_3Cl_2FeGeH_3NO_2{}^+$	$[CO(NO)FeClGe(CHCH_2)Cl]^+$	Fe:	Org.Verb.B1-133
$C_3Cl_2FeH_3N$	$NCCHCH_2FeCl_2$	Fe:	Org.Verb.B4-178, 182
$C_3Cl_2FeO_3$	$(CO)_3FeCl_2$	Fe:	Org.Verb.B3-121
$C_3Cl_2HN_3$	$C_3N_3HCl_2$	C:	MVol.D3-285
$C_3Cl_2HN_3O$	$C_3N_3Cl_2OH$	C:	MVol.D3-283/4
$C_3Cl_2H_2N_4$	$C_3N_3Cl_2NH_2$	C:	MVol.D3-284
$C_3Cl_2H_4O_2$	$ClCOOCH_2CH_2Cl$	C:	MVol.D3-101
$C_3Cl_2H_5NO_3S$	$ClSO_2NCCl(OC_2H_5)$	S:	S-N-Verb.1-127
$C_3Cl_2H_5NO_4S$	$ClSO_2NHCOOCH_2CH_2Cl$	S:	S-N-Verb.1-124/5
$C_3Cl_2H_5Ti$	$C_3H_5TiCl_2$	Ti:	Org.Verb.1-127
$C_3Cl_2H_6MnO$	$MnCl_2 \cdot (CH_3)_2CO$	Mn:	MVol.C5-80
		Mn:	MVol.D1-55
$C_3Cl_2H_6N_2O_2S$	$(CH_3)_2NSO_2NCCl_2$	S:	S-N-Verb.1-184
–	$ClSO_2NCCl[N(CH_3)_2]$	S:	S-N-Verb.1-126
$C_3Cl_2H_6O_3S_2{}^{2+}$	$[(CH_3)_2CO(SOCl)_2]^{2+}$	S:	SVol.1-52
$C_3Cl_2H_6Sn$	$(CH_3)(CH_2CH)SnCl_2$	Sn:	Org.Verb.6-196
$C_3Cl_2H_7NOSn$	$SnCl_2 \cdot OCHN(CH_3)_2$	Sn:	MVol.C5-43
$C_3Cl_2H_7NO_2S$	$SOCl_2 \cdot HCON(CH_3)_2$	S:	SVol.1-45
$C_3Cl_2H_7NO_2Sn$	$SnCl_2 \cdot CH_3CH(NH_2)COOH$	Sn:	MVol.C5-31/2
–	$SnCl_2 \cdot NH_2C_2H_4COOH$	Sn:	MVol.C5-31
$C_3Cl_2H_7NO_4S_2$	$(i\text{-}C_3H_7)N(SO_2Cl)_2$	S:	S-N-Verb.1-56
$C_3Cl_2H_7Ti$	$(i\text{-}C_3H_7)TiCl_2$	Ti:	Org.Verb.1-8
$C_3Cl_2H_8MnO$	$MnCl_2 \cdot C_3H_7OH$	Mn:	MVol.D1-33

$C_3Cl_2H_8OTi$	$CH_3Ti(OC_2H_5)Cl_2$	Ti: Org.Verb.1-59
$C_3Cl_2H_8Sn$	$(CH_3)(C_2H_5)SnCl_2$	Sn: Org.Verb.6-196
–	$(CH_3)_2(CH_2Cl)SnCl$	Sn: Org.Verb.5-191/2
$C_3Cl_2H_9NSn$	$SnCl_2 \cdot C_3H_7NH_2$	Sn: MVol.C5-11/2
$C_3Cl_2H_9Sn^-$	$[(CH_3)_3SnCl_2]^-$	Sn: Org.Verb.5-87
$C_3Cl_2H_{12}MnO_3$	$MnCl_2 \cdot 3\ CH_3OH$	Mn: MVol.D1-25/6
$C_3Cl_2H_{12}N_6O_5U$	$UO_2Cl_2 \cdot 3\ (NH_2)_2CO$	U: SVol.E1-85/6
–	$UO_2Cl_2 \cdot 3\ (NH_2)_2CO \cdot H_2O$	U: SVol.E1-83/4, 86
$C_3Cl_2H_{12}N_6O_8S_3Sn$	$Sn(ClO_4)_2 \cdot 3\ CS(NH_2)_2$	Sn: MVol.C5-55/6
$C_3Cl_2H_{12}N_6Se_3$	$[(NH_2)_2CSe]_3Cl_2 \cdot H_2O$	C: MVol.D6-246, 247/8
$C_3Cl_2H_{12}N_9O_5U$	$UO_2Cl_2 \cdot 3\ NH_2CONNH_2 \cdot 6\ H_2O$	U: SVol.E1-93, 95
$C_3Cl_2H_{14}MnO_4$	$MnCl_2(H_2O)(CH_3OH)_3$	Mn: MVol.C5-78
$C_3Cl_2H_{15}N_3O_2U$	$UO_2Cl_2 \cdot 3\ CH_3NH_2$	U: SVol.E1-18, 21
$C_3Cl_2N_3O^-$	$[C_3N_3Cl_2O]^-$	C: MVol.D3-284
$C_3Cl_3FH_2NO_3PS$	$ClP(O)[SCH_2C(O)Cl]ONCFCl$	F: PerFHalOrg.7-202
$C_3Cl_3FH_4NO_2PS$	$SPCl(ONCFCl)OC_2H_4Cl$	F: PerFHalOrg.7-201/2
$C_3Cl_3FH_4NO_3P$	$OPCl(ONCFCl)OCH_2CH_2Cl$	F: PerFHalOrg.7-201/2
$C_3Cl_3F_2H_2NO$	$CF_2ClCCl_2C(O)NH_2$	F: PerFHalOrg.7-18, 20/1, 45
$C_3Cl_3F_2N_3$	$NCFClNFCClNCCl$	F: PerFHalOrg.6-74/6, 97
$C_3Cl_3F_3H_4Sn$	$CF_3CH_2CH_2SnCl_3$	Sn: Org.Verb.6-254/5
$C_3Cl_3F_4N_3$	$NCFClNFCFClNFCCl$	F: PerFHalOrg.6-75/6, 97
$C_3Cl_3F_6NO$	$(CF_3)_2NOCCl_3$	F: PerFHalOrg.7-93/4, 109
$C_3Cl_3F_6N_3$	$ClN(CF_2NCl)_2CF_2$	F: PerFHalOrg.6-75/7, 97/8
–	$FN(CFClNF)_2CFCl$	F: PerFHalOrg.6-75/6, 97, 114
$C_3Cl_3F_7Si$	$C_3F_7SiCl_3$	F: PerFHalOrg.4-147
$C_3Cl_3F_9NOSb$	$NO[(CF_3)_3SbCl_3]$	F: PerFHalOrg.3-231
$C_3Cl_3F_9NP$	$CF_3PCl_3N(CF_3)_2$	F: PerFHalOrg.3-61, 66, 68
$C_3Cl_3F_9P^-$	$[(CF_3)_3PCl_3]$	F: PerFHalOrg.3-111
$C_3Cl_3F_9Sb^-$	$[(CF_3)_3SbCl_3]^-$	F: PerFHalOrg.3-230
$C_3Cl_3FeGeNO_4$	$(CO)_3Fe(NO)GeCl_3$	Fe: Org.Verb.B1-197/8
$C_3Cl_3FeNO_4Sn$	$(CO)_3Fe(NO)SnCl_3$	Fe: Org.Verb.B1-197/8
$C_3Cl_3HN_3O^-$	$[C_3N_3Cl_3OH]^-$	C: MVol.D3-284
$C_3Cl_3H_2N_3O$	$HC_3N_3Cl_3OH$	C: MVol.D3-284
$C_3Cl_3H_4NSn$	$(CH_2CH_2CN)SnCl_3$	Sn: MVol.C6-184
$C_3Cl_3H_5N_2O_2S$	$(NH_2)_2CS \cdot CCl_3COOH$	C: MVol.D6-112, 126
$C_3Cl_3H_5O_2Sn$	$HOCOCH_2CH_2SnCl_3$	Sn: Org.Verb.6-255
–	$SnCl_3(C_2H_5COO)$	Sn: MVol.C2-223
$C_3Cl_3H_5Sn$	$CH_2C(CH_3)SnCl_3$	Sn: Org.Verb.6-265
–	$CH_2CHCH_2SnCl_3$	Sn: Org.Verb.6-265
–	$(CH_2)_2CHSnCl_3$	Sn: Org.Verb.6-253
$C_3Cl_3H_5Ti$	$CH_2CHCH_2TiCl_3$	Ti: Org.Verb.1-16, 24, 127
$C_3Cl_3H_6O_3U^-$	$[UO_2Cl_3(CH_3COCH_3)]^-$	U: SVol.E1-79
$C_3Cl_3H_7NOSn^+$	$[SnCl_3(OCHN(CH_3)_2)]^+$	Sn: MVol.C6-41
$C_3Cl_3H_7O_2Ti$	$CH_3TiCl_3 \cdot C_2H_4O_2$	Ti: Org.Verb.1-36/7, 41
$C_3Cl_3H_7Sn$	$C_3H_7SnCl_3$	Sn: Org.Verb.6-234/6
–	$(i\text{-}C_3H_7)SnCl_3$	Sn: Org.Verb.6-236
$C_3Cl_3H_7Ti$	$C_3H_7TiCl_3$	Ti: Org.Verb.1-15, 23
$C_3Cl_3H_8N_3O_3Sn$	$SnCl_3(NHCHO) \cdot 2\ OCHNH_2$	Sn: MVol.C6-40

$C_3Cl_3H_8N_4O_2PS_2$	$[NS(O)Cl]_2[NPClNHC_3H_7]$	S:	S-N-Verb.1-30
$C_3Cl_3H_9N_6SiSn$	$SnCl_2(N_3)_2 \cdot (CH_3)_3SiCl$	Sn:	MVol.C6-130
–	$SnCl_3N_3 \cdot (CH_3)_3SiN_3$	Sn:	MVol.C6-130
$C_3Cl_3H_9OSSn$	$CH_3SnCl_3 \cdot (CH_3)_2SO$	Sn:	Org.Verb.6-222
$C_3Cl_3H_{10}MnN$	$(CH_3)_3NHMnCl_3$	Mn:	MVol.C5-191
–	$(CH_3)_3NHMnCl_3 \cdot 2\ H_2O$	Mn:	MVol.C5-191/3
–	$C_3H_7NH_3MnCl_3$	Mn:	MVol.C5-188
$C_3Cl_3H_{11}N_2SSn$	$CH_3SnCl_3 \cdot (CH_3)_2S(NH)_2$	Sn:	Org.Verb.6-222
$C_3Cl_3H_{12}IrN_6S_3$	$Ir(SC(NH_2)_2)_3Cl_3$	Ir:	SVol.2-168
$C_3Cl_3H_{12}MnO_3^-$	$[Mn(CH_3OH)_3Cl_3]^-$	Mn:	MVol.C5-77
		Mn:	MVol.D1-24
$C_3Cl_3H_{15}N_4Sn$	$C_3H_7SnCl_3 \cdot 2\ N_2H_4$	Sn:	Org.Verb.6-236
$C_3Cl_3N_3$	$C_3N_3Cl_3$		
	Chemical reactions	C:	MVol.D3-282/7
	Formation and preparation	C:	MVol.D3-273/6
	Molecule	C:	MVol.D3-276/9
	Physical properties	C:	MVol.D3-279/82
	Physiological action	C:	MVol.D3-273
	Purification	C:	MVol.D3-275
	Review	C:	MVol.D3-272/3
	Solubility	C:	MVol.D3-286/7
	Uses	C:	MVol.D3-275/6
$C_3Cl_3N_3O_3$	$(ClNCO)_3$	C:	MVol.D3-146/7
$C_3Cl_3NiO_3P$	$(CO)_3NiPCl_3$	Ni:	Org.Verb.1-161/2
$C_3Cl_3NiO_3Sb$	$(CO)_3NiSbCl_3$	Ni:	Org.Verb.1-187
$C_3Cl_4H_3NSn$	$SnCl_4 \cdot CH_2CHCN$	Sn:	MVol.C6-69/70
$C_3Cl_4H_4N_2OSn$	$SnCl_4 \cdot NCCH_2CONH_2$	Sn:	MVol.C6-76
$C_3Cl_4H_4OSn$	$ClC(O)CH_2CH_2SnCl_3$	Sn:	Org.Verb.6-255
$C_3Cl_4H_4O_4Sn$	$SnCl_4 \cdot CH_2(COOH)_2$	Sn:	MVol.C5-121
$C_3Cl_4H_{4.5}N_{1.5}U$	$[UCl_4(CH_3CN)_{1.5}]$	U:	SVol.E1-49
$C_3Cl_4H_5NSn$	$SnCl_4 \cdot C_2H_5CN$	Sn:	MVol.C6-68/9
$C_3Cl_4H_6N_2O_2Sn$	$SnCl_4 \cdot CH_3CONHCONH_2$	Sn:	MVol.C6-58/9
$C_3Cl_4H_6N_6Sn$	$SnCl_4 \cdot 3\ CNNH_2$	Sn:	MVol.C6-77/8
$C_3Cl_4H_6OSn$	$SnCl_4 \cdot CH_3COCH_3$	Sn:	MVol.C5-85/6
$C_3Cl_4H_6OTe$	$TeCl_4 \cdot (CH_3)_2CO$	Te:	SVol.B3-165/6
$C_3Cl_4H_6O_6Sn$	$SnCl_4 \cdot 3\ HCOOH$	Sn:	MVol.C5-117
$C_3Cl_4H_6Sn$	$(CH_2Cl)_3SnCl$	Sn:	Org.Verb.5-150
–	$(CH_3)_2(CCl_3)SnCl$	Sn:	Org.Verb.5-191/2
$C_3Cl_4H_7NOTe$	$TeCl_4 \cdot HCON(CH_3)_2$	Te:	SVol.B3-174
$C_3Cl_4H_7N_3O_2P_2S$	$[O_2S[NPCl_2]_2NC_3H_7]$	S:	S-N-Verb.1-35
$C_3Cl_4H_8N_2O_4Sn$	$SnCl_4 \cdot NH_2CH_2COOH \cdot CH_3NO_2$	Sn:	MVol.C6-11
$C_3Cl_4H_8OSn$	$SnCl_4 \cdot C_3H_7OH$	Sn:	MVol.C5-73
$C_3Cl_4H_8S_2Sn$	$SnCl_4 \cdot HSCH_2CH(SH)CH_3$	Sn:	MVol.C6-81
–	$SnCl_4 \cdot HSCH_2CH_2CH_2SH$	Sn:	MVol.C6-81
$C_3Cl_4H_8SiSn$	$Cl(CH_3)_2SiCH_2SnCl_3$	Sn:	Org.Verb.6-256
$C_3Cl_4H_9NOSn$	$SnCl_4 \cdot (CH_3)_3NO$	Sn:	MVol.C6-1
$C_3Cl_4H_9NSn$	$SnCl_4 \cdot (CH_3)_3N$	Sn:	MVol.C5-172/3
$C_3Cl_4H_9NU$	$UCl_4 \cdot C_3H_7NH_2$	U:	SVol.E1-18, 20
$C_3Cl_4H_9N_3SiSn$	$SnCl_4 \cdot (CH_3)_3SiN_3$	Sn:	MVol.C6-130
$C_3Cl_4H_9PSn$	$SnCl_4 \cdot P(CH_3)_3$	Sn:	MVol.C6-132

$C_3Cl_4H_9PSn$	$SnCl_4 \cdot P(CD_3)_3$	Sn: MVol.C6-132
$C_3Cl_4H_{12}MnN_2$	$NH_3(CH_2)_3NH_3MnCl_4$	Mn: MVol.C5-185/8
$C_3Cl_4H_{12}N_6O_3U$	$UCl_4 \cdot 3\ (NH_2)_2CO$	U: SVol.E1-82/3
–	$UCl_4 \cdot 3\ (NH_2)_2CO \cdot H_2O$	U: SVol.E1-82/3
$C_3Cl_4H_{12}N_6S_3U$	$UCl_4 \cdot 3\ (NH_2)_2CS$	U: SVol.E1-219
$C_3Cl_4NiO_3P_2$	$(CO)_3NiP_2Cl_4$	Ni: Org.Verb.1-185
$C_3Cl_4Ni_2O_3$	$Ni_2(CO)_3Cl_4$	Ni: Org.Verb.2-258
$C_3Cl_5F_3Si$	$CF_3C(SiCl_3)CCl_2$	F: PerFHalOrg.4-149/50, 156, 161
$C_3Cl_5H_9OPU^-$	$[UCl_5((CH_3)_3PO)]^-$	U: SVol.E1-156
$C_3Cl_6FeO_3P_2$	$(CO)_3Fe(PCl_3)_2$	Fe: Org.Verb.B1-149
$C_3Cl_6FeO_3Sb_2$	$(CO)_3Fe(SbCl_3)_2$	Fe: Org.Verb.B1-148/50, 156, 164
$C_3Cl_6O_3$	$CO(OCCl_3)_2$	C: MVol.D3-63/4
$C_3Cl_6S_3$	$(Cl_2CS)_3$	C: MVol.D6-172
$C_3Cl_7F_3Si$	$CF_3CCl(SiCl_3)CCl_3$	F: PerFHalOrg.4-149/50, 156, 161, 163
–	$CF_3CCl_2CCl_2SiCl_3$	F: PerFHalOrg.4-149/50, 156
C_3Cl_9OU	$UCl_5 \cdot Cl_2CCClCOCl$	U: SVol.E1-123/4
$C_3Cl_{14}H_6O_3S_2Sb_2$	$[(CH_3)_2CO(SOCl)_2][SbCl_6]_2$	S: SVol.1-54
$C_3Cl_{29}OU_5$	$5\ UCl_5 \cdot Cl_2CCClCOCl$	U: SVol.E1-123/4
$C_3CmH_5O_3^{2+}$	$Cm(CH_3CH(OH)COO)^{2+}$	Np: TrU.D1-154
$C_3CmN_3S_3$	$Cm(SCN)_3$	Np: TrU.D1-145
$C_3CoF_3H_2NO_3P$	$Co(NO)(CO)_2CF_3PH_2$	F: PerFHalOrg.3-27
$C_3CoH_6N_3S_6$	$Co[S_2CNH_2]_3$	C: MVol.D6-140
$C_3CrH_6N_3S_6$	$Cr[S_2CNH_2]_3$	C: MVol.D6-140
$C_3CsF_{12}P$	$Cs[(CF_3)_3PF_3]$	F: PerFHalOrg.3-134, 136, 137
$C_3CsH_3O_6Sn$	$CsSn(HCOO)_3$	Sn: MVol.C3-155
$C_3CsN_3S_3Sn$	$CsSn(SCN)_3$	Sn: MVol.C3-156
$C_3Cs_3DyO_9$	$Cs_3[Dy(CO_3)_3]$	Sc: MVol.B7-26
$C_3Cs_3ErO_9$	$Cs_3[Er(CO_3)_3]$	Sc: MVol.B7-26
$C_3Cs_4O_{11}U$	$Cs_4[UO_2(CO_3)_3]$	U: SVol.C2-263
$C_3CuH_5OS_2$	$Cu[S_2COC_2H_5]$	C: MVol.D4-256
$C_3CuH_5O_2S$	$Cu[SOCOC_2H_5]$	C: MVol.D4-237
$C_3CuH_5S_3$	$Cu[S_2CSC_2H_5]$	C: MVol.D4-269
$C_3CuH_8NS_4$	$[(CH_3)_2NH_2]CuCS_4$	C: MVol.D4-231
$C_3Cu_2K_2N_2S_3$	$Cu_2CS_3 \cdot 2\ KCN \cdot 2\ H_2O$	C: MVol.D4-224
$C_3DyK_3O_9$	$K_3[Dy(CO_3)_3]$	Sc: MVol.B7-26
$C_3DyNa_3O_9$	$Na_3[Dy(CO_3)_3]$	Sc: MVol.B7-26
C_3ErNaO_9	$Na[Er(CO_3)_3] \cdot 3.5\ H_2O$	Sc: MVol.B7-26
$C_3Er_2O_9$	$Er_2(CO_3)_3 \cdot x\ H_2O$	Sc: MVol.B7-26/7
$C_3Eu_2O_9$	$Eu_2(CO_3)_3$	Sc: MVol.B7-26
$C_3FH_2NO_4S$	$FSO_2NC(O)CH_2CO$	S: S-N-Verb.1-115
$C_3FH_3O_2$	$CH_2CFCOOH$	F: PerFHalOrg.4-89
$C_3FH_4N_5$	$C_3N_3F(NH_2)_2$	C: MVol.D3-271
$C_3FH_5O_2$	$FCOOC_2H_5$	C: MVol.D3-46
$C_3FH_6NO_2$	$(CH_3)_2CFNO_2$	F: PerFHalOrg.5-144
$C_3FH_6NO_2S$	$FSO_2NCHC_2H_5$	S: S-N-Verb.1-105

$C_3FH_6NO_3S$	$FSO_2NC(OCH_3)CH_3$	S:	S-N-Verb.1-105
$C_3FH_6NO_4S$	$FSO_2N(CH_3)COOCH_3$	S:	S-N-Verb.1-104
–	$FSO_2NHCOOC_2H_5$	S:	S-N-Verb.1-101
$C_3FH_6N_2O_2S^+$	$SN_2C_3H_6FO_2^+$	S:	S-N-Verb.1-194
$C_3FH_6N_3OS$	$(CH_3)_2NS(O)FNCN$	S:	S-N-Verb.1-195
$C_3FH_6N_3OS^+$	$(CH_3)_2NSOFNCN^+$	S:	S-N-Verb.1-195
$C_3FH_7N_2O_2S$	$FSO_2NC(NHCH_3)CH_3$	S:	S-N-Verb.1-105
$C_3FH_7N_2O_3S$	$FSO_2NHCON(CH_3)_2$	S:	S-N-Verb.1-126
C_3FH_8ISn	$(CH_3)_2(CH_2I)SnF$	Sn:	Org.Verb.5-40
$C_3FH_9NO_2SSn^+$	$(CH_3)_2SnNCH_3SO_2F^+$	S:	S-N-Verb.1-108
$C_3FH_9NSSi^+$	$SFNSi(CH_3)_3^+$	S:	S-N-Verb.1-201
$C_3FH_9N_2OS$	$(CH_3)_2NS(O)FNCH_3$	S:	S-N-Verb.1-193
$C_3FH_9N_2OS^+$	$(CH_3)_2NS(O)FNCH_3^+$	S:	S-N-Verb.1-193
C_3FH_9Sn	$(CH_3)_3SnF$	Sn:	Org.Verb.5-13/9
$C_3FH_{10}NO_2SSi$	$FSO_2NHSi(CH_3)_3$	S:	S-N-Verb.1-106
$C_3F_2H_2N_4$	$(NH_2)C_3F_2N_3$	F:	PerFHalOrg.6-78, 98, 111
$C_3F_2H_2O_4$	$CH_2(COOF)_2$	F:	PerFHalOrg.5-144
$C_3F_2H_3NOS^+$	$C_3H_3NSOF_2^+$	S:	S-N-Verb.1-157
$C_3F_2H_3NO_4$	$O_2NCF_2C(O)OCH_3$	F:	PerFHalOrg.7-191
$C_3F_2H_3N_3O_3$	NCF(OH)NFC(OH)NC(OH)	F:	PerFHalOrg.6-74/6
$C_3F_2H_4N_2O_2$	$NH_2C(O)CF_2C(O)NH_2$	F:	PerFHalOrg.7-18, 20, 47
$C_3F_2H_5NO_3S$	$FSO_2N(C_2H_5)C(O)F$	S:	S-N-Verb.1-104
$C_3F_2H_6NO_4P$	$CF_2NOPO(OCH_3)_2$	F:	PerFHalOrg.7-199
$C_3F_2H_7NO_3U$	$(UO_2F_2 \cdot HCON(CH_3)_2)_n$	U:	SVol.E1-100/1
$C_3F_2H_8N_2O_3U$	$[UO_2F_2 \cdot CH_3NHCONHCH_3]_n$	U:	SVol.E1-91/2, 94
$C_3F_2H_9NOSSi$	$(CH_3)_3SiNS(O)F_2$	S:	S-N-Verb.1-63/4
$C_3F_2H_9NOSSi^+$	$(CH_3)_3SiNS(O)F_2^+$	S:	S-N-Verb.1-64
$C_3F_2H_9NOSSn$	$(CH_3)_3SnNS(O)F_2$	S:	S-N-Verb.1-64
$C_3F_2H_9NO_4S_2Sn$	$(CH_3)_3SnN(SO_2F)_2$	S:	S-N-Verb.1-53
$C_3F_2H_9NSSi^+$	$SF_2NSi(CH_3)_3^+$	S:	S-N-Verb.1-201
$C_3F_2H_9N_2SSi^+$	$NSF_2[NSi(CH_3)_3]^+$	S:	S-N-Verb.1-201
$C_3F_2H_9N_2SSi_2^+$	$SF_2N_2Si_2(CH_3)_3^+$	S:	S-N-Verb.1-201
$C_3F_2H_9O_3P$	$(CH_3O)_3PF_2$	F:	PerFHalOrg.5-23
$C_3F_2H_9PS_2Sn$	$(CH_3)_3SnSPSF_2$	Sn:	Org.Verb.5-84
$C_3F_2H_9Sn^-$	$(CH_3)_3SnF_2^-$	Sn:	Org.Verb.5-19
$C_3F_2H_{10}N_4O_3S_3$	$[(CH_3)_3NH][NS(O)F]_2[NSO]$	S:	S-N-Verb.1-21
$C_3F_2H_{11}N_3O_3S$	$(CH_3)_3NO \cdot NH_2SO_2NF_2$	S:	S-N-Verb.1-176
$C_3F_2H_{12}N_6O_5U$	$UO_2F_2 \cdot 3\,(NH_2)_2CO$	U:	SVol.E1-83/5
$C_3F_3HN_2$	$CF_3C(NH)CN$	F:	PerFHalOrg.7-25/6, 52, 75
$C_3F_3HN_4O_2S$	$FSO_2NHC(NCF)_2N$	S:	S-N-Verb.1-103
$C_3F_3HO_2$	$CF_2CFCOOH$	F:	PerFHalOrg.4-90
$C_3F_3HO_3$	$CF_3C(O)COOH$	F:	PerFHalOrg.5-21
$C_3F_3H_2NO$	$CF_2CFC(O)NH_2$	F:	PerFHalOrg.7-18, 22, 46
$C_3F_3H_2NO_3$	$CF_3NHC(O)COOH$	F:	PerFHalOrg.7-23, 50, 63
$C_3F_3H_2N_3O$	$NH_2CNC(CF_3)NO$	F:	PerFHalOrg.5-66, 90, 101
$C_3F_3H_2N_3S$	$NC(CF_3)SC(NH_2)N$	F:	PerFHalOrg.5-67/8, 89, 100, 107/8, 110
–	$NH_2CSNC(CF_3)N$	F:	PerFHalOrg.5-67/8, 90, 101, 110

$C_3F_3H_2N_3Se$	$NC(CF_3)SeC(NH_2)N$	F:	PerFHalOrg.5-68, 100
$C_3F_3H_2OS_2^-$	$[S_2COCH_2CF_3]^-$	C:	MVol.D4-244
$C_3F_3H_3NO_2$	$CF_3N(O)COCH_3$	F:	PerFHalOrg.7-169
$C_3F_3H_3N_4$	$CH_3NNC(CF_3)NN$	F:	PerFHalOrg.5-103
–	$NC(CF_3)NC(NH_2)NH$	F:	PerFHalOrg.5-71/3, 92
–	$NN(CH_3)C(CF_3)NN$	F:	PerFHalOrg.5-103
$C_3F_3H_3N_4S$	$NC(CF_3)N(NH_2)C(SH)N$	F:	PerFHalOrg.5-71/3, 93, 100
$C_3F_3H_3O$	$(CF_3COCH_3)_n$	F:	PerFHalOrg.4-94
$C_3F_3H_4NO$	$[N(CF_3)OCH_2CH_2]_n$	F:	PerFHalOrg.7-161
$C_3F_3H_4N_3O$	$CF_3NNNHCOCH_3$	F:	PerFHalOrg.7-158/9
$C_3F_3H_5NO$	$CF_3N(O)C_2H_5$	F:	PerFHalOrg.7-169
$C_3F_3H_5NO_3P$	$CF_2NOPF(O)OC_2H_5$	F:	PerFHalOrg.7-199/200
$C_3F_3H_5N_2$	$CF_3NNC_2H_5$	F:	PerFHalOrg.7-159
$C_3F_3H_5N_2O$	$CF_3NNCH_2CH_2OH$	F:	PerFHalOrg.7-159
–	$C_2H_5NN(O)CF_3$	F:	PerFHalOrg.7-160/1
$C_3F_3H_5N_4O$	$CF_3CONHNHC(NH)NH_2$	F:	PerFHalOrg.7-6/7, 34
$C_3F_3H_6NOS$	$CF_3N(OH)SC_2H_5$	F:	PerFHalOrg.7-168
$C_3F_3H_6NO_2S$	$CF_3SO_2N(CH_3)_2$	S:	S-N-Verb.1-107
$C_3F_3H_6NO_2S_2$	$CF_3SO_2NS(CH_3)_2$	S:	S-N-Verb.1-107
$C_3F_3H_6NiP$	$CH_3CHCH_2NiPF_3$	Ni:	Org.Verb.1-356, 360
–	$C_3H_5Ni(PF_3)H$	Ni:	Org.Verb.2-12/3
$C_3F_3H_6OP$	$CF_3P(OCH_3)CH_3$	F:	PerFHalOrg.3-24
$C_3F_3H_6P$	$CF_3P(CH_3)_2$	F:	PerFHalOrg.3-173
$C_3F_3H_6PS_2$	$CF_3P(SCH_3)_2$	F:	PerFHalOrg.3-101, 107
$C_3F_3H_7N_2O$	$CF_3NO \cdot (CH_3)_2NH$	F:	PerFHalOrg.7-160
$C_3F_3H_7Sn$	$C_3H_7SnF_3$	Sn:	Org.Verb.5-51
$C_3F_3H_8N_2P$	$CF_3P(NHCH_3)_2$	F:	PerFHalOrg.3-102, 112
C_3F_3IMg	CF_3CCMgI	F:	PerFHalOrg.4-62/3, 74/6, 96, 106, 110
C_3F_3Li	CF_3CCLi	F:	PerFHalOrg.4-2, 17, 19, 33, 46/7
$C_3F_3N_3$	NCFNCFNCF	C:	MVol.D3-267/72
		F:	PerFHalOrg.6-77/8
$C_3F_3NiO_3P$	$(CO)_3NiPF_3$	Ni:	Org.Verb.1-161/2
C_3F_4HNO	$CF_3CFC(O)NH$	F:	PerFHalOrg.5-21
$C_3F_4HNO_2$	$CF_3C(O)NHC(O)F$	F:	PerFHalOrg.7-23, 50
$C_3F_4HNO_3$	$ONCF_2CF_2C(O)OH$	F:	PerFHalOrg.7-177, 184, 188, 192/3
$C_3F_4H_2NO_4S^+$	$FSO_2NHCOC(CF_3)OH^+$	S:	S-N-Verb.1-101
$C_3F_4H_2N_2O$	$CF_2NHCONHCF_2$	F:	PerFHalOrg.5-61, 84
$C_3F_4H_2N_2O_2$	$ONCF_2CF_2C(O)NH_2$	F:	PerFHalOrg.7-177, 184
$C_3F_4H_2N_2O_3$	$CF_3CF(NO_2)C(O)NH_2$	F:	PerFHalOrg.7-18/20, 46
$C_3F_4H_3NO_3$	$O_2NCF_2CF_2OCH_3$	F:	PerFHalOrg.7-191
$C_3F_4H_3N_2O_5P$	$NO_2CF_2CFNOPF(O)OCH_3$	F:	PerFHalOrg.7-199/200
$C_3F_4H_3N_3O_2$	$O_2NCF_2CF_2NNCH_3$	F:	PerFHalOrg.7-192
$C_3F_4H_3N_3O_3$	NCF(OH)NFCF(OH)NFC(OH)	F:	PerFHalOrg.6-75/6
$C_3F_4H_4N_2$	$CF_3NNCH_2CH_2F$	F:	PerFHalOrg.7-159
$C_3F_4H_6NP$	$CF_3P(F)N(CH_3)_2$	F:	PerFHalOrg.3-102, 112
$C_3F_4H_6N_2O_3S_2$	$CF_3SO_2NS(O)F[N(CH_3)_2]$	S:	S-N-Verb.1-70

$C_3F_4H_6N_2S_2$	$F_2SNCH_2CH(CH_3)NSF_2$	S:	SVol.2-78, 85
–	$F_2SN(CH_2)_3NSF_2$	S:	SVol.2-78, 85
$C_3F_4H_9NSn$	$SnF_4 \cdot (CH_3)_3N$	Sn:	MVol.C5-171
$C_3F_4H_9NTe$	$TeF_4 \cdot N(CH_3)_3$	Te:	SVol.B3-168
$C_3F_4H_9N_2OPSSi$	$(CH_3)_3SiNSF_2NP(O)F_2$	S:	S-N-Verb.1-201
$C_3F_4H_{10}MnN$	$NH_3(C_3H_7)MnF_4 \cdot H_2O$	Mn:	MVol.C4-212/3
$C_3F_4H_{14}N_6S_3Te$	$Te(SC(NH_2)_2)_3(HF_2)_2$	Te:	SVol.B3-142/3
$C_3F_4KNO_3$	$ONCF_2CF_2C(O)OK$	F:	PerFHalOrg.7-177, 184
$C_3F_4N_2O$	$NCCF_2CF_2NO$	F:	PerFHalOrg.7-175/6, 183
$C_3F_4N_2O_2S_2$	$FSO_2NC(NCS)CF_3$	S:	S-N-Verb.1-119
$C_3F_4N_2O_3S$	$FSO_2NC(NCO)CF_3$	S:	S-N-Verb.1-118
$C_3F_4N_2O_4$	$ONCF_2CF_2C(O)ONO$	F:	PerFHalOrg.7-177/8, 184
$C_3F_4N_2S$	$CFSNC(CF_3)N$	F:	PerFHalOrg.5-67/8
$C_3F_4N_4S$	$C_3N_3F_2(NSF_2)$	S:	SVol.2-82
$C_3F_5^+$	$C_3F_5^+$	F:	PerFHalOrg.3-216
C_3F_5H	$(CF_2)_2CHF$	F:	PerFHalOrg.5-17
$C_3F_5HO_2$	CF_3COOCF_2H	F:	PerFHalOrg.5-18
$C_3F_5H_2NO$	$CF_3CFCF(OH)NH$	F:	PerFHalOrg.5-21
–	$C_2F_5C(O)NH_2$	F:	PerFHalOrg.7-18/9, 22, 45, 62/3
–	$[N(CF_3)OCF_2CH_2]_n$	F:	PerFHalOrg.7-207/8
$C_3F_5H_2NO_3$	$CF_3C(OH)_2CF_2NO$	F:	PerFHalOrg.7-177, 184
$C_3F_5H_3N_2$	$C_2F_5C(NH)NH_2$	F:	PerFHalOrg.7-3/4, 30
$C_3F_5H_3N_2O$	$C_2F_5C(NOH)NH_2$	F:	PerFHalOrg.7-5, 32
$C_3F_5H_3O$	$CF_3CH_2OCF_2H$	F:	PerFHalOrg.5-19
$C_3F_5H_4OP$	$CF_3P(OCH_3)CF_2H$	F:	PerFHalOrg.3-24
$C_3F_5H_5OS$	$SF_5CH_2C(O)CH_3$	S:	SVol.2-288
$C_3F_5H_7NOSn^-$	$[SnF_5(OCHN(CH_3)_2)]^-$	Sn:	MVol.C6-40
$C_3F_5H_7NTe^+$	$TeF_5NC_2H_5CH_2^+$	Te:	SVol.B2-66
$C_3F_5H_8OSn^-$	$[SnF_5((CH_3)_2CHOH)]^-$	Sn:	MVol.C5-72
$C_3F_5H_9OSiTe$	$[(CH_3)_3Si][TeF_5O]$	Te:	SVol.B2-53, 63
$C_3F_5H_9OSnTe$	$(CH_3)_3SnTeF_5O$	Te:	SVol.B2-54
C_3F_5N	$CF_2NC(CF_3)$	F:	PerFHalOrg.5-2, 8, 19/20
–	CF_3CFNCF	F:	PerFHalOrg.5-2, 8, 19/21
–	CF_3NCCF_2	F:	PerFHalOrg.5-47
–	$[NCF(CF_3)CF]_n$	F:	PerFHalOrg.5-4, 10, 20
–	$[NCF_2C(CF_3)]_n$	F:	PerFHalOrg.5-4, 10, 20
C_3F_5NO	$CF_3NOCFCF$	F:	PerFHalOrg.5-33/4, 41
$C_3F_5NO_2$	$FC(O)NOCF_2CF_2$	F:	PerFHalOrg.5-30, 39, 46
–	$ONCF_2CF_2C(O)F$	F:	PerFHalOrg.7-177, 184
$C_3F_5N_3$	$NCF_2NFCFNCF$	F:	PerFHalOrg.6-74/6, 97
C_3F_6	CF_3CFCF_2	F:	PerFHalOrg.4-8, 68/9
$C_3F_6FeO_3P_2$	$(CO)_3Fe(PF_3)_2$	Fe:	Org.Verb.B1-148/50, 153, 161
C_3F_6HN	$CF_3CFNHCF_2$	F:	PerFHalOrg.5-2, 8
–	$(CF_3)_2CNH$	F:	PerFHalOrg.7-25/7, 51, 55/6, 64/75
C_3F_6HNO	CF_3NCF_2CFHO	F:	PerFHalOrg.7-161/2
–	$CF_3NCHFCF_2O$	F:	PerFHalOrg.7-161/2

C_3F_6HNO	$(CF_3)_2CNOH$	F:	PerFHalOrg.7-88/9, 105, 125, 141/3, 145
–	$[N(CF_3)OCF_2CHF]_n$	F:	PerFHalOrg.7-207
$C_3F_6HNO_2$	$(CF_3)_2CHNO_2$	F:	PerFHalOrg.7-141
$C_3F_6HNO_3S$	$C_2F_5C(O)NHSO_2F$	F:	PerFHalOrg.7-24/5, 51
$C_3F_6HN_2O_2$	$CF_3C(O)N(CF_3)NOH$	F:	PerFHalOrg.7-169
$C_3F_6H_2N_2$	$CF_3NNCH_2CF_3$	F:	PerFHalOrg.7-159
–	$(CF_3)_2C(NH)_2$	F:	PerFHalOrg.5-6, 12
$C_3F_6H_2N_2O$	$CF_3NHC(O)NHCF_3$	F:	PerFHalOrg.7-22/3, 49
$C_3F_6H_2N_2O_2S$	$C_2F_5C(NH_2)NSO_2F$	F:	PerFHalOrg.7-2, 29
		S:	S-N-Verb.1-119
$C_3F_6H_2N_4$	$(CF_3)_2C(N_3)NH_2$	F:	PerFHalOrg.7-2, 30
$C_3F_6H_3N$	$(CF_3)_2CHNH_2$	F:	PerFHalOrg.5-21
$C_3F_6H_3NO$	$(CF_3)_2NOCH_3$	F:	PerFHalOrg.7-144
$C_3F_6H_3NO_2$	$(CF_3)_2NOCH_2OH$	F:	PerFHalOrg.7-127
$C_3F_6H_3N_3O_3$	$FN(CF(OH)NF)_2CF(OH)$	F:	PerFHalOrg.6-75/6
$C_3F_6H_3OP$	$(CF_3)_2C(OH)PH_2$	F:	PerFHalOrg.3-15/6, 19, 24
$C_3F_6H_3O_2P$	$(CF_3)_2P(O)OCH_3$	F:	PerFHalOrg.3-53, 58
$C_3F_6H_3P$	$(CF_3)_2PCH_3$	F:	PerFHalOrg.3-101/2, 173
$C_3F_6H_3PS$	$(CF_3)_2PSCH_3$	F:	PerFHalOrg.3-101, 106
$C_3F_6H_3PS_2$	$(CF_3)_2P(S)SCH_3$	F:	PerFHalOrg.3-153, 155
$C_3F_6H_4NOP$	$(CF_3)_2P(O)N(H)CH_3$	F:	PerFHalOrg.3-58
–	$(CF_3)_2P(O)N(D)CH_3$	F:	PerFHalOrg.3-59
$C_3F_6H_4NP$	$(CF_3)_2PNHCH_3$	F:	PerFHalOrg.3-108, 113
–	$(CF_3)_2PNDCH_3$	F:	PerFHalOrg.3-108, 114
$C_3F_6H_4NPS$	$(CF_3)_2P(S)NHCH_3$	F:	PerFHalOrg.3-139, 153/4, 156
–	$(CF_3)_2P(S)ND(CH_3)$	F:	PerFHalOrg.3-156
$C_3F_6H_4N_2$	$(CF_3)_2C(NH_2)_2$	F:	PerFHalOrg.7-1/2, 28, 54, 57
$C_3F_6H_4OP_2$	$CF_3P(H)P(CF_3)OCH_3$	F:	PerFHalOrg.3-6
$C_3F_6H_4P_2$	$(CF_3)_2PP(H)CH_3$	F:	PerFHalOrg.3-101, 104
$C_3F_6H_5N_2P$	$(CF_3)_2PN(CH_3)NH_2$	F:	PerFHalOrg.3-109, 119
–	$(CF_3)_2PN(H)NHCH_3$	F:	PerFHalOrg.3-109, 118
$C_3F_6H_5N_3$	$(CF_3)_2C(NHNH_2)NH_2$	F:	PerFHalOrg.7-2/3, 30
$C_3F_6H_6N_2O$	$CH_3NH_2 \cdot (CF_3)_2NOH$	F:	PerFHalOrg.7-139
C_3F_6LiNO	$(CF_3)_2CNOLi$	F:	PerFHalOrg.7-88/9, 143
C_3F_6N	$(CF_3)_2CN$	F:	PerFHalOrg.7-56
C_3F_6NO	$(CF_3)_2CNO$	F:	PerFHalOrg.7-122
C_3F_6NOP	$(CF_3)_2PNCO$	F:	PerFHalOrg.3-60, 65
C_3F_6NP	$(CF_3)_2PCN$	F:	PerFHalOrg.3-164, 169, 173/4
C_3F_6NPS	$(CF_3)_2PNCS$	F:	PerFHalOrg.3-60, 65, 68/9
$C_3F_6N_2$	$CF_2NCF_2NCF_2$	F:	PerFHalOrg.5-15/6
–	$(CF_2)_2NNCF_2$	F:	PerFHalOrg.5-3, 9, 21/2
–	$(CF_3)_2CN_2$	F:	PerFHalOrg.5-5, 12, 21, 24/6

$C_3F_6N_2$	$(F)(C_2F_5)CN_2$	F: PerFHalOrg.5-5
–	$NCF(CF_3)CF_2N$	F: PerFHalOrg.5-49
–	$NCF_2CF_2CF_2N$	F: PerFHalOrg.5-60, 82, 97
$C_3F_6N_2O_2$	$CF_3CF(NO)CF_2NO$	F: PerFHalOrg.7-180
–	$(CF_3)_2CNONO$	F: PerFHalOrg.7-90/1, 106, 146
$C_3F_6N_2O_3$	$CF_3CF(NO)CF_2NO_2$	F: PerFHalOrg.7-180, 186, 189/92
C_3F_6O	CF_3CF_2COF	C: MVol.D3-46
–	$(CF_3)_2CO$	C: MVol.D3-45
$C_3F_6Se_3^+$	$(F_2CSe)_3^+$	C: MVol.D6-251
$C_3F_7^-$	$CF_3CF_2CF_2^-$	F: PerFHalOrg.4-68
$C_3F_7HN_2O_3$	$(NO_2CF_2CF_2)(CF_3)NOH$	F: PerFHalOrg.7-86/7, 104
$C_3F_7HO_2P^-$	$[n-C_3F_7P(O)(H)O]^-$	F: PerFHalOrg.3-49
$C_3F_7H_2N$	$(CF_3)_2CFNH_2$	F: PerFHalOrg.7-1/2, 28
–	$(CF_3CHF)(CF_3)NH$	F: PerFHalOrg.7-161
$C_3F_7H_2N_3$	$(CF_3)(NF_2CF_2)C(NH)_2$	F: PerFHalOrg.5-6/7, 12
$C_3F_7H_2O_2P$	$CF_3CF_2CF_2P(OH)_2$	F: PerFHalOrg.3-44/5, 49/50
$C_3F_7H_2O_3P$	$(n-C_3F_7)P(O)(OH)_2$	F: PerFHalOrg.3-41/3, 49/50
–	$(n-C_3F_7)P(O)(OH)_2 \cdot 2\ H_2O$	F: PerFHalOrg.3-45
C_3F_7IMg	$(n-C_3F_7)MgI$	F: PerFHalOrg.4-61/2, 68, 71, 92/6
$C_3F_7I_2P$	$C_3F_7PI_2$	F: PerFHalOrg.3-74/5, 76, 94, 100, 108
C_3F_7KO	$KOCF(CF_3)_2$	C: MVol.D3-134
C_3F_7Li	$CF_3CF_2CF_2Li$	F: PerFHalOrg.4-2, 8, 12/3, 18/9, 41/2
–	$(CF_3)_2CFLi$	F: PerFHalOrg.4-2, 8, 12/3, 18, 42
C_3F_7N	$(CF_2)_2C(NF_2)F$	F: PerFHalOrg.5-17
–	CF_3NCFCF_3	F: PerFHalOrg.5-47/8
C_3F_7NO	$CF_3NOCF_2CF_2$	F: PerFHalOrg.5-30/1, 40, 43/4, 46/8
–	$(CF_3)_2CFNO$	F: PerFHalOrg.7-176/7, 183, 189
–	$(CF_3)_2NCOF$	C: MVol.D3-46
–	C_3F_7NO	F: PerFHalOrg.7-176, 183, 189/91, 208/9
–	$[N(CF_3)OCF_2CF_2]_n$	F: PerFHalOrg.7-206/7, 212/4
$C_3F_7NO_2$	$CF_3OCF_2CF_2NO$	F: PerFHalOrg.7-179, 185
–	$(CF_3)_2NOC(O)F$	F: PerFHalOrg.7-94/5, 110, 126
$C_3F_7N_3$	NCF_2NFCF_2NFCF	F: PerFHalOrg.6-75/6, 97
$C_3F_7N_3O$	$FN(CF_2NF)_2C(O)$	F: PerFHalOrg.6-77, 98
$C_3F_7N_3O_4$	$(CF_3)(NO_2C_2F_4)NONO$	F: PerFHalOrg.7-89/90, 106
$C_3F_7Na_2O_3P$	$C_3F_7P(O)(ONa)_2$	F: PerFHalOrg.3-50
$C_3F_7P_2^+$	$C_3F_7P_2^+$	F: PerFHalOrg.3-179

$C_3F_8FeH_3O_3P_3$	$(CO)_2Fe(PF_3)_2PF_2OCH_3$	Fe:	Org.Verb.B1-97
C_3F_8HN	$(CF_3)_2CFN(H)F$	F:	PerFHalOrg.7-16/7, 43, 56
–	$C_2F_5N(H)CF_3$	F:	PerFHalOrg.7-13/4, 41
$C_3F_8H_2N_2S$	$(CF_3)_2C(NH_2)NSF_2$	F:	PerFHalOrg.7-2, 29
$C_3F_8H_2N_4S$	$SF_5N(CH_2)C(N_3)CF_3$	S:	S-N-Verb.1-148
$C_3F_8H_4S$	$CH_2FCF_2CH_2SF_5$	S:	SVol.2-200
$C_3F_8N_2$	$CF_2NFCF_2NFCF_2$	F:	PerFHalOrg.5-61/2, 84, 99
–	$NFCF_2CF_2CF_2NF$	F:	PerFHalOrg.5-60/1, 83
$C_3F_8N_2O$	$NF_2C(O)N(CF_3)_2$	C:	MVol.D3-136
$C_3F_8N_2O_2$	$(CF_3)_2NOC(O)NF_2$	F:	PerFHalOrg.7-94/5, 111/2
$C_3F_8N_2S_2$	$(CF_3)_2CNSNSF_2$	S:	SVol.2-78
$C_3F_8N_4O$	$NFCF(NF_2)NFC(O)NFCF_2$	F:	PerFHalOrg.6-77, 98
$C_3F_8P_3^+$	$C_3F_8P_3^+$	F:	PerFHalOrg.3-12
$C_3F_9FeH_5IP_3$	$C_3H_5Fe(PF_3)_3I$	Fe:	Org.Verb.B5-2/3
$C_3F_9FeH_5P_3$	$C_3H_5Fe(PF_3)_3$	Fe:	Org.Verb.B5-4
$C_3F_9Ge^+$	$GeF(CF_3)_2CF_2^+$	F:	PerFHalOrg.4-174
$C_3F_9HP_2$	$(CF_3)_2PP(H)CF_3$	F:	PerFHalOrg.3-16, 19
$C_3F_9H_2P$	$(CF_3)_3PH_2$	F:	PerFHalOrg.3-31
$C_3F_9H_2P_3$	$CF_3P(H)P(CF_3)P(H)CF_3$	F:	PerFHalOrg.3-16, 19, 23
$C_3F_9H_3NaO_3Sb$	$Na[Sb(CF_3)_3(OH)_3]$	F:	PerFHalOrg.3-231
$C_3F_9H_3O_3Sb^-$	$[Sb(CF_3)_3(OH)_3]^-$	F:	PerFHalOrg.3-226, 230/1
$C_3F_9H_4O_3Sb$	$H[(CF_3)_3Sb(OH)_3]$	F:	PerFHalOrg.3-226, 230
$C_3F_9H_6O_4Sb$	$[H_3O][Sb(CF_3)_3(OH)_3]$	F:	PerFHalOrg.3-230
C_3F_9ISi	$(CF_3)_3SiI$	F:	PerFHalOrg.4-147
C_3F_9ISn	$(CF_3)_3SnI$	F:	PerFHalOrg.4-181/2, 187
C_3F_9NO	$(CF_3)_2NOCF_3$	F:	PerFHalOrg.7-93/4, 109, 124/6
C_3F_9NS	$(CF_3)_2CFNSF_2$	S:	SVol.2-91
$C_3F_9N_3$	$FN(CF_2NF)_2CF_2$	F:	PerFHalOrg.6-75/6, 94/7
$C_3F_9N_5$	$(F_2NCF)_2NNC(NF_2)F$	F:	PerFHalOrg.5-3, 9
$C_3F_9N_5O$	$FN(CF(NF_2)NF)_2C(O)$	F:	PerFHalOrg.6-77, 98
C_3F_9NiOP	$(CF_3)_2PF_3Ni(CO)$	F:	PerFHalOrg.3-121
C_3F_9OP	$(CF_3)_3PO$	F:	PerFHalOrg.3-31/7
$C_3F_9OP_2^+$	$(CF_3)_2POP(CF_3)^+$	F:	PerFHalOrg.3-33
$C_3F_9O_3S$	$(CF_3O)_3S$	C:	MVol.D6-100
C_3F_9P	$(CF_3)_3P$	F:	PerFHalOrg.3-163, 165/6, 168, 172/4
–	$(n\text{-}C_3F_7)PF_2$	F:	PerFHalOrg.3-70/1, 84
C_3F_9PS	$(CF_3)_3PS$	F:	PerFHalOrg.3-137/8, 144, 151, 153
C_3F_9PSe	$(CF_3)_2PSeCF_3$	F:	PerFHalOrg.3-160
$C_3F_9P_2^+$	$(CF_3)_2PP(CF_3)^+$	F:	PerFHalOrg.3-179
$C_3F_9P_3S_5$	$[CF_3P(S)]_3S_2$	F:	PerFHalOrg.3-14
–	$(CF_3P(S)S)_2P(S)CF_3$	F:	PerFHalOrg.3-10, 13
$C_3F_9P_4S^+$	$(CF_3)_3P_4S^+$	F:	PerFHalOrg.3-12
C_3F_9Sb	$(CF_3)_3Sb$	F:	PerFHalOrg.3-227/9, 231
$C_3F_{10}FeH_6O_3P_4$	$COFeP_4F_{10}(OCH_3)_2$	Fe:	Org.Verb.B1-28/9

$C_3F_{10}N_4$	$NFCF(NF_2)NFCF_2NFCF_2$	F:	PerFHalOrg.6-77, 98
$C_3F_{10}Si$	$C_3F_7SiF_3$	F:	PerFHalOrg.4-147
$C_3F_{11}N_5$	$NFCF(NF_2)NFCF(NF_2)NFCF_2$	F:	PerFHalOrg.6-77, 98
$C_3F_{11}P$	$(CF_3)_3PF_2$	F:	PerFHalOrg.3-71, 76/7, 87, 98/9, 109
–	$(n-C_3F_7)PF_4$	F:	PerFHalOrg.3-71, 76/7, 86
$C_3F_{12}FeH_3NP_4$	$NCCHCH_2Fe(PF_3)_4$	Fe:	Org.Verb.B4-178, 182
$C_3F_{12}N_6$	$FN(CF(NF_2)NF)_2CF(NF_2)$	F:	PerFHalOrg.6-77, 98
$C_3F_{12}P^-$	$[(CF_3)_3PF_3]^-$	F:	PerFHalOrg.3-129, 134, 136/7
$C_3F_{12}Sb^-$	$[(CF_3)_3SbF_3]^-$	F:	PerFHalOrg.3-230
$C_3F_{13}NS$	$SF_5N(CF_3)C_2F_5$	S:	SVol.2-288
$C_3F_{13}NS_2$	$(CF_3)_2CFSF_4NSF_2$	S:	SVol.2-78
$C_3F_{16}FeOP_4$	$COFeP_4F_{10}(CF_3)_2$	Fe:	Org.Verb.B1-28/9
C_3FeHNO_4	$HFe(CO)_3NO$	Fe:	Org.Verb.B1-186
$C_3FeH_2^+$	$[FeC_3H_2]^+$	Fe:	Org.Verb.A1-93/4
$C_3FeH_3^+$	$[FeC_3H_3]^+$	Fe:	Org.Verb.A1-93/4
$C_3FeH_3I_2O_3P$	$(CO)_3Fe(PH_3)I_2$	Fe:	Org.Verb.B1-180/2
$C_3FeH_6N_2O_3$	$(CO)_3Fe(NH_3)_2$	Fe:	Org.Verb.B1-148
		Fe:	Org.Verb.C2-7
$C_3FeH_6N_2O_5S_2$	$[Fe(CH_3SO_2N)_2CO \cdot H_2O]_n$	Fe:	Org.Verb.C2-47
$C_3FeH_6N_4O^+$	$[(CO)Fe(CH_3NNNNCH_3)]^+$	Fe:	Org.Verb.B1-169
C_3FeH_9Li	$Li[Fe(CH_3)_3]$	Fe:	Org.Verb.B1-10
$C_3FeH_9NO_4P^+$	$[Fe(NO)P(OCH_3)_3]^+$	Fe:	Org.Verb.B1-47/8
$C_3FeH_9N_2O_5P^+$	$[Fe(NO)_2P(OCH_3)_3]^+$	Fe:	Org.Verb.B1-47/8
$C_3FeHgINO_4$	$(CO)_3Fe(NO)HgI$	Fe:	Org.Verb.B1-196/7
$C_3FeHg_2I_2O_3$	$(CO)_3FeHg_2I_2$	Fe:	Org.Verb.B1-192
$C_3FeI_4O_3P_8S_6$	$(CO)_3Fe(P_4S_3I_2)_2$	Fe:	Org.Verb.B3-124
C_3FeKNO_4	$K[Fe(CO)_3NO]$	Fe:	Org.Verb.B1-188
$C_3FeNNaO_4$	$Na[Fe(CO)_3NO]$	Fe:	Org.Verb.B1-186/8
$C_3FeNO_4^-$	$[Fe(CO)_3NO]^-$	Fe:	Org.Verb.B1-186
		Fe:	Org.Verb.B5-19/20
C_3FeNO_4Tl	$Tl[Fe(CO)_3NO]$	Fe:	Org.Verb.B1-188
$C_3FeN_2O_3$	$(CO)_3FeN_2$	Fe:	Org.Verb.B1-146
		Fe:	Org.Verb.B3-87
C_3FeO_3	$Fe(CO)_3$	Fe:	Org.Verb.B1-146/7
$C_3FeO_3^+$	$Fe(CO)_3^+$	Fe:	Org.Verb.B1-146
		Fe:	Org.Verb.B3-49, 78
$C_3FeO_3^{2+}$	$[Fe(CO)_3]^{2+}$	Fe:	Org.Verb.B3-78
$C_3FeO_3^-$	$[Fe(CO)_3]^-$	Fe:	Org.Verb.B3-78/9, 91
$C_3FeO_3P_2$	$[(CO)_3FeP_2]_n$	Fe:	Org.Verb.B3-108
$C_3FeO_{10}S_2$	$(CO)_3FeO(SO_3)_2$	Fe:	Org.Verb.B1-147
$C_3FeO_{15}P_8$	$(CO)_3Fe(P_4O_6)_2$	Fe:	Org.Verb.B1-148/50, 155, 163/4
$C_3FeO_{15}P_8^+$	$[(CO)_3Fe(P_4O_6)_2]^+$	Fe:	Org.Verb.B1-164
$C_3Fe_2H_{10}N_2O_6$	$Fe_2(NO)_2(CH_3O)_3OH$	Fe:	Org.Verb.B3-118
$C_3Fe_2O_3^+$	$[Fe_2(CO)_3]^+$	Fe:	Org.Verb.C2-26
$C_3GaH_6N_3S_6$	$Ga[S_2CNH_2]_3$	C:	MVol.D6-140
$C_3HNO_2S_2$	$C(O)SC(S)NHC(O)$	C:	MVol.D6-139

$C_3HN_3S_3$	$C_2S_3N_2 \cdot HCN$	C:	MVol.D6-42
–	$C_3S_3N_3H \cdot H_2O$	C:	MVol.D5-163
–	$H(SCN)_3$	C:	MVol.D6-42
–	$S_2(CN)_2 \cdot HNCS$	C:	MVol.D6-35
$C_3H_2N_4S_3$	$[NH(NCS)]_2CS$	C:	MVol.D6-160
$C_3H_2O_4Sn$	$Sn(CH_2(COO)_2)$	Sn:	MVol.C2-225
$C_3H_3KO_6Sn$	$KSn(HCOO)_3$	Sn:	MVol.C3-72/5
$C_3H_3NNiO_3$	$(CO)_3NiNH_3$	Ni:	Org.Verb.1-160
$C_3H_3NOS_2$	$CH_2SC(S)NHC(O)$	C:	MVol.D6-139
$C_3H_3NS_2$	$SC(SH)NCHCH$	C:	MVol.D6-139
$C_3H_3N_3OS_2$	$C_3H_3N_3(O)(S)_2$	C:	MVol.D6-46, 47/50
$C_3H_3N_3O_2S$	$C_3H_3N_3(O)_2(S)$	C:	MVol.D6-46, 47/50
$C_3H_3N_3O_3$	$C_3N_3(OH)_3$	C:	MVol.D3-283
$C_3H_3N_3S_3$	$C_3H_3N_3(S)_3$	C:	MVol.D6-46/50
–	$(HNCS)_3$	C:	MVol.D6-37
$C_3H_3N_3Se_3$	$C_3H_3N_3(Se)_3$	C:	MVol.D6-241
$C_3H_3NaO_6Sn$	$NaSn(HCOO)_3$	Sn:	MVol.C3-34
$C_3H_3NiO_3P$	$(CO)_3NiPH_3$	Ni:	Org.Verb.1-161
$C_3H_3O_6RbSn$	$RbSn(HCOO)_3$	Sn:	MVol.C3-140
$C_3H_3O_6Sn^-$	$Sn(HCOO)_3^-$	Sn:	MVol.C2-218
$C_3H_4N_2Na_2O_9U$	$Na_2[UO_2(O_2)C_2O_4((NH_2)_2CO)] \cdot n\ H_2O$	U:	SVol.E1-84, 91
$C_3H_4N_2O_6SU$	$UO_2(OH)_2 \cdot NHC(O)C(O)NHCS$	U:	SVol.E1-221
$C_3H_4N_2O_7U$	$UO_2(C_2O_4) \cdot (NH_2)_2CO$	U:	SVol.E1-84, 86, 89
–	$UO_2(C_2O_4) \cdot (NH_2)_2CO \cdot H_2O$	U:	SVol.E1-86, 89
$C_3H_4N_4OS$	$C_3H_2N_3(NH_2)(S)(O)$	C:	MVol.D6-43, 45/6
$C_3H_4N_4S_2$	$C_3H_2N_3(NH_2)(S)_2$	C:	MVol.D6-43, 44
$C_3H_4N_5NaS$	$C_3N_3(NH_2)_2SNa$	C:	MVol.D6-43
$C_3H_5KOS_2$	$K[SOCSC_2H_5]$	C:	MVol.D4-263
–	$K[S_2COC_2H_5]$	C:	MVol.D4-245, 251/3
$C_3H_5KOSe_2$	$K[Se_2COC_2H_5]$	C:	MVol.D6-219, 221
$C_3H_5KO_2S$	$K[SOCOC_2H_5]$	C:	MVol.D4-234/5, 237
–	$K[SOCOC_2D_5]$	C:	MVol.D4-234
$C_3H_5KS_3$	$K[S_2CSC_2H_5]$	C:	MVol.D4-267/9
$C_3H_5MoOS_4$	$MoS_2[S_2COC_2H_5]$	C:	MVol.D4-258
$C_3H_5MoO_2S_2$	$MoO[S_2COC_2H_5] \cdot C_5H_5N$	C:	MVol.D4-258
C_3H_5NSe	C_2H_5SeCN	C:	MVol.D6-227
$C_3H_5N_5S$	$C_3HN_3(NH_2)_2(S)$	C:	MVol.D6-43/4
$C_3H_5NaOSSe$	$Na[SSeCOC_2H_5]$	C:	MVol.D6-222
$C_3H_5NaOS_2$	$Na[S_2COC_2H_5]$	C:	MVol.D4-252
		C:	MVol.D6-170
–	$Na[S_2COC_2H_5] \cdot 2\ H_2O$	C:	MVol.D4-245
$C_3H_5NaO_2S$	$Na[SOCOC_2H_5]$	C:	MVol.D4-235
$C_3H_5NaS_3$	$Na[S_2CSC_2H_5]$	C:	MVol.D4-269
$C_3H_5NpO_5$	$NpO_2(CH_3CH(OH)COO)$	Np:	TrU.D1-155
$C_3H_5ORbS_2$	$Rb[S_2COC_2H_5]$	C:	MVol.D4-253
$C_3H_5OSSe^-$	$[SSeCOC_2H_5]^-$	C:	MVol.D6-222
$C_3H_5OS_2^-$	$[S_2COC_2H_5]^-$	C:	MVol.D4-244/5, 247/52
$C_3H_5OS_2Tl$	$Tl[S_2COC_2H_5]$	C:	MVol.D4-253
$C_3H_5OSe_2^-$	$[Se_2COC_2H_5]^-$	C:	MVol.D6-219/20
$C_3H_5O_2S^-$	$[SOCOC_2H_5]^-$	C:	MVol.D4-233/7

$C_3H_5O_2STl$	$Tl[SOCOC_2H_5]$	C:	MVol.D4-237
$C_3H_5O_2S_2^-$	$[S_2COC_2H_4OH]^-$	C:	MVol.D4-244, 250
$C_3H_5O_2S_2Sb$	$SbO[S_2COC_2H_5]$	C:	MVol.D4-254
$C_3H_5O_2Se^-$	$[OSeCOC_2H_5]^-$	C:	MVol.D6-221
$C_3H_5O_3Sn^+$	$Sn(OCOCH(OH)CH_3)^+$	Sn:	MVol.C2-225
$C_3H_5S_3^-$	$[S_2CSC_2H_5]^-$	C:	MVol.D4-268
$C_3H_5Sn^+$	$Sn(CH_2CHCH_2)^+$	Sn:	Org.Verb.1-134
$C_3H_6HgOS_2$	$CH_3Hg[S_2COCH_3]$	C:	MVol.D4-258
$C_3H_6InN_3S_6$	$In[S_2CNH_2]_3$	C:	MVol.D6-140
$C_3H_6IrN_5$	$Ir(CN)_3 \cdot 2\ NH_3 \cdot 1.5\ H_2O$	Ir:	SVol.2-181
$C_3H_6MoO_4$	$MoO_2(O(CH_2)_3O)$	Mo:	SVol.B1-141
$C_3H_6NO_2SSi^+$	$SONCOSi(CH_3)_2{}^+$	S:	S-N-Verb.1-25
$C_3H_6N_2O_4S$	$(NH_2)_2CS \cdot H_2C_2O_4$	C:	MVol.D6-126
$C_3H_6N_2O_4SSn$	$Sn(HCOO)_2 \cdot CS(NH_2)_2$	Sn:	MVol.C5-55
$C_3H_6N_2O_9U$	$UO_2(NO_3)_2 \cdot (CH_3)_2CO \cdot 2\ H_2O$	U:	SVol.E1-79/80
–	$UO_2(NO_3)_2 \cdot (CH_3)_2CO \cdot 3\ H_2O$	U:	SVol.E1-79/80
$C_3H_6N_3O_3S_2Si^+$	$[NSO_2][NSNCOSi(CH_3)_2]^+$	S:	S-N-Verb.1-25
$C_3H_6N_3O_5S_2^-$	$[NSO_2NCH_3CONCH_3SO_2]^-$	S:	S-N-Verb.1-24
$C_3H_6N_3S_6Sb$	$Sb[S_2CNH_2]_3$	C:	MVol.D6-140
C_3H_6Ni	C_3H_5NiH	Ni:	Org.Verb.2-2/3
$C_3H_6OS_2$	$SC(OC_2H_5)(SH)$	C:	MVol.D4-238/43
$C_3H_6OSe_2$	$SeC(OC_2H_5)(SeH)$	C:	MVol.D6-219
$C_3H_6O_2S$	$OC(OC_2H_5)(SH)$	C:	MVol.D4-232
–	$OC(SC_2H_5)(OH)$	C:	MVol.D4-232
$C_3H_6O_2Se$	$HOSeCOC_2H_5$	C:	MVol.D6-221
$C_3H_6O_3Sn$	$[(CH_3)_2SnCO_3]_x$	Sn:	Org.Verb.6-28
$C_3H_6O_7SU$	$UO_2SO_4 \cdot (CH_3)_2CO \cdot 2\ H_2O$	U:	SVol.E1-79/80
$C_3H_6SSe_2$	$SC(SeCH_3)_2$	C:	MVol.D6-215, 218
$C_3H_6S_2Se$	$SC(SCH_3)(SeCH_3)$	C:	MVol.D6-218
$C_3H_6S_3$	$SC(SCH_3)_2$	C:	MVol.D6-218
–	$SC(SC_2H_5)(SH)$	C:	MVol.D4-264/6
C_3H_7INiO	$C_3H_5Ni(H_2O)I$	Ni:	Org.Verb.2-280
$C_3H_7K_2NS_4$	$K_2CS_4 \cdot (CH_3)_2NH$	C:	MVol.D4-231
$C_3H_7NNa_2S_4$	$Na_2CS_4 \cdot (CH_3)_2NH$	C:	MVol.D4-231
–	$Na_2CS_4 \cdot C_2H_5NH_2$	C:	MVol.D4-231
$C_3H_7NOS_2$	$HNCS \cdot (CH_3)_2SO$	C:	MVol.D6-36
$C_3H_7NO_6SU$	$UO_2SO_3 \cdot HCON(CH_3)_2 \cdot n\ H_2O$	U:	SVol.E1-100/2
$C_3H_7NO_6Sn$	$NH_4Sn(HCOO)_3$	Sn:	MVol.C3-123/4
$C_3H_7NS_3$	$SC(SCH_2CH_2NH_2)(SH)$	C:	MVol.D4-264/5
$C_3H_7N_3O_5S_2$	$[CH_3NSO_2][HNSO_2][CH_3NCO]$	S:	S-N-Verb.1-23/4
$C_3H_7Sn^+$	$SnC_3H_7{}^+$	Sn:	Org.Verb.1-86
		Sn:	Org.Verb.5-24
$C_3H_{7.5}N_{1.5}O_{6.5}SU$	$UO_2SO_3 \cdot 1.5\ CH_3CONH_2 \cdot 0.5\ H_2O$	U:	SVol.E1-96/8
$C_3H_8N_2O_4S$	$C_2H_5OC(O)NHSO_2NH_2$	S:	S-N-Verb.1-167
$C_3H_8N_2O_6SU$	$UO_2SO_3 \cdot CO(NHCH_3)_2 \cdot 0.5\ H_2O$	U:	SVol.E1-93/5
$C_3H_8N_2O_9Sn_2$	$(NH_4)_2Sn_2(CO_3)_3 \cdot 3\ H_2O$	Sn:	MVol.C3-123
$C_3H_8N_2S_3Zn$	$Zn(C_2H_8N_2)CS_3$	C:	MVol.D4-224
$C_3H_8N_6O_2S$	$[NH_2C(O)NHNH]_2CS$	C:	MVol.D6-160
$C_3H_8N_6S$	$NH_2C(NH)NHC(NH)NHC(S)NH_2$	C:	MVol.D6-43
C_3H_8OSn	$[(CH_3)(C_2H_5)SnO]_x$	Sn:	Org.Verb.6-196

C_3H_8Ti	$CH_2Ti(CH_3)_2$	Ti:	Org.Verb.1-122
C_3H_9ITi	$(CH_3)_3TiI$	Ti:	Org.Verb.1-84
C_3H_9LiSn	$(CH_3)_3SnLi$	Sn:	Org.Verb.5-78
$C_3H_9NOSSi^+$	$(CH_3)_3SiNSO^+$	S:	S-N-Verb.1-198
$C_3H_9NOS_2$	$NH_4[S_2COC_2H_5]$	C:	MVol.D4-254
$C_3H_9NO_2S$	$HNCS \cdot 2\ CH_3OH$	C:	MVol.D6-35
–	$NH_4[SOCOC_2H_5]$	C:	MVol.D4-237
$C_3H_9NO_2Se$	$NH_4[OSeCOC_2H_5]$	C:	MVol.D6-221
$C_3H_9NO_3S$	$(CH_3)_3NSO_3$	S:	S-N-Verb.1-98
–	$NH_2SO_2OC_3H_7$	S:	S-N-Verb.1-97
$C_3H_9NO_4S_2Si$	$OSNSO_2OSi(CH_3)_3$	S:	S-N-Verb.1-98
$C_3H_9NSSi^+$	$(CH_3)_3SiNS^+$	S:	S-N-Verb.1-210, 211
$C_3H_9NS_7Sn$	$(CH_3)_3SnNS_7$	Sn:	Org.Verb.5-86
$C_3H_9N_2OSSi_2^+$	$(CH_3)_3SiNSONSi^+$	S:	S-N-Verb.1-198
$C_3H_9N_3O_2S$	$NH_2SO_2NHNC(CH_3)_2$	S:	S-N-Verb.1-222
$C_3H_9N_3O_3S$	$(CH_3)_2NSO_2NHCONH_2$	S:	S-N-Verb.1-171
$C_3H_9N_3O_3S_2Si^+$	$S_2N_3C_3H_9O_3Si^+$	S:	S-N-Verb.1-25
$C_3H_9N_3O_6S_3$	$[CH_3NSO_2]_3$	S:	S-N-Verb.1-9
C_3H_9NaSn	$(CH_3)_3SnNa$	Sn:	Org.Verb.5-78
$C_3H_9O_3Te^+$	$Te(OCH_3)_3{}^+$	Te:	SVol.B2-104
$C_3H_9Sn^+$	$HSn(CH_3)C_2H_5{}^+$	Sn:	Org.Verb.1-70/1
–	$Sn(CH_3)_3{}^+$	Sn:	Org.Verb.1-47, 50
		Sn:	Org.Verb.4-10
		Sn:	Org.Verb.5-84
–	$Sn(C_3H_7)H_2{}^+$	Sn:	Org.Verb.1-86
C_3H_9Ti	$(CH_3)_3Ti$	Ti:	Org.Verb.1-82
$C_3H_9Ti^+$	$[(CH_3)_3Ti]^+$	Ti:	Org.Verb.1-86
–	$[(CD_3)_3Ti]^+$	Ti:	Org.Verb.1-86
$C_3H_{10}N_2O_2S$	$(CH_3)_2NSO_2NHCH_3$	S:	S-N-Verb.1-170
$C_3H_{10}N_4O_8U$	$UO_2(NO_3)_2 \cdot NH_2(CH_2)_3NH_2$	U:	SVol.E1-23/5
$C_3H_{10}OSn$	$(CH_3)_3SnOH$	Sn:	Org.Verb.5-84
$C_3H_{10}O_5U$	$UO_2(OCH_3)_2 \cdot CH_3OH$	U:	SVol.E1-59, 63/4
$C_3H_{10}Sn$	$CH_3(C_2H_5)SnH_2$	Sn:	Org.Verb.4-119
–	$(CH_3)_3SnH$		
	Chemical reactions	Sn:	Org.Verb.4-10/8
	Formation and preparation	Sn:	Org.Verb.4-6/7
	Molecule	Sn:	Org.Verb.4-7/9
	Physical properties	Sn:	Org.Verb.4-10
	Spectra	Sn:	Org.Verb.4-7/9
–	$C_3H_7SnH_3$	Sn:	Org.Verb.4-125
–	$(i\text{-}C_3H_7)SnH_3$	Sn:	Org.Verb.4-125/6
$C_3H_{10}Sn^+$	$(CH_3)_3SnH^+$	Sn:	Org.Verb.4-10
$C_3H_{11}N_3O_4S_2$	$CH_3N(SO_2NHCH_3)_2$	S:	S-N-Verb.1-162
–	$NH_2SO_2NHSO_2NH(i\text{-}C_3H_7)$	S:	S-N-Verb.1-161
–	$NH_2SO_2NSO_2N(CH_3)_3$	S:	S-N-Verb.1-160
$C_3H_{11}PSn$	$(CH_3)_3SnPH_2$	Sn:	Org.Verb.5-85
$C_3H_{12}MnN_8O_9$	$Mn(NO_3)_2 \cdot 3\ CO(NH_2)_2 \cdot 3\ H_2O$	Mn:	MVol.C3-291
$C_3H_{12}N_4O_9U$	$(NH_4)_2[UO_2(O_2)C_2O_4((NH_2)_2CO)] \cdot n\ H_2O$	U:	SVol.E1-84, 91
$C_3H_{12}N_6O_9SU$	$UO_2SO_4 \cdot 3\ (NH_2)_2CO$	U:	SVol.E1-84/6, 89
$C_3H_{12}N_6S_3Te^{2+}$	$[Te(SC(NH_2)_2)_3]_2{}^{4+}$	Te:	SVol.B3-142/3

$C_3H_{12}N_6Se_3^{2+}$	$[(NH_2)_2CSe]_3^{2+}$	C:	MVol.D6-247
$C_3H_{12}N_8O_{11}U$	$UO_2(NO_3)_2 \cdot 3\ (NH_2)_2CO$	U:	SVol.E1-83, 85, 88/9
$C_3H_{12}SiSn$	$(CH_3)_3SnSiH_3$	Sn:	Org.Verb.5-84
$C_3H_{16}N_4O_{11}U$	$(NH_4)_4UO_2(CO_3)_3$	U:	SVol.C3-56
$C_3HgN_3Se_3^-$	$[Hg(SeCN)_3]^-$	C:	MVol.D6-227
C_3IIrO_3	$IrI(CO)_3$	Ir:	SVol.2-151
$C_3IN_3S_3$	$I(SCN)_3$	C:	MVol.D6-178
$C_3INaNiO_3$	$Na[(CO)_3NiI]$	Ni:	Org.Verb.1-191, 192
$C_3I_3IrO_3$	$IrI_3(CO)_3$	Ir:	SVol.2-151
$C_3I_3N_3$	$C_3N_3I_3$	C:	MVol.D3-288
$C_3I_3N_3O_3$	$(INCO)_3$	C:	MVol.D3-146, 147/8
$C_3I_3N_3S_3$	$(ISCN)_3$	C:	MVol.D6-178
$C_3I_6Ir_2O_3$	$Ir_2I_6(CO)_3$	Ir:	SVol.2-151
$C_3In_2S_9$	$In_2(CS_3)_3$	C:	MVol.D4-223
C_3IrN_3	$Ir(CN)_3$	Ir:	SVol.2-180
$C_3KN_3S_3$	$K(SCN)_3$	C:	MVol.D6-42
$C_3KN_3Se_3$	$K(SeCN)_3 \cdot 0.5\ H_2O$	C:	MVol.D6-232, 239
$C_3K_4O_{11}U$	$K_4[UO_2(CO_3)_3]$	U:	SVol.C2-156, 263
$C_3La_2O_9$	$La_2(CO_3)_3 \cdot 8\ H_2O$	Sc:	MVol.B7-27
$C_3Li_4O_{11}U$	$Li_4[UO_2(CO_3)_3]$	U:	SVol.C2-263
$C_3N_2NiO_3$	$(CO)_3NiN_2$	Ni:	Org.Verb.1-160
$C_3N_2OS_2$	$OC(NCS)_2$	C:	MVol.D3-46, 100
$C_3N_2O_3$	$CO(NCO)_2$	C:	MVol.D3-100, 141, 143, 147
$C_3N_3PuS_3$	$Pu(SCN)_3$	Np:	TrU.D1-144
$C_3N_3S^-$	$[S(CN)_3]^-$	C:	MVol.D5-221
$C_3N_3S_3^-$	$(SCN)_3^-$	C:	MVol.D5-211
$C_3N_3S_3Sn^-$	$Sn(SCN)_3^-$	Sn:	MVol.C2-228
$C_3N_3Se_3^-$	$(SeCN)_3^-$	C:	MVol.D6-232, 236
$C_3N_3Te^-$	$Te(CN)_3^-$	C:	MVol.D6-261
$C_3Na_4O_{11}U$	$Na_4[UO_2(CO_3)_3]$	U:	SVol.C2-247
$C_3Nd_2O_9$	$Nd_2(CO_3)_3$	Sc:	MVol.B7-26
C_3NiO_3	$Ni(CO)_3$	Ni:	Org.Verb.1-159
$C_3NiO_3^+$	$Ni(CO)_3^+$	Ni:	Org.Verb.1-215
$C_3NiO_3^-$	$Ni(CO)_3^-$	Ni:	Org.Verb.1-215
$C_3NiO_3^{2-}$	$[Ni(CO)_3]^{2-}$	Ni:	Org.Verb.1-228
$C_3NiO_9P_4$	$(CO)_3Ni(P_4O_6)$	Ni:	Org.Verb.1-185, 186
C_3OS	C_3OS	C:	MVol.D5-1
$C_3O_2S_3$	$C(O)SC(S)SC(O)$	C:	MVol.D4-215
$C_3O_{11}Rb_4U$	$Rb_4[UO_2(CO_3)_3]$	U:	SVol.C2-263
$C_3O_{11}U^{4-}$	$[UO_2(CO_3)_3]^{4-}$	U:	SVol.C2-283, 299
C_3Pu_2	Pu_2C_3	Np:	TrU.B2-25/6, 41
C_3S^+	$CCCS^+$	C:	MVol.D4-6
C_3S_2	C_3S_2	C:	MVol.D4-1/7
–	$(C_3S_2)_x$	C:	MVol.D4-6, 153
$C_3S_2^+$	$C_3S_2^+$	C:	MVol.D4-6
$C_3S_2^{2+}$	$C_3S_2^{2+}$	C:	MVol.D4-6
C_3S_5	$SCSC(S)C(S)S$	C:	MVol.D4-222
$C_{3.3}Cl_2H_{3.3}N_{1.1}Ni$	$(CH_2CHCN)_{1.1}NiCl_2$	Ni:	Org.Verb.1-339
$C_{3.4}H_{14.6}IN_{6.8}O_{6.4}U$	$UO_2(OH)I \cdot 3.4\ (NH_2)_2CO$	U:	SVol.E1-83, 87

$C_{3.4}H_{14.6}I_{1.96}N_{6.8}O_{6.4}U$	$UO_2(OH)I \cdot 3.4\ (NH_2)_2CO \cdot 0.96\ I$	U:	SVol.E1-83, 87
$C_{3.6}F_4H_{10.8}N_{1.2}Sn$	$SnF_4 \cdot 1.2\ (CH_3)_3N$	Sn:	MVol.C5-171
$C_4CaH_6O_4S_2$	$Ca[SOCOCH_3]_2$	C:	MVol.D4-237
C_4CaO_8SSn	$CaSnS(C_2O_4)_2$	Sn:	MVol.C3-179
$C_4CdF_{12}P_2S_4$	$Cd[S(S)P(CF_3)_2]_2$	F:	PerFHalOrg.3-159
$C_4CdFeH_6N_2O_4$	$Cd(NH_3)_2Fe(CO)_4$	Fe:	Org.Verb.B2-13, 27
C_4CdFeO_4	$CdFe(CO)_4$	Fe:	Org.Verb.B2-13
C_4CdO_8SSn	$CdSnS(C_2O_4)_2$	Sn:	MVol.C4-60
$C_4CeCl_5H_{16}N_2$	$CeCl_3 \cdot 2\ [(CH_3)_2NH_2]Cl$	Sc:	MVol.C5-213/5
$C_4Cl_{0.5}H_{16}N_{9.5}O_{10.5}U$	$UO_2(NO_3)_{1.5}Cl_{0.5} \cdot 4\ (NH_2)_2CO \cdot H_2O$	U:	SVol.E1-85
$C_4Cl_{0.5}H_{18}N_{9.5}O_{11.5}U$	$[UO_2((NH_2)_2CO)_4(H_2O)]Cl_{0.5}(NO_3)_{1.5}$	U:	SVol.E1-83, 85, 87
$C_4ClFH_6NO_4P$	$OP(ONCFCl)(OCH(CH_3)CH_2O)$	F:	PerFHalOrg.7-202
$C_4ClFH_7NO_3S$	$FSO_2NCCl(OC_3H_7)$	S:	S-N-Verb.1-118
$C_4ClF_2FeO_4P$	$(CO)_4FePF_2Cl$	Fe:	Org.Verb.B2-85, 93, 114
$C_4ClF_2HN_2O$	$NCClCFNC(OH)CF$	F:	PerFHalOrg.6-29/30, 50
$C_4ClF_2H_2N_3$	$NCFNC(NH_2)CClCF$	F:	PerFHalOrg.6-21/2, 42
–	$NC(NH_2)NCFCClCF$	F:	PerFHalOrg.6-21/2, 42
$C_4ClF_2H_3N_4$	$NCClCFNC(NHNH_2)CF$	F:	PerFHalOrg.6-28/30, 50, 71
$C_4ClF_2H_7NO_3P$	$CF_2NOPCl(O)OCH(CH_3)_2$	F:	PerFHalOrg.7-199/200
$C_4ClF_2H_{10}N_4O_2PS_2$	$[NS(O)F]_2[NPClN(C_2H_5)_2]$	S:	S-N-Verb.1-27
$C_4ClF_2H_{12}N_4O_2PS_2Si$	$[NS(O)F]_2[NPClN(CH_3)Si(CH_3)_3]$	S:	S-N-Verb.1-27
$C_4ClF_3HN_5O_2S$	$FSO_2NCClNHC(NCF)_2N$	S:	S-N-Verb.1-106
$C_4ClF_3H_2N_2$	$CF_3C(NH_2)CClCN$	F:	PerFHalOrg.7-3, 29
$C_4ClF_3H_3NO_2$	$CF_3N(OH)CHCHC(O)Cl$	F:	PerFHalOrg.7-164, 168
$C_4ClF_3H_5NO$	$CF_3NOCH_2CH(CH_2Cl)$	F:	PerFHalOrg.7-167
–	$CF_3N(OH)CHCHCH_2Cl$	F:	PerFHalOrg.7-164, 166/7
$C_4ClF_3H_5N_2O_5P$	$NO_2CF_2CFNOPCl(O)OC_2H_5$	F:	PerFHalOrg.7-199/200
$C_4ClF_3N_2$	$NCClCFNCFCF$	F:	PerFHalOrg.6-28/30, 48, 60, 70
–	$NCFNCFCClCF$	F:	PerFHalOrg.6-19/20, 34, 41, 58/9, 72
$C_4ClF_4H_7Sn$	$(CH_3)_2(CHF_2CF_2)SnCl$	Sn:	Org.Verb.5-192
C_4ClF_4I	$IC(CF_2CF_2)CCl$	F:	PerFHalOrg.4-107
C_4ClF_4Li	$CLiCF_2CF_2CCl$	F:	PerFHalOrg.4-3, 13, 33
$C_4ClF_5H_4S$	$SF_5CHCHCHCHCl$	S:	SVol.2-288
$C_4ClF_5H_6O_2S$	$SF_5CH_2CHClOC(O)CH_3$	S:	SVol.2-200
$C_4ClF_5N_2$	$NCFNCFCFClCF_2$	F:	PerFHalOrg.6-19/20, 34, 40
$C_4ClF_6HN_3OP$	$NP(OCl)NHC(CF_3)NC(CF_3)$	F:	PerFHalOrg.6-94, 110
$C_4ClF_6H_2NO$	$(CF_3)_2CClC(O)NH_2$	F:	PerFHalOrg.7-18, 20/2, 47
$C_4ClF_6H_6NOSi$	$(CF_3)_2NOSiCl(CH_3)_2$	F:	PerFHalOrg.7-127, 144
C_4ClF_6NO	$CF_2CFNOCFClCF_2$	F:	PerFHalOrg.5-31, 40, 47
–	$CF_3NOCClCCF_3$	F:	PerFHalOrg.5-33/4, 41
–	$CF_3NO \cdot CF_3CCCl$	F:	PerFHalOrg.7-168
$C_4ClF_6NO_2$	$ONCF_2CF_2CF_2C(O)Cl$	F:	PerFHalOrg.7-170/9, 185
$C_4ClF_6NO_3S$	$[NS(O)Cl][OC(CF_3)_2CO]$	F:	PerFHalOrg.5-57/8, 78/9
		S:	S-N-Verb.1-41
$C_4ClF_6O_{1.5}Si$	$[CF_3C(Cl)C(CF_3)SiO_{1.5}]_n$	F:	PerFHalOrg.4-147/8

C_4ClF_6P	$(CF_2CF)_2PCl$	F:	PerFHalOrg.3-73, 91
C_4ClF_7HNO	$C_3F_7C(Cl)NOH$	F:	PerFHalOrg.7-88, 105
$C_4ClF_7H_2N_2$	$(CF_2ClCF_2)(CF_3)C(NH)_2$	F:	PerFHalOrg.5-6
$C_4ClF_7H_4N_2$	$C_3F_7C(NH)NH_2 \cdot HCl$	F:	PerFHalOrg.7-3/4, 31
$C_4ClF_8H_4NS$	$SF_5N(CH_2)C(CH_2Cl)CF_3$	S:	S-N-Verb.1-148
C_4ClF_8NO	$(CF_2ClCF_2)(CF_3)NC(O)F$	F:	PerFHalOrg.3-109
–	$CF_2ClCF_2NOCF_2CF_2$	F:	PerFHalOrg.5-34, 42
–	$ClCF_2(CF_2)_3NO$	F:	PerFHalOrg.7-178, 184
–	$[N(CF_2CF_2Cl)OCF_2CF_2]_n$	F:	PerFHalOrg.7-208/9
$C_4ClF_8NO_2$	$CF_2ClCF_2(CF_3)NOC(O)F$	F:	PerFHalOrg.3-109
$C_4ClF_8N_2O_3$	$NO_2CF_2CF_2NOCF_2CF_2Cl$	F:	PerFHalOrg.7-103
$C_4ClF_8N_3O_4$	$NO_2CF_2CF_2NONOCF_2CF_2Cl$	F:	PerFHalOrg.7-106
$C_4ClF_9H_3N$	$(CF_3)_3CNH_2 \cdot HCl$	F:	PerFHalOrg.7-1/2, 28
$C_4ClF_{10}P$	$(C_2F_5)_2PCl$	F:	PerFHalOrg.3-72/3, 90
$C_4ClF_{11}N_2O$	$(CF_3)_2NOCF_2NCl(CF_3)$	F:	PerFHalOrg.7-96
$C_4ClF_{12}NS$	$SF_5NCClC_3F_7$	S:	S-N-Verb.1-148
$C_4ClF_{13}NP$	$(CF_3)_2PFClN(CF_3)_2$	F:	PerFHalOrg.3-61, 65, 68
$C_4ClFeGeH_6NO_3$	$(CO)_2(NO)FeClGe(CH_3)_2$	Fe:	Org.Verb.B1-133
$C_4ClFeIO_4$	$(CO)_4Fe(Cl)I$	Fe:	Org.Verb.B2-69
$C_4ClH_4MnO_2^+$	$[MnCl(CH_3COCHCO)]^+$	Mn:	MVol.D1-102
C_4ClH_7NiO	$CH_3CHCHCH_2NiClO$	Ni:	Org.Verb.2-290
$C_4ClH_8NO_2S$	$ClSO_2N(CH_2)_4$	S:	S-N-Verb.1-124
$C_4ClH_8NO_3S$	$ClSO_2N(CH_2CH_2)_2O$	S:	S-N-Verb.1-124
$C_4ClH_8NO_4S$	$ClSO_2NHCOOCH(CH_3)_2$	S:	S-N-Verb.1-124/5
$C_4ClH_9I_2MgSn$	$(CH_3)_3SnCI_2MgCl$	Sn:	Org.Verb.2-17/8
$C_4ClH_9IrN_2O_5^-$	$[Ir(NH_2CH_2COO)_2Cl(OH)]^-$	Ir:	SVol.2-81
$C_4ClH_9N_2NiS$	$C_3H_5Ni(SC(NH_2)_2)Cl$	Ni:	Org.Verb.2-10
$C_4ClH_9N_4OS$	$[NS(CH_3)_2][NCO][HNC(NH_2)]Cl$	S:	S-N-Verb.1-31
C_4ClH_9NiO	$CH_3CHCHCH_2Ni(H_2O)Cl$	Ni:	Org.Verb.2-290
$C_4ClH_9O_2Sn$	$(CH_3)_2ClSnOCOCH_3$	Sn:	Org.Verb.6-22, 25
C_4ClH_9Sn	$(CH_3)_2(CH_2CH)SnCl$	Sn:	Org.Verb.5-192, 197
$C_4ClH_9Sn^+$	$C_4H_9SnCl^+$	Sn:	Org.Verb.5-124, 138
$C_4ClH_{10}ISn$	$(CH_3)_3SnCHClI$	Sn:	Org.Verb.2-8
–	$(C_2H_5)_2SnClI$	Sn:	Org.Verb.5-104
$C_4ClH_{10}NO_2S$	$ClSO_2N(C_2H_5)_2$	S:	S-N-Verb.1-124
–	$ClSO_2NHC_4H_9$	S:	S-N-Verb.1-123
$C_4ClH_{10}OTi$	$CH_3Ti(OC_3H_7\text{-}i)Cl$	Ti:	Org.Verb.1-9
$C_4ClH_{10}O_3PS$	$(C_2H_5O)_2POSCl$	S:	SVol.2-238
$C_4ClH_{10}Sn^+$	$(C_2H_5)_2SnCl^+$	Sn:	Org.Verb.5-98
–	$C_4H_9SnHCl^+$	Sn:	Org.Verb.5-124, 138
$C_4ClH_{11}I_2SiSn$	$(CH_3)_3SiCH_2SnClI_2$	Sn:	Org.Verb.6-254
$C_4ClH_{11}MgSn$	$(CH_3)_3SnCH_2MgCl$	Sn:	Org.Verb.2-14
$C_4ClH_{11}O_2Sn$	$C_4H_9Sn(OH)_2Cl$	Sn:	Org.Verb.6-246
$C_4ClH_{11}Sn$	$(CH_3)_2(C_2H_5)SnCl$	Sn:	Org.Verb.5-192
–	$(CH_3)_3SnCH_2Cl$	Sn:	Org.Verb.2-6
–	$(C_2H_5)_2SnHCl$	Sn:	Org.Verb.6-289
–	$C_4H_9SnH_2Cl$	Sn:	Org.Verb.6-296
$C_4ClH_{12}LiTi$	$Li[(CH_3)_4TiCl] \cdot 4\ C_4H_8O$	Ti:	Org.Verb.1-120
$C_4ClH_{12}NO_4$	$(CH_3)_4NClO_4$ solid solutions		
	$(CH_3)_4NClO_4\text{-}(CH_3)_4NMnO_4$	Mn:	MVol.C2-201

$C_4ClH_{12}N_2O_4PS$	$(C_2H_5)_2NSO_2NHPClOOH$	S:	S-N-Verb.1-186
$C_4ClH_{12}N_4O_4PS_2$	$[SO_2(NCH_3)_2]_2PCl$	S:	S-N-Verb.1-42/3
$C_4ClH_{12}Ti^-$	$[(CH_3)_4TiCl]^-$	Ti:	Org.Verb.1-120
$C_4ClH_{13}O_2Sn_2$	$(CH_3)_2ClSnOSn(CH_3)_2OH$	Sn:	Org.Verb.6-23
$C_4ClH_{16}IIrN_4{}^+$	$[Ir(C_2H_4(NH_2)_2)_2ICl]^+$	Ir:	SVol.2-83, 84
$C_4ClH_{16}IIrN_5O_3$	$[Ir(C_2H_4(NH_2)_2)_2ICl]NO_3$	Ir:	SVol.2-84
$C_4ClH_{16}I_2IrN_4$	$[Ir(C_2H_4(NH_2)_2)_2I_2]Cl \cdot H_2O$	Ir:	SVol.2-84
$C_4ClH_{16}I_2IrN_4O_4$	$[Ir(C_2H_4(NH_2)_2)_2I_2]ClO_4$	Ir:	SVol.2-84
$C_4ClH_{16}IrN_4$	$[Ir(C_2H_4(NH_2)_2)_2]Cl$	Ir:	SVol.2-82/3
$C_4ClH_{16}IrN_4Na_2O_{10}S_2$	$[Ir(SO_3Na)_2(C_2H_4(NH_2)_2)_2]ClO_4$	Ir:	SVol.2-160
$C_4ClH_{16}IrN_4O_3S$	$Ir(SO_3)Cl(C_2H_4(NH_2)_2)_2 \cdot 3\ H_2O$	Ir:	SVol.2-157
$C_4ClH_{16}IrN_6O_5$	$[Ir(NO_2)(C_2H_4(NH_2)_2)_2Cl](NO_3)$	Ir:	SVol.2-62
$C_4ClH_{16}IrN_6O_8$	$[Ir(NO_2)_2(C_2H_4(NH_2)_2)_2]ClO_4$	Ir:	SVol.2-61
$C_4ClH_{16}IrN_7{}^+$	$[Ir(C_2H_4(NH_2)_2)_2(N_3)Cl]^+$	Ir:	SVol.2-83
$C_4ClH_{16}IrN_{10}$	$[Ir(C_2H_4(NH_2)_2)_2(N_3)_2]Cl$	Ir:	SVol.2-41
$C_4ClH_{17}IrN_4O^+$	$[Ir(C_2H_4(NH_2)_2)_2(OH)Cl]^+$	Ir:	SVol.2-83
$C_4ClH_{18}IrN_4O^{2+}$	$[Ir(C_2H_4(NH_2)_2)_2(H_2O)Cl]^{2+}$	Ir:	SVol.2-83
$C_4ClH_{18}MnO_5{}^+$	$MnCl(H_2O)(CH_3OH)_4{}^+$	Mn:	MVol.C5-78
$C_4Cl_2FFeO_4P$	$(CO)_4FePFCl_2$	Fe:	Org.Verb.B2-85, 93
$C_4Cl_2FH_7NO_3P$	$ClCFNOPCl(O)OCH(CH_3)_2$	F:	PerFHalOrg.7-199/200
–	$ClCFNOPCl(O)OC_3H_7$	F:	PerFHalOrg.7-199/200
$C_4Cl_2FH_7NO_4P$	$OP(OCH_3)(ONCFCl)OC_2H_4Cl$	F:	PerFHalOrg.7-201/2
$C_4Cl_2F_2H_6NO_3P$	$OPF(ONCFCl)OCH_2CHClCH_3$	F:	PerFHalOrg.7-201/2
$C_4Cl_2F_2N_2$	NCClCFNCClCF	F:	PerFHalOrg.6-28/30, 48, 60
–	NCClCFNCFCCl	F:	PerFHalOrg.6-28/30, 49
–	NCFCClCClCFN	F:	PerFHalOrg.6-14/5, 35, 70
–	NCFNCFCClCCl	F:	PerFHalOrg.6-19/20, 34, 41
$C_4Cl_2F_3H_5NOPS$	$C_2H_5SP(Cl)ONCFCF_2Cl$	F:	PerFHalOrg.7-200
$C_4Cl_2F_3H_5N_2$	$ClCF_2CFClNNC_2H_5$	F:	PerFHalOrg.7-192
$C_4Cl_2F_3N$	$CClCFNCF_2CCl$	F:	PerFHalOrg.5-51, 75
$C_4Cl_2F_4N_2$	$NCClNClCF_2CFCF$	F:	PerFHalOrg.6-19/20, 34, 40
–	$NCF_2CClNCF_2CCl$	F:	PerFHalOrg.6-28/30, 49
–	$NCF_2CF_2NCClCCl$	F:	PerFHalOrg.6-28/30, 49
$C_4Cl_2F_5H_2NO$	$CF_2ClCFClCF_2C(O)NH_2$	F:	PerFHalOrg.7-18/9, 46
$C_4Cl_2F_5NO$	$CF_3NOCClCCF_2Cl$	F:	PerFHalOrg.5-33/4, 41
$C_4Cl_2F_6H_6NP$	$(CF_3)_2PCl_2N(CH_3)_2$	F:	PerFHalOrg.3-109/10
$C_4Cl_2F_6N_2$	$(F)(CF_2ClCFClCF_2)CN_2$	F:	PerFHalOrg.5-5
$C_4Cl_2F_6N_3P$	$NPCl_2NC(CF_3)NC(CF_3)$	F:	PerFHalOrg.6-94, 110, 111/2
$C_4Cl_2F_6O_4Sn$	$SnCl_2(CF_3COO)_2$	Sn:	MVol.C2-223
$C_4Cl_2F_6Si$	$(CF_2CF)_2SiCl_2$	F:	PerFHalOrg.4-149/50
$C_4Cl_2F_7HNO_2P$	$C_3F_7C(O)NHP(O)Cl_2$	F:	PerFHalOrg.7-24/5, 51
$C_4Cl_2F_7N$	$CF_2ClCF_2NCFCF_2Cl$	F:	PerFHalOrg.3-109
–	$CF_3NCF_2CFClCFCl$	F:	PerFHalOrg.5-29, 37, 45, 48
$C_4Cl_2F_7NO$	$CF_2ClCFClNOCF_2CF_2$	F:	PerFHalOrg.5-34, 41

$C_4Cl_2F_7NO$	$CF_3CFClNOCFClCF_2$	F:	PerFHalOrg.5-34/5
–	$CF_3CFClNOCF_2CFCl$	F:	PerFHalOrg.5-34/5
–	$CF_3NOCCl(CF_2Cl)CF_2$	F:	PerFHalOrg.5-31/2, 40, 47
–	$CF_3NOCF_2CClCF_2Cl$	F:	PerFHalOrg.5-31/2, 40, 46/7
$C_4Cl_2F_8HNO$	$(ClCF_2CF_2)_2NOH$	F:	PerFHalOrg.7-86/7
$C_4Cl_2F_8NO$	$(ClCF_2CF_2)_2NO$	F:	PerFHalOrg.7-85/6, 103, 122, 145
$C_4Cl_2F_8N_2O_2$	$(ClCF_2CF_2)_2NONO$	F:	PerFHalOrg.7-90
$C_4Cl_2F_9N_2O_2P$	$CF_3N(OP(O)Cl_2)CF_2CFNCF_3$	F:	PerFHalOrg.7-100/1, 118, 126
$C_4Cl_2F_{10}Si$	$(C_2F_5)_2SiCl_2$	F:	PerFHalOrg.4-147
$C_4Cl_2F_{12}NP$	$(CF_3)_2PCl_2N(CF_3)_2$	F:	PerFHalOrg.3-61, 65, 68
$C_4Cl_2F_{14}O_2Si_4$	$[SiF_2OC(CF_3)ClSiF_2]_2$	F:	PerFHalOrg.4-147/8, 158
$C_4Cl_2FeGeH_3NO_3$	$(CO)_2(NO)FeClGe(CHCH_2)Cl$	Fe:	Org.Verb.B1-133
$C_4Cl_2FeGeH_3NO_3^+$	$[(CO)_2(NO)FeClGe(CHCH_2)Cl]^+$	Fe:	Org.Verb.B1-133
$C_4Cl_2FeHg_2O_4$	$(CO)_4Fe(HgCl)_2$	Fe:	Org.Verb.B2-153/6, 159
		Fe:	Org.Verb.B3-127/8
$C_4Cl_2FeO_4$	$(CO)_4FeCl_2$	Fe:	Org.Verb.B2-54/65
$C_4Cl_2FeO_4Si$	$(CO)_4FeSiCl_2$	Fe:	Org.Verb.B2-149
$C_4Cl_2FeO_5S$	$(CO)_4FeSOCl_2$	Fe:	Org.Verb.B3-121
$C_4Cl_2Fe_2N_6O_4$	$[(CO)_2Fe(N_3)_2Fe(CO)_2]Cl_2$	Fe:	Org.Verb.C1-2/3
$C_4Cl_2H_2Sn$	$(CHC)_2SnCl_2$	Sn:	Org.Verb.6-149
$C_4Cl_2H_3NiO_3P$	$(CO)_3NiP(CH_3)Cl_2$	Ni:	Org.Verb.1-164/5
$C_4Cl_2H_4N_2O_2U$	$UO_2Cl_2 \cdot N(CHCH)_2N \cdot H_2O$	U:	SVol.E1-46/8
$C_4Cl_2H_4O_6Pu$	$PuO_2(ClCH_2COO)_2$	Np:	TrU.D1-153
$C_4Cl_2H_6MoO_5$	$MoO_2Cl_2 \cdot (CH_3CO)_2O$	Mo:	SVol.B1-142
$C_4Cl_2H_6N_2O_8Sn$	$Sn(ClO_4)_2 \cdot 2\ CH_3CN$	Sn:	MVol.C5-44
$C_4Cl_2H_6Ni$	$(C_4H_6NiCl_2)_x$	Ni:	Org.Verb.2-398
$C_4Cl_2H_6O_4S_2Sn$	$SnCl_2(SCH_2COOH)_2$	Sn:	MVol.C6-89
$C_4Cl_2H_6O_4Sn$	$SnCl_2(CH_3COO)_2$	Sn:	MVol.C2-221/2
$C_4Cl_2H_6O_4Te$	$(HOCOCH_2)_2TeCl_2$	Te:	SVol.B2-98
$C_4Cl_2H_6O_5U$	$UO_2Cl_2 \cdot (CH_3CO)_2O$	U:	SVol.E1-123
$C_4Cl_2H_6Sn$	$(CH_2CH)_2SnCl_2$	Sn:	Org.Verb.6-142/3
–	$(CH_2CH)_2SnCl_2 \cdot 2\ H_2O$	Sn:	Org.Verb.6-145
$C_4Cl_2H_6Ti$	$(CH_2CH)_2TiCl_2$	Ti:	Org.Verb.1-73
$C_4Cl_2H_7NO_3S$	$ClSO_2NCCl(OC_3H_7)$	S:	S-N-Verb.1-127
$C_4Cl_2H_8MnO_2$	$MnCl_2 \cdot C_4H_8O_2$	Mn:	MVol.D1-152/3, 157
–	$MnCl_2 \cdot C_4H_8O_2 \cdot 2\ H_2O$	Mn:	MVol.D1-153
$C_4Cl_2H_8MnO_{10}$	$Mn(ClO_4)_2 \cdot C_4H_8O_2$	Mn:	MVol.D1-154
$C_4Cl_2H_8N_2O_8Sn$	$(NH_4)_2SnCl_2(C_2O_4)_2 \cdot H_2O$	Sn:	MVol.C3-125
$C_4Cl_2H_8OSSn$	$SnCl_2 \cdot S(CH_2CH_2)_2O$	Sn:	MVol.C5-52/3
$C_4Cl_2H_8OS_2Sn$	$SnCl_2(O(CH_2CH_2S)_2)$	Sn:	MVol.C6-85
$C_4Cl_2H_8O_2Sn$	$(CH_3)_2ClSnOCOCH_2Cl$	Sn:	Org.Verb.6-21
–	$SnCl_2 \cdot C_4H_8O_2$	Sn:	MVol.C5-10
$C_4Cl_2H_8O_4Sn^{2+}$	$[SnCl_2 \cdot 2\ CH_3COOH]^{2+}$	Sn:	MVol.C5-118
$C_4Cl_2H_8S_2Sn$	$SnCl_2 \cdot S(CH_2CH_2)_2S$	Sn:	MVol.C5-52/3
$C_4Cl_2H_8S_3Sn$	$SnCl_2(S(CH_2CH_2S)_2)$	Sn:	MVol.C6-85
$C_4Cl_2H_9LiSn$	$(CH_3)_3SnCCl_2Li$	Sn:	Org.Verb.2-16, 18

$C_4Cl_2H_9NOS_3$ ...... $[SOCl_2SCN]S(CH_3)_3$ .... S: SVol.1-46
$C_4Cl_2H_9NO_4Sn$ ..... $SnCl_2 \cdot NH_2CH_2COOH \cdot CH_3COOH$ ...... Sn: MVol.C5-30/1
$C_4Cl_2H_9Sn^+$ .... $C_4H_9SnCl_2^+$ .... Sn: Org.Verb.6-82
$C_4Cl_2H_9Ti$ .... $(i-C_4H_9)TiCl_2$ .... Ti: Org.Verb.1-8
$C_4Cl_2H_{10}MnO$ ...... $MnCl_2 \cdot C_4H_9OH$ .... Mn: MVol.D1-33
$C_4Cl_2H_{10}MnO_2$ ..... $MnCl_2 \cdot CH_3OCH_2CH_2OCH_3$ .... Mn: MVol.D1-140/1
$C_4Cl_2H_{10}N_2O_4U$ ..... $UO_2Cl_2 \cdot 2\ CH_3CONH_2 \cdot H_2O$ .... U: SVol.E1-96/8
$C_4Cl_2H_{10}N_2Sn$ ...... $SnCl_2 \cdot NH(CH_2CH_2)_2NH$ .... Sn: MVol.C5-23/4
$C_4Cl_2H_{10}N_4O_8Ti$ ..... $(NO_2CH_2)_2TiCl_2 \cdot 2\ CH_3NO_2$ .... Ti: Org.Verb.1-76
$C_4Cl_2H_{10}OSn$ .... $C_4H_9Sn(OH)Cl_2 \cdot H_2O$ .... Sn: Org.Verb.6-246
$C_4Cl_2H_{10}OTi$ .... $CH_3Ti(OC_3H_7-i)Cl_2$ .... Ti: Org.Verb.1-59
$C_4Cl_2H_{10}O_2Sn$ ...... $(CH_3OCH_2)_2SnCl_2$ .... Sn: Org.Verb.6-129
– .... $(HOCH_2CH_2)_2SnCl_2$ .... Sn: Org.Verb.6-129
$C_4Cl_2H_{10}O_2Te$ ...... $TeCl_2(OC_2H_5)_2$ .... Te: SVol.B2-103
$C_4Cl_2H_{10}O_3U$ .... $UO_2Cl_2 \cdot (C_2H_5)_2O$ .... U: SVol.E1-71, 75/6
– .... $UO_2Cl_2 \cdot (C_2H_5)_2O \cdot 2\ H_2O$ .... U: SVol.E1-71, 75/6
$C_4Cl_2H_{10}S_2Sn$ ...... $SnCl_2(SC_2H_5)_2$ .... Sn: MVol.C6-79/80
$C_4Cl_2H_{10}Sn$ .... $(CH_3)(i-C_3H_7)SnCl_2$ .... Sn: Org.Verb.6-196
– .... $(CH_3)_2Sn(CH_2Cl)_2$ .... Sn: Org.Verb.3-5/6
– .... $(CH_3)_3SnCHCl_2$ .... Sn: Org.Verb.2-7
– .... $(C_2H_5)_2SnCl_2$
Analysis .... Sn: Org.Verb.6-41/2
Chemical reactions .... Sn: Org.Verb.6-51/63
Formation and preparation .... Sn: Org.Verb.6-40/5
Molecule .... Sn: Org.Verb.6-45
Physical properties .... Sn: Org.Verb.6-49/50
Physiology .... Sn: Org.Verb.6-63/4
Polarography .... Sn: Org.Verb.6-51
Spectra .... Sn: Org.Verb.6-46/9
Structure .... Sn: Org.Verb.6-45
Uses .... Sn: Org.Verb.6-64/5
– .... $C_4H_9SnHCl_2$ .... Sn: Org.Verb.6-295
$C_4Cl_2H_{10}Ti$ .... $(C_2H_5)_2TiCl_2$ .... Ti: Org.Verb.1-72
$C_4Cl_2H_{11}ISiSn$ ...... $(CH_3)_3SiCH_2SnCl_2I$ .... Sn: Org.Verb.6-254
$C_4Cl_2H_{11}NSn$ .... $SnCl_2 \cdot (C_2H_5)_2NH$ .... Sn: MVol.C5-11/2
– .... $SnCl_2 \cdot C_4H_9NH_2$ .... Sn: MVol.C5-11/2
$C_4Cl_2H_{11}N_2O_3PS$ .... $(C_2H_5)_2NSO_2NHPOCl_2$ .... S: S N Verb.1-186/7
$C_4Cl_2H_{12}INOS$ ...... $SOCl_2 \cdot [(CH_3)_4N]I$ .... S: SVol.1-45/6
$C_4Cl_2H_{12}MnO_2$ ..... $MnCl_2 \cdot 2\ C_2H_5OH$ .... Mn: MVol.D1-29/30
$C_4Cl_2H_{12}MnO_4$ ..... $Mn(HOC_2H_4OH)_2Cl_2$ .... Mn: MVol.D1-35/6
– .... $Mn(HOC_2H_4OH)_2Cl_2 \cdot H_2O$ .... Mn: MVol.D1-36
$C_4Cl_2H_{12}N_2Sn$ ...... $SnCl_2 \cdot NH_2(CH_2)_4NH_2$ .... Sn: MVol.C5-14/5
$C_4Cl_2H_{12}NiP_2$ ...... $Ni(P(CH_3)_2)_2Cl_2$ .... Ni: Org.Verb.1-72
$C_4Cl_2H_{12}OSSn$ ..... $(CH_3)_2SnCl_2 \cdot (CH_3)_2SO$ .... Sn: Org.Verb.6-37
$C_4Cl_2H_{12}OSn_2$ ...... $[(CH_3)_2ClSn]_2O$ .... Sn: Org.Verb.5-191
Sn: Org.Verb.6-21, 23
– .... $(CH_3)_2SnO \cdot (CH_3)_2SnCl_2$ .... Sn: Org.Verb.6-21
$C_4Cl_2H_{12}O_2S_2Sn$ .... $SnCl_2 \cdot 2\ (CH_3)_2SO$ .... Sn: MVol.C5-45
$C_4Cl_2H_{12}O_4P_2Sn$ .... $[SnCl_2((CH_3)_2PO_2)_2]_n$ .... Sn: MVol.C6-151
$C_4Cl_2H_{12}O_4U$ .... $UO_2Cl_2 \cdot 2\ C_2H_5OH$ .... U: SVol.E1-58/9

$C_4Cl_2H_{12}O_6P_2Sn$	$SnCl_2(CH_3(CH_3O)P(O)O)_2$	Sn: MVol.C6-162
$C_4Cl_2H_{12}O_8P_2Sn$	$SnCl_2[(CH_3O)_2PO_2]_2$	Sn: MVol.C2-267
		Sn: MVol.C6-168/9
$C_4Cl_2H_{12}SSn_2$	$[(CH_3)_2ClSn]_2S$	Sn: Org.Verb.6-22
$C_4Cl_2H_{12}Sn_2$	$[(CH_3)_2ClSn]_2$	Sn: Org.Verb.6-15
$C_4Cl_2H_{13}O_3PSn$	$(CH_3)_2SnCl_2 \cdot (CH_3O)_2HPO$	Sn: Org.Verb.6-37
$C_4Cl_2H_{14}N_2O_2U$	$UO_2Cl_2 \cdot 2\ C_2H_5NH_2$	U: SVol.E1-18, 21
$C_4Cl_2H_{14}N_2SSn$	$(CH_3)_2SnCl_2 \cdot (CH_3)_2S(NH)_2$	Sn: Org.Verb.6-36
$C_4Cl_2H_{16}IIrN_4$	$[Ir(C_2H_4(NH_2)_2)_2Cl_2]I$	Ir: SVol.2-83
–	$[Ir(C_2H_4(NH_2)_2)_2ICl]Cl$	Ir: SVol.2-84
$C_4Cl_2H_{16}IIrN_4O_4$	$[Ir(C_2H_4(NH_2)_2)_2ICl]ClO_4$	Ir: SVol.2-84
$C_4Cl_2H_{16}IrN_4^+$	$[Ir(C_2H_4(NH_2)_2)_2Cl_2]^+$	Ir: SVol.2-83
$C_4Cl_2H_{16}IrN_5O_3$	$[Ir(C_2H_4(NH_2)_2)_2Cl_2]NO_3$	Ir: SVol.2-83
$C_4Cl_2H_{16}IrN_5O_6$	$[Ir(NO_2)(C_2H_4(NH_2)_2)_2Cl]ClO_4$	Ir: SVol.2-62
–	$[Ir(ONO)(C_2H_4(NH_2)_2)_2Cl]ClO_4$	Ir: SVol.2-62
$C_4Cl_2H_{16}IrN_8S_4^+$	$[Ir(SC(NH_2)_2)_4Cl_2]^+$	Ir: SVol.2-124
$C_4Cl_2H_{16}MnO_4$	$[MnCl_2(CH_3OH)_4]$	Mn: MVol.C5-77/8
		Mn: MVol.D1-24
–	$MnCl_2 \cdot 4\ CH_3OH$	Mn: MVol.D1-25
$C_4Cl_2H_{16}N_2Sn$	$(C_2H_5)_2SnCl_2 \cdot 2\ NH_3$	Sn: Org.Verb.6-61
$C_4Cl_2H_{16}N_4Sn$	$SnCl_2 \cdot 2\ C_2H_8N_2$	Sn: MVol.C5-14
$C_4Cl_2H_{16}N_8O_6U$	$UO_2Cl_2 \cdot 4\ (NH_2)_2CO$	U: SVol.E1-83/4, 86
$C_4Cl_2H_{16}N_8O_8S_4Te$	$Te(SC(NH_2)_2)_4(ClO_4)_2$	Te: SVol.B3-146
$C_4Cl_2H_{16}N_8S_4Te$	$Te(SC(NH_2)_2)_4Cl_2$	Te: SVol.B3-144/5
–	$Te(SC(NH_2)_2)_4Cl_2 \cdot 2\ H_2O$	Te: SVol.B3-145/6
$C_4Cl_2H_{16}N_8Se_4Te$	$Te(SeC(NH_2)_2)_4Cl_2$	Te: SVol.B3-162
$C_4Cl_2H_{18}N_4Sn$	$(C_2H_5)_2SnCl_2 \cdot 2\ N_2H_4$	Sn: Org.Verb.6-62
$C_4Cl_2H_{19}N_3Sn$	$(C_2H_5)_2SnCl_2 \cdot 3\ NH_3$	Sn: Org.Verb.6-61
$C_4Cl_2IrK_3O_8$	$K_3[Ir(C_2O_4)_2Cl_2]$	Ir: SVol.2-35
$C_4Cl_2K_2O_8Sn$	$K_2SnCl_2(C_2O_4)_2 \cdot H_2O$	Sn: MVol.C3-79
$C_4Cl_2N_4O$	$C_3N_3Cl_2NCO$	C: MVol.D3-285
$C_4Cl_2Na_2O_8Sn$	$Na_2SnCl_2(C_2O_4)_2$	Sn: MVol.C3-35
$C_4Cl_3FH_7N_2O_2P$	$OPCl(ONCFCl)N(CH_3)CH_2CH_2Cl$	F: PerFHalOrg.7-201/2
$C_4Cl_3FN_2$	NCClCClNCClCF	F: PerFHalOrg.6-28/30, 49
–	NCFNCClCClCCl	F: PerFHalOrg.6-19/20, 34, 41
$C_4Cl_3F_3H_5NOSi$	$CF_3N(OH)CHCHCH_2SiCl_3$	F: PerFHalOrg.7-166/7
$C_4Cl_3F_3N_2$	NCClNClCFClCFCF	F: PerFHalOrg.6-19/20, 34, 40
$C_4Cl_3F_3N_2O$	$CCl_3CNC(CF_3)NO$	F: PerFHalOrg.5-66, 90
$C_4Cl_3F_4H_2NO$	$CCl_3CF_2CF_2C(O)NH_2$	F: PerFHalOrg.7-18, 20/1, 46
$C_4Cl_3F_4N$	$CF_2CCl_2NCClCF_2$	F: PerFHalOrg.5-51, 75, 97
$C_4Cl_3F_4NO$	$CF_3NOCClCCFCl_2$	F: PerFHalOrg.5-33/4, 41
$C_4Cl_3F_6NO$	$CF_2ClCFClNOCFClCF_2$	F: PerFHalOrg.5-34, 41
$C_4Cl_3F_{10}N_2O_2P$	$Cl_3P(N(O)(CF_3)CF_2)_2 = Cl_3P(ON(CF_3)CF_2)_2$	F: PerFHalOrg.5-68/9, 90
$C_4Cl_3FeHO_4Si$	$(CO)_4Fe(SiCl_3)H$	Fe: Org.Verb.B2-138, 145
$C_4Cl_3FeHg_2O_4^-$	$[(CO)_4Fe(HgCl_2)HgCl]^-$	Fe: Org.Verb.B2-160
$C_4Cl_3FeO_4P$	$(CO)_4FePCl_3$	Fe: Org.Verb.B2-84/5, 93, 113/4

$C_4Cl_3FeO_4Si^-$	$[(CO)_4FeSiCl_3]^-$	Fe: Org.Verb.B2-147
$C_4Cl_3GaH_{12}Sn$	$(CH_3)_3SnCl \cdot CH_3GaCl_2$	Sn: Org.Verb.5-85
$C_4Cl_3H_3SSn$	$C_4H_3SSnCl_3$	Sn: Org.Verb.6-287
$C_4Cl_3H_6IrN_2$	$Ir(NCCH_3)_2Cl_3$	Ir: SVol.2-65
$C_4Cl_3H_7Ni$	$(CH_3CClCHCH_3)NiCl_2 \cdot H_2O$	Ni: Org.Verb.1-339
$C_4Cl_3H_7OSn$	$CH_3COCH_2CH_2SnCl_3$	Sn: Org.Verb.6-256
$C_4Cl_3H_7O_2Sn$	$CH_3OCOCH_2CH_2SnCl_3$	Sn: Org.Verb.6-254, 256
–	$HOCOCH_2CH(CH_3)SnCl_3$	Sn: Org.Verb.6-256
$C_4Cl_3H_7Sn$	$CH_3CHC(CH_3)SnCl_3$	Sn: Org.Verb.6-265
$C_4Cl_3H_8N_2O_3PS$	$O(CH_2CH_2)_2NSO_2NPCl_3$	S: S-N-Verb.1-188
$C_4Cl_3H_8O_2Sn^+$	$[SnCl_3(CH_3COOC_2H_5)]^+$	Sn: MVol.C5-130
$C_4Cl_3H_9OSn$	$CH_3SnCl_3 \cdot (CH_3)_2CO$	Sn: Org.Verb.6-222
$C_4Cl_3H_9Sn$	$(CH_3)_3SnCCl_3$	Sn: Org.Verb.2-10
–	$C_4H_9SnCl_3$	Sn: Org.Verb.6-236/49
–	$(i\text{-}C_4H_9)SnCl_3$	Sn: Org.Verb.6-250
–	$(s\text{-}C_4H_9)SnCl_3$	Sn: Org.Verb.6-250
–	$(t\text{-}C_4H_9)SnCl_3$	Sn: Org.Verb.6-250
$C_4Cl_3H_9Ti$	$(i\text{-}C_4H_9)TiCl_3$	Ti: Org.Verb.1-16, 24
$C_4Cl_3H_{10}NOSn$	$CH_3SnCl_3 \cdot HC(O)N(CH_3)_2$	Sn: Org.Verb.6-222
$C_4Cl_3H_{10}N_2O_2PS$	$(C_2H_5)_2NS(O)ClNP(O)Cl_2$	S: S-N-Verb.1-197
–	$(C_2H_5)_2NSO_2NPCl_3$	S: S-N-Verb.1-188
$C_4Cl_3H_{10}N_4O_2PS_2$	$[NS(O)Cl]_2[NPClNHC_4H_9]$	S: S-N-Verb.1-30
$C_4Cl_3H_{10}O_2PS_2Sn$	$SnCl_3((C_2H_5O)_2P(S)S)$	Sn: MVol.C6-175/6
$C_4Cl_3H_{11}OSiSn$	$(CH_3O)(CH_3)_2SiCH_2SnCl_3$	Sn: Org.Verb.6-256
$C_4Cl_3H_{11}OSn_2$	$(C_2H_5)Cl_2SnOSnCl(CH_3)_2$	Sn: Org.Verb.6-230
$C_4Cl_3H_{11}O_2Sn$	$(SnCl_3OC_2H_5 \cdot C_2H_5OH)_2$	Sn: MVol.C5-71
–	$SnCl_3OH \cdot (C_2H_5)_2O \cdot H_2O$	Sn: MVol.C5-150/1
$C_4Cl_3H_{11}SiSn$	$(CH_3)_3SiCH_2SnCl_3$	Sn: Org.Verb.6-254, 256
$C_4Cl_3H_{11}SiTi$	$(CH_3)_3SiCH_2TiCl_3$	Ti: Org.Verb.1-17, 25
$C_4Cl_3H_{12}MnN$	$(CH_3)_4NMnCl_3$	Mn: MVol.C5-194/204
–	$(CD_3)_4NMnCl_3$	Mn: MVol.C5-194, 198, 200
–	$C_4H_9NH_3MnCl_3$	Mn: MVol.C5-188
$C_4Cl_3H_{12}NOS$	$SOCl_2 \cdot [(CH_3)_4N]Cl$	S: SVol.1-45/6
$C_4Cl_3H_{12}NSn$	$[(CH_3)_4N]SnCl_3$	Sn: MVol.C3-95/6
$C_4Cl_3H_{12}N_2O_2PSSi$	$ClSO_2NPCl_2N(CH_3)Si(CH_3)_3$	S: S-N-Verb.1-132
$C_4Cl_3H_{16}IrN_4O_4$	$[Ir(C_2H_4(NH_2)_2)_2Cl_2]ClO_4$	Ir: SVol.2-83
$C_4Cl_3H_{16}IrN_8S_4$	$[Ir(SC(NH_2)_2)_4Cl_2]Cl$	Ir: SVol.2-168
$C_4Cl_3H_{18}IrN_4O_9$	$[Ir(C_2H_4(NH_2)_2)_2Cl(H_2O)](ClO_4)_2$	Ir: SVol.2-83
$C_4Cl_3H_{22}IrN_6O_{12}$	$[Ir(NH_3)_2(C_2H_4(NH_2)_2)_2](ClO_4)_3$	Ir: SVol.2-47
$C_4Cl_4F_5NO$	$CCl_3CF_2CFClCF_2NO$	F: PerFHalOrg.7-178/9, 184
$C_4Cl_4F_6Si$	$CF_3CClC(SiCl_3)CF_3$	F: PerFHalOrg.4-149/50
$C_4Cl_4FeO_4Sn$	$(CO)_4Fe(SnCl_3)Cl$	Fe: Org.Verb.B2-139, 146
$C_4Cl_4H_2SSn$	$C_4H_2ClSSnCl_3$	Sn: Org.Verb.6-287
$C_4Cl_4H_4N_2S_2Sn$	$SnCl_4 \cdot C_2H_4(SCN)_2$	Sn: MVol.C6-122
$C_4Cl_4H_4N_2Sn$	$SnCl_4 \cdot N(CHCH)_2N$	Sn: MVol.C5-225/6
–	$SnCl_4 \cdot NC(CH_2)_2CN$	Sn: MVol.C6-73/5
$C_4Cl_4H_4O_4Sn$	$SnCl_2(ClCH_2COO)_2$	Sn: MVol.C2-223
$C_4Cl_4H_4O_4Sn^{2-}$	$[SnCl_4(CH_2COO)_2]^{2-}$	Sn: MVol.C6-11
$C_4Cl_4H_4O_8Sn$	$SnCl_4 \cdot 2\ HOCOCOOH$	Sn: MVol.C5-121
$C_4Cl_4H_4Sn$	$(CHClCH)_2SnCl_2$	Sn: Org.Verb.6-147/8

$C_4Cl_4H_5NO_2Sn$	$SnCl_4 \cdot C_2H_4(CO)_2NH$	Sn: MVol.C6-53/4
$C_4Cl_4H_5NSn$	$SnCl_4 \cdot CH_2C(CH_3)CN$	Sn: MVol.C6-69/70
$C_4Cl_4H_6I_2O_4Sn$	$SnCl_4 \cdot 2\ CH_2ICOOH$	Sn: MVol.C5-119
$C_4Cl_4H_6N_2O_6Sn$	$(O_3NCH_2CHCl)_2SnCl_2$	Sn: Org.Verb.6-128
$C_4Cl_4H_6N_2Sn$	$(CH_3NCCl)_2SnCl_2$	Sn: Org.Verb.6-128
–	$SnCl_4 \cdot 2\ CH_3CN$	Sn: MVol.C6-64/6
–	$SnCl_4 \cdot 2\ CD_3CN$	Sn: MVol.C6-65, 67
$C_4Cl_4H_6N_2Te$	$[TeCl_3 \cdot 2\ CH_3CN]Cl$	Te: SVol.B3-175
$C_4Cl_4H_6N_2U$	$[UCl_4(CH_3CN)_2]$	U: SVol.E1-48/50
$C_4Cl_4H_6N_4O_2Sn$	$SnCl_4 \cdot C_4H_6N_4O_2$	Sn: MVol.C5-225
$C_4Cl_4H_6O_3Sn$	$SnCl_4 \cdot (CH_3CO)_2O$	Sn: MVol.C5-124/5
$C_4Cl_4H_6O_4Sn$	$SnCl_4 \cdot CH_3OCOCOOCH_3$	Sn: MVol.C5-139/40
–	$SnCl_4 \cdot C_2H_4(COOH)_2$	Sn: MVol.C5-121
$C_4Cl_4H_6O_4Sn^{2-}$	$[SnCl_4(CH_3COO)_2]^{2-}$	Sn: MVol.C5-125
$C_4Cl_4H_6O_4Sn_2$	$Sn_2Cl_4(CH_3COO)_2$	Sn: MVol.C2-222
$C_4Cl_4H_7NO_2Sn$	$SnCl_4 \cdot CH_3CONHCOCH_3$	Sn: MVol.C6-50/1
$C_4Cl_4H_7NSn$	$SnCl_4 \cdot C_3H_7CN$	Sn: MVol.C6-68/9
$C_4Cl_4H_8N_2O_2Sn$	$SnCl_4 \cdot CH_3CONHNHCOCH_3$	Sn: MVol.C6-59/60
–	$SnCl_4 \cdot NH_2COCH_2CH_2CONH_2$	Sn: MVol.C6-52
$C_4Cl_4H_8N_2O_4Sn^{2-}$	$[SnCl_4(NH_2CH_2COO)_2]^{2-}$	Sn: MVol.C6-10
$C_4Cl_4H_8N_4S_4Sn$	$SnCl_4 \cdot 2\ H_2NCSCSNH_2$	Sn: MVol.C6-116/7
$C_4Cl_4H_8OSn$	$SnCl_4 \cdot CH_3COC_2H_5$	Sn: MVol.C5-87
–	$SnCl_4 \cdot C_4H_8O$	Sn: MVol.C5-156
$C_4Cl_4H_8OTe$	$TeCl_4 \cdot CH_3COC_2H_5$	Te: SVol.B3-165/6
$C_4Cl_4H_8OU$	$UCl_4 \cdot (CH_2)_4O$	U: SVol.E1-70, 74/5
$C_4Cl_4H_8O_2SSn$	$SnCl_4 \cdot (CH_2)_4SO_2$	Sn: MVol.C6-111
$C_4Cl_4H_8O_2Sn$	$SnCl_4 \cdot CH_3COOC_2H_5$	Sn: MVol.C5-128/30
–	$SnCl_4 \cdot C_4H_8O_2$	Sn: MVol.C5-158
$C_4Cl_4H_8O_4Sn$	$SnCl_4 \cdot 2\ CH_3COOH$	Sn: MVol.C5-117/8
–	$SnCl_4 \cdot 2\ HCOOCH_3$	Sn: MVol.C5-126/7
$C_4Cl_4H_8SSn$	$SnCl_4 \cdot (CH_2)_4S$	Sn: MVol.C6-109/10
$C_4Cl_4H_8Sn$	$(CH_3CHCl)_2SnCl_2$	Sn: Org.Verb.6-129
–	$Sn(CH_2Cl)_4$	Sn: Org.Verb.1-122
$C_4Cl_4H_9NOSn$	$SnCl_4 \cdot NH(CH_2CH_2)_2O$	Sn: MVol.C5-245
$C_4Cl_4H_9N_3O_2P_2S$	$[O_2S[NPCl_2]_2NC_4H_9]$	S: S-N-Verb.1-35
$C_4Cl_4H_{10}N_2O_2Sn$	$SnCl_4 \cdot 2\ CH_3CONH_2$	Sn: MVol.C6-43
–	$SnCl_4 \cdot 2\ OCHNHCH_3$	Sn: MVol.C6-40
$C_4Cl_4H_{10}N_2O_2Te$	$TeCl_4 \cdot 2\ CH_3CONH_2$	Te: SVol.B3-174
–	$TeCl_4 \cdot 2\ HCONHCH_3$	Te: SVol.B3-174
$C_4Cl_4H_{10}N_2O_4Sn$	$SnCl_4 \cdot 2\ NH_2CH_2COOH$	Sn: MVol.C6-10
$C_4Cl_4H_{10}N_2Sn$	$SnCl_4 \cdot NH(CH_2CH_2)_2NH$	Sn: MVol.C5-227
$C_4Cl_4H_{10}N_2Te$	$TeCl_4 \cdot NH(CH_2CH_2)_2NH$	Te: SVol.B3-172
$C_4Cl_4H_{10}OSn$	$SnCl_4 \cdot (CH_3)_3COH$	Sn: MVol.C5-74
–	$SnCl_4 \cdot C_2H_5OC_2H_5$	Sn: MVol.C5-150/1
$C_4Cl_4H_{10}OTe$	$TeCl_4 \cdot (C_2H_5)_2O$	Te: SVol.B3-166
$C_4Cl_4H_{10}OU$	$UCl_4 \cdot (C_2H_5)_2O$	U: SVol.E1-70, 74/5
$C_4Cl_4H_{10}O_2Pa^-$	$[Pa(OC_2H_5)_2Cl_4]^-$	Pa: SVol.2-82
$C_4Cl_4H_{10}O_2SSn$	$SnCl_4 \cdot (C_2H_5)_2SO_2$	Sn: MVol.C6-103
$C_4Cl_4H_{10}O_2Sn$	$SnCl_4 \cdot C_2H_4(OCH_3)_2$	Sn: MVol.C5-154
$C_4Cl_4H_{10}O_5Sn$	$H_3O[HSnCl_4(CH_3COO)_2]$	Sn: MVol.C5-118

$C_4Cl_4H_{10}S_2Sn$	$SnCl_4 \cdot C_2H_4(SCH_3)_2$	Sn: MVol.C6-96
$C_4Cl_4H_{10}Se_2Sn$	$SnCl_4 \cdot CH_3Se(CH_2)_2SeCH_3$	Sn: MVol.C6-123
$C_4Cl_4H_{11}NSn_2$	$(CH_3SnCl_2)_2NC_2H_5$	Sn: Org.Verb.6-218
–	$2\ SnCl_2 \cdot (C_2H_5)_2NH$	Sn: MVol.C5-11/2
$C_4Cl_4H_{11}O_4PSn$	$SnCl_4 \cdot (C_2H_5O)_2P(O)OH$	Sn: MVol.C6-168/9
$C_4Cl_4H_{12}N_2O_2Sn$	$SnCl_4 \cdot (CH_3)_3N \cdot CH_3NO_2$	Sn: MVol.C5-172/3
$C_4Cl_4H_{12}N_4O_2P_2S$	$SO_2[NPCl_2N(CH_3)_2]_2$	S: S-N-Verb.1-192
–	$SO_2[NPCl_2NHC_2H_5]_2$	S: S-N-Verb.1-192
$C_4Cl_4H_{12}N_4O_2Sn$	$SnCl_4 \cdot 2\ CH_3CONHNH_2$	Sn: MVol.C6-59/60
$C_4Cl_4H_{12}N_4O_4Sn$	$SnCl_4 \cdot 4\ OCHNH_2$	Sn: MVol.C6-40
$C_4Cl_4H_{12}N_4Sn$	$SnCl_4 \cdot 2\ CH_3NNCH_3$	Sn: MVol.C6-31
$C_4Cl_4H_{12}O_2S_2Sn$	$SnCl_4 \cdot 2\ (CH_3)_2SO$	Sn: MVol.C6-100/1
–	$SnCl_4 \cdot 2\ (CD_3)_2SO$	Sn: MVol.C6-101
$C_4Cl_4H_{12}O_2S_2Te$	$TeCl_4 \cdot 2\ (CH_3)_2SO$	Te: SVol.B3-176
$C_4Cl_4H_{12}O_2Se_2Sn$	$SnCl_4 \cdot 2\ (CH_3)_2SeO$	Sn: MVol.C6-123
$C_4Cl_4H_{12}O_2Sn$	$SnCl_4 \cdot 2\ CH_3OCH_3$	Sn: MVol.C5-148
–	$SnCl_4 \cdot 2\ C_2H_5OH$	Sn: MVol.C5-70
$C_4Cl_4H_{12}O_2Te$	$TeCl_4 \cdot 2\ C_2H_5OH$	Te: SVol.B3-165
$C_4Cl_4H_{12}O_4Sn$	$SnCl_4 \cdot 2\ C_2H_4(OH)_2$	Sn: MVol.C5-76
$C_4Cl_4H_{12}O_6S_2Sn$	$SnCl_4 \cdot 2\ (CH_3O)_2SO$	Sn: MVol.C6-102
$C_4Cl_4H_{12}P_2S_2Sn$	$SnCl_4 \cdot (CH_3)_2PSPS(CH_3)_2$	Sn: MVol.C6-148
$C_4Cl_4H_{12}S_2Sn$	$SnCl_4 \cdot 2\ CH_3SCH_3$	Sn: MVol.C6-92
–	$SnCl_4 \cdot 2\ C_2H_5SH$	Sn: MVol.C6-79/80
$C_4Cl_4H_{14}MnN_2$	$NH_3(CH_2)_4NH_3MnCl_4$	Mn: MVol.C5-185/8
$C_4Cl_4H_{14}N_2Sn$	$SnCl_4 \cdot 2\ (CH_3)_2NH$	Sn: MVol.C5-170/1
–	$SnCl_4 \cdot 2\ C_2H_5NH_2$	Sn: MVol.C5-169
$C_4Cl_4H_{14}N_2U$	$UCl_4 \cdot 2\ C_2H_5NH_2$	U: SVol.E1-18, 20
$C_4Cl_4H_{14}O_4P_2Sn$	$SnCl_4 \cdot 2\ (CH_3)_2P(O)OH$	Sn: MVol.C6-150/1
$C_4Cl_4H_{14}O_6P_2Sn$	$SnCl_4 \cdot 2\ (CH_3)(CH_3O)P(O)OH$	Sn: MVol.C6-161
–	$SnCl_4 \cdot 2\ HP(O)(OCH_3)_2$	Sn: MVol.C6-160/1
$C_4Cl_4H_{14}O_8P_2Sn$	$SnCl_4 \cdot 2\ (CH_3O)_2P(O)OH$	Sn: MVol.C6-168/9
$C_4Cl_4H_{16}MnN_2$	$[(CH_3)_2NH_2]_2MnCl_4$	Mn: MVol.C5-189
–	$(C_2H_5NH_3)_2MnCl_4$	Mn: MVol.C5-182/5
$C_4Cl_4H_{16}N_4Sn$	$SnCl_4 \cdot 2\ C_2H_8N_2$	Sn: MVol.C5-184
$C_4Cl_4H_{16}N_4Sn_2$	$[Cl_2SnClSn(NHC_2H_4NH_3)_2]Cl$	Sn: MVol.C5-14
$C_4Cl_4H_{16}N_4Te$	$TeCl_4 \cdot 2\ C_2H_8N_2$	Te: SVol.B3-168
$C_4Cl_4H_{16}N_8O_4U$	$UCl_4 \cdot 4\ (NH_2)_2CO \cdot 3\ H_2O$	U: SVol.E1-82/3
$C_4Cl_4H_{16}N_8Se_4Sn$	$SnCl_4 \cdot 4\ CSe(NH_2)_2$	Sn: MVol.C6-125
$C_4Cl_4H_{16}O_4Sn$	$SnCl_4 \cdot 4\ CH_3OH$	Sn: MVol.C5-68/9
$C_4Cl_4H_{16}O_4U$	$UCl_4 \cdot 4\ CH_3OH$	U: SVol.E1-58/9
$C_4Cl_4H_{17}IrN_4$	$[Ir(C_2H_4(NH_2)_2)_2Cl_2]Cl \cdot HCl \cdot 2\ H_2O$	Ir: SVol.2-83
$C_4Cl_4H_{20}N_4Sn$	$SnCl_4 \cdot 4\ CH_3NH_2$	Sn: MVol.C5-168
$C_4Cl_4H_{22}N_6Te$	$TeCl_4 \cdot 2\ C_2H_8N_2 \cdot 2\ NH_3$	Te: SVol.B3-168
$C_4Cl_4Hg_2N_4Sn$	$SnCl_4 \cdot 2\ Hg(CN)_2$	Sn: MVol.C4-97/8
$C_4Cl_4N_2$	$N(CClCCl)_2N$	F: PerFHalOrg.6-57
$C_4Cl_4N_4$	$(ClCN)_4$	C: MVol.D3-203
$C_4Cl_6FeO_4Sb$	$(CO)_4FeCl_2 \cdot SbCl_3$	Fe: Org.Verb.B2-54
$C_4Cl_5H_4IrK_2N_2$	$K_2[Ir(CHCHNCHCHN)Cl_5]$	Ir: SVol.2-78
–	$K_2[Ir(CHCHNCHCHN)Cl_5] \cdot 0.5\ H_2O$	Ir: SVol.2-78
–	$K_2[Ir(CHCHNNCHCH)Cl_5]$	Ir: SVol.2-79

$C_4Cl_5H_4IrK_2N_2$	$K_2[Ir(CHNCHNCHCH)Cl_5] \cdot H_2O$	Ir:	SVol.2-80
$C_4Cl_5H_4IrN_2^{2-}$	$[Ir(CHCHNCHCHN)Cl_5]^{2-}$	Ir:	SVol.2-78, 138
–	$[Ir(CHCHNNCHCH)Cl_5]^{2-}$	Ir:	SVol.2-79, 138
–	$[Ir(CHNCHNCHCH)Cl_5]^{2-}$	Ir:	SVol.2-80, 138
$C_4Cl_5H_4IrN_2Na_2$	$Na_2[Ir(CHCHNNCHCH)Cl_5] \cdot H_2O$	Ir:	SVol.2-79
$C_4Cl_5H_6IrN_2$	$H[IrH(CHCHNCHCHN)Cl_5] \cdot 3\ H_2O$	Ir:	SVol.2-78
–	$H[IrH(CHCHNNCHCH)Cl_5] \cdot 3\ H_2O$	Ir:	SVol.2-79
–	$H_2[Ir(CHNCHNCHCH)Cl_5] \cdot n\ H_2O$	Ir:	SVol.2-80
$C_4Cl_5H_6NO_2Sn$	$SnCl_4 \cdot CH_3CONHCOCH_2Cl$	Sn:	MVol.C6-50/1
$C_4Cl_5H_6N_2U$	$UCl_5 \cdot 2\ CH_3CN$	U:	SVol.C9-37
$C_4Cl_5H_7O_2Sn$	$SnCl_4 \cdot CH_2ClCOOC_2H_5$	Sn:	MVol.C5-133/4
$C_4Cl_5H_8OU$	$UCl_5 \cdot (CH_2)_4O$	U:	SVol.E1-70, 75
$C_4Cl_5H_8O_2Sn^-$	$[SnCl_5(CH_3COOC_2H_5)]^-$	Sn:	MVol.C5-130
$C_4Cl_5H_8O_2U$	$UCl_5 \cdot C_4H_8O_2$	U:	SVol.E1-71, 75
$C_4Cl_5H_9NOU^-$	$[UCl_5(CH_3CON(CH_3)_2)]^-$	U:	SVol.E1-103
$C_4Cl_5H_9Sn^{2-}$	$[C_4H_9SnCl_5]^{2-}$	Sn:	Org.Verb.6-246
$C_4Cl_5H_{10}OSn^-$	$[SnCl_5 \cdot C_4H_9OH]^-$	Sn:	MVol.C5-72
$C_4Cl_5H_{10}OU$	$UCl_5 \cdot (C_2H_5)_2O$	U:	SVol.E1-70, 74/5
$C_4Cl_5H_{12}IrN_4$	$(NH_4)_2[Ir(C_4H_4N_2)Cl_5]$	Ir:	SVol.2-78
–	$(NH_4)_2[Ir(C_4H_4N_2)Cl_5] \cdot H_2O$	Ir:	SVol.2-79
$C_4Cl_5H_{12}NSn$	$[(CH_3)_4N]SnCl_5$	Sn:	MVol.C3-101
$C_4Cl_5H_{12}N_4O_4PS_2Ti$	$[SO_2(NCH_3)_2]_2PCl \cdot TiCl_4$	S:	S-N-Verb.1-43
$C_4Cl_5H_{15}IrN_7$	$[(H_2N)_3C]_2[Ir(NCCH_3)Cl_5]$	Ir:	SVol.2-65
$C_4Cl_5H_{16}LaN_2$	$LaCl_3 \cdot 2\ [(CH_3)_2NH_2]Cl$	Sc:	MVol.C5-213/5
$C_4Cl_5H_{16}MnN_2$	$[(CH_3)_2NH_2]_2MnCl_5 \cdot C_2H_5OH$	Mn:	MVol.C5-210
$C_4Cl_5H_{16}MnN_3$	$(NH_3C_2H_4)_2NH_2ClMnCl_4$	Mn:	MVol.C5-204
$C_4Cl_5H_{16}N_2Nd$	$NdCl_3 \cdot 2\ [(CH_3)_2NH_2]Cl$	Sc:	MVol.C5-213/5
$C_4Cl_5H_{16}N_2Pr$	$PrCl_3 \cdot 2\ [(CH_3)_2NH_2]Cl$	Sc:	MVol.C5-213/5
$C_4Cl_5H_{18}IrN_4$	$[(C_2H_4(NH_2)_2)H_2Cl][Ir(C_2H_4(NH_2)_2)Cl_4]$	Ir:	SVol.2-85
$C_4Cl_5H_{20}IrN_6$	$[Ir(NH_2Cl)_2(C_2H_4(NH_2)_2)_2]Cl_3$	Ir:	SVol.2-63
$C_4Cl_5H_{23}N_6Ni_2$	$CH_2CClCHCH_2 \cdot 2\ NiCl_2 \cdot 6\ NH_3$	Ni:	Org.Verb.2-332
$C_4Cl_6FeGe_2O_4$	$(CO)_4Fe(GeCl_3)_2$	Fe:	Org.Verb.B2-140
$C_4Cl_6FeHg_4O_4$	$(CO)_4Fe(HgCl \cdot HgCl_2)_2$	Fe:	Org.Verb.B3-127/8
$C_4Cl_6FeO_4SiSn$	$(CO)_4Fe(SnCl_3)(SiCl_3)$	Fe:	Org.Verb.B2-141
$C_4Cl_6FeO_4Si_2$	$(CO)_4Fe(SiCl_3)_2$	Fe:	Org.Verb.B2-136, 139, 146/7
$C_4Cl_6FeO_4Sn_2$	$(CO)_4Fe(SnCl_3)_2$	Fe:	Org.Verb.B2-141, 148
$C_4Cl_6H_4O_3Te_2$	$(Cl_3TeCH_2CO)_2O$	Te:	SVol.B2-98
$C_4Cl_6H_6O_2Sn$	$SnCl_4 \cdot CHCl_2COOC_2H_5$	Sn:	MVol.C5-133/4
–	$SnCl_4 \cdot 2\ CH_3COCl$	Sn:	MVol.C5-125/6
$C_4Cl_6H_6O_4Sn$	$SnCl_4 \cdot 2\ CH_2ClCOOH$	Sn:	MVol.C5-119
$C_4Cl_6H_7O_2Sn^-$	$[SnCl_5(CH_2ClCOOC_2H_5)]^-$	Sn:	MVol.C5-134
$C_4Cl_6H_8N_2Sn$	$[CH_3CNH]_2[SnCl_6]$	Sn:	MVol.C6-66
$C_4Cl_6H_8OSn$	$SnCl_4 \cdot (ClC_2H_4)_2O$	Sn:	MVol.C5-151
$C_4Cl_6H_{10}O_2Sn$	$SnCl_4 \cdot 2\ CH_3OCH_2Cl$	Sn:	MVol.C5-148
$C_4Cl_6H_{10}Si_2Sn$	$[CH_3SiCl_2CH_2]_2SnCl_2$	Sn:	Org.Verb.6-129
$C_4Cl_6H_{12}NPa$	$[N(CH_3)_4]PaCl_6$	Pa:	SVol.2-58
$C_4Cl_6H_{12}NU$	$[N(CH_3)_4]UCl_6$	U:	SVol.C9-60/3
$C_4Cl_6H_{12}O_2P_2Sn$	$SnCl_4 \cdot 2\ (CH_3)_2P(O)Cl$	Sn:	MVol.C6-156
$C_4Cl_6H_{12}P_2S_2Sn$	$SnCl_4 \cdot 2\ (CH_3)_2P(S)Cl$	Sn:	MVol.C6-160

$C_4Cl_6H_{14}O_2Sn$	$[(CH_3)_2OH]_2[SnCl_6] \cdot 2\ (CH_3)_2O$	Sn:	MVol.C3-107
$C_4Cl_6H_{14}S_2Sn$	$(C_2H_5SH_2)_2[SnCl_6]$	Sn:	MVol.C6-80
$C_4Cl_6H_{16}IrN_2$	$((CH_3)_2NH_2)_2[IrCl_6]$	Ir:	SVol.2-127
–	$(C_2H_5NH_3)_2[IrCl_6]$	Ir:	SVol.2-126/7
$C_4Cl_6H_{16}N_2Sb$	$[C_2H_5NH_3]_2[SbCl_6]$ solid solutions $[C_2H_5NH_3]_2[(Sb,Sn)Cl_6]$	Sn:	MVol.C3-104
$C_4Cl_6H_{16}N_2Sn$	$[(CH_3)_2NH_2]_2[SnCl_6]$	Sn:	MVol.C3-104/5
–	$[C_2H_5NH_3]_2[SnCl_6]$	Sn:	MVol.C3-103/4
–	$[C_2H_5NH_3]_2[SnCl_6]$ solid solutions $[C_2H_5NH_3]_2[(Sn,Sb)Cl_6]$	Sn:	MVol.C3-104
$C_4Cl_6H_{16}N_2Te$	$[C_2H_5NH_3]_2TeCl_6$	Te:	SVol.B2-137/8
$C_4Cl_6NiS_2$	$(Ni(SCCl_2CCl)_2)_n$	Ni:	Org.Verb.1-233
$C_4Cl_6O_5Pa^+$	$PaO(Cl_3CCOO)_2{}^+$	Pa:	SVol.2-167
$C_4Cl_8H_9PSn$	$[(CH_3)_3CPCl_3][SnCl_5]$	Sn:	MVol.C2-260
$C_4Cl_8H_{10}N_2Sn$	$[CH_3C(Cl)NH_2]_2[SnCl_6]$	Sn:	MVol.C6-66
$C_4Cl_8H_{12}NOPSn$	$[(CH_3)_4N][POCl_2][SnCl_6]$	Sn:	MVol.C2-264
$C_4Cl_8H_{12}N_2O_2P_2Sn$	$SnCl_4 \cdot 2\ Cl_2P(O)N(CH_3)_2$	Sn:	MVol.C6-173
$C_4Cl_{10}H_{10}SnTi_2$	$(C_2H_5)_2SnCl_2 \cdot 2\ TiCl_4$	Sn:	Org.Verb.6-61
$C_4Cl_{11}H_{12}N_4O_4PS_2Sb_2$	$[SO_2(NCH_3)_2]_2PCl \cdot 2\ SbCl_5$	S:	S-N-Verb.1-43
$C_4CmH_5NO_4{}^+$	$CmNH(CH_2COO)_2{}^+$	Np:	TrU.D1-162
$C_4CmH_6O_4{}^+$	$Cm(CH_3COO)_2{}^+$	Np:	TrU.D1-153
$C_4CmH_6O_6{}^+$	$Cm(HOCH_2COO)_2{}^+$	Np:	TrU.D1-153
$C_4CmH_7O_3{}^{2+}$	$Cm((CH_3)_2C(OH)COO)^{2+}$	Np:	TrU.D1-154
$C_4CmO_8{}^-$	$Cm(C_2O_4)_2{}^-$	Np:	TrU.D1-158
$C_4CoF_6HNO_3P$	$Co(CO)_2(NO) \cdot HP(CF_3)_2$	F:	PerFHalOrg.3-27
$C_4CoF_{12}N_2O_2$	$[(CF_3)_2NO]_2Co$	F:	PerFHalOrg.7-102
$C_4CoF_{12}P_2S_4$	$Co[S(S)P(CF_3)_2]_2$	F:	PerFHalOrg.3-159
$C_4CoN_4Se_4{}^{2-}$	$[Co(NCSe)_4]^{2-}$	C:	MVol.D6-227
$C_4Co_2H_{10}N_2O_2S_7$	$Co_2(NH_3)_2C_2S_7 \cdot CH_3COOH \cdot 4\ H_2O$	C:	MVol.D4-225
$C_4CrF_{12}P_2S_4$	$Cr[S(S)P(CF_3)_2]_2$	F:	PerFHalOrg.3-154
$C_4CrH_{10}N_2O_8U$	$UO_2CrO_4 \cdot 2\ CH_3CONH_2$	U:	SVol.E1-96/8
–	$[UO_2CrO_4(CH_3CONH_2)_2]_n$	U:	SVol.E1-98
$C_4CrH_{12}O_5Sn_2$	$[(CH_3)_2SnCrO_4 \cdot (CH_3)_2SnO]_x$	Sn:	Org.Verb.6-29
$C_4CrH_{12}O_8S_2U$	$UO_2CrO_4 \cdot 2\ (CH_3)_2SO$	U:	SVol.E1-204, 208
$C_4CsF_{12}IN_2O_2$	$CsI[ON(CF_3)_2]_2$	F:	PerFHalOrg.7-102
$C_4CsF_{13}H_2N_2O_2$	$[(CF_3)_2NOH]_2 \cdot CsF$	F:	PerFHalOrg.7-86/7, 104, 140/1
$C_4CuH_7OS_2$	$Cu[S_2COC_3H_7]$	C:	MVol.D4-256
$C_4CuH_7S_3$	$Cu[S_2CSC_3H_7]$	C:	MVol.D4-269
$C_4CuH_{10}NS_4$	$[(CH_3)_3NH]CuCS_4$	C:	MVol.D4-231
$C_4Cu_2FeH_6N_2O_4$	$(CO)_4FeCu_2(NH_3)_2$	Fe:	Org.Verb.B2-173
$C_4FH_3N_2O_2$	$NHC(O)NHC(O)CHCF$	F:	PerFHalOrg.6-57, 72
$C_4FH_4N_5O_2$	$NCFNC(NH_2)C(NO_2)C(NH_2)$	F:	PerFHalOrg.6-21/2, 42
$C_4FH_5O_2$	$CH_3CHCFCOOH$	F:	PerFHalOrg.4-89
$C_4FH_7N_2O_5S$	$HOCOCH_2NHS(O)FNCH_2COOH$	S:	S-N-Verb.1-192/3
$C_4FH_8NO_2S$	$FSO_2NCHCH(CH_3)_2$	S:	S-N-Verb.1-105
$C_4FH_8NO_3S$	$FSO_2NC(OC_2H_5)CH_3$	S:	S-N-Verb.1-105
$C_4FH_8NO_3Sn$	$[SnF(HOC_2H_4NHCH_2COO)]$	Sn:	MVol.C5-31
$C_4FH_8NO_4S$	$FSO_2N(CH_3)COOC_2H_5$	S:	S-N-Verb.1-104

$C_4FH_9N_2O_2S$	$(CH_3)_2NC(CH_3)NSO_2F$	S:	S-N-Verb.1-115
–	$FSO_2NC(NHC_2H_5)CH_3$	S:	S-N-Verb.1-105
$C_4FH_9N_2O_2SSi$	$FSO_2NCNSi(CH_3)_3$	S:	S-N-Verb.1-106
$C_4FH_9N_2O_3S$	$FSO_2N(CH_3)C(O)N(CH_3)_2$	S:	S-N-Verb.1-104
–	$FSO_2NC[N(CH_3)_2]OCH_3$	S:	S-N-Verb.1-119
$C_4FH_{10}ISn$	$(C_2H_5)_2SnFI$	Sn:	Org.Verb.5-22
$C_4FH_{10}NO_2S$	$FSO_2N(C_2H_5)_2$	S:	S-N-Verb.1-104
$C_4FH_{10}Sn^+$	$(C_2H_5)_2SnF^+$	Sn:	Org.Verb.5-20
$C_4FH_{11}Sn$	$(CH_3)_3SnCH_2F$	Sn:	Org.Verb.2-6
$C_4FH_{12}NO_2SSi$	$FSO_2N(CH_3)Si(CH_3)_3$	S:	S-N-Verb.1-106
$C_4FH_{12}NO_3S$	$[(C_2H_5)_2NH_2]SO_3F$	S:	S-N-Verb.1-104
$C_4FH_{12}N_3O_3S_2$	$[(CH_3)_2N]_2S(O)NSO_2F$	S:	S-N-Verb.1-207
$C_4FH_{12}N_4O_4PS_2$	$[SO_2(NCH_3)_2]_2PF$	S:	S-N-Verb.1-42
$C_4FH_{12}PS_2Sn$	$(CH_3)_3SnSP(S)FCH_3$	Sn:	Org.Verb.5-84
$C_4F_2FeIO_4P$	$(CO)_4FePF_2I$	Fe:	Org.Verb.B2-82
$C_4F_2FeN_3O_4P$	$(CO)_4FePF_2N_3$	Fe:	Org.Verb.B2-85, 93
$C_4F_2FeO_{10}S_2$	$(CO)_4Fe(SO_3F)_2$	Fe:	Org.Verb.B3-120
$C_4F_2H_2IN_3$	$NCFNC(NH_2)CFCl$	F:	PerFHalOrg.6-20/1, 42
$C_4F_2H_2N_2$	$C_4F_2H_2N_2$	F:	PerFHalOrg.6-69/70
$C_4F_2H_2N_2O_2$	NHCOCFCFCONH	F:	PerFHalOrg.6-15, 35
$C_4F_2H_2N_2O_3$	$NHC(O)NHC(O)CF_2C(O)$	F:	PerFHalOrg.6-21/2, 43
$C_4F_2H_3NOS^+$	$C_4H_3NSOF_2{}^+$	S:	S-N-Verb.1-157
$C_4F_2H_4N_4$	$NCFNC(NH_2)CFC(NH_2)$	F:	PerFHalOrg.6-20/1, 41
–	$NC(NH_2)CFNCFC(NH_2)$	F:	PerFHalOrg.6-29/30, 50/1
$C_4F_2H_6Sn$	$(CH_2CH)_2SnF_2$	Sn:	Org.Verb.5-49
$C_4F_2H_7NOS^+$	$C_4H_7NSOF_2{}^+$	S:	S-N-Verb.1-157
$C_4F_2H_7NO_3S$	$FSO_2N(C_3H_7)C(O)F$	S:	S-N-Verb.1-104
$C_4F_2H_8O$	$(CH_3)_2CHOCF_2H$	F:	PerFHalOrg.5-19
$C_4F_2H_9NOS^+$	$C_4H_9NSOF_2{}^+$	S:	S-N-Verb.1-157
$C_4F_2H_9NO_3Sn$	$SnF_2 \cdot HOC_2H_4NHCH_2COOH$	Sn:	MVol.C5-31
$C_4F_2H_9NO_3U$	$[UO_2F_2 \cdot CH_3CON(CH_3)_2]_n$	U:	SVol.E1-103/4, 106
$C_4F_2H_{10}N_2O_3S_2$	$(C_2H_5)_2NS(O)FNSO_2F$	S:	S-N-Verb.1-196
$C_4F_2H_{10}N_2S$	$NSF_2[N(C_2H_5)_2]$	S:	S-N-Verb.1-204
$C_4F_2H_{10}N_4O_3S_3$	$[NS(O)F]_2[NS(O)N(C_2H_5)_2]$	S:	S-N-Verb.1-14
$C_4F_2H_{10}O_6S_2Sn$	$(C_2H_5)_2Sn(SO_3F)_2$	Sn:	Org.Verb.6-56
$C_4F_2H_{10}Sn$	$(CH_3)_3SnCHF_2$	Sn:	Org.Verb.2-7
–	$(C_2H_5)_2SnF_2$	Sn:	Org.Verb.5-45/6
$C_4F_2H_{12}N_2O_4S_2$	$[(CH_3)_4N]N(SO_2F)_2$	S:	S-N-Verb.1-50, 53
$C_4F_2H_{12}N_4O_4S_3$	$[(CH_3)_4N][SO_2(NS(O)F)_2N]$	S:	S-N-Verb.1-11
–	$(FSO_2N)_2S[N(CH_3)_2]_2$	S:	S-N-Verb.1-212
$C_4F_2H_{13}N_5O_3S_3$	$[NS(O)F]_2[NS(O)NHC_2H_5] \cdot C_2H_5NH_2$	S:	S-N-Verb.1-14
$C_4F_2H_{16}N_8S_4Te$	$Te(SC(NH_2)_2)_4F_2 \cdot 2\ H_2O$	Te:	SVol.B3-144
$C_4F_2I_2N_2$	NCFNCICFCl	F:	PerFHalOrg.6-19/20, 42
$C_4F_2N_6O_{10}$	$(NO_2)_2CFCNON(O)CCF(NO_2)_2$	F:	PerFHalOrg.5-66/7, 89
$C_4F_2N_8$	$NC(N_3)CFNC(N_3)CF$	F:	PerFHalOrg.6-29/30, 51
$C_4F_3FeO_4P$	$(CO)_4FePF_3$	Fe:	Org.Verb.B2-84, 93, 112/3
$C_4F_3GeH_{10}P$	$(CH_3)_3GeP(H)CF_3$	F:	PerFHalOrg.3-25
$C_4F_3HN_2$	$C_4F_3HN_2$	F:	PerFHalOrg.6-69/70

$C_4F_3HN_2O$ .... NCFNC(OH)CFCF .... F: PerFHalOrg.6-20/1, 41, 58, 60
– .... NC(OH)CFCFCFN .... F: PerFHalOrg.6-14/5, 34
– .... NC(OH)CFNCFCF .... F: PerFHalOrg.6-28/30, 50, 60
– .... NHNCFCFCFC(O) .... F: PerFHalOrg.6-60, 72
$C_4F_3H_2N_3$ .... $NCFC(NH_2)CFCFN$ .... F: PerFHalOrg.6-14/5, 34/5, 72
– .... $NCFNC(NH_2)CFCF$ .... F: PerFHalOrg.6-20/1, 41
– .... $NC(NH_2)CFNCFCF$ .... F: PerFHalOrg.6-28/30, 50
$C_4F_3H_2N_3O_2$ .... $(CF_3)(OH)_2C_3N_3$ .... F: PerFHalOrg.6-79/80, 100, 111
$C_4F_3H_2N_5O_3S$ ...... $FSO_2NHCONHC(NCF)_2N$ .... S: S-N-Verb.1-102
$C_4F_3H_2NiO_3P$ .... $(CO)_3NiP(CF_3)H_2$ .... Ni: Org.Verb.1-164, 165
$C_4F_3H_3N_4$ .... $NC(NHNH_2)CFNCFCF$ .... F: PerFHalOrg.6-28/30, 50
$C_4F_3H_4NO_3$ .... $CF_3N(OH)CHCHCOOH$ .... F: PerFHalOrg.7-164, 168
$C_4F_3H_4N_5$ .... $(CF_3)(NH_2)_2C_3N_3$ .... F: PerFHalOrg.6-78/80, 100, 112, 115
$C_4F_3H_5N_2$ .... $CF_3NNCH_2CHCH_2$ .... F: PerFHalOrg.7-159
$C_4F_3H_5O_2$ .... $CF_3COOC_2H_5$ .... F: PerFHalOrg.6-112
$C_4F_3H_6NO$ .... $CF_3N(OH)CH_2CHCH_2$ .... F: PerFHalOrg.7-164, 166/7
$C_4F_3H_6N_2O_6P$ ...... $NO_2CF_2CFNOPO(OCH_3)_2$ .... F: PerFHalOrg.7-199, 200
$C_4F_3H_7NO$ .... $CF_3N(O)C_3H_7$ .... F: PerFHalOrg.7-169
$C_4F_3H_7N_2$ .... $CF_3NNCH(CH_3)_2$ .... F: PerFHalOrg.7-159
– .... $CF_3NNC_3H_7$ .... F: PerFHalOrg.7-158/9
$C_4F_3H_7N_2O$ .... $(CH_3)_2CHNN(O)CF_3$ .... F: PerFHalOrg.7-160/1
$C_4F_3H_7N_2O_2$ .... $CF_3C(NH)NH_2 \cdot CH_3COOH$ .... F: PerFHalOrg.7-57
$C_4F_3H_7O$ .... $CF_3(CH_3)_2COH$ .... F: PerFHalOrg.4-44
$C_4F_3H_7Si$ .... $CF_2CFSi(CH_3)_2H$ .... F: PerFHalOrg.4-72
$C_4F_3H_7Sn$ .... $(CH_3)_2(CF_2CF)SnH$ .... Sn: Org.Verb.4-88
$C_4F_3H_9IP$ .... $[CF_3P(CH_3)_3]I$ .... F: PerFHalOrg.3-173
$C_4F_3H_9NO_2SSn^+$ .... $(CH_3)_2SnNCH_3SO_2CF_3{}^+$ .... S: S-N-Verb.1-108
$C_4F_3H_9P_2$ .... $(CH_3)_3PPCF_3$ .... F: PerFHalOrg.3-6, 7
$C_4F_3H_9Sn$ .... $(CH_3)_2(CHF_2CHF)SnH$ .... Sn: Org.Verb.4-89
– .... $(CH_3)_3SnCF_3$ .... Sn: Org.Verb.2-8/9
– .... $C_4H_9SnF_3$ .... Sn: Org.Verb.5-51/2
$C_4F_3H_{10}NOS$ .... $(C_2H_5)_2NS(O)F_3$ .... S: S-N-Verb.1-157
$C_4F_3H_{10}NOSi$ .... $CF_3N(H)OSi(CH_3)_3$ .... F: PerFHalOrg.7-169
$C_4F_3H_{10}NO_2SSi$ ..... $CF_3SO_2NHSi(CH_3)_3$ .... S: S-N-Verb.1-107
$C_4F_3H_{10}NO_3Sn$ ..... $[HOC_2H_4NH_2CH_2COOH][SnF_3]$ .... Sn: MVol.C5-31
$C_4F_3H_{10}PSi$ .... $(CH_3)_3SiP(H)CF_3$ .... F: PerFHalOrg.3-25
$C_4F_3H_{10}PSn$ .... $(CH_3)_3SnP(H)CF_3$ .... F: PerFHalOrg.3-25
$C_4F_3H_{10}Sn^-$ .... $[(C_2H_5)_2SnF_3]^-$ .... Sn: Org.Verb.5-46
$C_4F_3IN_2$ .... NCFNCFCFCI .... F: PerFHalOrg.6-19/20, 42, 58, 70
$C_4F_3N_3OS$ .... $NC(CF_3)SC(NCO)N$ .... F: PerFHalOrg.5-67/8, 107
$C_4F_3N_3O_2$ .... $NCFNCFC(NO_2)CF$ .... F: PerFHalOrg.6-21/2, 42
$C_4F_4FeH_2O_4$ ...... $(CHF_2COO)_2Fe$ .... Fe: Org.Verb.B3-176/7
$C_4F_4HNO_2$ .... $HN(COCF_2)_2$ .... F: PerFHalOrg.5-51/2, 75
$C_4F_4H_2N_2$ .... $CF_3C(NH_2)CFCN$ .... F: PerFHalOrg.7-3, 28

$C_4F_4H_2N_2OS$	$CF(CF_3)CONC(NH_2)S$	F:	PerFHalOrg.5-56/8, 78
$C_4F_4H_2N_2O_3$	$NO_2CF_2CFCFC(O)NH_2$	F:	PerFHalOrg.7-18, 20/1, 46
$C_4F_4H_2O_5$	$O(CF_2COOH)_2$	F:	PerFHalOrg.6-12
$C_4F_4H_3NO_3$	$HOCO(CF_2)_2C(O)NH_2$	F:	PerFHalOrg.7-18, 21, 47
$C_4F_4H_3N_3$	$CF_2C(NH_2)NC(NH)CF_2$	F:	PerFHalOrg.5-51, 97
$C_4F_4H_3N_5O_4$	$NO_2CF_2C(NH)NC(NH_2)CF_2NO_2$	F:	PerFHalOrg.7-3/4, 31
$C_4F_4H_4N_2O_2$	$NH_2C(O)(CF_2)_2C(O)NH_2$	F:	PerFHalOrg.7-18, 20, 22, 47
$C_4F_4H_4N_2O_2S$	$NH_2C(O)CF_2SCF_2C(O)NH_2$	F:	PerFHalOrg.7-18, 20, 47
$C_4F_4H_4N_2O_3$	$NH_2C(O)CF_2OCF_2C(O)NH_2$	F:	PerFHalOrg.7-18, 20, 47
$C_4F_4H_5N_2O_5P$	$NO_2CF_2CFNOPF(O)OC_2H_5$	F:	PerFHalOrg.7-199/200
$C_4F_4H_5N_3O_2$	$O_2NCF_2CF_2NNC_2H_5$	F:	PerFHalOrg.7-192
$C_4F_4H_6N_2O_2S$	$(CH_3)_2NS(O)FNC(O)CF_3$	S:	S-N-Verb.1-194
–	$FSO_2NC[N(CH_3)_2]CF_3$	S:	S-N-Verb.1-118
$C_4F_4H_6N_2O_2S^+$	$(CH_3)_2NSOFNCOCF_3{}^+$	S:	S-N-Verb.1-194
$C_4F_4H_6N_2O_3$	$NH_4OCO(CF_2)_2C(O)NH_2$	F:	PerFHalOrg.7-18, 21, 47
$C_4F_4H_6N_2Sn$	$SnF_4 \cdot 2\ CH_3CN$	Sn:	MVol.C6-64
$C_4F_4H_7NO_2S$	$CF_3CF(CH_3)SO_2NHCH_3$	F:	PerFHalOrg.7-193
$C_4F_4H_7N_3O_2S$	$CF_3CF(SO_2NHCH_3)NNCH_3$	F:	PerFHalOrg.7-193
$C_4F_4H_8N_2S_2$	$F_2SNCH_2C(CH_3)_2NSF_2$	S:	SVol.2-78, 85
$C_4F_4H_8OSn$	$SnF_4 \cdot C_4H_8O$	Sn:	MVol.C5-156
$C_4F_4H_8O_2Te$	$TeF_4 \cdot C_4H_8O_2$	Te:	SVol.B3-166
$C_4F_4H_8Sn$	$(CH_3)_2(CHF_2CF_2)SnH$	Sn:	Org.Verb.4-88/9
$C_4F_4H_9NO_3Sn_2$	$2\ SnF_2 \cdot HOC_2H_4NHCH_2COOH$	Sn:	MVol.C5-31
$C_4F_4H_{10}K_2Sn$	$(C_2H_5)_2SnF_2 \cdot 2\ KF$	Sn:	Org.Verb.5-46
$C_4F_4H_{10}O_2Sn$	$SnF_4 \cdot C_2H_4(OCH_3)_2$	Sn:	MVol.C5-153/4
$C_4F_4H_{10}O_4P_2Sn$	$(C_2H_5)_2Sn(PO_2F_2)_2$	Sn:	Org.Verb.6-56
$C_4F_4H_{10}Sn^{2-}$	$[(C_2H_5)_2SnF_4]^{2-}$	Sn:	Org.Verb.5-46
$C_4F_4H_{12}MnN$	$N(CH_3)_4MnF_4 \cdot 2\ H_2O$	Mn:	MVol.C4-212/3
–	$NH_2(C_2H_5)_2MnF_4 \cdot 2\ H_2O$	Mn:	MVol.C4-212/3
$C_4F_4H_{12}N_2Te$	$[(CH_3)_2N]_2TeF_4$	Te:	SVol.B2-65
$C_4F_4H_{12}N_3S_2{}^+$	$SF_4NS(N(CH_3)_2)_2{}^+$	S:	S-N-Verb.1-141
$C_4F_4H_{12}O_2S_2Sn$	$SnF_4 \cdot 2\ (CH_3)_2SO$	Sn:	MVol.C6-100
$C_4F_4H_{12}O_2Sn$	$SnF_4 \cdot 2\ C_2H_5OH$	Sn:	MVol.C5-70
$C_4F_4H_{16}N_8O_8U_2$	$[UO_2F_2 \cdot 2\ (NH_2)_2CO]_2$	U:	SVol.E1-83/4, 87
$C_4F_4H_{18}N_8S_4Te$	$Te(SC(NH_2)_2)_4(HF_2)_2$	Te:	SVol.B3-144
$C_4F_4Li_2$	$LiCCF_2CF_2CLi$	F:	PerFHalOrg.4-3, 13, 26, 31, 33
$C_4F_4N_2$	$C_4F_4N_2$	F:	PerFHalOrg.6-32/3, 53, 55/6
–	NCFCFCFCFN	F:	PerFHalOrg.6-14/5, 33/4, 55/60, 69/70, 72
–	NCFCFNCFCF	F:	PerFHalOrg.6-28/30, 33/4, 48, 57/60, 69/70
–	NCFNCFCFCF	F:	PerFHalOrg.6-19/20, 33/4, 40, 57/60, 69
$C_4F_4N_2O_2$	$ONCF_2CF_2C(O)CN$	F:	PerFHalOrg.7-177, 184
$C_4F_4N_2O_3$	$ONCF_2CF_2C(O)NCO$	F:	PerFHalOrg.7-177, 184
$C_4F_{4.6}H_{12}NS_{0.9}$	$[(CH_3)_4N]F \cdot 0.9\ SF_4$	S:	SVol.2-75

$C_4F_5FeH_3O_4P_2$	$(CO)_3FeP_2F_5(OCH_3)$	Fe:	Org.Verb.B1-161
$C_4F_5HN_2$	$CF_3CFNHC(CN)(F)$	F:	PerFHalOrg.5-2/3, 9
–	$C_2F_5C(NH)CN$	F:	PerFHalOrg.7-25/6, 52
$C_4F_5HN_4$	$(CF_3NH)C_3F_2N_3$	F:	PerFHalOrg.6-78, 99
$C_4F_5H_3N_4$	$NC(C_2F_5)NC(NH_2)NH$	F:	PerFHalOrg.5-71/3, 92/3
$C_4F_5H_3N_4S$	$NC(C_2F_5)N(NH_2)C(SH)N$	F:	PerFHalOrg.5-71/3, 93, 100
$C_4F_5H_3O$	$(CF_2)_2C(OCH_3)F$	F:	PerFHalOrg.5-17
$C_4F_5H_3O_2$	$CH_3O(CF_2)_2CFO$	C:	MVol.D3-46
$C_4F_5H_4N_3O_4$	$NO_2CF_2C(NH)NH_2 \cdot CF_3COOH$	F:	PerFHalOrg.7-3/4, 30
$C_4F_5H_5N_4O$	$C_2F_5CONHNHC(NH)NH_2$	F:	PerFHalOrg.7-6/7, 34
$C_4F_5H_6OP$	$CF_3P(OC_2H_5)CHF_2$	F:	PerFHalOrg.3-24
$C_4F_5H_7N_2$	$(CH_3)_2NCF_2(CF_3)NH$	F:	PerFHalOrg.7-58
$C_4F_5H_7O_3Te$	$TeF_5OCH_2COOC_2H_5$	Te:	SVol.B2-53
$C_4F_5H_{10}NTe$	$(C_2H_5)_2NTeF_5$	Te:	SVol.B2-65/6
$C_4F_5H_{10}NTe^+$	$TeF_5N(C_2H_5)_2{}^+$	Te:	SVol.B2-66
$C_4F_5H_{10}N_2S_2{}^+$	$SF_5NSN(C_2H_5)_2{}^+$	S:	S-N-Verb.1-142, 144
$C_4F_5H_{12}NS$	$[(CH_3)_4N]SF_5$	S:	SVol.2-211
$C_4F_5H_{12}NTe$	$[(CH_3)_4N]TeF_5$	Te:	SVol.B2-15
$C_4F_5H_{12}N_3S_2$	$SF_5NS[N(CH_3)_2]_2$	S:	S-N-Verb.1-141/2
$C_4F_5H_{12}N_3S_2{}^+$	$SF_5NS(N(CH_3)_2)_2{}^+$	S:	S-N-Verb.1-141
$C_4F_5H_{16}MnN_2$	$(NH_3C_2H_5)_2MnF_5$	Mn:	MVol.C4-212/3
C_4F_5IMg	CF_3CF_2CCMgI	F:	PerFHalOrg.4-62/3, 75/6
C_4F_5Li	$CFCF_2CF_2CLi$	F:	PerFHalOrg.4-3, 44
C_4F_5N	$(CF_2)_2C(CN)F$	F:	PerFHalOrg.5-17
$C_4F_5N_3$	$(CF_3)C_3F_2N_3$	F:	PerFHalOrg.6-78/80, 99, 111
$C_4F_6FeNOS_2$	$C_4F_6S_2FeNO$	Fe:	Org.Verb.C1-64
$C_4F_6FeO_4$	$(CF_3COO)_2Fe$	Fe:	Org.Verb.B3-176/7
C_4F_6HNO	$(CF_2)_3CNOH$	F:	PerFHalOrg.7-88/9, 105
$C_4F_6HNO_2$	$[CF_3C(O)]_2NH$	F:	PerFHalOrg.7-24, 50, 55, 63
$C_4F_6HNO_3$	$ONCF_2CF_2CF_2C(O)OH$	F:	PerFHalOrg.7-178/9, 185, 192/3
$C_4F_6HNO_4S$	$[HNSO_2][OC(CF_3)_2CO]$	F:	PerFHalOrg.5-57/8, 79/80
		S:	S-N-Verb.1-40
$C_4F_6HN_3$	$NC(CF_3)NHC(CF_3)N$	F:	PerFHalOrg.5-71/2, 92, 97, 99
$C_4F_6HN_3O_2S$	$[NSO_2][NCCF_3][HNCCF_3]$	F:	PerFHalOrg.6-94, 110
		S:	S-N-Verb.1-31
$C_4F_6HNiO_2P$	$[(CF_3)_2PHNi(CO)_2]_x$	Ni:	Org.Verb.2-395
$C_4F_6H_2INO$	$CF_2ICF_2CF_2C(O)NH_2$	F:	PerFHalOrg.7-18/20, 46
$C_4F_6H_2N_2$	$(CF_3)_2C(CN)NH_2$	F:	PerFHalOrg.7-2/3, 30
$C_4F_6H_2N_2O$	$CF_3CFNHCONHCF_2$	F:	PerFHalOrg.5-61/2, 84
–	$(CF_3)_2C(NCO)NH_2$	F:	PerFHalOrg.7-2, 30
$C_4F_6H_2N_2O_2$	$CF_3C(NH_2)NOC(O)CF_3$	F:	PerFHalOrg.7-5/6, 54
–	$CF_3C(NOH)C(NOH)CF_3$	F:	PerFHalOrg.7-89, 105
–	$CF_3NHC(O)C(O)NHCF_3$	F:	PerFHalOrg.7-22/3, 49
$C_4F_6H_2N_2O_3$	$CF_3N(OH)C(O)C(O)NHCF_3$	F:	PerFHalOrg.7-22/3, 49

$C_4F_6H_2N_2O_4$	$CF_3N(OH)COCON(OH)CF_3$	F:	PerFHalOrg.7-87, 104
$C_4F_6H_2O$	$CF_3CH(OH)CFCF_2$	F:	PerFHalOrg.4-89
$C_4F_6H_2O_2$	$(CF_3)_2CHC(O)OH$	F:	PerFHalOrg.7-190
$C_4F_6H_3IN_2$	$I(CF_2)_3C(NH)NH_2$	F:	PerFHalOrg.7-3/4, 31
$C_4F_6H_3NO$	$(CF_3)_2CNOCH_3$	F:	PerFHalOrg.7-143
$C_4F_6H_3NO_2$	$(CF_3)_2NOC(O)CH_3$	F:	PerFHalOrg.7-129/30, 136, 140
$C_4F_6H_3NO_3$	$CF_3NHC(O)C(OH)_2CF_3$	F:	PerFHalOrg.7-22/3, 49
–	$(CF_3)_2NOC(O)OCH_3$	F:	PerFHalOrg.7-126
$C_4F_6H_3N_3$	$CF_3C(NH)NC(NH_2)CF_3$	F:	PerFHalOrg.7-7, 34, 58
$C_4F_6H_3N_3O$	$CF_3C(NH_2)NNHCOCF_3$	F:	PerFHalOrg.7-6/7, 33
$C_4F_6H_3O_2P$	$(CF_3)_2POC(O)CH_3$	F:	PerFHalOrg.3-101, 103
$C_4F_6H_4N_2O$	$(CF_3)_2C(NH_2)C(O)NH_2$	F:	PerFHalOrg.7-2, 29
$C_4F_6H_4N_4$	$CF_3NNCH_2CH_2NNCF_3$	F:	PerFHalOrg.7-158/9
$C_4F_6H_4NiO_2P_2$	$(CO)_2Ni(P(CF_3)H_2)_2$	Ni:	Org.Verb.1-127, 128
$C_4F_6H_5NO$	$(CF_3)_2C(NH_2)OCH_3$	F:	PerFHalOrg.7-65
–	$(CF_3)_2NOCH_2CH_3$	F:	PerFHalOrg.7-129/30
$C_4F_6H_5NO_2$	$(CF_3)_2CNOH \cdot CH_3OH$	F:	PerFHalOrg.7-88/9
$C_4F_6H_5OP$	$(CF_3)_2POC_2H_5$	F:	PerFHalOrg.3-101, 103
$C_4F_6H_5P$	$(CF_3)_2PC_2H_5$	F:	PerFHalOrg.3-102
$C_4F_6H_6NOP$	$(CF_3)_2P(O)N(CH_3)_2$	F:	PerFHalOrg.3-35, 36
$C_4F_6H_6NP$	$(CF_3)_2PN(CH_3)_2$	F:	PerFHalOrg.3-108, 114
–	$(CF_3)_2PNHC_2H_5$	F:	PerFHalOrg.3-108, 115
$C_4F_6H_6NPS$	$(CF_3)_2P(S)N(CH_3)_2$	F:	PerFHalOrg.3-153/5
$C_4F_6H_6N_2O$	$(CF_3)_2NON(CH_3)_2$	F:	PerFHalOrg.7-127, 128
$C_4F_6H_6P_2$	$(CF_3)_2PP(CH_3)_2$	F:	PerFHalOrg.3-101, 104, 182
$C_4F_6H_7N_2P$	$(CF_3)_2PN(H)N(CH_3)_2$	F:	PerFHalOrg.3-109, 118
$C_4F_6H_8NOPS$	$[(CH_3)_2NH_2][(CF_3)_2P(S)O]$	F:	PerFHalOrg.3-153
$C_4F_6H_8NO_2P$	$[(CH_3)_2NH_2][(CF_3)_2P(O)O]$	F:	PerFHalOrg.3-153
$C_4F_6H_8N_2O$	$(CH_3)_2NH \cdot (CF_3)_2NOH$	F:	PerFHalOrg.7-139
–	$C_2H_5NH_2 \cdot (CF_3)_2NOH$	F:	PerFHalOrg.7-139
$C_4F_6H_8N_2U$	$UF_4 \cdot 2\ HF \cdot 2\ CH_3CN$	U:	SVol.E1-48/50
$C_4F_6H_{10}N_2S_2$	$SF_5NSF[N(C_2H_5)_2]$	S:	S-N-Verb.1-144
$C_4F_6H_{10}N_2S_2^+$	$SF_5NSFN(C_2H_5)_2^+$	S:	S-N-Verb.1-144
$C_4F_6H_{10}NiO_2P_2$	$Ni(PF_3)_2(CH_3OC_2H_4OCH_3)$	Ni:	Org.Verb.1-124
$C_4F_6H_{16}IrN_{10}P$	$[Ir(C_2H_4(NH_2)_2)_2(N_3)_2]PF_6$	Ir:	SVol.2-41
C_4F_6Hg	$(CF_2CF)_2Hg$	F:	PerFHalOrg.4-33, 110
$C_4F_6KNO_3$	$ONCF_2CF_2CF_2C(O)OK$	F:	PerFHalOrg.7-178/9, 185
$C_4F_6N_2$	NCF_2CF_2NCFCF	F:	PerFHalOrg.6-28/30, 49
$C_4F_6N_2O$	$CF_3CNC(CF_3)NO$	F:	PerFHalOrg.5-66, 88
–	$(CF_3)_2C(CN)NO$	F:	PerFHalOrg.7-177, 185
–	$NC(CF_3)OC(CF_3)N$	F:	PerFHalOrg.5-64/5, 87
$C_4F_6N_2O_2$	$CF_3CNON(O)CCF_3$	F:	PerFHalOrg.5-66/7, 89
–	$C_4F_6N_2O_2$	F:	PerFHalOrg.6-123, 138
$C_4F_6N_2O_4$	$ONCF_2CF_2CF_2C(O)ONO$	F:	PerFHalOrg.7-178/9
$C_4F_6NiS_2$	$NiS_2C_2(CF_3)_2$	Ni:	Org.Verb.1-233
C_4F_6OSi	$[(CF_2CF)_2SiO]_n$	F:	PerFHalOrg.4-147/8
$C_4F_6O_3$	$[CF_3C(O)]_2O$	F:	PerFHalOrg.4-179

C_4F_7HINO	$C_3F_7C(O)NHI$	F:	PerFHalOrg.7-18, 22, 56/7
$C_4F_7HN_2O_3$	$C_3F_7C(NO_2)NOH$	F:	PerFHalOrg.7-88/9
$C_4F_7HN_4$	$NHN_3C(C_3F_7)$	F:	PerFHalOrg.5-73/4, 93
C_4F_7HO	C_3F_7CHO	F:	PerFHalOrg.4-104
$C_4F_7HO_2$	C_3F_7COOH	F:	PerFHalOrg.4-95
$C_4F_7H_2NO$	$(CF_3)_2CHNHCOF$	F:	PerFHalOrg.5-21
–	$C_3F_7C(O)NH_2$	F:	PerFHalOrg.7-18/21, 46, 57, 63
–	$[N(C_3F_7)OCH_2]_n$	F:	PerFHalOrg.7-209
$C_4F_7H_2NO_2$	$CF_3OCF_2CF_2C(O)NH_2$	F:	PerFHalOrg.7-18, 21
–	$C_3F_7NHCOOH$	F:	PerFHalOrg.7-22/3, 49
$C_4F_7H_2NO_4S$	$(CF_3)_2C(OH)C(O)NHSO_2F$	F:	PerFHalOrg.7-24/5, 51
		S:	S-N-Verb.1-101
$C_4F_7H_2NO_4S^+$	$FSO_2NHCOC(CF_3)_2OH^+$	S:	S-N-Verb.1-101
$C_4F_7H_3N_2$	$C_3F_7C(NH)NH_2$	F:	PerFHalOrg.7-3/4, 30, 57/8
$C_4F_7H_3N_2O$	$CF_3OCF_2CF_2C(NH)NH_2$	F:	PerFHalOrg.7-3/4, 57/8
–	$C_3F_7C(NOH)NH_2$	F:	PerFHalOrg.7-5, 32, 58
$C_4F_7H_3O$	$CF_3CF_2CF_2CH_2OH$	F:	PerFHalOrg.4-92
$C_4F_7H_3O_2$	$C_3F_7CH(OH)_2$	F:	PerFHalOrg.4-95
$C_4F_7H_4N_3$	$C_3F_7C(NH_2)NNH_2$	F:	PerFHalOrg.7-6/7, 33, 58
$C_4F_7H_4P$	$(CF_3)_2PCH_2CH_2F$	F:	PerFHalOrg.3-29
C_4F_7N	$CF_2CF_2NCFCF_2$	F:	PerFHalOrg.5-51, 75, 97
–	$(CF_3)_2CNCF$	F:	PerFHalOrg.5-2, 8, 21
C_4F_7NO	$CF_2CFNOCF_2CF_2$	F:	PerFHalOrg.5-31, 40
–	$(CF_2)_3CFNO$	F:	PerFHalOrg.7-178/9
–	$CF_3NOCF_2CCF_2$	F:	PerFHalOrg.5-31/2, 40, 44, 47/9
–	$[N(CFCF_2)OCF_2CF_2]_n$	F:	PerFHalOrg.7-208/9
–	$OCF_2CFNCF_2CF_2$	F:	PerFHalOrg.6-1, 6, 11/2
$C_4F_7NO_2$	$CF_3NOCF(COF)CF_2$	F:	PerFHalOrg.5-30, 39/40
–	$(CF_3)_2C(NO)C(O)F$	F:	PerFHalOrg.7-178/9, 187, 190/1
–	$FC(O)CF_2NOCF_2CF_2$	F:	PerFHalOrg.5-35, 42
–	$ONCF_2CF_2CF_2C(O)F$	F:	PerFHalOrg.7-178/9, 185, 189/90
$C_4F_7NO_3S$	$[NS(O)F][OC(CF_3)_2CO]$	F:	PerFHalOrg.5-57/8, 79, 98
		S:	S-N-Verb.1-40/1
$C_4F_7N_3S$	$NNNC(C_3F_7)S$	F:	PerFHalOrg.5-72/3, 93, 97
C_4F_7P	$(CF_2CF)_2PF$	F:	PerFHalOrg.3-71, 85
C_4F_8	$(CF_2)_2CFCF_3$	F:	PerFHalOrg.5-17
–	$CF_3CFCFCF_3$	F:	PerFHalOrg.4-161
C_4F_8HN	$(CF_3)_2CNHCF_2$	F:	PerFHalOrg.5-2, 8
C_4F_8HNO	$O(CF_2CF_2)_2NH$	F:	PerFHalOrg.6-1, 6
$C_4F_8H_2N_2$	$(CF_3)(C_2F_5)C(NH)_2$	F:	PerFHalOrg.5-6
$C_4F_8H_2N_2O$	$(CF_3)_2NCF_2C(O)NH_2$	F:	PerFHalOrg.7-17, 19/20, 45

$C_4F_8H_3O_2P$	$HCF_2(CF_2)_3P(OH)_2$	F:	PerFHalOrg.3-13
$C_4F_8H_3O_3P$	$HCF_2(CF_2)_3P(O)(OH)_2$	F:	PerFHalOrg.3-13
$C_4F_8H_3P$	$(CF_3)_2PCH_2CHF_2$	F:	PerFHalOrg.3-28
–	$(HCF_2CF_2)_2PH$	F:	PerFHalOrg.3-16
$C_4F_8H_5N_2P$	$(CF_3)_2C(NH_2)NP(CH_3)F_2$	F:	PerFHalOrg.7-57
$C_4F_8H_6S$	$C_2H_5CF_2CHFSF_5$	S:	SVol.2-200
$C_4F_8H_9N_2O_3P$	$HCF_2(CF_2)_3P(O)(ONH_4)_2$	F:	PerFHalOrg.3-13
$C_4F_8H_9PSn$	$[(CH_3)_3Sn][CF_3PF_5]$	F:	PerFHalOrg.3-134
C_4F_8IP	$PI(CF_2)_4$	F:	PerFHalOrg.3-9, 12
$C_4F_8I_2P_2$	$IP(CF_2CF_2)_2PI$	F:	PerFHalOrg.3-9, 12, 13
$C_4F_8Li_2$	$LiCF_2CF_2CF_2CF_2Li$	F:	PerFHalOrg.4-3, 42
$C_4F_8NO_2$	$ON(CF_2CF_2)_2O$	F:	PerFHalOrg.6-2/3, 7
$C_4F_8N_2$	$CFNCF_2N(CF_3)CF_2$	F:	PerFHalOrg.5-62, 85
–	$CF_2NCFN(CF_3)CF_2$	F:	PerFHalOrg.5-62, 85
–	$CF_2N(CF_2)_2NCF_2$	F:	PerFHalOrg.5-15/6
–	$(CF_2)_2NN(CF_2)_2$	F:	PerFHalOrg.5-3, 9/10
$C_4F_8N_2O$	$CF_3NOC(NCF_3)CF_2$	F:	PerFHalOrg.5-31/2, 40, 47/8
$C_4F_8N_2O_2$	$[N(CF_2CF_2NO)OCF_2CF_2]_n$	F:	PerFHalOrg.7-208/9
$C_4F_8N_2O_3$	$C_4N_2O_3F_8$	F:	PerFHalOrg.7-207
–	$[N(CF_2CF_2NO_2)OCF_2CF_2]_n$	F:	PerFHalOrg.7-208/9
–	$NO_2CF_2CF_2NOCF_2CF_2$	F:	PerFHalOrg.5-35, 42, 48
–	$O(CF_2CF_2)_2NNO_2$	F:	PerFHalOrg.6-1, 6
–	$O(CF_2CF_2)_2NONO$	F:	PerFHalOrg.6-2/3, 8
$C_4F_8NiO_2P$	$(CF_3)_2PF_2Ni(CO)_2$	F:	PerFHalOrg.3-121
$C_4F_8Se_3^+$	$[(F_2CSe)_3CF_2]^+$	C:	MVol.D6-251
$C_4F_9FeH_6OP_3$	$(CH_2CHC(O)CH_3)Fe(PF_3)_3$	Fe:	Org.Verb.B5-112/3
–	$(CH_3CHCHCHO)Fe(PF_3)_3$	Fe:	Org.Verb.B5-112/3
$C_4F_9FeH_9O_4P_4$	$COFeP_4F_9(OCH_3)_3$	Fe:	Org.Verb.B1-28/9
$C_4F_9H_2N$	$(CF_3)_3CNH_2$	F:	PerFHalOrg.7-1/2, 28, 57
$C_4F_9H_2NO$	$(CF_3)_3CN(H)OH$	F:	PerFHalOrg.7-16/7, 43
$C_4F_9H_2O_3P$	$C_4F_9P(O)(OH)_2$	F:	PerFHalOrg.3-43, 50
$C_4F_9H_2P$	$(CF_3)_2PCFHCHF_2$	F:	PerFHalOrg.3-28
–	$(CF_3)_2PCF_2CH_2F$	F:	PerFHalOrg.3-28
$C_4F_9I_2P$	$(n\text{-}C_4F_9)PI_2$	F:	PerFHalOrg.3-76, 94
C_4F_9K	$(CF_3)_3CK$	F:	PerFHalOrg.4-1
C_4F_9N	$(CF_2)_4NF$	F:	PerFHalOrg.5-51, 74
–	$CF_3N(CF_2)_3$	F:	PerFHalOrg.5-29, 37, 45/6, 48
–	$(n\text{-}C_3F_7)NCF_2$	F:	PerFHalOrg.5-46
C_4F_9NO	$CF_2N(CF_3)CF_2OCF_2$	F:	PerFHalOrg.5-54/5, 77, 96, 98
–	$CF_3NOCF(CF_3)CF_2$	F:	PerFHalOrg.5-32/3, 41, 47
–	$CF_3NOCF_2CFCF_3$	F:	PerFHalOrg.5-32/3, 41, 47
–	$(CF_3)_3CNO$	F:	PerFHalOrg.7-178, 184, 189, 191/3, 195, 200
–	$C_2F_5CF(CF_3)NO$	F:	PerFHalOrg.7-178, 184
–	$C_2F_5NOCF_2CF_2$	F:	PerFHalOrg.5-30/1, 40

C_4F_9NO	C_4F_9NO	F:	PerFHalOrg.7-178, 184
–	$[N(CF_3)OCF_2CF(CF_3)]_n$	F:	PerFHalOrg.7-207/8, 214
–	$[N(C_2F_5)OCF_2CF_2]_n$	F:	PerFHalOrg.7-208
–	$O(CF_2CF_2)_2NF$	F:	PerFHalOrg.6-1, 6, 11/2
C_4F_9NOS	$CF_3NOCF(SCF_3)CF_2$	F:	PerFHalOrg.5-33/4
–	$CF_3NOCF_2CF(SCF_3)$	F:	PerFHalOrg.5-33/4
$C_4F_9NO_2$	$CF_3NOCF(OCF_3)CF_2$	F:	PerFHalOrg.5-33/4, 41
–	$CF_3NOCF_2CFOCF_3$	F:	PerFHalOrg.5-33/4, 41
–	$(CF_3)_2NOC(O)CF_3$	F:	PerFHalOrg.7-94/5, 111, 124
–	$C_2F_5OCF_2CF_2NO$	F:	PerFHalOrg.7-179, 185
$C_4F_9O_2P$	$(CF_3)_2POC(O)CF_3$	F:	PerFHalOrg.3-31, 34, 36
C_4F_9P	$(CF_2CF)_2PF_3$	F:	PerFHalOrg.3-71, 87
$C_4F_{10}HN$	$(CF_3)N(H)CF(CF_3)_2$	F:	PerFHalOrg.7-13/4, 41
–	$(C_2F_5)_2NH$	F:	PerFHalOrg.7-13/4, 41
$C_4F_{10}HN_2Sb$	$[N(CFCF)_2NH][SbF_6]$	F:	PerFHalOrg.6-29/30, 51
$C_4F_{10}HP$	$(CF_3)_2PCF_2CF_2H$	F:	PerFHalOrg.3-28
$C_4F_{10}H_2N_2$	$CF_3NHCF_2CF_2NHCF_3$	F:	PerFHalOrg.7-13/4, 41
$C_4F_{10}H_2N_2O_2$	$CF_3N(OH)CF_2CF_2N(OH)CF_3$	F:	PerFHalOrg.7-87, 104
$C_4F_{10}H_3PS$	$(CF_3)_3P(F)SCH_3$	F:	PerFHalOrg.3-110
$C_4F_{10}IP$	$(C_2F_5)_2PI$	F:	PerFHalOrg.3-74/5, 93
$C_4F_{10}N_2$	$CF_3NCF_2CF_2NCF_3$	F:	PerFHalOrg.5-36, 42, 47
–	$NFCF_2CF_2NFCF_2CF_2$	F:	PerFHalOrg.6-29/30, 51
$C_4F_{10}N_2O$	$CF_2N(CF_3)ON(CF_3)CF_2$	F:	PerFHalOrg.5-66/7, 89, 97, 99
–	$(CF_3)_2NOCFNCF_3$	F:	PerFHalOrg.7-96
$C_4F_{10}N_2O_2$	$CF_3N(O)CF_2CF_2N(O)CF_3$	F:	PerFHalOrg.7-86, 103
–	$(CF_3)_2NOCF_2CF_2NO$	F:	PerFHalOrg.7-95/6, 113
–	$[N(CF_3)OCF_2CF_2ON(CF_3)]_n$	F:	PerFHalOrg.7-213
–	$[ON(CF_3)CF_2CF_2N(CF_3)O]_n$	F:	PerFHalOrg.7-213
$C_4F_{10}N_4O_4$	$CF_3N(ONO)CF_2CF_2N(ONO)CF_3$	F:	PerFHalOrg.7-90/1, 106
$C_4F_{10}NiO_2P_2$	$(CO)_2Ni(P(CF_3)F_2)_2$	Ni:	Org.Verb.1-127, 128
$C_4F_{10}P_2$	$CF_3P(CF_2)_2PCF_3$	F:	PerFHalOrg.3-9/14
$C_4F_{11}Ge^+$	$Ge(CF_3)_3CF_2^+$	F:	PerFHalOrg.4-174
$C_4F_{11}HN_2O$	$(CF_3)_2NOCF_2N(H)CF_3$	F:	PerFHalOrg.7-13/4, 96
$C_4F_{11}NS$	$F_4SCF_2CF_2NCFCF_2$	F:	PerFHalOrg.6-4/5
$C_4F_{11}P$	$(C_2F_5)_2PF$	F:	PerFHalOrg.3-70/1
$C_4F_{11}P_2^+$	$(CF_3)_2PP(CF_3)CF_2^+$	F:	PerFHalOrg.3-179
$C_4F_{11}P_4S^+$	$(CF_3)_3(CF_2)P_4S^+$	F:	PerFHalOrg.3-12
$C_4F_{12}FeH_5NP_4$	$NCCHCH(CH_3)Fe(PF_3)_4$	Fe:	Org.Verb.B4-179, 182
$C_4F_{12}FeH_6O_2P_4$	$CH_3OC(O)CHCH_2Fe(PF_3)_4$	Fe:	Org.Verb.B4-178, 182
$C_4F_{12}FeP_2S_4$	$Fe[S(S)P(CF_3)_2]_2$	F:	PerFHalOrg.3-159
$C_4F_{12}Fe_2N_4O_4P_2$	$Fe_2(NO)_4[P(CF_3)_2]_2$	F:	PerFHalOrg.3-27
$C_4F_{12}Ge$	$(CF_3)_4Ge$	F:	PerFHalOrg.4-168, 173/4, 177
$C_4F_{12}HNP_2$	$[(CF_3)_2P]_2NH$	F:	PerFHalOrg.3-60, 64, 69
$C_4F_{12}HNP_2S_2$	$[(CF_3)_2P(S)]_2NH$	F:	PerFHalOrg.3-138/9, 153
$C_4F_{12}H_2P_4$	$CF_3P(H)P(CF_3)P(CF_3)P(H)CF_3$	F:	PerFHalOrg.3-16, 20
$C_4F_{12}H_4N_2P_2$	$[(CF_3)_2P]_2NH \cdot NH_3$	F:	PerFHalOrg.3-60, 64, 69
$C_4F_{12}HgN_2$	$Hg[N(CF_3)_2]_2$	C:	MVol.D3-214

$C_4F_{12}HgN_2O_2$	$[(CF_3)_2NO]_2Hg$	F: PerFHalOrg.7-102
$C_4F_{12}HgO_4P_2$	$Hg[O(O)P(CF_3)_2]_2$	F: PerFHalOrg.3-50
$C_4F_{12}HgP_2$	$[(CF_3)_2P]_2Hg$	F: PerFHalOrg.3-186
$C_4F_{12}HgP_2S_4$	$Hg[S(S)P(CF_3)_2]_2$	F: PerFHalOrg.3-138/9, 159
$C_4F_{12}MnP_2S_4$	$Mn[S(S)P(CF_3)_2]_2$	F: PerFHalOrg.3-159
$C_4F_{12}NNaP_2$	$NaN[P(CF_3)_2]_2$	F: PerFHalOrg.3-60, 68
$C_4F_{12}NOP$	$(CF_3)_2NOP(CF_3)_2$	F: PerFHalOrg.7-100/1
$C_4F_{12}NP$	$(CF_3)_2PN(CF_3)_2$	F: PerFHalOrg.3-61, 65, 68/9
$C_4F_{12}N_2O$	$(CF_3)_2NON(CF_3)_2$	F: PerFHalOrg.7-91, 106, 124/5, 143/4
$C_4F_{12}N_2O_2$	$(CF_3)_2NOON(CF_3)_2$	F: PerFHalOrg.7-89/90, 106, 124/5
$C_4F_{12}N_2O_2Pb$	$[(CF_3)_2NO]_2Pb$	F: PerFHalOrg.7-102
$C_4F_{12}N_2O_2S$	$[(CF_3)_2NO]_2S$	F: PerFHalOrg.7-91/3, 107
$C_4F_{12}N_2O_2S_2$	$[(CF_3)_2NO]_2S_2$	F: PerFHalOrg.7-91/3, 107
$C_4F_{12}N_2O_2Sn$	$[(CF_3)_2NO]_2Sn$	F: PerFHalOrg.7-102
$C_4F_{12}N_2O_3S$	$[(CF_3)_2NO]_2SO$	F: PerFHalOrg.7-91/3, 107, 126
$C_4F_{12}N_2O_4S$	$[(CF_3)_2NO]_2SO_2$	F: PerFHalOrg.7-92/3, 107, 126
$C_4F_{12}NiP_2S_4$	$Ni[S(S)P(CF_3)_2]_2$	F: PerFHalOrg.3-159
$C_4F_{12}OP_2$	$(CF_3)_2POP(CF_3)_2$	F: PerFHalOrg.3-32/9
$C_4F_{12}OP_2S_2$	$(CF_3)_2P(S)O(S)P(CF_3)_2$	F: PerFHalOrg.3-139/40, 143, 146, 151/3
$C_4F_{12}OP_2S_4V$	$VO[S(S)P(CF_3)_2]_2$	F: PerFHalOrg.3-154
$C_4F_{12}O_3P_2$	$(CF_3)_2P(O)OP(O)(CF_3)_2$	F: PerFHalOrg.3-32/3, 35, 37
$C_4F_{12}O_3P_2Si$	$[(CF_3)_2PO]_2SiO$	F: PerFHalOrg.3-41
$C_4F_{12}P_2$	$(CF_3)_2PP(CF_3)_2$	F: PerFHalOrg.3-176/85
$C_4F_{12}P_2^+$	$(CF_3)_2PP(CF_3)_2^+$	F: PerFHalOrg.3-179
$C_4F_{12}P_2PdS_4$	$Pd[S(S)P(CF_3)_2]_2$	F: PerFHalOrg.3-159
$C_4F_{12}P_2S$	$(CF_3)_2PSP(CF_3)_2$	F: PerFHalOrg.3-137/8, 142, 144, 151/4
$C_4F_{12}P_2S_2$	$(CF_3)_2P(S)SP(CF_3)_2$	F: PerFHalOrg.3-137/8, 141, 144, 151/3
$C_4F_{12}P_2S_3$	$(CF_3)_2PS_3P(CF_3)_2$	F: PerFHalOrg.3-138
$C_4F_{12}P_2S_4$	$(CF_3)_2P(S)SS(S)P(CF_3)_2$	F: PerFHalOrg.3-137/8, 141, 144, 151/3, 159
$C_4F_{12}P_2S_4Zn$	$Zn[S(S)P(CF_3)_2]_2$	F: PerFHalOrg.3-154, 159
$C_4F_{12}P_2Se$	$(CF_3)_2PSeP(CF_3)_2$	F: PerFHalOrg.3-160
$C_4F_{12}P_4$	$(CF_3P)_4$	F: PerFHalOrg.3-1/8
$C_4F_{12}P_4S$	$S(PCF_3)_4$	F: PerFHalOrg.3-10/4
$C_4F_{12}P_4S^+$	$(CF_3)_4P_4S^+$	F: PerFHalOrg.3-12
$C_4F_{12}Pb$	$Pb(CF_3)_4$	F: PerFHalOrg.4-190
$C_4F_{12}Sb_2$	$(CF_3)_2SbSb(CF_3)_2$	F: PerFHalOrg.3-227, 229/31
$C_4F_{12}Si$	$C_4F_9SiF_3$	F: PerFHalOrg.4-147
$C_4F_{12}Sn$	$(CF_3)_4Sn$	F: PerFHalOrg.4-182, 187
		Sn: Org.Verb.1-123

$C_4F_{13}H_2KN_2O_2$	$[(CF_3)_2NOH]_2 \cdot KF$	F:	PerFHalOrg.7-86/7
$C_4F_{13}NS$	$F_4S(CF_2CF_2)_2NF$	F:	PerFHalOrg.6-4/5
–	$SF_5NCFC_3F_7$	S:	S-N-Verb.1-147
$C_4F_{13}N_3O_2S$	$[(CF_3)_2NO]_2S(F)N$	F:	PerFHalOrg.7-92/3, 108
$C_4F_{13}P$	$(C_2F_5)_2PF_3$	F:	PerFHalOrg.3-71, 76/7, 86
$C_4F_{14}P^-$	$[(CF_3)_4PF_2]^-$	F:	PerFHalOrg.3-129, 134, 137
$C_4F_{15}N_2O_2P$	$[(CF_3)_2NO]_2PF_3$	F:	PerFHalOrg.7-100/1, 117, 145
$C_4F_{15}NiOP_3$	$CONi(P(CF_3)F_2)_3$	Ni:	Org.Verb.1-114
$C_4F_{16}N_2O_2S$	$[(CF_3)_2NO]_2SF_4$	F:	PerFHalOrg.7-91/3, 107
$C_4F_{18}FeOP_4$	$COFeP_4F_9(CF_3)_3$	Fe:	Org.Verb.B1-28/9
$C_4F_{20}NiP_4$	$Ni(CF_3PF_2)_4$	F:	PerFHalOrg.3-110/1, 120
		Ni:	Org.Verb.1-235, 238
$C_4F_{20}P_4Pt$	$Pt(CF_3PF_2)_4$	F:	PerFHalOrg.3-110/1, 121
$C_4FeGeH_4O_4$	$(CO)_4Fe(GeH_3)H$	Fe:	Org.Verb.B2-138, 145
$C_4FeGeI_4O_4$	$(CO)_4Fe(GeI_3)I$	Fe:	Org.Verb.B2-138, 146
$C_4FeGe_2H_6O_4$	$(CO)_4Fe(GeH_3)_2$	Fe:	Org.Verb.B2-140, 147
$C_4FeGe_2I_6O_4$	$(CO)_4Fe(GeI_3)_2$	Fe:	Org.Verb.B2-140, 147/8
C_4FeHKO_4	$K[HFe(CO)_4]$	Fe:	Org.Verb.B2-17/8, 32, 36/46
		Fe:	Org.Verb.B3-115
$C_4FeHNaO_4$	$Na[HFe(CO)_4]$	Fe:	Org.Verb.B2-15, 17, 32, 35, 38, 41/6
$C_4FeHO_4^-$	$[HFe(CO)_4]^-$	Fe:	Org.Verb.B2-13/7, 25/48
		Fe:	Org.Verb.B3-67, 114/6, 157, 175
–	$[DFe(CO)_4]^-$	Fe:	Org.Verb.B3-114
$C_4FeHO_4Zn^+$	$[ZnHFe(CO)_4]^+$	Fe:	Org.Verb.B2-27
$C_4FeH_2O_4$	$H_2Fe(CO)_4$	Fe:	Org.Verb.B2-7/13
		Fe:	Org.Verb.B3-114/6
$C_4FeH_2O_6Zn_2$	$Zn_2Fe(CO)_4(OH)_2$	Fe:	Org.Verb.B2-27
$C_4FeH_3NO_4$	$(CO)_4FeNH_3$	Fe:	Org.Verb.B2-71/3
$C_4FeH_3O_4P$	$(CO)_4FePH_3$	Fe:	Org.Verb.B2-86, 103
$C_4FeH_3O_7Zn_2^-$	$[Zn_2Fe(CO)_4(OH)_3]^-$	Fe:	Org.Verb.B2-27
$C_4FeH_4N_6$	$Fe(N_3C_2H_2)_2$	Fe:	Org.Verb.B3-193
$C_4FeH_4O_4$	$Fe(CO)_3 \cdot HOCH_3$	Fe:	Org.Verb.B3-174
$C_4FeH_4O_4Si$	$(CO)_4Fe(SiH_3)H$	Fe:	Org.Verb.B2-138, 145
$C_4FeH_6N_4O_2$	$(CH_3NC)_2Fe(NO)_2$	Fe:	Org.Verb.B4-11
$C_4FeH_6N_4O_2^+$	$[(CO)_2Fe(CH_3NNNNCH_3)]^+$	Fe:	Org.Verb.B1-169
$C_4FeH_6O_2S_4$	$Fe[S_2COCH_3]_2 \cdot 2\ C_5H_5N$	C:	MVol.D4-259
$C_4FeH_6O_4$	$(CH_3COO)_2Fe$	Fe:	Org.Verb.B3-176
$C_4FeH_6O_4Si_2$	$(CO)_4Fe(SiH_3)_2$	Fe:	Org.Verb.B2-139, 146
$C_4FeH_8N_2O_4$	$[NH_4]_2[Fe(CO)_4]$	Fe:	Org.Verb.B2-14, 22
$C_4FeH_9N_2O_6P$	$COFe(NO)_2P(OCH_3)_3$	Fe:	Org.Verb.B1-42/3, 45, 47/8
$C_4FeH_9N_3O_4Zn$	$(CO)_4FeZn(NH_3)_3$	Fe:	Org.Verb.B2-173
$C_4FeH_{12}Li_2$	$Li_2[Fe(CH_3)_4]$	Fe:	Org.Verb.B1-10
$C_4FeH_{13}N_2P^+$	$[HFeP(N(CH_3)_2)_2]^+$	Fe:	Org.Verb.B1-164

C_4FeHgO_4 .... $HgFe(CO)_4$ .... Fe: Org.Verb.B2-13, 29
Fe: Org.Verb.B3-126, 141/2
– .... $[(CO)_4FeHg]_n$ .... Fe: Org.Verb.B2-148, 154
Fe: Org.Verb.B3-142
$C_4FeHg_2I_2O_4$ .... $(CO)_4Fe(HgI)_2$ .... Fe: Org.Verb.B2-153/6, 159
$C_4FeHg_2O_8S$ .... $(CO)_4FeHg \cdot HgSO_4$ .... Fe: Org.Verb.B2-154
$C_4FeINaO_4$ .... $Na[(CO)_4FeI]$ .... Fe: Org.Verb.B2-51
C_4FeIO_4 .... $(CO)_4FeI$ .... Fe: Org.Verb.B2-50
$C_4FeI_2O_4$ .... $(CO)_4FeI_2$ .... Fe: Org.Verb.B2-54/67
– .... $(CO)_4Fe^{129}I_2$ .... Fe: Org.Verb.B2-55
$C_4FeI_2O_4P_4S_3$ .... $(CO)_4FeP_4S_3I_2$ .... Fe: Org.Verb.B2-99, 120
$C_4FeI_4O_4Sn$ .... $(CO)_4Fe(SnI_3)I$ .... Fe: Org.Verb.B2-139
$C_4FeI_6O_4$ .... $(CO)_4Fe(I \cdot I_2)_2$ .... Fe: Org.Verb.B2-64
$C_4FeK_2O_4$ .... $K_2[Fe(CO)_4]$ .... Fe: Org.Verb.B2-22, 28, 31, 36/42
Fe: Org.Verb.B3-134
C_4FeKrO_4 .... $Fe(CO)_4Kr$ .... Fe: Org.Verb.B2-3
C_4FeMnO_4Re .... $ReFeMn(CO)_4$ .... Fe: Org.Verb.B3-144
$C_4FeN_2O_2S_2$ .... $(CO)_2Fe(SCN)_2$ .... Fe: Org.Verb.B3-121
$C_4FeN_2O_4$ .... $[Fe(CO)_4N_2]$ .... Fe: Org.Verb.B2-1, 3
Fe: Org.Verb.B3-87
$C_4FeN_4O_4^{2-}$ .... $[Fe(NCO)_4]^{2-}$ .... Fe: Org.Verb.B3-157
$C_4FeNaO_4^-$ .... $[NaFe(CO)_4]^-$ .... Fe: Org.Verb.B2-21, 34
$C_4FeNa_2O_4$ .... $Na_2[Fe(CO)_4]$ .... Fe: Org.Verb.B2-14, 20, 25/45
C_4FeO_4 .... $Fe(CO)_4$ .... Fe: Org.Verb.B2-1/5
$C_4FeO_4^+$ .... $Fe(CO)_4^+$ .... Fe: Org.Verb.B3-49, 78
$C_4FeO_4^{2+}$ .... $[Fe(CO)_4]^{2+}$ .... Fe: Org.Verb.B3-78
$C_4FeO_4^-$ .... $[Fe(CO)_4]^-$ .... Fe: Org.Verb.B3-78, 86, 91
$C_4FeO_4^{2-}$ .... $[Fe(CO)_4]^{2-}$ .... Fe: Org.Verb.B2-13/7, 25/48
C_4FeO_4Pb .... $(CO)_4FePb$ .... Fe: Org.Verb.B2-136
C_4FeO_4Sb .... $(CO)_4FeSb$ .... Fe: Org.Verb.B2-153
C_4FeO_4Xe .... $[Fe(CO)_4Xe]$ .... Fe: Org.Verb.B2-1
Fe: Org.Verb.B3-87
C_4FeO_4Zn .... $ZnFe(CO)_4$ .... Fe: Org.Verb.B2-27, 173
$C_4FeO_4Zn_2^{2+}$ .... $[Zn_2Fe(CO)_4]^{2+}$ .... Fe: Org.Verb.B2-27
C_4FeO_6S .... $(CO)_4FeSO_2$ .... Fe: Org.Verb.C2-27
$C_4FeO_{10}P_4$ .... $(CO)_4FeP_4O_6$ .... Fe: Org.Verb.B2-96, 117
$C_4FeO_{11}P_4$ .... $(CO)_4FeP_4O_7$ .... Fe: Org.Verb.B2-96, 117
$C_4Fe_2H_{12}N_4O_4P_2$ .... $(NO)_2Fe(P(CH_3)_2)_2Fe(NO)_2$ .... Fe: Org.Verb.B1-143
$C_4Fe_2N_6O_4^{2+}$ .... $[(CO)_2Fe(N_3)_2Fe(CO)_2]^{2+}$ .... Fe: Org.Verb.C1-2/3
$C_4Fe_2O_4^+$ .... $[Fe_2(CO)_4]^+$ .... Fe: Org.Verb.C2-26
C_4HIrO_4 .... $HIr(CO)_4$ .... Ir: SVol.2-113
C_4HO_9Pa .... $PaO(C_2O_4)(HC_2O_4)$ .... Pa: SVol.2-170
– .... $Pa(C_2O_4)_2OH \cdot 6\ H_2O$ .... Pa: SVol.2-82
$C_4H_2K_2O_7Sn$ .... $K_2SnO(C_4H_2O_6) \cdot 5\ H_2O$ .... Sn: MVol.C3-80
$C_4H_2N_4S_4$ .... $S_2(CN)_2 \cdot 2\ HNCS$ .... C: MVol.D6-35
$C_4H_2Na_2O_7Sn$ .... $Na_2SnOC_4H_2O_6 \cdot 5\ H_2O$ .... Sn: MVol.C3-35
C_4H_2Ni .... $(HCC)_2Ni$ .... Ni: Org.Verb.1-80
$C_4H_2O_2S$ .... $[SC(O)CHCHC(O)]_n$ (n = 4 to 5) .... C: MVol.D4-215

$C_4H_2O_2S_3$	$CH_2C(O)SC(S)SC(O)$	C: MVol.D4-215
$C_4H_2O_7Sn^{2-}$	$SnO(OCOCH(O)CH(O)COO)^{2-}$	Sn: MVol.C2-226
$C_4H_2O_8Pa^{3+}$	$Pa(HC_2O_4)_2{}^{3+}$	Pa: SVol.2-170
$C_4H_2O_9Sn_2{}^{2-}$	$Sn_2O_3(OCOCH(O)CH(O)COO)^{2-}$	Sn: MVol.C2-226
$C_4H_2O_{11}Pa^{3-}$	$PaO(OH)_2(C_2O_4)_2{}^{3-}$	Pa: SVol.2-170
$C_4H_3KO_7Sn$	$KSnO(C_4H_3O_6)$	Sn: MVol.C3-80
$C_4H_3NSe_8$	$C_4H_3Se_8N$	C: MVol.D6-248
$C_4H_3N_3S_3Sn$	$CH_3Sn(NCS)_3$	Sn: Org.Verb.6-219/20
$C_4H_3O_6Sn^-$	$Sn(OCOCH(O)CH(OH)COO)^-$	Sn: MVol.C2-225/6
$C_4H_4I_4N_2Sn$	$SnI_4 \cdot N(CHCH)_2N$	Sn: MVol.C5-225/6
$C_4H_4K_2O_8Sn$	$K_2Sn(HCOO)_4 \cdot 5\ H_2O$	Sn: MVol.C3-72
$C_4H_4MnO_2{}^+$	$[Mn(CH_3COCHCO)]^+$	Mn: MVol.D1-92/3
$C_4H_4MnO_6$	$[MnC_4O_4(H_2O)_2]$	Mn: MVol.D1-127/8
$C_4H_4N_2NiO_3S$	$(CO)_3NiSC(NH_2)_2$	Ni: Org.Verb.1-160
$C_4H_4Na_2O_8Sn$	$Na_2Sn(HCOO)_4 \cdot 5\ H_2O$	Sn: MVol.C3-34
$C_4H_4NpO_8{}^-$	$NpO_2(C_4H_4O_6)^-$	Np: TrU.D1-157, 159
$C_4H_4O_2S_4{}^{2-}$	$[S_2COC_2H_4OCS_2]^{2-}$	C: MVol.D4-250
$C_4H_4O_5Sn$	$Sn(OCOCH_2CH(OH)COO)$	Sn: MVol.C2-225
$C_4H_4O_6Pu^+$	$Pu(C_4H_4O_6)^+$	Np: TrU.D1-159
$C_4H_4O_6Ra$	$RaC_4H_4O_6$	Ra: SVol.2-371/2
$C_4H_4O_6Sn$	$Sn(OCOCHOHCHOHCOO)$	Sn: MVol.C2-225/6
$C_4H_4O_8Pa$	$Pa(HCOO)_4$	Pa: SVol.2-81
$C_4H_4O_8Sn$	$Sn(HCOO)_4$	Sn: MVol.C2-219
$C_4H_4Se_9$	$C_4H_4Se_9$	C: MVol.D6-248
$C_4H_5I_5N_2Te$	$[C_4H_4N_2H]TeI_5$	Te: SVol.B3-65, 69/70
$C_4H_5MnO^+$	$[Mn(CH_2COCHCH_2)]^+$	Mn: MVol.D1-92/3
$C_4H_5NNpO_6{}^-$	$(NpO_2)NH(CH_2COO)_2{}^-$	Np: TrU.D1-162
$C_4H_5NOS_2$	$CH(CH_3)SC(S)NHC(O)$	C: MVol.D6-139
$C_4H_5NO_6Pu^-$	$(PuO_2)NH(CH_2COO)_2{}^-$	Np: TrU.D1-162
$C_4H_5NO_7U$	$UO_2C_2O_4 \cdot CH_3CONH_2$	U: SVol.E1-96/8
$C_4H_5NS_2$	$SC(S)NHC(CH_3)CH$	C: MVol.D6-139
$C_4H_5NpO_8$	$NpO_2(C_4H_5O_6)$	Np: TrU.D1-157, 159
$C_4H_5OS_2{}^-$	$[S_2COC_3H_5]^-$	C: MVol.D4-244
$C_4H_5O_6Pa^{4+}$	$Pa(HC_4H_4O_6)^{4+}$	Pa: SVol.2-172
$C_4H_5O_6Pu^{2+}$	$Pu(C_4H_5O_6)^{2+}$	Np: TrU.D1-159
$C_4H_6HgO_2S_4$	$Hg[S_2COCH_3]_2$	C: MVol.D4-258
$C_4H_6HgO_4S_2$	$Hg[SOCOCH_3]_2$	C: MVol.D4-237
$C_4H_6I_4N_2Te$	$[(CH_3CN)_2TeI_3]I$	Te: SVol.B3-64
$C_4H_6MgO_4S_2$	$Mg[SOCOCH_3]_2$	C: MVol.D4-237
$C_4H_6Mo_2O_4S_6$	$Mo_2O_2S_2[S_2COCH_3]_2$	C: MVol.D4-258
$C_4H_6NNpO_6$	$(NpO_2)NH(CH_2COOH)(CH_2COO)$	Np: TrU.D1-162
$C_4H_6N_2NiO_2$	$(Ni(O_2)(NCCH_3)_2)_x$	Ni: Org.Verb.1-328
$C_4H_6N_2O_2Sn$	$(CH_3)_2Sn(CNO)_2$	Sn: Org.Verb.6-28
–	$Sn(CH_3C(NO)C(NO)CH_3)$	Sn: MVol.C5-41
$C_4H_6N_2O_7U$	$UO_2C_2O_4 \cdot NH_2CONH(CH_3)$	U: SVol.E1-93, 95
$C_4H_6N_4O_8U$	$[UO_2(NO_3)_2(CH_3CN)_2]$	U: SVol.E1-48/50
$C_4H_6N_8Ni$	$(CH_3N_4C)_2Ni$	Ni: Org.Verb.1-87, 88
$C_4H_6N_{10}O_{25}Pa_2$	$Pa_2O(NO_3)_8 \cdot 2\ CH_3CN$	Pa: SVol.2-35
$C_4H_6Na_2O_6S_2Sn$	$Na_2[Sn(OH)_2(SCH_2COO)_2] \cdot 6\ H_2O$	Sn: MVol.C6-89
$C_4H_6NiO_2S_4$	$Ni[S_2COCH_3]_2$	C: MVol.D4-247/8, 260

$C_4H_6NpO_6$	$NpO_2(CH_3COO)_2$	Np: TrU.D1-153
$C_4H_6NpO_6^-$	$NpO_2(CH_3COO)_2^-$	Np: TrU.D1-153
$C_4H_6OS_2$	$SC(OC_3H_5)(SH)$	C: MVol.D4-243
C_4H_6OSn	$[(CH_2CH)_2SnO]_x$	Sn: Org.Verb.6-144
$C_4H_6O_2PbS_4$	$Pb[S_2COCH_3]_2$	C: MVol.D4-254
$C_4H_6O_2PdS_4$	$Pd[S_2COCH_3]_2$	C: MVol.D4-248, 261
$C_4H_6O_2PtS_4$	$Pt[S_2COCH_3]_2$	C: MVol.D4-247/8, 261
–	$Pt[S_2COCD_3]_2$	C: MVol.D4-247
$C_4H_6O_2S_4$	$CH_3OC(S)SSC(S)OCH_3$	C: MVol.D4-251
$C_4H_6O_2S_4Se$	$Se[S_2COCH_3]_2$	C: MVol.D4-245, 255
$C_4H_6O_2S_4Sn$	$Sn[S_2COCH_3]_2$	C: MVol.D4-245, 253
$C_4H_6O_2S_4Te$	$Te[S_2COCH_3]_2$	C: MVol.D4-245, 255
		Te: SVol.B3-134/5
$C_4H_6O_2S_4Zn$	$Zn[S_2COCH_3]_2$	C: MVol.D4-257
$C_4H_6O_2S_5$	$S[S_2COCH_3]_2$	C: MVol.D4-245, 254
$C_4H_6O_4Pu^+$	$Pu(CH_3COO)_2^+$	Np: TrU.D1-152
$C_4H_6O_4Pu^{2+}$	$Pu(CH_3COO)_2^{2+}$	Np: TrU.D1-153
$C_4H_6O_4Ra$	$Ra(CH_3COO)_2$	Ra: SVol.2-371, 377
–	$Ra(CH_3COO)_2$ systems	
	$Ra(CH_3COO)_2$-$Ba(CH_3COO)_2$-H_2O	Ra: SVol.2-371
$C_4H_6O_4S_2$	$CH_3OC(O)SS(O)COCH_3$	C: MVol.D4-236/7
$C_4H_6O_4S_2Te$	$[Te(SCH_2COOH)_2]$	Te: SVol.B3-134
$C_4H_6O_4S_4U$	$UO_2[S_2COCH_3]_2$	C: MVol.D4-261
$C_4H_6O_4Sn$	$Sn(CH_3COO)_2$	Sn: MVol.C2-219/20
–	$(CH_3)_2Sn(C_2O_4) \cdot H_2O$	Sn: Org.Verb.6-25
$C_4H_6O_6Pu$	$PuO_2(CH_3COO)_2$	Np: TrU.D1-153
$C_4H_6O_7SU$	$UO_2(C_2O_4) \cdot (CH_3)_2SO$	U: SVol.E1-204, 208
$C_4H_6O_8Pu$	$PuO_2(HOCH_2COO)_2$	Np: TrU.D1-153
$C_4H_6O_8SU$	$UO_2SO_4 \cdot O(CH_2)_3C(O) \cdot 2\ H_2O$	U: SVol.E1-119/20
$C_4H_7KOS_2$	$K[S_2COC_3H_7]$	C: MVol.D4-252/3
$C_4H_7KOSe_2$	$K[Se_2COC_3H_7]$	C: MVol.D6-221
$C_4H_7KO_2S$	$K[SOCOC_3H_7]$	C: MVol.D4-237
$C_4H_7KS_3$	$K[S_2CSC_3H_7]$	C: MVol.D4-269
$C_4H_7NOS_2$	$SC(S)NHC(OH)(CH_3)CH_2$	C: MVol.D6-139
$C_4H_7N_3O_4S$	$[CH_3NSO_2][CH_3NCO][HNCO]$	S: S-N-Verb.1-30
$C_4H_7NaOS_2$	$Na[S_2COC_3H_7]$	C: MVol.D4-252
$C_4H_7NaO_2S$	$Na[SOCOC_3H_7]$	C: MVol.D4-237
$C_4H_7NaS_3$	$Na[S_2CSC_3H_7]$	C: MVol.D4-267, 269
$C_4H_7NpO_5$	$NpO_2(C_2H_5CH(OH)COO)$	Np: TrU.D1-154
$C_4H_7OS_2^-$	$[S_2COC_3H_7]^-$	C: MVol.D4-244/5, 251/2
$C_4H_7OS_2Tl$	$Tl[S_2COC_3H_7]$	C: MVol.D4-253
$C_4H_7O_2S^-$	$[SOCOC_3H_7]^-$	C: MVol.D4-236/9
$C_4H_7O_2S_2^-$	$[S_2COCH_2CH(OH)CH_3]^-$	C: MVol.D4-244
$C_4H_7S_3^-$	$[S_2CSC_3H_7]^-$	C: MVol.D4-268
$C_4H_7Sn^+$	$(CH_2CH)_2SnH^+$	Sn: Org.Verb.3-41
$C_4H_8HgOS_2$	$CH_3Hg[S_2COC_2H_5]$	C: MVol.D4-258
$C_4H_8I_4O_2S_2Sn$	$SnI_4 \cdot SO(CH_2CH_2)_2SO$	Sn: MVol.C6-114
$C_4H_8I_4O_2Sn$	$SnI_4 \cdot C_4H_8O_2$	Sn: MVol.C5-159/60
$C_4H_8I_4Sn$	$Sn(CH_2I)_4$	Sn: Org.Verb.1-122
$C_4H_8MnN_2O_2S_2$	$Mn(NCS)_2 \cdot 2\ CH_3OH$	Mn: MVol.D1-26

$C_4H_8MnN_3O_4$	$Mn(NO)_3 \cdot C_4H_8O$	Mn: MVol.D1-142
$C_4H_8N_2NpO_6^-$	$NpO_2(NH_2CH_2COO)_2^-$	Np: TrU.D1-155
$C_4H_8N_2O_2Sn$	$Sn(CH_3CHNO)_2$	Sn: MVol.C5-41
$C_4H_8N_2O_8Sn$	$(NH_4)_2Sn(C_2O_4)_2 \cdot H_2O$	Sn: MVol.C3-125
$C_4H_8N_2O_9U$	$UO_2(NO_3)_2 \cdot CH_3COC_2H_5 \cdot 3\ H_2O$	U: SVol.E1-79/80
$C_4H_8N_4OS$	$[NS(CH_3)_2][NCO][HNC(NH)]$	S: S-N-Verb.1-31
$C_4H_8N_6O_4S_2U$	$UO_2(NCS)_2 \cdot 2\ (NH_2)_2CO$	U: SVol.E1-83, 85, 87/8
–	$UO_2(NCS)_2 \cdot 2\ (NH_2)_2CO \cdot H_2O$	U: SVol.E1-83, 85/8
$C_4H_8N_6S_4Te$	$Te(SC(NH_2)_2)_2(SCN)_2$	Te: SVol.B3-142
$C_4H_8OS_2$	$SC(OC_3H_7)(SH)$	C: MVol.D4-239/40, 243
$C_4H_8O_2SSn$	$(CH_3)_2Sn(SCH_2COO)$	Sn: Org.Verb.6-30
$C_4H_8O_2Sn$	$[(CH_3)_2SnCH_2COO]_x$	Sn: Org.Verb.6-28
$C_4H_8O_4Sn$	$(CH_3)_2Sn(OC(O)H)_2$	Sn: Org.Verb.6-27
–	$Sn[(OCH_2)_2]_2$	Sn: Org.Verb.1-88
$C_4H_8O_6S_2Sn$	$Sn(OH)_2(SCH_2COOH)_2$	Sn: MVol.C6-89
$C_4H_8O_6Th$	$Th(OH)_2(CH_3COO)_2 \cdot 3\ H_2O$	Th: SVol.C1-249
$C_4H_8S_2Sn$	$(CH_3)_2Sn(SCH)_2$	Sn: Org.Verb.6-30
$C_4H_8S_3Sn$	$[Sn(S(CH_2CH_2S)_2)]$	Sn: MVol.C5-46
$C_4H_8S_4Sn$	$Sn(SCH_2CH_2S)_2$	Sn: MVol.C6-82
$C_4H_9IN_2NiS$	$C_3H_5Ni(SC(NH_2)_2)I$	Ni: Org.Verb.2-10
C_4H_9INiO	$CH_3CHCHCH_2Ni(H_2O)I$	Ni: Org.Verb.2-294
$C_4H_9I_3Sn$	$(CH_3)_3SnCl_3$	Sn: Org.Verb.2-11
C_4H_9NOSi	$(CH_3)_3SiNCO$	C: MVol.D3-142
$C_4H_9NO_3SSi^+$	$SO_2NCOSi(CH_3)_3^+$	S: S-N-Verb.1-25
$C_4H_9NO_5U$	$UO(CH_3COO)_2 \cdot NH_3$	U: SVol.E1-12
$C_4H_9N_2NaO_4S$	$Na[(CH_3)_2NSO_2NCOOCH_3]$	S: S-N-Verb.1-171
$C_4H_9N_2O_3P$	$(CH_3O)_3PNCN$	F: PerFHalOrg.5-23
$C_4H_9N_3O_3S_2Si$	$[NSO_2][NSNCOSi(CH_3)_3]$	S: S-N-Verb.1-25/6
$C_4H_9N_3O_3S_2Si^+$	$[NSO_2][NSNCOSi(CH_3)_3]^+$	S: S-N-Verb.1-25
$C_4H_9N_3O_4S_2Si$	$[NSO_2][NSO][NCOSi(CH_3)_3]$	S: S-N-Verb.1-24/5
$C_4H_9N_3O_5S_2$	$[CH_3NSO_2]_2[CH_3NCO]$	S: S-N-Verb.1-24
$C_4H_9N_3O_8U$	$U(C_2O_4)_2 \cdot 3\ NH_3 \cdot H_2O$	U: SVol.E1-12, 15
$C_4H_9N_5O_4S$	$[NS(CH_3)_2][NCO][HNC(NH_2)]NO_3$	S: S-N-Verb.1-31
$C_4H_9O_{1.5}Sn$	$[C_4H_9SnO_{1.5}]_x$	Sn: Org.Verb.6-245
$C_4H_9S_{1.5}Sn$	$[C_4H_9SnS_{1.5}]_x$	Sn: Org.Verb.6-247
$C_4H_9Sn^+$	$C_4H_9Sn^+$	Sn: Org.Verb.5-30
$C_4H_{10}I_2MnO$	$MnI_2 \cdot (C_2H_5)_2O$	Mn: MVol.D1-140
$C_4H_{10}I_2Sn$	$(CH_3)_2Sn(CH_2I)_2$	Sn: Org.Verb.3-6
–	$(CH_3)_3SnCHI_2$	Sn: Org.Verb.2-8
$C_4H_{10}I_2SnZn$	$(CH_3)_3SnCHIZnI$	Sn: Org.Verb.2-16
$C_4H_{10}I_4S_2Sn$	$SnI_4 \cdot C_2H_4(SCH_3)_2$	Sn: MVol.C6-96
$C_4H_{10}I_8S_2Sn_2$	$2\ SnI_4 \cdot C_2H_4(SCH_3)_2$	Sn: MVol.C6-97
$C_4H_{10}MoO_6$	$MoO_2(OCH_2CH_2OH)_2$	Mo: SVol.B1-141
$C_4H_{10}N_2O_2Sn$	$Sn(NHCH_2CH_2O)_2$	Sn: MVol.C6-2
$C_4H_{10}N_2O_3S$	$O(CH_2CH_2)_2NSO_2NH_2$	S: S-N-Verb.1-168
$C_4H_{10}N_2O_4S$	$(CH_3)_2NSO_2NHCOOCH_3$	S: S-N-Verb.1-171
$C_4H_{10}N_2O_8SU$	$UO_2SO_4 \cdot 2\ CH_3CONH_2$	U: SVol.E1-96/8
$C_4H_{10}N_2O_8U$	$U(HCOO)_4 \cdot 2\ NH_3$	U: SVol.E1-12, 15
$C_4H_{10}N_2O_9U$	$UO_2(NO_3)_2 \cdot (C_2H_5)_2O \cdot 3\ H_2O$	U: SVol.E1-71, 76
–	$UO_2(NO_3)_2 \cdot C_4H_9OH$	U: SVol.E1-59, 63

$C_4H_{10}N_2O_9U$	$UO_2(NO_3)_2 \cdot C_4H_9OH \cdot 2\ H_2O$	U:	SVol.E1-59, 63
$C_4H_{10}N_2Sn$	$(CH_3)_3SnCHN_2$	Sn:	Org.Verb.2-13
$C_4H_{10}N_4O_2S$	$C_4H_9NHSO_2N_3$	S:	S-N-Verb.1-214
$C_4H_{10}N_4O_8U$	$UO_2(OH)_2 \cdot 2\ NH_2COCONH_2$	U:	SVol.E1-113/4
$C_4H_{10}N_4O_{10}U$	$UO_2(NO_3)_2 \cdot 2\ CH_3CONH_2$	U:	SVol.E1-96/8
$C_4H_{10}N_6O_5S_2U$	$[UO_2(NCS)_2((NH_2)_2CO)_2(H_2O)]$	U:	SVol.E1-87/8
$C_4H_{10}NiO_2S_2$	$Ni(SCH_2CH_2OH)_2$	Ni:	Org.Verb.1-160
$C_4H_{10}Ni_2$	$C_2H_5NiNiC_2H_5$	Ni:	Org.Verb.2-268
$C_4H_{10}OSn$	$[(C_2H_5)_2SnO]_x$	Sn:	Org.Verb.6-54, 56, 58
$C_4H_{10}O_2Sn$	$[C_4H_9Sn(OH)O]_x$	Sn:	Org.Verb.6-246
$C_4H_{10}S_2Sn$	$(CH_3)_2Sn(SCH_2)_2$	Sn:	Org.Verb.6-21, 25
$C_4H_{10}Sn$	$[(C_2H_5)_2Sn]_x$	Sn:	Org.Verb.6-53
$C_4H_{11}ISn$	$(CH_3)_3SnCH_2I$	Sn:	Org.Verb.2-7
$C_4H_{11}I_3SiSn$	$(CH_3)_3SiCH_2SnI_3$	Sn:	Org.Verb.6-254
$C_4H_{11}NO_3S$	$NH_2SO_2OC_4H_9$	S:	S-N-Verb.1-97
$C_4H_{11}N_2NaO_3S$	$Na[SO_3NC_4H_9NH_2]$	S:	S-N-Verb.1-219
$C_4H_{11}N_2SSi^+$	$(CH_3)_3SiNSNCH_2^+$	S:	S-N-Verb.1-211
$C_4H_{11}N_3O_9U$	$UO_2(NO_3)_2 \cdot [(CH_3)_2(C_2H_4OH)N]$	U:	SVol.E1-20, 23
$C_4H_{11}NiP^+$	$Ni(CH_2P(CH_3)_3)^+$	Ni:	Org.Verb.1-182
$C_4H_{11}Sn^+$	$(CH_3)_2SnC_2H_5^+$	Sn:	Org.Verb.2-33
–	$Sn(C_2H_5)_2H^+$	Sn:	Org.Verb.1-70/1
$C_4H_{11}Sn^-$	$(CH_3)_3SnCH_2^-$	Sn:	Org.Verb.1-47
$C_4H_{12}I_2NiP_2$	$Ni(P(CH_3)_2)_2I_2$	Ni:	Org.Verb.1-72
$C_4H_{12}I_2Ti$	$(CH_3)_4Ti \cdot I_2$	Ti:	Org.Verb.1-116, 119
$C_4H_{12}I_3NSn$	$[(CH_3)_4N]SnI_3$	Sn:	MVol.C3-119/20
$C_4H_{12}I_4O_2S_2Sn$	$SnI_4 \cdot 2\ (CH_3)_2SO$	Sn:	MVol.C6-100/1
$C_4H_{12}I_4O_2S_2Te$	$TeI_4 \cdot 2\ (CH_3)_2SO$	Te:	SVol.B3-176
$C_4H_{12}I_4O_2Se_2Sn$	$SnI_4 \cdot 2\ (CH_3)_2SeO$	Sn:	MVol.C6-123
$C_4H_{12}I_4O_2Te$	$TeI_4 \cdot 2\ C_2H_5OH$	Te:	SVol.B3-165
$C_4H_{12}Li_2Ni$	$Li_2[Ni(CH_3)_4] \cdot 2\ C_4H_8O$	Ni:	Org.Verb.1-104
$C_4H_{12}MnNO_4$	$(CH_3)_4NMnO_4$	Mn:	MVol.C2-37, 45, 201
–	$(CH_3)_4NMnO_4$ solid solutions $(CH_3)_4NMnO_4$-$(CH_3)_4NClO_4$	Mn:	MVol.C2-201
$C_4H_{12}MnN_2O_{10}$	$Mn(HOC_2H_4OH)_2(NO_3)_2$	Mn:	MVol.D1-35
$C_4H_{12}N_2OSSi_2^+$	$(CH_3)_3SiNSONSiCH_3^+$	S:	S-N-Verb.1-198
$C_4H_{12}N_2O_2S$	$CH_3NHSO_2NHC_3H_7$	S:	S-N-Verb.1-170
–	$(CH_3)_2NSO_2N(CH_3)_2$	S:	S-N-Verb.1-172
–	$C_2H_5NHSO_2NHC_2H_5$	S:	S-N-Verb.1-169
–	$(C_2H_5)_2NSO_2NH_2$	S:	S-N-Verb.1-168
$C_4H_{12}N_2O_2S_3$	$SO_2[NS(CH_3)_2]_2$	S:	S-N-Verb.1-182
$C_4H_{12}N_2O_2S_4Zn$	$Zn[S_2COCH_3]_2(NH_3)_2$	C:	MVol.D4-257
$C_4H_{12}N_2O_{10}S_2U$	$UO_2(NO_3)_2 \cdot 2\ (CH_3)_2SO$	U:	SVol.E1-204, 207/8
$C_4H_{12}N_2O_{10}U$	$UO_2(NO_3)_2 \cdot 2\ C_2H_5OH$	U:	SVol.E1-58/9, 63
$C_4H_{12}N_3NaO_4S_2$	$Na[N(SO_2N(CH_3)_2)_2]$	S:	S-N-Verb.1-162
$C_4H_{12}N_4O_3S$	$[(CH_3)_4N][SO_3N_3]$	S:	S-N-Verb.1-213
$C_4H_{12}N_4O_4S_2$	$NH_2SO_2N(CH_2CH_2)_2NSO_2NH_2$	S:	S-N-Verb.1-168
$C_4H_{12}N_4O_4S_3$	$S[NSO_2N(CH_3)_2]_2$	S:	S-N-Verb.1-181
$C_4H_{12}N_4O_8SU$	$UO_2SO_4 \cdot 2\ NH_2CONHCH_3$	U:	SVol.E1-93, 95
$C_4H_{12}N_4O_8S_4$	$[CH_3NSO_2]_4$	S:	S-N-Verb.1-4
$C_4H_{12}N_6O_{10}U$	$UO_2(NO_3)_2 \cdot 2\ NH_2CONHCH_3$	U:	SVol.E1-92, 95

$C_4H_{12}N_7O_{18}Pa$	$[N(CH_3)_4]Pa(NO_3)_6$	Pa: SVol.2-35
$C_4H_{12}OSn$	$(CH_3)_3SnCH_2OH$	Sn: Org.Verb.2-15
$C_4H_{12}O_3Sn$	$C_4H_9Sn(OH)_3 \cdot x\ H_2O$	Sn: Org.Verb.6-246
$C_4H_{12}O_4P_2S_4Te$	$Te((CH_3O)_2P(S)S)_2$	Te: SVol.B3-164/5
$C_4H_{12}O_4Te$	$Te(OCH_3)_4$	Te: SVol.B2-104
$C_4H_{12}O_6S_2Sn$	$(CH_3)_2Sn(OSO_2CH_3)_2$	Sn: Org.Verb.6-28
$C_4H_{12}O_7S_3U$	$UO_2SO_3 \cdot 2\ (CH_3)_2SO \cdot 0.5\ H_2O$	U: SVol.E1-204, 208
$C_4H_{12}O_8S_3U$	$UO_2SO_4 \cdot 2\ (CH_3)_2SO$	U: SVol.E1-204, 208
$C_4H_{12}P_2S_4Te$	$Te((CH_3)_2P(S)S)_2$	Te: SVol.B3-162
$C_4H_{12}S_2Sn$	$(CH_3)_2Sn(SCH_3)_2$	Sn: Org.Verb.6-31
$C_4H_{12}S_3Sn$	$CH_3Sn(SCH_3)_3$	Sn: Org.Verb.6-218
$C_4H_{12}SeSn$	$(CH_3)_3SnSeCH_3$	Sn: Org.Verb.5-81, 84
$C_4H_{12}Se_2Sn$	$(CH_3)_2Sn(SeCH_3)_2$	Sn: Org.Verb.6-21
$C_4H_{12}Se_3Sn$	$CH_3Sn(SeCH_3)_3$	Sn: Org.Verb.6-218
$C_4H_{12}Sn$	$(CH_3)(i\text{-}C_3H_7)SnH_2$	Sn: Org.Verb.4-119
–	$(CH_3)_2(C_2H_5)SnH$	Sn: Org.Verb.4-89
–	$(C_2H_5)_2SnH_2$	Sn: Org.Verb.4-98/102
–	$C_4H_9SnH_3$	Sn: Org.Verb.4-126
–	$(i\text{-}C_4H_9)SnH_3$	Sn: Org.Verb.4-126
–	$Sn(CH_3)_4$	
	Chemical reactions	Sn: Org.Verb.1-46/56
	Formation	Sn: Org.Verb.1-30/5
	Molecule	Sn: Org.Verb.1-35/42
	Physical properties	Sn: Org.Verb.1-42/5
	Physiology	Sn: Org.Verb.1-56
	Preparation	Sn: Org.Verb.1-30/5
	Purity testing	Sn: Org.Verb.1-32
	Spectra	Sn: Org.Verb.1-35/42
	Uses	Sn: Org.Verb.1-56/7
–	$Sn(CH_3)_4$ solutions	
	$Sn(CH_3)_4\text{-}CH_3OH\text{-}H_2O$	Sn: Org.Verb.1-56
$C_4H_{12}Sn^+$	$Sn(CH_3)_4^+$	Sn: Org.Verb.1-47
$C_4H_{12}Ti$	$(CH_3)_4Ti$	Ti: Org.Verb.1-88/9, 96/7
–	$(CD_3)_4Ti$	Ti: Org.Verb.1-90, 97
$C_4H_{13}N_2O_5PS$	$(C_2H_5)_2NSO_2NHPO(OH)_2$	S: S-N-Verb.1-185
$C_4H_{13}N_3O_2S$	$(C_2H_5)_2NSO_2NHNH_2$	S: S-N-Verb.1-222
$C_4H_{13}N_3O_4S_2$	$HN(SO_2N(CH_3)_2)_2$	S: S-N-Verb.1-101
$C_4H_{14}I_4N_2Sn$	$SnI_4 \cdot 2\ (CH_3)_2NH$	Sn: MVol.C5-170
$C_4H_{14}N_4Ni$	$(CHC)_2Ni(NH_3)_4$	Ni: Org.Verb.1-81/2
$C_4H_{14}N_4O_4S_2$	$(CH_3)_2NSO_2NHNHSO_2N(CH_3)_2$	S: S-N-Verb.1-220
$C_4H_{14}N_4O_4S_6Te$	$Te(SC(NH_2)_2)_2(S_2O_2CH_3)_2$	Te: SVol.B3-142
$C_4H_{14}Na_2O_6Sn$	$Na_2[Sn(OH)_2(OCH_3)_4]$	Sn: MVol.C3-15
$C_4H_{15}N_4O_6PS_2$	$(CH_3NHSO_2NCH_3)_2POOH$	S: S-N-Verb.1-185/6
$C_4H_{16}I_{0.5}N_{9.5}O_{10.5}U$	$UO_2(NO_3)_{1.5}I_{0.5} \cdot 4\ (NH_2)_2CO \cdot H_2O$	U: SVol.E1-85
$C_4H_{16}I_{1.1}N_{8.9}O_{8.7}U$	$UO_2(NO_3)_{0.9}I_{1.1} \cdot 4\ (NH_2)_2CO \cdot H_2O$	U: SVol.E1-85
$C_4H_{16}I_{1.7}N_{8.3}O_{6.9}U$	$UO_2(NO_3)_{0.3}I_{1.7} \cdot 4\ (NH_2)_2CO \cdot H_2O$	U: SVol.E1-85
$C_4H_{16}I_2IrN_4^+$	$[Ir(C_2H_4(NH_2)_2)_2I_2]^+$	Ir: SVol.2-83, 84
$C_4H_{16}I_2N_8O_6U$	$UO_2I_2 \cdot 4\ (NH_2)_2CO \cdot H_2O$	U: SVol.E1-83, 86/7
$C_4H_{16}I_3IrN_4$	$[Ir(C_2H_4(NH_2)_2)_2I_2]I$	Ir: SVol.2-84

$C_4H_{16}I_4N_4Sn$	$SnI_4 \cdot 2\ C_2H_8N_2$	Sn:	MVol.C5-184
$C_4H_{16}IrN_4NaO_6S_2$	$Na[Ir(C_2H_4(NH_2)_2)_2(SO_3)_2]$	Ir:	SVol.2-83
$C_4H_{16}IrN_6O_4^+$	$[Ir(C_2H_4(NH_2)_2)_2(NO_2)_2]^+$	Ir:	SVol.2-83
$C_4H_{16}IrN_{10}^+$	$[Ir(C_2H_4(NH_2)_2)_2(N_3)_2]^+$	Ir:	SVol.2-83
$C_4H_{16}MnN_{10}O_{10}$	$Mn(NO_3)_2 \cdot 4\ CO(NH_2)_2 \cdot 2\ H_2O$	Mn:	MVol.C3-294/5
$C_4H_{16}N_2O_{13}U_4$	$[(CH_3)_2NH_2]_2O \cdot 4\ UO_3 \cdot n\ H_2O$	U:	SVol.C3-65/6
$C_4H_{16}N_2O_{25}U_8$	$[(CH_3)_2NH_2]_2O \cdot 8\ UO_3 \cdot n\ H_2O$	U:	SVol.C3-65/6
$C_4H_{16}N_2O_{37}U_{12}$	$[NH_2(CH_3)_2]_2U_{12}O_{37}$	U:	SVol.C3-65
$C_4H_{16}N_2O_{49}U_{16}$	$[(CH_3)_2NH_2]_2O \cdot 16\ UO_3 \cdot n\ H_2O$	U:	SVol.C3-65/6
$C_4H_{16}N_4O_4S_2$	$NH_4[N(SO_2N(CH_3)_2)_2]$	S:	S-N-Verb.1-162
$C_4H_{16}N_6O_8U$	$UO_2(NO_3)_2 \cdot 2\ C_2H_8N_2$	U:	SVol.E1-23/4
$C_4H_{16}N_8O_2OsS_4^{2+}$	$[OsO_2((NH_2)_2CS)_4]^{2+}$	C:	MVol.D6-108
$C_4H_{16}N_8O_4S_5Te$	$Te(SC(NH_2)_2)_4SO_4 \cdot 2\ H_2O$	Te:	SVol.B3-146
$C_4H_{16}N_8O_6OsS_5$	$[OsO_2((NH_2)_2CS)_4]SO_4$	C:	MVol.D6-108
$C_4H_{16}N_8O_{10}SU$	$UO_2SO_4 \cdot 4\ (NH_2)_2CO$	U:	SVol.E1-84/6, 89
$C_4H_{16}N_8O_{12}S_2U$	$U(SO_4)_2 \cdot 4\ (NH_2)_2CO$	U:	SVol.E1-82/3
–	$U(SO_4)_2 \cdot 4\ (NH_2)_2CO \cdot 4\ H_2O$	U:	SVol.E1-82/3
$C_4H_{16}N_8S_4Te^{2+}$	$Te(SC(NH_2)_2)_4^{2+}$	Te:	SVol.B3-140, 144/5
$C_4H_{16}N_8Se_4Te^{2+}$	$[Te(SeC(NH_2)_2)_4]^{2+}$	Te:	SVol.B3-162
$C_4H_{16}N_{10}O_6S_4Te$	$Te(SC(NH_2)_2)_4(NO_3)_2$	Te:	SVol.B3-143
–	$Te(SC(NH_2)_2)_4(NO_3)_2 \cdot 2\ H_2O$	Te:	SVol.B3-143
$C_4H_{16}N_{10}O_{12}U$	$UO_2(NO_3)_2 \cdot 4\ (NH_2)_2CO \cdot H_2O$	U:	SVol.E1-83, 85/6, 88/9
–	$UO_2(NO_3)_2 \cdot 4\ (NH_2)_2CO \cdot 2\ H_2O$	U:	SVol.E1-83, 88/9
$C_4H_{17}IN_8S_4$	$4\ (NH_2)_2CS \cdot HI$	C:	MVol.D6-125
$C_4H_{18}IN_9O_{10}U$	$[UO_2((NH_2)_2CO)_4(H_2O)]I(NO_3)$	U:	SVol.E1-83, 87
$C_4H_{18}I_{1.4}N_{8.6}O_{8.8}U$	$[UO_2((NH_2)_2CO)_4(H_2O)]I_{1.4}(NO_3)_{0.6}$	U:	SVol.E1-83, 87
$C_4H_{18}I_2N_8O_7U$	$[UO_2((NH_2)_2CO)_4(H_2O)]I_2$	U:	SVol.E1-83, 87
$C_4H_{18}N_4O_{14}S_3U$	$(NH_4)_2[U(SO_4)_3(CH_3CONH_2)_2]$	U:	SVol.E1-96/8
–	$(NH_4)_2[U(SO_4)_3(CH_3CONH_2)_2] \cdot 4\ H_2O$	U:	SVol.E1-96/8
$C_4H_{18}N_8O_7U^{2+}$	$[UO_2((NH_2)_2CO)_4(H_2O)]^{2+}$	U:	SVol.E1-82
$C_4H_{18}N_{10}O_{13}U$	$[UO_2((NH_2)_2CO)_4(H_2O)](NO_3)_2$	U:	SVol.E1-83, 87
–	$[UO_2((NH_2)_2CO)_4(H_2O)](NO_3)_2 \cdot H_2O$	U:	SVol.E1-88
$C_4H_{24}N_6NiO_2S_4$	$Ni[S_2COCH_3]_2(NH_3)_6$	C:	MVol.D4-260
$C_4H_{24}N_{12}O_8U$	$(CH_6N_3)_4UO_8$	U:	SVol.C3-66
$C_4HgN_4Se_4^{2-}$	$[Hg(SeCN)_4]^{2-}$	C:	MVol.D6-227
$C_4I_6IrK_2O_4$	$K_2[IrI_6(CO)_4]$	Ir:	SVol.2-151
$C_4I_6Ir_2O_4$	$(IrI_3(CO)_2)_2$	Ir:	SVol.2-151
$C_4IrK_3N_2O_{12}$	$K_3[Ir(C_2O_4)_2(NO_2)_2]$	Ir:	SVol.2-35, 61
$C_4IrN_2O_{12}Tl_3$	$Tl_3[Ir(NO_2)_2(C_2O_4)_2]$	Ir:	SVol.2-61
$C_4IrO_8^-$	$[Ir(C_2O_4)_2]^-$	Ir:	SVol.2-36
$C_4Ir_3O_8^{5+}$	$[Ir_3(C_2O_4)_2]^{5+}$	Ir:	SVol.2-36
C_4KNNiO_3	$K[(CO)_3NiCN]$	Ni:	Org.Verb.1-191, 192
$C_4K_2N_2NiO_2$	$K_2[(CO)_2Ni(CN)_2]$	Ni:	Org.Verb.1-157/8
$C_4K_2N_4S_4Sn$	$K_2Sn(SCN)_4 \cdot CH_3OH$	Sn:	MVol.C3-80
$C_4K_2O_8SSn$	$K_2SnS(C_2O_4)_2$	Sn:	MVol.C3-79
–	$K_2SnS(C_2O_4)_2 \cdot 2\ H_2O$	Sn:	MVol.C3-79/80
$C_4K_2O_8Sn$	$K_2Sn(C_2O_4)_2$	Sn:	MVol.C3-78
–	$K_2Sn(C_2O_4)_2 \cdot H_2O$	Sn:	MVol.C3-78
$C_4Li_2O_8SSn$	$Li_2SnS(C_2O_4)_2$	Sn:	MVol.C3-7
C_4MgO_8SSn	$MgSnS(C_2O_4)_2$	Sn:	MVol.C3-168

C_4MnO_4 .... MnC_4O_4 .... Mn: MVol.D1-127
$C_4MnO_8^-$ .... $[Mn(C_2O_4)_2]^-$ .... Mn: MVol.C2-82/3
$C_4NNaNiO_3$ .... $Na[(CO)_3NiCN]$ .... Ni: Org.Verb.1-191, 192
$C_4N_4O_4Sn$ .... $Sn(NCO)_4$ .... Sn: MVol.C2-227
$C_4N_4PbSe_4$ .... $Pb(SeCN)_4$ .... C: MVol.D6-229
C_4N_4PtRa .... $RaPt(CN)_4$ .... Ra: SVol.2-376
$C_4N_4S_4Sn$ .... $Sn(SCN)_4$ .... Sn: MVol.C2-229
$C_4N_4S_4Sn^{2-}$ .... $Sn(SCN)_4^{2-}$ .... Sn: MVol.C2-228
C_4N_4Sn .... $Sn(CN)_4$ .... Sn: MVol.C2-227
C_4NiO_4 .... $Ni(CO)_4$
Chemical reactions .... Ni: Org.Verb.1-223/51
Decomposition .... Ni: Org.Verb.1-223/6
Detection .... Ni: Org.Verb.1-299/301
Determination .... Ni: Org.Verb.1-299/301
Dissociation .... Ni: Org.Verb.1-224/6
Exchange reactions with CO .... Ni: Org.Verb.1-226
Formation .... Ni: Org.Verb.1-202/8
Hazards .... Ni: Org.Verb.1-297/9
Molecule .... Ni: Org.Verb.1-216/22
Physical properties .... Ni: Org.Verb.1-208/16
Preparation .... Ni: Org.Verb.1-195/202
Purification .... Ni: Org.Verb.1-205/6
Recovery .... Ni: Org.Verb.1-205/6
Removal .... Ni: Org.Verb.1-205/6
Safety precautions .... Ni: Org.Verb.1-297/9
Solubility .... Ni: Org.Verb.1-223
Solvent power .... Ni: Org.Verb.1-223
Toxicity .... Ni: Org.Verb.1-287/97
Uses .... Ni: Org.Verb.1-251/87
– .... $Ni(C^{18}O)_4$ .... Ni: Org.Verb.1-196
– .... $Ni(^{13}CO)_4$ .... Ni: Org.Verb.1-196
– .... $Ni(^{14}CO)_4$ .... Ni: Org.Verb.1-196
– .... $Ni(CO)_4$ systems
$Ni(CO)_4$-$Co_2(CO)_8$-$Co_4(CO)_{12}$-$Fe(CO)_5$. Fe: Org.Verb.B3-64
$Ni(CO)_4$-$Co_2(CO)_8$-$Fe(CO)_5$ .... Fe: Org.Verb.B3-64
$Ni(CO)_4$-$Fe(CO)_5$ .... Fe: Org.Verb.B3-64
$C_4NiO_4^+$ .... $Ni(CO)_4^+$ .... Ni: Org.Verb.1-215
C_4NiS_4 .... $Ni(CS)_4$ .... C: MVol.D4-28
C_4NpO_8 .... $Np(C_2O_4)_2 \cdot 6\,H_2O$
Solubility .... Np: TrU.D1-157
$C_4NpO_{10}^{2-}$ .... $NpO_2(C_2O_4)_2^{2-}$ .... Np: TrU.D1-158
$C_4NpO_{10}^{3-}$ .... $NpO_2(C_2O_4)_2^{3-}$ .... Np: TrU.D1-28, 157, 158
$C_4O_8Pa^+$ .... $Pa(C_2O_4)_2^+$ .... Pa: SVol.2-170
C_4O_8Pu .... $Pu(C_2O_4)_2 \cdot 6\,H_2O$
Solubility .... Np: TrU.D1-157, 158
$C_4O_8Pu^-$ .... $Pu(C_2O_4)_2^-$ .... Np: TrU.D1-156, 158
$C_4O_8SSn^{2-}$ .... $SnS(C_2O_4)_2^{2-}$ .... Sn: MVol.C2-227/8
C_4O_8SSnSr .... $SrSnS(C_2O_4)_2$ .... Sn: MVol.C3-190
C_4O_8Sn .... $Sn(C_2O_4)_2$ .... Sn: MVol.C2-224/5

C_4O_8Sn	$Sn(C_2O_4)_2$ systems	
	$Sn(C_2O_4)_2$-$K_2C_2O_4$-H_2O	Sn: MVol.C3-78
$C_4O_8Sn^{2-}$	$Sn(C_2O_4)_2^{2-}$	Sn: MVol.C2-223/4
$C_4O_9Pa^-$	$PaO(C_2O_4)_2^-$	Pa: SVol.2-169/70
$C_4O_{10}Pu^{2-}$	$PuO_2(C_2O_4)_2^{2-}$	Np: TrU.D1-157/8
$C_4O_{10}Pu^{3-}$	$PuO_2(C_2O_4)_2^{3-}$	Np: TrU.D1-157/8
$C_{4.4}F_6H_{9.2}N_{0.8}P_2$	$(CF_3)_2PPH_2 \cdot 0.8\ N(CH_3)_3$	F: PerFHalOrg.3-26
$C_{4.5}H_{10.5}I_4N_{1.5}O_{1.5}Te$	$TeI_4 \cdot 1.5\ HCON(CH_3)_2$	Te: SVol.B3-174
$C_{4.8}H_{10.6}IMnO_{1.2}$	$MnHI \cdot 1.2\ C_4H_8O$	Mn: MVol.D1-143
$C_5CeCl_4H_6N$	$CeCl_3 \cdot [C_5H_5NH]Cl \cdot 2\ H_2O$	
	$= [C_5H_5NH][CeCl_4(H_2O)_2]$	Sc: MVol.C5-213/5, 223
$C_5CeCl_4H_{10}NO_2$	$[C_5H_5NH][CeCl_4(H_2O)_2]$	
	$= CeCl_3 \cdot [C_5H_5NH]Cl \cdot 2\ H_2O$	Sc: MVol.C5-213/5, 223
$C_5CeH_{20}I_3N_{10}O_5$	$CeI_3 \cdot 5\ CO(NH_2)_2$	Sc: MVol.C6-186
$C_5CeO_{15}^{6-}$	$[Ce(CO_3)_5]^{6-}$	Sc: MVol.B7-76
$C_5ClCoH_{18}IrN_{11}$	$[Co(NH_3)_6][Ir(CN)_5Cl]$	Ir: SVol.2-184
$C_5ClCuH_{12}Sn$	$(CH_3)_3SnCHCH_2 \cdot CuCl$	Sn: Org.Verb.2-77
C_5ClFHI_2NO	HOC_5FClI_2N	F: PerFHalOrg.5-182/3
		F: PerFHalOrg.6-13
$C_5ClFH_8NO_4P$	$OP(ONCFCl)(OCH(CH_3))_2$	F: PerFHalOrg.7-202
$C_5ClFH_{10}NO_2PS_2$	$(C_2H_5S)_2P(O)ONCFCl$	F: PerFHalOrg.7-200
$C_5ClFH_{10}NO_4P$	$ClCFNOPO(OC_2H_5)_2$	F: PerFHalOrg.7-199/200
$C_5ClFH_{10}N_2O_2S$	$FSO_2NCCl[N(C_2H_5)_2]$	S: S-N-Verb.1-118
$C_5ClF_2H_3N_2O$	$NCFCClNCFC(OCH_3)$	F: PerFHalOrg.6-58, 63
–	$NCFNCFCClC(OCH_3)$	F: PerFHalOrg.6-58, 62
–	$NC(OCH_3)NCFCClCF$	F: PerFHalOrg.6-58, 62
$C_5ClF_2H_{14}N_4O_2PS_2Si$	$[NS(O)F]_2[NPClN(C_2H_5)Si(CH_3)_3]$	S: S-N-Verb.1-27
$C_5ClF_2I_2N$	$C_5F_2ClI_2N$	F: PerFHalOrg.5-118/9
$C_5ClF_2N_3$	$NCFNCClC(CN)CF$	F: PerFHalOrg.6-21/2, 42
$C_5ClF_3FeO_6S$	$(CO)_4Fe(SO_2CF_3)Cl$	Fe: Org.Verb.B2-69
C_5ClF_3HNO	HOC_5F_3ClN	F: PerFHalOrg.5-181/3, 199
		F: PerFHalOrg.6-13
C_5ClF_3HNS	HSC_5F_3ClN	F: PerFHalOrg.5-192, 206, 216
$C_5ClF_3HN_2Na$	$NaNHC_5F_3ClN$	F: PerFHalOrg.5-215/6
$C_5ClF_3H_2N_2$	$NH_2C_5F_3ClN$	F: PerFHalOrg.5-185/6, 201
$C_5ClF_3H_2N_4O_2$	$NCClNC(NH_2)C(NO_2)C(CF_3)$	F: PerFHalOrg.6-26/7, 47, 58/9
$C_5ClF_3H_3N_3$	$NC_5F_3ClNHNH_2$	F: PerFHalOrg.5-189, 203
$C_5ClF_3H_3N_3S$	$NC(CF_3)NC(SH)C(NH_2)CCl$	F: PerFHalOrg.6-25, 47, 71
–	$NCClNC(SH)C(NH_2)C(CF_3)$	F: PerFHalOrg.6-26, 47, 71
$C_5ClF_3H_4N_4$	$NC(CF_3)NC(NH_2)C(NH_2)CCl$	F: PerFHalOrg.6-25, 47, 71
–	$NCClNC(NH_2)C(NH_2)C(CF_3)$	F: PerFHalOrg.6-26, 47, 70/1
$C_5ClF_3H_5NO$	$CF_3N(O)CH_2CHCHCHCl$	F: PerFHalOrg.7-168
$C_5ClF_3H_5N_5$	$NCClNC(NHNH_2)C(NH_2)C(CF_3)$	F: PerFHalOrg.6-26/7, 47, 71
$C_5ClF_3H_8NO_3P$	$FClCNOPF_2(OCH(CH_3))_2$	F: PerFHalOrg.7-200, 202
$C_5ClF_3H_9P$	$CF_3PCl(n\text{-}C_4H_9)$	F: PerFHalOrg.3-101, 105

$C_5ClF_3H_{10}NP$	$CF_3PClN(C_2H_5)_2$	F:	PerFHalOrg.3-108, 113
$C_5ClF_3H_{10}Sn$	$(CH_3)_3SnCF_2CHClF$	Sn:	Org.Verb.2-35/6
C_5ClF_3IN	C_5F_3ClIN	F:	PerFHalOrg.5-118/9, 129
$C_5ClF_3N_4$	$N_3C_5F_3ClN$	F:	PerFHalOrg.5-184/5, 200/1
$C_5ClF_4H_6NO$	$CF_3CFClNOCH(CH_3)CH_2$	F:	PerFHalOrg.7-194
C_5ClF_4MgN	NC_5F_4MgCl	F:	PerFHalOrg.5-115, 134
C_5ClF_4N	ClC_5F_4N	F:	PerFHalOrg.5-116/7, 124/6, 134, 138, 141
C_5ClF_4NS	$ClSC_5F_4N$	F:	PerFHalOrg.5-191/2
$C_5ClF_5H_3NOS$	$SC(CF_2Cl)(CF_3)NHC(O)CH_2$	F:	PerFHalOrg.7-66/7
$C_5ClF_5H_4O$	$CH_3(CF_3)C(OH)CClCF_2$	F:	PerFHalOrg.4-46
$C_5ClF_5H_{12}SSi$	$(CH_3)_3SiC_2H_3ClSF_5$	S:	SVol.2-289
$C_5ClF_6H_2O_3P$	$C_5F_6(Cl)PO(OH)_2$	F:	PerFHalOrg.3-44, 46
C_5ClF_6Li	$LiC(CF_2)_3CCl$	F:	PerFHalOrg.4-3, 13, 27, 34
C_5ClF_6N	$NCF_2CFCClCF_2CF$	F:	PerFHalOrg.5-120/1, 131
–	$N(CF_2)_2CClCFCF$	F:	PerFHalOrg.5-120/1, 131
C_5ClF_6NO	$NCO(CF_2)_3CCl$	F:	PerFHalOrg.5-180/1, 198
$C_5ClF_6N_3$	$(CF_3CFCl)C_3F_2N_3$	F:	PerFHalOrg.6-78/80, 100
$C_5ClF_7H_2N_2O_2$	$C_3F_7C(NH_2)NOC(O)Cl$	F:	PerFHalOrg.7-5/6, 32
$C_5ClF_7N_2S$	$CClNC(C_3F_7)NS$	F:	PerFHalOrg.5-67/8
C_5ClF_8HO	$(CF_3)_2C(OH)CClCF_2$	F:	PerFHalOrg.4-46
C_5ClF_8N	$NCF_2CFCl(CF_2)_2CF$	F:	PerFHalOrg.5-119/21, 130
–	$N(CF_2)_2CFClCF_2CF$	F:	PerFHalOrg.5-119/21, 130
–	$N(CF_2)_3CFClCF$	F:	PerFHalOrg.5-119/21, 130
C_5ClF_8NO	$ON(CF_2Cl)CF_2CFCFCF_2$	F:	PerFHalOrg.6-4/5, 9
$C_5ClF_9H_6NP$	$(CF_3)_3PClN(CH_3)_2$	F:	PerFHalOrg.3-109, 110
$C_5ClF_9N_2O_2$	$(CF_3)_2CNOCF_2CFClNO$	F:	PerFHalOrg.7-96/7, 114
–	$[N(CF_3)OCF_2CF(NO)CFClCF_2]_n$	F:	PerFHalOrg.7-208
$C_5ClF_{10}N$	$FN(CF_2CFClCF_2CF_2CF_2)$	F:	PerFHalOrg.5-120/1, 131
$C_5ClFeHO_5$	$[(CO)_5FeH]Cl$	Fe:	Org.Verb.B3-252
$C_5ClFeH_3O_6S$	$CH_3SO_2Fe(CO)_4Cl$	Fe:	Org.Verb.B3-187
$C_5ClFeH_4^+$	$[C_5H_4FeCl]^+$	Fe:	Org.Verb.A3-152
$C_5ClFeH_4NO_3$	$CHClCHCH_2Fe(CO)_2NO$	Fe:	Org.Verb.B5-14/5, 21
–	$CH_2CClCH_2Fe(CO)_2NO$	Fe:	Org.Verb.B5-14/6, 22/3
C_5ClFeH_5	$[C_5H_5Fe]Cl$	Fe:	Org.Verb.A1-182
$C_5ClFeH_5^+$	$[C_5H_5FeCl]^+$	Fe:	Org.Verb.A1-385
		Fe:	Org.Verb.A2-242
$C_5ClFeH_9NO_2$	$Cl(CO)FeN(O)C(CH_3)_3$	Fe:	Org.Verb.B3-171
$C_5ClFeNO_6$	$[(CO)_5FeNO]Cl$	Fe:	Org.Verb.B3-255
$C_5ClFeO_5^+$	$[(CO)_5FeCl]^+$	Fe:	Org.Verb.B3-107, 252
$C_5ClH_5NOS^+$	$(C_5H_5N \cdot SOCl)^+$	S:	SVol.1-52, 57
C_5ClH_5Ti	C_5H_5TiCl	Ti:	Org.Verb.1-132
$C_5ClH_5Ti^+$	$[C_5H_5TiCl]^+$	Ti:	Org.Verb.1-181, 184
$C_5ClH_6NiO_3P$	$(CO)_3NiP(CH_3)_2Cl$	Ni:	Org.Verb.1-164, 165
$C_5ClH_7MnO_2^+$	$[MnCl(CH(COCH_3)_2)]^+$	Mn:	MVol.D1-102

$C_5ClH_7O_2$	$(CH_3)_2CCClCOOH$	F:	PerFHalOrg.4-45
$C_5ClH_7O_2Sn$	$[SnCl(CH(COCH_3)_2)]$	Sn:	MVol.C5-9
$C_5ClH_8NO_3S$	$ClSO_2NHCOCH_2C(CH_3)CH_2$	S:	S-N-Verb.1-126
C_5ClH_8NSn	$(CH_3)_2(NCCHCH)SnCl$	Sn:	Org.Verb.5-193
C_5ClH_9Sn	$(CH_2CH)_2(CH_3)SnCl$	Sn:	Org.Verb.5-222
–	$(CH_3)_3SnCCCl$	Sn:	Org.Verb.2-111/2
$C_5ClH_{10}NO_2S$	$ClSO_2N(CH_2)_5$	S:	S-N-Verb.1-124
$C_5ClH_{11}MgNi$	$C_3H_5NiCH_3 \cdot Mg(CH_3)Cl \cdot C_4H_8O$	Ni:	Org.Verb.2-285
$C_5ClH_{11}N_2NiS$	$CH_3C_3H_4Ni(SC(NH_2)_2)Cl$	Ni:	Org.Verb.2-10
$C_5ClH_{11}Sn$	$(CH_2)_4SnClCH_3$	Sn:	Org.Verb.5-229
–	$(CH_3)_2(CH_2CHCH_2)SnCl$	Sn:	Org.Verb.5-193
$C_5ClH_{12}NO_2S$	$ClSO_2NHC_5H_{11}$	S:	S-N-Verb.1-123
$C_5ClH_{12}NSn$	$(CH_3)_3SnCl \cdot CH_3CN$	Sn:	Org.Verb.5-87
$C_5ClH_{12}N_4O_3PS$	$[SO_2(NCH_3)_2]PCl[(NCH_3)_2CO]$	S:	S-N-Verb.1-44
$C_5ClH_{13}N_4NiS_2$	$C_3H_5Ni(SC(NH_2)_2)_2Cl$	Ni:	Org.Verb.2-21/2
$C_5ClH_{13}OSn$	$(CH_3)_2[HO(CH_2)_3]SnCl$	Sn:	Org.Verb.5-193, 197
$C_5ClH_{13}O_2Ti$	$CH_3Ti(OC_2H_5)_2Cl$	Ti:	Org.Verb.1-59
$C_5ClH_{13}Sn$	$(CH_3)_2(C_3H_7)SnCl$	Sn:	Org.Verb.5-193
–	$(CH_3)_2Sn(C_2H_5)(CH_2Cl)$	Sn:	Org.Verb.3-76
–	$(C_2H_5)_2(CH_3)SnCl$	Sn:	Org.Verb.5-201
$C_5ClH_{15}N_2O_2S$	$[(CH_3)_2NSO_2N(CH_3)_3]Cl$	S:	S-N-Verb.1-172
$C_5ClH_{15}OSSn$	$(CH_3)_3SnCl \cdot (CH_3)_2SO$	Sn:	Org.Verb.5-87
$C_5ClH_{16}IrN_5S^+$	$[Ir(C_2H_4(NH_2)_2)_2(NCS)Cl]^+$	Ir:	SVol.2-83
$C_5ClH_{16}IrN_6O_3S$	$[Ir(NCS)Cl(C_2H_4(NH_2)_2)_2]NO_3$	Ir:	SVol.2-162
$C_5ClH_{16}NSn$	$(CH_3)_2NH \cdot (CH_3)_3SnCl$	Sn:	Org.Verb.5-84
$C_5ClH_{17}IrN_5^{2+}$	$[Ir(C_5H_5N)(NH_3)_4Cl]^{2+}$	Ir:	SVol.2-72
$C_5ClH_{17}N_2SSn$	$(CH_3)_3SnCl \cdot (CH_3)_2S(NH)_2$	Sn:	Org.Verb.5-87
$C_5ClH_{17}N_2Sn$	$(CH_3)_3SnCl \cdot C_2H_8N_2$	Sn:	Org.Verb.5-87
$C_5ClH_{19}N_5Ti$	$C_5H_5Ti(NH_2)Cl \cdot 4\ NH_3$	Ti:	Org.Verb.1-189/90
$C_5ClH_{20}I_3N_{10}O_5U$	$UI_3Cl \cdot 5\ (NH_2)_2CO \cdot 2\ H_2O$	U:	SVol.E1-82/3
$C_5ClH_{20}IrN_{10}S_5^{2+}$	$[Ir(SC(NH_2)_2)_5Cl]^{2+}$	Ir:	SVol.2-124, 168
$C_5ClH_{20}MnO_5^+$	$Mn(CH_3OH)_5Cl^+$	Mn:	MVol.C5-77/8
		Mn:	MVol.D1-24
$C_5ClH_{20}MnO_9^+$	$[MnClO_4(CH_3OH)_5]^+$	Mn:	MVol.D1-24/5
$C_5ClH_{20}N_{11}O_{10}U$	$[UO_2((NH_2)_2CO)_5] \cdot Cl(NO_3)$	U:	SVol.E1-83, 85, 87
$C_5ClIrK_3N_5$	$K_3[Ir(CN)_5Cl]$	Ir:	SVol.2-184
$C_5ClIrN_5^{3-}$	$[Ir(CN)_5Cl]^{3-}$	Ir:	SVol.2-183
$C_5Cl_2FH_2NO_2$	$(HO)_2C_5FCl_2N$	F:	PerFHalOrg.5-181/3, 199
		F:	PerFHalOrg.6-13
$C_5Cl_2FH_4N_3O_2S$	$(NH_2)(NH_2SO_2)C_5FCl_2N$	F:	PerFHalOrg.5-192, 206
$C_5Cl_2FH_6N_5$	$(NH_2NH)_2C_5FCl_2N$	F:	PerFHalOrg.5-189, 204
$C_5Cl_2FH_7O$	$(CH_3)_2C(OH)CClCFCl$	F:	PerFHalOrg.4-46
$C_5Cl_2FH_9NOPS$	$C_4H_9SP(Cl)ONCFCl$	F:	PerFHalOrg.7-200
$C_5Cl_2FN_3$	NCClNCClC(CN)CF	F:	PerFHalOrg.6-21/2, 43
–	NCFNCClC(CN)CCl	F:	PerFHalOrg.6-21/2, 43
$C_5Cl_2F_2HNO$	HNC(O)CClCFCClCF	F:	PerFHalOrg.5-181
–	$HOC_5F_2Cl_2N$	F:	PerFHalOrg.5-181/3, 199, 212
		F:	PerFHalOrg.6-13
$C_5Cl_2F_2HNO_2$	$ONC_5F_2Cl_2(OH)$	F:	PerFHalOrg.5-184, 200

$C_5Cl_2F_2HNS$	$HSC_5F_2Cl_2N$	F:	PerFHalOrg.5-192, 206, 216/7
$C_5Cl_2F_2HN_2Na$	$NaNHC_5F_2Cl_2N$	F:	PerFHalOrg.5-187/9, 215/6
$C_5Cl_2F_2H_2N_2$	$NH_2C_5F_2Cl_2N$	F:	PerFHalOrg.5-185/6, 201, 214/6
		F:	PerFHalOrg.6-14
$C_5Cl_2F_2H_2N_2O$	$(HONH)C_5F_2Cl_2N$	F:	PerFHalOrg.5-189, 204
$C_5Cl_2F_2H_2N_2O_2S$	$(NH_2SO_2)C_5Cl_2F_2N$	F:	PerFHalOrg.5-192, 206
$C_5Cl_2F_2H_3N_3$	$NC_5F_2Cl_2NHNH_2$	F:	PerFHalOrg.5-189, 203
$C_5Cl_2F_2H_8NO_3P$	$OPF(ONCFCl)OCH_2CCl(CH_3)_2$	F:	PerFHalOrg.7-201/2
$C_5Cl_2F_2H_9N_2OPS$	$SPF(ONCFCl)N(C_2H_5)CH_2CH_2Cl$	F:	PerFHalOrg.7-201/2
$C_5Cl_2F_2H_9N_2O_2P$	$OPF(ONCFCl)N(C_2H_5)CH_2CH_2Cl$	F:	PerFHalOrg.7-201/2
$C_5Cl_2F_2N_4$	$N(CFCCl)_2CN_3$	F:	PerFHalOrg.5-184/5, 201
$C_5Cl_2F_3H_2N_3$	$NC(CF_3)NCClC(NH_2)CCl$	F:	PerFHalOrg.6-25, 47
–	$NCClNCClC(NH_2)C(CF_3)$	F:	PerFHalOrg.6-26/7, 47, 59
$C_5Cl_2F_3N$	$C_5F_3Cl_2N$	F:	PerFHalOrg.5-116/7, 124/5, 127, 134/6, 138/43
		F:	PerFHalOrg.6-13/4
$C_5Cl_2F_3N_3O_2$	$NC(CF_3)NCClC(NO_2)CCl$	F:	PerFHalOrg.6-25, 47
–	$NCClNCClC(NO_2)C(CF_3)$	F:	PerFHalOrg.6-26/7, 47, 60, 71
$C_5Cl_2F_4H_2N_2O_3$	$OC(O)NHC(O)NHC(CF_2Cl)_2$	F:	PerFHalOrg.6-27, 48
$C_5Cl_2F_4H_3NOS$	$SC(CF_2Cl)_2NHC(O)CH_2$	F:	PerFHalOrg.7-66/7
$C_5Cl_2F_4H_3NO_2$	$OC(CF_2Cl)_2NHC(O)CH_2$	F:	PerFHalOrg.7-66, 68
$C_5Cl_2F_4H_4O$	$CH_3(CF_3)C(OH)CClCFCl$	F:	PerFHalOrg.4-46
$C_5Cl_2F_4N_2$	$Cl_2NC_5F_4N$	F:	PerFHalOrg.5-185/6, 201
$C_5Cl_2F_5N$	$N(CF_2)_2(CCl)_2CF$	F:	PerFHalOrg.5-120/1, 131
$C_5Cl_2F_5N_3$	$(CF_2Cl)_2C_3FN_3$	F:	PerFHalOrg.6-112
$C_5Cl_2F_7HO$	$CF_2CFC(CF_2Cl)_2OH$	F:	PerFHalOrg.4-45
$C_5Cl_2F_7H_5Si$	$(C_3F_7)(C_2H_5)SiCl_2$	F:	PerFHalOrg.4-18
$C_5Cl_2F_7N$	$NCF_2CFClCF_2CFClCF$	F:	PerFHalOrg.5-119/21
–	$NCF_2(CFCl)_2CF_2CF$	F:	PerFHalOrg.5-119/21
–	$N(CF_2)_2(CFCl)_2CF$	F:	PerFHalOrg.5-119/21
$C_5Cl_2F_7NO$	$CFClCF_2N(CF_3)COCFCl$	F:	PerFHalOrg.5-54, 77, 98
$C_5Cl_2F_9HNSb$	$[C_5F_3Cl_2NH][SbF_6]$	F:	PerFHalOrg.5-119, 129
$C_5Cl_2F_9NO$	$[N(CF_3)OCF_2CF(CFClCF_2Cl)]_n$	F:	PerFHalOrg.7-207/8, 214
–	$ON(CF_3)CF_2(CFCl)_2CF_2$	F:	PerFHalOrg.6-4/5, 9/10, 11
$C_5Cl_2F_{11}NO$	$ClCF_2CF_2N(CF_3)OCF_2CF_2Cl$	F:	PerFHalOrg.7-95, 112
$C_5Cl_2F_{13}P$	$(n\text{-}C_3F_7)P(CF_3)_2Cl_2$	F:	PerFHalOrg.3-74, 92, 100
$C_5Cl_2FeHgO_5$	$(CO)_5FeHgCl_2$	Fe:	Org.Verb.B3-250
$C_5Cl_2FeO_5$	$(CO)_4Fe(COCl)Cl$	Fe:	Org.Verb.B2-176/7
–	$[(CO)_5FeCl]Cl$	Fe:	Org.Verb.B3-105/7, 252
$C_5Cl_2H_2N_8$	$(NH_2)(N_3)_2C_5Cl_2N$	F:	PerFHalOrg.5-215
$C_5Cl_2H_4Ti$	$C_5H_4TiCl_2$	Ti:	Org.Verb.1-122
$C_5Cl_2H_5NOSn$	$SnCl_2 \cdot C_5H_5NO$	Sn:	MVol.C5-19
–	$SnOCl_2 \cdot C_5H_5N$	Sn:	MVol.C5-197

$C_5Cl_2H_5NO_2U$	$UO_2Cl_2 \cdot C_5H_5N \cdot H_2O$	U: SVol.E1-28, 34
$C_5Cl_2H_5NSn$	$SnCl_2 \cdot C_5H_5N$	Sn: MVol.C5-17
$C_5Cl_2H_5O_{12}P_8Ti$	$C_5H_5Ti(P_4O_6)_2Cl_2$	Ti: Org.Verb.1-132/3
$C_5Cl_2H_5Ti$	$C_5H_5TiCl_2 \cdot 1.5\ C_4H_8O$	Ti: Org.Verb.1-135/6
$C_5Cl_2H_5Ti^+$	$[C_5H_5TiCl_2]^+$	Ti: Org.Verb.1-145, 196
$C_5Cl_2H_6N_2O_2U$	$UO_2Cl_2 \cdot CH_3C_4H_3N_2 \cdot H_2O$	U: SVol.E1-46/8
$C_5Cl_2H_6N_4$	$C_3N_3Cl_2N(CH_3)_2$	C: MVol.D3-285
$C_5Cl_2H_6O_2Te$	$Cl_2Te(CH_2CO)_2CH_2$	Te: SVol.B2-98
$C_5Cl_2H_7MnO_2$	$[MnCl_2(CH(COCH_3)_2)]$	Mn: MVol.D1-104
$C_5Cl_2H_7NO_3U$	$(C_5H_5NH)[UO_2Cl_2(OH)]$	U: SVol.E1-34
$C_5Cl_2H_7NTi$	$C_5H_5Ti(NH_2)Cl_2$	Ti: Org.Verb.1-189
$C_5Cl_2H_9MnO_3$	$[MnCl_2(CH(COCH_3)_2)H_2O]$	Mn: MVol.D1-104/5
$C_5Cl_2H_9NNiO$	$(t\text{-}C_4H_9NCl_2NiCO)_x$	Ni: Org.Verb.2-394
$C_5Cl_2H_{10}N_2O_2S$	$(C_2H_5)_2NSO_2NCCl_2$	S: S-N-Verb.1-184
–	$ClSO_2NCCl[N(C_2H_5)_2]$	S: S-N-Verb.1-126
$C_5Cl_2H_{10}O_2Sn$	$(CH_3)(CH_3OCOCH_2CH_2)SnCl_2$	Sn: Org.Verb.6-195/6
$C_5Cl_2H_{10}Sn$	$CH_2(CH_2CH_2)_2SnCl_2$	Sn: Org.Verb.6-206
$C_5Cl_2H_{11}NO_4Sn$	$SnCl_2 \cdot NH_2CH(CH_3)COOH \cdot CH_3COOH$	Sn: MVol.C5-31/2
–	$SnCl_2 \cdot NH_2(CH_2)_2COOH \cdot CH_3COOH$	Sn: MVol.C5-31/2
$C_5Cl_2H_{11}NSn$	$SnCl_2 \cdot C_5H_{10}NH$	Sn: MVol.C5-15/6
$C_5Cl_2H_{12}N_2SSn$	$SnCl_2 \cdot CS(N(CH_3)_2)_2$	Sn: MVol.C5-56/7
$C_5Cl_2H_{12}N_2STe$	$Te((CH_3)_2NCSN(CH_3)_2)Cl_2$	Te: SVol.B3-147
$C_5Cl_2H_{12}OSn$	$(CH_3)_2SnCl_2 \cdot (CH_3)_2CO$	Sn: Org.Verb.6-36, 38
–	$C_2H_5SnCl_2(OC_3H_7)$	Sn: Org.Verb.6-230
–	$(C_4H_9)(CH_3O)SnCl_2$	Sn: Org.Verb.6-245
$C_5Cl_2H_{12}OTi$	$CH_3Ti(OC_4H_9\text{-}t)Cl_2$	Ti: Org.Verb.1-59
$C_5Cl_2H_{12}SSn$	$(C_4H_9)(CH_3S)SnCl_2$	Sn: Org.Verb.6-245
$C_5Cl_2H_{12}Sn$	$(CH_2Cl)_2Sn(CH_3)(C_2H_5)$	Sn: Org.Verb.3-92/3
–	$(CH_3)(C_4H_9)SnCl_2$	Sn: Org.Verb.6-195/6
–	$(CH_3)_2[Cl(CH_2)_3]SnCl$	Sn: Org.Verb.5-193, 197
–	$(CH_3)_3SnCCl_2CH_3$	Sn: Org.Verb.2-35/6
–	$(C_2H_5)(C_3H_7)SnCl_2$	Sn: Org.Verb.6-198
–	$(C_2H_5)_2(CH_2Cl)SnCl$	Sn: Org.Verb.5-201, 204
$C_5Cl_2H_{13}NOSn$	$(CH_3)_2SnCl_2 \cdot HCON(CH_3)_2$	Sn: Org.Verb.6-36
$C_5Cl_2H_{13}NSn$	$SnCl_2 \cdot C_5H_{11}NH_2$	Sn: MVol.C5-11/2
$C_5Cl_2H_{14}N_2SSn$	$(CH_3)_2SnCl_2 \cdot (CH_3NH)_2CS$	Sn: Org.Verb.6-36
$C_5Cl_2H_{14}SiSn$	$(CH_3)_3SnCH_2SiCl_2CH_3$	Sn: Org.Verb.2-15
$C_5Cl_2H_{15}N_5O_2P_2S$	$[O_2S(NPClN(CH_3)_2)_2NCH_3]$	S: S-N-Verb.1-34
$C_5Cl_2H_{15}OPSn$	$(CH_3)_3SnCl \cdot (CH_3)_2POCl$	Sn: Org.Verb.5-87
$C_5Cl_2H_{16}IrN_5O_4S$	$[Ir(NCS)Cl(C_2H_4(NH_2)_2)_2]ClO_4$	Ir: SVol.2-163
$C_5Cl_2H_{20}N_{10}O_{15}U$	$UO_2(ClO_4)_2 \cdot 5\ (NH_2)_2CO$	U: SVol.E1-83, 89
$C_5Cl_2H_{24}IrN_5O_{10}$	$[Ir(NH_3)_5(C(CH_3)_3COO)](ClO_4)_2$	Ir: SVol.2-32
C_5Cl_3FHNO	HOC_5FCl_3N	F: PerFHalOrg.5-181/3, 199
		F: PerFHalOrg.6-13
C_5Cl_3FHNS	HSC_5FCl_3N	F: PerFHalOrg.5-192, 206, 216/7
$C_5Cl_3FHN_2Na$	$NaNHC_5FCl_3N$	F: PerFHalOrg.5-215/6
$C_5Cl_3FH_2N_2$	$NH_2C_5FCl_3N$	F: PerFHalOrg.5-185/6, 202, 215
		F: PerFHalOrg.6-13

$C_5Cl_3FH_3N_3$	$(NH_2NH)C_5FCl_3N$	F: PerFHalOrg.5-189, 204
$C_5Cl_3FH_8NO_3P$	$OPCl(ONCFCl)OCH_2CCl(CH_3)_2$	F: PerFHalOrg.7-201/2
$C_5Cl_3FH_9N_2OPS$	$SPCl(ONCFCl)N(C_2H_5)CH_2CH_2Cl$	F: PerFHalOrg.7-201/2
$C_5Cl_3FN_2O$	NOC_5FCl_3N	F: PerFHalOrg.5-190/1
$C_5Cl_3FN_2O_2$	$NO_2C_5FCl_3N$	F: PerFHalOrg.5-190/1, 205
$C_5Cl_3F_2N$	$C_5F_2Cl_3N$	F: PerFHalOrg.5-116/7, 123/5, 127/8, 136, 141
		F: PerFHalOrg.6-13/4
$C_5Cl_3F_2NO_2S$	$(SO_2Cl)C_5F_2Cl_2N$	F: PerFHalOrg.5-192, 206
$C_5Cl_3F_3H_2O_4$	$HOCO(CFCl)_3COOH$	F: PerFHalOrg.5-133
$C_5Cl_3F_3N_3P$	$NPCl_2NCClC(CN)C(CF_3)$	F: PerFHalOrg.6-27/8, 48
$C_5Cl_3F_4N_3O$	$N(CN)C(CF_2Cl)_2OCClN$	F: PerFHalOrg.5-65/6, 88
$C_5Cl_3F_6H_2NO$	$CCl_3CF_2CF_2CF_2C(O)NH_2$	F: PerFHalOrg.7-18, 20/1, 47
$C_5Cl_3F_6N$	$NCCl_2(CF_2)_3CCl$	F: PerFHalOrg.5-120/1, 130, 133/4
$C_5Cl_3F_6OP$	$C_5F_6(Cl)POCl_2$	F: PerFHalOrg.3-52, 56, 57
$C_5Cl_3F_8NO$	$ClCF_2CFClCF_2CFClCF_2NO$	F: PerFHalOrg.7-179, 185
$C_5Cl_3H_5NO_2U^-$	$[UO_2Cl_3(C_5H_5N)]^-$	U: SVol.E1-34
$C_5Cl_3H_5Sn$	$C_5H_5SnCl_3$	Sn: Org.Verb.6-264, 266
$C_5Cl_3H_5Ti$	$C_5H_5TiCl_3$	Ti: Org.Verb.1-137/9, 142/8
$C_5Cl_3H_6MnN$	$C_5H_5NHMnCl_3$	Mn: MVol.C5-207/8
–	$C_5H_5NHMnCl_3 \cdot H_2O$	Mn: MVol.C5-208
$C_5Cl_3H_6NSn$	$[C_5H_6N]SnCl_3$	Sn: MVol.C3-95
$C_5Cl_3H_7O_2Te$	$(CH_3C(OH)CHCOCH_2)TeCl_3$	Te: SVol.B2-98
$C_5Cl_3H_9N_2Ti$	$CH_3TiCl_3 \cdot 2\ CH_3CN$	Ti: Org.Verb.1-30/1, 39
$C_5Cl_3H_9O_2Sn$	$CH_3OCOCH(CH_3)CH_2SnCl_3$	Sn: Org.Verb.6-256
–	$CH_3OCOCH_2CH(CH_3)SnCl_3$	Sn: Org.Verb.6-256
–	$C_2H_5OCOCH_2CH_2SnCl_3$	Sn: Org.Verb.6-256
$C_5Cl_3H_9SSiSn$	$(CH_3)_3SiC(CS)SnCl_3$	Sn: Org.Verb.6-266
$C_5Cl_3H_9Sn$	$(CH_3)_3SnCClCCl_2$	Sn: Org.Verb.2-77
$C_5Cl_3H_{11}OSn$	$C_2H_5SnCl_3 \cdot (CH_3)_2CO$	Sn: Org.Verb.6-232
$C_5Cl_3H_{11}OTi$	$CH_3TiCl_3 \cdot C_4H_8O$	Ti: Org.Verb.1-30/1, 38
$C_5Cl_3H_{11}O_2SiSn$	$(CH_3COO)(CH_3)_2SiCH_2SnCl_3$	Sn: Org.Verb.6-256
$C_5Cl_3H_{11}O_2Sn$	$CH_3SnCl_3 \cdot C_4H_8O_2$	Sn: Org.Verb.6-222
$C_5Cl_3H_{11}O_2Ti$	$CH_3TiCl_3 \cdot C_4H_8O_2$	Ti: Org.Verb.1-30/1, 39
$C_5Cl_3H_{11}S_2Ti$	$CH_3TiCl_3 \cdot C_4H_8S_2$	Ti: Org.Verb.1-30/1, 39
$C_5Cl_3H_{11}Sn$	$C_5H_{11}SnCl_3$	Sn: Org.Verb.6-250
–	$(i\text{-}C_5H_{11})SnCl_3$	Sn: Org.Verb.6-250
$C_5Cl_3H_{11}Ti$	$(CH_3)_3CCH_2TiCl_3$	Ti: Org.Verb.1-16, 24
–	$C_5H_{11}TiCl_3$	Ti: Org.Verb.1-16, 24
$C_5Cl_3H_{12}MnN$	$C_5H_{10}NH_2MnCl_3$	Mn: MVol.C5-204/5
$C_5Cl_3H_{13}OSTi$	$CH_3TiCl_3 \cdot CH_3OCH_2CH_2SCH_3$	Ti: Org.Verb.1-32/3, 40
$C_5Cl_3H_{13}OSiSn$	$(C_2H_5O)(CH_3)_2SiCH_2SnCl_3$	Sn: Org.Verb.6-257
$C_5Cl_3H_{13}O_2Sn$	$CH_3SnCl_3 \cdot CH_3OCH_2CH_2OCH_3$	Sn: Org.Verb.6-222
$C_5Cl_3H_{13}O_2Ti$	$CH_3TiCl_3 \cdot CH_3OCH_2CH_2OCH_3$	Ti: Org.Verb.1-32/3, 40
$C_5Cl_3H_{13}S_2Ti$	$CH_3TiCl_3 \cdot CH_3SCH_2CH_2SCH_3$	Ti: Org.Verb.1-32/3, 40
$C_5Cl_3H_{14}MnN$	$C_5H_{11}NH_3MnCl_3$	Mn: MVol.C5-188
$C_5Cl_3H_{14}OPSn$	$C_2H_5SnCl_3 \cdot (CH_3)_3PO$	Sn: Org.Verb.6-232
$C_5Cl_3H_{15}O_2S_2Sn$	$CH_3SnCl_3 \cdot 2\ (CH_3)_2SO$	Sn: Org.Verb.6-26

$C_5Cl_3H_{15}S_2Ti$	$CH_3TiCl_3 \cdot 2\ S(CH_3)_2$	Ti:	Org.Verb.1-30/1, 39
$C_5Cl_3H_{20}IrN_{10}S_5$	$[Ir(SC(NH_2)_2)_5Cl]Cl_2$	Ir:	SVol.2-168
$C_5Cl_3H_{20}LaO_5$	$LaCl_3(CH_3OH)_5$	Sc:	MVol.B7-9
C_5Cl_4FN	C_5FCl_4N	F:	PerFHalOrg.5-116/7, 124/5, 128, 143
		F:	PerFHalOrg.6-13/4
$C_5Cl_4FNO_2S$	$(SO_2Cl)C_5FCl_3N$	F:	PerFHalOrg.5-192, 206
$C_5Cl_4F_5N$	$N(CFCl)_4CF$	F:	PerFHalOrg.5-119/21, 130, 133/4
$C_5Cl_4F_6H_2N_2O_4P_2$	$C_3F_6[C(O)NHP(O)Cl_2]_2$	F:	PerFHalOrg.7-17, 55, 63
$C_5Cl_4F_6NO_2P$	$NCO(CF_2)_3C(OPCl_4)$	F:	PerFHalOrg.5-180/1, 198
$C_5Cl_4F_6O_2P_2$	$C_5F_6(POCl_2)_2$	F:	PerFHalOrg.3-52, 55, 57
$C_5Cl_4F_9NO$	$ClCF_2CFClN(CF_3)OCF_2CFCl_2$	F:	PerFHalOrg.7-95, 112
$C_5Cl_4FeHg_2O_5$	$(CO)_5FeHgCl_2 \cdot HgCl_2$	Fe:	Org.Verb.B3-127/8, 250
$C_5Cl_4FeO_6S$	$CCl_3SO_2Fe(CO)_4Cl$	Fe:	Org.Verb.B3-187
$C_5Cl_4HN_2Na$	$NaNHC_5Cl_4N$	F:	PerFHalOrg.5-188
$C_5Cl_4H_5NOTe$	$TeCl_4 \cdot C_5H_5NO$	Te:	SVol.B3-171
$C_5Cl_4H_5NO_2S_2$	$2\ SOCl_2 \cdot C_5H_5N$	S:	SVol.1-45
$C_5Cl_4H_5NSn$	$SnCl_4 \cdot C_5H_5N$	Sn:	MVol.C5-197
$C_5Cl_4H_5NSn_2$	$2\ SnCl_2 \cdot C_5H_5N$	Sn:	MVol.C5-18
$C_5Cl_4H_5NTe$	$TeCl_4 \cdot C_5H_5N$	Te:	SVol.B3-169
$C_5Cl_4H_6MnN$	$C_5H_5NHMnCl_4$	Mn:	MVol.C5-212
$C_5Cl_4H_6NNd$	$(Nd,Pr)Cl_3 \cdot [C_5H_5NH]Cl \cdot x\ C_2H_5OH$	Sc:	MVol.C5-213/5, 217
$C_5Cl_4H_6NPr$	$(Nd,Pr)Cl_3 \cdot [C_5H_5NH]Cl \cdot x\ C_2H_5OH$	Sc:	MVol.C5-213/5, 217
$C_5Cl_4H_6N_2Sn$	$SnCl_4 \cdot NC(CH_2)_3CN$	Sn:	MVol.C6-73/4
$C_5Cl_4H_8O_2Sn$	$SnCl_4 \cdot CH_2(COCH_3)_2$	Sn:	MVol.C5-101
–	$SnCl_4 \cdot CH_2C(CH_3)COOCH_3$	Sn:	MVol.C5-137/8
$C_5Cl_4H_8O_4Sn$	$SnCl_4 \cdot CH_2(COOCH_3)_2$	Sn:	MVol.C5-140/1
–	$SnCl_4 \cdot C_3H_6(COOH)_2$	Sn:	MVol.C5-121
$C_5Cl_4H_9NO_2Sn$	$SnCl_4 \cdot CH_3CONHCOC_2H_5$	Sn:	MVol.C6-50/1
$C_5Cl_4H_{10}OSn$	$SnCl_4 \cdot CH_3C_4H_7O$	Sn:	MVol.C5-157
$C_5Cl_4H_{10}O_3Sn$	$SnCl_4 \cdot OC(OC_2H_5)_2$	Sn:	MVol.C5-138/9
$C_5Cl_4H_{10}Sn$	$(CH_3CHCl)_2(CH_2Cl)SnCl$	Sn:	Org.Verb.5-222
$C_5Cl_4H_{11}NSn$	$SnCl_4 \cdot (CH_2)_5NH \cdot n\ CH_3COOH$ (n = 1, 2)	Sn:	MVol.C5-193
$C_5Cl_4H_{12}OSn$	$SnCl_4 \cdot C_5H_{11}OH$	Sn:	MVol.C5-73
$C_5Cl_4H_{12}Se_2Sn$	$SnCl_4 \cdot CH_3Se(CH_2)_3SeCH_3$	Sn:	MVol.C6-123
$C_5Cl_4H_{16}MnN_2$	$NH_3(CH_2)_5NH_3MnCl_4$	Mn:	MVol.C5-185/8
$C_5Cl_4H_{20}N_{10}S_5Sn_2$	$2\ SnCl_2 \cdot 5\ CS(NH_2)_2 \cdot 2\ H_2O$	Sn:	MVol.C5-55
$C_5Cl_5Cs_2H_5IrN$	$Cs_2[Ir(C_5H_5N)Cl_5]$	Ir:	SVol.2-72
$C_5Cl_5H_5IrK_2N$	$K_2[Ir(C_5H_5N)Cl_5]$	Ir:	SVol.2-72
$C_5Cl_5H_5NSn^-$	$[SnCl_5(C_5H_5N)]^-$	Sn:	MVol.C5-196
$C_5Cl_5H_6Mn_2N$	$C_5H_5NHMn_2Cl_5$	Mn:	MVol.C5-208
$C_5Cl_5H_6NTe$	$[C_5H_5NH]TeCl_5$	Te:	SVol.B2-104, 108, 132
$C_5Cl_5H_{12}OSn^-$	$[SnCl_5 \cdot C_5H_{11}OH]^-$	Sn:	MVol.C5-72
$C_5Cl_5H_{15}O_2P_2Sn$	$CH_3SnCl_3 \cdot 2\ (CH_3)_2P(O)Cl$	Sn:	Org.Verb.6-222
$C_5Cl_5N_5S_5$	$(ClSCN)_5$	C:	MVol.D6-177/8
$C_5Cl_6CoIrN_5^{5-}$	$[Cl_5IrClCo(CN)_5]^{5-}$	Ir:	SVol.2-121, 137
$C_5Cl_6H_8N_2Sn$	$SnCl_4 \cdot 2\ CH_3CN \cdot CH_2Cl_2$	Sn:	MVol.C6-66
$C_5Cl_7H_{12}PSn$	$[((CH_3)_3C)(CH_3)PCl_2][SnCl_5]$	Sn:	MVol.C2-261

$C_5Cl_8H_5NO_2Se_2Sn$	$[C_5H_5NSeOCl][SeOCl][SnCl_6]$	Sn:	MVol.C2-126
$C_5Cl_8H_7NO_2Sn_2$	$2\ SnCl_4 \cdot NCCH_2COOC_2H_5$	Sn:	MVol.C6-75
$C_5Cl_8Ni_2O_5P_4$	$Cl_2PPCl_2Ni(CO)_2Cl_2PPCl_2Ni(CO)_3$	Ni:	Org.Verb.2-262
$C_5Cl_{16}H_5NO_4Se_4Sn_2$	$[C_5H_5NSeOCl][SeOCl]_3[SnCl_6]_2$	Sn:	MVol.C2-126
$C_5CmH_9O_4$	$Cm(OH)_2(CH_3COCHCOCH_3)$	Np:	TrU.D2-242
$C_5CoF_9NO_3P$	$Co(NO)(CO)_2 \cdot (CF_3)_3P$	F:	PerFHalOrg.3-174
$C_5CoH_6KNO_2S_4$	$K[Co(CN)(S_2COCH_3)_2] \cdot H_2O$	C:	MVol.D4-260
$C_5CoH_{18}IrN_{11}$	$[Co(NH_3)_6][Ir(CN)_5I]$	Ir:	SVol.2-184
$C_5Cr_2H_5NO_9U$	$UO_2Cr_2O_7 \cdot C_5H_5N$	U:	SVol.E1-29
$C_5CsH_8N_7O_4S_3U$	$Cs[UO_2(NCS)_3((NH_2)_2CO)_2]$	U:	SVol.E1-84, 86, 88
$C_5CsN_5O_3$	$N(C(NCO)N)_2COCs$	C:	MVol.D3-211/2
$C_5Cs_2H_4N_2O_{11}U$	$Cs_2[UO_2(C_2O_4)_2((NH_2)_2CO)] \cdot H_2O$	U:	SVol.E1-84, 90
C_5CuF_4N	CuC_5F_4N	F:	PerFHalOrg.5-115/6
$C_5CuH_9OS_2$	$Cu[S_2COC_4H_9]$	C:	MVol.D4-256
$C_5CuH_9O_2S$	$Cu[SOCOC_4H_9]$	C:	MVol.D4-237
$C_5CuH_9S_3$	$Cu[S_2CSC_4H_9]$	C:	MVol.D4-269
$C_5CuH_{12}NS_4$	$[(CH_3)_4N]CuCS_4$	C:	MVol.D4-231
C_5F^+	C_5F^+	F:	PerFHalOrg.3-172, 216
C_5FFeHO_5	$[(CO)_5FeH]F$	Fe:	Org.Verb.B3-251
C_5FFeHO_8S	$[(CO)_5FeH]SO_3F$	Fe:	Org.Verb.B3-251
$C_5FFeH_5{}^+$	$[C_5H_5FeF]^+$	Fe:	Org.Verb.A2-287
$C_5FH_3N_2O_4$	$NHC(O)NHC(O)CFC(COOH)$	F:	PerFHalOrg.6-21/2, 43, 72
–	$NHC(O)NHC(O)CFC(COOH) \cdot H_2O$	F:	PerFHalOrg.6-21/2, 43
$C_5FH_4N_5$	$NC(NH_2)NC(NH_2)C(CN)CF$	F:	PerFHalOrg.6-21/2, 43
$C_5FH_6NO_2$	$CH_2ON(COF)CH_2CHCH$	F:	PerFHalOrg.7-194/5
$C_5FH_7O_2$	$(CH_3)_2CCFCOOH$	F:	PerFHalOrg.4-45, 89
$C_5FH_{10}N$	$(CH_2)_5NF$	F:	PerFHalOrg.5-143/4
$C_5FH_{10}N_3OS$	$(C_2H_5)_2NS(O)FNCN$	S:	S-N-Verb.1-196
$C_5FH_{11}N_2O_3S$	$FSO_2NHC(O)N(C_2H_5)_2$	S:	S-N-Verb.1-102
$C_5FH_{11}Sn$	$(CH_3)_3SnCFCH_2$	Sn:	Org.Verb.2-77, 84
–	$(CH_3)_3SnCHCHF$	Sn:	Org.Verb.2-77, 84
$C_5FH_{12}N_4O_3PS$	$[SO_2(NCH_3)_2]PF[(NCH_3)_2CO]$	S:	S-N-Verb.1-44
$C_5FH_{13}N_2OS$	$(C_2H_5)_2NS(O)FNCH_3$	S:	S-N-Verb.1-193
$C_5F_2FeH_2O_5$	$[(CO)_5FeH]HF_2$	Fe:	Org.Verb.B3-251/2
$C_5F_2FeH_3O_5P$	$(CO)_4FePF_2(OCH_3)$	Fe:	Org.Verb.B2-82, 113
$C_5F_2FeNO_4PS$	$(CO)_4FePF_2SCN$	Fe:	Org.Verb.B2-85, 93
$C_5F_2HI_2NO$	$HOC_5F_2I_2N$	F:	PerFHalOrg.5-182/3
		F:	PerFHalOrg.6-13
$C_5F_2H_3IN_2O$	$NCFNC(OCH_3)CFCl$	F:	PerFHalOrg.6-58, 62
$C_5F_2H_4N_4O_2$	$(NH_2)_2(NO_2)C_5F_2N$	F:	PerFHalOrg.5-190/1, 205
$C_5F_2H_4Ti$	$C_5H_4TiF_2$	Ti:	Org.Verb.1-122
$C_5F_2H_5NSn$	$SnF_2 \cdot C_5H_5N$	Sn:	MVol.C5-17
$C_5F_2H_6$	$CH(CHCH_2)CF_2CH_2$	F:	PerFHalOrg.5-16
$C_5F_2H_6NNiO_3P$	$(CO)_3NiP(N(CH_3)_2)F_2$	Ni:	Org.Verb.1-162
$C_5F_2H_6N_4$	$(NH_2)_3C_5F_2N$	F:	PerFHalOrg.5-185/6, 202
$C_5F_2H_8$	$(CH_3)_2CCF_2CH_2$	F:	PerFHalOrg.5-16
–	$(CH_3CH)_2CF_2$	F:	PerFHalOrg.5-17
$C_5F_2H_9NO_4Te$	$[C_5H_5NH][TeF_2(OH)_3O] \cdot 2\ C_5H_5N \cdot 7\ HF$	Te:	SVol.B2-47/8
$C_5F_2H_{10}NO_4P$	$CF_2NOPO(OC_2H_5)_2$	F:	PerFHalOrg.7-199

$C_5F_2H_{10}N_2S$	$NSF_2NC_5H_{10}$	S:	S-N-Verb.1-204
$C_5F_2H_{10}N_4O_3S_3$	$[NS(O)F]_2[NS(O)NC_5H_{10}]$	S:	S-N-Verb.1-14
$C_5F_2H_{10}O_3S$	$C_4H_9SO_3CHF_2$	F:	PerFHalOrg.5-19
$C_5F_2H_{10}Sn$	$(CH_3)_3SnCFCHF$	Sn:	Org.Verb.2-77, 84
–	$(CH_3)_3SnCHCF_2$	Sn:	Org.Verb.2-77
$C_5F_2H_{12}N_2O_3U$	$[UO_2F_2 \cdot (CH_3)_2NCON(CH_3)_2]_n$	U:	SVol.E1-91/2, 94
$C_5F_2H_{12}Sn$	$(CH_3)_3SnCH_2CHF_2$	Sn:	Org.Verb.2-35/6
$C_5F_2H_{15}N_2OP$	$[(CH_3)_2N]_2PF_2(OCH_3)$	F:	PerFHalOrg.5-23
$C_5F_2H_{15}N_2SSi_2^+$	$SF_2N_2Si_2(CH_3)_5^+$	S:	S-N-Verb.1-201
$C_5F_2H_{15}N_5O_2P_2S$	$[SO_2[NPFN(CH_3)_2]_2NCH_3]$	S:	S-N-Verb.1-33
$C_5F_2MnNO_9S_2$	$Mn(CO)_5N(SO_2F)_2$	S:	S-N-Verb.1-54
$C_5F_2NO_6ReS$	$Re(CO)_5[NS(O)F_2]$	S:	S-N-Verb.1-61
$C_5F_3^+$	$C_5F_3^+$	F:	PerFHalOrg.3-172
$C_5F_3FeIO_4$	$(CO)_4Fe(CF_3)I$	Fe:	Org.Verb.B2-177/8, 180
C_5F_3HINO	HOC_5F_3IN	F:	PerFHalOrg.5-182/3, 199
$C_5F_3H_2IN_2$	$NH_2C_5F_3IN$	F:	PerFHalOrg.5-185/6, 201
$C_5F_3H_2N$	$C_5H_2F_3N$	F:	PerFHalOrg.5-134
$C_5F_3H_2NO_2$	$(HO)_2C_5F_3N$	F:	PerFHalOrg.5-181/2, 198
$C_5F_3H_2N_3O_2$	$(NO_2)(NH_2)C_5F_3N$	F:	PerFHalOrg.5-190/1, 205
$C_5F_3H_2N_3O_3$	$NC(CF_3)NC(OH)C(NO)C(OH)$	F:	PerFHalOrg.6-25/6, 47
$C_5F_3H_2N_3O_4$	$NC(CF_3)NC(OH)C(NO_2)C(OH)$	F:	PerFHalOrg.6-25, 47
$C_5F_3H_3N_2$	$NCFCFNCFC(CH_3)$	F:	PerFHalOrg.6-60, 68
–	$NCFNCFC(CH_3)CF$	F:	PerFHalOrg.6-55
–	$NH_2C_5HF_3N$	F:	PerFHalOrg.5-216
$C_5F_3H_3N_2O$	$CH_3NNCFCFCFC(O)$	F:	PerFHalOrg.6-60
–	$NCFCFCFC(OCH_3)N$	F:	PerFHalOrg.6-57/8, 61
–	$NCFCFCFC(O)N(CH_3)$	F:	PerFHalOrg.6-58, 60, 61, 67
–	$NCFCFNCFC(OCH_3)$	F:	PerFHalOrg.6-58, 63
–	$NCFC(OCH_3)CFCFN$	F:	PerFHalOrg.6-57, 60
–	$NCFNC(OCH_3)CFCF$	F:	PerFHalOrg.6-58, 62
$C_5F_3H_3N_4O_2$	$NC(CF_3)NC(NH_2)C(NO)C(OH)$	F:	PerFHalOrg.6-25/6, 47
$C_5F_3H_3N_4O_3$	$NC(CF_3)NC(NH_2)C(NO_2)C(OH)$	F:	PerFHalOrg.6-25/6, 47
$C_5F_3H_3N_4S$	$(CH_3)(CF_3)C_3N_4S$	F:	PerFHalOrg.5-100
$C_5F_3H_4N_3$	$NCFNC(NHCH_3)CFCF$	F:	PerFHalOrg.6-59, 66
–	$(NH_2)_2C_5F_3N$	F:	PerFHalOrg.5-185/6, 202
$C_5F_3H_4N_3O_2$	$NC(CF_3)NC(OH)C(NH_2)C(OH)$	F:	PerFHalOrg.6-25/6, 47
$C_5F_3H_4N_3S_2$	$NC(SH)NC(SH)C(NH_2)C(CF_3)$	F:	PerFHalOrg.6-26, 47, 70
$C_5F_3H_4N_5O_2$	$NC(CF_3)NC(NH_2)C(NO_2)C(NH_2)$	F:	PerFHalOrg.6-25, 47
–	$NC(NH_2)NC(NH_2)C(NO_2)C(CF_3)$	F:	PerFHalOrg.6-26/7, 47
$C_5F_3H_5N_2OS$	$CF_3N(OH)CHCHCH_2NCS$	F:	PerFHalOrg.7-166/7
$C_5F_3H_5N_4O$	$NC(CF_3)NC(NH_2)C(NH_2)C(OH)$	F:	PerFHalOrg.6-25/6, 47
$C_5F_3H_5N_4S$	$NC(SH)NC(NH_2)C(NH_2)C(CF_3)$	F:	PerFHalOrg.6-26/7, 47, 69/70
$C_5F_3H_5Ti$	$C_5H_5TiF_3$	Ti:	Org.Verb.1-137/8
$C_5F_3H_6NO$	$CH_2ON(CF_3)CH_2CHCH$	F:	PerFHalOrg.7-194
–	$OCH_2CHCHCH_2NCF_3$	F:	PerFHalOrg.7-162, 165, 168
$C_5F_3H_6NO_2$	$CF_3C(O)NOCH(CH_3)CH_2$	F:	PerFHalOrg.7-194
–	$CF_3C(O)N(OH)CH_2CHCH_2$	F:	PerFHalOrg.7-194

$C_5F_3H_6NO_3$	$CF_3NOCH(COOCH_3)CH_2$	F:	PerFHalOrg.7-165, 168
–	$CF_3NOCH(OCOCH_3)CH_2$	F:	PerFHalOrg.7-165, 167/8
–	$CF_3N(OH)CHCHCOOCH_3$	F:	PerFHalOrg.7-163, 168
–	$CF_3N(OH)CHCHOC(O)CH_3$	F:	PerFHalOrg.7-164, 168
$C_5F_3H_6N_3O$	$CH_3NCC(NCH_3)ON(CF_3)$	F:	PerFHalOrg.7-164, 166, 170
$C_5F_3H_6N_5$	$NC(CF_3)NC(NH_2)C(NH_2)C(NH_2)$	F:	PerFHalOrg.6-25, 47
–	$NC(NH_2)NC(NH_2)C(NH_2)C(CF_3)$	F:	PerFHalOrg.6-26/7, 47, 70/1
$C_5F_3H_6O_4P$	$CF_3P[OC(O)CH_3]_2$	F:	PerFHalOrg.3-101, 103
$C_5F_3H_7N_2O_2$	$CF_3NNCH_2COOC_2H_5$	F:	PerFHalOrg.7-160
–	$CF_3NNC_3H_6COOH$	F:	PerFHalOrg.7-160
$C_5F_3H_7O_2$	$CF_3COOC_3H_7$	F:	PerFHalOrg.6-112
$C_5F_3H_8NO$	$CF_3N(OH)CH_2C(CH_3)CH_2$	F:	PerFHalOrg.7-164, 166/7
$C_5F_3H_9NO$	$CF_3N(O)C_4H_9$	F:	PerFHalOrg.7-169
–	$CH_2CH(CH_3)CH_2N(OH)CF_3$	F:	PerFHalOrg.7-162, 165
$C_5F_3H_9N_2$	$CF_3NNCH(CH_3)C_2H_5$	F:	PerFHalOrg.7-159
–	$CF_3NNC_4H_9$	F:	PerFHalOrg.7-158/9
$C_5F_3H_9N_2O_2SSi$	$CF_3SO_2NCNSi(CH_3)_3$	S:	S-N-Verb.1-108
$C_5F_3H_9Si$	$CF_2CFSi(CH_3)_3$	F:	PerFHalOrg.4-19
$C_5F_3H_9Sn$	$(CH_3)_3SnCFCF_2$	F:	PerFHalOrg.4-28
		Sn:	Org.Verb.2-70/1
$C_5F_3H_{10}INP$	$CF_3PIN(C_2H_5)_2$	F:	PerFHalOrg.3-108, 113
$C_5F_3H_{10}NOS$	$C_5H_{10}NS(O)F_3$	S:	S-N-Verb.1-157/8
$C_5F_3H_{10}NO_2S$	$CF_3SO_2N(C_2H_5)_2$	S:	S-N-Verb.1-107
$C_5F_3H_{10}N_3$	$CF_3NNN(C_2H_5)_2$	F:	PerFHalOrg.7-158/9
$C_5F_3H_{10}P$	$CF_3P(C_2H_5)_2$	F:	PerFHalOrg.3-102
$C_5F_3H_{11}N_2O$	$2\ CF_3NO \cdot 2\ (C_2H_5)_2NH$	F:	PerFHalOrg.7-160
$C_5F_3H_{11}Sn$	$(CH_3)_3SnCF_2CH_2F$	Sn:	Org.Verb.2-35/6
–	$(CH_3)_3SnCHFCHF_2$	Sn:	Org.Verb.2-35/6
–	$(CH_3)_3SnCH_2CF_3$	Sn:	Org.Verb.2-36
–	$C_5H_{11}SnF_3$	Sn:	Org.Verb.5-52
$C_5F_3H_{12}NO_2SSn$	$CF_3SO_2N(CH_3)Sn(CH_3)_3$	S:	S-N-Verb.1-108
$C_5F_3H_{12}NP_2$	$(CH_3)_2NP(CF_3)P(CH_3)_2$	F:	PerFHalOrg.3-7
–	$(CH_3)_2N(PCF_3)P(CH_3)_2$	F:	PerFHalOrg.3-7
–	$(CH_3)_2NP(PCF_3)(CH_3)_2$	F:	PerFHalOrg.3-7
$C_5F_3H_{12}N_2OP$	$CF_3P(O)[N(CH_3)_2]_2$	F:	PerFHalOrg.3-57, 59
$C_5F_3H_{12}N_2P$	$CF_3P[N(CH_3)_2]_2$	F:	PerFHalOrg.3-102, 112/3
$C_5F_3H_{12}N_2PS$	$CF_3P(S)[N(CH_3)_2]_2$	F:	PerFHalOrg.3-153, 155
$C_5F_3H_{12}N_3O_2$	$[(C_2H_5)_2NH_2]ON(NO)CF_3$	F:	PerFHalOrg.7-145
$C_5F_3H_{12}P_3$	$CF_3P[P(CH_3)_2]_2$	F:	PerFHalOrg.3-7/8
$C_5F_3H_{14}N_4P$	$CF_3P[N(H)N(CH_3)_2]_2$	F:	PerFHalOrg.3-109, 119
$C_5F_3I_2N$	$C_5F_3I_2N$	F:	PerFHalOrg.5-118/9
$C_5F_3Li_2N$	$Li_2C_5F_3N$	F:	PerFHalOrg.4-7, 14
C_5F_3N	$(C_5F_3N)_n$	F:	PerFHalOrg.5-115/6, 133
$C_5F_3N_3$	$NCFNCFC(CN)CF$	F:	PerFHalOrg.6-21/2, 42, 58
$C_5F_4FeH_4O_5P_2$	$(CO)_3Fe(F_2POCH_2CH_2OPF_2)$	Fe:	Org.Verb.B3-205
$C_5F_4FeH_6O_5P_2$	$(CO)_3FeP_2F_4(OCH_3)_2$	Fe:	Org.Verb.B1-161
C_5F_4HN	C_5HF_4N	F:	PerFHalOrg.5-134, 143

C_5F_4HNO	HOC_5F_4N	F:	PerFHalOrg.5-181/3, 198, 211/2
		F:	PerFHalOrg.6-13/4
$C_5F_4HNO_3S$	$NC_5F_4SO_3H$	F:	PerFHalOrg.5-191/2
C_5F_4HNS	HSC_5F_4N	F:	PerFHalOrg.5-191/2
$C_5F_4HN_2Na$	$NaNHC_5F_4N$	F:	PerFHalOrg.5-188, 215/6
$C_5F_4H_2N_2$	$NH_2C_5F_4N$	F:	PerFHalOrg.5-185/6, 201, 214/6
$C_5F_4H_3MnO_2^-$	$[Mn(CF_3COCHCOCH_2)F]^-$	Mn:	MVol.D1-108/9
$C_5F_4H_3N_3$	$NC_5F_4NHNH_2$	F:	PerFHalOrg.5-189, 203, 216
$C_5F_4H_5NO$	$CF_3CFC(OC_2H_5)N$	F:	PerFHalOrg.5-21
$C_5F_4H_5NTe$	$TeF_4 \cdot C_5H_5N$	Te:	SVol.B3-169
$C_5F_4H_6MnN$	$C_5H_5NHMnF_4 \cdot H_2O$	Mn:	MVol.C4-212/3
$C_5F_4H_7NO_2Te$	$[C_5H_5NH][TeF_4(OH)O]$	Te:	SVol.B2-47/8
$C_5F_4H_{10}Si$	$CF_3CFHSi(CH_3)_3$	F:	PerFHalOrg.4-162
$C_5F_4H_{10}Sn$	$(CH_3)_3SnCF_2CHF_2$	Sn:	Org.Verb.2-35/6
C_5F_4IMgN	NC_5F_4MgI	F:	PerFHalOrg.4-65/6, 77, 97/8, 106
		F:	PerFHalOrg.5-115/6, 143
C_5F_4IN	C_5F_4IN	F:	PerFHalOrg.5-118/9, 134
C_5F_4LiN	LiC_5F_4N	F:	PerFHalOrg.4-7, 14, 30, 39
		F:	PerFHalOrg.5-115/6, 142
C_5F_4NNaO	$NaOC_5F_4N$	F:	PerFHalOrg.5-181/2, 198
$C_5F_4N_2O_2$	$NO_2C_5F_4N$	F:	PerFHalOrg.5-190/1, 205, 214
$C_5F_4N_4$	$N(CFCF)_2CN_3$	F:	PerFHalOrg.5-184/5, 200, 212/3
$C_5F_5FeO_4P$	$(CO)_4FeP(CF_3)F_2$	Fe:	Org.Verb.B2-82
$C_5F_5H_2N_3O_2$	$(C_2F_5)(OH)_2C_3N_3$	F:	PerFHalOrg.6-79/80, 100
$C_5F_5H_4MnO_2^-$	$[MnF_2(CF_3COCHCOCH_3)]^-$	Mn:	MVol.D1-108/9
$C_5F_5H_4N_5$	$(C_2F_5)(NH_2)_2C_3N_3$	F:	PerFHalOrg.6-79/80, 101, 112
$C_5F_5H_6NOTe$	$[C_5H_5NH]TeF_5O$	Te:	SVol.B2-63
$C_5F_5H_6NTe$	$[C_5H_5NH]TeF_5$	Te:	SVol.B2-15
$C_5F_5H_7O$	$CF_3CF_2C(OH)(CH_3)_2$	F:	PerFHalOrg.4-90
$C_5F_5H_9P_2$	$C_2F_5PP(CH_3)_3$	F:	PerFHalOrg.3-7
$C_5F_5H_9Sn$	$(CH_3)_3SnC_2F_5$	Sn:	Org.Verb.2-33
$C_5F_5H_{12}NOSnTe$	$(CH_3)_3Sn(CH_3CN)TeF_5O$	Te:	SVol.B2-55
$C_5F_5H_{13}O_2Te$	$[C_2H_5O(CH_3)C_2H_5][TeF_5O]$	Te:	SVol.B2-53, 54
C_5F_5N	C_5F_5N	F:	PerFHalOrg.5-115/6, 122/5, 134/7, 141/5
$C_5F_5Si^+$	$C_5F_5Si^+$	F:	PerFHalOrg.4-153
$C_5F_6FeHO_5P$	$[(CO)_5FeH]PF_6$	Fe:	Org.Verb.B3-252
$C_5F_6FeHO_5Sb$	$[(CO)_5FeH]SbF_6$	Fe:	Org.Verb.B3-251
$C_5F_6FeHO_5Ta$	$[(CO)_5FeH]TaF_6$	Fe:	Org.Verb.B3-251
$C_5F_6GeH_9P$	$(CH_3)_3GeP(CF_3)_2$	F:	PerFHalOrg.3-24/5
$C_5F_6HNO_2$	$HN(COCF_2)_2CF_2$	F:	PerFHalOrg.5-180/1, 198, 211

$C_5F_6HN_3OS$	$NC(CF_3)SC(NHCOCF_3)N$	F:	PerFHalOrg.5-67/8, 90
$C_5F_6HNiO_3P$	$(CO)_3NiP(CF_3)_2H$	F:	PerFHalOrg.3-27
		Ni:	Org.Verb.1-165, 167
$C_5F_6H_2N_2$	$C_2F_5C(NH_2)CFCN$	F:	PerFHalOrg.7-3
$C_5F_6H_2N_2OS$	$(CF_3)_2CCONC(NH_2)S$	F:	PerFHalOrg.5-56/8, 78
$C_5F_6H_2N_2O_3$	$OC(O)NHC(O)NHC(CF_3)_2$	F:	PerFHalOrg.6-27, 48, 60
$C_5F_6H_2N_4S$	$CF_3NCNHC(NH)SCNCF_3$	F:	PerFHalOrg.5-57/8, 78, 110
$C_5F_6H_2N_8$	$NHNNNC(CF_2)_3CNNNNH$	F:	PerFHalOrg.5-73/4, 93
$C_5F_6H_2O_2$	$C_2F_5CHCFC(O)OH$	F:	PerFHalOrg.4-44
$C_5F_6H_3NO$	$(CF_3)_2CNOCHCH_2$	F:	PerFHalOrg.7-141
$C_5F_6H_3NOS$	$SC(CF_3)_2NHC(O)CH_2$	F:	PerFHalOrg.7-66
$C_5F_6H_3NO_2$	$OC(CF_3)_2NHC(O)CH_2$	F:	PerFHalOrg.7-66, 68
$C_5F_6H_3NO_3$	$(CF_3)_2CNOC(O)OCH_3$	F:	PerFHalOrg.7-141/2
–	$HOCO(CF_2)_3C(O)NH_2$	F:	PerFHalOrg.7-18, 21/2, 47
$C_5F_6H_3NO_4S$	$[CH_3NSO_2][OC(CF_3)_2CO]$	S:	S-N-Verb.1-40
$C_5F_6H_3N_2O_4S$	$(CF_3)_2CC(O)N(CH_3)S(ONO)O$	F:	PerFHalOrg.5-103
$C_5F_6H_3N_3$	$NC(NH)(CF_2)_3C(NH_2)$	F:	PerFHalOrg.5-185/6, 202
$C_5F_6H_3N_3O_2$	$HN(C(O)NH)_2C(CF_3)_2$	F:	PerFHalOrg.6-75/6, 98
$C_5F_6H_3OP$	$(CF_3)_2P(O)CCCH_3$	F:	PerFHalOrg.3-101
$C_5F_6H_4N_2O_2$	$NH_2C(O)(CF_2)_3C(O)NH_2$	F:	PerFHalOrg.7-18, 20, 48
$C_5F_6H_4N_2O_3$	$NH_2C(O)CF(CF_3)OCF_2C(O)NH_2$	F:	PerFHalOrg.7-18, 20, 48
$C_5F_6H_4O$	$CF_2CFC(OH)(CH_3)CF_3$	F:	PerFHalOrg.4-44, 89
$C_5F_6H_4O_6P_2$	$C_5F_6(PO(OH)_2)_2$	F:	PerFHalOrg.3-43/4, 46, 50
$C_5F_6H_5NO$	$(CF_3)_2CNOC_2H_5$	F:	PerFHalOrg.7-143
–	$CH_2CHCH_2ON(CF_3)_2$	F:	PerFHalOrg.7-140
$C_5F_6H_5NO_2$	$(CF_3)_2NOC(O)C_2H_5$	F:	PerFHalOrg.7-136, 140/1
$C_5F_6H_5N_3O$	$NH_2CO(CF_2)_3CNHNH_2$	F:	PerFHalOrg.7-4/5, 31
$C_5F_6H_5NiSb$	$[C_5H_5Ni][SbF_6]$	Ni:	Org.Verb.2-135
$C_5F_6H_6N_2O_3$	$NH_4OCO(CF_2)_3C(O)NH_2$	F:	PerFHalOrg.7-18, 21, 47
$C_5F_6H_6N_2O_8P_2$	$C_3F_6[C(O)NHP(O)(OH)_2]_2$	F:	PerFHalOrg.7-17
$C_5F_6H_6N_4$	$NH_2CNH(CF_2)_3CNHNH_2$	F:	PerFHalOrg.7-4/5
$C_5F_6H_6N_4O$	$NH_2(NH)CCF(CF_3)O(CF_2)C(NH)NH_2$	F:	PerFHalOrg.7-5
$C_5F_6H_6N_4O_2$	$NH_2(NOH)C(CF_2)_3C(NOH)NH_2$	F:	PerFHalOrg.7-5/6, 32
$C_5F_6H_7NO$	$(CF_3)_2C(NH_2)OC_2H_5$	F:	PerFHalOrg.7-65
–	$(CF_3)_2NOCH(CH_3)_2$	F:	PerFHalOrg.7-129/30
$C_5F_6H_7O_3P$	$(CF_3)_2C(OH)P(CH_2OH)_2$	F:	PerFHalOrg.3-24
$C_5F_6H_7P$	$(CF_3)_2P(CH_2)_2CH_3$	F:	PerFHalOrg.3-28
$C_5F_6H_8IrO_2P_2$	$Ir(CH_3COCH_2COCH_3)(PF_3)_2$	Ir:	SVol.2-194
$C_5F_6H_8N_6$	$(NH_2NC(NH_2)CF_2)_2CF_2$	F:	PerFHalOrg.7-6/7, 33
$C_5F_6H_9NOSi$	$(CF_3)_2NOSi(CH_3)_3$	F:	PerFHalOrg.7-127, 140, 144, 169/70
$C_5F_6H_9N_2P$	$(CF_3)_2PN(CH_3)N(CH_3)_2$	F:	PerFHalOrg.3-109, 118
$C_5F_6H_9N_3O_2$	$2\ CF_3NO \cdot (CH_3)_3N$	F:	PerFHalOrg.7-160
$C_5F_6H_9OPSSi$	$(CF_3)_2P(S)OSi(CH_3)_3$	F:	PerFHalOrg.3-57, 58, 153, 156
$C_5F_6H_9OPSi$	$(CF_3)_2POSi(CH_3)_3$	F:	PerFHalOrg.3-37, 38, 102
$C_5F_6H_9O_2PSi$	$(CF_3)_2P(O)OSi(CH_3)_3$	F:	PerFHalOrg.3-57, 58

$C_5F_6H_9PSSi$	$(CF_3)_2PSSi(CH_3)_3$	F:	PerFHalOrg.3-102, 107, 153, 157
$C_5F_6H_9PS_2Si$	$(CF_3)_2P(S)SSi(CH_3)_3$	F:	PerFHalOrg.3-153, 157
$C_5F_6H_9PSi$	$(CH_3)_3SiP(CF_3)_2$	F:	PerFHalOrg.3-15, 24/5, 102
$C_5F_6H_9PSn$	$(CH_3)_3SnP(CF_3)_2$	F:	PerFHalOrg.3-24/5
$C_5F_6H_{10}NOP$	$(CF_3)_2POH \cdot (CH_3)_3N$	F:	PerFHalOrg.3-50
$C_5F_6H_{10}NPS$	$[(CH_3)_3NH][(CF_3)_2PS]$	F:	PerFHalOrg.3-153, 157
$C_5F_6H_{10}NPS_2$	$[(CH_3)_3NH][(CF_3)_2P(S)S]$	F:	PerFHalOrg.3-153
$C_5F_6Li_2$	$LiC(CF_2)_3CLi$	F:	PerFHalOrg.4-3, 13, 26, 31, 34
$C_5F_6N_2$	$NC(CF_3)NCFCFCF$	F:	PerFHalOrg.6-22, 43, 58, 60, 72
–	$NCFNC(CF_3)CFCF$	F:	PerFHalOrg.6-22, 43, 58, 60
–	$NCFNCFC(CF_3)CF$	F:	PerFHalOrg.6-22, 43, 58, 60
–	$(C_5F_6N_2)_m$	F:	PerFHalOrg.6-120
$C_5F_6N_4S$	$(CF_3)_2C_3N_4S$	F:	PerFHalOrg.6-123/4, 138
$C_5F_7FeH_6IN_2O_2$	$(CO)_2Fe(C_3F_7)(NH_3)_2I$	Fe:	Org.Verb.B1-121/2, 126
$C_5F_7FeH_9IN_3O_2$	$[(CO)_2Fe(C_3F_7)(NH_3)_3]I$	Fe:	Org.Verb.B1-121, 124, 128
$C_5F_7HMnO_2^-$	$[Mn(CF_3COCHCOCF_3)F]^-$	Mn:	MVol.D1-110, 111
$C_5F_7HN_2$	$CF_3CF_2CF_2C(NH)CN$	F:	PerFHalOrg.7-25/6, 52
$C_5F_7HN_2O_2$	$C(O)NHC(C_3F_7)NO$	F:	PerFHalOrg.5-66, 89
$C_5F_7H_2NO$	$CF_3N(O)CF_2CHCHCF_2$	F:	PerFHalOrg.7-168
$C_5F_7H_2NO_5$	$HOCOCF_2ON(CF_3)CF_2COOH$	F:	PerFHalOrg.6-12
$C_5F_7H_3N_4$	$NC(C_3F_7)NC(NH_2)NH$	F:	PerFHalOrg.5-71/3, 93
$C_5F_7H_3N_4O$	$NHC(NH_2)NC(CF_2CF_2OCF_3)N$	F:	PerFHalOrg.5-71/3, 93
$C_5F_7H_3N_4S$	$NC(C_3F_7)N(NH_2)C(SH)N$	F:	PerFHalOrg.5-71/3, 93, 100
$C_5F_7H_3O$	$C_3F_7C(O)CH_3$	F:	PerFHalOrg.4-93, 95
$C_5F_7H_5N_4O$	$C_3F_7CONHNHC(NH)NH_2$	F:	PerFHalOrg.7-6/7, 34
$C_5F_7H_5N_4O_2$	$CF_3OCF_2CF_2CONHNHC(NH)NH_2$	F:	PerFHalOrg.7-6/7, 34
$C_5F_7H_5O$	$(CF_3)_2CFCH(OH)CH_3$	F:	PerFHalOrg.4-91
–	$C_3F_7CH(OH)CH_3$	F:	PerFHalOrg.4-91, 92, 95
$C_5F_7H_{13}NP$	$[CF_3PF_4H]NH_2(C_2H_5)_2$	F:	PerFHalOrg.3-102
C_5F_7IMg	$(CF_3)_2CFCCMgI$	F:	PerFHalOrg.4-62/3, 76
C_5F_7Li	$FCCF_2CF_2CF_2CLi$	F:	PerFHalOrg.4-3, 44
C_5F_7N	$N(CF_2)_2(CF)_2CF$	F:	PerFHalOrg.5-120/1, 130
C_5F_7NO	$CFCF_2N(CF_3)COCF$	F:	PerFHalOrg.5-54, 76
$C_5F_7N_3$	$(CF_3)_2C_3FN_3$	F:	PerFHalOrg.6-78/80, 99, 111
–	$(C_2F_5)C_3F_2N_3$	F:	PerFHalOrg.6-78/80, 99
$C_5F_8FeH_{12}O_5P_4$	$COFeP_4F_8(OCH_3)_4$	Fe:	Org.Verb.B1-28/9
$C_5F_8HMnO_2^-$	$[Mn(CF_3COCHCOCF_3)F_2]^-$	Mn:	MVol.D1-110, 111
$C_5F_8H_2N_4O$	$(CF_2)_2NNHC(O)NHN(CF_2)_2$	F:	PerFHalOrg.5-4, 10
$C_5F_8H_2O$	$CF_2CFCH(OH)C_2F_5$	F:	PerFHalOrg.4-44
$C_5F_8H_3NO$	$N(CF_3)OCF_2CH_2CHFCF_2$	F:	PerFHalOrg.6-12
$C_5F_8N_2S$	$CFNC(C_3F_7)NS$	F:	PerFHalOrg.5-67/8

$C_5F_8N_4$	$[(CF_3)_2N]C_3F_2N_3$	F:	PerFHalOrg.6-78, 99
$C_5F_9FeH_8O_2P_3$	$(CH_2C(CH_3)COOCH_3)Fe(PF_3)_3$	Fe:	Org.Verb.B5-112/3
$C_5F_9HN_2O$	$(CF_3)_2NC(COH)NCF_3$	F:	PerFHalOrg.7-163, 165
–	$(CF_3)_2NCOCHNCF_3$	F:	PerFHalOrg.7-162, 165
C_5F_9HO	$CF_2CFC(CF_3)_2OH$	F:	PerFHalOrg.4-45, 90
$C_5F_9HO_2$	$(n-C_3F_7)COOCF_2H$	F:	PerFHalOrg.5-18
$C_5F_9H_2NO$	$CF_3CF_2CF(CF_3)C(O)NH_2$	F:	PerFHalOrg.7-18/9, 46
–	$(CF_3)_3CC(O)NH_2$	F:	PerFHalOrg.7-18, 20/1, 47
–	$N(CF_3)OCF_2CHFCHFCF_2$	F:	PerFHalOrg.6-12
$C_5F_9H_3N_2$	$C_4F_9C(NH)NH_2$	F:	PerFHalOrg.7-57
$C_5F_9H_4NO$	$(CF_3)_3CN(OH)CH_3$	F:	PerFHalOrg.7-192
$C_5F_9H_4N_3$	$C_4F_9C(NH_2)NNH_2$	F:	PerFHalOrg.7-6/7, 33
$C_5F_9H_4P$	$(CF_3)_2PCH_2CH_2CF_3$	F:	PerFHalOrg.3-185
C_5F_9N	$N(CF_2)_4CF$	F:	PerFHalOrg.5-119/21, 129/30, 133/4, 144
C_5F_9NO	$CF_2CF_2N(CF_3)COCF_2$	F:	PerFHalOrg.5-54, 76, 97/8
–	$ON(CF_3)CF_2CFCFCF_2$	F:	PerFHalOrg.6-4/5, 8/9, 11/2
		F:	PerFHalOrg.7-168
–	$[N(CF_3)OCF_2CF(CFCF_2)]_n$	F:	PerFHalOrg.7-207/8, 213/4
$C_5F_9NO_2$	$(CF_3)_2CNOC(O)CF_3$	F:	PerFHalOrg.7-96/7, 114
–	$O(CF_2CF_2)_2NC(O)F$	F:	PerFHalOrg.6-2, 7
$C_5F_9NO_3$	$(CF_3)_2C(NO)OC(O)CF_3$	F:	PerFHalOrg.7-179, 187
$C_5F_9NO_3S$	$[NS(O)CF_3][OC(CF_3)_2CO]$	F:	PerFHalOrg.5-57/8, 79
		S:	S-N-Verb.1-41
$C_5F_9NS_2$	$SC(CF_3)NC(CF_3)_2S$	F:	PerFHalOrg.5-57/8, 80
$C_5F_9O_4P$	$CF_3P[OC(O)CF_3]_2$	F:	PerFHalOrg.3-31/2, 34, 36
$C_5F_{10}FeO_3P_2$	$(CO)_3FeP_2F_4(CF_3)_2$	Fe:	Org.Verb.B1-149
$C_5F_{10}HN$	$(C_2F_5)_2CNH$	F:	PerFHalOrg.7-25/6, 52, 66/7, 75
–	$HN(CF_2)_5$	F:	PerFHalOrg.5-120/1, 131
$C_5F_{10}HNO$	$C_4F_9NHC(O)F$	F:	PerFHalOrg.7-22/3
–	$HON(CF_2CF_2)_2CF_2$	F:	PerFHalOrg.5-184, 200
$C_5F_{10}H_2O$	$CF_3CF_2CF_2CH(OH)CF_3$	F:	PerFHalOrg.4-94, 95
$C_5F_{10}H_2O_2$	$(C_3F_7)(CF_3)C(OH)_2$	F:	PerFHalOrg.4-105
$C_5F_{10}NO$	$ON(CF_2)_5$	F:	PerFHalOrg.5-180/1, 198
$C_5F_{10}NP$	$(C_2F_5)_2PCN$	F:	PerFHalOrg.3-164, 169
$C_5F_{10}N_2$	$CF_2N(CF_2)_3NCF_2$	F:	PerFHalOrg.5-15
–	$(C_2F_5)_2CN_2$	F:	PerFHalOrg.5-5
$C_5F_{10}N_2O_2$	$NO_2N(CF_2)_5$	F:	PerFHalOrg.5-191, 205
–	$ONON(CF_2CF_2)_2CF_2$	F:	PerFHalOrg.5-184, 200
$C_5F_{10}N_4$	$NNC(CF_3)N(CF(CF_3)_2)N$	F:	PerFHalOrg.5-73/4, 93
$C_5F_{10}O$	$C_3F_7C(O)CF_3$	F:	PerFHalOrg.4-93, 95
$C_5F_{11}HNSb$	$[C_5F_5NH][SbF_6]$	F:	PerFHalOrg.5-119, 129
$C_5F_{11}N$	$CF_2NFCF_2CF(CF_3)CF_2$	F:	PerFHalOrg.5-52/3, 76
–	$(CF_2)_4NCF_3$	F:	PerFHalOrg.5-52/3, 75, 96/7

$C_5F_{11}N$	$FN(CF_2)_5$	F:	PerFHalOrg.5-120/1, 124, 131, 133/4, 143/5
$C_5F_{11}NO$	$CF_2N(C_2F_5)CF_2OCF_2$	F:	PerFHalOrg.5-54/5, 77, 96/8
–	$(CF_3)_2CFNOCF_2CF_2$	F:	PerFHalOrg.5-30/1, 40
–	$C_3F_7NOCF_2CF_2$	F:	PerFHalOrg.5-30/1, 40, 43/4, 46
–	$[N(C_3F_7)OCF_2CF_2]_n$	F:	PerFHalOrg.7-208/9, 214
–	$O(CF_2CF_2)_2NCF_3$	F:	PerFHalOrg.6-1/2, 6, 14
–	$ON(CF_3)(CF_2)_4$	F:	PerFHalOrg.6-4/5, 9, 11
$C_5F_{11}NO_2$	$C_3F_7OCF_2CF_2NO$	F:	PerFHalOrg.7-179, 185
$C_5F_{11}P$	$CF_3P(CF_2)_4$	F:	PerFHalOrg.3-9, 11
$C_5F_{12}FeN_2O_3P_2S$	$COFe(NO)_2(CF_3)_2PSP(CF_3)_2$	Fe:	Org.Verb.B1-42/4, 47
$C_5F_{12}FeN_2O_3P_2Se$	$COFe(NO)_2(CF_3)_2PSeP(CF_3)_2$	Fe:	Org.Verb.B1-42/4, 47
$C_5F_{12}FeN_2O_4P_2$	$COFe(NO)_2(CF_3)_2POP(CF_3)_2$	Fe:	Org.Verb.B1-42/4, 47
$C_5F_{12}FeO_5Sb_2Sn$	$[(CO)_5FeSn][SbF_6]_2$	Fe:	Org.Verb.B3-250
$C_5F_{12}HN$	$(CF_3)_2CFN(H)C_2F_5$	F:	PerFHalOrg.7-13/4
$C_5F_{12}H_3NP_2$	$[(CF_3)_2P]_2NCH_3$	F:	PerFHalOrg.3-108, 114
$C_5F_{12}H_3P_3$	$(CF_3)_2PP(CH_3)P(CF_3)_2$	F:	PerFHalOrg.3-101, 104
$C_5F_{12}H_6IrNP_4$	$(C_5H_5NH)[Ir(PF_3)_4]$	Ir:	SVol.2-193
$C_5F_{12}N_2$	$CF_3NCF_2CF_2CF_2NCF_3$	F:	PerFHalOrg.5-60/1, 83
$C_5F_{12}N_2O_3$	$[(CF_3)_2NO]_2CO$	F:	PerFHalOrg.7-94/5, 110
$C_5F_{13}OP$	$(CF_3)_2(n\text{-}C_3F_7)PO$	F:	PerFHalOrg.3-31, 34
$C_5F_{13}P$	$(CF_3)_2PC_3F_7$	F:	PerFHalOrg.3-163, 168, 173
$C_5F_{14}N_2O$	$(CF_3)_2NOCF_2N(CF_3)_2$	F:	PerFHalOrg.7-94/5, 112
$C_5F_{14}N_2O_2$	$[(CF_3)_2NO]_2CF_2$	F:	PerFHalOrg.7-94, 109
$C_5F_{15}N_2P$	$CF_3P[N(CF_3)_2]_2$	F:	PerFHalOrg.3-61, 65, 68
$C_5F_{15}P_3$	$(CF_3)_2PP(CF_3)P(CF_3)_2$	F:	PerFHalOrg.3-176, 180
$C_5F_{15}P_5$	$(CF_3P)_5$	F:	PerFHalOrg.3-1/8
$C_5F_{20}FeOP_4$	$COFeP_4F_8(CF_3)_4$	Fe:	Org.Verb.B1-28/9
C_5FeHKO_5	$K[HC(O)Fe(CO)_4]$	Fe:	Org.Verb.B4-156, 167
$C_5FeHLiO_5$	$Li[HC(O)Fe(CO)_4]$	Fe:	Org.Verb.B4-156, 166/7
C_5FeHNO_8	$[(CO)_5FeH]NO_3$	Fe:	Org.Verb.B3-251
$C_5FeHNaO_5$	$Na[HC(O)Fe(CO)_4]$	Fe:	Org.Verb.B4-149, 158
$C_5FeHO_5^+$	$[(CO)_5FeH]^+$	Fe:	Org.Verb.B3-251
$C_5FeHO_5^-$	$[HC(O)Fe(CO)_4]^-$	Fe:	Org.Verb.B4-149
$C_5FeH_3I_2NO_3$	$CH_3NCFe(CO)_3I_2$	Fe:	Org.Verb.B4-2/3
$C_5FeH_3K_2N_5O$	$K_2[COFe(NH_3)(CN)_4] \cdot 1.5\ H_2O$	Fe:	Org.Verb.B1-33
$C_5FeH_3NO_5$	$(CO)_5FeNH_3$	Fe:	Org.Verb.B3-113, 250
$C_5FeH_4N_4O_3$	$(NH_2NC)_2Fe(CO)_3$	Fe:	Org.Verb.B4-18/9
$C_5FeH_4O_4$	$[Fe(CO)_4CH_4]$	Fe:	Org.Verb.B2-1
		Fe:	Org.Verb.B3-87
$C_5FeH_4O_7$	$(CO)_5Fe(H_2O)_2$	Fe:	Org.Verb.B3-249
$C_5FeH_5^+$	$[FeC_5H_5]^+$	Fe:	Org.Verb.A1-93/4
–	$[FeC_5H_4D]^+$	Fe:	Org.Verb.A2-17
$C_5FeH_5NO_3$	$C_3H_5Fe(CO)_2NO$	Fe:	Org.Verb.B5-14/5, 18/21
$C_5FeH_5NaO_5$	$Na[HFe(CO)_4] \cdot CH_3OH$	Fe:	Org.Verb.B2-17
$C_5FeH_5O^+$	$[FeC_5H_5O]^+$	Fe:	Org.Verb.A2-109
$C_5FeH_5O_4P$	$(CO)_4FeP(CH_3)H_2$	Fe:	Org.Verb.B2-88, 108

$C_5FeH_6^+$	$[FeC_5H_6]^+$	Fe: Org.Verb.A1-264
$C_5FeH_6N_2O_5$	$(CO)_5Fe(NH_3)_2$	Fe: Org.Verb.B3-250
$C_5FeH_6N_4O_3$	$(CO)_3Fe(CH_3NNNNCH_3)$	Fe: Org.Verb.B1-168/70
$C_5FeH_6N_4O_3^+$	$[(CO)_3Fe(CH_3NNNNCH_3)]^+$	Fe: Org.Verb.B1-169
$C_5FeH_6O^+$	$[C_5H_5FeOH]^+$	Fe: Org.Verb.A2-3, 17, 28, 71, 79/80
		Fe: Org.Verb.A3-57
$C_5FeH_9N_3O_4$	$(CH_3NH)(C_2H_5O)CFe(NO)_2CO$	Fe: Org.Verb.B4-109/11
$C_5FeH_{20}NO_6^{2+}$	$[Fe(CH_3OH)_5NO]^{2+}$	Fe: Org.Verb.B1-141
$C_5FeIO_5^+$	$[(CO)_5FeI]^+$	Fe: Org.Verb.B3-107, 252, 254
$C_5FeI_2O_5$	$[(CO)_5FeI]I$	Fe: Org.Verb.B3-105/7, 254
$C_5FeI_4O_5$	$Fe(CO)_5 \cdot 2\ I_2$	Fe: Org.Verb.B3-107
C_5FeKNO_4	$K[(CO)_4FeCN]$	Fe: Org.Verb.B2-50
C_5FeNO_4	$(CO)_4FeCN$	Fe: Org.Verb.C2-107
$C_5FeNO_4^-$	$[(CO)_4FeCN]^-$	Fe: Org.Verb.B2-51/2
$C_5FeNO_4S^-$	$[(CO)_4FeNCS]^-$	Fe: Org.Verb.B2-52
$C_5FeO_4S_2$	$(CO)_4FeSCS$	Fe: Org.Verb.B2-129
C_5FeO_5	$Fe(CO)_5$	
	Analysis	Fe: Org.Verb.B3-243/7
	Chemical reactions	
	with elements	Fe: Org.Verb.B3-94/110
	with inorganic compounds	Fe: Org.Verb.B3-110/31
	with organic compounds	Fe: Org.Verb.B3-154/206
	with organoelement and analogous compounds	Fe: Org.Verb.B3-132/53
	with polymers	Fe: Org.Verb.B3-206/7
	Decomposition	Fe: Org.Verb.B3-67/91
	Electrical properties	Fe: Org.Verb.B3-26/8, 47/50
	Electrochemical behavior	Fe: Org.Verb.B3-66/7
	Exchange reactions	Fe: Org.Verb.B3-92/4
	Formation	Fe: Org.Verb.B3-1/17, 19/22
	General properties	Fe: Org.Verb.B3-22
	Ionization energies	Fe: Org.Verb.B3-47/50
	Magnetic properties	Fe: Org.Verb.B3-26/8
	Mechanical properties	Fe: Org.Verb.B3-22/3
	Molecule	Fe: Org.Verb.B3-51/61
	Nuclear magnetic resonance	Fe: Org.Verb.B3-28/31
	Optical properties	Fe: Org.Verb.B3-26/8
	Physiological properties	Fe: Org.Verb.B3-247/9
	Preparation	Fe: Org.Verb.B3-1/13, 18
	Purification	Fe: Org.Verb.B3-18/9
	Removal from gases and liquids	Fe: Org.Verb.B3-13/7
	Solubility	Fe: Org.Verb.B3-62/6
	Spectra	Fe: Org.Verb.B3-28/51
	Storage	Fe: Org.Verb.B3-18/9
	Thermal properties	Fe: Org.Verb.B3-22/6
	Undesired formation	Fe: Org.Verb.B3-13/4
	Uses	Fe: Org.Verb.B3-207/43

C_5FeO_5	$Fe(CO)_4(C^{18}O)$	Fe:	Org.Verb.B3-86, 92
–	$Fe(CO)_3(C^{18}O)_2$	Fe:	Org.Verb.B3-86
–	$Fe(C^{18}O)_5$	Fe:	Org.Verb.B3-18, 93/4
–	$Fe(^{12}CO)_4(^{13}CO)$	Fe:	Org.Verb.B3-81, 92
–	$Fe(^{13}CO)_5$	Fe:	Org.Verb.B3-18, 93/4
–	$Fe(^{13}C^{17}O)_5$	Fe:	Org.Verb.B3-18
–	$^{54}Fe(CO)_5$	Fe:	Org.Verb.B3-18
–	$^{59}Fe(CO)_5$	Fe:	Org.Verb.B3-18
–	$(CO)_5Fe \cdot 2\ H_2O$	Fe:	Org.Verb.B3-175, 249
–	$Fe(CO)_5$ solutions		
	$Fe(CO)_5$-HF	Fe:	Org.Verb.B3-62/3
	$Fe(CO)_5$-$Mo(CO)_6$	Fe:	Org.Verb.B3-64
	$Fe(CO)_5$-organic solvents	Fe:	Org.Verb.B3-63, 249
–	$Fe(CO)_5$ systems		
	$Fe(CO)_5$-$C_5H_5Mn(CO)_3$	Fe:	Org.Verb.B3-64
	$Fe(CO)_5$-$Co_2(CO)_8$	Fe:	Org.Verb.B3-64
	$Fe(CO)_5$-$Co_2(CO)_8$-$Co_4(CO)_{12}$-$Ni(CO)_4$	Fe:	Org.Verb.B3-64
	$Fe(CO)_5$-$Co_2(CO)_8$-$Ni(CO)_4$	Fe:	Org.Verb.B3-64
	$Fe(CO)_5$-$Co_4(CO)_{12}$	Fe:	Org.Verb.B3-64
	$Fe(CO)_5$-$Cr(CO)_6$	Fe:	Org.Verb.B3-64
	$Fe(CO)_5$-$Mn_2(CO)_{10}$	Fe:	Org.Verb.B3-64
	$Fe(CO)_5$-$Ni(CO)_4$	Fe:	Org.Verb.B3-64
	$Fe(CO)_5$-$W(CO)_6$	Fe:	Org.Verb.B3-64
$C_5FeO_5^+$	$[Fe(CO)_5]^+$	Fe:	Org.Verb.B3-49, 66, 77/8
		Fe:	Org.Verb.C2-26
$C_5FeO_5^-$	$[Fe(CO)_5]^-$	Fe:	Org.Verb.B3-67
$C_5Fe_2H_{12}N_8O$	$[Fe(NH_3)_3][COFe(NH_3)(CN)_4]$	Fe:	Org.Verb.B1-33
$C_5Fe_2O_5^+$	$[Fe_2(CO)_5]^+$	Fe:	Org.Verb.C2-26
$C_5GeH_{32}MnN_{15}O_{40}W_{11}$			
	$(CH_6N_3)_5H_2GeMnW_{11}O_{40} \cdot 6\ H_2O$	Mn:	MVol.C3-226
$C_5HIrK_3N_5$	$K_3[IrH(CN)_5]$	Ir:	SVol.2-183
$C_5HIrN_5^{3-}$	$[Ir(CN)_5H]^{3-}$	Ir:	SVol.2-182
$C_5HIrN_5O^{3-}$	$[Ir(CN)_5OH]^{3-}$	Ir:	SVol.2-184
$C_5HN_5PaS_5$	$H[Pa(SCN)_5 \cdot 3\ (C_4H_9)_3PO_4]$	Pa:	SVol.2-152/3
$C_5HN_5S_6$	$HS_6C_5N_5$	C:	MVol.D6-42
$C_5H_2IrN_5O^{2-}$	$[Ir(CN)_5H_2O]^{2-}$	Ir:	SVol.2-182, 183
$C_5H_3LiNiO_4$	$Li[(CO)_3NiC(O)CH_3]$	Ni:	Org.Verb.1-193, 194
$C_5H_3NNiO_3$	$CH_3NCNi(CO)_3$	Ni:	Org.Verb.1-302/3
$C_5H_3NpO_4S$	$NpO_2(SC_4H_3COO)$	Np:	TrU.D1-156
$C_5H_3NpO_5$	$NpO_2(OC_4H_3COO)$	Np:	TrU.D1-156
$C_5H_3O_4PuS^+$	$PuO_2(SC_4H_3COO)^+$	Np:	TrU.D1-156
$C_5H_3O_5Pu^+$	$PuO_2(OC_4H_3COO)^+$	Np:	TrU.D1-156
$C_5H_4K_2N_2O_{11}U$	$K_2[UO_2(C_2O_4)_2((NH_2)_2CO)] \cdot H_2O$	U:	SVol.E1-84, 86, 90
$C_5H_4K_2N_4OS_4Sn$	$K_2Sn(SCN)_4 \cdot CH_3OH$	Sn:	MVol.C3-80
$C_5H_4LiNNiO_4$	$Li[(CO)_3NiC(O)N(H)CH_3]$	Ni:	Org.Verb.1-193, 194
$C_5H_4N_2Na_2O_{11}U$	$Na_2[UO_2(C_2O_4)_2((NH_2)_2CO)] \cdot 3\ H_2O$	U:	SVol.E1-84, 86, 90
$C_5H_4N_2O_{11}Rb_2U$	$Rb_2[UO_2(C_2O_4)_2((NH_2)_2CO)] \cdot H_2O$	U:	SVol.E1-84, 86, 90
–	$Rb_2[UO_2(C_2O_4)_2((NH_2)_2CO)] \cdot 3\ H_2O$	U:	SVol.E1-90
$C_5H_4N_2O_{11}U^{2-}$	$[UO_2(C_2O_4)_2((NH_2)_2CO)]^{2-}$	U:	SVol.E1-90

$C_5H_4N_6O_2S_3$	$[NH_2C(O)N(NCS)]_2CS$	C:	MVol.D6-160
$C_5H_4N_6S_5$	$[NH_2C(S)N(NCS)]_2CS$	C:	MVol.D6-160
$C_5H_5I_2NOSn$	$SnOI_2 \cdot C_5H_5N$	Sn:	MVol.C5-200
$C_5H_5I_2Ti$	$C_5H_5TiI_2 \cdot C_4H_8O$	Ti:	Org.Verb.1-135/6
$C_5H_5I_3Ti$	$C_5H_5TiI_3$	Ti:	Org.Verb.1-139/40, 149
$C_5H_5KN_2Ni$	$K[C_3H_5Ni(CN)_2]$	Ni:	Org.Verb.1-108
$C_5H_5Mn_2NO_{12}S_3$	$Mn_2(SO_4)_3 \cdot C_5H_5N$	Mn:	MVol.C6-173
$C_5H_5MoNO_3$	$MoO_3 \cdot C_5H_5N$	Mo:	SVol.B1-142
C_5H_5NNiO	C_5H_5NiNO	Ni:	Org.Verb.2-108/12, 113
–	$C_5H_5Ni^{15}NO$	Ni:	Org.Verb.2-108/9
–	C_5D_5NiNO	Ni:	Org.Verb.2-108/9
$C_5H_5NNiO^+$	$[C_5H_5NiNO]^+$	Ni:	Org.Verb.2-111
$C_5H_5NNiO^-$	$[C_5H_5NiNO]^-$	Ni:	Org.Verb.2-111
$C_5H_5NO_3S$	$C_5H_5N \cdot SO_3$	S:	S-N-Verb.1-23
$C_5H_5NO_3U$	$UO_3 \cdot C_5H_5N$	U:	SVol.E1-38
$C_5H_5NO_4SSn$	$SnSO_4 \cdot C_5H_5N$	Sn:	MVol.C5-17
$C_5H_5N_3O_8U$	$UO_2(NO_3)_2 \cdot C_5H_5N \cdot H_2O$	U:	SVol.E1-29, 34
$C_5H_5Ni^+$	$[C_5H_5Ni]^+$	Ni:	Org.Verb.2-177/8, 181, 183, 201
–	$[C_5H_5Ni(H_2O)_n]^+$	Ni:	Org.Verb.2-118
$C_5H_5OPa^+$	$PaC_5H_5O^+$	Pa:	SVol.2-99
$C_5H_5OTi^+$	$[C_5H_5TiO]^+$	Ti:	Org.Verb.1-160, 181, 184
$C_5H_5O_{10}Sn_2^-$	$Sn_2(HCOO)_5^-$	Sn:	MVol.C2-218
C_5H_5Ti	C_5H_5Ti	Ti:	Org.Verb.1-132
$C_5H_6I_2NiO$	$CH_2IC_3H_4Ni(CO)I$	Ni:	Org.Verb.2-42
$C_5H_6I_3NSn$	$[C_5H_6N]SnI_3$	Sn:	MVol.C3-119/20
$C_5H_6MnO_8$	$[MnC_5O_5(H_2O)_3]_n$	Mn:	MVol.D1-132/3
$C_5H_6N_2O_5U$	$UO_2(NCO)_2 \cdot (CH_3)_2CO$	U:	SVol.E1-79/80
$C_5H_6N_2O_{12}U^{2-}$	$[UO_2(C_2O_4)_2((NH_2)_2CO)(H_2O)]^{2-}$	U:	SVol.E1-90
$C_5H_7MnO_2^+$	$[Mn(CH(COCH_3)_2)]^+$	Mn:	MVol.D1-65/6, 69, 92
$C_5H_7NNpO_6^-$	$[(NpO_2)N(CH_3)(CH_2COO)_2]^-$	Np:	TrU.D1-162
$C_5H_7NO_8U$	$[UO_2(C_2O_4)(NH_2COOC_2H_5)]_n$	U:	SVol.E1-121, 123
$C_5H_7N_3O_4S_2$	$NH_2SO_2NSO_2NC_5H_5$	S:	S-N-Verb.1-160
$C_5H_7N_3O_9U$	$UO_2(NO_3)_2 \cdot NH_2(CH)_4CHO$	U:	SVol.E1-81
$C_5H_7NpO_2^{3+}$	$Np(CH_3COCHCOCH_3)^{3+}$	Np:	TrU.D2-242
$C_5H_7O_2Pu^{3+}$	$Pu(CH_3COCHCOCH_3)^{3+}$	Np:	TrU.D1-167/8
$C_5H_7O_3Pa^{2+}$	$PaO((CH_3CO)_2CH)^{2+}$	Pa:	SVol.2-177/8
$C_5H_7O_4Pa$	$PaO_2((CH_3CO)_2CH)$	Pa:	SVol.2-177/8
$C_5H_8NNpO_6$	$(NpO_2)N(CH_3)(CH_2COOH)(CH_2COO)$	Np:	TrU.D1-162
$C_5H_8N_2O_7U$	$UO_2C_2O_4 \cdot CO(NHCH_3)_2$	U:	SVol.E1-93, 95
–	$UO_2C_2O_4 \cdot NH_2CONH(C_2H_5)$	U:	SVol.E1-93, 95
$C_5H_8N_7O_4S_3U^-$	$[UO_2(NCS)_3((NH_2)_2CO)_2]^-$	U:	SVol.E1-88
$C_5H_8O_3Pa^{2+}$	$Pa(OH)((CH_3CO)_2CH)^{2+}$	Pa:	SVol.2-176
$C_5H_9KOS_2$	$K[S_2COC_4H_9]$	C:	MVol.D4-249, 252/3
$C_5H_9KOSe_2$	$K[Se_2COC_4H_9]$	C:	MVol.D6-221
$C_5H_9KO_2S$	$K[SOCOC_4H_9]$	C:	MVol.D4-237
$C_5H_9KS_3$	$K[S_2CSC_4H_9]$	C:	MVol.D4-267, 269
C_5H_9NNiO	$[NiOCNC_4H_9\text{-}t]_x$	Ni:	Org.Verb.2-396
C_5H_9NOS	$HNCS \cdot C_4H_8O \cdot n\ H_2O$	C:	MVol.D6-30
$C_5H_9NaOS_2$	$Na[S_2COC_4H_9]$	C:	MVol.D4-253

$C_5H_9NaS_3$	$Na[S_2CSC_4H_9]$	C:	MVol.D4-269
$C_5H_9NpO_5$	$NpO_2(CH_3CH_2CH_2CH(OH)COO)$	Np:	TrU.D1-154
$C_5H_9OS_2^-$	$[S_2COC_4H_9]^-$	C:	MVol.D4-244/5, 249/52
$C_5H_9OS_2Tl$	$Tl[S_2COC_4H_9]$	C:	MVol.D4-253
$C_5H_9O_2S^-$	$[SOCOC_4H_9]^-$	C:	MVol.D4-236/7
$C_5H_9O_2S_2^-$	$[S_2COC_2H_4OC_2H_5]^-$	C:	MVol.D4-244
$C_5H_9O_2S_2Sb$	$SbO[S_2COC_4H_9]$	C:	MVol.D4-254
$C_5H_9O_3S_2^-$	$[S_2COC_2H_4OC_2H_4OH]^-$	C:	MVol.D4-244
$C_5H_9S_3^-$	$[S_2CSC_4H_9]^-$	C:	MVol.D4-268
$C_5H_{10}HgOS_2$	$C_2H_5Hg[S_2COC_2H_5]$	C:	MVol.D4-258
$C_5H_{10}IrN_5O_3$	$(H_3O)_3[IrH(CN)_5]$	Ir:	SVol.2-183
$C_5H_{10}MoOS_4$	$(C_2H_5S)MoS[S_2COC_2H_5]$	C:	MVol.D4-258
$C_5H_{10}N_2O_4SSn$	$Sn(CH_3COO)_2 \cdot CS(NH_2)_2$	Sn:	MVol.C5-55
$C_5H_{10}N_2O_6SU$	$UO_2(CH_3COO)_2 \cdot (NH_2)_2CS$	U:	SVol.E1-220/1
$C_5H_{10}N_2O_7U$	$UO_2(CH_3COO)_2 \cdot (NH_2)_2CO$	U:	SVol.E1-84, 86, 89
$C_5H_{10}N_4O_4S_2$	$[C_5H_5NH][N(SO_2NH_2)_2]$	S:	S-N-Verb.1-160
$C_5H_{10}Ni$	$C_3H_5Ni(C_2H_4)H$	Ni:	Org.Verb.2-32
$C_5H_{10}NiS_4$	$[Ni(S_2CSC_2H_5)(SC_2H_5)]_2$	C:	MVol.D4-266, 270
$C_5H_{10}OS_2$	$SC(OC_4H_9)(SH)$	C:	MVol.D4-239/43
$C_5H_{10}OSe_2$	$CH_3[Se_2COC_3H_7]$	C:	MVol.D6-220
$C_5H_{10}O_2SSn$	$(CH_3)_2Sn(SCH(CH_3)COO)$	Sn:	Org.Verb.6-30
$C_5H_{10}PdS_4$	$[Pd(S_2CSC_2H_5)(SC_2H_5)]_3$	C:	MVol.D4-266, 270
$C_5H_{10}S_3$	$SC(SC_4H_9)(SH)$	C:	MVol.D4-264
$C_5H_{11}NOS$	$(C_2H_5)_2O \cdot HNCS$	C:	MVol.D6-30, 33, 35
–	$HNCS \cdot C_4H_9OH \cdot H_2O$	C:	MVol.D6-30/1
$C_5H_{11}NO_7U$	$[UO_2(CH_3COO)(OH)(NH_2COOC_2H_5)]_n$	U:	SVol.E1-121, 123
$C_5H_{11}NSSn$	$(CH_3)_3SnCH_2SCN$	Sn:	Org.Verb.2-15
$C_5H_{11}NS_3$	$SC(SCH_2CH_2N(CH_3)_2)(SH)$	C:	MVol.D4-264/5
$C_5H_{11}N_3O_4S$	$O(CH_2CH_2)_2NSO_2NHCONH_2$	S:	S-N-Verb.1-171
$C_5H_{11}N_5O_6S_3$	$SO_2(NNHC_5H_5SO_2NH_2)NHSO_2NH_2 \cdot H_2O$	S:	S-N-Verb.1-179
$C_5H_{11}NiOP^+$	$(CO)Ni(CH_2P(CH_3)_3)^+$	Ni:	Org.Verb.1-182
$C_5H_{12}N_2O_2S$	$(CH_2)_5NSO_2NH_2$	S:	S-N-Verb.1-168
$C_5H_{12}N_2O_6SU$	$UO_2SO_3 \cdot (CH_3)_2NCON(CH_3)_2$	U:	SVol.E1-93/5
$C_5H_{12}N_2S$	$[N(CH_3)_4]SCN$	C:	MVol.D6-173
$C_5H_{12}N_2STe^{2+}$	$[Te((CH_3)_2NCSN(CH_3)_2)]^{2+}$	Te:	SVol.B3-147
$C_5H_{12}N_2Te$	$(CH_3)_4NTeCN$	C:	MVol.D6-259/60
$C_5H_{12}N_4O_{11}U$	$(NH_4)_2[UO_2(C_2O_4)_2((NH_2)_2CO)]$	U:	SVol.E1-84, 90
–	$(NH_4)_2[UO_2(C_2O_4)_2((NH_2)_2CO)] \cdot H_2O$	U:	SVol.E1-84, 90
$C_5H_{12}N_6O_9U$	$UO_2(C_2O_4) \cdot 3\ (NH_2)_2CO$	U:	SVol.E1-84, 86, 89
$C_5H_{12}N_8O_5S_2U$	$UO_2(NCS)_2 \cdot 3\ (NH_2)_2CO$	U:	SVol.E1-83, 85/8
$C_5H_{12}OSn$	$(CH_3)_2(CH_2CH)SnOCH_3$	Sn:	Org.Verb.5-197
$C_5H_{12}Sn$	$(CH_3)_3SnCHCH_2$	Sn:	Org.Verb.2-69/70
$C_5H_{13}ISn$	$(CH_3)_3SnCHICH_3$	Sn:	Org.Verb.2-35/6
$C_5H_{13}NO_2S$	$[(CH_3)_3NH][SOCOCH_3]$	C:	MVol.D4-237
–	$HNCS \cdot 2\ C_2H_5OH$	C:	MVol.D6-35
$C_5H_{13}N_3NiS_3$	$Ni(NH(C_2H_4NH_2)_2)CS_3$	C:	MVol.D4-225
$C_5H_{13}N_3O_2S$	$(CH_2)_5NSO_2NHNH_2$	S:	S-N-Verb.1-222
$C_5H_{13}N_3O_3S$	$(CH_3)_2NSO_2NHCON(CH_3)_2$	S:	S-N-Verb.1-171
–	$(C_2H_5)_2NSO_2NHCONH_2$	S:	S-N-Verb.1-171
$C_5H_{13}N_3O_4S_2$	$NH_2SO_2NHSO_2NC_5H_{10}$	S:	S-N-Verb.1-161

$C_5H_{13}Sn^+$	$(C_2H_5)_2SnCH_3^+$	Sn:	Org.Verb.1-70/1
$C_5H_{14}OSn$	$(CH_3)_3SnCH_2OCH_3$	Sn:	Org.Verb.2-11
$C_5H_{14}O_3Sn$	$C_2H_5Sn(OCH_3)_3$	Sn:	Org.Verb.6-230/1
$C_5H_{14}Sn$	$(CH_3)_2(i\text{-}C_3H_7)SnH$	Sn:	Org.Verb.4-89
–	$(CH_3)_3SnC_2H_5$	Sn:	Org.Verb.2-31/3
$C_5H_{15}LiTi$	$Li[(CH_3)_5Ti]$	Ti:	Org.Verb.1-121
$C_5H_{15}NSn$	$(CH_3)_3SnCH_2NHCH_3$	Sn:	Org.Verb.2-12
–	$(CH_3)_3SnN(CH_3)_2$	Sn:	Org.Verb.5-85
$C_5H_{15}N_2OSSi_2^+$	$(CH_3)_3SiNSONSi(CH_3)_2^+$	S:	S-N-Verb.1-198
$C_5H_{15}N_2O_{3.5}S$	$(CH_3)_3NNHSO_3 \cdot 0.5\ (CH_3)_3COH$	S:	S-N-Verb.1-219
$C_5H_{15}N_2SSi_2^+$	$S[NSi(CH_3)_3]NSi(CH_3)_2^+$	S:	S-N-Verb.1-207
$C_5H_{15}N_3O_2S_2Si$	$(CH_3)_2NSO_2NSNSi(CH_3)_3$	S:	S-N-Verb.1-181
$C_5H_{15}N_3O_{10}U$	$[UO_2(NO_3)_2(HOC_2H_4N(CH_3)_3OH)]_2$	U:	SVol.E1-23
$C_5H_{15}N_3SSi^+$	$(CH_3)_3SiN_2SN(CH_3)_2^+$	S:	S-N-Verb.1-210
$C_5H_{15}N_4SSi^+$	$(CH_3)_3SiN_3SN(CH_3)_2^+$	S:	S-N-Verb.1-210
$C_5H_{15}O_5Te^-$	$Te(OCH_3)_5^-$	Te:	SVol.B2-104
$C_5H_{15}Ti^-$	$[(CH_3)_5Ti]^-$	Ti:	Org.Verb.1-120/1
$C_5H_{16}N_2O_2SSi$	$(CH_3)_2NSO_2NHSi(CH_3)_3$	S:	S-N-Verb.1-171/2
$C_5H_{16}N_8O_9U$	$(CN_3H_6)_2[UO_2(O_2)C_2O_4((NH_2)_2CO)] \cdot n\ H_2O$	U:	SVol.E1-84, 91
$C_5H_{19}N_7O_2P_2S$	$[SO_2[NP(NHCH_3)_2]_2NCH_3]$	S:	S-N-Verb.1-33
$C_5H_{20}IN_{11}O_{10}U$	$[UO_2((NH_2)_2CO)_5]I(NO_3)$	U:	SVol.E1-83, 85, 87
$C_5H_{20}I_2N_{10}O_7U$	$UO_2I_2 \cdot 5\ (NH_2)_2CO$	U:	SVol.E1-85
–	$UO_2I_2 \cdot 5\ (NH_2)_2CO \cdot H_2O$	U:	SVol.E1-83, 86/7
$C_5H_{20}I_3LaN_{10}O_5$	$LaI_3 \cdot 5\ CO(NH_2)_2$	Sc:	MVol.C6-184
$C_5H_{20}I_3N_{10}NdO_5$	$NdI_3 \cdot 5\ CO(NH_2)_2$	Sc:	MVol.C6-191
–	$NdI_3 \cdot 5\ CO(NH_2)_2$ solid solutions $(Nd,Pr)I_3 \cdot 5\ CO(NH_2)_2$	Sc:	MVol.C6-191
$C_5H_{20}I_3N_{10}O_5Pr$	$PrI_3 \cdot 5\ CO(NH_2)_2$	Sc:	MVol.C6-188
–	$PrI_3 \cdot 5\ CO(NH_2)_2$ solid solutions $(Pr,Nd)I_3 \cdot 5\ CO(NH_2)_2$	Sc:	MVol.C6-191
$C_5H_{20}I_3N_{10}O_5Sm$	$SmI_3 \cdot 5\ CO(NH_2)_2$	Sc:	MVol.C6-193
$C_5H_{20}I_4N_{10}O_5U$	$UI_4 \cdot 5\ (NH_2)_2CO \cdot 2\ H_2O$	U:	SVol.E1-82/3
$C_5H_{20}I_{11}N_{10}NdO_5$	$NdI_{11} \cdot 5\ CO(NH_2)_2$ solid solutions $(Nd,Pr)I_{11} \cdot 5\ CO(NH_2)_2$	Sc:	MVol.C6-191
$C_5H_{20}I_{11}N_{10}O_5Pr$	$PrI_{11} \cdot 5\ CO(NH_2)_2$ solid solutions $(Pr,Nd)I_{11} \cdot 5\ CO(NH_2)_2$	Sc:	MVol.C6 191
$C_5H_{20}MnNO_8^+$	$[MnNO_3(CH_3OH)_5]^+$	Mn:	MVol.D1-24/5
$C_5H_{20}N_{10}O_4S_6Sn$	$SnSO_4 \cdot 5\ CS(NH_2)_2$	Sn:	MVol.C5-55
$C_5H_{20}N_{10}O_7U^{2+}$	$[UO_2((NH_2)_2CO)_5]^{2+}$	U:	SVol.E1-82
$C_5H_{20}N_{12}O_{13}U$	$UO_2(NO_3)_2 \cdot 5\ (NH_2)_2CO$	U:	SVol.E1-83, 85, 88/9
–	$UO_2(NO_3)_2 \cdot 5\ (NH_2)_2CO \cdot H_2O$	U:	SVol.E1-83, 85/6, 88/9
$C_5H_{24}N_{12}O_6S_7Sn$	$Sn(SO_3NH_2)_2 \cdot 5\ CS(NH_2)_2 \cdot H_2O$	Sn:	MVol.C5-55
$C_5H_{30}MnN_{15}O_{37}W_{11}$	$(CH_6N_3)_5MnW_{11}O_{37} \cdot 8.5\ H_2O$	Mn:	MVol.C3-223/4
$C_5H_{48}Ir_2N_{21}NaO_{18}S_6$	$Na(C(NH_2)_3)_5[Ir(SO_3)_3(NH_3)_3]_2$	Ir:	SVol.2-158
$C_5IIrK_3N_5$	$K_3[Ir(CN)_5I]$	Ir:	SVol.2-184
$C_5IIrN_5^{3-}$	$[Ir(CN)_5I]^{3-}$	Ir:	SVol.2-183
$C_5IrK_3N_5$	$K_3[Ir(CN)_5]$	Ir:	SVol.2-181
$C_5IrN_5^{3-}$	$[Ir(CN)_5]^{3-}$	Ir:	SVol.2-181
$C_5KN_5O_3$	$N(C(NCO)N)_2COK$	C:	MVol.D3-211/2

$C_5K_8O_{21}U_2$	$K_8[(UO_2)_2(O_2)(CO_3)_5]$	U:	SVol.C3-44
$C_5Li_6O_{19}U_2$	$Li_6[(UO_2)_2(CO_3)_5]$	U:	SVol.C2-263
$C_5Na_6O_{19}U_2$	$Na_6[(UO_2)_2(CO_3)_5]$	U:	SVol.C2-247
$C_5Na_6O_{28}U_5$	$Na_6[(UO_3)_3(UO_2)_2(CO_3)_5]$	U:	SVol.C2-247
$C_6CaCl_6H_{24}N_{12}O_6Sn$	$[Ca(CO(NH_2)_2)_6][SnCl_6]$	Sn:	MVol.C3-176
C_6CaF_6	C_6F_5CaF	F:	PerFHalOrg.4-60
$C_6CaFeKN_5O$	$KCa[COFe(CN)_5] \cdot 5\ H_2O$	Fe:	Org.Verb.B1-25
$C_6CaH_{10}O_4S_2$	$Ca[SOCOC_2H_5]_2$	C:	MVol.D4-237
$C_6CaN_6S_6Sn$	$Ca[Sn(SCN)_6] \cdot 7\ H_2O$	Sn:	MVol.C3-179
$C_6CdFeH_8N_2O_4$	$(C_2H_8N_2)CdFe(CO)_4$	Fe:	Org.Verb.B3-126
$C_6CdFe_2H_6N_4O_8$	$[Cd(NH_3)_2][Fe(CO)_3NO]_2$	Fe:	Org.Verb.B1-189
$C_6CdFe_2N_2O_8$	$Cd[Fe(CO)_3NO]_2$	Fe:	Org.Verb.B1-196
		Fe:	Org.Verb.C1-62
$C_6CdH_{10}O_2S_4$	$Cd[S_2COC_2H_5]_2$	C:	MVol.D4-245, 258
$C_6CdH_{10}O_2Se_4$	$Cd[Se_2COC_2H_5]_2$	C:	MVol.D6-221
$C_6CeCsFeN_6$	$CsCe[Fe(CN)_6]$	Sc:	MVol.B7-30
$C_6CeFeH_4N_7$	$NH_4Ce[Fe(CN)_6]$	Sc:	MVol.B7-30
$C_6CeFeKN_6$	$KCe[Fe(CN)_6]$	Sc:	MVol.B7-30
C_6CeFeN_6Rb	$RbCe[Fe(CN)_6]$	Sc:	MVol.B7-30
C_6CeFeN_6Tl	$TlCe[Fe(CN)_6]$	Sc:	MVol.B7-30
$C_6CeH_{33}O_{21}S_3$	$[Ce(H_2O)_9](C_2H_5SO_4)_3$	Sc:	MVol.B1-41
$C_6CfH_6NO_6$	$Cf[N(CH_2COO)_3]$	Np:	TrU.D1-163
$C_6CfO_{12}{}^{3-}$	$Cf(C_2O_4)_3{}^{3-}$	Np:	TrU.D1-158
$C_6ClF_2FeHO_4$	$F_2CCHClFe(CO)_4$	Fe:	Org.Verb.B4-315/6
$C_6ClF_2H_6N_3$	$NCClCFNCFC(N(CH_3)_2)$	F:	PerFHalOrg.6-60, 67
$C_6ClF_3FeO_4$	$CF_2CFClFe(CO)_4$	Fe:	Org.Verb.B4-320, 322/4
$C_6ClF_3HN_3S$	$(CF_3)C_5HClN_3S$	F:	PerFHalOrg.6-71
$C_6ClF_3H_2N_4$	$(CF_3)C_5H_2ClN_4$	F:	PerFHalOrg.6-71
$C_6ClF_3H_4OS$	$(ClC_6H_4O)SF_3$	S:	SVol.2-83
$C_6ClF_3H_5NO_2$	$CF_3C(O)NC_4H_5ClO$	F:	PerFHalOrg.7-194/5
$C_6ClF_3H_6N_4$	$(CH_3NH)(NH_2)(CF_3)C_4ClN_2$	F:	PerFHalOrg.6-59
$C_6ClF_3H_8N_2O_3$	$O_2NCF_2CFClNOC(CH_3)_2CH_2$	F:	PerFHalOrg.7-194
$C_6ClF_3H_{10}NO_4P$	$CF_2ClCFNOPO(OC_2H_5)_2$	F:	PerFHalOrg.7-199
–	$CF_3CClNOPO(OC_2H_5)_2$	F:	PerFHalOrg.7-199
$C_6ClF_3H_{11}NO_2$	$CF_2ClC(F)NOH \cdot (C_2H_5)_2O$	F:	PerFHalOrg.7-88, 105
–	$CF_3CClNOH \cdot (C_2H_5)_2O$	F:	PerFHalOrg.7-88, 105
$C_6ClF_3N_2$	$NC_5F_3Cl(CN)$	F:	PerFHalOrg.5-195/7, 209/10
		F:	PerFHalOrg.6-13
$C_6ClF_3N_4O_2S$	$NCClNC(CF_3)C(NO_2)C(SCN)$	F:	PerFHalOrg.6-71
$C_6ClF_4H_2N$	$ClC_6F_4NH_2$	F:	PerFHalOrg.7-9, 36
$C_6ClF_4H_4NO$	$HOC_6F_4NH_2 \cdot HCl$	F:	PerFHalOrg.7-8/9, 36
$C_6ClF_4H_8NO$	$CF_3CFClNOC(CH_3)_2CH_2$	F:	PerFHalOrg.7-194
C_6ClF_4NO	$NC_5F_4(COCl)$	F:	PerFHalOrg.5-194, 208, 217
–	NOC_6F_4Cl	F:	PerFHalOrg.7-180
C_6ClF_5	C_6F_5Cl	F:	PerFHalOrg.4-136
$C_6ClF_5H_3N$	$C_6F_5NH_2 \cdot HCl$	F:	PerFHalOrg.7-8, 36
$C_6ClF_5H_{18}N_2S$	$SF_5Cl \cdot 2\ (CH_3)_3N$	S:	SVol.2-288
C_6ClF_5Hg	C_6F_5HgCl	F:	PerFHalOrg.4-110

C_6ClF_5Mg	C_6F_5MgCl	F: PerFHalOrg.4-63/4, 67/8, 70/1, 79/81, 98/9, 107, 110/1
$C_6ClF_5N_2$	C_6F_5NNCl	F: PerFHalOrg.7-61/2
C_6ClF_5Ni	$(C_6F_5NiCl)_x$	Ni: Org.Verb.2-394
$C_6ClF_6H_6N$	$CH_3C(Cl)CHC(CF_3)_2NH_2$	F: PerFHalOrg.7-69, 73
$C_6ClF_6H_{10}NP_2$	$(CF_3)_2PNP(C_2H_5)_2Cl$	F: PerFHalOrg.3-69
$C_6ClF_7H_8OSi$	$CH_3(C_3F_7)Si(OC_2H_5)Cl$	F: PerFHalOrg.4-18
C_6ClF_7Si	$C_6F_5SiF_2Cl$	F: PerFHalOrg.4-149
$C_6ClF_8H_2N_3$	$(CF_3)_2(CF_2Cl)C_3H_2N_3$	F: PerFHalOrg.6-113
C_6ClF_8Li	$ClC(CF_2)_4CLi$	F: PerFHalOrg.4-3, 13, 27, 34
C_6ClF_8NO	$FC(CFCF)(CFCl)(N(CF_3)O)CF$	F: PerFHalOrg.6-4/5, 10
–	$FC(CFCF)(CF_2)(N(Cl)O)C(CF_3)$	F: PerFHalOrg.6-4/5, 10
$C_6ClF_8N_3$	$(CF_3)_2(CF_2Cl)C_3N_3$	F: PerFHalOrg.6-81/2, 101, 113
$C_6ClF_{11}H_4N_2$	$CF_3(CF_2)_4C(NH)NH_2 \cdot HCl$	F: PerFHalOrg.7-3/4, 31
$C_6ClF_{12}NO$	$ClCF_2(CF_2)_5NO$	F: PerFHalOrg.7-179, 186
$C_6ClF_{14}P$	$(n\text{-}C_3F_7)_2PCl$	F: PerFHalOrg.3-72/3, 90, 99/101
$C_6ClF_{15}N_2O_2$	$(CF_3)_2NOCFClCF_2ON(CF_3)_2$	F: PerFHalOrg.7-95/6, 112
$C_6ClF_{15}N_2O_4S$	$(CF_3)_2NOCF_2CF(SO_2Cl)ON(CF_3)_2$	F: PerFHalOrg.7-97/8, 115
$C_6ClF_{15}Si$	$(C_2F_5)_3SiCl$	F: PerFHalOrg.4-147
$C_6ClF_{18}N_2O_2P$	$[(CF_3)_2NO]_2P(CF_3)_2Cl$	F: PerFHalOrg.7-100/1
$C_6ClFeH_3O_4$	$ClCHCH_2Fe(CO)_4$	Fe: Org.Verb.B4-233, 239/40
$C_6ClFeH_4O^+$	$[FeC_5H_4COCl]^+$	Fe: Org.Verb.A3-152
$C_6ClFeH_5O_3$	$C_3H_5Fe(CO)_3Cl$	Fe: Org.Verb.B5-35/7, 48/50, 54/8
$C_6ClFeH_5O_7$	$[C_3H_5Fe(CO)_3]ClO_4$	Fe: Org.Verb.B5-82/5
$C_6ClFeH_6^+$	$[ClFeC_6H_6]^+$	Fe: Org.Verb.A2-245
$C_6ClFeH_6NO_3$	$(CH_3)(Cl)C_3H_3Fe(CO)_2NO$	Fe: Org.Verb.B5-14/5, 17, 23
$C_6ClFeH_6O_4P$	$(CO)_4FeP(CH_3)_2Cl$	Fe: Org.Verb.B2-94, 115
$C_6ClH_3MnO_2$	$[Mn(ClC_6H_3O_2)]$	Mn: MVol.D1-45
C_6ClH_3Sn	$(CHC)_3SnCl$	Sn: Org.Verb.5-157
$C_6ClH_4S_{1.5}Sn$	$[ClC_6H_4SnS_{1.5}]_x$	Sn: Org.Verb.6-283
$C_6ClH_5N_4O_2S$	$ClC_6H_4NHSO_2N_3$	S: S-N-Verb.1-214
$C_6ClH_5NiO_3$	$CH_2CHCH_2C(O)Ni(CO)_2Cl$	Ni: Org.Verb.2-275
$C_6ClH_5O_2Sn$	$(ClC_6H_4SnOOH)_x$	Sn: Org.Verb.6-283
C_6ClH_5Ti	C_6H_5TiCl	Ti: Org.Verb.1-7
C_6ClH_6NNiO	$[C_6H_6Ni(NO)]Cl$	Ni: Org.Verb.2-221, 225
$C_6ClH_7N_2O_2S$	$ClC_6H_4NHSO_2NH_2$	S: S-N-Verb.1-167
$C_6ClH_7O_2Sn$	$(C_6H_5)ClSn(OH)_2$	Sn: Org.Verb.6-275/6
$C_6ClH_9Na_2O_4S_2Sn$	$C_2H_5SnCl(SCH_2COONa)_2$	Sn: Org.Verb.6-230/1
$C_6ClH_9O_2SSn$	$(CH_2CH)_2ClSnOS(O)CHCH_2$	Sn: Org.Verb.5-156
$C_6ClH_9O_6Te$	$(HOCOCH_2)_3TeCl$	Te: SVol.B2-98
C_6ClH_9Sn	$(CH_2CH)_3SnCl$	Sn: Org.Verb.5-156
$C_6ClH_{10}NOS_2Sn$	$Sn(O)Cl(SC(S)N(CH_2)_5) \cdot 2\ H_2O$	Sn: MVol.C6-107
$C_6ClH_{11}N_2NiS$	$C_3H_5Ni[SC(NHCH_2)_2]Cl$	Ni: Org.Verb.2-275
$C_6ClH_{12}IrN_3O_6^-$	$[Ir(NH_2CH_2COO)(NH_2CH_2COO)_2Cl]^-$	Ir: SVol.2-81
$C_6ClH_{12}NO_2S$	$ClSO_2NHC_6H_{11}$	S: S-N-Verb.1-123

$C_6ClH_{12}N_3O_5S_2$	$ClSO_2NHCONHSO_2N(CH_2)_5$	S:	S-N-Verb.1-125
$C_6ClH_{13}IrN_3O_6$	$Ir(NH_2CH_2COO)_2(NH_2CH_2COOH)Cl$	Ir:	SVol.2-81
$C_6ClH_{13}OSn$	$(CH_3)_2(CH_3COCH_2CH_2)SnCl$	Sn:	Org.Verb.5-193, 197
$C_6ClH_{13}O_2Sn$	$(CH_3)_2(CH_3OCOCH_2CH_2)SnCl$	Sn:	Org.Verb.5-193, 197
–	$(CH_3)_3Sn(OCOCH_2CH_2Cl)$	Sn:	Org.Verb.5-84
$C_6ClH_{13}Sn$	$CH_2(CH_2CH_2)_2Sn(CH_3)Cl$	Sn:	Org.Verb.5-229
$C_6ClH_{14}NO_2S$	$ClSO_2N(C_3H_7)_2$	S:	S-N-Verb.1-124
–	$ClSO_2NHC_6H_{13}$	S:	S-N-Verb.1-123
$C_6ClH_{14}NiP$	$C_3H_5Ni(P(CH_3)_3)Cl$	Ni:	Org.Verb.2-12/4
$C_6ClH_{14}Sn^+$	$(C_3H_7)_2SnCl^+$	Sn:	Org.Verb.5-112, 116
$C_6ClH_{15}NOS^+$	$[(C_2H_5)_3NSOCl]^+$	S:	SVol.1-52
$C_6ClH_{15}OSn$	$(CH_3)_3SnCl \cdot (CH_3)_2CO$	Sn:	Org.Verb.5-87
–	$(C_2H_5)_2(CH_3OCH_2)SnCl$	Sn:	Org.Verb.5-201, 204
$C_6ClH_{15}O_2Sn$	$(C_4H_9)(CH_3O)_2SnCl$	Sn:	Org.Verb.6-245
$C_6ClH_{15}S_2Sn$	$(C_4H_9)(CH_3S)_2SnCl$	Sn:	Org.Verb.6-245
$C_6ClH_{15}Sn$	$(CH_3)_2(C_4H_9)SnCl$	Sn:	Org.Verb.5-193, 197
–	$(C_2H_5)_3SnCl$		
	Analysis	Sn:	Org.Verb.5-92
	Chemical reactions	Sn:	Org.Verb.5-101/7
	Formation and preparation	Sn:	Org.Verb.5-89/94
	Molecule	Sn:	Org.Verb.5-95
	Physical properties	Sn:	Org.Verb.5-98/100
	Physiological action	Sn:	Org.Verb.5-107/9
	Polarography	Sn:	Org.Verb.5-101
	Spectra	Sn:	Org.Verb.5-95/8
	Uses	Sn:	Org.Verb.5-109/10
–	$(C_3H_7)_2SnHCl$	Sn:	Org.Verb.6-289
–	$(i\text{-}C_3H_7)_2SnHCl$	Sn:	Org.Verb.6-289
$C_6ClH_{15}Sn^+$	$(C_2H_5)_3SnCl^+$	Sn:	Org.Verb.5-98
$C_6ClH_{15}SnZn$	$(C_2H_5)_3SnZnCl$	Sn:	Org.Verb.5-102
$C_6ClH_{16}IrN_4O_6$	$(NH_4)[IrCl(NH_2CH_2COO)(NH_2CH_2COO)_2]$	Ir:	SVol.2-81
$C_6ClH_{16}NSn$	$(CH_3)_2ClSnN(C_2H_5)_2$	Sn:	Org.Verb.6-22
$C_6ClH_{17}O_2Sn_2$	$(CH_3)_2ClSnOSn(CH_3)_2OC_2H_5$	Sn:	Org.Verb.6-23
$C_6ClH_{17}SiSn$	$(CH_3)_2Sn(CH_2Cl)(CH_2SiH(CH_3)_2)$	Sn:	Org.Verb.3-82
–	$(CH_3)_3SnCH_2SiCl(CH_3)_2$	Sn:	Org.Verb.2-14/5
$C_6ClH_{18}NSn$	$[(CH_3)_3SnCH_2NH(CH_3)_2]Cl$	Sn:	Org.Verb.2-14/5
$C_6ClH_{18}OPSSn$	$(CH_3)_3SnCl \cdot (CH_3)_2P(O)SCH_3$	Sn:	Org.Verb.5-87
$C_6ClH_{18}O_2PSn$	$(CH_3)_3SnCl \cdot (CH_3)_2P(O)OCH_3$	Sn:	Org.Verb.5-87
$C_6ClH_{19}OSn_2$	$(CH_3)_3SnCl \cdot (CH_3)_3SnOH \cdot H_2O$	Sn:	Org.Verb.5-82
$C_6ClH_{21}IrN_{15}$	$[Ir((NH_2C(NH))_2NH)_3]Cl$	Ir:	SVol.2-94
$C_6ClH_{21}N_2Sn$	$(C_2H_5)_3SnCl \cdot 2\ NH_3$	Sn:	Org.Verb.5-107
$C_6ClH_{21}O_6Sn_3^{2+}$	$[(C_2H_5Sn)_3(OH)_6Cl]^{2+}$	Sn:	Org.Verb.6-230
$C_6Cl_2Cu_2H_{12}Sn$	$(CH_3)_2Sn(CHCH_2)_2 \cdot 2\ CuCl$	Sn:	Org.Verb.3-7, 16
$C_6Cl_2FH_2N_3$	$(NH_2)(CN)C_5FCl_2N$	F:	PerFHalOrg.5-196/7, 210
$C_6Cl_2FH_5N_2O$	$(NH_2)(CH_3O)C_5FCl_2N$	F:	PerFHalOrg.5-214
$C_6Cl_2FH_5N_2S$	$(NH_2)(CH_3S)C_5FCl_2N$	F:	PerFHalOrg.5-214/5
$C_6Cl_2FH_6N_3$	$NCClC(N(CH_3)_2)NCClCF$	F:	PerFHalOrg.6-60, 66
–	$(NH_2)(CH_3NH)C_5FCl_2N$	F:	PerFHalOrg.5-214/5
$C_6Cl_2F_2FeO_4$	$CF_2CCl_2Fe(CO)_4$	Fe:	Org.Verb.B4-320, 322/4
$C_6Cl_2F_2H_2N_2O$	$NH_2C(O)C_5F_2Cl_2N$	F:	PerFHalOrg.5-194/5, 208

$C_6Cl_2F_2H_2N_2O_2$	$NH_2COOC_5F_2Cl_2N$	F:	PerFHalOrg.5-143
$C_6Cl_2F_2H_3NO$	$CH_3N(CFCCl)_2C(O)$	F:	PerFHalOrg.5-212
–	$CH_3OC_5Cl_2F_2N$	F:	PerFHalOrg.5-135, 212
$C_6Cl_2F_2H_3NS$	$CH_3SC_5F_2Cl_2N$	F:	PerFHalOrg.5-217
$C_6Cl_2F_2H_3N_3$	$NC_5F_2Cl_2(C(NH)NH_2)$	F:	PerFHalOrg.5-194/5, 208
$C_6Cl_2F_2H_3N_3O$	$(NH_2CONH)C_5F_2Cl_2N$	F:	PerFHalOrg.5-189, 204
$C_6Cl_2F_2N_2$	$NC_5F_2Cl_2(CN)$	F:	PerFHalOrg.5-195/7, 209
		F:	PerFHalOrg.6-13
$C_6Cl_2F_3HN_4$	$CF_3C_5Cl_2HN_4$	F:	PerFHalOrg.6-127, 139, 162
$C_6Cl_2F_3H_2N$	$Cl_2C_6F_3NH_2$	F:	PerFHalOrg.7-9, 36
$C_6Cl_2F_3Li$	$Cl_2C_6F_3Li$	F:	PerFHalOrg.4-26
$C_6Cl_2F_4HNO$	$(HO)(CF_3)C_5FCl_2N$	F:	PerFHalOrg.5-183/4
$C_6Cl_2F_4H_2N_2$	$(CF_3)(NH_2)C_5FCl_2N$	F:	PerFHalOrg.5-186
$C_6Cl_2F_4H_3NO_3S$	$C_5F_3Cl_2NCH_3OSO_2F$	F:	PerFHalOrg.5-143
$C_6Cl_2F_4H_5NOS$	$SC(CF_2Cl)_2NHC(O)CH(CH_3)$	F:	PerFHalOrg.7-66/7
–	$SC(CF_2Cl)_2NHC(O)CH_2CH_2$	F:	PerFHalOrg.7-66, 68
$C_6Cl_2F_4Mg$	ClC_6F_4MgCl	F:	PerFHalOrg.4-63, 65, 77/8
$C_6Cl_2F_5In$	$C_6F_5InCl_2 \cdot C_4H_8O_2$	F:	PerFHalOrg.4-128/9
$C_6Cl_2F_5NS$	$C_6F_5NSCl_2$	S:	SVol.2-237
$C_6Cl_2F_5P$	$C_6F_5PCl_2$	F:	PerFHalOrg.3-124/9
$C_6Cl_2F_5PS$	$C_6F_5P(S)Cl_2$	F:	PerFHalOrg.3-141, 148
$C_6Cl_2F_6H_8Sn$	$(CF_3CH_2CH_2)_2SnCl_2$	Sn:	Org.Verb.6-129, 138
$C_6Cl_2F_6Si$	$C_6F_5SiFCl_2$	F:	PerFHalOrg.4-149, 156, 161
$C_6Cl_2F_7H_2N_3$	$(CF_2Cl)_2(CF_3)C_3H_2N_3$	F:	PerFHalOrg.6-113
$C_6Cl_2F_7N_3$	$(CF_3)(CF_2Cl)_2C_3N_3$	F:	PerFHalOrg.6-81/2, 102, 113
$C_6Cl_2F_7P$	$C_6F_5PF_2Cl_2$	F:	PerFHalOrg.3-123, 125
$C_6Cl_2F_{14}N_2O_2S$	$ClS[N(CF_3)OCF_2CF_2]_2Cl$	F:	PerFHalOrg.7-92/3, 108, 126
$C_6Cl_2F_{14}P_2$	$(CF_3)_2PCCl_2CF_2P(CF_3)_2$	F:	PerFHalOrg.3-164, 169
$C_6Cl_2F_{14}Si$	$(C_3F_7)_2SiCl_2$	F:	PerFHalOrg.4-147
$C_6Cl_2FeH_2O_4$	$ClCHCHClFe(CO)_4$	Fe:	Org.Verb.B4-268/9, 278/80
–	$Cl_2CCH_2Fe(CO)_4$	Fe:	Org.Verb.B4-258, 260
$C_6Cl_2FeH_4O_3$	$ClC_3H_4Fe(CO)_3Cl$	Fe:	Org.Verb.B5-44, 51/2, 60, 62/3
$C_6Cl_2FeO_6$	$Fe(CO)_6Cl_2$	Fe:	Org.Verb.B3-256
$C_6Cl_2Fe_2GeH_{10}N_4O_4$	$(CH_2CHCH_2Fe(NO)_2)_2GeCl_2$	Fe:	Org.Verb.C2-156, 165
$C_6Cl_2Fe_2GeO_6S_2$	$(CO)_3FeS_2(GeCl_2)Fe(CO)_3$	Fe:	Org.Verb.C1-23/4
$C_6Cl_2Fe_2H_{10}N_4O_4Pb$	$(CH_2CHCH_2Fe(NO)_2)_2PbCl_2$	Fe:	Org.Verb.C2-156, 165
$C_6Cl_2Fe_2H_{10}N_4O_4Sn$	$(C_3H_5Fe(NO)_2)_2SnCl_2$	Fe:	Org.Verb.C2-156, 165/6
$C_6Cl_2Fe_2N_2O_8Sn$	$NO(CO)_3FeSnCl_2Fe(CO)_3NO$	Fe:	Org.Verb.C1-67
$C_6Cl_2H_5NSTi$	$C_5H_5Ti(NCS)Cl_2$	Ti:	Org.Verb.1-141, 152
$C_6Cl_2H_6N_2O_2$	$NCClC(OCH_3)NCClC(OCH_3)$	F:	PerFHalOrg.6-58/9, 64
$C_6Cl_2H_6SSn$	$(CH_2CH)(C_4H_3S)SnCl_2$	Sn:	Org.Verb.6-202
$C_6Cl_2H_7NSn$	$SnCl_2 \cdot CH_3C_5H_4N$	Sn:	MVol.C5-18
–	$SnCl_2 \cdot C_6H_5NH_2$	Sn:	MVol.C5-12/3

$C_6Cl_2H_8N_2O_2U$	$UO_2Cl_2 \cdot (CH_3)_2C_4H_2N_2 \cdot H_2O$	U:	SVol.E1-46/8
–	$UO_2Cl_2 \cdot C_6H_4(NH_2)_2$	U:	SVol.E1-23/5
$C_6Cl_2H_8N_2Sn$	$(CH_2CH_2CN)_2SnCl_2$	Sn:	MVol.C6-184
		Sn:	Org.Verb.6-129
$C_6Cl_2H_8NiPd$	$(ClNiC_3H_4C_3H_4PdCl)_x$	Ni:	Org.Verb.2-399
$C_6Cl_2H_8OTi$	$C_5H_5Ti(OCH_3)Cl_2$	Ti:	Org.Verb.1-180/1
$C_6Cl_2H_8STi$	$C_5H_5Ti(SCH_3)Cl_2$	Ti:	Org.Verb.1-189
$C_6Cl_2H_8Sn$	$(CH_3)(C_5H_5)SnCl_2$	Sn:	Org.Verb.6-195/6
$C_6Cl_2H_9N_3O_{10}U$	$[UO_2(CH_3CN)_3](ClO_4)_2$	U:	SVol.E1-48/50
$C_6Cl_2H_{10}N_2O_8S_2$	$C_2H_4(CH_2OCONHSO_2Cl)_2$	S:	S-N-Verb.1-125
$C_6Cl_2H_{10}Ni_2$	$(C_3H_5NiCl)_2$	Ni:	Org.Verb.2-274/6
$C_6Cl_2H_{10}O_2Pd_2S_4$	$[PdCl(S_2COC_2H_5)]_2$	C:	MVol.D4-261
$C_6Cl_2H_{10}O_2Sn$	$[CH_3C(O)CH_2]_2SnCl_2$	Sn:	Org.Verb.6-130
$C_6Cl_2H_{10}O_3Sn$	$[SnCl_2(CH(COCH_3)_2)OCH_3]$	Sn:	MVol.C5-106
$C_6Cl_2H_{10}O_4Sn$	$(CH_3)_2Sn(OCOCH_2Cl)_2$	Sn:	Org.Verb.6-21
–	$(HOCOCH_2CH_2)_2SnCl_2$	Sn:	Org.Verb.6-129/30
–	$SnCl_2(C_2H_5COO)_2$	Sn:	MVol.C2-223
$C_6Cl_2H_{10}Sn$	$(CH_2CHCH_2)_2SnCl_2$	Sn:	Org.Verb.6-148, 150
–	$[(CH_2)_2CH]_2SnCl_2$	Sn:	Org.Verb.6-127
–	$[CH_3C(CH_2)]_2SnCl_2$	Sn:	Org.Verb.6-148
–	$(CH_3CHCH)_2SnCl_2$	Sn:	Org.Verb.6-148
$C_6Cl_2H_{10}Ti$	$(C_3H_5)_2TiCl_2$	Ti:	Org.Verb.1-127
$C_6Cl_2H_{11}NNi$	$(C_4H_5N(CH_3)_2)NiCl_2 \cdot 2\ H_2O$		
	$= [Ni(C_4H_5N(CH_3)_2)(H_2O)_2]Cl_2$	Ni:	Org.Verb.1-339
$C_6Cl_2H_{12}HgMnN_2O_2S$	$MnHg(NCS)_2Cl_2 \cdot 2\ C_2H_5OH$	Mn:	MVol.D1-32
$C_6Cl_2H_{12}MnO_{1.5}$	$MnCl_2 \cdot 1.5\ C_4H_8O$	Mn:	MVol.D1-142
$C_6Cl_2H_{12}N_2O_2Sn$	$[H_2NC(O)CH_2CH_2]_2SnCl_2$	Sn:	Org.Verb.6-130, 138
$C_6Cl_2H_{12}N_2S_4Sn$	$[SnCl_2((CH_3)_2NCSS)_2]$	Sn:	MVol.C6-117/8
$C_6Cl_2H_{12}N_2Se_4Sn$	$[SnCl_2((CH_3)_2NCSeSe)_2]$	Sn:	MVol.C6-124/5
$C_6Cl_2H_{12}N_4O_4S_2Sn$	$Sn(CH_2ClCOO)_2 \cdot 2\ CS(NH_2)_2$	Sn:	MVol.C5-55
$C_6Cl_2H_{12}O_2Sn$	$(CH_3)(C_2H_5OCOCH_2CH_2)SnCl_2$	Sn:	Org.Verb.6-197
$C_6Cl_2H_{12}Sn$	$(CH_3)_3SnCHCCl_2CH_2$	Sn:	Org.Verb.2-58/9
–	$(C_4H_9)(CH_2CH)SnCl_2$	Sn:	Org.Verb.6-199, 204
$C_6Cl_2H_{13}NO_2Sn$	$SnCl_2 \cdot (CH_3)_2C_3H_4(NH_2)COOH$	Sn:	MVol.C5-31/2
$C_6Cl_2H_{14}IrN_3O_6$	$Ir(NH_2CH_2COO)(NH_2CH_2COOH)_2Cl_2$	Ir:	SVol.2-81
$C_6Cl_2H_{14}N_2O_2S$	$SO_2[N(C_3H_7)Cl]_2$	S:	S-N-Verb.1-177
–	$SO_2[N(i\text{-}C_3H_7)Cl]_2$	S:	S-N-Verb.1-177
$C_6Cl_2H_{14}N_2O_2Sn$	$SnCl_2 \cdot 2\ OCHN(CH_3)_2$	Sn:	MVol.C5-43
$C_6Cl_2H_{14}N_2O_4Sn$	$SnCl_2 \cdot 2\ CH_3CH(NH_2)COOH$	Sn:	MVol.C5-31/2
–	$SnCl_2 \cdot 2\ NH_2(CH_2)_2COOH$	Sn:	MVol.C5-31
$C_6Cl_2H_{14}OSSn$	$(CH_3)_2SnCl_2 \cdot (CH_2)_4SO$	Sn:	Org.Verb.6-37
$C_6Cl_2H_{14}OSn$	$(C_2H_5)[(CH_3)_3CO]SnCl_2$	Sn:	Org.Verb.6-55
$C_6Cl_2H_{14}O_2Sn$	$(CH_3)_2SnCl_2 \cdot O(CH_2CH_2)_2O$	Sn:	Org.Verb.6-37
–	$(C_2H_5OCH_2)_2SnCl_2$	Sn:	Org.Verb.6-130
–	$(HOCH_2CH_2CH_2)_2SnCl_2$	Sn:	Org.Verb.6-130
$C_6Cl_2H_{14}S_2Sn$	$SnCl_2(SC_3H_7)_2$	Sn:	MVol.C6-79/80
$C_6Cl_2H_{14}Sn$	$(CH_3)(C_5H_{11})SnCl_2$	Sn:	Org.Verb.6-195, 197
–	$(C_2H_5)(C_4H_9)SnCl_2$	Sn:	Org.Verb.6-198, 204
–	$(C_3H_7)_2SnCl_2$	Sn:	Org.Verb.6-65/70
–	$(i\text{-}C_3H_7)_2SnCl_2$	Sn:	Org.Verb.6-70

$C_6Cl_2H_{15}NNiO_2$	$[Ni(C_4H_5N(CH_3)_2)(H_2O)_2]Cl_2$ = $(C_4H_5N(CH_3)_2)NiCl_2 \cdot 2\ H_2O$	Ni: Org.Verb.1-339
$C_6Cl_2H_{15}NSn$	$SnCl_2 \cdot (C_3H_7)_2NH$	Sn: MVol.C5-11/2
$C_6Cl_2H_{15}OPS_2Sn$	$C_4H_9Sn[SP(S)(CH_3)(OCH_3)]Cl_2$	Sn: Org.Verb.6-247
$C_6Cl_2H_{15}O_2PS_2Sn$	$C_4H_9Sn[SP(S)(OCH_3)_2]Cl_2$	Sn: Org.Verb.6-247
$C_6Cl_2H_{16}IrN_2$	$Ir(C_2H_4(N(CH_3)_2)_2)Cl_2$	Ir: SVol.2-86
$C_6Cl_2H_{16}N_4O_4U$	$UO_2Cl_2 \cdot 2\ CO(NHCH_3)_2$	U: SVol.E1-92/3
$C_6Cl_2H_{16}OSSn$	$(CH_3)_2SnCl_2 \cdot (C_2H_5)_2SO$	Sn: Org.Verb.6-37
$C_6Cl_2H_{16}OSi_2Sn$	$O(Si(CH_3)_2CH_2)_2SnCl_2$	Sn: Org.Verb.6-207
$C_6Cl_2H_{16}OSn_2$	$(C_2H_5)_2SnO \cdot (CH_3)_2SnCl_2$	Sn: Org.Verb.6-21
$C_6Cl_2H_{16}O_2Sn$	$(CH_3)_2SnCl_2 \cdot C_2H_4(OCH_3)_2$	Sn: Org.Verb.6-37
$C_6Cl_2H_{16}SiSn$	$(CH_3)_2Sn(CH_2Cl)(CH_2SiCl(CH_3)_2)$	Sn: Org.Verb.3-81
$C_6Cl_2H_{17}NSn_2$	$[(CH_3)_2ClSn]_2NC_2H_5$	Sn: Org.Verb.6-22
$C_6Cl_2H_{17}N_5O_2P_2S$	$[O_2S[NPClN(CH_3)_2]_2NC_2H_5]$	S: S-N-Verb.1-35
$C_6Cl_2H_{17}O_3PSn$	$(CH_3)_2SnCl_2 \cdot (C_2H_5O)_2HPO$	Sn: Org.Verb.6-37
$C_6Cl_2H_{18}IrN_4^+$	$[Ir((NH_2C_2H_4NH)_2C_2H_4)Cl_2]^+$	Ir: SVol.2-95, 124
$C_6Cl_2H_{18}MnO_3$	$MnCl_2 \cdot 3\ C_2H_5OH$	Mn: MVol.C5-78
		Mn: MVol.D1-27, 28
$C_6Cl_2H_{18}N_2O_2U$	$UO_2Cl_2 \cdot 2\ C_3H_7NH_2$	U: SVol.E1-18, 21
$C_6Cl_2H_{18}N_2Sn$	$SnCl_2 \cdot 2\ C_3H_7NH_2$	Sn: MVol.C5-11/2
$C_6Cl_2H_{18}N_3OPSn$	$SnCl_2 \cdot OP(N(CH_3)_2)_3$	Sn: MVol.C5-61
$C_6Cl_2H_{18}N_4O_8S_4$	$[(NS(CH_3)_2NS(CH_3))_2](ClO_4)_2$	S: S-N-Verb.1-5
$C_6Cl_2H_{18}N_6O_4P_2S_2$	$[SO_2(NCH_3)_2PClNCH_3]_2$	S: S-N-Verb.1-44/5
$C_6Cl_2H_{18}N_6O_5U$	$UO_2Cl_2 \cdot 3\ CH_3CONHNH_2 \cdot n\ H_2O$	U: SVol.E1-114/6
$C_6Cl_2H_{18}NiP_2$	$Ni(P(CH_3)_3)_2Cl_2$	Ni: Org.Verb.1-11
$C_6Cl_2H_{18}O_2S_2Sn$	$(CH_3)_2SnCl_2 \cdot 2\ (CH_3)_2SO$	Sn: Org.Verb.6-37
$C_6Cl_2H_{18}O_2Se_2Sn$	$(CH_3)_2SnCl_2 \cdot 2\ (CH_3)_2SeO$	Sn: Org.Verb.6-37
$C_6Cl_2H_{18}O_2Sn_3$	$[(CH_3)_2SnO]_2 \cdot (CH_3)_2SnCl_2$	Sn: Org.Verb.6-21
$C_6Cl_2H_{18}O_4P_2U$	$UO_2Cl_2 \cdot 2\ (CH_3)_3PO$	U: SVol.E1-148, 156
$C_6Cl_2H_{18}O_4S_4$	$SOCl_2 \cdot 3\ (CH_3)_2SO$	S: SVol.1-45
$C_6Cl_2H_{18}O_5S_3U$	$UO_2Cl_2 \cdot 3\ (CH_3)_2SO$	U: SVol.E1-203, 207
$C_6Cl_2H_{18}S_2Sn_3$	$(CH_3)_2Sn[SSnCl(CH_3)_2]_2$	Sn: Org.Verb.6-22
$C_6Cl_2H_{21}N_3O_2U$	$UO_2Cl_2 \cdot 3\ C_2H_5NH_2$	U: SVol.E1-18, 21
$C_6Cl_2H_{22}N_4Sn$	$(C_3H_7)_2SnCl_2 \cdot 2\ N_2H_4$	Sn: Org.Verb.6-69
$C_6Cl_2H_{24}MnO_{14}$	$[Mn(CH_3OH)_6](ClO_4)_2$	Mn: MVol.D1-25
$C_6Cl_2H_{24}NdO_6^+$	$NdCl_2(CH_3OH)_6^+$	Sc: MVol.B7-9
$C_6Cl_2H_{26}N_4Sn$	$(C_3H_7)_2SnCl_2 \cdot 4\ NH_3$	Sn: Org.Verb.6-69
$C_6Cl_3CsH_6O_6Sn$	$CsSn(CH_2ClCOO)_3$	Sn: MVol.C3-155
$C_6Cl_3FH_4Sn$	$FC_6H_4SnCl_3$	Sn: Org.Verb.6-283/4
$C_6Cl_3FH_{11}N_2O_2P$	$OPCl(ONCFCl)N(C_2H_5)CH_2CHClCH_3$	F: PerFHalOrg.7-201/2
$C_6Cl_3FN_2$	$NC_5FCl_3(CN)$	F: PerFHalOrg.5-196/7, 209/10
		F: PerFHalOrg.6-13
$C_6Cl_3F_3HNO$	$(HO)(CF_3)C_5Cl_3N$	F: PerFHalOrg.5-183/4, 199
$C_6Cl_3F_3HNS$	$(HS)(CF_3)C_5Cl_3N$	F: PerFHalOrg.5-192, 206
$C_6Cl_3F_3H_2N_2$	$(CF_3)(NH_2)C_5Cl_3N$	F: PerFHalOrg.5-186, 202
$C_6Cl_3F_3Mg$	$Cl_2C_6F_3MgCl$	F: PerFHalOrg.4-63, 65, 98
$C_6Cl_3F_3N_2O$	$(NO)(CF_3)C_5Cl_3N$	F: PerFHalOrg.5-190/1, 205
$C_6Cl_3F_3N_2O_2$	$(NO_2)(CF_3)C_5Cl_3N$	F: PerFHalOrg.5-190/1, 205
$C_6Cl_3F_4N$	$Cl_3CC_5F_4N$	F: PerFHalOrg.5-156/7, 169

$C_6Cl_3F_4P$	$ClC_6F_4PCl_2$	F:	PerFHalOrg.4-109
$C_6Cl_3F_5Si$	$C_6F_5SiCl_3$	F:	PerFHalOrg.4-149/50, 156, 162
$C_6Cl_3F_5Sn$	$C_6F_5SnCl_3$	F:	PerFHalOrg.4-181/2, 186, 189
		Sn:	Org.Verb.6-282/3
$C_6Cl_3F_5Ti$	$C_6F_5TiCl_3$	F:	PerFHalOrg.4-35
		Ti:	Org.Verb.1-17, 24/5
$C_6Cl_3F_6H_2N_3$	$(CF_2Cl)_3C_3H_2N_3$	F:	PerFHalOrg.6-113
$C_6Cl_3F_6N_3$	$(CF_2Cl)_3C_3N_3$	F:	PerFHalOrg.6-81/2, 96, 102, 113
$C_6Cl_3F_{12}H_3N_2OSi$	$Cl_3SiCH[ON(CF_3)_2]CH_2N(CF_3)_2$	F:	PerFHalOrg.7-143
$C_6Cl_3F_{12}H_3N_2O_2Si$	$Cl_3SiCH[ON(CF_3)_2]CH_2ON(CF_3)_2$	F:	PerFHalOrg.7-127/8
$C_6Cl_3F_{14}NP_2$	$(C_3F_7)_2PNPCl_3$	F:	PerFHalOrg.3-61, 65
$C_6Cl_3F_{14}P$	$(n\text{-}C_3F_7)_2PCl_3$	F:	PerFHalOrg.3-73, 91, 99, 100
$C_6Cl_3FeGeH_5O_3$	$C_3H_5Fe(CO)_3GeCl_3$	Fe:	Org.Verb.B5-67
$C_6Cl_3FeHO_4$	$Cl_2CCHClFe(CO)_4$	Fe:	Org.Verb.B4-315/6
$C_6Cl_3FeH_5O_3Si$	$C_3H_5Fe(CO)_3SiCl_3$	Fe:	Org.Verb.B5-67
$C_6Cl_3FeH_{10}N_2$	$(C_2H_5NC)_2FeCl_3$	Fe:	Org.Verb.B4-12/3
$C_6Cl_3GeH_5NiO$	$C_5H_5Ni(CO)GeCl_3$	Ni:	Org.Verb.2-164/6
$C_6Cl_3H_4ISn$	$IC_6H_4SnCl_3$	Sn:	Org.Verb.6-284
$C_6Cl_3H_4NO_2Sn$	$NO_2C_6H_4SnCl_3$	Sn:	Org.Verb.6-284
$C_6Cl_3H_5N_2O$	$(NH_2)(CH_3O)C_5Cl_3N$	F:	PerFHalOrg.5-214/5
$C_6Cl_3H_5N_3O_2PS_2$	$[NS(O)Cl][NS(O)C_6H_5][NPCl_2]$	S:	S-N-Verb.1-30
$C_6Cl_3H_5NiOSi$	$C_5H_5Ni(CO)SiCl_3$	Ni:	Org.Verb.2-164/6
$C_6Cl_3H_5NiOSn$	$C_5H_5Ni(CO)SnCl_3$	Ni:	Org.Verb.2-164/6
$C_6Cl_3H_5Sn$	$C_6H_5SnCl_3$	Sn:	Org.Verb.6-267/81
$C_6Cl_3H_5Ti$	$C_6H_5TiCl_3$	Ti:	Org.Verb.1-17, 24
$C_6Cl_3H_6KO_6Sn$	$KSn(CH_2ClCOO)_3$	Sn:	MVol.C3-77/8
$C_6Cl_3H_6O_6RbSn$	$RbSn(CH_2ClCOO)_3$	Sn:	MVol.C3-140
$C_6Cl_3H_6O_8Pu^-$	$PuO_2(ClCH_2COO)_3^-$	Np:	TrU.D1-153
$C_6Cl_3H_7Ti$	$CH_3C_5H_4TiCl_3$	Ti:	Org.Verb.1-140, 149/50
$C_6Cl_3H_8MnN$	$[CH_3C_5H_4NH]MnCl_3$	Mn:	MVol.C5-208/9
–	$C_6H_5NH_3MnCl_3$	Mn:	MVol.C5-205/6
$C_6Cl_3H_9O_3Sn$	$SnCl_3(CH_3COCHCOOC_2H_5)$	Sn:	MVol.C5-137
$C_6Cl_3H_9Sn$	$C_6H_9SnCl_3$	Sn:	Org.Verb.6-266
$C_6Cl_3H_{10}NO_6Sn$	$NH_4Sn(CH_2ClCOO)_3$	Sn:	MVol.C3-125
$C_6Cl_3H_{10}N_3Sn$	$SnCl_3N_2H_3 \cdot CH_3C_5H_4N$	Sn:	MVol.C5-202
$C_6Cl_3H_{11}OSn$	$CH_3COCH_2C(CH_3)_2SnCl_3$	Sn:	Org.Verb.6-257
$C_6Cl_3H_{11}O_4U$	$UCl_3(CH_3COO) \cdot CH_3COOC_2H_5$	U:	SVol.E1-120/1
$C_6Cl_3H_{11}Sn$	$(CH_2)_5CHSnCl_3$	Sn:	Org.Verb.6-253
–	$(CH_3)(Cl(CH_2)_3CHCH)SnCl_2$	Sn:	Org.Verb.6-197
–	$(CH_3)_2CHCH_2C(CH_2)SnCl_3$	Sn:	Org.Verb.6-265
$C_6Cl_3H_{13}N_4Sn$	$C_6H_5SnCl_3 \cdot 2\ N_2H_4$	Sn:	Org.Verb.6-279
$C_6Cl_3H_{13}OSn$	$C_3H_7OCH(CH_3)CH_2SnCl_3$	Sn:	Org.Verb.6-257
$C_6Cl_3H_{13}OTi$	$CH_3TiCl_3 \cdot CH_3OCH_2CH_2CHCH_2$	Ti:	Org.Verb.1-36/7, 41, 124
$C_6Cl_3H_{13}O_2Ti$	$CH_2CHTiCl_3 \cdot CH_3OCH_2CH_2OCH_3$	Ti:	Org.Verb.1-36/7, 41
$C_6Cl_3H_{13}O_3Te$	$TeCl_3OC_2H_5 \cdot CH_3COOC_2H_5$	Te:	SVol.B2-105
$C_6Cl_3H_{13}Sn$	$(CH_2Cl)_2(C_4H_9)SnCl$	Sn:	Org.Verb.5-222

$C_6Cl_3H_{13}Sn$	$C_6H_{13}SnCl_3$	Sn:	Org.Verb.6-250
$C_6Cl_3H_{14}N_2O_2PS$	$(C_3H_7)_2NSO_2NPCl_3$	S:	S-N-Verb.1-188
$C_6Cl_3H_{15}IrN_3S_3$	$Ir(S(NH_2)CCH_3)_3Cl_3$	Ir:	SVol.2-168
$C_6Cl_3H_{15}IrSe_3$	$IrCl_3(SeC_2H_5)_3$	Ir:	SVol.2-174
$C_6Cl_3H_{15}OSn_2$	$Cl_2(C_4H_9)SnOSn(CH_3)_2Cl$	Sn:	Org.Verb.6-245
$C_6Cl_3H_{15}O_2SiSn$	$(CH_3)(C_2H_5O)_2SiCH_2SnCl_3$	Sn:	Org.Verb.6-257
$C_6Cl_3H_{16}NOTi$	$CH_3TiCl_3 \cdot CH_3OCH_2CH_2N(CH_3)_2$	Ti:	Org.Verb.1-32/3, 40
$C_6Cl_3H_{16}NSTi$	$CH_3TiCl_3 \cdot (CH_3)_2NCH_2CH_2SCH_3$	Ti:	Org.Verb.1-32/3, 40
$C_6Cl_3H_{18}IrN_4$	$[Ir((NH_2C_2H_4NH)_2C_2H_4)Cl_2]Cl$	Ir:	SVol.2-95
$C_6Cl_3H_{18}IrS$	$Ir(S(CH_3)_2)_3Cl_3$	Ir:	SVol.2-164/5
$C_6Cl_3H_{24}IrN_6$	$[Ir(C_2H_4(NH_2)_2)_3]Cl_3$	Ir:	SVol.2-82
$C_6Cl_3H_{24}IrN_{12}S_6$	$[Ir(SC(NH_2)_2)_6]Cl_3$	Ir:	SVol.2-168
$C_6Cl_4FH_2N$	$Cl_4C_6FNH_2$	F:	PerFHalOrg.7-9, 36/7
$C_6Cl_4F_2H_3NO_3S$	$C_5FCl_4NCH_3OSO_2F$	F:	PerFHalOrg.5-143
$C_6Cl_4F_3N$	$CF_3C_5Cl_4N$	F:	PerFHalOrg.5-177
$C_6Cl_4F_5P$	$C_6F_5PCl_4$	F:	PerFHalOrg.3-124
$C_6Cl_4F_6H_3N_3$	$C_2F_3Cl_2C(NH)NC(NH_2)C_2F_3Cl_2$	F:	PerFHalOrg.7-7, 34
$C_6Cl_4F_6MoO_4P_2$	$Mo(CO)_4(CF_3PCl_2)_2$	F:	PerFHalOrg.3-110/1, 120
$C_6Cl_4F_6Si_2$	$C_6F_4(SiFCl_2)_2$	F:	PerFHalOrg.4-149, 161
$C_6Cl_4F_{12}H_2N_2O_2Si$	$Cl_3SiCH(ON(CF_3)_2)CHClON(CF_3)_2$	F:	PerFHalOrg.7-127/8
$C_6Cl_4F_{12}N_2O_2$	$(CF_3)_2NOCCl_2CCl_2ON(CF_3)_2$	F:	PerFHalOrg.7-95/6, 113
$C_6Cl_4FeO_4$	$Cl_2CCCl_2Fe(CO)_4$	Fe:	Org.Verb.B4-321, 325
$C_6Cl_4H_3NO_4Sn$	$SnCl_4 \cdot NO_2C_6H_3O_2$	Sn:	MVol.C5-111/2
$C_6Cl_4H_4N_4Sn$	$SnCl_4 \cdot 2\ NCCH_2CN$	Sn:	MVol.C6-73/5
$C_6Cl_4H_4O_2Sn$	$SnCl_4 \cdot C_6H_4O_2$	Sn:	MVol.C5-111/2
$C_6Cl_4H_4Sn$	$ClC_6H_4SnCl_3$	Sn:	Org.Verb.6-283/4
$C_6Cl_4H_5N_3O_4Sn$	$SnCl_4 \cdot (NO_2)_2C_6H_3NH_2$	Sn:	MVol.C5-178
$C_6Cl_4H_6INSn$	$SnCl_4 \cdot IC_6H_4NH_2$	Sn:	MVol.C5-180
$C_6Cl_4H_6N_2O_2Sn$	$SnCl_4 \cdot NO_2C_6H_4NH_2$	Sn:	MVol.C5-177
$C_6Cl_4H_6N_2S_2Sn$	$SnCl_4 \cdot NCC(SCH_3)C(SCH_3)CN$	Sn:	MVol.C6-99
$C_6Cl_4H_6N_2Sn$	$SnCl_4 \cdot 2\ CH_2CHCN$	Sn:	MVol.C6-69/70
$C_6Cl_4H_6N_6Sn$	$SnCl_4 \cdot 2\ C_3H_3N_3$	Sn:	MVol.C5-229
$C_6Cl_4H_6OSn$	$SnCl_4 \cdot C_6H_5OH$	Sn:	MVol.C5-76
$C_6Cl_4H_6Sn$	$(CHClCH)_3SnCl$	Sn:	Org.Verb.5-156/7
$C_6Cl_4H_7NSn_2$	$2\ SnCl_2 \cdot CH_3C_5H_4N$	Sn:	MVol.C5-18
$C_6Cl_4H_8N_2Sn$	$SnCl_4 \cdot C_6H_4(NH_2)_2$	Sn:	MVol.C5-186/7
–	$SnCl_4 \cdot C_6H_5NHNH_2$	Sn:	MVol.C5-189
–	$SnCl_4 \cdot NC(CH_2)_4CN$	Sn:	MVol.C6-73/5
$C_6Cl_4H_8N_4O_2Sn$	$SnCl_4 \cdot 2\ NCCH_2CONH_2$	Sn:	MVol.C6-75
$C_6Cl_4H_8N_4Sn$	$SnCl_4 \cdot 2\ C_3H_4N_2$	Sn:	MVol.C5-223
$C_6Cl_4H_8O_4Sn$	$SnCl_4 \cdot C_2H_2(COOCH_3)_2$	Sn:	MVol.C5-143/4
$C_6Cl_4H_8O_6Sn$	$SnCl_4 \cdot 2\ (CH_2O)_2CO$	Sn:	MVol.C5-155
$C_6Cl_4H_8O_7Sn$	$SnCl_4 \cdot HOC_3H_4(COOH)_3$	Sn:	MVol.C5-121
$C_6Cl_4H_9NO_2Sn$	$2\ SnCl_4 \cdot 2\ NCC_2H_4COOC_2H_5$	Sn:	MVol.C6-75
$C_6Cl_4H_9N_3Sn$	$SnCl_4 \cdot 3\ CH_3CN$	Sn:	MVol.C6-64/5
$C_6Cl_4H_{10}N_2S_2Sn$	$SnCl_4 \cdot 2\ C_2H_5SCN$	Sn:	MVol.C6-122
$C_6Cl_4H_{10}N_2Sn$	$SnCl_4 \cdot 2\ C_2H_5CN$	Sn:	MVol.C6-68/9
$C_6Cl_4H_{10}N_6O_2Sn$	$SnCl_4 \cdot 2\ CH_2CNCONHNH_2$	Sn:	MVol.C6-59/60
$C_6Cl_4H_{10}NiTi$	$C_3H_5NiCl \cdot C_3H_5TiCl_3$	Ni:	Org.Verb.2-5
$C_6Cl_4H_{10}Ni_3$	$(C_3H_5NiCl)_2 \cdot NiCl_2$	Ni:	Org.Verb.2-275

$C_6Cl_4H_{10}O_2Sn$	$SnCl_4 \cdot (CH_3CO)_2CHCH_3$	Sn:	MVol.C5-108
$C_6Cl_4H_{10}O_3Sn$	$SnCl_4 \cdot CH_3COCH_2COOC_2H_5$	Sn:	MVol.C5-137
–	$SnCl_4 \cdot (C_2H_5CO)_2O$	Sn:	MVol.C5-125
$C_6Cl_4H_{10}O_4Sn$	$SnCl_4 \cdot C_2H_4(COOCH_3)_2$	Sn:	MVol.C5-141/3
–	$SnCl_4 \cdot C_2H_4(OCOCH_3)_2$	Sn:	MVol.C5-133
–	$SnCl_4 \cdot C_2H_5OCOCOOC_2H_5$	Sn:	MVol.C5-139/40
–	$SnCl_4 \cdot C_4H_8(COOH)_2$	Sn:	MVol.C5-121
$C_6Cl_4H_{10}SSn$	$SnCl_4 \cdot (C_3H_5)_2S$	Sn:	MVol.C6-93/4
$C_6Cl_4H_{11}NO_2Sn$	$SnCl_4 \cdot CH_3CONHCOC_3H_7$	Sn:	MVol.C6-50/1
$C_6Cl_4H_{12}NO_3PSn$	$SnCl_4 \cdot CNCH_2P(O)(OC_2H_5)_2$	Sn:	MVol.C6-164/5
$C_6Cl_4H_{12}N_2Sn$	$SnCl_4 \cdot (CH_3)_2CNNC(CH_3)_2$	Sn:	MVol.C6-30/1
–	$SnCl_4 \cdot (C_2H_5)_2NCH_2CN$	Sn:	MVol.C6-73
$C_6Cl_4H_{12}N_4O_2Sn$	$SnCl_4 \cdot 2\ CO(NHCH_2)_2$	Sn:	MVol.C5-224/5
$C_6Cl_4H_{12}N_4S_2Sn$	$SnCl_4 \cdot 2\ CH_2NHC(SH)NCH_2$	Sn:	MVol.C6-104/5
$C_6Cl_4H_{12}N_4Sn$	$SnCl_4 \cdot 2\ CNN(CH_3)_2$	Sn:	MVol.C6-77/8
$C_6Cl_4H_{12}OSn$	$SnCl_4 \cdot (CH_3)_2C_4H_6O$	Sn:	MVol.C5-157
$C_6Cl_4H_{12}O_2Sn$	$SnCl_4 \cdot 2\ CH_3COCH_3$	Sn:	MVol.C5-85/6
$C_6Cl_4H_{12}O_2Te$	$[TeCl_3((CH_3)_2CO)_2]Cl$	Te:	SVol.B3-165/6
$C_6Cl_4H_{12}O_4Sn$	$SnCl_4 \cdot 2\ CH_3COOCH_3$	Sn:	MVol.C5-128
–	$SnCl_4 \cdot 2\ C_2H_5COOH$	Sn:	MVol.C5-120/1
–	$SnCl_4 \cdot 2\ HCOOC_2H_5$	Sn:	MVol.C5-127
$C_6Cl_4H_{12}O_6Sn$	$SnCl_4 \cdot 3\ CH_3COOH$	Sn:	MVol.C5-117/8
$C_6Cl_4H_{12}SSn$	$SnCl_4 \cdot S(CH(CH_3)CH_2)_2$	Sn:	MVol.C6-110
$C_6Cl_4H_{12}Sn$	$(CH_2Cl)_2Sn(CHClCH_3)_2$	Sn:	Org.Verb.3-69
–	$(CH_3CHCl)_3SnCl$	Sn:	Org.Verb.5-150
$C_6Cl_4H_{13}O_3U$	$U(OC_2H_5)Cl_4 \cdot CH_3COOC_2H_5$	U:	SVol.E1-120/2
$C_6Cl_4H_{14}MnN_2$	$[HN(C_2H_4)_3NH]MnCl_4$	Mn:	MVol.C5-204/5
$C_6Cl_4H_{14}MnN_2O_2^{2-}$	$[MnCl_4(HCON(CH_3)_2)_2]^{2-}$	Mn:	MVol.C5-87
$C_6Cl_4H_{14}N_2O_2Sn$	$SnCl_4 \cdot 2\ CH_3CONHCH_3$	Sn:	MVol.C6-43
–	$SnCl_4 \cdot 2\ OCHN(CH_3)_2$	Sn:	MVol.C6-41/2
$C_6Cl_4H_{14}N_2O_2Te$	$TeCl_4 \cdot 2\ CH_3CONHCH_3$	Te:	SVol.B3-174
–	$TeCl_4 \cdot 2\ HCON(CH_3)_2$	Te:	SVol.B3-174
$C_6Cl_4H_{14}N_2O_4Sn$	$SnCl_4 \cdot 2\ CH_3CH(NH_2)COOH$	Sn:	MVol.C6-12
–	$SnCl_4 \cdot 2\ NH_2CH_2CH_2COOH$	Sn:	MVol.C6-12
$C_6Cl_4H_{14}N_2Sn$	$[((CH_3)_2N)_2CCH_2SnCl_3]Cl$	Sn:	Org.Verb.6-257
$C_6Cl_4H_{14}OSn$	$SnCl_4 \cdot (C_3H_7)_2O$	Sn:	MVol.C5-151, 152
$C_6Cl_4H_{14}O_2SSn$	$SnCl_4 \cdot (n-C_3H_7)_2SO_2$	Sn:	MVol.C6-103
$C_6Cl_4H_{14}O_2Sn$	$SnCl_4 \cdot C_2H_4(OC_2H_5)_2$	Sn:	MVol.C5-154/5
$C_6Cl_4H_{14}O_3Sn$	$SnCl_4 \cdot (CH_3OC_2H_4)_2O$	Sn:	MVol.C5-154/5
$C_6Cl_4H_{14}O_3U$	$UCl_4 \cdot O(C_2H_4OCH_3)_2$	U:	SVol.E1-70, 74/5
$C_6Cl_4H_{14}SSn$	$SnCl_4 \cdot (C_3H_7)_2S$	Sn:	MVol.C6-93/4
$C_6Cl_4H_{14}STe$	$TeCl_4 \cdot (C_3H_7)_2S$	Te:	SVol.B3-175/6
$C_6Cl_4H_{14}S_2Sn$	$SnCl_4 \cdot C_2H_5S(CH_2)_2SC_2H_5$	Sn:	MVol.C6-97
$C_6Cl_4H_{15}NSn$	$SnCl_4 \cdot (C_2H_5)_3N$	Sn:	MVol.C5-173/4
$C_6Cl_4H_{15}NSn_2$	$2\ SnCl_2 \cdot (C_3H_7)_2NH$	Sn:	MVol.C5-11/2
$C_6Cl_4H_{15}NU$	$UCl_4 \cdot (C_2H_5)_3N$	U:	SVol.E1-20, 22/3
$C_6Cl_4H_{15}O_3PSn$	$(CH_3)_2SnCl_2 \cdot (ClCH_2CH_2O)_2HPO$	Sn:	Org.Verb.6-37
$C_6Cl_4H_{15}O_4PSn$	$SnCl_4 \cdot (C_2H_5O)_3PO$	Sn:	MVol.C6-170
$C_6Cl_4H_{16}IrN_3O_6$	$(NH_3CH_2COOH)[Ir(NH_2CH_2COOH)_2Cl_4]$	Ir:	SVol.2-81
$C_6Cl_4H_{16}N_2Sn$	$SnCl_4 \cdot C_2H_4(N(CH_3)_2)_2$	Sn:	MVol.C5-184

$C_6Cl_4H_{16}N_2Te$	$TeCl_4 \cdot (CH_3)_2NCH_2CH_2N(CH_3)_2$	Te: SVol.B3-168
$C_6Cl_4H_{16}N_4O_2P_2S$	$SO_2[NPCl_2NHC_3H_7]_2$	S: S-N-Verb.1-192
$C_6Cl_4H_{16}N_4O_2Sn$	$SnCl_4 \cdot 2\ (CH_3)_2NCONH_2$	Sn: MVol.C6-56
–	$SnCl_4 \cdot 2\ (CH_3NH)_2CO$	Sn: MVol.C6-56
$C_6Cl_4H_{16}O_2Sn$	$SnCl_4 \cdot 2\ C_3H_7OH$	Sn: MVol.C5-72/3
$C_6Cl_4H_{16}O_2Sn_2$	$(C_2H_5)_2SnCl_2 \cdot (C_2H_5)Cl_2SnOOH$	Sn: Org.Verb.6-52
$C_6Cl_4H_{16}O_4Sn$	$SnCl_4 \cdot 2\ CH_2(OCH_3)_2$	Sn: MVol.C5-148
$C_6Cl_4H_{16}P_2U$	$UCl_4 \cdot (CH_3)_2PC_2H_4P(CH_3)_2$	U: SVol.E1-51
$C_6Cl_4H_{16}S_2Sn$	$SnCl_4 \cdot 2\ C_3H_7SH$	Sn: MVol.C6-79/80
$C_6Cl_4H_{16}Si_2Sn$	$[(CH_3)_2ClSiCH_2]_2SnCl_2$	Sn: Org.Verb.6-130, 138
–	$(CH_3)_2SiClCH_2Si(CH_3)_2CH_2SnCl_3$	Sn: Org.Verb.6-257
$C_6Cl_4H_{16}Si_2Ti$	$Cl(CH_3)_2SiCH_2Si(CH_3)_2CH_2TiCl_3$	Ti: Org.Verb.1-17, 25
$C_6Cl_4H_{18}IrP_2$	$Ir(P(CH_3)_3)_2Cl_4$	Ir: SVol.2-202/3
$C_6Cl_4H_{18}MnO_8P_2^{2-}$	$[MnCl_4(PO(OCH_3)_3)_2]^{2-}$	Mn: MVol.C5-80
$C_6Cl_4H_{18}Mn_2O_3$	$2\ MnCl_2 \cdot 3\ C_2H_5OH$	Mn: MVol.D1-30/2
$C_6Cl_4H_{18}N_2Sn$	$SnCl_4 \cdot 2\ (CH_3)_3N$	Sn: MVol.C5-171
$C_6Cl_4H_{18}N_2Sn_2$	$[Sn_2Cl_3(C_3H_7NH_2)_2]Cl$	Sn: MVol.C5-11/2
$C_6Cl_4H_{18}N_2U$	$UCl_4 \cdot 2\ (CH_3)_3N$	U: SVol.E1-20, 22/3
$C_6Cl_4H_{18}N_3P_3Sn$	$SnCl_4 \cdot NP(CH_3)_2NP(CH_3)_2NP(CH_3)_2$	Sn: MVol.C6-177
$C_6Cl_4H_{18}O_2P_2S_2Sn$	$SnCl_4 \cdot 2\ (CH_3)_2P(O)SCH_3$	Sn: MVol.C6-158/60
–	$SnCl_4 \cdot 2\ (CH_3)_2P(S)OCH_3$	Sn: MVol.C6-158/60
$C_6Cl_4H_{18}O_2P_2Sn$	$(CH_3)_2SnCl_2 \cdot 2\ (CH_3)_2ClPO$	Sn: Org.Verb.6-37
–	$SnCl_4 \cdot 2\ (CH_3)_3PO$	Sn: MVol.C6-143/4
$C_6Cl_4H_{18}O_2P_2U$	$UCl_4 \cdot 2\ (CH_3)_3PO$	U: SVol.E1-147/8, 153, 156
$C_6Cl_4H_{18}O_3PaS_3$	$PaCl_4 \cdot 3\ (CH_3)_2SO$	Pa: SVol.2-91/2
$C_6Cl_4H_{18}O_3S_3Sn_2$	$2\ SnCl_2 \cdot 3\ (CH_3)_2SO$	Sn: MVol.C5-45
$C_6Cl_4H_{18}O_3S_3U$	$UCl_4 \cdot 3\ (CH_3)_2SO$	U: SVol.E1-3, 202/3, 206/7
$C_6Cl_4H_{18}O_3Sn$	$SnCl_4 \cdot 3\ C_2H_5OH$	Sn: MVol.C5-70
$C_6Cl_4H_{18}O_4P_2S_4Sn$	$SnCl_4 \cdot 2\ (CH_3O)(CH_3S)_2PO$	Sn: MVol.C6-176
$C_6Cl_4H_{18}O_4P_2Sn$	$SnCl_4 \cdot 2\ (CH_3)_2P(O)OCH_3$	Sn: MVol.C6-153/4
–	$SnCl_4 \cdot 2\ C_2H_5(H)P(O)OCH_3$	Sn: MVol.C6-151/2
$C_6Cl_4H_{18}O_6P_2S_2Sn$	$SnCl_4 \cdot 2\ (CH_3O)_2(CH_3S)PO$	Sn: MVol.C6-174
$C_6Cl_4H_{18}O_6P_2Sn$	$SnCl_4 \cdot 2\ CH_3P(O)(OCH_3)_2$	Sn: MVol.C6-162/3
–	$SnCl_4 \cdot 2\ P(OCH_3)_3$	Sn: MVol.C6-141/2
$C_6Cl_4H_{18}O_8P_2Sn$	$SnCl_4 \cdot 2\ (CH_3O)_3PO$	Sn: MVol.C6-169/70
$C_6Cl_4H_{18}P_2S_2Sn$	$SnCl_4 \cdot 2\ (CH_3)_3PS$	Sn: MVol.C6-148
$C_6Cl_4H_{18}P_2Sn$	$SnCl_4 \cdot 2\ P(CH_3)_3$	Sn: MVol.C6-132
–	$SnCl_4 \cdot 2\ P(CD_3)_3$	Sn: MVol.C6-132
$C_6Cl_4H_{19}NSi_2Sn$	$SnCl_4 \cdot [(CH_3)_3Si]_2NH$	Sn: MVol.C6-130
$C_6Cl_4H_{20}MnN_2$	$[(CH_3)_3NH]_2MnCl_4$	Mn: MVol.C5-190
–	$(C_3H_7NH_3)_2MnCl_4$	Mn: MVol.C5-182/5
$C_6Cl_4H_{20}N_2O_2U$	$[N(CH_3)_3H]_2UO_2Cl_4$	U: SVol.C9-93, 101
$C_6Cl_4H_{20}O_2P_2U$	$[P(CH_3)_3H]_2UO_2Cl_4$	U: SVol.C9-59
$C_6Cl_4H_{21}N_3Si_3Sn$	$SnCl_4 \cdot NH(Si(CH_3)_2NH)_2Si(CH_3)_2$	Sn: MVol.C6-131
$C_6Cl_4H_{21}N_3U$	$UCl_4 \cdot 3\ (CH_3)_2NH$	U: SVol.E1-19, 22
$C_6Cl_4H_{24}N_6Sn$	$SnCl_4 \cdot 3\ C_2H_8N_2$	Sn: MVol.C5-183
$C_6Cl_4H_{24}N_{12}S_6Te$	$Te(SC(NH_2)_2)_6Cl_4$	Te: SVol.B3-179
$C_6Cl_4O_4Sn$	$[SnCl_2(C_6Cl_2O_4)]$	Sn: MVol.C5-111/2
$C_6Cl_5F_8NO$	$Cl_3C(CF_2CFCl)_2CF_2NO$	F: PerFHalOrg.7-179, 186
$C_6Cl_5H_4ISn$	$Cl_2IC_6H_4SnCl_3$	Sn: Org.Verb.6-284

$C_6Cl_5H_5N_2O_2Sn$	$SnCl_4 \cdot NO_2C_6H_3(Cl)NH_2$	Sn: MVol.C5-177
$C_6Cl_5H_5OPSn^+$	$[SnCl_3(C_6H_5POCl_2)]^+$	Sn: MVol.C6-167
$C_6Cl_5H_7OSn$	$SnCl_4 \cdot ClC_5H_7CO$	Sn: MVol.C5-94
$C_6Cl_5H_{14}OU$	$UCl_5 \cdot (C_3H_7)_2O$	U: SVol.E1-70, 75
$C_6Cl_5H_{15}O_3SiSn$	$SnCl_4 \cdot ClSi(OC_2H_5)_3$	Sn: MVol.C6-128/9
$C_6Cl_5H_{17}IrN_3O_6$	$(NH_3CH_2COOH)_2[Ir(NH_2CH_2COOH)Cl_5]$	Ir: SVol.2-81
$C_6Cl_5H_{17}O_2P_2Sn$	$C_2H_5SnCl_3 \cdot 2\ Cl(CH_3)_2PO$	Sn: Org.Verb.6-232
$C_6Cl_5H_{18}N_3OPU$	$UCl_5 \cdot [(CH_3)_2N]_3PO$	U: SVol.E1-162
$C_6Cl_5H_{20}MnN_2$	$[(CH_3)_3NH]_2MnCl_5$	Mn: MVol.C5-210
$C_6Cl_5H_{22}MnN_2O$	$[(CH_3)_2NH_2]_2MnCl_5 \cdot C_2H_5OH$	Mn: MVol.C5-210
$C_6Cl_6F_3N_3$	$(CFCl_2)_3C_3N_3$	F: PerFHalOrg.6-81/2, 102, 112
$C_6Cl_6F_4Si_2$	$C_6F_4(SiCl_3)_2$	F: PerFHalOrg.4-149/50, 159
$C_6Cl_6FeO_2S_2$	$[Fe(CO)_2(SCClCCl_2)_2]_n$	Fe: Org.Verb.B3-186
$C_6Cl_6H_3KO_6Sn$	$KSn(CHCl_2COO)_3$	Sn: MVol.C3-78
$C_6Cl_6H_4N_2O_2Sn$	$SnCl_4 \cdot NO_2C_6H_2Cl_2NH_2$	Sn: MVol.C5-178
$C_6Cl_6H_8N_6Sn$	$[C_3H_3N_3 \cdot H]_2SnCl_6$	Sn: MVol.C5-229
$C_6Cl_6H_{10}Ni_2Ti$	$(C_3H_5NiCl)_2 \cdot TiCl_4$	Ni: Org.Verb.2-275, 285
$C_6Cl_6H_{10}O_4Sn$	$SnCl_4 \cdot 2\ CH_2ClCOOCH_3$	Sn: MVol.C5-133/4
$C_6Cl_6H_{18}O_2Sn$	$[(CH_3)_3O]_2[SnCl_6]$	Sn: MVol.C3-107
$C_6Cl_6H_{18}S_2Sn$	$(C_3H_7SH_2)_2[SnCl_6]$	Sn: MVol.C6-80
$C_6Cl_6H_{20}IrN_2$	$((CH_3)_3NH)_2[IrCl_6]$	Ir: SVol.2-127
$C_6Cl_6H_{20}N_2Sn$	$[(CH_3)_3NH]_2[SnCl_6]$	Sn: MVol.C3-105
–	$[C_3H_7NH_3]_2[SnCl_6]$	Sn: MVol.C3-103/4
–	$[i\text{-}C_3H_7NH_3]_2[SnCl_6]$	Sn: Org.Verb.5-151
$C_6Cl_6H_{20}N_2Te$	$[(CH_3)_3NH]_2TeCl_6$	Te: SVol.B2-137/8
$C_6Cl_6H_{20}P_2U$	$(P(CH_3)_3H)_2UCl_6$	U: SVol.C9-58/9
$C_6Cl_6H_{24}Ir_2N_6$	$[Ir(C_2H_4(NH_2)_2)_2Cl_2][Ir(C_2H_4(NH_2)_2)Cl_4]$	Ir: SVol.2-85
$C_6Cl_6H_{24}Ir_2N_8S_2$	$[Ir(SC(NH_2)_2)_2(C_2H_4(NH_2)_2)_2][IrCl_6] \cdot 5\ H_2O$	Ir: SVol.2-168
$C_6Cl_6H_{24}MgN_{12}O_6Sn$	$[Mg(CO(NH_2)_2)_6][SnCl_6]$	Sn: MVol.C3-166
$C_6Cl_6N_6S_6$	$(ClSCN)_6$	C: MVol.D6-177/8
$C_6Cl_7H_5OPSn^-$	$[SnCl_5 \cdot C_6H_5POCl_2]^-$	Sn: MVol.C6-167
$C_6Cl_7H_{15}NOSSb$	$[(C_2H_5)_3NSOCl][SbCl_6]$	S: SVol.1-54
$C_6Cl_7H_{15}Si_3Sn$	$(CH_3SiCl_2CH_2)_3SnCl$	Sn: Org.Verb.5-154
$C_6Cl_9F_2N_3$	$NCF(CCl_3)NFC(CCl_3)NC(CCl_3)$	F: PerFHalOrg.6-75/6, 97
$C_6Cl_9F_4N_3$	$NCF(CCl_3)NFCF(CCl_3)NFC(CCl_3)$	F: PerFHalOrg.6-75/6
$C_6Cl_9F_6N_3$	$FN(CF(CCl_3)NF)_2CF(CCl_3)$	F: PerFHalOrg.6-75/6
$C_6Cl_9H_{15}SnTi_2$	$(C_2H_5)_3SnCl \cdot 2\ TiCl_4$	Sn: Org.Verb.5-106
$C_6Cl_9KO_6Sn$	$KSn(CCl_3COO)_3$	Sn: MVol.C3-78
$C_6Cl_9O_7Pa$	$PaO(Cl_3CCOO)_3$	Pa: SVol.2-167
$C_6Cl_{10}H_2Ni_2$	$(Cl_2CCHCCl_2NiCl)_2$	Ni: Org.Verb.2-318
$C_6CmH_6NO_6$	$Cm(N(CH_2COO)_3)$	Np: TrU.D1-163
$C_6CmH_6O_8^-$	$Cm(OH)(C_6H_5O_7)^-$	Np: TrU.D1-157, 159
$C_6CmH_7O_9^{2-}$	$Cm(OH)_2(C_6H_5O_7)^{2-}$	Np: TrU.D1-157, 159
$C_6CmH_8NO_7S$	$Cm(OSO_2(CH_2)_2N(CH_2COO)_2)$	Np: TrU.D1-163
$C_6CmH_9NO_5^+$	$CmN(C_2H_4OH)(CH_2COO)_2^+$	Np: TrU.D1-162
$C_6CmH_9NO_7S^+$	$Cm(OSO_2(CH_2)_2N(CH_2COO)_2H)^+$	Np: TrU.D1-163
$C_6CmH_{10}O_6^+$	$Cm(CH_3CH(OH)COO)_2^+$	Np: TrU.D1-154
$C_6CmO_{12}^{3-}$	$Cm(C_2O_4)_3^{3-}$	Np: TrU.D1-158

$C_6CoF_{18}P_3S_6$	$Co[S(S)P(CF_3)_2]_3$	F:	PerFHalOrg.3-159
$C_6CoH_9O_3S_6$	$Co[S_2COCH_3]_3$	C:	MVol.D4-259
$C_6CoH_{10}O_2S_4$	$Co[S_2COC_2H_5]_2 \cdot NC_5H_4C_5H_4N$	C:	MVol.D4-259
–	$Co[S_2COC_2H_5]_2 \cdot n\ C_{12}H_8N_2$	C:	MVol.D4-259
$C_6CoHg_2N_6Se_6$	$Co(SeCN)_2 \cdot 2\ Hg(SeCN)_2$	C:	MVol.D6-229
$C_6CrF_6H_4O_4P_2$	$Cr(CO)_4(CF_3PH_2)_2$	F:	PerFHalOrg.3-27
$C_6CrH_9O_3S_6$	$Cr[S_2COCH_3]_3$	C:	MVol.D4-258
$C_6CrH_9S_9$	$Cr[S_2CSCH_3]_3$	C:	MVol.D4-269
$C_6CrH_{14}N_2O_8U$	$UO_2CrO_4 \cdot 2\ HCON(CH_3)_2$	U:	SVol.E1-100/2
C_6CrO_6	$Cr(CO)_6$ systems		
	$Cr(CO)_6$-$Fe(CO)_5$	Fe:	Org.Verb.B3-64
$C_6Cr_2H_{18}N_4O_7S_4$	$[(NS(CH_3)_2NS(CH_3))_2]Cr_2O_7$	S:	S-N-Verb.1-5
$C_6CsFeGdN_6$	$CsGd[Fe(CN)_6]$	Sc:	MVol.B7-30
$C_6CsFeLaN_6$	$CsLa[Fe(CN)_6]$	Sc:	MVol.B7-30
C_6CsFeN_6Nd	$CsNd[Fe(CN)_6]$	Sc:	MVol.B7-30
C_6CsFeN_6Pr	$CsPr[Fe(CN)_6]$	Sc:	MVol.B7-30
C_6CsFeN_6Sm	$CsSm[Fe(CN)_6]$	Sc:	MVol.B7-30
–	$CsSm[Fe(CN)_6] \cdot 4\ H_2O$	Sc:	MVol.B7-30
C_6CsFeN_6Y	$CsY[Fe(CN)_6]$	Sc:	MVol.B7-30
$C_6CsH_9O_6Sn$	$CsSn(CH_3COO)_3$	Sn:	MVol.C3-155
$C_6Cs_2IrLiN_6$	$LiCs_2[Ir(CN)_6]$	Ir:	SVol.2-181/2
$C_6Cs_3IrN_6$	$Cs_3[Ir(CN)_6]$	Ir:	SVol.2-182
C_6CuF_5	CuC_6F_5	F:	PerFHalOrg.4-110, 111
$C_6CuH_{11}OS_2$	$Cu[S_2COC_5H_{11}]$	C:	MVol.D4-256
$C_6CuH_{11}S_3$	$Cu[S_2CSC_5H_{11}]$	C:	MVol.D4-269
$C_6CuH_{16}N_6Se_2$	$Cu(NH_2C_2H_4NH_2)_2(SeCN)_2$	C:	MVol.D6-230
$C_6CuHg_2N_6Se_6$	$Cu(SeCN)_2 \cdot 2\ Hg(SeCN)_2$	C:	MVol.D6-229
$C_6FFeH_4O^+$	$[C_5H_4COFeF]^+$	Fe:	Org.Verb.A2-287
$C_6FFeH_6O_6P$	$(CO)_4FePF(OCH_3)_2$	Fe:	Org.Verb.B2-82, 113
$C_6FFeN_2O_4PS_2$	$(CO)_4FePF(SCN)_2$	Fe:	Org.Verb.B2-85, 93
C_6FH_7Ti	$CH_3C_5H_4TiF$	Ti:	Org.Verb.1-133
C_6FH_9Sn	$(CH_2CH)_3SnF$	Sn:	Org.Verb.5-37
$C_6FH_{10}NO_3S_2$	$FSO_2NHS(O)CHC(CH_3)C(CH_3)CH_2$	S:	S-N-Verb.1-106
$C_6FH_{10}N_3OS$	$C_5H_{10}NS(O)FNCN$	S:	S-N-Verb.1-196
$C_6FH_{13}N_2OS$	$C_5H_{10}NS(O)FNCH_3$	S:	S-N-Verb.1-193
$C_6FH_{13}N_2O_2S$	$FSO_2NC(N(C_2H_5)_2)CH_3$	S:	S-N-Verb.1-105
$C_6FH_{13}N_2O_3S$	$FSO_2N(CH_3)C(O)N(C_2H_5)_2$	S:	S-N-Verb.1-104
–	$FSO_2NC[N(C_2H_5)_2]OCH_3$	S:	S-N-Verb.1-119
$C_6FH_{13}N_2O_3Sn$	$[SnF(NH_2C_2H_4N(C_2H_4OH)CH_2COO)]$	Sn:	MVol.C5-31
$C_6FH_{15}OSn$	$(C_2H_5)_2Sn(F)OC_2H_5$	Sn:	Org.Verb.5-46
$C_6FH_{15}Sn$	$(C_2H_5)_3SnF$	Sn:	Org.Verb.5-19/22
$C_6FH_{18}N_2SSi_2{}^+$	$SF[NSi(CH_3)_3]_2{}^+$	S:	S-N-Verb.1-201
$C_6F_2{}^+$	$C_6F_2{}^+$	F:	PerFHalOrg.3-172
$C_6F_2FeH_2O_4$	$CH_2CF_2Fe(CO)_4$	Fe:	Org.Verb.B4-259, 264
$C_6F_2FeH_5O_5P$	$(CO)_4FePF_2(OC_2H_5)$	Fe:	Org.Verb.B2-82, 113
$C_6F_2H_4O_3$	$OC_4H_3COOCHF_2$	F:	PerFHalOrg.5-18
$C_6F_2H_5LiN_4O_3S_3$	$[NS(O)F]_2[NS(O)NLiC_6H_5]$	S:	S-N-Verb.1-14
$C_6F_2H_5NOS$	$C_6H_5NS(O)F_2$	S:	S-N-Verb.1-63
–	$NSF_2(OC_6H_5)$	S:	S-N-Verb.1-155
$C_6F_2H_5NS$	$C_6H_5NSF_2$	S:	SVol.2-85

$C_6F_2H_5N_3O_3S_3$	$[NS(O)F]_2[NS(O)C_6H_5]$	S:	S-N-Verb.1-14
$C_6F_2H_6N_2O_2$	$NCFC(OCH_3)C(OCH_3)CFN$	F:	PerFHalOrg.6-57, 61
–	$NCFNC(OCH_3)CFC(OCH_3)$	F:	PerFHalOrg.6-58, 62
–	$NC(OCH_3)CFCFC(OCH_3)N$	F:	PerFHalOrg.6-57/8, 61
–	$NC(OCH_3)C(OCH_3)NCFCF$	F:	PerFHalOrg.6-58, 63
$C_6F_2H_6N_4O_2$	$(NH_2)_3(NO_2)C_6F_2$	F:	PerFHalOrg.7-11/2, 40
$C_6F_2H_6N_4O_3S_3$	$[NS(O)F]_2[NS(O)NHC_6H_5]$	S:	S-N-Verb.1-14
$C_6F_2H_8N_2O_4$	$O_2NCF_2C(O)NOC(CH_3)_2CH_2$	F:	PerFHalOrg.7-194
–	$O_2NCF_2C(O)N(OH)CH_2C(CH_3)CH_2$	F:	PerFHalOrg.7-194
$C_6F_2H_8N_4$	$FC(NCNHCH_3)_2CF$	F:	PerFHalOrg.6-59, 66
$C_6F_2H_{10}Ni_2$	$(C_3H_5NiF)_2$	Ni:	Org.Verb.2-274
$C_6F_2H_{10}O_2$	$(n\text{-}C_4H_9)COOCF_2H$	F:	PerFHalOrg.5-18
$C_6F_2H_{10}Sn$	$(CH_2CHCH_2)_2SnF_2$	Sn:	Org.Verb.5-49/50
–	$(CH_3)_2Sn(CHCHF)_2$	Sn:	Org.Verb.3-7, 16
$C_6F_2H_{12}NiO_2P_2$	$(CO)_2Ni(P(CH_3)_2F)_2$	Ni:	Org.Verb.1-236
$C_6F_2H_{14}N_2O_3Sn$	$SnF_2 \cdot NH_2C_2H_4N(C_2H_4OH)CH_2COOH$	Sn:	MVol.C5-31
$C_6F_2H_{14}O_6S_2Sn$	$(C_3H_7)_2Sn(SO_3F)_2$	Sn:	Org.Verb.6-68
$C_6F_2H_{14}Sn$	$(C_3H_7)_2SnF_2$	Sn:	Org.Verb.5-46
–	$(i\text{-}C_3H_7)_2SnF_2$	Sn:	Org.Verb.5-49
$C_6F_2H_{16}OSi_2Sn$	$O(Si(CH_3)_2CH_2)_2SnF_2$	Sn:	Org.Verb.6-207
$C_6F_2H_{18}N_2SSi_2$	$SF_2[NSi(CH_3)_3]_2$	S:	S-N-Verb.1-201
$C_6F_2H_{18}N_2SSi_2^+$	$SF_2[NSi(CH_3)_3]_2^+$	S:	S-N-Verb.1-201
$C_6F_2H_{18}N_3P$	$[(CH_3)_2N]_3PF_2$	F:	PerFHalOrg.5-23
$C_6F_2H_{19}N_3O_3S_4$	$[(CH_3)_2S(O)N]_3S[HF_2]$	S:	SVol.2-78
$C_6F_3^+$	$C_6F_3^+$	F:	PerFHalOrg.3-172, 216
$C_6F_3FeHO_4$	$F_2CCHFFe(CO)_4$	Fe:	Org.Verb.B4-315/6
$C_6F_3FeH_9O_6P_2$	$(CO)_3FeP_2F_3(OCH_3)_3$	Fe:	Org.Verb.B1-161
$C_6F_3FeIO_4$	$(CO)_4Fe(CFCF_2)I$	Fe:	Org.Verb.B2-179, 181
$C_6F_3GeH_9$	$CF_3CCGe(CH_3)_3$	F:	PerFHalOrg.4-74
$C_6F_3H_3INO$	$(I)(CH_3O)C_5F_3N$	F:	PerFHalOrg.5-134/5
$C_6F_3H_3N_2O_3$	$(NO_2)(CH_3O)C_5F_3N$	F:	PerFHalOrg.5-214
$C_6F_3H_4NO_3S$	$(NO_2C_6H_4O)SF_3$	S:	SVol.2-83
$C_6F_3H_4N_3$	$CF_3NNC_5H_4N$	F:	PerFHalOrg.7-160
$C_6F_3H_4N_3OS_2$	$NC(SH)NC(SH)C(NHCHO)C(CF_3)$	F:	PerFHalOrg.6-70
$C_6F_3H_4N_3O_2$	$(NH_2)_2(NO_2)C_6F_3$	F:	PerFHalOrg.7-11/3, 39, 55
$C_6F_3H_5NOSSi^+$	$C_6H_5SiF(NSOF_2)^+$	S:	S-N-Verb.1-83
$C_6F_3H_5NO_2PS$	$FSO_2NPF_2C_6H_5$	S:	S-N-Verb.1-120
$C_6F_3H_5N_2O$	$NCFCFCFC(O)N(C_2H_5)$	F:	PerFHalOrg.6-58, 62
–	$NCFCFNCFC(OC_2H_5)$	F:	PerFHalOrg.6-63
–	$(NH_2)(CH_3O)C_5F_3N$	F:	PerFHalOrg.5-214/5
$C_6F_3H_5N_2O_2$	$NCFCFNCFC(OCH_2CH_2OH)$	F:	PerFHalOrg.6-58, 64
$C_6F_3H_5N_4O_3$	$NC(OCH_3)NC(NH_2)C(NO_2)C(CF_3)$	F:	PerFHalOrg.6-58
$C_6F_3H_5N_4S$	$(CF_3)(C_2H_5)C_3N_4S$	F:	PerFHalOrg.5-100
$C_6F_3H_5OS$	$(C_6H_5O)SF_3$	S:	SVol.2-83
$C_6F_3H_5Sn$	$C_6H_5SnF_3$	Sn:	Org.Verb.5-52/3
$C_6F_3H_6KO_6Sn$	$KSn(CH_2FCOO)_3$	Sn:	MVol.C3-77
$C_6F_3H_6N_3$	$NCFCFNCFC(N(CH_3)_2)$	F:	PerFHalOrg.6-60, 67
–	$NCFNC(N(CH_3)_2)CFCF$	F:	PerFHalOrg.6-59, 66
–	$(NH_2)_3C_6F_3$	F:	PerFHalOrg.7-12/3, 40

$C_6F_3H_6NiO_3P$	$(CO)_3NiP(CF_3)(CH_3)_2$	Ni:	Org.Verb.1-165
$C_6F_3H_7N_4OS$	$SNC(CF_3)NC(NHCON(CH_3)_2)$	F:	PerFHalOrg.5-101
$C_6F_3H_7N_4S$	$NC(SCH_3)NC(NH_2)C(NH_2)C(CF_3)$	F:	PerFHalOrg.6-70
$C_6F_3H_7O$	$CF_3CCC(CH_3)_2OH$	F:	PerFHalOrg.4-46, 96
–	$CF_3CC(OH)CHC_2H_5$	F:	PerFHalOrg.4-47
$C_6F_3H_8NO_2$	$CF_3C(O)NOC(CH_3)_2CH_2$	F:	PerFHalOrg.7-194
–	$CF_3C(O)N(OH)CH_2C(CH_3)CH_2$	F:	PerFHalOrg.7-194
–	$CF_3N(OH)CH_2C(CH_2)C(O)CH_3$	F:	PerFHalOrg.7-166/7
$C_6F_3H_8NO_3$	$CF_3N(OH)CHC(CH_3)COOCH_3$	F:	PerFHalOrg.7-163, 168
–	$CF_3N(OH)CH_2C(CH_2)C(O)OCH_3$	F:	PerFHalOrg.7-165, 168
$C_6F_3H_8O_2P$	$CF_3(CH_2CCH)P(O)OC_2H_5$	F:	PerFHalOrg.3-101
$C_6F_3H_9Si$	$(CH_3)_3SiCCCF_3$	F:	PerFHalOrg.4-74
$C_6F_3H_9Sn$	$(CH_3)_3SnCCCF_3$	Sn:	Org.Verb.2-111/2
$C_6F_3H_{10}NO_6Sn$	$NH_4Sn(CH_2FCOO)_3$	Sn:	MVol.C3-124/5
$C_6F_3H_{10}N_2O_6P$	$NO_2CF_2CFNOPO(OC_2H_5)_2$	F:	PerFHalOrg.7-199
$C_6F_3H_{11}N_2O$	$CF_3C(NOH)N(C_2H_5)_2$	F:	PerFHalOrg.7-146
$C_6F_3H_{11}N_2O_4$	$NO_2CF_2CFNOH \cdot (C_2H_5)_2O$	F:	PerFHalOrg.7-88, 105
$C_6F_3H_{11}OSi$	$CF_2CFSi(OC_2H_5)(CH_3)_2$	F:	PerFHalOrg.4-19
$C_6F_3H_{11}Sn$	$(CH_3)_3SnC(CF_3)CH_2$	Sn:	Org.Verb.2-77
–	$(CH_3)_3SnCHCHCF_3$	Sn:	Org.Verb.2-77
$C_6F_3H_{13}Sn$	$(CH_3)_3SnCH_2CH_2CF_3$	Sn:	Org.Verb.2-43
–	$C_6H_{13}SnF_3$	Sn:	Org.Verb.5-52
$C_6F_3H_{15}N_2O_3Sn$	$[NH_2C_2H_4NH(C_2H_4OH)CH_2COOH][SnF_3]$	Sn:	MVol.C5-31
$C_6F_3Li_3$	$Li_3C_6F_3$	F:	PerFHalOrg.4-5/6, 25
$C_6F_3N_5$	$(CF_3)(NC)_2C_3N_3$	F:	PerFHalOrg.6-79/80, 101
C_6F_4	C_6F_4	F:	PerFHalOrg.4-70
$C_6F_4^+$	$C_6F_4^+$	F:	PerFHalOrg.3-172
$C_6F_4FeO_4$	$CF_2CF_2Fe(CO)_4$	Fe:	Org.Verb.B4-320, 322/4
$C_6F_4Fe_2O_6P_2$	$(CO)_3Fe(PF_2)_2Fe(CO)_3$	Fe:	Org.Verb.C1-164/6, 170
C_6F_4HNO	$CHOC_5F_4N$	F:	PerFHalOrg.5-142, 217
$C_6F_4HNO_2$	NC_5F_4COOH	F:	PerFHalOrg.5-192/3, 206/7, 211, 217
$C_6F_4HN_3$	C_6F_4NHNN	F:	PerFHalOrg.6-128, 140, 160
$C_6F_4HN_3O$	$C_6F_4N(OH)NN$	F:	PerFHalOrg.6-128, 140
$C_6F_4H_2IN$	$IC_6F_4NH_2$	F:	PerFHalOrg.7-9, 37
$C_6F_4H_2LiN$	$NH_2C_6F_4Li$	F:	PerFHalOrg.4-5/6, 13, 14
$C_6F_4H_2N_2O$	$NH_2C(O)C_5F_4N$	F:	PerFHalOrg.5-194, 208
$C_6F_4H_2N_2O_2$	$NO_2C_6F_4NH_2$	F:	PerFHalOrg.7-9, 36, 55, 60/1
$C_6F_4H_3N$	$CH_3C_5F_4N$	F:	PerFHalOrg.5-141
–	$C_6F_4HNH_2$	F:	PerFHalOrg.7-55, 57
$C_6F_4H_3NO$	$CH_3OC_5F_4N$	F:	PerFHalOrg.5-135, 211/2
–	$HOC_6F_4NH_2$	F:	PerFHalOrg.7-8/9, 36, 60
$C_6F_4H_4N_2$	$(NH_2)_2C_6F_4$	F:	PerFHalOrg.7-11/2, 39, 55, 61
$C_6F_4H_4N_2S_2$	$F_2SNC_6H_4NSF_2$	S:	SVol.2-78, 85
$C_6F_4H_4OS$	$(FC_6H_4O)SF_3$	S:	SVol.2-83
$C_6F_4H_4O_2$	$CH_3C(CF_3)CCFCOOH$	F:	PerFHalOrg.4-89

$C_6F_4H_5N_2O_2S_2Si^+$	$C_6H_5Si(NSOF_2)_2^+$	S:	S-N-Verb.1-83
$C_6F_4H_5N_3$	$NH_2NHC_6F_4NH_2$	F:	PerFHalOrg.7-9, 36, 59
$C_6F_4H_6$	$CH_2CF_2CHCHCF_2CH_2$	F:	PerFHalOrg.5-16
$C_6F_4H_6N_2O_3$	$CH_2ON(CF_2CF_2NO_2)CH_2CHCH$	F:	PerFHalOrg.7-194
$C_6F_4H_7NSn$	$SnF_4 \cdot CH_3C_5H_4N$	Sn:	MVol.C5-201
$C_6F_4H_8Sn$	$(CH_3)_2Sn(CFCHF)_2$	Sn:	Org.Verb.3-6, 16
–	$(CH_3)_2Sn(CHCF_2)_2$	Sn:	Org.Verb.3-7, 16
$C_6F_4H_{10}NO_4P$	$CF_3CFNOPO(OC_2H_5)_2$	F:	PerFHalOrg.7-199
$C_6F_4H_{10}N_2O_2S$	$(C_2H_5)_2NS(O)FNC(O)CF_3$	S:	S-N-Verb.1-194
–	$FSO_2NC[N(C_2H_5)_2]CF_3$	S:	S-N-Verb.1-118
$C_6F_4H_{11}NO_2$	$CF_3CFNOH \cdot (C_2H_5)_2O$	F:	PerFHalOrg.7-88, 105
$C_6F_4H_{12}N_2NiO_2P_2$	$(CO)_2Ni(P(N(CH_3)_2)F_2)_2$	Ni:	Org.Verb.1-126
$C_6F_4H_{13}N_3O_2S$	$CF_3CF[N(CH_3)_2]SO_2NHN(CH_3)_2$	F:	PerFHalOrg.7-193
$C_6F_4H_{14}N_2O_2Sn$	$SnF_4 \cdot 2\ (CH_3)_2CNOH$	Sn:	MVol.C6-36
–	$SnF_4 \cdot 2\ OCHN(CH_3)_2$	Sn:	MVol.C6-41
$C_6F_4H_{14}N_2O_3Sn_2$	$2\ SnF_2 \cdot NH_2C_2H_4N(C_2H_4OH)CH_2COOH$	Sn:	MVol.C5-31
$C_6F_4H_{14}O_4P_2Sn$	$(C_3H_7)_2Sn(PO_2F_2)_2$	Sn:	Org.Verb.6-68
$C_6F_4H_{16}N_2Sn$	$SnF_4 \cdot C_2H_4(N(CH_3)_2)_2$	Sn:	MVol.C5-184
$C_6F_4H_{16}N_2Te$	$(CH_3)_2N(C_2H_5)_2NTeF_4$	Te:	SVol.B2-66
$C_6F_4H_{18}N_2Sn$	$SnF_4 \cdot 2\ (CH_3)_3N$	Sn:	MVol.C5-171
$C_6F_4H_{18}O_2P_2Sn$	$SnF_4 \cdot 2\ (CH_3)_3PO$	Sn:	MVol.C6-143/4
C_6F_4ILi	IC_6F_4Li	F:	PerFHalOrg.4-5/6
C_6F_4INO	NOC_6F_4I	F:	PerFHalOrg.7-180
$C_6F_4Li_2$	$Li_2C_6F_4$	F:	PerFHalOrg.4-5, 13, 15, 24/5, 29, 39
$C_6F_4Li_2O$	$(LiO)C_6F_4Li$	F:	PerFHalOrg.4-6
$C_6F_4Li_2S$	LiC_6F_4SLi	F:	PerFHalOrg.4-5/7, 13
$C_6F_4NO_2Tl$	NC_5F_4COOTl	F:	PerFHalOrg.5-192/3, 207
$C_6F_4N_2$	NC_5F_4CN	F:	PerFHalOrg.5-195/7, 209, 217
$C_6F_4O_2$	$C_6F_4O_2$	F:	PerFHalOrg.7-60
$C_6F_5^+$	$C_6F_5^+$	F:	PerFHalOrg.3-172
$C_6F_5Fe^+$	$[FeC_6F_5]^+$	Fe:	Org.Verb.A2-287
$C_6F_5FeIO_4$	$(CO)_4Fe(C_2F_5)I$	Fe:	Org.Verb.B2-177/8, 180
$C_6F_5GeH_3$	$C_6F_5GeH_3$	F:	PerFHalOrg.4-165, 174
$C_6F_5GeO_{1.5}$	$(C_6F_5GeO_{1.5})_m$	F:	PerFHalOrg.4-166
C_6F_5HNNa	C_6F_5NHNa	F:	PerFHalOrg.7-60
$C_6F_5HN_2$	C_6F_5NNH	F:	PerFHalOrg.7-61/2
$C_6F_5HN_4$	$CF_3C_5F_2HN_4$	F:	PerFHalOrg.6-127, 139, 162
$C_6F_5HNaO_2P$	$C_6F_5PH(O)ONa$	F:	PerFHalOrg.3-46
$C_6F_5HO_2Sn$	$C_6F_5Sn(O)OH$	F:	PerFHalOrg.4-181, 185, 188
–	$(C_6F_5SnOOH)_x$	Sn:	Org.Verb.6-283
$C_6F_5H_2N$	$C_6F_5NH_2$	F:	PerFHalOrg.7-8/9, 27, 35/6, 53/5, 58/62, 76
$C_6F_5H_2O_2P$	$C_6F_5P(OH)_2$	F:	PerFHalOrg.3-46, 48
$C_6F_5H_2P$	$C_6F_5PH_2$	F:	PerFHalOrg.3-15, 18, 24/5
$C_6F_5H_3N_2O$	$NCFNC(OCH_3)CFC(CF_3)$	F:	PerFHalOrg.6-63

$C_6F_5H_3N_2O$	$NCFNCFCCF_3C(OCH_3)$	F:	PerFHalOrg.6-58, 63
–	$NC(OCH_3)NCFC(CF_3)CF$	F:	PerFHalOrg.6-58, 63
$C_6F_5H_5O$	$FC(CF_2)_2CCH(OH)CH_3$	F:	PerFHalOrg.4-44
$C_6F_5H_5S$	$C_6H_5SF_5$	S:	SVol.2-200
$C_6F_5H_7NiO_4$	$(C_4H_7O_2F_2NiO_2CCF_3)_n$	Ni:	Org.Verb.2-296
C_6F_5I	C_6F_5I	F:	PerFHalOrg.4-38, 107, 111
C_6F_5IMg	C_6F_5MgI	F:	PerFHalOrg.4-63/5, 67/8, 71, 99, 108/9, 112
$C_6F_5IN_2$	C_6F_5NNI	F:	PerFHalOrg.7-61/2
$C_6F_5I_2P$	$C_6F_5PI_2$	F:	PerFHalOrg.3-125, 128
$C_6F_5I_3Tl^-$	$[C_6F_5TlI_3]^-$	F:	PerFHalOrg.4-137
C_6F_5Li	C_6F_5Li	F:	PerFHalOrg.4-5, 9/12, 14/7, 19/24, 28, 30/1, 34/9, 47/53
C_6F_5NO	C_6F_5NO	F:	PerFHalOrg.7-180, 181, 186/95, 209
–	$NC_5F_4(COF)$	F:	PerFHalOrg.5-194, 208
$C_6F_5O_{1.5}Si$	$(C_6F_5SiO_{1.5})_x$	F:	PerFHalOrg.4-147/8, 154, 161/2
$C_6F_5P^+$	$C_6F_5P^+$	F:	PerFHalOrg.3-172, 179
$C_6F_5Pb^+$	$C_6F_5Pb^+$	F:	PerFHalOrg.4-191
$C_6F_5Sn^+$	$C_6F_5Sn^+$	F:	PerFHalOrg.4-185
		Sn:	Org.Verb.1-168
$C_6F_6FeHO_4P$	$(CO)_4FeP(CF_3)_2H$	Fe:	Org.Verb.B2-88, 108/9
$C_6F_6FeH_{10}P_2$	$(C_3H_5)_2Fe(PF_3)_2$	Fe:	Org.Verb.B5-107
$C_6F_6GeH_9N$	$(CH_3)_3Ge[NC(CF_3)_2]$	F:	PerFHalOrg.7-64
$C_6F_6H_2N_2$	$(CF_3)(NH_2)C_5F_3N$	F:	PerFHalOrg.5-186, 202
$C_6F_6H_3KO_6Sn$	$KSn(CHF_2COO)_3$	Sn:	MVol.C3-77
$C_6F_6H_3NiO_3P$	$(CO)_3NiP(CF_3)_2CH_3$	Ni:	Org.Verb.1-165, 168
$C_6F_6H_4MoO_4P_2$	$Mo(CO)_4(CF_3PH_2)_2$	F:	PerFHalOrg.3-27
$C_6F_6H_4O$	$CF_3CCC(OH)(CF_3)CH_3$	F:	PerFHalOrg.4-46
$C_6F_6H_5N$	$CH_2CCHC(CF_3)_2NH_2$	F:	PerFHalOrg.7-65
$C_6F_6H_5NOS$	$SC(CF_3)_2NHC(O)CH(CH_3)$	F:	PerFHalOrg.7-66
–	$SC(CF_3)_2NHC(O)CH_2CH_2$	F:	PerFHalOrg.7-66, 68
$C_6F_6H_5NO_3$	$(CF_3)_2C(NO)C(O)OC_2H_5$	F:	PerFHalOrg.7-141/2, 191
$C_6F_6H_6Si$	$(CF_2CF)_2Si(CH_3)_2$	F:	PerFHalOrg.4-19
$C_6F_6H_6Sn$	$(CH_3)_2Sn(CFCF_2)_2$	Sn:	Org.Verb.3-5/6
$C_6F_6H_7N$	$CH_2CHCH_2C(CF_3)_2NH_2$	F:	PerFHalOrg.7-65, 72, 76
$C_6F_6H_7NO$	$(CF_3)_2NOCH_2C(CH_3)CH_2$	F:	PerFHalOrg.7-133, 134
$C_6F_6H_7NO_2$	$(CF_3)_2CNOH \cdot OC(CH_3)_2$	F:	PerFHalOrg.7-89, 141
$C_6F_6H_8N_2O_2$	$CF_3N(OH)CH_2C(CH_3)CH_2ON(CF_3)$	F:	PerFHalOrg.7-164, 167
–	$(CF_3)_2CNOH \cdot HC(O)N(CH_3)_2$	F:	PerFHalOrg.7-141
$C_6F_6H_8N_4$	$CF_3NN(CH_2)_4NNCF_3$	F:	PerFHalOrg.7-159
$C_6F_6H_9NO$	$(CF_3)_2NOC(CH_3)_3$	F:	PerFHalOrg.7-129/30, 133
–	$(CF_3)_2NOCH_2CH(CH_3)_2$	F:	PerFHalOrg.7-131
$C_6F_6H_9NOSi$	$(CF_3)_2CNOSi(CH_3)_3$	F:	PerFHalOrg.7-143
$C_6F_6H_9NSi$	$(CH_3)_3SiNC(CF_3)_2$	F:	PerFHalOrg.7-64
$C_6F_6H_9NSn$	$(CH_3)_3SnNC(CF_3)_2$	F:	PerFHalOrg.7-64
		Sn:	Org.Verb.5-84

$C_6F_6H_9P$	$(CF_3)_2PCH(CH_3)CH_2CH_3$	F:	PerFHalOrg.3-101, 105
–	$(CF_3)_2PCH_2CH(CH_3)_2$	F:	PerFHalOrg.3-102
–	$(CF_3)_2P(n\text{-}C_4H_9)$	F:	PerFHalOrg.3-28, 101, 104
$C_6F_6H_9PS$	$(CF_3)_2PSC(CH_3)_3$	F:	PerFHalOrg.3-101, 107
$C_6F_6H_{10}NO_3P$	$(CF_3)_2NOP(OC_2H_5)_2$	F:	PerFHalOrg.7-126
$C_6F_6H_{10}NP$	$(CF_3)_2PNHC(CH_3)_3$	F:	PerFHalOrg.3-108, 115
$C_6F_6H_{10}O_6S_2Sn$	$(C_2H_5)_2Sn(SO_3CF_3)_2$	Sn:	Org.Verb.6-56
$C_6F_6H_{10}Sn$	$(CH_3)_3SnCF(CF_3)CHF_2$	Sn:	Org.Verb.2-43, 46
–	$(CH_3)_3SnCF_2CHFCF_3$	Sn:	Org.Verb.2-42/3
–	$(CH_3)_3SnCH(CF_3)_2$	Sn:	Org.Verb.2-43, 46
$C_6F_6H_{11}NOSi$	$(CH_3)_3SiCH_2ON(CF_3)_2$	F:	PerFHalOrg.7-127
$C_6F_6H_{11}NO_4Si$	$(CF_3)_2NOCH_2OSi(OCH_3)_2CH_3$	F:	PerFHalOrg.7-127
–	$(CF_3)_2NOCH_2Si(OCH_3)_3$	F:	PerFHalOrg.7-127
$C_6F_6H_{11}N_3O_2$	$2\ CF_3NO \cdot (C_2H_5)_2NH$	F:	PerFHalOrg.7-160
$C_6F_6H_{12}NOPS$	$[(CH_3)_4N][(CF_3)_2P(S)O]$	F:	PerFHalOrg.3-153
$C_6F_6H_{12}N_2P_2$	$(CF_3)_2PNP(C_2H_5)_2NH_2$	F:	PerFHalOrg.3-69
$C_6F_6H_{12}N_3OP$	$(CF_3)_2NOP[N(CH_3)_2]_2$	F:	PerFHalOrg.7-128
$C_6F_6H_{16}N_2Te$	$TeF_6 \cdot (CH_3)_2NCH_2CH_2N(CH_3)_2$	Te:	SVol.B3-183
$C_6F_6H_{18}N_2Te$	$TeF_6 \cdot 2\ N(CH_3)_3$	Te:	SVol.B3-183
C_6F_6Mg	C_6F_5MgF	F:	PerFHalOrg.4-63/4, 71, 79
$C_6F_6NOS^+$	$C_6F_5NSOF^+$	S:	S-N-Verb.1-63
$C_6F_6N_2$	C_6F_5NNF	F:	PerFHalOrg.7-61/2
C_6F_6Ni	$(NiC_6F_6)_x$	Ni:	Org.Verb.2-400
$C_6F_6P^+$	$C_6F_5PF^+$	F:	PerFHalOrg.3-172
$C_6F_6Pb^+$	$C_6F_5PbF^+$	F:	PerFHalOrg.4-191
$C_6F_6Si^+$	$C_6F_5SiF^+$	F:	PerFHalOrg.4-153
$C_6F_7FeO_4P$	$(CO)_4FeP(CF_3)_2F$	Fe:	Org.Verb.B2-82
$C_6F_7Ge^-$	$C_6F_5GeF_2^-$	F:	PerFHalOrg.4-174
$C_6F_7HMnO_4P$	$HMn(CO)_4P(CF_3)_2F$	F:	PerFHalOrg.3-110/1, 120
$C_6F_7HO_2$	$HOC(CF_2)_3C(O)CF$	F:	PerFHalOrg.7-57
$C_6F_7H_2NO$	$CF_2C(O)CFC(NH_2)CF_2CF_2$	F:	PerFHalOrg.7-8, 35
$C_6F_7H_3N_2$	$CF_2C(NH)CFC(NH_2)CF_2CF_2$	F:	PerFHalOrg.7-8, 35
–	$CF_3CC(NH)CF_2CF_2C(NH_2)$	F:	PerFHalOrg.7-8, 35
$C_6F_7H_4N_5$	$(C_3F_7)(NH_2)_2C_3N_3$	F:	PerFHalOrg.6-79/80, 101, 112
$C_6F_7H_5O$	$C_3F_7C(O)C_2H_5$	F:	PerFHalOrg.4-93, 96
$C_6F_7H_7N_2O_2$	$C_3F_7C(NH)NH_2 \cdot CH_3COOH$	F:	PerFHalOrg.7-57
$C_6F_7H_7O$	$(CF_3)_2CFCH(OH)C_2H_5$	F:	PerFHalOrg.4-42, 90
–	$(CF_3)_2CFC(OH)(CH_3)_2$	F:	PerFHalOrg.4-42
–	$C_3F_7CH(OH)C_2H_5$	F:	PerFHalOrg.4-41, 94, 96
–	$C_3F_7C(OH)(CH_3)_2$	F:	PerFHalOrg.4-41, 92
$C_6F_7H_7Sn$	$(CH_3)_2Sn(CF_2CHF_2)(CFCF_2)$	Sn:	Org.Verb.3-82
$C_6F_7H_9Si$	$CF_3CF_2CF_2Si(CH_3)_3$	F:	PerFHalOrg.4-18
–	$(CF_3)_2CFSi(CH_3)_3$	F:	PerFHalOrg.4-18
$C_6F_7H_9Sn$	$(CH_3)_3SnCF(CF_3)_2$	Sn:	Org.Verb.2-43, 46
$C_6F_7I_2N_3$	$(CF_2I)_2(CF_3)C_3N_3$	F:	PerFHalOrg.6-81/2
C_6F_7N	$CF_3C_5F_4N$	F:	PerFHalOrg.5-148/9, 159, 177

C_6F_7NOS	$C_6F_5NS(O)F_2$	S:	S-N-Verb.1-63
$C_6F_7NOS^+$	$C_6F_5NSOF_2{}^+$	S:	S-N-Verb.1-63
C_6F_7OP	$C_6F_5P(O)F_2$	F:	PerFHalOrg.3-52, 55
C_6F_7P	$C_6F_5PF_2$	F:	PerFHalOrg.3-123, 125/6, 128
C_6F_7PS	$C_6F_5P(S)F_2$	F:	PerFHalOrg.3-141, 148
$C_6F_7Sn^+$	$C_6F_5SnF_2{}^+$	F:	PerFHalOrg.4-184/5
		Sn:	Org.Verb.1-168
C_6F_8	$C(CFCF_2)_2CF_2$	F:	PerFHalOrg.4-12
C_6F_8Ge	$C_6F_5GeF_3$	F:	PerFHalOrg.4-166/7, 176
$C_6F_8H_2N_2$	$C_3F_7C(NH_2)CFCN$	F:	PerFHalOrg.7-3
$C_6F_8H_4N_2O_2$	$NH_2C(O)(CF_2)_4C(O)NH_2$	F:	PerFHalOrg.7-18, 20, 48
$C_6F_8H_5N_3O$	$NH_2CO(CF_2)_4CNHNH_2$	F:	PerFHalOrg.7-4/5, 31
$C_6F_8H_5SSb$	$C_6H_5SbF_4 \cdot SF_4$	S:	SVol.2-86
$C_6F_8H_6N_4$	$NH_2CNH(CF_2)_4CNHNH_2$	F:	PerFHalOrg.7-4/5
$C_6F_8H_6N_4O$	$NH_2(NH)CCF(CF_3)O(CF_2)_2C(NH)NH_2$	F:	PerFHalOrg.7-5
$C_6F_8H_6N_4O_2$	$NH_2(NOH)C(CF_2)_4C(NOH)NH_2$	F:	PerFHalOrg.7-5/6, 32
$C_6F_8H_8N_6$	$(CF_2)_4(C(NH_2)NNH_2)_2$	F:	PerFHalOrg.7-6/7, 33
$C_6F_8H_8Sn$	$(CH_3)_2Sn(CF_2CHF_2)_2$	Sn:	Org.Verb.3-7, 16
$C_6F_8H_{10}Si_2$	$C_2F_4(CF_2SiH_2CH_3)_2$	F:	PerFHalOrg.4-87
$C_6F_8H_{16}N_2Te_2$	$2\ TeF_4 \cdot (CH_3)_2NCH_2CH_2N(CH_3)_2$	Te:	SVol.B3-168
$C_6F_8H_{18}N_2Te_2$	$[TeF_3 \cdot 2\ N(CH_3)_3][TeF_5]$	Te:	SVol.B3-168
$C_6F_8N_2$	$C_2F_5C_4F_3N_2$	F:	PerFHalOrg.6-32/3, 53, 55/6
–	$NC(C_2F_5)CFNCFCF$	F:	PerFHalOrg.6-30/2, 51
–	$NCFC(C_2F_5)CFCFN$	F:	PerFHalOrg.6-15/6, 35, 55
–	$NCFNC(C_2F_5)CFCF$	F:	PerFHalOrg.6-22, 43/4
–	$NCFNCFC(C_2F_5)CF$	F:	PerFHalOrg.6-22, 44
$C_6F_8N_2O_5$	$[ONCF_2CF_2C(O)]_2O$	F:	PerFHalOrg.7-177/8, 184
$C_6F_8N_4S$	$(C_2F_5)(CF_3)C_3N_4S$	F:	PerFHalOrg.6-123/4, 138
C_6F_8Si	$C_6F_5SiF_3$	F:	PerFHalOrg.4-148, 155, 161
$C_6F_9FeH_{10}P_3$	$C_2H_5CHCHCHCH_2Fe(PF_3)_3$	Fe:	Org.Verb.B1-8
C_6F_9HO	$(CF_2CF)_2C(OH)CF_3$	F:	PerFHalOrg.4-45
–	$CF_3CCC(CF_3)_2OH$	F:	PerFHalOrg.4-47
$C_6F_9H_2N_3$	$CF_3CH(NCCF_3)_2NH$	F:	PerFHalOrg.6-113
–	$NHCNC(CF_3)NHC(CF_3)_2$	F:	PerFHalOrg.5-63/4, 86
$C_6F_9H_3N_2O$	$(CF_3)_2NCHC(CH_2)N(CF_3)O$	F:	PerFHalOrg.7-163, 165
–	$(CF_3)_2NCOC(CH_3)NCF_3$	F:	PerFHalOrg.7-163, 165
$C_6F_9H_3N_2O_2$	$[N(CF_3)OCH_2CF_2N(CF_3)OCHCF]_n$	F:	PerFHalOrg.7-161
$C_6F_9H_3N_4$	$HN(C(CF_3)N)_2C(CF_3)NH_2$	F:	PerFHalOrg.6-75/6, 98
$C_6F_9H_6N_3O_3$	$CF_3N(OH)CH_2C_2H_2NO(N(OH)CF_3)CF_3$	F:	PerFHalOrg.7-164, 167
$C_6F_9H_9O_3P$	$(CF_3)_3P(OCH_3)_3$	F:	PerFHalOrg.3-110
C_6F_9I	$IC(CF_2)_4CF$	F:	PerFHalOrg.4-14
$C_6F_9KO_6Sn$	$KSn(CF_3COO)_3$	Sn:	MVol.C3-77
C_6F_9Li	$CF(CF_2)_4CLi$	F:	PerFHalOrg.4-3, 9, 11/2, 14/5, 44
$C_6F_9N_3$	$(CF_3)_3C_3N_3$	F:	PerFHalOrg.6-81/2, 95/6, 101, 112/3, 115

$C_6F_9N_3$	$(i\text{-}C_3F_7)C_3F_2N_3$	F:	PerFHalOrg.6-78/80, 99
$C_6F_9NiO_3P$	$(CO)_3NiP(CF_3)_3$	Ni:	Org.Verb.1-165, 168
C_6F_9P	$(CF_2CF)_3P$	F:	PerFHalOrg.3-163, 169, 173
–	$C_6F_5PF_4$	F:	PerFHalOrg.3-123, 126
		F:	PerFHalOrg.4-109
C_6F_9Sb	$(CF_2CF)_3Sb$	F:	PerFHalOrg.3-227, 229/31
$C_6F_{10}HN_3$	$NC(C_2F_5)NHC(C_2F_5)N$	F:	PerFHalOrg.5-71/2, 92, 97
$C_6F_{10}H_2N_2O_2$	$C_2F_5C(NH_2)NOC(O)C_2F_5$	F:	PerFHalOrg.7-5/6, 32
–	$C_2F_5C(NOH)C(NOH)C_2F_5$	F:	PerFHalOrg.7-89, 105
$C_6F_{10}H_2N_4$	$(CF_3)(C_3F_7)C_2H_2N_4$	F:	PerFHalOrg.6-122
–	$NC(CF_3)N(NH_2)C(C_3F_7)N$	F:	PerFHalOrg.5-71/2, 92
–	$NC(C_2F_5)N(NH_2)C(C_2F_5)N$	F:	PerFHalOrg.5-71/2, 92, 101
$C_6F_{10}H_3N_3$	$CF_3C(NH)NC(NH_2)C_3F_7$	F:	PerFHalOrg.7-7, 34, 58
–	$C_2F_5C(NH)NC(NH_2)C_2F_5$	F:	PerFHalOrg.7-7, 34, 58
$C_6F_{10}H_3N_3O$	$C_2F_5C(NH_2)NNHC(O)C_2F_5$	F:	PerFHalOrg.7-6/7, 33
$C_6F_{10}H_4HgN_4$	$(C_2F_5C(NH)NH)_2Hg$	F:	PerFHalOrg.7-3, 30
$C_6F_{10}H_4N_2O_2$	$CF_3C(NH)NH_2 \cdot C_3F_7COOH$	F:	PerFHalOrg.7-3/4
–	$C_3F_7C(NH)NH_2 \cdot CF_3COOH$	F:	PerFHalOrg.7-3/4
$C_6F_{10}H_4N_2O_3$	$CH_3C(O)ON(CF_3)CF_2CF_2N(OH)CF_3$	F:	PerFHalOrg.7-169
$C_6F_{10}H_5N_2O_4P$	$(CF_2N(O)CF_3)_2P(O)OC_2H_5$	F:	PerFHalOrg.5-99
$C_6F_{10}H_6Sn$	$(CH_3)_2Sn(C_2F_5)_2$	Sn:	Org.Verb.3-6
$C_6F_{10}MoO_4P_2$	$Mo(CO)_4(CF_3PF_2)_2$	F:	PerFHalOrg.3-110/1, 121
$C_6F_{10}N_2$	$(CF_2)_5CN_2$	F:	PerFHalOrg.5-5
$C_6F_{10}N_2O$	$CF_3CNC(C_3F_7)NO$	F:	PerFHalOrg.5-66, 88
–	$C_2F_5CNC(C_2F_5)NO$	F:	PerFHalOrg.5-66, 88
–	$NC(C_2F_5)OC(C_2F_5)N$	F:	PerFHalOrg.5-64/5, 87
$C_6F_{10}N_2O_2$	$CF_3NCOCF_2CF_2CONCF_3$	F:	PerFHalOrg.6-15, 35, 55
–	$C_2F_5CNON(O)CC_2F_5$	F:	PerFHalOrg.5-66/7, 89
$C_6F_{10}N_2O_3$	$CF_3NC(O)CF_2OCF_2C(O)NCF_3$	F:	PerFHalOrg.6-165
–	$ON(C_2F_4NO_2)CF_2CFCFCF_2$	F:	PerFHalOrg.6-4/5, 9
$C_6F_{10}N_2S$	$NC(C_2F_5)SC(C_2F_5)N$	F:	PerFHalOrg.5-67/8, 89
$C_6F_{10}N_3P_3$	$(C_6F_5)F_5P_3N_3$	F:	PerFHalOrg.3-62, 67
$C_6F_{10}N_4$	$(CF_3)(C_3F_7)C_2N_4$	F:	PerFHalOrg.6-122
$C_6F_{10}Si_2$	$C_6F_4(SiF_3)_2$	F:	PerFHalOrg.4-148, 159, 161
$C_6F_{11}H_2NO$	$C_5F_{11}C(O)NH_2$	F:	PerFHalOrg.7-18, 20/1, 47, 76
$C_6F_{11}H_3N_2$	$CF_3(CF_2)_4C(NH)NH_2$	F:	PerFHalOrg.7-3/4, 31
$C_6F_{11}H_4N_3$	$C_5F_{11}C(NH_2)NNH_2$	F:	PerFHalOrg.7-6/7, 33
$C_6F_{11}NO$	$O(CF_2CF_2)_2NCFCF_2$	F:	PerFHalOrg.6-1/2, 6, 12
$C_6F_{11}N_3$	$NCF(CF_3)NFC(CF_3)NC(CF_3)$	F:	PerFHalOrg.6-75/6
$C_6F_{11}P$	$(CF_3)_2PC_4F_5$	F:	PerFHalOrg.3-164
$C_6F_{12}Fe$	$(CF_3CFCF_2)_2Fe$	Fe:	Org.Verb.B4-356
$C_6F_{12}FeH_5NOP_2S_4$	$Fe[S(S)P(CF_3)_2]_2 \cdot CH_3C(O)NH_2$	F:	PerFHalOrg.3-159
$C_6F_{12}FeH_{10}P_4$	$(CH_2CHCH_2)_2Fe(PF_3)_4$	Fe:	Org.Verb.B1-8
$C_6F_{12}HNO$	$(CF_3)_2CNOCF_2CHFCF_3$	F:	PerFHalOrg.7-141

$C_6F_{12}HN_3S$	$[(CF_3)_2CN]_2SNH$	S:	SVol.2-91
$C_6F_{12}HP$	$(CF_3)_2PC(CF_3)CHCF_3$	F:	PerFHalOrg.3-29
$C_6F_{12}H_2NiO_2P_2$	$(CO)_2Ni(P(CF_3)_2H)_2$	Ni:	Org.Verb.1-127, 128
$C_6F_{12}H_2O$	$C_3F_7CH(OH)C_2F_5$	F:	PerFHalOrg.4-94, 96
$C_6F_{12}H_2P_2$	$(CF_3)_2PCHCHP(CF_3)_2$	F:	PerFHalOrg.3-184
$C_6F_{12}H_4N_2O_2$	$(CF_3)_2NOCH_2CH_2ON(CF_3)_2$	F:	PerFHalOrg.7-129, 133
$C_6F_{12}H_4P_2$	$(CF_3)_2PCH_2CH_2P(CF_3)_2$	F:	PerFHalOrg.3-182
$C_6F_{12}H_6N_2O_2Si$	$(CH_3)_2Si[ON(CF_3)_2]_2$	F:	PerFHalOrg.7-127, 144
$C_6F_{12}H_9PSn$	$[(CH_3)_3Sn][(CF_3)_3PF_3]$	F:	PerFHalOrg.3-134, 136
$C_6F_{12}IrMnO_6P_4$	$[Mn(CO)_6][Ir(PF_3)_4]$	Ir:	SVol.2-193
$C_6F_{12}N_2$	$CF_2N(CF_2)_4NCF_2$	F:	PerFHalOrg.5-15
–	$CF_3CNCF_2N(CF_3)CFCF_3$	F:	PerFHalOrg.5-62, 85
–	$(CF_3)_2CNNC(CF_3)_2$	F:	PerFHalOrg.5-21
–	$(F)(C_5F_{11})CN_2$	F:	PerFHalOrg.5-5
–	$N(CF_2CF_2)_3N$	F:	PerFHalOrg.6-32/3, 54
$C_6F_{12}N_2O_2$	$(CF_3)_2CNOCON(CF_3)_2$	F:	PerFHalOrg.7-141/2
–	$C_6N_2O_2F_{12}$	F:	PerFHalOrg.7-207
$C_6F_{12}N_2O_3S$	$[(CF_3)_2CNO]_2SO$	F:	PerFHalOrg.7-92/3, 107
$C_6F_{12}N_2O_4$	$(CF_3)_2NOC(O)C(O)ON(CF_3)_2$	F:	PerFHalOrg.7-96, 114
$C_6F_{12}N_2S$	$NC(CF_3)_2SC(CF_3)_2N$	F:	PerFHalOrg.5-67/8
$C_6F_{12}N_4P_4$	$(C_6F_5)F_7P_4N_4$	F:	PerFHalOrg.3-62, 67
$C_6F_{12}N_9$	$C_6F_{12}N_9$	F:	PerFHalOrg.6-137, 150
$C_6F_{12}O$	$C_3F_7COC_2F_5$	F:	PerFHalOrg.4-94, 96
$C_6F_{12}P_2$	$C_2P_2(CF_3)_4$	F:	PerFHalOrg.3-8/14
$C_6F_{12}Si_3$	$C_6F_{12}Si_3$	F:	PerFHalOrg.4-148
$C_6F_{13}H_2O_3P$	$C_6F_{13}P(O)(OH)_2$	F:	PerFHalOrg.3-43, 46, 50
$C_6F_{13}H_3P_2$	$(CF_3)_2PCFHCH_2P(CF_3)_2$	F:	PerFHalOrg.3-183
$C_6F_{13}IMg$	$(n\text{-}C_6F_{13})MgI$	F:	PerFHalOrg.4-68
$C_6F_{13}I_2P$	$C_6F_{13}PI_2$	F:	PerFHalOrg.3-76, 94
$C_6F_{13}N$	$CF_3N(CF_2)_5$	F:	PerFHalOrg.5-157/8, 170, 175
–	$FNC_5F_9CF_3$	F:	PerFHalOrg.5-150/1, 161, 174/5
$C_6F_{13}NO$	$CF_3(CF_2)_5NO$	F:	PerFHalOrg.7-179, 185
–	$O(CF_2CF_2)_2NC_2F_5$	F:	PerFHalOrg.6-1/2, 6
$C_6F_{13}NO_2$	$(CF_3)_2NOC(O)C_3F_7$	F:	PerFHalOrg.7-94/5, 111
–	$[N(CF_3)OCF_2CF(OCF_3)CF_2CF_2]_n$	F:	PerFHalOrg.7-208
$C_6F_{13}N_3$	$NCF(CF_3)NFCF(CF_3)NFC(CF_3)$	F:	PerFHalOrg.6-75/6, 97
$C_6F_{14}HO_2P$	$(CF_3CF_2CF_2)_2P(O)OH$	F:	PerFHalOrg.3-43, 45, 49/50
$C_6F_{14}HP$	$(C_3F_7)_2PH$	F:	PerFHalOrg.3-14/5, 18
$C_6F_{14}H_2NP$	$(n\text{-}C_3F_7)_2PNH_2$	F:	PerFHalOrg.3-57, 64, 68/9
$C_6F_{14}H_2N_2O_2$	$(CF_3)_2NOCH_2CF_2ON(CF_3)_2$	F:	PerFHalOrg.7-134
$C_6F_{14}H_2P_2$	$(CF_3)_2PCF_2CH_2P(CF_3)_2$	F:	PerFHalOrg.3-184
$C_6F_{14}IP$	$(C_3F_7)_2PI$	F:	PerFHalOrg.3-74/5, 76, 94, 100, 108
$C_6F_{14}NP$	$[(C_3F_7)_2PN]_x$	F:	PerFHalOrg.3-62
$C_6F_{14}N_2O$	$CF_2N(CF_3)CF_2OCF(N(CF_3)_2)$	F:	PerFHalOrg.5-55, 77
$C_6F_{14}N_2O_2$	$CF_2CF_2ON(CF_3)CF_2CF_2N(CF_3)O$	F:	PerFHalOrg.6-166/7

$C_6F_{14}N_2S$ .... $[(CF_3)_2CFN]_2S$ .... S: SVol.2-91
$C_6F_{14}N_5P_5$ .... $(C_6F_5)F_9P_5N_5$ .... F: PerFHalOrg.3-62, 67
$C_6F_{14}NiO_2P_2$ .... $(CO)_2Ni(P(CF_3)_2F)_2$ .... Ni: Org.Verb.1-127, 128
$C_6F_{14}O_2P^-$ .... $[(n-C_3F_7)_2P(O)O]^-$ .... F: PerFHalOrg.3-49
$C_6F_{14}Si_4$ .... $C_6F_{14}Si_4$ .... F: PerFHalOrg.4-148
$C_6F_{15}HN_2O$ .... $(CF_3)_2NOCF_2CHFN(CF_3)_2$ .... F: PerFHalOrg.7-143/4
– .... $(CF_3)_2NOCHFCF_2N(CF_3)_2$ .... F: PerFHalOrg.7-143/4
$C_6F_{15}HN_2O_2$ .... $(CF_3)_2NOCHFCF_2ON(CF_3)_2$ .... F: PerFHalOrg.7-134
$C_6F_{15}MoO_3P_3$ .... $Mo(CO)_3(CF_3PF_2)_3$ .... F: PerFHalOrg.3-110/1, 121
$C_6F_{15}N_3$ .... $FN(CF(CF_3)NF)_2CF(CF_3)$ .... F: PerFHalOrg.6-75/6, 97
$C_6F_{15}N_3O_2$ .... $(CF_3)_2NOCFNCF_2ON(CF_3)_2$ .... F: PerFHalOrg.7-96
– .... $[(CF_3)_2NO]_2CNCF_3$ .... F: PerFHalOrg.7-96
$C_6F_{15}P$ .... $(n-C_3F_7)_2PF$ .... F: PerFHalOrg.3-70/1, 85, 121
$C_6F_{15}P_3$ .... $(C_2F_5P)_3$ .... F: PerFHalOrg.3-1, 7
$C_6F_{16}N_2$ .... $(CF_3)_2NCF_2CF_2N(CF_3)_2$ .... F: PerFHalOrg.7-76
$C_6F_{16}N_2O_2$ .... $(CF_3)_2NOCF_2CF_2ON(CF_3)_2$ .... F: PerFHalOrg.7-95/6, 112, 124/5, 145
$C_6F_{16}N_2O_4S$ .... $(CF_3)_2NOCF_2CF(SO_2F)ON(CF_3)_2$ .... F: PerFHalOrg.7-97/8, 115
$C_6F_{16}N_6P_6$ .... $(C_6F_5)F_{11}P_6N_6$ .... F: PerFHalOrg.3-62, 67
$C_6F_{16}P_2$ .... $(CF_3)_2PCF_2CF_2P(CF_3)_2$ .... F: PerFHalOrg.3-164, 168, 174
$C_6F_{17}H_3N_2O_2S$ .... $(CF_3)_2NOCH(SF_5)CH_2ON(CF_3)_2$ .... F: PerFHalOrg.7-135
$C_6F_{17}P$ .... $(n-C_3F_7)_2PF_3$ .... F: PerFHalOrg.3-71, 76/7, 86
$C_6F_{18}FeP_3S_6$ .... $Fe[S(S)P(CF_3)_2]_3$ .... F: PerFHalOrg.3-159
$C_6F_{18}HO_3P_3Si$ .... $[(CF_3)_2PO]_3SiH$ .... F: PerFHalOrg.3-40, 42
$C_6F_{18}NP_3$ .... $[(CF_3)_2P]_3N$ .... F: PerFHalOrg.3-60, 62/4, 68/9
$C_6F_{18}N_3O_3PS$ .... $[(CF_3)_2NO]_3PS$ .... F: PerFHalOrg.7-100/1, 119
$C_6F_{18}N_3O_3Sb$ .... $[(CF_3)_2NO]_3Sb$ .... F: PerFHalOrg.7-101/2, 119
$C_6F_{18}N_3O_4P$ .... $[(CF_3)_2NO]_3PO$ .... F: PerFHalOrg.7-100/1, 119
$C_6F_{18}N_3P_3$ .... $[(CF_3)_2PN]_3$ .... F: PerFHalOrg.3-61/2, 66, 68/9
$C_6F_{18}N_6O_3S_3$ .... $[(CF_3)_2NO]_3S_3N_3$ .... F: PerFHalOrg.7-92/3, 108
$C_6F_{18}N_6O_6S_3$ .... $[(CF_3)_2NO]_3O_3S_3N_3$ .... F: PerFHalOrg.7-92/3, 108, 145
$C_6F_{18}N_7P_7$ .... $(C_6F_5)F_{13}P_7N_7$ .... F: PerFHalOrg.3-62, 67
$C_6F_{18}P_3S_6V$ .... $V[S(S)P(CF_3)_2]_3$ .... F: PerFHalOrg.3-154
$C_6F_{20}N_3O_3P$ .... $[(CF_3)_2NO]_3PF_2$ .... F: PerFHalOrg.7-100/1, 117
$C_6F_{20}N_8P_8$ .... $(C_6F_5)F_{15}P_8N_8$ .... F: PerFHalOrg.3-62, 67
$C_6FeGdKN_6$ .... $KGd[Fe(CN)_6]$ .... Sc: MVol.B7-30
C_6FeGdN_6Na .... $NaGd[Fe(CN)_6]$ .... Sc: MVol.B7-30
C_6FeGdN_6Rb .... $RbGd[Fe(CN)_6]$ .... Sc: MVol.B7-30
$C_6FeGeH_9NO_4$ .... $(CO)_3Fe(NO)Ge(CH_3)_3$ .... Fe: Org.Verb.B1-197
$C_6FeH_2I_2O_4$ .... $ICHCHIFe(CO)_4$ .... Fe: Org.Verb.B4-269, 278/80
$C_6FeH_2NO_4Tl$ .... $Tl[(CO)_4FeCH_2CN]$ .... Fe: Org.Verb.B2-173
$C_6FeH_3KO_5$ .... $K[CH_3C(O)Fe(CO)_4]$ .... Fe: Org.Verb.B4-156, 167
$C_6FeH_3LiO_5$ .... $Li[CH_3C(O)Fe(CO)_4]$ .... Fe: Org.Verb.B4-149, 159
$C_6FeH_3NO_4$ .... $CH_3NCFe(CO)_4$ .... Fe: Org.Verb.B4-5/6

$C_6FeH_3NO_4$ .... $(CO)_4Fe(CH_3CN)$ .... Fe: Org.Verb.B2-71, 76
$C_6FeH_3N_5O$ .... $H_3[COFe(CN)_5] \cdot H_2O$ .... Fe: Org.Verb.B1-24
$C_6FeH_3NaO_5$ .... $Na[CH_3C(O)Fe(CO)_4]$ .... Fe: Org.Verb.B4-149, 159
$C_6FeH_3NaO_6$ .... $(CO)_5Fe(NaOCH_3)$ .... Fe: Org.Verb.B3-249/50
$C_6FeH_3O_3S^+$ .... $[C_3H_3SFe(CO)_3]^+$ .... Fe: Org.Verb.B5-152
$C_6FeH_3O_5^-$ .... $[CH_3C(O)Fe(CO)_4]^-$ .... Fe: Org.Verb.B4-149, 156, 159, 167
$C_6FeH_3O_6^-$ .... $[CH_3OC(O)Fe(CO)_4]^-$ .... Fe: Org.Verb.B4-154, 162/3
$C_6FeH_4KNO_3$ .... $K[(CO)_3FeCH_2CH_2CN]$ .... Fe: Org.Verb.B1-205
$C_6FeH_4LaN_7$ .... $NH_4La[Fe(CN)_6]$ .... Sc: MVol.B7-30
$C_6FeH_4LiNO_3$ ...... $Li[(CO)_3FeCH_2CH_2CN]$ .... Fe: Org.Verb.B1-205
$C_6FeH_4NNaO_3$ ...... $Na[(CO)_3FeCH_2CH_2CN]$ .... Fe: Org.Verb.B1-205
$C_6FeH_4N_2O_3$ ...... $NCC_3H_4Fe(CO)_2NO$ .... Fe: Org.Verb.B5-14/5, 21
$C_6FeH_4N_7Nd$ .... $NH_4Nd[Fe(CN)_6]$ .... Sc: MVol.B7-30
$C_6FeH_4N_7Pr$ .... $NH_4Pr[Fe(CN)_6]$ .... Sc: MVol.B7-30
$C_6FeH_4N_7Sm$ ...... $NH_4Sm[Fe(CN)_6]$ .... Sc: MVol.B7-30
$C_6FeH_4O_3S$ .... $(CH_2CHCHS)Fe(CO)_3$ .... Fe: Org.Verb.B5-122, 135, 148
$C_6FeH_4O_4$ .... $HC(O)CHCH_2Fe(CO)_3$ .... Fe: Org.Verb.B4-244
– .... $H_2CCH_2Fe(CO)_4$ .... Fe: Org.Verb.B4-228/30
$C_6FeH_4O_4S$ .... $(CH_2CHCHSO)Fe(CO)_3$ .... Fe: Org.Verb.B5-135, 148/9
$C_6FeH_4O_5$ .... $HOCHCH_2Fe(CO)_4$ .... Fe: Org.Verb.B4-234, 240/1
$C_6FeH_5^+$ .... $[FeC_6H_5]^+$ .... Fe: Org.Verb.A1-93/4, 306
C_6FeH_5I .... C_6H_5FeI .... Fe: Org.Verb.B1-11
$C_6FeH_5IO_3$ .... $C_3H_5Fe(CO)_3I$ .... Fe: Org.Verb.B2-66
Fe: Org.Verb.B5-35/9, 47/50, 54/8
Fe: Org.Verb.C2-164/5
$C_6FeH_5NO_5$ .... $C_3H_5Fe(CO)_3ONO$ .... Fe: Org.Verb.B5-48, 51, 59
$C_6FeH_5NO_6$ .... $C_3H_5Fe(CO)_3ONO_2$ .... Fe: Org.Verb.B5-39, 48, 51, 59
$C_6FeH_5NaO_3$ .... $Na[C_3H_5Fe(CO)_3]$ .... Fe: Org.Verb.B5-72
$C_6FeH_5O_2^+$ .... $[FeC_5H_4COOH]^+$ .... Fe: Org.Verb.A3-57
$C_6FeH_5O_3$ .... $[C_3H_5Fe(CO)_3]$ .... Fe: Org.Verb.B5-69/71
$C_6FeH_5O_3^+$ .... $[C_3H_5Fe(CO)_3]^+$ .... Fe: Org.Verb.B5-82/5
$C_6FeH_5O_3^-$ .... $[C_3H_5Fe(CO)_3]^-$ .... Fe: Org.Verb.B5-72/3
$C_6FeH_5O_5^-$ .... $[HOCH_2CH_2C(O)Fe(CO)_3]^-$ .... Fe: Org.Verb.B4-109
$C_6FeH_6^+$ .... $[FeC_5H_4CH_2]^+$ .... Fe: Org.Verb.A1-346
– .... $[FeC_6H_6]^+$ .... Fe: Org.Verb.A1-93/4, 257
Fe: Org.Verb.A2-17
$C_6FeH_6Hg_2O_4$ ...... $(CO)_4Fe(HgCH_3)_2$ .... Fe: Org.Verb.B2-153, 156, 159/60
$C_6FeH_6INO_3S_2$ ..... $(CO)_3Fe(S_2CN(CH_3)_2)I$ .... Fe: Org.Verb.B1-184
$C_6FeH_6I_2N_2O_2$ ...... $(CH_3NC)_2Fe(CO)_2I_2$ .... Fe: Org.Verb.B4-16/7
$C_6FeH_6LiO_4P$ ...... $(CO)_4FeP(CH_3)_2Li$ .... Fe: Org.Verb.B2-108
$C_6FeH_6N_4$ .... $Fe(N_2C_3H_3)_2$ .... Fe: Org.Verb.B3-193
$C_6FeH_6N_6O_2$ .... $[CH_3NCFe(CN)_4NO_2]H_3 \cdot H_2O$ .... Fe: Org.Verb.B4-9
$C_6FeH_6O_2^+$ .... $[C_5H_5FeCOOH]^+$ .... Fe: Org.Verb.A3-57
$C_6FeH_6O_5S$ .... $(CO)_4FeOS(CH_3)_2$ .... Fe: Org.Verb.B2-128/31
$C_6FeH_7^+$ .... $[FeC_6H_7]^+$ .... Fe: Org.Verb.A1-264, 269, 273

$C_6FeH_7NO_3$ .... $CH_2C(CH_3)CH_2Fe(CO)_2NO$ .... Fe: Org.Verb.B5-14/5, 17, 22/3
– .... $CH_3CHCHCH_2Fe(CO)_2NO$ .... Fe: Org.Verb.B5-14/6, 22
$C_6FeH_7NaO_5$ .... $Na[HFe(CO)_4] \cdot C_2H_5OH$ .... Fe: Org.Verb.B2-17
$C_6FeH_7O^+$ .... $[FeC_5H_4CH_2OH]^+$ .... Fe: Org.Verb.A2-17
– .... $[FeC_5H_4OCH_3]^+$ .... Fe: Org.Verb.A2-109
$C_6FeH_7O_4P$ .... $(CO)_4FeP(CH_3)_2H$ .... Fe: Org.Verb.B2-88, 108
$C_6FeH_8^+$ .... $[C_5H_5FeCH_3]^+$ .... Fe: Org.Verb.A2-109, 208
$C_6FeH_8N_2O_3$ .... $NH_4[(CO)_3FeCH_2CH_2CN]$ .... Fe: Org.Verb.B1-205
$C_6FeH_8NiP^+$ .... $NiPCH_3FeC_5H_5^+$ .... Ni: Org.Verb.1-167
$C_6FeH_8O^+$ .... $[C_5H_5FeOCH_3]^+$ .... Fe: Org.Verb.A3-109
$C_6FeH_9I_2O_3P$ .... $(CO)_3Fe(P(CH_3)_3)I_2$ .... Fe: Org.Verb.B1-180/2
$C_6FeH_9I_2O_6P$ .... $(CO)_3Fe(P(OCH_3)_3)I_2$ .... Fe: Org.Verb.B1-180/3
$C_6FeH_9NO_4Pb$ ...... $(CO)_3Fe(NO)Pb(CH_3)_3$ .... Fe: Org.Verb.B1-197/8
$C_6FeH_9NO_4Si$ ...... $(CO)_3Fe(NO)Si(CH_3)_3$ .... Fe: Org.Verb.B1-197
$C_6FeH_9NO_4Sn$ ...... $(CO)_3Fe(NO)Sn(CH_3)_3$ .... Fe: Org.Verb.B1-197
$C_6FeH_9N_2O_6P$ ...... $COFe(NO)_2P(OCH_2)_3CCH_3$ .... Fe: Org.Verb.B1-42/3, 45
$C_6FeH_9O_3S_6$ .... $Fe[S_2COCH_3]_3$ .... C: MVol.D4-259
$C_6FeH_{10}N_2O_2$ ...... $(H_2NC(O)CHCH_2)_2Fe$ .... Fe: Org.Verb.B4-356
$C_6FeH_{10}N_2O_4S_4$ ..... $(NO)_2Fe[S_2COC_2H_5]_2$ .... C: MVol.D4-259
$C_6FeH_{10}N_4O_2$ ...... $(CH_3NC)_2Fe(H_2O)_2(CN)_2$ .... Fe: Org.Verb.B4-28
– .... $(C_2H_5NC)_2Fe(NO)_2$ .... Fe: Org.Verb.B4-11/3
$C_6FeH_{10}N_4O_3$ ...... $(CH_2NCH_3)_2CFe(NO)_2CO$ .... Fe: Org.Verb.B4-109/11
$C_6FeH_{10}O_2S_4$ .... $Fe[S_2COC_2H_5]_2$ .... C: MVol.D4-259
$C_6FeH_{11}^+$ .... $[FeC_6H_{11}]^+$ .... Fe: Org.Verb.A1-94
$C_6FeH_{11}NO_4S$ ...... $(CH_3)_2NCH_2CCCH_3FeSO_4 \cdot 2\ H_2O$ .... Fe: Org.Verb.B4-328
$C_6FeH_{11}N_3O_4$ ...... $((CH_3)_2N)(C_2H_5O)CFe(NO)_2CO$ .... Fe: Org.Verb.B4-109/11
$C_6FeH_{12}N_2S_4$ .... $Fe(S_2CN(CH_3)_2)_2$ .... Fe: Org.Verb.B3-163
$C_6FeH_{12}N_4S_2^+$ ...... $[Fe(SC(NHCH_2)_2)_2]^+$ .... Fe: Org.Verb.B1-165
$C_6FeH_{15}NO_4P^+$ ..... $[Fe(NO)P(OC_2H_5)_3]^+$ .... Fe: Org.Verb.B1-47/8
$C_6FeH_{15}N_2O_5P^+$ .... $[Fe(NO)_2P(OC_2H_5)_3]^+$ .... Fe: Org.Verb.B1-47/8
$C_6FeH_{15}O_3P^+$ ...... $[FeP(OC_2H_5)_3]^+$ .... Fe: Org.Verb.B1-48
$C_6FeH_{18}N_3P^+$ ...... $[FeP(N(CH_3)_2)_3]^+$ .... Fe: Org.Verb.B1-164
$C_6FeKLaN_6$ .... $KLa[Fe(CN)_6]$ .... Sc: MVol.B7-30
C_6FeKN_6Nd .... $KNd[Fe(CN)_6]$ .... Sc: MVol.B7-30
C_6FeKN_6Pr .... $KPr[Fe(CN)_6]$ .... Sc: MVol.B7-30
C_6FeKN_6Sm .... $KSm[Fe(CN)_6]$ .... Sc: MVol.B7-30
– .... $KSm[Fe(CN)_6] \cdot 4\ H_2O$ .... Sc: MVol.B7-30
$C_6FeK_3N_5O$ .... $K_3[COFe(CN)_5]$ .... Fe: Org.Verb.B1-25/7
– .... $K_3[COFe(CN)_5] \cdot n\ H_2O$ .... Fe: Org.Verb.B1-25
$C_6FeLaLiN_6$ .... $LiLa[Fe(CN)_6]$ .... Sc: MVol.B7-30
C_6FeLaN_6Na .... $NaLa[Fe(CN)_6]$ .... Sc: MVol.B7-30
C_6FeLaN_6Rb .... $RbLa[Fe(CN)_6]$ .... Sc: MVol.B7-30
C_6FeLaN_6Tl .... $TlLa[Fe(CN)_6]$ .... Sc: MVol.B7-30
C_6FeLiN_6Nd .... $LiNd[Fe(CN)_6]$ .... Sc: MVol.B7-30
C_6FeLiN_6Pr .... $LiPr[Fe(CN)_6]$ .... Sc: MVol.B7-30
C_6FeLiN_6Sc .... $LiSc[Fe(CN)_6]$ .... Sc: MVol.B7-30
C_6FeLiN_6Sm .... $LiSm[Fe(CN)_6]$ .... Sc: MVol.B7-30
$C_6FeN_2O_4S_2$ .... $(CO)_4Fe(SCN)_2$ .... Fe: Org.Verb.B2-54
Fe: Org.Verb.B3-121

$C_6FeN_5Na_3O$	$Na_3[COFe(CN)_5]$	Fe: Org.Verb.B1-25/7
–	$Na_3[COFe(CN)_5] \cdot n\ H_2O$	Fe: Org.Verb.B1-25
$C_6FeN_5O^{3-}$	$[COFe(CN)_5]^{3-}$	Fe: Org.Verb.B1-24/7
C_6FeN_6NaNd	$NaNd[Fe(CN)_6]$	Sc: MVol.B7-30
C_6FeN_6NaPr	$NaPr[Fe(CN)_6]$	Sc: MVol.B7-30
C_6FeN_6NaSc	$NaSc[Fe(CN)_6]$	Sc: MVol.B7-30
C_6FeN_6NaSm	$NaSm[Fe(CN)_6]$	Sc: MVol.B7-30
–	$NaSm[Fe(CN)_6] \cdot 3\ H_2O$	Sc: MVol.B7-30
C_6FeN_6NdRb	$RbNd[Fe(CN)_6]$	Sc: MVol.B7-30
C_6FeN_6PrRb	$RbPr[Fe(CN)_6]$	Sc: MVol.B7-30
$C_6FeN_6Ra_2$	$Ra_2[Fe(CN)_6]$	Ra: SVol.2-375
C_6FeN_6RbSm	$RbSm[Fe(CN)_6]$	Sc: MVol.B7-30
–	$RbSm[Fe(CN)_6] \cdot 4\ H_2O$	Sc: MVol.B7-30
C_6FeN_6RbY	$RbY[Fe(CN)_6]$	Sc: MVol.B7-30
$C_6FeO_6^{2+}$	$[Fe(CO)_6]^{2+}$	Fe: Org.Verb.B3-256
$C_6Fe_2GeH_{10}I_2N_4O_4$	$(CH_2CHCH_2Fe(NO)_2)_2GeI_2$	Fe: Org.Verb.C2-156, 165
$C_6Fe_2HO_6Tl$	$(CO)_3FeTl \cdot HFe(CO)_3$	Fe: Org.Verb.C1-249
$C_6Fe_2H_2N_2O_6$	$(CO)_3Fe(NH)_2Fe(CO)_3$	Fe: Org.Verb.C1-134
$C_6Fe_2H_2O_6S_2$	$(CO)_3Fe(SH)_2Fe(CO)_3$	Fe: Org.Verb.C1-80/1, 84
$C_6Fe_2H_4N_2O_6$	$(CO)_3Fe(NH_2)_2Fe(CO)_3$	Fe: Org.Verb.C1-130, 134/5
$C_6Fe_2H_4O_6P_2$	$(CO)_3Fe(PH_2)_2Fe(CO)_3$	Fe: Org.Verb.C1-165
$C_6Fe_2H_4O_6P_2^-$	$[(CO)_3Fe(PH_2)_2Fe(CO)_3]^-$	Fe: Org.Verb.C1-165
$C_6Fe_2H_4O_6P_2^{2-}$	$[(CO)_3Fe(PH_2)_2Fe(CO)_3]^{2-}$	Fe: Org.Verb.C1-165
$C_6Fe_2H_9N_7O_2$	$Fe_2(CN)_4(CO)_2(NH_3)_3$	Fe: Org.Verb.B3-113
$C_6Fe_2H_{10}I_2N_4O_4Pb$	$(CH_2CHCH_2Fe(NO)_2)_2PbI_2$	Fe: Org.Verb.C2-156, 165
$C_6Fe_2H_{10}I_2N_4O_4Sn$	$(CH_2CHCH_2Fe(NO)_2)_2SnI_2$	Fe: Org.Verb.C2-156, 165
$C_6Fe_2H_{12}N_4O_6P_2$	$(NO)_2(CO)Fe(P(CH_3)_2)_2Fe(CO)(NO)_2$	Fe: Org.Verb.B1-143
		Fe: Org.Verb.C1-67, 69/70
$C_6Fe_2HgN_2O_8$	$[(CO)_3FeNO]_2Hg$	Fe: Org.Verb.C1-63/4
$C_6Fe_2I_2O_6$	$(CO)_3FeI_2Fe(CO)_3$	Fe: Org.Verb.C1-73/4
$C_6Fe_2I_4O_6Si$	$(CO)_3Fe(SiI_4)Fe(CO)_3$	Fe: Org.Verb.C1-5
$C_6Fe_2O_6^+$	$[Fe_2(CO)_6]^+$	Fe: Org.Verb.C2-26
$C_6Fe_2O_6^-$	$[Fe_2(CO)_6]^-$	Fe: Org.Verb.B3-79
$C_6Fe_2O_6S_2$	$(CO)_3FeS_2Fe(CO)_3$	Fe: Org.Verb.C1-74/6
$C_6Fe_2O_6Se_2$	$(CO)_3FeSe_2Fe(CO)_3$	Fe: Org.Verb.C1-74, 76/7
$C_6Fe_3I_5O_6S_2$	$Fe_3(CO)_6S_2I_5$	Fe: Org.Verb.C1-18/20
$C_6GaH_9O_3S_6$	$Ga[S_2COCH_3]_3$	C: MVol.D4-253
$C_6GeH_5I_3NiO$	$C_5H_5Ni(CO)GeI_3$	Ni: Org.Verb.2-164/6
$C_6HNi_2O_6^-$	$[HNi_2(CO)_6]^-$	Ni: Org.Verb.2-265
$C_6HO_{13}Pa^{2-}$	$PaO(HC_2O_4)(C_2O_4)_2^{2-}$	Pa: SVol.2-170
$C_6H_2MnN_3O_7^+$	$[Mn(OC_6H_2(NO_2)_3)]^+$	Mn: MVol.D1-40/1
$C_6H_2O_{14}Pa^{3-}$	$PaO(OH)(HC_2O_4)(C_2O_4)_2^{3-}$	Pa: SVol.2-170
$C_6H_3IrN_6$	$H_3[Ir(CN)_6]$	Ir: SVol.2-183
$C_6H_3IrN_6S_9$	$Ir(S_3N_2C_2H)_3$	Ir: SVol.2-171
$C_6H_3MnNO_4$	$[Mn(NO_2C_6H_3O_2)]$	Mn: MVol.D1-45
$C_6H_3MnN_2O_5^+$	$[Mn(OC_6H_3(NO_2)_2)]^+$	Mn: MVol.D1-40/1
$C_6H_3MnO_4S_3$	$Mn(CO)_4[S_2CSCH_3]$	C: MVol.D4-269
$C_6H_3MnO_5S_2$	$(CO)_4Mn[S_2COCH_3]$	C: MVol.D4-259
$C_6H_3O_4ReS_3$	$Re(CO)_4[S_2CSCH_3]$	C: MVol.D4-269
$C_6H_3O_5ReS_2$	$(CO)_4Re[S_2COCH_3]$	C: MVol.D4-259

$C_6H_3O_{12}Pa^{2+}$	$Pa(HC_2O_4)_3^{2+}$	Pa: SVol.2-170
$C_6H_4IS_{1.5}Sn$	$(IC_6H_4SnS_{1.5})_x$	Sn: Org.Verb.6-284
$C_6H_4I_2Ni^+$	$NiI_2C_6H_4^+$	Ni: Org.Verb.2-398
$C_6H_4MnNO_3^+$	$[Mn(OC_6H_4NO_2)]^+$	Mn: MVol.D1-40/1
$C_6H_4MnO_2$	$Mn(OC_6H_4O)$	Mn: MVol.D1-43/4
$C_6H_4MnO_4$	$Mn(C_2(CHO)_4) \cdot 2\ H_2O$	Mn: MVol.D1-54
$C_6H_4MnO_5S$	$[Mn(O_2C_6H_3SO_3H)]$	Mn: MVol.D1-46
$C_6H_4MnO_8S_2$	$[Mn(O_2C_6H_2(SO_3H)_2)]$	Mn: MVol.D1-46
$C_6H_4NNpO_2^{3+}$	$Np(NC_5H_4COO)^{3+}$	Np: TrU.D1-155
$C_6H_4NNpO_3^{3+}$	$Np(ONC_5H_4COO)^{3+}$	Np: TrU.D1-155
$C_6H_4NNpO_4$	$NpO_2(NC_5H_4COO)$	Np: TrU.D1-155
$C_6H_4NNpO_5$	$NpO_2(ONC_5H_4COO)$	Np: TrU.D1-155
$C_6H_4NO_4Pu^+$	$PuO_2(NC_5H_4COO)^+$	Np: TrU.D1-155
$C_6H_4NO_5Pu^+$	$PuO_2(ONC_5H_4COO)^+$	Np: TrU.D1-155
$C_6H_4N_2Ni$	$C_4H_4Ni(CN)_2$	Ni: Org.Verb.2-324
$C_6H_4N_6S_6$	$(C_3H_2N_3S_2)_2S_2$	C: MVol.D6-50/1
$C_6H_4NiO_2$	$C_4H_4Ni(CO)_2$	Ni: Org.Verb.2-324
$C_6H_4O_2Sn^{2+}$	$[Sn(C_6H_4O_2)]^{2+}$	Sn: MVol.C5-77
$C_6H_4O_2Sn_2^{6+}$	$[Sn_2(C_6H_4O_2)]^{6+}$	Sn: MVol.C5-77
$C_6H_4O_4S_2$	$CH_2C(O)SC(O)CH_2C(O)SC(O)$	C: MVol.D4-215
$C_6H_4O_8Sn^{2-}$	$Sn(CH_2(COO)_2)_2^{2-}$	Sn: MVol.C2-225
C_6H_5INiO	$C_5H_5Ni(CO)I$	Ni: Org.Verb.2-163/4
$C_6H_5IO_2Sn$	$(IC_6H_4SnOOH)_x$	Sn: Org.Verb.6-284
$C_6H_5I_3Sn$	$C_6H_5SnI_3$	Sn: Org.Verb.6-275/6
C_6H_5KNiO	$K(C_5H_5NiCO)$	Ni: Org.Verb.2-163
$C_6H_5MnO_2^+$	$[Mn(OC_6H_4OH)]^+$	Mn: MVol.D1-43/4
$C_6H_5MnO_3^+$	$[Mn(CH_3C_5H_2O_3)]^+$	Mn: MVol.D1-150
$C_6H_5MnO_4^+$	$[Mn(C_5H_2O_3CH_2OH)]^+$	Mn: MVol.D1-150/1
$C_6H_5MnO_8S_2^+$	$[Mn(OC_6H_2(SO_3H)_2OH)]^+$	Mn: MVol.D1-46
$C_6H_5NNiO_3$	$C_2H_5NCNi(CO)_3$	Ni: Org.Verb.1-303
$C_6H_5NO_4Pu^{2+}$	$PuO_2(NC_5H_4COOH)^{2+}$	Np: TrU.D1-155
$C_6H_5NO_5Sn$	$SnO_2 \cdot NO_2C_6H_4OH \cdot 2\ H_2O$	Sn: Org.Verb.1-158
$C_6H_5NPtS_3$	$Pt(C_5H_5N)CS_3$	C: MVol.D4-226
$C_6H_5NiO^+$	$C_5H_5Ni(CO)^+$	Ni: Org.Verb.2-345
$C_6H_5NiO^-$	$[C_5H_5NiCO]^-$	Ni: Org.Verb.2-163
$C_6H_5NpO_9^{2-}$	$NpO_2(C_6H_5O_7)^{2-}$	Np: TrU.D1-157, 159
$C_6H_5O_7Pu$	$Pu(C_6H_5O_7)$	Np: TrU.D1-159
$C_6H_5O_7Pu^+$	$Pu(C_6H_5O_7)^+$	Np: TrU.D1-159
C_6H_5Sn	$(C_6H_5Sn)_x$	Sn: Org.Verb.6-273
$C_6H_5Sn^+$	$Sn(C_6H_5)^+$	Sn: Org.Verb.1-155
		Sn: Org.Verb.2-355, 386
$C_6H_6I_3NSSn$	$[SnI_3(SCH_2C_5H_4N)]_2$	Sn: MVol.C6-105
$C_6H_6K_2O_{12}Sn$	$K_2[Sn(HCOO)_6] \cdot 5\ H_2O$	Sn: MVol.C3-75
$C_6H_6LiNNiO_4$	$Li[(CO)_3NiC(O)N(CH_3)_2]$	Ni: Org.Verb.1-193, 194
$C_6H_6MnO_3$	$[MnOH(OC_6H_4OH)]$	Mn: MVol.D1-43/4
$C_6H_6NNpO_6^+$	$Np(N(CH_2COO)_3)^+$	Np: TrU.D1-163
$C_6H_6NNpO_8^{2-}$	$NpO_2(N(CH_2COO)_3)^{2-}$	Np: TrU.D1-163
$C_6H_6NO_8Pu^{2-}$	$PuO_2(N(CH_2COO)_3)^{2-}$	Np: TrU.D1-163
$C_6H_6N_2Ni$	$Ni(CH_2CHCN)_2$	Ni: Org.Verb.1-388/90
$C_6H_6N_2NiO_2$	$(CH_3NC)_2Ni(CO)_2$	Ni: Org.Verb.1-308, 309

$C_6H_6N_2S_2Sn$ .... $(CH_3)_2Sn(SC(CN))_2$ .... Sn: Org.Verb.6-30
$C_6H_6N_4O_2S$ .... $C_6H_5NHSO_2N_3$ .... S: S-N-Verb.1-214
$C_6H_6N_6Sn$ .... $(CH_3)_2Sn[N(CN)_2]_2$ .... Sn: Org.Verb.6-31
$C_6H_6Na_2O_{12}Sn$ ..... $Na_2[Sn(HCOO)_6] \cdot 5\ H_2O$ .... Sn: MVol.C3-34
C_6H_6Ni .... $(CH_3CC)_2Ni$ .... Ni: Org.Verb.1-81/2
$C_6H_6NpO_9^-$ .... $NpO_2(C_6H_6O_7)^-$ .... Np: TrU.D1-157, 159
$C_6H_6O_2Sn$ .... $[C_6H_5SnOOH]_x$ .... Sn: Org.Verb.6-275/6
$C_6H_6O_7Sn$ .... $Sn(HOC_3H_4(COO)_2COOH)$ .... Sn: MVol.C2-226
$C_6H_6O_9Pa^-$ .... $PaO(OH)(C_6H_5O_7)^-$ .... Pa: SVol.2-173/4
$C_6H_6O_{12}Sn_2Sr$ ...... $Sr[Sn(HCOO)_3]_2$ .... Sn: MVol.C3-190
$C_6H_7I_2NSn$ .... $SnI_2 \cdot CH_3C_5H_4N$ .... Sn: MVol.C5-18
$C_6H_7MnO_5^-$ .... $[Mn(C_6H_7O_5)]^-$ .... Mn: MVol.D1-147
$C_6H_7MnO_6^+$ ...... $[Mn(C_4HO_2(OH)_2CHOHCH_2O)]^+$ .... Mn: MVol.D1-145
C_6H_7NNiO .... $CH_3C_5H_4NiNO$ .... Ni: Org.Verb.2-112/3
$C_6H_7NNpO_8^-$ .... $NpO_2(N(CH_2COO)_2(CH_2COOH))^-$ .... Np: TrU.D1-163
$C_6H_7NO_2S$ .... $C_6H_5SO_2NH_2$ .... S: S-N-Verb.1-123
$C_6H_7N_2NaO_3S$ ..... $Na[SO_3NC_6H_5NH_2]$ .... S: S-N-Verb.1-219
$C_6H_7N_3O_4S$ .... $NO_2C_6H_4NHSO_2NH_2$ .... S: S-N-Verb.1-167
$C_6H_7N_3O_8U$ .... $UO_2(NO_3)_2 \cdot CH_3C_5H_4N \cdot H_2O$ .... U: SVol.E1-29, 34
$C_6H_7O_9Pa$ .... $PaO(OH)(HC_6H_5O_7)$ .... Pa: SVol.2-173/4
$C_6H_7O_{10}Pa^{2-}$ .... $PaO(OH)_2(C_6H_5O_7)^{2-}$ .... Pa: SVol.2-173
$C_6H_7Sn^+$ .... $Sn(C_6H_5)H_2^+$ .... Sn: Org.Verb.3-25
$C_6H_8I_3NSn$ .... $[C_6H_5NH_3]SnI_3$ .... Sn: MVol.C3-119/20
$C_6H_8I_4N_2Sn$ .... $SnI_4 \cdot NC_5H_4CH_2NH_2$ .... Sn: MVol.C5-203
$C_6H_8K_2O_4S_2$ .... $K_2[SOCO(CH_2)_4OCOS]$ .... C: MVol.D4-237
$C_6H_8MnN_2O_2S_2$ ..... $Mn(NCS)_2 \cdot C_4H_8O_2$ .... Mn: MVol.D1-155
$C_6H_8MnN_2O_2Se_2$ .... $Mn(NCSe)_2 \cdot C_4H_8O_2$ .... Mn: MVol.D1-156
$C_6H_8MnO_4$ .... $[Mn(CH(COCH_3)_2)HCOO]$ .... Mn: MVol.D1-75
$C_6H_8N_2O_2S$ .... $C_6H_5NHSO_2NH_2$ .... S: S-N-Verb.1-167
$C_6H_8N_2O_6SU$ .... $UO_2SO_4 \cdot C_6H_5HNNH_2$ .... U: SVol.E1-26/7
$C_6H_8N_4Na_2O_{12}U$ .... $Na_2[UO_2(C_2O_4)_2((NH_2)_2CO)] \cdot (NH_2)_2CO \cdot 3\ H_2O$
U: SVol.E1-84, 86, 90
$C_6H_8N_4O_{10}U$ .... $U(C_2O_4)_2 \cdot 2\ (NH_2)_2CO$ .... U: SVol.E1-82/4
$C_6H_8N_{10}S_2$ .... $(C_3N_3(NH_2)_2)_2S_2$ .... C: MVol.D6-44
C_6H_8NI .... $C_5H_5NICH_3$ .... Ni: Org.Verb.2-207
$C_6H_8NiO_2$ .... $Ni(CH_2CHCHO)_2$ .... Ni: Org.Verb.1-387/8
$C_6H_8O_4S_2^{2-}$ .... $[SOCO(CH_2)_4OCOS]^{2-}$ .... C: MVol.D4-236
$C_6H_8O_8Sn$ .... $SnC_6H_8O_8$ .... Sn: MVol.C2-226
$C_6H_8O_9Pa^+$ .... $PaO(OH)(H_2C_6H_5O_7)^+$ .... Pa: SVol.2-173/4
$C_6H_8S_2Sn$ .... $(CH_2CH)_2Sn(SCHCHS)$ .... Sn: Org.Verb.6-145
C_6H_8Sn .... $(CH_3)_2Sn(CCH)_2$ .... Sn: Org.Verb.3-6, 15
– .... $C_6H_5SnH_3$ .... Sn: Org.Verb.4-128
$C_6H_9InO_3S_6$ .... $In[S_2COCH_3]_3$ .... C: MVol.D4-253
$C_6H_9KO_6Sn$ .... $KSn(CH_3COO)_3$ .... Sn: MVol.C3-75/6
– .... $KSn(CH_3COO)_3 \cdot 2\ H_2O$ .... Sn: MVol.C3-77
$C_6H_9MnO_3^+$ ...... $[Mn(CH_3OCH_2COCHCOCH_3)]^+$ .... Mn: MVol.D1-112
$C_6H_9MnO_8^-$ .... $[Mn(C_4HO_2(OH)_2CHOHCH_2O)(OH)_2]^-$ ...... Mn: MVol.D1-145
$C_6H_9NNpO_5^{2+}$ ...... $NpN(C_2H_4OH)(CH_2COO)_2^{2+}$ .... Np: TrU.D1-162
$C_6H_9NNpO_7^-$ .... $(NpO_2)N(C_2H_4OH)(CH_2COO)_2^-$ .... Np: TrU.D1-162
$C_6H_9N_2O_2Sn$ .... $[C_2H_5Sn(NCOCH_2NCOCH_2)]_x$ .... Sn: Org.Verb.6-230/1

$C_6H_9N_3O_4S_2$	$NH_2SO_2NHSO_2NHC_6H_5$	S:	S-N-Verb.1-161
$C_6H_9NaO_6Sn$	$NaSn(CH_3COO)_3 \cdot 2\ H_2O$	Sn:	MVol.C3-34
$C_6H_9NaO_8Pu$	$NaPuO_2(CH_3COO)_3$	Np:	TrU.D1-152
$C_6H_9NiO_3P$	$(CO)_3NiP(CH_3)_3$	Ni:	Org.Verb.1-165, 167
$C_6H_9NiO_3PS_3$	$(CO)_3NiP(SCH_3)_3$	Ni:	Org.Verb.1-182
$C_6H_9NiO_6P$	$(CO)_3NiP(OCH_3)_3$	Ni:	Org.Verb.1-177, 179
$C_6H_9NpO_8^-$	$NpO_2(CH_3COO)_3^-$	Np:	TrU.D1-153
$C_6H_9O_3S_6Sb$	$Sb[S_2COCH_3]_3$	C:	MVol.D4-254
$C_6H_9O_3S_6Tl$	$Tl[S_2COCH_3]_3$	C:	MVol.D4-253
$C_6H_9O_6Pu$	$Pu(CH_3COO)_3$	Np:	TrU.D1-152
$C_6H_9O_6Pu^+$	$Pu(CH_3COO)_3^+$	Np:	TrU.D1-153
$C_6H_9O_6RbSn$	$RbSn(CH_3COO)_3$	Sn:	MVol.C3-140
$C_6H_9O_6Sn^-$	$Sn(CH_3COO)_3^-$	Sn:	MVol.C2-219/20
$C_6H_9O_8Pu^-$	$PuO_2(CH_3COO)_3^-$	Np:	TrU.D1-153
$C_6H_9Sn^+$	$Sn(C_3H_4)(CH_2CHCH_2)^+$	Sn:	Org.Verb.1-134
$C_6H_{10}HgO_2S_4$	$Hg[S_2COC_2H_5]_2$	C:	MVol.D4-258
$C_6H_{10}HgO_4S_2$	$Hg[SOCOC_2H_5]_2$	C:	MVol.D4-237
$C_6H_{10}Hg_2O_2S_4$	$Hg_2[S_2COC_2H_5]_2$	C:	MVol.D4-258
$C_6H_{10}Hg_{2.5}O_{3.5}S_4$	$Hg[S_2COC_2H_5]_2 \cdot 1.5\ HgO \cdot H_2O$	C:	MVol.D4-258
$C_6H_{10}I_2Ni_2$	$(C_3H_5NiI)_2$	Ni:	Org.Verb.2-280/1
$C_6H_{10}I_2O_3Sn$	$[SnI_2(CH(COCH_3)_2)OCH_3]$	Sn:	MVol.C5-106
$C_6H_{10}I_4Ni_3$	$2\ C_3H_5NiI \cdot NiI_2$	Ni:	Org.Verb.2-280
$C_6H_{10}MnNa_2O_6$	$Na_2[Mn(OCH(CH_2O)_2)_2]$	Mn:	MVol.D1-38
$C_6H_{10}MnO_2S_4$	$Mn[S_2COC_2H_5]_2$	C:	MVol.D4-258
$C_6H_{10}MnO_6$	$Mn(OCHCH(O)(CHOH)_4)$	Mn:	MVol.D1-39
$C_6H_{10}Mo_2O_4S_6$	$Mo_2O_2S_2[S_2COC_2H_5]_2$	C:	MVol.D4-258
$C_6H_{10}NNpO_7$	$(NpO_2)N(C_2H_4OH)(CH_2COOH)(CH_2COO)$	Np:	TrU.D1-162
$C_6H_{10}NNpO_8^{2-}$	$NpO_2(OH)N(C_2H_4OH)(CH_2COO)_2^{2-}$	Np:	TrU.D1-162
$C_6H_{10}N_2O_{12}S_3Sn$	$[C_6H_4(NH_3)_2]Sn(SO_4)_3$	Sn:	MVol.C3-122
$C_6H_{10}N_4O_4S_2U$	$UO_2(NCS)_2 \cdot 2\ CH_3CONH_2 \cdot H_2O$	U:	SVol.E1-96/8
$C_6H_{10}Ni$	$Ni(CH_2CHCH_2)_2$		
	Chemical reactions	Ni:	Org.Verb.2-55/9
	Formation and preparation	Ni:	Org.Verb.2-46/50
	Physical properties	Ni:	Org.Verb.2-50/4
	Solubility	Ni:	Org.Verb.2-55
–	$(CH_3CHCH)_2Ni$	Ni:	Org.Verb.1-70, 76
$C_6H_{10}NiO_2S_4$	$Ni[S_2COC_2H_5]_2$	C:	MVol.D4-245, 247/8, 260
$C_6H_{10}NiO_2Se_4$	$Ni[Se_2COC_2H_5]_2$	C:	MVol.D6-219, 221
$C_6H_{10}NiO_4S_2$	$Ni[SOCOC_2H_5]_2 \cdot 2\ C_5H_5N$	C:	MVol.D4-237
$C_6H_{10}NiS_6$	$Ni[S_2CSC_2H_5]_2$	C:	MVol.D4-267, 270
$C_6H_{10}NpO_8^-$	$NpO_2(CH_3CH(OH)COO)_2^-$	Np:	TrU.D1-155
$C_6H_{10}OSn$	$(CH_2CH)_3SnOH$	Sn:	Org.Verb.5-156
$C_6H_{10}O_2PbS_4$	$Pb[S_2COC_2H_5]_2$	C:	MVol.D4-245, 254
$C_6H_{10}O_2PdS_4$	$Pd[S_2COC_2H_5]_2$	C:	MVol.D4-261
$C_6H_{10}O_2PdSe_4$	$Pd[Se_2COC_2H_5]_2$	C:	MVol.D6-221
$C_6H_{10}O_2PtS_4$	$Pt[S_2COC_2H_5]_2$	C:	MVol.D4-261
$C_6H_{10}O_2S_3$	$[C_2H_5O(S)C]_2S$	C:	MVol.D6-171
$C_6H_{10}O_2S_4$	$C_2H_5OC(S)SSC(S)OC_2H_5$	C:	MVol.D4-251
$C_6H_{10}O_2S_4Se$	$Se[S_2COC_2H_5]_2$	C:	MVol.D4-255
$C_6H_{10}O_2S_4Sn$	$Sn[S_2COC_2H_5]_2$	C:	MVol.D4-253

$C_6H_{10}O_2S_4Te$	$Te[S_2COC_2H_5]_2$	C: MVol.D4-245, 255
		Te: SVol.B3-135/6
$C_6H_{10}O_2S_4Zn$	$Zn[S_2COC_2H_5]_2$	C: MVol.D4-245, 257
$C_6H_{10}O_2S_5$	$S[S_2COC_2H_5]_2$	C: MVol.D4-254
$C_6H_{10}O_2S_6$	$S_2[S_2COC_2H_5]_2$	C: MVol.D4-255
$C_6H_{10}O_2S_7$	$S_3[S_2COC_2H_5]_2$	C: MVol.D4-255
$C_6H_{10}O_2S_8$	$S_4[S_2COC_2H_5]_2$	C: MVol.D4-255
$C_6H_{10}O_2Se_3$	$C_2H_5OC(Se)SeC(Se)OC_2H_5$	C: MVol.D6-220
$C_6H_{10}O_2Se_4$	$[Se_2COC_2H_5]_2$	C: MVol.D6-220
$C_6H_{10}O_4PbS_2$	$Pb[SOCOC_2H_5]_2$	C: MVol.D4-237
$C_6H_{10}O_4S_2$	$C_2H_5OC(O)SS(O)COC_2H_5$	C: MVol.D4-236/7
$C_6H_{10}O_4S_2Te$	$Te(SCH_2CH_2COOH)_2$	Te: SVol.B3-134
$C_6H_{10}O_4S_3$	$C_2H_5OC(O)S_3(O)COC_2H_5$	C: MVol.D4-236/7
$C_6H_{10}O_4S_4$	$C_2H_5OC(O)S_4(O)COC_2H_5$	C: MVol.D4-236/7
$C_6H_{10}O_4S_4U$	$UO_2[S_2COC_2H_5]_2$	C: MVol.D4-261
$C_6H_{10}O_4S_5$	$C_2H_5OC(O)S_5(O)COC_2H_5$	C: MVol.D4-236/7
$C_6H_{10}O_4S_6$	$C_2H_5OC(O)S_6(O)COC_2H_5$	C: MVol.D4-236/7
$C_6H_{10}O_4S_7$	$C_2H_5OC(O)S_7(O)COC_2H_5$	C: MVol.D4-236/7
$C_6H_{10}O_4S_8$	$C_2H_5OC(O)S_8(O)COC_2H_5$	C: MVol.D4-236/7
$C_6H_{10}O_4Sn$	$(CH_3)_2Sn(OCOCH_2)_2$	Sn: Org.Verb.6-28
$C_6H_{10}O_6S_6Sn^{4-}$	$[Sn(SCH_2CH(S)CH_2SO_3)_2]^{4-}$	Sn: MVol.C5-49/50
$C_6H_{10}O_6Sn$	$(CH_3)_2Sn(OCO(CHOH)_2COO)$	Sn: Org.Verb.6-25
–	$Sn(OCOCH(OH)CH_3)_2$	Sn: MVol.C2-225
$C_6H_{10}S_5$	$S[(S)CSC_2H_5]_2$	C: MVol.D4-268
$C_6H_{10}S_7$	$S[S_2CSC_2H_5]_2$	C: MVol.D4-269
$C_6H_{10}S_8$	$S_2[S_2CSC_2H_5]_2$	C: MVol.D4-269
$C_6H_{10}Sn^+$	$Sn(CH_2CHCH_2)_2{}^+$	Sn: Org.Verb.1-134
$C_6H_{11}IN_2NiS$	$C_3H_5Ni(SC(NHCH_2)_2)I$	Ni: Org.Verb.2-281
$C_6H_{11}KOS_2$	$K[S_2COC_5H_{11}]$	C: MVol.D4-249, 252/3
$C_6H_{11}KOSe_2$	$K[Se_2COC_5H_{11}]$	C: MVol.D6-221
$C_6H_{11}KO_2S$	$K[SOCOC_5H_{11}]$	C: MVol.D4-237
$C_6H_{11}MnO_9P$	$[Mn(C_5H_5O(OH)_4CH_2OPO_3)]$	Mn: MVol.D1-147
$C_6H_{11}NO_8U$	$UO_2(CH_3COO)_2 \cdot NH_2CH_2COOH$	U: SVol.E1-18, 21
$C_6H_{11}NSn$	$(CH_3)_3SnC(CN)CH_2$	Sn: Org.Verb.2-77, 84
–	$(CH_3)_3SnCHCHCN$	Sn: Org.Verb.2-77, 84
$C_6H_{11}N_3O_8U$	$[UO_2(CH_3COO)_2(NH(CONH_2)_2)] \cdot 2\ H_2O$	U: SVol.E1-93, 95
$C_6H_{11}NaOS_2$	$Na[S_2COC_5H_{11}]$	C: MVol.D4-253
$C_6H_{11}NiO_2P^+$	$(CO)_2Ni(CH_2P(CH_3)_3)^+$	Ni: Org.Verb.1-182
$C_6H_{11}NpO_5$	$NpO_2(CH_3CH_2CH_2CH_2CH(OH)COO)$	Np: TrU.D1-154
$C_6H_{11}OS_2^-$	$[S_2COC_5H_{11}]^-$	C: MVol.D4-249, 251/2
$C_6H_{11}OS_2Tl$	$Tl[S_2COC_5H_{11}]$	C: MVol.D4-253
$C_6H_{11}O_2S^-$	$[SOCOC_5H_{11}]^-$	C: MVol.D4-236/7
$C_6H_{12}HgOS_2$	$C_3H_7Hg[S_2COC_2H_5]$	C: MVol.D4-258
$C_6H_{12}I_2N_2Se_4Sn$	$[SnI_2((CH_3)_2NCSeSe)_2]$	Sn: MVol.C6-124/5
$C_6H_{12}I_2N_4S_2Te$	$Te(SC(NHCH_2)_2)_2I_2$	Te: SVol.B3-153/4
$C_6H_{12}I_4N_2S_4Sn$	$SnI_4 \cdot (CH_3)_2NCSSSCSN(CH_3)_2$	Sn: MVol.C6-118
$C_6H_{12}I_4N_4O_2Sn$	$SnI_4 \cdot 2\ CO(NHCH_2)_2$	Sn: MVol.C5-224/5
$C_6H_{12}I_4N_4S_2Sn$	$SnI_4 \cdot 2\ CH_2NHC(SH)NCH_2$	Sn: MVol.C6-104/5
$C_6H_{12}I_4O_2Te$	$[TeI_3((CH_3)_2CO)_2]I$	Te: SVol.B3-165/6
$C_6H_{12}IrN_3O_{12}$	$(NH_4)_3[Ir(C_2O_4)_3]$	Ir: SVol.2-35

$C_6H_{12}IrN_9$ .... $(NH_4)_3[Ir(CN)_6]$ .... Ir: SVol.2-182
$C_6H_{12}MnN_2O_2S_2$ .... $Mn(NCS)_2 \cdot 2\ C_2H_5OH$ .... Mn: MVol.D1-30
$C_6H_{12}MnN_3O_3S_3^-$ ... $[Mn(SCN)_3(CH_3OH)_3]^-$ .... Mn: MVol.D1-24
$C_6H_{12}NNaO_3S$ ...... $Na[C_6H_{11}NHSO_3]$ .... S: S-N-Verb.1-97
$C_6H_{12}N_2O_2S_4Sn$ ..... $Sn(NCS)_2 \cdot 2\ (CH_3)_2SO$ .... Sn: MVol.C5-45
$C_6H_{12}N_2O_2Sn$ ...... $Sn((CH_3)_2CNO)_2$ .... Sn: MVol.C5-41
$C_6H_{12}N_2O_6S$ .... $SO_2(NHCOOC_2H_5)_2$ .... S: S-N-Verb.1-183
$C_6H_{12}N_2O_9U$ .... $UO_2(NO_3)_2 \cdot CH_3COC_4H_9$ .... U: SVol.E1-79/80
– .... $UO_2(NO_3)_2 \cdot CH_3COC_4H_9 \cdot 2\ H_2O$ .... U: SVol.E1-79/80
$C_6H_{12}N_2O_{10}U$ .... $UO_2(NO_3)_2 \cdot 2\ (CH_3)_2CO$ .... U: SVol.E1-79/80
$C_6H_{12}N_2O_{12}U$ .... $UO_2(NO_3)_2 \cdot 2\ CH_3COOCH_3$ .... U: SVol.E1-120/2
$C_6H_{12}N_2S_4Se$ .... $Se(S_2CN(CH_3)_2)_2$ .... Se: SVol.A1-193
$C_6H_{12}N_2S_4Te$ .... $Te((CH_3)_2NCSS)_2$ .... Te: SVol.B3-137
$C_6H_{12}N_2Sn$ .... $(CH_3)_3SnC_3H_3N_2$ .... Sn: Org.Verb.2-153
$C_6H_{12}N_4O_2S$ .... $C_6H_{11}NHSO_2N_3$ .... S: S-N-Verb.1-214
$C_6H_{12}N_4O_4S_2$ .... $[C_6H_5NH_3][N(SO_2NH_2)_2]$ .... S: S-N-Verb.1-160
$C_6H_{12}N_4O_5S_2U$ ..... $[UO_2(NCS)_2(CH_3CONH_2)_2(H_2O)]$ .... U: SVol.E1-97/8
$C_6H_{12}N_4O_6S$ .... $SO_2(NCO)_2 \cdot 2\ (CH_3)_2NO$ .... S: S-N-Verb.1-183
$C_6H_{12}N_4S_2Te^{2+}$ ..... $[Te(SC(NHCH_2)_2)_2]^{2+}$ .... Te: SVol.B3-153/7
$C_6H_{12}N_6O_8U$ .... $UO_2(NO_3)_2 \cdot (CH_2)_6N_4$ .... U: SVol.E1-23/6
$C_6H_{12}N_6O_{10}U$ .... $UO_2(NO_3)_2 \cdot 2\ CO(NHCH_2)_2$ .... U: SVol.E1-92, 95
$C_6H_{12}N_{10}NiO_4$ ...... $(CH_3N_4C)_2Ni \cdot 2\ CH_3NO_2$ .... Ni: Org.Verb.1-87
$C_6H_{12}Ni$ .... $Ni(C_2H_4)_3$ .... Ni: Org.Verb.1-417/8
$C_6H_{12}OS_2$ .... $SC(OC_5H_{11})(SH)$ .... C: MVol.D4-238/43
$C_6H_{12}O_2SSn$ .... $(C_2H_5)_2Sn(SCH_2C(O)O)$ .... Sn: Org.Verb.6-59
$C_6H_{12}O_2S_2$ .... $SC(OC_5H_{10}OH)(SH)$ .... C: MVol.D4-241
$C_6H_{12}O_4Sn$ .... $(CH_3)_2Sn(OCOCH_3)_2$ .... Sn: Org.Verb.6-22
$C_6H_{12}O_5S_2Sn$ ...... $C_2H_5Sn(OH)(SCH_2COOH)_2$ .... Sn: Org.Verb.6-230/1
$C_6H_{12}O_6S_6Sn$ ...... $[Sn(SCH_2CH(SH)CH_2SO_3)_2]$ .... Sn: MVol.C5-49/50
$C_6H_{12}O_7SU$ .... $UO_2(CH_3COO)_2 \cdot (CH_3)_2SO$ .... U: SVol.E1-204, 208
$C_6H_{12}O_8S_2U$ .... $UO_2(C_2O_4) \cdot 2\ (CH_3)_2SO$ .... U: SVol.E1-208
$C_6H_{12}S_2Sn$ .... $(C_2H_5)_2Sn(SCHCHS)$ .... Sn: Org.Verb.6-59
$C_6H_{12}S_4Sn$ .... $Sn(SCH_2CH(S)CH_3)_2$ .... Sn: MVol.C6-83
– .... $Sn(SCH_2CH_2CH_2S)_2$ .... Sn: MVol.C6-84
$C_6H_{12}Sn$ .... $(CH_3)_2Sn(CHCH_2)_2$ .... Sn: Org.Verb.3-7, 16
– .... $(CH_3)_3SnCCCH_3$ .... Sn: Org.Verb.2-112, 116
– .... $(CH_3)_3SnCHCCH_2$ .... Sn: Org.Verb.2-71
– .... $(CH_3)_3SnCDCCH_2$ .... Sn: Org.Verb.2-71
– .... $(CH_3)_3SnCH_2CCH$ .... Sn: Org.Verb.2-111/2
– .... $(CH_3)_3SnCH_2CCD$ .... Sn: Org.Verb.2-111
$C_6H_{13}IMnO_{1.5}$ ...... $MnHI \cdot 1.5\ C_4H_8O$ .... Mn: MVol.D1-143
$C_6H_{13}MnO_2^{2+}$ ...... $[Mn(OC(CH_3)_2C(CH_3)_2OH)]^{2+}$ .... Mn: MVol.D1-36/7
$C_6H_{13}NO_6Sn$ .... $NH_4Sn(CH_3COO)_3$ .... Sn: MVol.C3-124
– .... $NH_4Sn(CH_3COO)_3 \cdot 2\ H_2O$ .... Sn: MVol.C3-124
$C_6H_{13}NSn$ .... $(CH_3)_3SnCH(CN)CH_3$ .... Sn: Org.Verb.2-36
– .... $(CH_3)_3SnCH_2CH_2CN$ .... Sn: Org.Verb.2-35/6
$C_6H_{13}N_3O_3S$ .... $C_5H_{10}NSO_2NHCONH_2$ .... S: S-N-Verb.1-171
$C_6H_{13}Ni$ .... $C_3H_5Ni(C_2H_5)CH_3$ .... Ni: Org.Verb.2-285
$C_6H_{13}Sn^+$ .... $(C_4H_9)(CH_2CH)SnH^+$ .... Sn: Org.Verb.3-41
$C_6H_{13}Ti^+$ .... $[((CH_3)_3CCH_2)TiCH_2]^+$ .... Ti: Org.Verb.1-99

$C_6H_{14}I_4N_2O_2Sn$	$SnI_4 \cdot 2\ OCHN(CH_3)_2$	Sn: MVol.C6-41/2
$C_6H_{14}I_4N_2O_2Te$	$TeI_4 \cdot 2\ HCON(CH_3)_2$	Te: SVol.B3-174
$C_6H_{14}Li_2Ni$	$C_2H_4Ni(LiC_2H_5)_2 \cdot (C_2H_5)_2O$	Ni: Org.Verb.1-340/1
$C_6H_{14}MnO_2{}^{3+}$	$[Mn(HOC(CH_3)_2C(CH_3)_2OH)]^{3+}$	Mn: MVol.D1-36/7
$C_6H_{14}MoO_6$	$MoO_2(OCH_2CHOHCH_3)_2$	Mo: SVol.B1-141
$C_6H_{14}N_2O_2Sn^{2+}$	$[Sn(OCHN(CH_3)_2)_2]^{2+}$	Sn: MVol.C5-43
$C_6H_{14}N_2O_8SU$	$UO_2SO_4 \cdot 2\ HCON(CH_3)_2$	U: SVol.E1-100/2
$C_6H_{14}N_2O_9U$	$UO_2(NO_3)_2 \cdot (C_3H_7)_2O \cdot 3\ H_2O$	U: SVol.E1-71, 76
$C_6H_{14}N_2O_{10}U$	$UO_2(NO_3)_2 \cdot C_2H_4(OC_2H_5)_2 \cdot 3\ H_2O$	U: SVol.E1-71, 76
$C_6H_{14}N_2O_{12}Sn$	$(NH_4)_2[Sn(HCOO)_6] \cdot 5\ H_2O$	Sn: MVol.C3-124
$C_6H_{14}N_4O_4S$	$SO_2[N(C_3H_7)NO]_2$	S: S-N-Verb.1-175
$C_6H_{14}N_4O_6S$	$SO_2[N(C_3H_7)NO_2]_2$	S: S-N-Verb.1-176
$C_6H_{14}N_4O_6S_2U$	$UO_2(CH_3COO)_2 \cdot 2\ (NH_2)_2CS$	U: SVol.E1-220/1
$C_6H_{14}N_4O_8U$	$UO_2(CH_3COO)_2 \cdot 2\ (NH_2)_2CO$	U: SVol.E1-84, 89
$C_6H_{14}N_4O_{10}U$	$UO_2(NO_3)_2 \cdot 2\ HCON(CH_3)_2$	U: SVol.E1-100/1
$C_6H_{14}N_4O_{12}U$	$UO_2(NO_3)_2 \cdot 2\ NH_2COOC_2H_5$	U: SVol.E1-120/3
$C_6H_{14}N_{10}O_4S_3U$	$CN_3H_6[UO_2(NCS)_3((NH_2)_2CO)_2]$	U: SVol.E1-84, 86, 88, 90
$C_6H_{14}NiP^+$	$(C_3H_5)Ni(P(CH_3)_3)^+$	Ni: Org.Verb.2-13/4
$C_6H_{14}OS_2Sn$	$(CH_3)_3Sn[S_2COC_2H_5]$	C: MVol.D4-253
$C_6H_{14}OSn$	$(CH_3)_3SnCH_2COCH_3$	Sn: Org.Verb.2-20
–	$[(C_3H_7)_2SnO]_x$	Sn: Org.Verb.6-68
$C_6H_{14}O_2Sn$	$(CH_3)_3SnCH_2OOCCH_3$	Sn: Org.Verb.2-11/2
–	$(C_6H_{13}SnO_2H)_x$	Sn: Org.Verb.6-250
$C_6H_{14}O_8U$	$UO_2(CH_3COO)_2 \cdot 2\ CH_3OH$	U: SVol.E1-59, 63
$C_6H_{14}Sn$	$(CH_2)_5CHSnH_3$	Sn: Org.Verb.4-127
–	$(CH_3)_2(CH_3CHCHCH_2)SnH$	Sn: Org.Verb.4-89/90
–	$(CH_3)_2Sn(CH_2)_4$	Sn: Org.Verb.6-18/9
–	$[(CH_3)_2Sn(CH_2)_4]_x$	Sn: Org.Verb.6-18/9
–	$(CH_3)_3SnC(CH_3)CH_2$	Sn: Org.Verb.2-74
–	$(CH_3)_3SnCHCHCH_3$	Sn: Org.Verb.2-74
	$(CH_3)_3SnCHCH_2CH_2$	Sn: Org.Verb.2-57
–	$(CH_3)_3SnCH_2CHCH_2$	Sn: Org.Verb.2-71/3
–	$[(C_3H_7)_2Sn]_x$	Sn: Org.Verb.6-67
$C_6H_{15}IrN_3O_7{}^+$	$[Ir(NH_2CH_2COO)_2(NH_2CH_2COOH)H_2O]^+$	Ir: SVol.2-81
$C_6H_{15}NO_2S$	$[(CH_3)_3NH][SOCOC_2H_5]$	C: MVol.D4-237
–	$[(CH_3)_4N][SOCOCH_3]$	C: MVol.D4-237
$C_6H_{15}NO_3S$	$C_3H_7NHSO_2OC_3H_7$	S: S-N-Verb.1-97
–	$C_4H_9NHSO_2OC_2H_5$	S: S-N-Verb.1-97
–	$(i\text{-}C_4H_9)NHSO_2OC_2H_5$	S: S-N-Verb.1-97
$C_6H_{15}NSn$	$(CH_3)_3SnCH_2N(CH_2)_2$	Sn: Org.Verb.2-12
$C_6H_{15}N_3O_4S_2$	$NH_2SO_2NHSO_2NHC_6H_{11}$	S: S-N-Verb.1-161
$C_6H_{15}N_3O_6S_3$	$[C_2H_5NSO_2]_3$	S: S-N-Verb.1-9
$C_6H_{15}N_3O_9SU$	$UO_2SO_4 \cdot 3\ CH_3CONH_2 \cdot H_2O$	U: SVol.E1-96/8
$C_6H_{15}N_3O_{12}SU$	$UO_2SO_4 \cdot 3\ NH_2CH_2COOH \cdot 2\ H_2O$	U: SVol.E1-18, 21, 25
$C_6H_{15}N_4OP$	$[(CH_3)_2N]_2P(NCN)OCH_3$	F: PerFHalOrg.5-23
$C_6H_{15}Sn^+$	$Sn(C_2H_5)_3{}^+$	Sn: Org.Verb.1-70/1
		Sn: Org.Verb.2-157
–	$Sn(C_3H_7)_2H^+$	Sn: Org.Verb.1-86
$C_6H_{16}I_4N_2Sn$	$SnI_4 \cdot C_2H_4(N(CH_3)_2)_2$	Sn: MVol.C5-184
$C_6H_{16}IrN_7O_3S_2$	$[Ir(NCS)_2(C_2H_4(NH_2)_2)_2]NO_3$	Ir: SVol.2-162

$C_6H_{16}MnN_2O_4S_2$	$[Mn(SCN)_2(CH_3OH)_4]$	Mn:	MVol.D1-24
$C_6H_{16}N_2O_2PtS_4$	$Pt[S_2COC_2H_5]_2(NH_3)_2$	C:	MVol.D4-261
$C_6H_{16}N_2O_2S$	$C_2H_5NHSO_2NHC_4H_9$	S:	S-N-Verb.1-170
–	$(C_2H_5)_2NSO_2N(CH_3)_2$	S:	S-N-Verb.1-173
–	$C_3H_7NHSO_2NHC_3H_7$	S:	S-N-Verb.1-169
–	$(C_3H_7)_2NSO_2NH_2$	S:	S-N-Verb.1-168
–	$(i\text{-}C_3H_7NH)_2SO_2$	S:	S-N-Verb.1-169
$C_6H_{16}N_2O_2S_4Zn$	$Zn[S_2COC_2H_5]_2(NH_3)_2$	C:	MVol.D4-257
$C_6H_{16}N_2O_{10}S_3$	$(C_2H_5OSO_2N(OCH_3))_2SO_2$	S:	S-N-Verb.1-23
$C_6H_{16}N_6O_{10}U$	$UO_2(NO_3)_2 \cdot 2\ CO(NHCH_3)_2$	U:	SVol.E1-92, 95
–	$UO_2(NO_3)_2 \cdot 2\ NH_2CONHC_2H_5$	U:	SVol.E1-92, 95
$C_6H_{16}N_{10}S_6Te$	$Te(SC(NH_2)_2)_4(SCN)_2$	Te:	SVol.B3-146
$C_6H_{16}OSSi_2Sn$	$[O(Si(CH_3)_2CH_2)_2SnS]_x$	Sn:	Org.Verb.6-207
$C_6H_{16}OSn$	$(CH_3)_3SnCH_2CH_2CH_2OH$	Sn:	Org.Verb.2-43, 46
–	$(CH_3)_3SnCH_2OC_2H_5$	Sn:	Org.Verb.2-15
–	$(C_2H_5)_3SnOH$	Sn:	Org.Verb.5-105
$C_6H_{16}O_2Si_2Sn$	$[O(Si(CH_3)_2CH_2)_2SnO]_x$	Sn:	Org.Verb.6-207
$C_6H_{16}O_3Si_2Sn$	$[(OSi(CH_3)_2CH_2)_2SnO]_x$	Sn:	Org.Verb.6-138
$C_6H_{16}SSn$	$(CH_3)_3SnSC_3H_7$	Sn:	Org.Verb.5-81
$C_6H_{16}Sn$	$(CH_3)_2Sn(C_2H_5)_2$	Sn:	Org.Verb.3-3
–	$(CH_3)_3SnC_3H_7$	Sn:	Org.Verb.2-41/2
–	$(CH_3)_3Sn(i\text{-}C_3H_7)$	Sn:	Org.Verb.2-42
–	$(C_2H_5)(C_4H_9)SnH_2$	Sn:	Org.Verb.4-119
–	$(C_2H_5)_3SnH$		
	Chemical reactions	Sn:	Org.Verb.4-22/36
	Formation and preparation	Sn:	Org.Verb.4-19
	Molecule	Sn:	Org.Verb.4-20/1
	Physical properties	Sn:	Org.Verb.4-21/2
	Spectra	Sn:	Org.Verb.4-20/1
	Uses	Sn:	Org.Verb.4-36
–	$(C_2H_5)_3SnD$	Sn:	Org.Verb.4-19/23, 31
–	$(C_3H_7)_2SnH_2$	Sn:	Org.Verb.4-102/3
–	$(i\text{-}C_3H_7)_2SnH_2$	Sn:	Org.Verb.4-103/4
$C_6H_{17}IrN_6OS_2$	$[Ir(NCS)_2(C_2H_4(NH_2)_2)_2]OH$	Ir:	SVol.2-162
$C_6H_{17}KOSiSn$	$(CH_3)_3SnCH_2Si(CH_3)_2OK$	Sn:	Org.Verb.2-14/5
$C_6H_{17}NSn$	$(CH_3)_3SnCH_2N(CH_3)_2$	Sn:	Org.Verb.2-12
$C_6H_{17}N_3O_2S$	$(C_3H_7)_2NSO_2NHNH_2$	S:	S-N-Verb.1-222
$C_6H_{17}N_3O_4S_2$	$NH_2SO_2NSO_2N(C_2H_5)_3$	S:	S-N-Verb.1-160
$C_6H_{17}N_9O_{10}U$	$UO_2(OH)_2 \cdot 3\ NH(CONH_2)_2$	U:	SVol.E1-93, 95
$C_6H_{18}I_2N_4S_4$	$[(NS(CH_3)_2NS(CH_3))_2]I_2$	S:	S-N-Verb.1-5
$C_6H_{18}I_4N_2Sn$	$SnI_4 \cdot 2\ (CH_3)_3N$	Sn:	MVol.C5-172/3
$C_6H_{18}I_4O_2P_2Sn$	$SnI_4 \cdot 2\ (CH_3)_3PO$	Sn:	MVol.C6-143/4
$C_6H_{18}I_4O_6P_2Sn$	$SnI_4 \cdot 2\ CH_3P(O)(OCH_3)_2$	Sn:	MVol.C6-162/3
–	$SnI_4 \cdot 2\ P(OCH_3)_3$	Sn:	MVol.C6-141/2
$C_6H_{18}MnO_{16}P_4^-$	$[Mn(C_2(OH)_2(CH_3)_4)(H_2P_2O_7)_2]^-$	Mn:	MVol.D1-37
$C_6H_{18}N_2OSSi_2$	$SO[NSi(CH_3)_3]_2$	S:	S N Verb.1 197/8
$C_6H_{18}N_2OSSi_2^+$	$SO[NSi(CH_3)_3]_2^+$	S:	S-N-Verb.1-198
$C_6H_{18}N_2O_{10}P_2U$	$UO_2(NO_3)_2 \cdot 2\ (CH_3)_3PO$	U:	SVol.E1-148, 156
$C_6H_{18}N_2SSi_2^+$	$[(CH_3)_3SiN]_2S^+$	S:	S-N-Verb.1-211
$C_6H_{18}N_3SSiSn^+$	$(CH_3)_3SiNSN_2Sn(CH_3)_3^+$	S:	S-N-Verb.1-211

$C_6H_{18}N_4Ni$	$(CH_3CC)_2Ni(NH_3)_4$	Ni: Org.Verb.1-81/2
$C_6H_{18}N_4O_4S_2$	$NH_2SO_2NH(CH_2)_6NHSO_2NH_2$	S: S-N-Verb.1-168
$C_6H_{18}N_4O_6S_3$	$SO_2[N(CH_3)SO_2N(CH_3)_2]_2$	S: S-N-Verb.1-180
$C_6H_{18}N_4O_{10}U$	$UO_2(NO_3)_2 \cdot 2\ (CH_3)_3NO$	U: SVol.E1-131/2, 135
$C_6H_{18}N_4O_{14}P_2U$	$U(NO_3)_4 \cdot 2\ (CH_3)_3PO$	U: SVol.E1-156
$C_6H_{18}N_4O_{15}S_3U$	$U(NO_3)_4 \cdot 3\ (CH_3)_2SO$	U: SVol.E1-203, 207
$C_6H_{18}N_6O_2Sn_3$	$(CH_3)_2Sn[OSn(CH_3)_2N_3]_2$	Sn: Org.Verb.6-32
$C_6H_{18}N_6O_6S_4$	$[(NS(CH_3)_2NS(CH_3))_2](NO_3)_2$	S: S-N-Verb.1-5
$C_6H_{18}N_8Ni_2$	$(NC)Ni(NH_3)_3C_4Ni(NH_3)_3(CN) \cdot HCN$	Ni: Org.Verb.2-269
$C_6H_{18}N_8O_{11}U$	$UO_2(NO_3)_2 \cdot 3\ CH_3CONHNH_2 \cdot n\ H_2O$	U: SVol.E1-114/6
$C_6H_{18}SSn_2$	$[(CH_3)_3Sn]_2S$	Sn: Org.Verb.5-84
$C_6H_{18}S_3Sn_3$	$[(CH_3)_2SnS]_3$	Sn: Org.Verb.6-25
–	$(C_2H_5SnHS)_3$	Sn: Org.Verb.6-230/1
$C_6H_{18}SiSn$	$(CH_3)_3SnCH_2SiH(CH_3)_2$	Sn: Org.Verb.2-14/5
$C_6H_{18}Sn_2$	$[(CH_3)_3Sn]_2$	Sn: Org.Verb.5-78
$C_6H_{19}IrK_2N_6$	$K_2[Ir(H_2NCH_2CH_2NH)(HNCH_2CH_2NH)_2]$	Ir: SVol.2-86
$C_6H_{19}NO_3SSi_2$	$(CH_3)_3SiNHSO_2OSi(CH_3)_3$	S: S-N-Verb.1-97
$C_6H_{19}N_3SSi_2$	$(CH_3)_2S[NSi(CH_3)_2]_2NH$	S: S-N-Verb.1-32
$C_6H_{20}MnNO_5S^+$	$[Mn(SCN)(CH_3OH)_5]^+$	Mn: MVol.D1-24
$C_6H_{20}N_2O_2SSi_2$	$(CH_3)_3SiNHSO_2NHSi(CH_3)_3$	S: S-N-Verb.1-170
$C_6H_{20}N_2O_{13}U_4$	$[(CH_3)_3NH]_2O \cdot 4\ UO_3 \cdot n\ H_2O$	U: SVol.C3-65/6
$C_6H_{20}N_2O_{25}U_8$	$[(CH_3)_3NH]_2O \cdot 8\ UO_3 \cdot n\ H_2O$	U: SVol.C3-65/6
$C_6H_{20}N_2O_{49}U_{16}$	$[(CH_3)_3NH]_2O \cdot 16\ UO_3 \cdot n\ H_2O$	U: SVol.C3-65/6
$C_6H_{20}N_4O_4S_2$	$[(CH_3)_2NH_2][N(SO_2N(CH_3)_2)_2]$	S: S-N-Verb.1-162
$C_6H_{21}IrN_{15}^{3+}$	$[Ir((NH_2CNH)_2NH)_3]^{3+}$	Ir: SVol.2-94, 124
$C_6H_{21}IrN_{16}O_3$	$[Ir((NH_2C(NH))_2NH)_3]NO_3$	Ir: SVol.2-94
$C_6H_{22}IIrN_6$	$[Ir(C_2H_4(NH_2)_2)(H_2NCH_2CH_2NH)_2]I$	Ir: SVol.2-86
$C_6H_{23}IrP_2$	$Ir(P(CH_3)_3)_2H_5$	Ir: SVol.2-195
$C_6H_{24}I_3IrN_6$	$[Ir(C_2H_4(NH_2)_2)_3]I_3$	Ir: SVol.2-82
–	$[Ir(C_2H_4(NH_2)_2)_3]I_3 \cdot H_2O$	Ir: SVol.2-82
$C_6H_{24}I_4N_6Sn$	$SnI_4 \cdot 3\ C_2H_8N_2$	Sn: MVol.C5-183
$C_6H_{24}IrN_6^{2+}$	$[Ir(C_2H_4(NH_2)_2)_3]^{2+}$	Ir: SVol.2-81
$C_6H_{24}IrN_6^{3+}$	$[Ir(C_2H_4(NH_2)_2)_3]^{3+}$	Ir: SVol.2-82
$C_6H_{24}IrN_9O_6$	$[Ir(C_2H_4(NH_2)_2)_3](NO_2)_3$	Ir: SVol.2-82
$C_6H_{24}IrN_9O_9$	$[Ir(C_2H_4(NH_2)_2)_3](NO_3)_3$	Ir: SVol.2-82
$C_6H_{24}IrN_{12}S_6^{3+}$	$[Ir(SC(NH_2)_2)_6]^{3+}$	Ir: SVol.2-124, 168
$C_6H_{24}MnN_2O_{12}$	$[Mn(CH_3OH)_6](NO_3)_2$	Mn: MVol.D1-25
$C_6H_{24}MnO_6^{2+}$	$Mn(CH_3OH)_6^{2+}$	Mn: MVol.C5-77
		Mn: MVol.D1-24/5
$C_6H_{24}N_{12}OsS_6^{3+}$	$[Os((NH_2)_2CS)_6]^{3+}$	C: MVol.D6-108
$C_6H_{24}N_{12}S_6Te_2^{4+}$	$[Te(SC(NH_2)_2)_3]_2^{4+}$	Te: SVol.B3-142/3
$C_6H_{24}N_{14}O_{14}U$	$UO_2(NO_3)_2 \cdot 6\ (NH_2)_2CO$	U: SVol.E1-83, 88/9
$C_6H_{25}O_{10}Sn_3^-$	$[(C_2H_5Sn)_3(OH)_{10}]^-$	Sn: Org.Verb.6-230
$C_6H_{26}I_4N_{12}O_{12}U_2$	$[UO_2(OH)((NH_2)_2CO)_3]_2I_4$	U: SVol.E1-87
$C_6H_{28}N_6NiO_2S_4$	$Ni[S_2COC_2H_5]_2(NH_3)_6$	C: MVol.D4-260
$C_6H_{29}O_6S_8Sn_3^{5-}$	$[(C_2H_5Sn)_3(OH)_6(SH)_8]^{5-}$	Sn: Org.Verb.6-230/1
$C_6H_{33}O_{21}PrS_3$	$[Pr(H_2O)_9](C_2H_5SO_4)_3$	Sc: MVol.B1-23/4
$C_6Hg_2N_6NiSe_6$	$Ni(SeCN)_2 \cdot 2\ Hg(SeCN)_2$	C: MVol.D6-229
C_6InIrN_6	$In[Ir(CN)_6]$	Ir: SVol.2-183
$C_6IrK_2O_{12}$	$K_2[Ir(C_2O_4)_3]$	Ir: SVol.2-35

$C_6IrK_3N_6$	$K_3[Ir(CN)_6]$	Ir:	SVol.2-181/2
$C_6IrK_3O_{12}$	$K_3[Ir(C_2O_4)_3] \cdot 2\ H_2O$	Ir:	SVol.2-33/4
–	$K_3[Ir(C_2O_4)_3] \cdot 4.5\ H_2O$	Ir:	SVol.2-33/5
$C_6IrLi_3N_6$	$Li_3[Ir(CN)_6]$	Ir:	SVol.2-182
$C_6IrLi_3O_{12}$	$Li_3[Ir(C_2O_4)_3]$	Ir:	SVol.2-33
$C_6IrN_6^{3-}$	$[Ir(CN)_6]^{3-}$	Ir:	SVol.2-181/2
$C_6IrN_6Na_3S_6$	$Na_3[Ir(SCN)_6]$	Ir:	SVol.2-161
$C_6IrN_6S_6^{3-}$	$[Ir(SCN)_6]^{3-}$	Ir:	SVol.2-124, 161
$C_6IrNa_3O_{12}$	$Na_3[Ir(C_2O_4)_3]$	Ir:	SVol.2-33
$C_6IrO_{12}^{2-}$	$[Ir(C_2O_4)_3]^{2-}$	Ir:	SVol.2-35
$C_6IrO_{12}^{3-}$	$[Ir(C_2O_4)_3]^{3-}$	Ir:	SVol.2-34/5
$C_6K_2N_6S_6Sn$	$K_2[Sn(SCN)_6] \cdot 4\ H_2O$	Sn:	MVol.C3-80
$C_6MgN_6S_6Sn$	$Mg[Sn(SCN)_6] \cdot 6\ H_2O$	Sn:	MVol.C3-168
C_6MnO_6	$[Mn(C_6O_6)]$	Mn:	MVol.D1-133
$C_6MnO_{12}^{3-}$	$[Mn(C_2O_4)_3]^{3-}$	Mn:	MVol.C2-82/3
$C_6N_6Na_2S_6Sn$	$Na_2[Sn(SCN)_6] \cdot 6\ H_2O$	Sn:	MVol.C3-35
$C_6N_6O_6Sn^{2-}$	$[Sn(NCO)_6]^{2-}$	Sn:	MVol.C2-227
$C_6N_6S_6SnSr$	$Sr[Sn(SCN)_6] \cdot 12\ H_2O$	Sn:	MVol.C3-190
$C_6Ni_2O_6$	$Ni_2(CO)_6$	Ni:	Org.Verb.2-262
$C_6Ni_2O_6^-$	$[Ni_2(CO)_6]^-$	Ni:	Org.Verb.1-215
$C_6Ni_2O_6^{2-}$	$[Ni_2(CO)_6]^{2-}$	Ni:	Org.Verb.2-265
$C_6Ni_2O_{12}P_4$	$(CO)_3NiP_4O_6Ni(CO)_3$	Ni:	Org.Verb.2-264
$C_6NpO_{12}^{2-}$	$Np(C_2O_4)_3^{2-}$	Np:	TrU.D1-158
C_6O_6Ra	$Ra(C_6O_6)$	Ra:	SVol.2-372
C_6O_6W	$W(CO)_6$ systems		
	$W(CO)_6$-$Fe(CO)_5$	Fe:	Org.Verb.B3-64
$C_6O_{12}Pa^-$	$Pa(C_2O_4)_3^-$	Pa:	SVol.2-169/70
$C_6O_{12}Pu^{2-}$	$Pu(C_2O_4)_3^{2-}$	Np:	TrU.D1-158
$C_6O_{12}Pu^{3-}$	$Pu(C_2O_4)_3^{3-}$	Np:	TrU.D1-158
$C_6O_{12}Pu_2$	$Pu_2(C_2O_4)_3 \cdot 9\ H_2O$		
	Solubility	Np:	TrU.D1-156
$C_6O_{12}Sn^{4-}$	$Sn(C_2O_4)_3^{4-}$	Sn:	MVol.C2-223/4
$C_{6.6}ClH_{6.6}N_{2.2}Ni$	$(NCCHCH_2)_{2.2}NiCl$	Ni:	Org.Verb.1-386
$C_7CdF_3NO_4$	$NC_5F_3(COO)_2Cd$	F:	PerFHalOrg.5-193, 207
$C_7CfH_8NO_6$	$Cf(N(CH_2CH_2COO)(CH_2COO)_2)$	Np:	TrU.D1-164
$C_7CfH_9NO_6^+$	$Cf(N(CH_2CH_2COO)(CH_2COO)_2H)^+$	Np:	TrU.D1-164
$C_7ClFH_{12}NO_4P$	$OP(ONCFCl)(OC(CH_3)_2)_2$	F:	PerFHalOrg.7-202
$C_7ClF_2H_7N_4$	$NCFC(NHNC(CH_3)_2)NCFCCl$	F:	PerFHalOrg.6-71
$C_7ClF_3H_2N_4$	$(CF_3)C_6H_2ClN_4$	F:	PerFHalOrg.6-70/1
$C_7ClF_3H_3NOS$	$(CH_3COS)C_5F_3ClN$	F:	PerFHalOrg.5-141
$C_7ClF_3H_4N_4O$	$(CF_3)(HO)C_6H_3ClN_4$	F:	PerFHalOrg.6-70
$C_7ClF_3H_6N_2$	$(CH_3)_2NC_5F_3ClN$	F:	PerFHalOrg.5-138
$C_7ClF_3H_6N_4O_2$	$(C_2H_5NH)(NO_2)(CF_3)C_4ClN_2$	F:	PerFHalOrg.6-60
$C_7ClF_3H_8N_4$	$(C_2H_5NH)(NH_2)(CF_3)C_4ClN_2$	F:	PerFHalOrg.6-59
$C_7ClF_4HO_2$	ClC_6F_4COOH	F:	PerFHalOrg.4-98
$C_7ClF_4H_8NO$	$CF_3CFClNC_4H_5(CH_3)O$	F:	PerFHalOrg.7-194/5
C_7ClF_5HNO	$C_6F_5C(Cl)NOH$	F:	PerFHalOrg.7-88, 105
$C_7ClF_5H_4Si$	$C_6F_5(CH_3)SiHCl$	F:	PerFHalOrg.4-80
$C_7ClF_6HN_4$	$(CF_3)_2C_5HClN_4$	F:	PerFHalOrg.6-126/7, 139

$C_7ClF_6H_9N_2O$	$(CH_3)_3CNCClON(CF_3)_2$	F:	PerFHalOrg.7-144
$C_7ClF_7H_{10}Si$	$C_3F_7Si(C_2H_5)_2Cl$	F:	PerFHalOrg.4-19
C_7ClF_7Mg	$CF_3C_6F_4MgCl$	F:	PerFHalOrg.4-63, 65
C_7ClF_8N	$CF_3CFClC_5F_4N$	F:	PerFHalOrg.5-149/50, 161, 178
$C_7ClF_{10}H_3N_2O$	$(NH_2)(CF_3)_2(CHF_2)(CClF_2)C_3NO$	F:	PerFHalOrg.7-65/6
$C_7ClF_{10}N_3$	$(C_2F_5)_2C_3ClN_3$	F:	PerFHalOrg.6-79/80, 100, 112
$C_7ClF_{11}H_2N_2O$	$(CF_3)_2CC(NH_2)NC(CF_3)(CF_2Cl)O$	F:	PerFHalOrg.5-56, 78
$C_7ClF_{11}H_3N_3$	$NH_2CNC(CF_2Cl)(CF_3)NHC(CF_3)_2$	F:	PerFHalOrg.5-63/4, 86
–	$NH_2CNC(CF_3)_2NHC(CF_3)CF_2Cl$	F:	PerFHalOrg.5-63/4, 86
$C_7ClF_{12}HN_2$	$ClCNC(CF_3)_2NHC(CF_3)_2$	F:	PerFHalOrg.5-63/4, 87
$C_7ClF_{12}NO$	$(CF_3)_2CCClNC(CF_3)_2O$	F:	PerFHalOrg.5-56, 78
$C_7ClF_{12}NO_2$	$NC(CF_3)_2OC(CF_3)_2OCCl$	F:	PerFHalOrg.6-4/5, 10
$C_7ClF_{14}H_3Si$	$CH_3(C_3F_7)_2SiCl$	F:	PerFHalOrg.4-18
$C_7ClF_{15}N_2O_2$	$CF_3C(ON(CF_3)_2)C(ON(CF_3)_2)Cl$	F:	PerFHalOrg.7-97/8, 115
$C_7ClFeH_3O_5$	$ClC(O)CHCH_2Fe(CO)_4$	Fe:	Org.Verb.B4-236, 247/8
$C_7ClFeH_5O_2$	$C_5H_5Fe(CO)_2Cl$	Fe:	Org.Verb.A2-245
$C_7ClFeH_5O_4$	$(ClCHCHC(CH_3)O)Fe(CO)_3$	Fe:	Org.Verb.B5-133, 146
$C_7ClFeH_5O_4Pd$	$(CO)_4FePd(C_3H_5)Cl$	Fe:	Org.Verb.B2-162
$C_7ClFeH_6O^+$	$[FeC_5H_4COCH_2Cl]^+$	Fe:	Org.Verb.A2-245
$C_7ClFeH_7O_3$	$CH_3C_3H_4Fe(CO)_3Cl$	Fe:	Org.Verb.B5-36, 40/1, 44/5, 48, 51, 53, 60/1, 63
$C_7ClFeH_7O_4$	$CH_3OC_3H_4Fe(CO)_3Cl$	Fe:	Org.Verb.B5-53, 63
$C_7ClFeH_7O_7$	$[CH_3C_3H_4Fe(CO)_3]ClO_4$	Fe:	Org.Verb.B5-82/6
$C_7ClFeH_{13}NO_5P$	$ClC_3H_4Fe(CO)(NO)P(OCH_3)_3$	Fe:	Org.Verb.B5-108/11
$C_7ClFe_2GeH_{13}N_4O_4$	$(C_3H_5Fe(NO)_2)_2Ge(Cl)CH_3$	Fe:	Org.Verb.C2-165
$C_7ClFe_2H_{13}N_4O_4Pb$	$(C_3H_5Fe(NO)_2)_2Pb(Cl)CH_3$	Fe:	Org.Verb.C2-165
$C_7ClFe_2H_{13}N_4O_4Sn$	$(C_3H_5Fe(NO)_2)_2Sn(Cl)CH_3$	Fe:	Org.Verb.C2-165
$C_7ClH_3MnNO_4{}^+$	$[Mn(OC_6H_2Cl(NO_2)CHO)]^+$	Mn:	MVol.D1-53/4
$C_7ClH_4MnO_2{}^+$	$[Mn(OC_6H_3(Cl)CHO)]^+$	Mn:	MVol.D1-53/4
$C_7ClH_5N_2O_4S$	$NO_2C_6H_4CHNSO_2Cl$	S:	S-N-Verb.1-126
$C_7ClH_5N_2O_6S$	$ClSO_2NHCOOC_6H_4NO_2$	S:	S-N-Verb.1-124/5
C_7ClH_5O	C_6H_5COCl	C:	MVol.D3-101
$C_7ClH_6MnO_5Sn$	$(CH_3)_2ClSnMn(CO)_5$	Sn:	Org.Verb.6-33
$C_7ClH_6NO_2S$	$ClSO_2NCHC_6H_5$	S:	S-N-Verb.1-126
$C_7ClH_6NO_3S_2$	$ClSO_2NHCOSC_6H_5$	S:	S-N-Verb.1-124/5
$C_7ClH_6NO_4S$	$ClSO_2NHCOOC_6H_5$	S:	S-N-Verb.1-124/5
C_7ClH_6NiO	$[C_6H_6Ni(CO)]Cl$	Ni:	Org.Verb.2-221, 225
$C_7ClH_7N_2O_5S_2$	$ClSO_2NHCONHSO_2C_6H_5$	S:	S-N-Verb.1-125
C_7ClH_7Ni	$C_6H_5CH_2NiCl$	Ni:	Org.Verb.1-16
$C_7ClH_7NiO_2S$	$CH_3C_6H_4SO_2NiCl$	Ni:	Org.Verb.1-233
$C_7ClH_7Sn^+$	$C_6H_5CH_2SnCl^+$	Sn:	Org.Verb.6-119
$C_7ClH_8NiO_5P$	$(CO)_3NiP(C_2H_2O_2(CH_3)_2)Cl$	Ni:	Org.Verb.1-178, 179
$C_7ClH_9MnO_4$	$[Mn(CH(COCH_3)_2)CH_2ClCOO]$	Mn:	MVol.D1-75
C_7ClH_9Ni	$[C_6H_6Ni(CH_3)]Cl$	Ni:	Org.Verb.2-221, 225
$C_7ClH_{10}NiO_3Sb$	$(CO)_3NiSb(C_2H_5)_2Cl$	Ni:	Org.Verb.1-188
$C_7ClH_{11}S_2Ti$	$C_5H_5Ti(SCH_3)_2Cl$	Ti:	Org.Verb.1-189
$C_7ClH_{11}Sn$	$(CH_3)_2(C_5H_5)SnCl$	Sn:	Org.Verb.5-193, 197

$C_7ClH_{12}NO_4S$	$ClSO_2NHCOOC_6H_{11}$	S:	S-N-Verb.1-124/5
$C_7ClH_{13}Sn$	$CH_2(CH_2CH_2)_2Sn(CHCH_2)Cl$	Sn:	Org.Verb.5-230
$C_7ClH_{15}MgNi$	$(CH_3)_2C_3H_3Ni(CH_3)ClMgCH_3 \cdot (C_2H_5)_2O$	Ni:	Org.Verb.2-315
$C_7ClH_{15}N_2NiS$	$C_3H_5Ni[SC(NH_2)(NHCH(CH_3)_2)]Cl$	Ni:	Org.Verb.2-275
$C_7ClH_{15}OSn$	$(CH_3)_2[CH_3CO(CH_2)_3]SnCl$	Sn:	Org.Verb.5-194, 198
$C_7ClH_{15}O_2Sn$	$(CH_3)_2(C_2H_5OCOCH_2CH_2)SnCl$	Sn:	Org.Verb.6-197
–	$(C_2H_5)_2(CH_3OCOCH_2)SnCl$	Sn:	Org.Verb.5-201, 204
$C_7ClH_{15}Sn$	$(CH_3)(C_4H_9)(CH_2CH)SnCl$	Sn:	Org.Verb.5-225
$C_7ClH_{16}NS_2Sn$	$(CH_3)_2ClSnSSCN(C_2H_5)_2$	Sn:	Org.Verb.6-22
$C_7ClH_{17}OSn$	$(C_2H_5)_2(C_2H_5OCH_2)SnCl$	Sn:	Org.Verb.5-201, 204
$C_7ClH_{17}O_2Ti$	$CH_3Ti(OC_3H_7\text{-i})_2Cl$	Ti:	Org.Verb.1-59/60
$C_7ClH_{17}O_3Ti$	$CH_3OTi(OC_3H_7\text{-i})_2Cl$	Ti:	Org.Verb.1-60
$C_7ClH_{17}Sn$	$(CH_3)(C_2H_5)(C_4H_9)SnCl$	Sn:	Org.Verb.5-225
–	$(CH_3)_2(C_5H_{11})SnCl$	Sn:	Org.Verb.5-194, 198
–	$(C_2H_5)_2(C_3H_7)SnCl$	Sn:	Org.Verb.5-201
–	$(C_2H_5)_3SnCH_2Cl$	Sn:	Org.Verb.2-163/4
$C_7ClH_{18}NOSn$	$(CH_3)_3SnCl \cdot CH_3CON(CH_3)_2$	Sn:	Org.Verb.5-87
$C_7ClH_{18}NSSn$	$(CH_3)_3SnCl \cdot CH_3CSN(CH_3)_2$	Sn:	Org.Verb.5-87
$C_7ClH_{19}N_2O_2S$	$[(C_2H_5)_2NSO_2N(CH_3)_3]Cl$	S:	S-N-Verb.1-172
$C_7ClH_{19}OSi_2Sn$	$O(Si(CH_3)_2CH_2)_2Sn(CH_3)Cl$	Sn:	Org.Verb.5-230
$C_7ClH_{20}NSn_2$	$(CH_3)_3SnCl \cdot (CH_3)_2SnNC_2H_5$	Sn:	Org.Verb.5-82
$C_7ClH_{21}NiP_2$	$CH_3Ni(P(CH_3)_3)_2Cl$	Ni:	Org.Verb.1-11/2, 19
$C_7ClH_{23}O_6P_2Sn$	$(CH_3)_3SnCl \cdot 2\ (CH_3O)_2HPO$	Sn:	Org.Verb.5-87
$C_7Cl_2FH_6NO_2$	$(Cl)_2(CH_3O)_2C_5FN$	F:	PerFHalOrg.5-135/6
$C_7Cl_2FH_7N_2O$	$(CH_3NH)(CH_3O)C_5FCl_2N$	F:	PerFHalOrg.5-212
–	$(NH_2)(C_2H_5O)C_5FCl_2N$	F:	PerFHalOrg.5-214
$C_7Cl_2FH_8N_3$	$(NH_2)((CH_3)_2N)C_5FCl_2N$	F:	PerFHalOrg.5-214/5
–	$(NH_2)(C_2H_5NH)C_5FCl_2N$	F:	PerFHalOrg.5-214/5
$C_7Cl_2FH_{13}NO_3P$	$OP(t\text{-}C_4H_9)(ONCFCl)OC_2H_4Cl$	F:	PerFHalOrg.7-201/2
$C_7Cl_2FH_{14}N_2O_2PS$	$(C_2H_5)_2NPO(ONCFCl)SC_2H_4Cl$	F:	PerFHalOrg.7-201/2
$C_7Cl_2FH_{14}N_2O_3P$	$(C_2H_5)_2NPO(ONCFCl)OC_2H_4Cl$	F:	PerFHalOrg.7-201/2
$C_7Cl_2F_2H_3NO$	$(CH_3CO)C_5F_2Cl_2N$	F:	PerFHalOrg.5-142
$C_7Cl_2F_2H_3NO_2$	$CH_3COOC_5F_2Cl_2N$	F:	PerFHalOrg.5-212
$C_7Cl_2F_2H_6NO_3PS$	$NC_5F_2Cl_2(OPS(OCH_3)_2)$	F:	PerFHalOrg.5-212
$C_7Cl_2F_2H_6N_2$	$(CH_3)_2NC_5F_2Cl_2N$	F:	PerFHalOrg.5-138
$C_7Cl_2F_2H_7N_3$	$N(CFCCl)_2CNH(CH_2)_2NH_2$	F:	PerFHalOrg.5-140
$C_7Cl_2F_3HO_2$	$Cl_2C_6F_3COOH$	F:	PerFHalOrg.4-98
$C_7Cl_2F_4H_4Si$	$C_6F_4ClSiH(CH_3)Cl$	F:	PerFHalOrg.4-78
$C_7Cl_2F_4H_5NO_3S$	$SC(CF_2Cl)_2NHC(O)CH(CH_2COOH)$	F:	PerFHalOrg.7-66/7
$C_7Cl_2F_4H_7NOS$	$SC(CF_2Cl)_2NHC(O)C(CH_3)_2$	F:	PerFHalOrg.7-66/7
$C_7Cl_2F_5HN_4$	$C_2F_5C_5Cl_2HN_4$	F:	PerFHalOrg.6-127, 139, 162
$C_7Cl_2F_7N$	$C_2F_5C_5F_2Cl_2N$	F:	PerFHalOrg.5-149/50
$C_7Cl_2F_7P$	$CF_3C_6F_4PCl_2$	F:	PerFHalOrg.4-109
$C_7Cl_2F_9N_3$	$(CF_3CFCl)_2C_3FN_3$	F:	PerFHalOrg.6-78/80, 100
$C_7Cl_2F_{10}HNO_2$	$CF_3C(CF_2Cl)CONHC(CF_3)(CF_2Cl)O$	F:	PerFHalOrg.5-55/6, 77
$C_7Cl_2F_{10}H_3N_3$	$NH_2CNC(CF_2Cl)_2NHC(CF_3)_2$	F:	PerFHalOrg.5-63/4, 85
–	$NH_2CNCCF_3(CF_2Cl)NHC(CF_3)CF_2Cl$	F:	PerFHalOrg.5-63/4, 86
–	$NH_2CNC(CF_3)_2NHC(CF_2Cl)_2$	F:	PerFHalOrg.5-63/4, 85

$C_7Cl_2F_{11}NO$	$(CF_3)C_6F_8Cl_2NO$	F:	PerFHalOrg.6-130, 142, 158
$C_7Cl_2F_{12}H_6N_2OSi$	$CH_3SiCl_2CH[ON(CF_3)_2]CH_2N(CF_3)_2$	F:	PerFHalOrg.7-143
$C_7Cl_2F_{12}H_6N_2O_2Si$	$CH_3SiCl_2CH[ON(CF_3)_2]CH_2ON(CF_3)_2$	F:	PerFHalOrg.7-127/8
$C_7Cl_2F_{17}P$	$(n\text{-}C_3F_7)_2(CF_3)PCl_2$	F:	PerFHalOrg.3-74, 92, 100
$C_7Cl_2FeH_7O_2S$	$CH_3C_6H_4SO_2FeCl_2$	Fe:	Org.Verb.B3-187
$C_7Cl_2FeH_9Hg_2NO_4$	$(CO)_3Fe(HN(CH_2CH_2)_2O)(HgCl)_2$	Fe:	Org.Verb.B1-192/4
$C_7Cl_2H_5NO_4S$	$ClSO_2NHCOOC_6H_4Cl$	S:	S-N-Verb.1-124/5
$C_7Cl_2H_6MnO$	$MnCl_2 \cdot CO(CH)_6$	Mn:	MVol.D1-59
$C_7Cl_2H_6N_2SSn$	$SnCl_2 \cdot C_7H_6N_2S$	Sn:	MVol.C5-50
$C_7Cl_2H_7Sn^+$	$C_6H_5CH_2SnCl_2{}^+$	Sn:	Org.Verb.6-119
$C_7Cl_2H_8MnO_2$	$MnCl_2 \cdot O(CCH_3CH)_2CO$	Mn:	MVol.D1-148
$C_7Cl_2H_8MnO_4$	$[Mn(CH(COCH_3)_2)CHCl_2COO]$	Mn:	MVol.D1-75
$C_7Cl_2H_8N_2O_2$	$(NH_2)(CH_3O)_2C_5Cl_2N$	F:	PerFHalOrg.5-214/5
$C_7Cl_2H_8SSn$	$(CH_2CH)(CH_3C_4H_2S)SnCl_2$	Sn:	Org.Verb.6-203
$C_7Cl_2H_8Sn$	$(CH_2CH)(C_5H_5)SnCl_2$	Sn:	Org.Verb.6-202
–	$(CH_3)(C_6H_5)SnCl_2$	Sn:	Org.Verb.6-195, 197
$C_7Cl_2H_9NO_2U$	$UO_2Cl_2 \cdot (C_6H_5)(CH_3)NH$	U:	SVol.E1-19, 22
$C_7Cl_2H_9NiO_4P$	$(CO)_3NiP(OC_4H_9\text{-}n)Cl_2$	Ni:	Org.Verb.1-178
$C_7Cl_2H_{10}OTi$	$C_5H_5Ti(OC_2H_5)Cl_2$	Ti:	Org.Verb.1-180/2
$C_7Cl_2H_{10}STi$	$C_5H_5Ti(SC_2H_5)Cl_2$	Ti:	Org.Verb.1-189
$C_7Cl_2H_{10}Sn$	$(C_2H_5)(C_5H_5)SnCl_2$	Sn:	Org.Verb.6-198
$C_7Cl_2H_{11}NNiO$	$(C_6H_{11}NCl_2NiCO)_x$	Ni:	Org.Verb.2-394
$C_7Cl_2H_{11}NOSn$	$(CH_3)_2SnCl_2 \cdot C_5H_5NO$	Sn:	Org.Verb.6-36
$C_7Cl_2H_{11}NSn$	$(CH_3)_2SnCl_2 \cdot C_5H_5N$	Sn:	Org.Verb.6-35, 38
$C_7Cl_2H_{12}Sn$	$(CH_2CH)(C_5H_9)SnCl_2$	Sn:	Org.Verb.6-203
$C_7Cl_2H_{14}Sn$	$(CH_3)_2[Cl(CH_2)_3CHCH]SnCl$	Sn:	Org.Verb.5-194, 197/8
–	$(CH_3)_3SnCH_2CH(CCl_2)CH_2$	Sn:	Org.Verb.2-26, 29
–	$(C_4H_9)(CH_2CHCH_2)SnCl_2$	Sn:	Org.Verb.6-199
$C_7Cl_2H_{15}NNiO$	$(t\text{-}C_4H_9NC)Ni(C_2H_5OH)Cl_2$	Ni:	Org.Verb.1-304, 307
$C_7Cl_2H_{16}OSn$	$(C_2H_5)_2SnCl_2 \cdot (CH_3)_2CO$	Sn:	Org.Verb.6-61/2
–	$(C_4H_9)(i\text{-}C_3H_7O)SnCl_2$	Sn:	Org.Verb.6-245
$C_7Cl_2H_{16}Sn$	$(CH_3)(C_6H_{13})SnCl_2$	Sn:	Org.Verb.6-195, 197
–	$(CH_3)_3SnCH_2CH(CH_3)CHCl_2$	Sn:	Org.Verb.2-50/1
–	$(C_2H_5)[(CH_3)_3CCH_2]SnCl_2$	Sn:	Org.Verb.6-198
–	$(C_2H_5)_3SnCHCl_2$	Sn:	Org.Verb.2-164
–	$(C_3H_7)(C_4H_9)SnCl_2$	Sn:	Org.Verb.6-198
$C_7Cl_2H_{18}SiSn$	$(CH_3)_3SnCCl_2Si(CH_3)_3$	Sn:	Org.Verb.2-17/8
$C_7Cl_2H_{19}OPSn$	$(C_2H_5)_2SnCl_2 \cdot (CH_3)_3PO$	Sn:	Org.Verb.6-61
$C_7Cl_2H_{19}O_2PSn$	$(CH_3)_2SnCl_2 \cdot (C_2H_5)(C_3H_7O)HPO$	Sn:	Org.Verb.6-37
$C_7Cl_3CsH_5IrNO_4$	$Cs[Ir(C_2O_4)(C_5H_5N)Cl_3]$	Ir:	SVol.2-36
$C_7Cl_3Cs_2H_5IrNO_4$	$Cs_2[Ir(C_2O_4)(C_5H_5N)Cl_3]$	Ir:	SVol.2-36
$C_7Cl_3F_3H_3N$	$(CF_3)(CH_3)C_5Cl_3N$	F:	PerFHalOrg.5-177
$C_7Cl_3F_3H_4Sn$	$CF_3C_6H_4SnCl_3$	Sn:	Org.Verb.6-283, 285
$C_7Cl_3F_{10}N$	$NCClCF_2CF(C_2F_5)CF_2CCl_2$	F:	PerFHalOrg.5-150/1, 162
$C_7Cl_3FeGeH_{10}NO_4$	$[N(CH_3)_3H][(CO)_4FeGeCl_3]$	Fe:	Org.Verb.B2-136
$C_7Cl_3H_4NO_2S$	$Cl_2C_6H_3CHNSO_2Cl$	S:	S-N-Verb.1-126
$C_7Cl_3H_4NO_4S$	$ClSO_2NHCOOC_6H_3Cl_2$	S:	S-N-Verb.1-124/5
$C_7Cl_3H_5IrKNO_4$	$K[Ir(C_2O_4)(C_5H_5N)Cl_3]$	Ir:	SVol.2-36

$C_7Cl_3H_5IrK_2NO_4$	$K_2[Ir(C_2O_4)(C_5H_5N)Cl_3]$	Ir:	SVol.2-36
$C_7Cl_3H_5IrNO_4Rb$	$Rb[Ir(C_2O_4)(C_5H_5N)Cl_3]$	Ir:	SVol.2-36
$C_7Cl_3H_5IrNO_4Rb_2$	$Rb_2[Ir(C_2O_4)(C_5H_5N)Cl_3]$	Ir:	SVol.2-36
$C_7Cl_3H_5IrNO_4Tl$	$Tl[Ir(C_2O_4)(C_5H_5N)Cl_3]$	Ir:	SVol.2-36
$C_7Cl_3H_5MnO$	$MnCl_2 \cdot C_6H_5COCl$	Mn:	MVol.C5-81
$C_7Cl_3H_5O_3Sn$	$SnCl_3(OC_6H_4)COOH$	Sn:	MVol.C5-123/4
$C_7Cl_3H_6NO_2Sn$	$(CH_3)(NO_2)C_6H_3SnCl_3$	Sn:	Org.Verb.6-286
–	$[SnCl_3(OC_6H_4CONH_2)]$	Sn:	MVol.C6-48
$C_7Cl_3H_6NO_3Sn$	$(CH_3O)(NO_2)C_6H_3SnCl_3$	Sn:	Org.Verb.6-286
$C_7Cl_3H_7MnO_4$	$[Mn(CH(COCH_3)_2)CCl_3COO]$	Mn:	MVol.D1-75
$C_7Cl_3H_7N_2O$	$(NH_2)(C_2H_5O)C_5Cl_3N$	F:	PerFHalOrg.5-214/5
$C_7Cl_3H_7OSn$	$CH_3OC_6H_4SnCl_3$	Sn:	Org.Verb.6-283, 285
$C_7Cl_3H_7Sn$	$CH_3C_6H_4SnCl_3$	Sn:	Org.Verb.6-282
–	$(C_6H_5)(CH_2Cl)SnCl_2$	Sn:	Org.Verb.6-200
–	$C_6H_5CH_2SnCl_3$	Sn:	Org.Verb.6-253
$C_7Cl_3H_7Ti$	$C_6H_5CH_2TiCl_3$	Ti:	Org.Verb.1-16, 24
$C_7Cl_3H_8NO_2Sn$	$CH_3SnCl_3 \cdot C_6H_5NO_2$	Sn:	Org.Verb.6-222
$C_7Cl_3H_8N_3$	$(NH_2)((CH_3)_2N)C_5Cl_3N$	F:	PerFHalOrg.5-214/5
$C_7Cl_3H_9Ti$	$C_2H_5C_5H_4TiCl_3$	Ti:	Org.Verb.1-140, 150
$C_7Cl_3H_{10}MnN$	$CH_3C_6H_4NH_3MnCl_3$	Mn:	MVol.C5-205/6
$C_7Cl_3H_{10}NTi$	$CH_3TiCl_3 \cdot NC_5H_4CH_3$	Ti:	Org.Verb.1-30/1, 40
$C_7Cl_3H_{10}N_3O_2Sn$	$CH_3SnCl_3 \cdot (NH_2)_2C_6H_3NO_2$	Sn:	Org.Verb.6-222
$C_7Cl_3H_{11}O_2PtS_4$	$Pt[S_2COC_2H_5]_2 \cdot HCCl_3$	C:	MVol.D4-261
$C_7Cl_3H_{13}O_4U$	$UCl_3(CH_3COO) \cdot CH_3COOC_3H_7$	U:	SVol.E1-120/1
$C_7Cl_3H_{14}O_2Sn^+$	$[SnCl_3(C_4H_9COOC_2H_5)]^+$	Sn:	MVol.C5-136/7
$C_7Cl_3H_{15}OSn$	$C_4H_9SnCl_3 \cdot (CH_3)_2CO$	Sn:	Org.Verb.6-247/8
$C_7Cl_3H_{15}Sn$	$(C_2H_5)_3SnCCl_3$	Sn:	Org.Verb.2-164
–	$C_7H_{15}SnCl_3$	Sn:	Org.Verb.6-250/1
$C_7Cl_3H_{17}OSiSn$	$(C_4H_9O)(CH_3)_2SiCH_2SnCl_3$	Sn:	Org.Verb.6-258
$C_7Cl_3H_{17}O_3SiSn$	$(C_2H_5O)_3SiCH_2SnCl_3$	Sn:	Org.Verb.6-258
$C_7Cl_3H_{17}SiSn$	$(C_2H_5)_3SiCH_2SnCl_3$	Sn:	Org.Verb.6-258
–	$(C_4H_9)(CH_3)_2SiCH_2SnCl_3$	Sn:	Org.Verb.6-258
$C_7Cl_3H_{19}NPTi$	$CH_3TiCl_3 \cdot (CH_3)_2PCH_2CH_2N(CH_3)_2$	Ti:	Org.Verb.1-32/3
$C_7Cl_3H_{19}N_2Ti$	$CH_3TiCl_3 \cdot (CH_3)_2NCH_2CH_2N(CH_3)_2$	Ti:	Org.Verb.1-32/3, 40
$C_7Cl_3H_{19}P_2Ti$	$CH_3TiCl_3 \cdot (CH_3)_2PCH_2CH_2P(CH_3)_2$	Ti:	Org.Verb.1-32/3
$C_7Cl_3H_{19}Si_2Sn$	$[(CH_3)_3Si]_2CHSnCl_3$	Sn:	Org.Verb.6-258
$C_7Cl_3H_{20}IrN_6O_{12}$	$[Ir(NCC_6H_5)(NH_3)_5](ClO_4)_3$	Ir:	SVol.2-65
$C_7Cl_3H_{21}N_3OPSn$	$CH_3SnCl_3 \cdot [(CH_3)_2N]_3PO$	Sn:	Org.Verb.6-222
$C_7Cl_3H_{21}O_2P_2S_2Sn$	$CH_3SnCl_3 \cdot 2\ (CH_3)_2P(O)SCH_3$	Sn:	Org.Verb.6-222/3
$C_7Cl_3H_{21}O_4P_2Sn$	$CH_3SnCl_3 \cdot 2\ (CH_3)_2P(O)OCH_3$	Sn:	Org.Verb.6-222
$C_7Cl_4FH_4NSn$	$SnCl_4 \cdot FC_6H_4CN$	Sn:	MVol.C6-71/2
$C_7Cl_4F_8HNO_2$	$(CF_2Cl)_2CCONHC(CF_2Cl)_2O$	F:	PerFHalOrg.5-55/6, 77
$C_7Cl_4F_8H_3N_3$	$NH_2CNC(CF_2Cl)_2NHC(CF_2Cl)_2$	F:	PerFHalOrg.5-62/4, 85
$C_7Cl_4F_{11}NO$	$ClCF_2CFCl(CF_2CFCl)_2CF_2NO$	F:	PerFHalOrg.7-179, 185
$C_7Cl_4H_4N_2O_2Sn$	$SnCl_4 \cdot NO_2C_6H_4CN$	Sn:	MVol.C6-71/2
$C_7Cl_4H_5NSn$	$SnCl_4 \cdot C_6H_5CN$	Sn:	MVol.C6-71
$C_7Cl_4H_6OSn$	$SnCl_4 \cdot C_6H_5CHO$	Sn:	MVol.C5-79/80
$C_7Cl_4H_6O_2Sn$	$SnCl_4 \cdot C_6H_5COOH$	Sn:	MVol.C5-121/3
$C_7Cl_4H_6O_6SSn$	$SnCl_4 \cdot HOC_6H_3(SO_3H)COOH$	Sn:	MVol.C5-123/4
$C_7Cl_4H_6Sn$	$ClC_6H_3(CH_3)SnCl_3$	Sn:	Org.Verb.6-286

$C_7Cl_4H_7NOSn$	$SnCl_4 \cdot CH_3COC_5H_4N$	Sn:	MVol.C5-205
$C_7Cl_4H_7NO_2Sn$	$SnCl_4 \cdot NH_2C_6H_4COOH$	Sn:	MVol.C6-14
$C_7Cl_4H_7NO_3Sn$	$SnCl_4 \cdot NO_2C_6H_4OCH_3$	Sn:	MVol.C5-152/3
$C_7Cl_4H_8N_2O_2Sn$	$SnCl_4 \cdot NO_2C_6H_3(CH_3)NH_2$	Sn:	MVol.C5-178
$C_7Cl_4H_8OSn$	$SnCl_4 \cdot C_6H_5OCH_3$	Sn:	MVol.C5-152
$C_7Cl_4H_8Sn$	$ClC_7H_8SnCl_3$	Sn:	Org.Verb.6-257
$C_7Cl_4H_9NO_2Sn$	$SnCl_4 \cdot C_5H_5N \cdot CH_3COOH$	Sn:	MVol.C5-197
$C_7Cl_4H_{10}N_2Sn$	$SnCl_4 \cdot NC(CH_2)_5CN$	Sn:	MVol.C6-73/5
$C_7Cl_4H_{12}O_2Sn$	$SnCl_4 \cdot (CH_3)_2C(COCH_3)_2$	Sn:	MVol.C5-108
$C_7Cl_4H_{12}O_4Sn$	$SnCl_4 \cdot CH_2(COOC_2H_5)_2$	Sn:	MVol.C5-140/1
–	$SnCl_4 \cdot C_3H_6(COOCH_3)_2$	Sn:	MVol.C5-141/3
$C_7Cl_4H_{13}NO_2Sn$	$SnCl_4 \cdot CH_3CONHCOC_4H_9$	Sn:	MVol.C6-50/1
$C_7Cl_4H_{14}O_2Sn$	$SnCl_4 \cdot (CH_3)_3CCOOC_2H_5$	Sn:	MVol.C5-136/7
–	$SnCl_4 \cdot C_4H_9COOC_2H_5$	Sn:	MVol.C5-136
$C_7Cl_4H_{15}NO_2Sn$	$SnCl_4 \cdot CH_3COOH \cdot (CH_2)_5NH$	Sn:	MVol.C5-193
$C_7Cl_4H_{15}O_4PSn$	$SnCl_4 \cdot (CH_3)_3CCOP(O)(OCH_3)_2$	Sn:	MVol.C6-164
$C_7Cl_4H_{16}O_2Sn$	$SnCl_4 \cdot C_3H_6(OC_2H_5)_2$	Sn:	MVol.C5-154/5
$C_7Cl_4H_{18}NO_2PSn$	$SnCl_4 \cdot (CH_3)_2CHOPH(O)N(C_2H_5)_2$	Sn:	MVol.C6-167
$C_7Cl_4H_{21}MnN_3$	$[((CH_3)_2N)_2CH](CH_3)_2NH_2MnCl_4$	Mn:	MVol.C5-204/5
$C_7Cl_4H_{28}N_{14}O_7U$	$UCl_4 \cdot 7\ (NH_2)_2CO \cdot 3\ H_2O$	U:	SVol.E1-82/3
$C_7Cl_5H_4NSn$	$SnCl_4 \cdot ClC_6H_4CN$	Sn:	MVol.C6-71/2
$C_7Cl_5H_5OSn$	$SnCl_4 \cdot ClC_6H_4CHO$	Sn:	MVol.C5-81
$C_7Cl_5H_9OSn$	$SnCl_4 \cdot ClCHC_6H_8O$	Sn:	MVol.C5-94
$C_7Cl_5H_{14}O_2Sn^-$	$[SnCl_5(C_4H_9COOC_2H_5)]^-$	Sn:	MVol.C5-136/7
$C_7Cl_6FeH_7O_3Sb$	$[CH_3C_3H_4Fe(CO)_3]SbCl_6$	Fe:	Org.Verb.B5-82/3
$C_7Cl_7H_9N_4Sn$	$SnCl_4 \cdot 2\ C_3H_4N_2 \cdot CHCl_3$	Sn:	MVol.C5-223
$C_7Cl_8F_3N_3$	$(CCl_3)_2(CF_2ClCFCl)C_3N_3$	F:	PerFHalOrg.6-82/3, 102
$C_7Cl_8H_8OSn_2$	$2\ SnCl_4 \cdot C_6H_5OCH_3$	Sn:	MVol.C5-152
$C_7Cl_8H_{14}O_2Sn_2$	$2\ SnCl_4 \cdot C_4H_9COOC_2H_5$	Sn:	MVol.C5-136/7
$C_7Cl_9Fe_2O_7Si_2Sn^-$	$[(CO)_7Fe_2Si_2SnCl_9]^-$	Fe:	Org.Verb.C1-60
$C_7CmH_8NO_6$	$Cm(N(CH_2CH_2COO)(CH_2COO)_2)$	Np:	TrU.D1-164
$C_7CmH_9NO_6^+$	$Cm(N(CH_2CH_2COO)(CH_2COO)_2H)^+$	Np:	TrU.D1-164
$C_7CoF_{18}NO_2P_2$	$Co(NO)(CO) \cdot [P(CF_3)_3]_2$	F:	PerFHalOrg.3-174
$C_7CoH_9O_4Sn$	$(CH_3)_3SnCo(CO)_4$	Sn:	Org.Verb.5-82, 86
$C_7CoH_{24}N_6S_3$	$Co(C_2H_8N_2)_3CS_3$	C:	MVol.D4-225
$C_7CrH_6O_5Sn$	$(CH_3)_2SnCr(CO)_5$	Sn:	Org.Verb.6-32
$C_7CsF_{12}NO_2$	$(CF_3)_2CCONCsC(CF_3)_2O$	F:	PerFHalOrg.5-55/6, 110
$C_7CuFeH_3N_6$	$[CH_3NCFe(CN)_5]Cu \cdot x\ H_2O$	Fe:	Org.Verb.B4-10
$C_7CuH_8NOS_2$	$Cu[S_2COCH_3](C_5H_5N)$	C:	MVol.D4-257
$C_7CuH_8NO_2S$	$Cu[SOCOCH_3] \cdot C_5H_5N$	C:	MVol.D4-237
$C_7CuH_{13}S_3$	$Cu[S_2CSC_6H_{13}]$	C:	MVol.D4-269
$C_7FFeH_7O_6S$	$CH_3C_3H_4Fe(CO)_3OSO_2F$	Fe:	Org.Verb.B5-41, 52, 61/2
$C_7FH_5O_2$	$FCOOC_6H_5$	C:	MVol.D3-46
$C_7FH_6NO_2S$	$FSO_2NCHC_6H_5$	S:	S-N-Verb.1-105
$C_7FH_6N_3O_2$	$(CH_3O)_2(NC)C_4FN_2$	F:	PerFHalOrg.6-58, 63
$C_7FH_7N_2O_3S$	$FSO_2NHC(O)NHC_6H_5$	S:	S-N-Verb.1-102
$C_7FH_8NO_2S$	$FSO_2N(CH_3)C_6H_5$	S:	S-N-Verb.1-104
$C_7FH_9N_2O_3$	$NCFC(OCH_3)C(OCH_3)C(OCH_3)N$	F:	PerFHalOrg.6-57, 61
–	$NC(OCH_3)NC(OCH_3)CFC(OCH_3)$	F:	PerFHalOrg.6-58, 62
$C_7FH_{17}Sn$	$(C_2H_5)_2(C_3H_7)SnF$	Sn:	Org.Verb.5-40/1

$C_7FH_{18}N_4O_3PS$	$[(CH_3)_2N]_3PCONSO_2F$	S:	S-N-Verb.1-115
$C_7F_2FeH_5NO_3S$	$(C_5H_5)Fe(CO)_2[NS(O)F_2]$	S:	S-N-Verb.1-62
$C_7F_2FeH_{12}O_7P_2$	$(CO)_3FeP_2F_2(OCH_3)_4$	Fe:	Org.Verb.B1-161
$C_7F_2H_3MnN_2O_9S_2$	$[Mn(CO)_5(CH_3CN)]N(SO_2F)_2$	S:	S-N-Verb.1-54
$C_7F_2H_6INO_2$	$(I)(CH_3O)_2C_5F_2N$	F:	PerFHalOrg.5-134/5
$C_7F_2H_8N_2O_4$	$NO_2CF_2C(O)NC_4H_5(CH_3)O$	F:	PerFHalOrg.7-194/5
$C_7F_2H_9NiO_3P$	$(CO)_3NiP(C_4H_9\text{-}t)F_2$	Ni:	Org.Verb.1-166, 168
$C_7F_2H_{10}NNiO_3P$	$(CO)_3NiP(N(C_2H_5)_2)F_2$	Ni:	Org.Verb.1-162
$C_7F_2H_{10}OTi$	$C_5H_5Ti(OC_2H_5)F_2$	Ti:	Org.Verb.1-180/2
$C_7F_2H_{10}O_4$	$HOCO(CH_2)_4COOCHF_2$	F:	PerFHalOrg.5-19
$C_7F_2H_{14}Sn$	$(CH_3)_3SnCH_2CH(CH_2)CF_2$	Sn:	Org.Verb.2-26
$C_7F_3FeH_5O_3S$	$C_3H_5Fe(CO)_3SCF_3$	Fe:	Org.Verb.B5-51, 59/60
$C_7F_3Ge_2H_{18}P$	$[(CH_3)_3Ge]_2PCF_3$	F:	PerFHalOrg.3-25
$C_7F_3H_2NO_4$	$NC_5F_3(COOH)_2$	F:	PerFHalOrg.5-193, 207
		F:	PerFHalOrg.6-159
$C_7F_3H_4N_3O_2$	$CF_3NNC_6H_4NO_2$	F:	PerFHalOrg.7-158/9
$C_7F_3H_4N_5$	$(CF_3)(NH_2)C_6H_2N_4$	F:	PerFHalOrg.6-70/1
$C_7F_3H_5NO$	$CF_3N(O)C_6H_5$	F:	PerFHalOrg.7-169
$C_7F_3H_5N_2$	$CF_3NNC_6H_5$	F:	PerFHalOrg.7-158/9, 170
$C_7F_3H_5N_2O$	$CF_3NN(O)C_6H_5$	F:	PerFHalOrg.7-58
–	$C_6H_5NN(O)CF_3$	F:	PerFHalOrg.7-160/1
$C_7F_3H_5N_2O_3$	$(NO_2)(CH_3O)C_6F_3NH_2$	F:	PerFHalOrg.7-60
$C_7F_3H_5N_2O_3S$	$CF_3NNC_6H_4SO_3H$	F:	PerFHalOrg.7-160
$C_7F_3H_5N_4OS$	$(CF_3)(OH)(S)C_6H_4N_4$	F:	PerFHalOrg.6-70
$C_7F_3H_5NiO$	$C_5H_5Ni(CO)CF_3$	Ni:	Org.Verb.2-167
$C_7F_3H_5NiOS$	$C_5H_5Ni(CO)SCF_3$	Ni:	Org.Verb.2-164
$C_7F_3H_5NiO_3S$	$C_5H_5Ni(CO)SO_2CF_3$	Ni:	Org.Verb.2-164
$C_7F_3H_6N$	$(CH_3)_2C_5F_3N$	F:	PerFHalOrg.5-141
$C_7F_3H_6NO_2$	$(CH_3O)_2C_5F_3N$	F:	PerFHalOrg.5-135
$C_7F_3H_6N_3$	$CF_3NNNHC_6H_5$	F:	PerFHalOrg.7-158/9
$C_7F_3H_6N_3O_2$	$(NO_2)(CH_3NH)C_6F_3NH_2$	F:	PerFHalOrg.7-60
$C_7F_3H_6O_2P$	$CF_3(CH_2CCH)P(O)OCH_2CCH$	F:	PerFHalOrg.3-101
–	$CF_3P(OCH_2CCH)_2$	F:	PerFHalOrg.3-101
$C_7F_3H_7MnO_4$	$[Mn(CH(COCH_3)_2)CF_3COO]$	Mn:	MVol.D1-75
$C_7F_3H_7N_2O$	$(NH_2)(C_2H_5O)C_5F_3N$	F:	PerFHalOrg.5-214/5
$C_7F_3H_7N_4O_3$	$NC(OC_2H_5)NC(NH_2)C(NO_2)C(CF_3)$	F:	PerFHalOrg.6-58
$C_7F_3H_7OS$	$(CH_3C_6H_4O)SF_3$	S:	SVol.2-83
$C_7F_3H_8NO_2$	$CF_3C(O)NC_4H_5(CH_3)O$	F:	PerFHalOrg.7-194/5
$C_7F_3H_8N_5O_2$	$NC(N(CH_3)_2)NC(NH_2)C(NO_2)C(CF_3)$	F:	PerFHalOrg.6-59
$C_7F_3H_9NO_3P$	$CF_3P(O)(OH)_2 \cdot C_6H_5NH_2$	F:	PerFHalOrg.3-50
$C_7F_3H_{10}NO$	$CF_3NCH_2C(CH_3)C(CH_3)CH_2O$	F:	PerFHalOrg.7-168
–	$CF_3N(O)CH_2C(CH_3)C(CH_3)CH_2$	F:	PerFHalOrg.7-168
$C_7F_3H_{10}N_3O_5$	$C_2H_5C(O)ONN(OCOC_2H_5)ON(CF_3)$	F:	PerFHalOrg.7-164, 170
$C_7F_3H_{10}N_5O_3$	$(CF_3)(N(CH_2OH)_2)(NHCH_2OH)C_3N_3$	F:	PerFHalOrg.6-112
$C_7F_3H_{11}NO_2$	$(CF_3NOCHCHOC_4H_9)_n$	F:	PerFHalOrg.7-166/7
$C_7F_3H_{11}N_2$	$CF_3NNC_6H_{11}$	F:	PerFHalOrg.7-158/9
$C_7F_3H_{18}PSn_2$	$[(CH_3)_3Sn]_2PCF_3$	F:	PerFHalOrg.3-25
$C_7F_3HgNO_4$	$C_5F_3N(COO)_2Hg$	F:	PerFHalOrg.5-193, 207
$C_7F_3NO_4Zn$	$NC_5F_3(COO)_2Zn$	F:	PerFHalOrg.5-193, 207

$C_7F_4HNO_3$	$NOC_6F_4C(O)OH$	F:	PerFHalOrg.7-180, 187, 190, 198/9, 209
$C_7F_4HSi^+$	$(C_6F_3)SiHCF^+$	F:	PerFHalOrg.4-152
$C_7F_4H_2N_2$	$CNC_6F_4NH_2$	F:	PerFHalOrg.7-10, 37
–	C_6F_4NHCHN	F:	PerFHalOrg.7-61
$C_7F_4H_2N_2O_3$	$NO_2C_6F_4C(O)NH_2$	F:	PerFHalOrg.7-19, 22, 48
$C_7F_4H_3NO_2$	$CH_3OC(O)C_5F_4N$	F:	PerFHalOrg.5-217
–	$CH_3OC_6F_4NO$	F:	PerFHalOrg.7-192
–	$NH_2C_6F_4COOH$	F:	PerFHalOrg.7-9/10, 37, 55, 61/2
$C_7F_4H_3N_3$	$C_6F_4NC(NH_2)NH$	F:	PerFHalOrg.6-124/5, 139
$C_7F_4H_4N_2O$	$NH_2C_6F_4C(O)NH_2$	F:	PerFHalOrg.7-9/10, 37, 57
$C_7F_4H_5NO$	$CH_3OC_6F_4NH_2$	F:	PerFHalOrg.7-59, 60
$C_7F_4H_5NO_2$	$(HOC_2H_4O)C_5F_4N$	F:	PerFHalOrg.5-136
$C_7F_4H_5NS_2$	$HSCH_2CH_2SC_5F_4N$	F:	PerFHalOrg.5-141
$C_7F_4H_6N_2$	$CH_3NHC_6F_4NH_2$	F:	PerFHalOrg.7-59
–	$(CH_3)_2NC_5F_4N$	F:	PerFHalOrg.5-136
$C_7F_4H_6N_2O$	$(HOC_2H_4NH)C_5F_4N$	F:	PerFHalOrg.5-137
$C_7F_4H_6N_2OS$	$[(CH_3)_2S(O)N]C_5F_4N$	F:	PerFHalOrg.5-212
$C_7F_4H_7N_3$	$(NH_2CH_2CH_2NH)C_5F_4N$	F:	PerFHalOrg.5-137
$C_7F_4H_8N_2O_2S_2Si$	$(C_6H_5)(CH_3)Si[NS(O)F_2]_2$	S:	S-N-Verb.1-83
$C_7F_4H_8N_2O_2S_2Si^+$	$CH_3(C_6H_5)Si(NSOF_2)_2{}^+$	S:	S-N-Verb.1-83
$C_7F_4H_9NSn$	$SnF_4 \cdot (CH_3)_2C_5H_3N$	Sn:	MVol.C5-202
$C_7F_4H_{10}N_2O_2S$	$C_5H_{10}NS(O)FNC(O)CF_3$	S:	S-N-Verb.1-194
$C_7F_4H_{14}Sn$	$(CH_3)_3SnCH_2CF_2CH_2CHF_2$	Sn:	Org.Verb.2-50
$C_7F_4Li_2O_2$	LiC_6F_4COOLi	F:	PerFHalOrg.4-6, 13, 14
$C_7F_4N_2O_2$	$[N_2C_6F_4COO]$	F:	PerFHalOrg.7-62
$C_7F_5FeHO_4$	$CF_3CFCHFFe(CO)_4$	Fe:	Org.Verb.B4-315/6
–	$F_2CCH(CF_3)Fe(CO)_4$	Fe:	Org.Verb.B4-315/6
$C_7F_5FeH_6O_5SSb$	$(CO)_3FeOS(OSbF_5)C_4H_6$	Fe:	Org.Verb.B5-89
$C_7F_5FeIO_4$	$(CO)_4Fe(CFCFCF_3)I$	Fe:	Org.Verb.B2-177, 179, 181/2
$C_7F_5GeH_9$	$CF_3CF_2CCGe(CH_3)_3$	F:	PerFHalOrg.4-75, 76
C_7F_5HO	$C_6F_5C(O)H$	F:	PerFHalOrg.4-47, 98, 99, 108
$C_7F_5HO_2$	C_6F_5COOH	F:	PerFHalOrg.4-98/9, 102, 104
$C_7F_5HSi^+$	$(C_6F_3)SiHCF_2{}^+$	F:	PerFHalOrg.4-152
$C_7F_5H_2NO$	$C_6F_5C(O)NH_2$	F:	PerFHalOrg.7-19/21, 48
–	$C_6F_5NHC(O)H$	F:	PerFHalOrg.7-60
$C_7F_5H_3$	$C_6F_5CH_3$	F:	PerFHalOrg.4-108
$C_7F_5H_3Hg$	$CH_3HgC_6F_5$	F:	PerFHalOrg.4-110, 111
$C_7F_5H_3N_2O$	$C_6F_5NHC(O)NH_2$	F:	PerFHalOrg.7-24, 50, 56/7
$C_7F_5H_3N_2S$	$C_6F_5NHC(S)NH_2$	F:	PerFHalOrg.7-24, 50
$C_7F_5H_3S$	$C_6F_5SCH_3$	F:	PerFHalOrg.4-31
$C_7F_5H_4N$	$C_6F_5NHCH_3$	F:	PerFHalOrg.7-59
$C_7F_5H_9Sn$	$(CH_3)_3SnCCCF_2CF_3$	Sn:	Org.Verb.2-112, 116

$C_7F_5H_9Sn$	$(CH_3)_3SnCCF_2CCF_3$	Sn:	Org.Verb.2-97/8
$C_7F_5H_{21}O_3S_2SnTe$	$[(CH_3)_3Sn((CH_3)_2SO)_2][TeF_5O]$	Te:	SVol.B2-55
$C_7F_5NS^+$	$[C_6F_5C(N)S]^+$	F:	PerFHalOrg.5-95
$C_7F_6FeO_4$	$CF_2CFCF_3Fe(CO)_4$	Fe:	Org.Verb.B4-320, 322/4
$C_7F_6HSi^+$	$(C_6F_3)SiHCF_3{}^+$	F:	PerFHalOrg.4-152
$C_7F_6H_2N_4$	$NC(CF_3)NC(NH_2)C(CN)C(CF_3)$	F:	PerFHalOrg.6-25, 47
$C_7F_6H_3NO$	$(CH_3O)(CF_3)C_5F_3N$	F:	PerFHalOrg.5-177
$C_7F_6H_4N_2$	$(NH_2)_2(CF_3)C_6F_3$	F:	PerFHalOrg.7-11/2
$C_7F_6H_4OS$	$(CF_3C_6H_4O)SF_3$	S:	SVol.2-83
$C_7F_6H_5NO_3S$	$SC(CF_3)_2NHC(O)CH(CH_2COOH)$	F:	PerFHalOrg.7-66/7
$C_7F_6H_5NiOS_4$	$[Ni(S_2COC_2H_5)(S_2C_2(CF_3)_2)]_2$	C:	MVol.D4-260
$C_7F_6H_5NiO_3P$	$(CO)_3NiP(C_2H_5)(CF_3)_2$	Ni:	Org.Verb.1-165, 168
$C_7F_6H_6$	$CH_3CC(CF_3)_2CCH_3$	F:	PerFHalOrg.5-24
$C_7F_6H_6N_2O_2$	$CH_3OCOCH(CN)C(CF_3)_2NH_2$	F:	PerFHalOrg.7-65, 73
$C_7F_6H_6N_8$	$CF_2(CF_2CNNHC(NH_2)N)_2$	F:	PerFHalOrg.5-72, 93
$C_7F_6H_7NOS$	$SC(CF_3)_2NHC(O)C(CH_3)_2$	F:	PerFHalOrg.7-66
$C_7F_6H_7P$	$(CF_3)_2PC_5H_7$	F:	PerFHalOrg.3-26
–	$(CF_3)_2PC_5H_5D_2$	F:	PerFHalOrg.3-27
$C_7F_6H_8$	$(CF_3)_2CC(CH_3)CH_2CH_3$	F:	PerFHalOrg.5-25
–	$(CF_3)_2C(CHCH_3)_2$	F:	PerFHalOrg.5-25
–	$CH_3CHC(CH_3)CH(CF_3)_2$	F:	PerFHalOrg.5-25
–	$CH_3CHCHCH_2CH(CF_3)_2$	F:	PerFHalOrg.5-25
$C_7F_6H_9N$	$CH_3CHCHCH_2C(CF_3)_2NH_2$	F:	PerFHalOrg.7-65, 72, 76
–	$CH_3CH_2CH(CH_3)NC(CF_3)_2$	F:	PerFHalOrg.5-26
–	$CH_3CH_2CH_2CH_2NC(CF_3)_2$	F:	PerFHalOrg.5-26
$C_7F_6H_9NO$	$CH_2CHC(CH_3)_2ON(CF_3)_2$	F:	PerFHalOrg.7-134/5
–	$(CH_3)_2CCHCH_2ON(CF_3)_2$	F:	PerFHalOrg.7-134/5
$C_7F_6H_9NO_2$	$(CF_3)_2NOCOC(CH_3)_3$	F:	PerFHalOrg.7-129/30, 136
$C_7F_6H_9N_3O_2SSi$	$(CH_3)_3SiNS[NC(O)CF_3]_2$	S:	S-N-Verb.1-206
$C_7F_6H_{10}$	$CH_3CH_2CH(CH_3)CH(CF_3)_2$	F:	PerFHalOrg.5-26
–	$CH_3CH_2CH_2CH_2CH(CF_3)_2$	F:	PerFHalOrg.5-26
$C_7F_6H_{10}NO_3P$	$(CF_3)_2CNOP(O)(CH_3)OCH(CH_3)_2$	F:	PerFHalOrg.7-141/2
$C_7F_6H_{10}NO_4P$	$(CF_3)_2CNOP(O)(OC_2H_5)_2$	F:	PerFHalOrg.7-142, 199
$C_7F_6H_{10}N_2O_2S$	$FSO_2NC[N(C_2H_5)_2]C_2F_5$	S:	S-N-Verb.1-119
$C_7F_6H_{10}N_8O_2$	$(CF_2)_3(C(O)NHNHC(NH)NH_2)_2$	F:	PerFHalOrg.7-6/7, 34
$C_7F_6H_{10}Sn$	$(CH_3)_3SnC(CF_3)CHCF_3$	Sn:	Org.Verb.2-77, 84
–	$(CH_3)_3SnCF(CF_2)_2CHF$	Sn:	Org.Verb.2-60, 67
$C_7F_6H_{11}NO$	$(CF_3)_2NOC(CH_3)_2CH_2CH_3$	F:	PerFHalOrg.7-129, 131
–	$(CF_3)_2NOCH_2C(CH_3)_3$	F:	PerFHalOrg.7-129/31
$C_7F_6H_{11}NO_2$	$(CF_3)_2CNOH \cdot (C_2H_5)_2O$	F:	PerFHalOrg.7-88/9
$C_7F_6H_{12}Sn$	$(CH_3)_3SnCHFCF_2CHFCHF_2$	Sn:	Org.Verb.2-50
$C_7F_6H_{18}N_3NiOP_3$	$CONi(P(N(CH_3)_2)F_2)_3$	Ni:	Org.Verb.1-113
$C_7F_6MnNO_6$	$(CF_3)_2NOMn(CO)_5$	F:	PerFHalOrg.7-127
$C_7F_6MoNO_4PS$	$[(CF_3)_2PNCS]Mo(CO)_4$	F:	PerFHalOrg.3-69
$C_7F_7FeHO_4$	$(CO)_4Fe(C_3F_7)H$	Fe:	Org.Verb.B2-177
$C_7F_7FeIO_4$	$(CO)_4Fe(C_3F_7)I$	Fe:	Org.Verb.B2-177/8, 180/1
$C_7F_7H_2N$	$CF_3C_6F_4NH_2$	F:	PerFHalOrg.7-10, 37, 55
$C_7F_7H_4N_3$	$(NH_2)_2(CF_3)_2C_5FN$	F:	PerFHalOrg.5-186, 202
$C_7F_7H_5O$	$FC(CF_2)_3CCH(OH)CH_3$	F:	PerFHalOrg.4-44

$C_7F_7H_7O$	$C_3F_7COC_3H_7$	F:	PerFHalOrg.4-93, 96
$C_7F_7H_9O$	$CF_3(CF_2)_2CH(OH)C_3H_7$	F:	PerFHalOrg.4-94
–	$(CF_3)_2CFC(CH_3)(OH)CH_2CH_3$	F:	PerFHalOrg.4-91
–	$(CF_3)_2CFCH(OH)C_3H_7$	F:	PerFHalOrg.4-91
–	$C_3F_7CH(OH)C_3H_7$	F:	PerFHalOrg.4-91, 96
$C_7F_7H_{10}INP$	$C_3F_7PIN(C_2H_5)_2$	F:	PerFHalOrg.3-108, 113
$C_7F_7H_{12}N_2P$	$C_3F_7P[N(CH_3)_2]_2$	F:	PerFHalOrg.3-108, 113
C_7F_7Li	$CF_3C_6F_4Li$	F:	PerFHalOrg.4-6, 53
$C_7F_8H_2N_2$	$(CF_3)_2(NH_2)C_5F_2N$	F:	PerFHalOrg.5-186, 202
$C_7F_8H_{10}Sn$	$(CH_3)_3SnCF_2CF_2CF_2CHF_2$	Sn:	Org.Verb.2-50
$C_7F_8I_2Mg$	$IC(CFCF)(CF_2)(CF_2CF_2)CMgI$	F:	PerFHalOrg.4-66
C_7F_8NO	$(C_6F_5)(CF_3)NO$	F:	PerFHalOrg.7-86, 122/3
$C_7F_9HMnO_4P$	$H[Mn(CO)_4P(CF_3)_3]$	F:	PerFHalOrg.3-174
$C_7F_9HN_4$	$H(CF_3)_3C_4N_4$	F:	PerFHalOrg.6-123, 138, 151, 156, 159/60
$C_7F_9H_5$	$CF_3CHC(CH_3)CH(CF_3)_2$	F:	PerFHalOrg.5-25
$C_7F_9H_5N_2O_2$	$CF_3NNC(CF_3)_2COOC_2H_5$	F:	PerFHalOrg.7-160
$C_7F_9H_{11}NNiP_3$	$C_6H_{11}NCNi(PF_3)_3$	Ni:	Org.Verb.1-303
$C_7F_9H_{12}N_2P$	$(CF_3)_3P[N(CH_3)_2]_2$	F:	PerFHalOrg.3-109
C_7F_9I	$FC(CF_2CF_2)(CF_2)(CFCF)CI$	F:	PerFHalOrg.4-69
C_7F_9IMg	$FC(CFCF)(CF_2)(CF_2CF_2)CMgI$	F:	PerFHalOrg.4-66, 71, 97
C_7F_9Li	$FC(CF_2)(CFCF)(CF_2CF_2)CLi$	F:	PerFHalOrg.4-4, 13, 43/4
C_7F_9N	$(CF_3)_2C_5F_3N$	F:	PerFHalOrg.5-148/9, 159, 177
–	$C_2F_5C_5F_4N$	F:	PerFHalOrg.5-149/50, 159
$C_7F_9NO_4$	$[N(CF_3)OCF_2CF_2(C_4F_2O_3)]_n$	F:	PerFHalOrg.7-209/10
$C_7F_9N_3$	$[(CF_2)_3C_3N_3(CF_3)]_n$	F:	PerFHalOrg.6-121
$C_7F_9N_4Na$	$Na(CF_3)_3C_4N_4$	F:	PerFHalOrg.6-123
$C_7F_{10}H_2N_2$	$C_4F_9C(NH_2)CFCN$	F:	PerFHalOrg.7-3
$C_7F_{10}H_3NOS$	$SC(C_2F_5)_2NHC(O)CH_2$	F:	PerFHalOrg.7-66/7
$C_7F_{10}H_4N_2O$	$(NH_2)(CF_3)_2(CHF_2)_2C_3NO$	F:	PerFHalOrg.7-65/6
$C_7F_{10}H_4N_2O_2$	$NH_2CO(CF_2)_2CF(CF_3)CF_2C(O)NH_2$	F:	PerFHalOrg.7-18, 20, 48
$C_7F_{10}H_4N_2O_3$	$NH_2C(O)CF(CF_3)O(CF_2)_3C(O)NH_2$	F:	PerFHalOrg.7-18, 20, 48
$C_7F_{10}H_4N_2O_4$	$NH_2COCF(CF_3)O(CF_2)_2OCF_2CONH_2$	F:	PerFHalOrg.7-19/20
$C_7F_{10}H_6N_4O$	$NH_2(NH)CCF(CF_3)O(CF_2)_3C(NH)NH_2$	F:	PerFHalOrg.7-5
$C_7F_{10}H_6N_4O_2$	$NH_2(NH)CCF(CF_3)O(CF_2)_2(OCF_2)C(NH)NH_2$	F:	PerFHalOrg.7-5
$C_7F_{10}H_9N_2O_5P$	$2\ CF_3NO \cdot C_2F_4 \cdot P(OCH_3)_3$	F:	PerFHalOrg.5-100
$C_7F_{10}ILi$	$IC(CF_2)(CF_2CF_2)(CFLiCF_2)CF$	F:	PerFHalOrg.4-4, 9, 15
$C_7F_{10}I_2$	$IC(CF_2)(CF_2CF_2)_2CI$	F:	PerFHalOrg.4-14
$C_7F_{10}Li_2$	$LiC(CF_2)(CF_2CF_2)_2CLi$	F:	PerFHalOrg.4-4, 14
$C_7F_{10}N_2$	$(i\text{-}C_3F_7)C_4F_3N_2$	F:	PerFHalOrg.6-32/3, 53, 55/6
–	$NC(i\text{-}C_3F_7)CFNCFCF$	F:	PerFHalOrg.6-30/2, 51
–	$NCFC(i\text{-}C_3F_7)CFCFN$	F:	PerFHalOrg.6-16, 36
–	$NCFNC(i\text{-}C_3F_7)CFCF$	F:	PerFHalOrg.6-23, 44
–	$NCFNCFC(i\text{-}C_3F_7)CF$	F:	PerFHalOrg.6-23, 45
$C_7F_{10}N_2O$	$(CF_3)_2NOC_5F_4N$	F:	PerFHalOrg.7-99/100
$C_7F_{10}N_4S$	$(C_2F_5)_2C_3N_4S$	F:	PerFHalOrg.6-123/4, 138
$C_7F_{11}H$	$FC(CF_2)(CF_2CF_2)_2CH$	F:	PerFHalOrg.4-43

$C_7F_{11}H_2NO$	$C_6F_{11}C(O)NH_2$	F:	PerFHalOrg.7-19/21, 48
$C_7F_{11}H_3N_2O$	$(NH_2)(CF_3)_3(CHF_2)C_3NO$	F:	PerFHalOrg.7-65/6
$C_7F_{11}I$	$IC(CF_2)(CF_2CF_2)_2CF$	F:	PerFHalOrg.4-14
$C_7F_{11}IMg$	$FC(CF_2)(CF_2CF_2)_2CMgI$	F:	PerFHalOrg.4-66, 69, 71
$C_7F_{11}Li$	$LiC(CF_2)(CF_2CF_2)_2CF$	F:	PerFHalOrg.4-4, 9, 11/7, 33, 43
$C_7F_{11}NO$	$CF_3C_6F_8NO$	F:	PerFHalOrg.6-130, 141
$C_7F_{11}N_3$	$(C_2F_5)_2C_3FN_3$	F:	PerFHalOrg.6-78/80, 99/100
$C_7F_{12}GeO_6$	$CF_3Ge(OCOCF_3)_3$	F:	PerFHalOrg.4-166, 175, 179
$C_7F_{12}HNO$	$(CF_3)_2CNC(O)CH(CF_3)_2$	F:	PerFHalOrg.7-65
$C_7F_{12}HNO_2$	$(CF_3)_2CCONHC(CF_3)_2O$	F:	PerFHalOrg.5-55/6, 77, 101, 110
$C_7F_{12}HNO_5$	$O(CF_2CF_2)_2NOCF_2OCF_2COOH$	F:	PerFHalOrg.6-2/3
$C_7F_{12}HN_2O$	$C(O)C(CF_3)_2NHC(CF_3)_2N$	F:	PerFHalOrg.5-99
$C_7F_{12}HN_3Na_2$	$NaNHCNC(CF_3)_2NNaC(CF_3)_2$	F:	PerFHalOrg.5-64
$C_7F_{12}HN_3O_3$	$(CF_3)_2CC(NHNO_2)NC(CF_3)_2O$	F:	PerFHalOrg.5-56, 78
$C_7F_{12}H_2N_2O$	$(CF_3)_2CC(NH_2)NC(CF_3)_2O$	F:	PerFHalOrg.5-56, 78, 98/9, 101/3
–	$C(O)NHC(CF_3)_2NHC(CF_3)_2$	F:	PerFHalOrg.5-62, 84, 103
$C_7F_{12}H_2N_2O_4$	$O(CF_2CF_2)_2NOCF_2OCF_2CONH_2$	F:	PerFHalOrg.6-2/3, 8
$C_7F_{12}H_2N_3Na$	$NaNHCNC(CF_3)_2NHC(CF_3)_2$	F:	PerFHalOrg.5-64
$C_7F_{12}H_2N_4O_2$	$NO_2NHCNC(CF_3)_2NHC(CF_3)_2$	F:	PerFHalOrg.5-63/4, 86, 99, 108
$C_7F_{12}H_3N_3$	$C_2F_5C(NH)NC(NH_2)C_3F_7$	F:	PerFHalOrg.7-7, 34, 58
–	$NH_2CNC(CF_3)_2NHC(CF_3)_2$	F:	PerFHalOrg.5-62/4, 85, 96/9, 103/6, 110
$C_7F_{12}H_3N_3O$	$(CF_3)_2CC(NHNH_2)NC(CF_3)_2O$	F:	PerFHalOrg.5-56, 78
–	$C_2F_5C(NH)NC(NH_2)(CF_2)_2OCF_3$	F:	PerFHalOrg.7-7, 35
$C_7F_{12}H_4N_2O$	$(CF_3)_2CHNHCONHCH(CF_3)_2$	F:	PerFHalOrg.5-21
$C_7F_{12}H_4N_4$	$NH_2NHCNC(CF_3)_2NHC(CF_3)_2$	F:	PerFHalOrg.5-63/4, 87, 100/1
$C_7F_{12}H_5NO$	$(CF_3)_2C(OH)CH_2NHC(CF_3)_2H$	F:	PerFHalOrg.5-101
$C_7F_{12}H_6N_2O_2$	$(CF_3)_2NOCH(CH_3)CH_2ON(CF_3)_2$	F:	PerFHalOrg.7-129, 131
$C_7F_{12}H_6P_2$	$(CF_3)_2PCH_2CH(CH_3)P(CF_3)_2$	F:	PerFHalOrg.3-183
$C_7F_{12}H_9NOP_2$	$(CH_3)_3N \cdot (CF_3)_2POP(CF_3)_2$	F:	PerFHalOrg.3-37
$C_7F_{12}H_9N_2O_2Sb$	$[(CF_3)_2NO]_2Sb(CH_3)_3$	F:	PerFHalOrg.7-128
$C_7F_{12}H_9N_2O_5P$	$(CH_3O)_3P[ON(CF_3)_2]_2$	F:	PerFHalOrg.7-128
$C_7F_{12}H_{10}N_2P_2$	$[(CF_3)_2P]_2NH \cdot N(CH_3)_3$	F:	PerFHalOrg.3-69
$C_7F_{12}H_{11}N_3O_2$	$(CH_3)_3N \cdot [(CF_3)_2NOH]_2$	F:	PerFHalOrg.7-139
$C_7F_{12}KNO_2$	$(CF_3)_2CCONKC(CF_3)_2O$	F:	PerFHalOrg.5-55/6, 110
$C_7F_{12}NNaO_2$	$(CF_3)_2CCONNaC(CF_3)_2O$	F:	PerFHalOrg.5-55/6, 77, 101/2, 110
$C_7F_{12}NNiO_3P$	$(CO)_3NiP(CF_3)_2N(CF_3)_2$	Ni:	Org.Verb.1-165, 167
$C_7F_{12}NO_2$	$C(O)C(CF_3)_2OC(CF_3)_2N$	F:	PerFHalOrg.5-99
$C_7F_{12}N_2$	$NC(CF_3)_2C(CF_3)C(CF_3)N$	F:	PerFHalOrg.5-61, 84, 97
$C_7F_{12}N_2O$	$NC(CF_3)_2COC(CF_3)_2N$	F:	PerFHalOrg.5-60, 82
$C_7F_{12}N_2O_2$	$CF_3NC(O)(CF_2)_3C(O)NCF_3$	F:	PerFHalOrg.6-165
–	$NC(CF_3)_2COC(CF_3)_2N(O)$	F:	PerFHalOrg.5-60, 83

$C_7F_{12}N_2O_3$	$(CF_3)_2C(NO)C(O)ONC(CF_3)_2$	F:	PerFHalOrg.7-97, 115
–	$[(CF_3)_2CNO]_2CO$	F:	PerFHalOrg.7-94/5, 110
$C_7F_{12}N_2S$	$NC(C(CF_3)_2)SC(CF_3)_2N$	F:	PerFHalOrg.5-67/8
$C_7F_{13}H_2NO$	$C_6F_{13}C(O)NH_2$	F:	PerFHalOrg.7-18, 76
$C_7F_{13}N$	$CF_2CFN(CF_2)_5$	F:	PerFHalOrg.5-157/8, 170, 175
–	$NC(CF_3)(CF_2)_3CF(CF_3)$	F:	PerFHalOrg.5-150/1, 162
$C_7F_{13}NO$	$(CF_2)_4NCF_2CF_2COF$	F:	PerFHalOrg.5-52/3
$C_7F_{13}N_5$	$[(CF_3)_2N]_2C_3FN_3$	F:	PerFHalOrg.6-78, 99
C_7F_{14}	$C_5F_{11}CFCF_2$	F:	PerFHalOrg.4-9
$C_7F_{14}FeO_3P_2$	$(CO)_3FeP_2F_2(CF_3)_4$	Fe:	Org.Verb.B1-149
$C_7F_{14}HNO$	$(CF_3)_2CNOCF_2CH(CF_3)_2$	F:	PerFHalOrg.7-141
$C_7F_{14}H_2O$	$(n-C_3F_7)_2CHOH$	F:	PerFHalOrg.4-42, 94
$C_7F_{14}H_9PSn$	$[(CH_3)_3Sn][(CF_3)_4PF_2]$	F:	PerFHalOrg.3-134, 137
$C_7F_{14}N_2O$	$N(CF_2)_4C(ON(CF_3)_2)$	F:	PerFHalOrg.5-184
		F:	PerFHalOrg.7-99/100
$C_7F_{14}O$	$(C_3F_7)_2CO$	F:	PerFHalOrg.4-41, 42, 94, 96
$C_7F_{15}HN_2O_2$	$CF_3C(O)CH[N(CF_3)_2]ON(CF_3)_2$	F:	PerFHalOrg.7-135/6
$C_7F_{15}HN_2O_3$	$((CF_3)_2NO)_2CHCOCF_3$	F:	PerFHalOrg.7-135/6
$C_7F_{15}Li$	$(n-C_7F_{15})Li$	F:	PerFHalOrg.4-2, 8/9, 13, 42
$C_7F_{15}N$	$CF_2CF(C_2F_5)NFCF(CF_3)CF_2$	F:	PerFHalOrg.5-52/3, 76
–	$(CF_2)_4NCF(CF_3)_2$	F:	PerFHalOrg.5-52/3, 75
–	$(CF_2)_4NCF_2CF_2CF_3$	F:	PerFHalOrg.5-52/3, 75
–	$C_2F_5N(CF_2)_5$	F:	PerFHalOrg.5-157/8, 170, 175
–	$FNC_5F_8(CF_3)_2$	F:	PerFHalOrg.5-150/1, 161/2, 176
–	$FNC_5F_9C_2F_5$	F:	PerFHalOrg.5-150/1, 162
$C_7F_{15}NO$	$(CF_3)_2CNOC(CF_3)_3$	F:	PerFHalOrg.7-96/7, 114, 145
–	$O(CF_2CF_2)_2NCF(CF_3)_2$	F:	PerFHalOrg.6-1/2, 6
–	$O(CF_2CF_2)_2NCF_2CF_2CF_3$	F:	PerFHalOrg.6-1/2, 6
$C_7F_{15}NO_2$	$C_5F_{11}OCF_2CF_2NO$	F:	PerFHalOrg.7-179, 185
$C_7F_{15}P_3$	$C_2P_3(CF_3)_5$	F:	PerFHalOrg.3-8/14
$C_7F_{16}N_2O$	$(CF_3)_2NOCF_2C(N(CF_3)_2)CF_2$	F:	PerFHalOrg.7-96/7, 114
$C_7F_{16}N_2O_2$	$CF(CF_3)CF_2ON(CF_3)CF_2CF_2N(CF_3)O$	F:	PerFHalOrg.6-166/7
–	$(CF_3)_2NOC(N(CF_3)_2)(CF_3)COF$	F:	PerFHalOrg.7-97/8, 116
$C_7F_{16}N_2O_3$	$[(CF_3)_2NO]_2C(CF_3)COF$	F:	PerFHalOrg.7-97/8, 116
–	$[(CF_3)_2NOCF_2]_2CO$	F:	PerFHalOrg.7-96/7, 114
$C_7F_{17}P$	$CF_3P(C_3F_7)_2$	F:	PerFHalOrg.3-163, 168, 173
$C_7F_{18}N_2O_2$	$(CF_3)_2NOCF_2CF(CF_3)ON(CF_3)_2$	F:	PerFHalOrg.7-95/6, 113, 145
$C_7F_{21}N_2O_2P$	$[(CF_3)_2NO]_2P(CF_3)_3$	F:	PerFHalOrg.7-100/1
$C_7F_{21}NiOP_3$	$Ni(CO)[(CF_3)_2PF]_3$	F:	PerFHalOrg.3-121
		Ni:	Org.Verb.1-114
$C_7FeGeH_9NO_4$	$(CO)_3Fe(NO)Ge(CH_3)_2CHCH_2$	Fe:	Org.Verb.B1-197/8
$C_7FeH_2O_7$	$OC(OCH)_2Fe(CO)_4$	Fe:	Org.Verb.B4-302, 304

$C_7FeH_3LiO_5$	$Li[CH_2CHC(O)Fe(CO)_4]$	Fe: Org.Verb.B3-132
$C_7FeH_3NO_4$	$(CO)_4Fe(NCCHCH_2)$	Fe: Org.Verb.B2-71, 76, 80
		Fe: Org.Verb.B4-233, 235, 242/3
$C_7FeH_3N_6^{2-}$	$[CH_3NCFe(CN)_5]^{2-}$	Fe: Org.Verb.B4-10
$C_7FeH_3N_6Ni$	$[CH_3NCFe(CN)_5]Ni \cdot x\ H_2O$	Fe: Org.Verb.B4-10
$C_7FeH_3N_6Zn$	$[CH_3NCFe(CN)_5]Zn \cdot x\ H_2O$	Fe: Org.Verb.B4-10
$C_7FeH_3O_5^-$	$[CH_2CHC(O)Fe(CO)_4]^-$	Fe: Org.Verb.B4-150
$C_7FeH_3O_6^-$	$[OC(O)CHCH_2Fe(CO)_4]^-$	Fe: Org.Verb.B4-245/6
$C_7FeH_4N_2O_4$	$(CO)_4FeN_2C_3H_4$	Fe: Org.Verb.B2-71/2, 74
$C_7FeH_4O_3$	$C_4H_4Fe(CO)_3$	Fe: Org.Verb.B2-31
$C_7FeH_4O_5$	$HC(O)CHCH_2Fe(CO)_4$	Fe: Org.Verb.B4-233, 235, 243/4
$C_7FeH_4O_6$	$HOC(O)CHCH_2Fe(CO)_4$	Fe: Org.Verb.B4-233, 235/6, 244/6
$C_7FeH_4O_6S$	$(SO_2CHCHCH_2)Fe(CO)_4$	Fe: Org.Verb.B4-303, 306
$C_7FeH_5IO_2$	$C_5H_5Fe(CO)_2I$	Fe: Org.Verb.B2-66
$C_7FeH_5IO_4$	$ICHCHCOCH_3Fe(CO)_3$	Fe: Org.Verb.B4-289
$C_7FeH_5KNO_5$	$K[NH_2COCHCH_2Fe(CO)_4]$	Fe: Org.Verb.B4-248
$C_7FeH_5KO_6$	$K[HOCH_2CH_2C(O)Fe(CO)_4]$	Fe: Org.Verb.B4-153, 162
$C_7FeH_5NO_4$	$(CH_2N(CH_3)C)Fe(CO)_4$	Fe: Org.Verb.B4-128, 131, 135
–	$C_2H_5NCFe(CO)_4$	Fe: Org.Verb.B4-5/6
$C_7FeH_5NO_5$	$NH_2C(O)CHCH_2Fe(CO)_4$	Fe: Org.Verb.B4-233, 236/7, 248/9
$C_7FeH_5NO_5^-$	$[NH_2COCHCH_2Fe(CO)_4]^-$	Fe: Org.Verb.B4-248
$C_7FeH_5N_6$	$[CH_3NCFe(CN)_5]H_2$	Fe: Org.Verb.B4-10
$C_7FeH_5NaO_4$	$Na[CH(O)C_3H_4Fe(CO)_3]$	Fe: Org.Verb.B5-72, 75
$C_7FeH_5NaO_5$	$Na[C_2H_5C(O)Fe(CO)_4]$	Fe: Org.Verb.B4-150, 159
$C_7FeH_5NaO_6$	$(CO)_5Fe(NaOC_2H_5)$	Fe: Org.Verb.B3-249/50
$C_7FeH_5O_4^+$	$[C_3H_5Fe(CO)_4]^+$	Fe: Org.Verb.B5-99, 102/4
$C_7FeH_5O_4^-$	$[CH(O)C_3H_4Fe(CO)_3]^-$	Fe: Org.Verb.B5-73, 75
$C_7FeH_5O_5^-$	$[C_2H_5C(O)Fe(CO)_4]^-$	Fe: Org.Verb.B4-150, 159
$C_7FeH_5O_6^-$	$[C_2H_5OC(O)Fe(CO)_4]^-$	Fe: Org.Verb.B4-154, 163
–	$[HOCH_2CH_2C(O)Fe(CO)_4]^-$	Fe: Org.Verb.B4-153, 162
$C_7FeH_6LiNO_5$	$Li[(CH_3)_2NC(O)Fe(CO)_4] \cdot 2\ C_4H_8O$	Fe: Org.Verb.B4-155, 163
$C_7FeH_6NO_5^-$	$[(CH_3)_2NC(O)Fe(CO)_4]^-$	Fe: Org.Verb.B4-155, 163/4
$C_7FeH_6N_2O_3$	$(CH_3NC)_2Fe(CO)_3$	Fe: Org.Verb.B4-18/9
$C_7FeH_6N_2O_4S$	$(CO)_4FeSC(NHCH_2)_2$	Fe: Org.Verb.B2-128/30
$C_7FeH_6N_4O_4$	$(NN(CH_3)NN(CH_3)C)Fe(CO)_4$	Fe: Org.Verb.B4-128/9
–	$(NNCH_3)_2CFe(CO)_4$	Fe: Org.Verb.B4-128/9, 132
$C_7FeH_6N_6O$	$(CH_3NC)_2Fe(CN)_3NO$	Fe: Org.Verb.B4-11
$C_7FeH_6O_4$	$CH_3CHCH_2Fe(CO)_4$	Fe: Org.Verb.B4-233, 237, 249/50
$C_7FeH_6O_4S$	$(CHCH_2)_2S(O)Fe(CO)_3$	Fe: Org.Verb.B4-213, 220
$C_7FeH_6O_5$	$CH_3OC(O)CHCH_2Fe(CO)_3$	Fe: Org.Verb.B4-194, 195
		Fe: Org.Verb.B5-131, 140/1
–	$(HO)_2C_4H_4Fe(CO)_3$	Fe: Org.Verb.B5-76
$C_7FeH_6O_7^{2-}$	$[(CO)_3Fe(COOCH_3)_2]^{2-}$	Fe: Org.Verb.B3-175
$C_7FeH_6O_7P_2$	$(CO)_4FeP(CH_2O)_3P$	Fe: Org.Verb.B2-86, 104

$C_7FeH_6O_7P_2$ .... $(CO)_4FeP(OCH_2)_3P$ .... Fe: Org.Verb.B2-96, 117
$C_7FeH_7^+$ .... $[FeC_7H_7]^+$ .... Fe: Org.Verb.A1-264, 269
$C_7FeH_7IO_3$ .... $CH_3C_3H_4Fe(CO)_3I$ .... Fe: Org.Verb.B5-45, 48, 52/3, 60/1, 63
$C_7FeH_7IO_4$ .... $CH_3OC_3H_4Fe(CO)_3I$ .... Fe: Org.Verb.B5-53, 63
C_7FeH_7N .... $[Fe(C_7H_7N)]_n$ .... Fe: Org.Verb.B3-206
$C_7FeH_7NO_3$ .... $C_5H_7Fe(CO)_2NO$ .... Fe: Org.Verb.B5-14/5, 17, 23
$C_7FeH_7NO_4$ .... $(CH_3)_2NCHFe(CO)_4$ .... Fe: Org.Verb.B4-118/9
$C_7FeH_7NO_5$ .... $CH_3OOCC_3H_4Fe(CO)_2NO$ .... Fe: Org.Verb.B5-14/6
$C_7FeH_7NO_6$ .... $CH_3C_3H_4Fe(CO)_3ONO_2$ .... Fe: Org.Verb.B5-41, 45, 48, 52/3, 61/3
– .... $NH_4[OC(O)CHCH_2Fe(CO)_4]$ .... Fe: Org.Verb.B4-245/6
$C_7FeH_7O^+$ .... $[FeC_5H_4COCH_3]^+$ .... Fe: Org.Verb.A2-208, 245
– .... $[FeC_6H_4OCH_3]^+$ .... Fe: Org.Verb.A2-296
$C_7FeH_7O_3$ .... $[CH_3C_3H_4Fe(CO)_3]$ .... Fe: Org.Verb.B5-69, 71
$C_7FeH_7O_3^+$ .... $[CH_3C_3H_4Fe(CO)_3]^+$ .... Fe: Org.Verb.B5-82/6
– .... $[C_4H_6Fe(CO)_3H]^+$ .... Fe: Org.Verb.B5-62
$C_7FeH_7O_4^-$ .... $[(CO)_4FeC_3H_7]^-$ .... Fe: Org.Verb.B2-176
$C_7FeH_7O_5^-$ .... $[CH_3CH(OH)CH_2C(O)Fe(CO)_3]^-$ .... Fe: Org.Verb.B4-109
$C_7FeH_8^+$ .... $[FeC_7H_8]^+$ .... Fe: Org.Verb.A1-264, 269
$C_7FeH_8NNaO_6$ ...... $Na[Fe(CO)_3NO] \cdot C_4H_8O_2$ .... Fe: Org.Verb.B1-188
$C_7FeH_8NO_4^+$ .... $[(CO)_4FeCH_2N(CH_3)_2]^+$ .... Fe: Org.Verb.B2-188
$C_7FeH_8N_2O_4$ .... $(CH_3NH)_2CFe(CO)_4$ .... Fe: Org.Verb.B4-119/20
– .... $(CH_3ND)_2CFe(CO)_4$ .... Fe: Org.Verb.B4-120
$C_7FeH_8N_2O_5$ .... $(CO)_5Fe(C_2H_8N_2)$ .... Fe: Org.Verb.B3-249
$C_7FeH_8N_4O_4$ .... $[(NNCH_3)_2CH][HFe(CO)_4]$ .... Fe: Org.Verb.B4-119, 132
$C_7FeH_8N_6O_2$ .... $[C_2H_5NCFe(CN)_4NO_2]H_3 \cdot 2\ H_2O$ .... Fe: Org.Verb.B4-10
$C_7FeH_8O_4$ .... $(CH_3)(C_2H_5O)CFe(CO)_3$ .... Fe: Org.Verb.B4-116
$C_7FeH_8O_7S$ .... $[CH_3C_3H_4Fe(CO)_3]HSO_4$ .... Fe: Org.Verb.B5-84, 86
$C_7FeH_9NO_3$ .... $(CH_3)_2C_3H_3Fe(CO)_2NO$ .... Fe: Org.Verb.B5-17, 23
$C_7FeH_9NO_4$ .... $(CO)_4FeN(CH_3)_3$ .... Fe: Org.Verb.B2-70
$C_7FeH_9O_4P$ .... $(CO)_4FeP(CH_3)_3$ .... Fe: Org.Verb.B2-84, 86, 103/4
$C_7FeH_9O_4Sb$ .... $(CO)_4FeSb(CH_3)_3$ .... Fe: Org.Verb.B2-83/4, 101, 103, 122
$C_7FeH_9O_7P$ .... $(CO)_4FeP(OCH_3)_3$ .... Fe: Org.Verb.B2-84/5, 95, 115/6
$C_7FeH_{10}O_4Sn$ ...... $(CH_3)_3SnFe(CO)_4H$ .... Fe: Org.Verb.B3-140
$C_7FeH_{11}NiP^+$ ...... $NiP(CH_3)_2Fe(C_5H_5)^+$ .... Ni: Org.Verb.1-167
$C_7FeH_{12}N_2O_4$ ...... $[N(CH_3)_4][Fe(CO)_3NO]$ .... Fe: Org.Verb.B1-190/1
$C_7FeH_{12}N_4OS_2^+$ .... $[(CO)Fe(SC(NHCH_2)_2)_2]^+$ .... Fe: Org.Verb.B1-165
$C_7FeH_{13}NO_4Si$ ...... $[SiH_3 \cdot N(CH_3)_3][HFe(CO)_4]$ .... Fe: Org.Verb.B2-20
$C_7FeH_{14}NO_5P$ ...... $C_3H_5Fe(CO)NOP(OCH_3)_3$ .... Fe: Org.Verb.B5-6/9, 20/1
$C_7FeH_{14}O_3Si_2$ ...... $(CO)_3Fe(H)(Si(CH_3)_2H)Si(CH_3)_2$ .... Fe: Org.Verb.C2-52
$C_7FeH_{15}N_2O_6P$ ..... $COFe(NO)_2P(OC_2H_5)_3$ .... Fe: Org.Verb.B1-42/3, 45, 47/8
$C_7FeN_2O_5$ .... $(CO)_5Fe(CN)_2$ .... Fe: Org.Verb.B3-254
$C_7FeN_2O_5S_2$ .... $(CO)_5Fe(SCN)_2$ .... Fe: Org.Verb.B3-254/5
$C_7FeNiO_{13}P_4$ .... $(CO)_3Ni(P_4O_6) \cdot Fe(CO)_4$ .... Ni: Org.Verb.1-185
$C_7Fe_2GeH_{13}IN_4O_4$ ... $(C_3H_5Fe(NO)_2)_2Ge(I)CH_3$ .... Fe: Org.Verb.C2-165

$C_7Fe_2H_2O_6S_2$	$(CO)_3Fe(SCH_2S)Fe(CO)_3$	Fe: Org.Verb.C2-44
$C_7Fe_2H_{10}N_2O_5S_2$	$(CO)_3Fe_2(NO)_2(SC_2H_5)_2$	Fe: Org.Verb.C1-220
$C_7Fe_2H_{13}IN_4O_4Pb$	$(C_3H_5Fe(NO)_2)_2Pb(I)CH_3$	Fe: Org.Verb.C2-165
$C_7Fe_2H_{13}IN_4O_4Sn$	$(C_3H_5Fe(NO)_2)_2Sn(I)CH_3$	Fe: Org.Verb.C2-165
$C_7Fe_2N_2O_7$	$(CO)_4FeFe(CO)_3N_2$	Fe: Org.Verb.C2-1
$C_7Fe_2O_7^+$	$[Fe_2(CO)_7]^+$	Fe: Org.Verb.C2-26
$C_7H_3IrK_2N_6$	$K_2[Ir(CN)_5(NCCH_3)]$	Ir: SVol.2-184
$C_7H_3IrN_6^{2-}$	$[Ir(CN)_5(CH_3CN)]^{2-}$	Ir: SVol.2-183
$C_7H_3NO_4Sn$	$Sn(NC_5H_3(COO)_2)$	Sn: MVol.C5-19
$C_7H_3O_5ReS_3$	$Re(CO)_5[S_2CSCH_3]$	C: MVol.D4-269
$C_7H_4IMnO_2^+$	$[Mn(OC_6H_3(I)CHO)]^+$	Mn: MVol.D1-53/4
$C_7H_4MnNO_4^+$	$[Mn(OC_6H_3(NO_2)CHO)]^+$	Mn: MVol.D1-53/4
$C_7H_4N_6O_4S_3$	$[NH_2C(O)C(O)N(NCS)]_2CS$	C: MVol.D6-160
C_7H_5HgN	$(C_6H_5)HgCN$	C: MVol.D6-235
C_7H_5HgNSe	$C_6H_5HgSeCN$	C: MVol.D6-239
$C_7H_5MnO_2^+$	$[Mn(OC(CH)_5CO)]^+$	Mn: MVol.D1-59
–	$[Mn(OC_6H_4CHO)]^+$	Mn: MVol.D1-52/3
$C_7H_5MnO_3^+$	$[Mn(OC_6H_3(OH)CHO)]^+$	Mn: MVol.D1-53/4
$C_7H_5MnO_5S^+$	$[Mn(OC_6H_3(SO_3H)CHO)]^+$	Mn: MVol.D1-53/4
$C_7H_5MnO_5S_2$	$(CO)_4Mn[S_2COC_2H_5]$	C: MVol.D4-259
$C_7H_5NO_2Pu^{2+}$	$Pu(C_6H_5CONO)^{2+}$	Np: TrU.D1-168/9
$C_7H_5NO_2Sn$	$Sn(OC_6H_4CHNO)$	Sn: MVol.C5-41/2
$C_7H_5NO_3S$	$C_6H_5C(O)NSO_2$	S: S-N-Verb.1-59
C_7H_5NSe	$(C_6H_5)SeCN$	C: MVol.D6-235, 239
$C_7H_5N_3O_5S$	$SO_2(NCO)_2 \cdot C_5H_5NO$	S: S-N-Verb.1-183
$C_7H_5O_5ReS_2$	$(CO)_4Re[S_2COC_2H_5]$	C: MVol.D4-259
C_7H_6INiO	$[C_6H_6Ni(CO)]I$	Ni: Org.Verb.2-221, 225
$C_7H_6NNpO_4$	$NpO_2(NC_5H_3(CH_3)COO)$	Np: TrU.D1-155
–	$NpO_2(NC_5H_4CH_2COO)$	Np: TrU.D1-156
$C_7H_6NO_2Pu^{3+}$	$Pu(C_6H_5CONHO)^{3+}$	Np: TrU.D1-168/9
$C_7H_6NO_4Pu^+$	$PuO_2(NC_5H_3(CH_3)COO)^+$	Np: TrU.D1-155
$C_7H_6N_2NiO$	$[C_6H_6Ni(NO)]CN$	Ni: Org.Verb.2-221
$C_7H_6N_2Se$	$H_2NC_6H_4SeCN$	C: MVol.D6-239
C_7H_6Sn	$(CHC)_3SnCH_3$	Sn: Org.Verb.2-351/2
$C_7H_7NNpO_4^{2+}$	$NpO_2(NC_5H_3(CH_3)COOH)^{2+}$	Np: TrU.D1-155
$C_7H_7NO_4Pu^{2+}$	$PuO_2(NC_5H_3(CH_3)COOH)^{2+}$	Np: TrU.D1-155
$C_7H_7NO_4Sn$	$Sn(HCOO)_2 \cdot C_5H_5N$	Sn: MVol.C5-17
$C_7H_7NO_9U$	$U(C_2O_4)_2 \cdot HCON(CH_3)_2 \cdot 2\ H_2O$	U: SVol.E1-99/100
$C_7H_7S_{1.5}Sn$	$(CH_3C_6H_4SnS_{1.5})_x$	Sn: Org.Verb.6-282
$C_7H_8I_4N_2SSn$	$SnI_4 \cdot C_5H_3N(CH_3)CSNH_2$	Sn: MVol.C6-106
$C_7H_8NNpO_8^{2-}$	$NpO_2(N(CH_2CH_2COO)(CH_2COO)_2)^{2-}$	Np: TrU.D1-164
$C_7H_8N_2O_4S$	$C_6H_5OC(O)NHSO_2NH_2$	S: S-N-Verb.1-167
$C_7H_8N_2O_8U$	$UO_2(OCOC_6H_4OH)(NO_3) \cdot NH_3$	U: SVol.E1-15
$C_7H_8N_4O_2S$	$CH_3C_6H_4NHSO_2N_3$	S: S-N-Verb.1-214
C_7H_8Ni	$(NiC_7H_8)_x$	Ni: Org.Verb.2-399
C_7H_8NiO	$C_5H_5Ni(CO)CH_3$	Ni: Org.Verb.2-167
$C_7H_8NiO_2S$	$(C_5H_5NiSCH_2COOH)_n$	Ni: Org.Verb.2-338
$C_7H_8NiS_3$	$(C_5H_5)Ni[S_2CSCH_3]$	C: MVol.D4-270
		Ni: Org.Verb.2-115
$C_7H_8O_2Sn$	$(CH_3C_6H_4SnOOH)_x$	Sn: Org.Verb.6-282

$C_7H_8O_3Sn$	$SnO_2 \cdot CH_3C_6H_4OH \cdot 3\ H_2O$	Sn:	Org.Verb.1-158
C_7H_8Ti	$(CH_3)(C_6H_5)Ti$	Ti:	Org.Verb.1-68
$C_7H_9NNiO_4$	$CH_3(H)N(C_2H_5O)CNi(CO)_3$	Ni:	Org.Verb.1-334
–	$(CO)_3Ni(O(CH_2CH_2)_2NH)$	Ni:	Org.Verb.1-160
$C_7H_9NNpO_8^-$	$NpO_2(N(CH_2CH_2COO)(CH_2COO)_2H)^-$	Np:	TrU.D1-164
$C_7H_9NNpO_9^{3-}$	$NpO_2(OH)(N(CH_2CH_2COO)(CH_2COO)_2)^{3-}$	Np:	TrU.D1-164
$C_7H_9NO_2S$	$CH_3C_6H_4SO_2NH_2$	S:	S-N-Verb.1-123
$C_7H_9N_2NaO_3S$	$Na[SO_3N(C_6H_4CH_3)NH_2]$	S:	S-N-Verb.1-219
$C_7H_9N_3NiO$	$(CH_3NC)_3NiCO$	Ni:	Org.Verb.1-323, 324
$C_7H_9N_3O_3S$	$NH_2SO_2NHNCHC_6H_4OH$	S:	S-N-Verb.1-222
$C_7H_9Sn^+$	$Sn(C_6H_5)(CH_3)H^+$	Sn:	Org.Verb.3-25
$C_7H_{10}I_2OTi$	$C_5H_5Ti(OC_2H_5)I_2$	Ti:	Org.Verb.1-180/2
$C_7H_{10}MnO_4$	$[Mn(CH(COCH_3)_2)CH_3COO]$	Mn:	MVol.D1-75
$C_7H_{10}N_2O_2S$	$CH_3C_6H_4NHSO_2NH_2$	S:	S-N-Verb.1-167
–	$C_6H_5CH_2NHSO_2NH_2$	S:	S-N-Verb.1-167
$C_7H_{10}N_2O_3S$	$CH_3OC_6H_4NHSO_2NH_2$	S:	S-N-Verb.1-167
$C_7H_{10}N_2Sn$	$(CH_3)_3SnC(CN)CHCN$	Sn:	Org.Verb.2-78
$C_7H_{10}N_{12}S$	$C_6H_9N_{11} \cdot HNCS$	C:	MVol.D6-92
$C_7H_{10}NiO$	$(C_3H_5)_2NiCO$	Ni:	Org.Verb.2-55
$C_7H_{10}NiOS$	$(C_5H_5NiSCH_2CH_2OH)_n$	Ni:	Org.Verb.2-338
$C_7H_{10}NiO_3S$	$(CO)_3NiS(C_2H_5)_2$	Ni:	Org.Verb.1-160
$C_7H_{10}OTi^+$	$[C_5H_5TiOC_2H_5]^+$	Ti:	Org.Verb.1-184
$C_7H_{10}O_4OsSn$	$(CH_3)_3SnOsH(CO)_4$	Sn:	Org.Verb.5-86
$C_7H_{11}MnO_2^+$	$[Mn(C_2H_5COCHCOC_2H_5)]^+$	Mn:	MVol.D1-125
$C_7H_{11}N_3O_9U$	$UO_2(NO_3)_2 \cdot C_5H_5N \cdot C_2H_5OH$	U:	SVol.E1-29, 34
$C_7H_{11}NiO_3P$	$(CO)_3Ni(CH_2P(CH_3)_3)$	Ni:	Org.Verb.1-182, 183
–	$(CO)_3NiP(C_2H_5)(CH_3)_2$	Ni:	Org.Verb.1-165, 168
$C_7H_{11}OS_2^-$	$[S_2COC_6H_{11}]^-$	C:	MVol.D4-252
$C_7H_{11}Ti$	$C_5H_5Ti(CH_3)_2$	Ti:	Org.Verb.1-208
$C_7H_{12}NNiO_3P$	$(CO)_3NiP(CH_3)_2N(CH_3)_2$	Ni:	Org.Verb.1-161, 165
$C_7H_{12}N_2Ni$	$(C_3H_5)_2NiCH_2N_2$	Ni:	Org.Verb.2-59
$C_7H_{12}N_2O_7U$	$UO_2C_2O_4 \cdot CO(NHC_2H_5)_2$	U:	SVol.E1-93, 95
–	$UO_2C_2O_4 \cdot (CH_3)_2NCON(CH_3)_2$	U:	SVol.E1-93, 95
$C_7H_{12}N_4S_3Sn$	$[(CH_3)_4N]Sn(SCN)_3$	Sn:	MVol.C3-126
$C_7H_{12}N_6O_{11}U$	$U(C_2O_4)_2 \cdot 3\ (NH_2)_2CO$	U:	SVol.E1-82/4
–	$U(C_2O_4)_2 \cdot 3\ (NH_2)_2CO \cdot H_2O$	U:	SVol.E1-82/4
$C_7H_{12}Ni$	$C_3H_5NiC_3H_4CH_3$	Ni:	Org.Verb.2-60
$C_7H_{12}OS_2$	$SC(OC_6H_{11})(SH)$	C:	MVol.D4-243
$C_7H_{12}O_2S_4$	$C_5H_{10}(OCSSH)_2$	C:	MVol.D4-241
$C_7H_{12}O_2Sn$	$(CH_3)_3SnCCCOOCH_3$	Sn:	Org.Verb.2-112
$C_7H_{12}SSn$	$(CH_3)_3SnC_4H_3S$	Sn:	Org.Verb.2-153, 156
$C_7H_{12}Sn$	$(CH_2CH)_3SnCH_3$	Sn:	Org.Verb.2-347
–	$(CH_3)_3SnCCCHCH_2$	Sn:	Org.Verb.2-112, 116/7
$C_7H_{13}KOS_2$	$K[S_2COC_6H_{13}]$	C:	MVol.D4-253
$C_7H_{13}KOSe_2$	$K[Se_2COC_6H_{13}]$	C:	MVol.D6-221
$C_7H_{13}KO_2S$	$K[SOCOC_6H_{13}]$	C:	MVol.D4-237
$C_7H_{13}NNiS_2$	$CH_3C_3H_4NiSC(S)N(CH_3)_2$	Ni:	Org.Verb.2-9
$C_7H_{13}NO_7U$	$UO_2(CH_3COO)_2 \cdot HCON(CH_3)_2$	U:	SVol.E1-100, 102
$C_7H_{13}OS_2^-$	$[S_2COC_6H_{13}]^-$	C:	MVol.D4-251
$C_7H_{13}O_2S^-$	$[SOCOC_6H_{13}]^-$	C:	MVol.D4-236/7

$C_7H_{13}O_4S_2^-$	$[S_2CO(C_2H_4O)_2C_2H_4OH]^-$	C:	MVol.D4-244
$C_7H_{14}HgOS_2$	$C_4H_9Hg[S_2COC_2H_5]$	C:	MVol.D4-258
$C_7H_{14}NO_3S_2^-$	$[S_2COC_2H_4N(C_2H_4OH)_2]^-$	C:	MVol.D4-244
$C_7H_{14}N_2O_2Sn$	$(CH_3)_3SnC(N_2)COOC_2H_5$	Sn:	Org.Verb.2-21
$C_7H_{14}N_4O_{11}U$	$[C_2H_4(NH_3)_2][UO_2(C_2O_4)_2((NH_2)_2CO)]$	U:	SVol.E1-84, 90
$C_7H_{14}NiS_4$	$[Ni(S_2CSC_3H_7)(SC_3H_7)]_2$	C:	MVol.D4-270
$C_7H_{14}OS_2$	$SC(OC_6H_{13})(SH)$	C:	MVol.D4-239, 241
$C_7H_{14}OSn$	$(CH_3)_3SnCCOC_2H_5$	Sn:	Org.Verb.2-113, 116
$C_7H_{14}O_2SSn$	$(C_2H_5)_2Sn(SCH(CH_3)C(O)O)$	Sn:	Org.Verb.6-59
$C_7H_{14}O_2Sn$	$(CH_3)_3SnC(CH_2)COOCH_3$	Sn:	Org.Verb.2-78, 85
–	$(CH_3)_3SnCHCHCOOCH_3$	Sn:	Org.Verb.2-78, 84/5
–	$(CH_3)_3Sn(OCOC(CH_3)CH_2)$	Sn:	Org.Verb.5-84
$C_7H_{14}Sn$	$(CH_3)_2Sn(CHCH_2)(CHCH_2CH_2)$	Sn:	Org.Verb.3-82
–	$(CH_3)_3SnCHCCHCH_3$	Sn:	Org.Verb.2-78, 84
–	$(CH_3)_3SnCH_2CCCH_3$	Sn:	Org.Verb.2-113
$C_7H_{15}IN_2NiS$	$C_3H_5Ni(SC(NH_2)(NHCH(CH_3)_2))I$	Ni:	Org.Verb.2-281
$C_7H_{15}NO_4Sn$	$CH_3OSnN(CH_2CH_2O)_3$	Sn:	MVol.C6-3
$C_7H_{15}NO_7U$	$UO_2(CH_3COO)_2 \cdot (CH_3)_3NO$	U:	SVol.E1-131/2
–	$UO_2(CH_3COO)_2 \cdot (CH_3)_3NO \cdot H_2O$	U:	SVol.E1-131/2, 135
$C_7H_{15}NSn$	$(C_2H_5)_2(CNCH_2CH_2)SnH$	Sn:	Org.Verb.4-90
$C_7H_{15}O_2Ti$	$C_3H_5Ti(OC_2H_5)_2$	Ti:	Org.Verb.1-127
$C_7H_{16}MnN_5O_4$	$[MnOH(CH(COCH_3)_2)(C_2H_6N_5)H_2O]$	Mn:	MVol.D1-106/7
$C_7H_{16}N_2Se$	$((CH_3)_2CHNH)_2CSe$	C:	MVol.D6-201
$C_7H_{16}N_8O_{11}U$	$(CN_3H_6)_2[UO_2(C_2O_4)_2((NH_2)_2CO)]$	U:	SVol.E1-84, 86, 90
$C_7H_{16}OSn$	$(CH_3)_3SnCHCHOC_2H_5$	Sn:	Org.Verb.2-78
–	$(CH_3)_3SnCH_2CH_2CH(O)CH_2$	Sn:	Org.Verb.2-37
–	$(CH_3)_3SnCH_2CH_2COCH_3$	Sn:	Org.Verb.2-34
$C_7H_{16}OTi$	$(i\text{-}C_3H_7)Ti(OC_4H_9)$	Ti:	Org.Verb.1-45
$C_7H_{16}O_2Sn$	$(CH_3)_3SnCH_2CH_2COOCH_3$	Sn:	Org.Verb.2-34
$C_7H_{16}Sn$	$(CH_3)_2Sn(CH_2)_5$	Sn:	Org.Verb.3-105
–	$(CH_3)_3SnC(CH_3)CHCH_3$	Sn:	Org.Verb.2-78, 85
–	$(CH_3)_3SnC(C_2H_5)CH_2$	Sn:	Org.Verb.2-78, 85/6
–	$(CH_3)_3SnCHC(CH_3)_2$	Sn:	Org.Verb.2-78, 85
–	$(CH_3)_3SnCHCH_2CHCH_3$	Sn:	Org.Verb.2-58/9
–	$(CH_3)_3SnCH(CH_2)_3$	Sn:	Org.Verb.2-60
–	$(CH_3)_3SnCH(CH_3)CHCH_2$	Sn:	Org.Verb.2-78, 85
–	$(CH_3)_3SnCH_2C(CH_3)CH_2$	Sn:	Org.Verb.2-78, 85
–	$(CH_3)_3SnCH_2CHCHCH_3$	Sn:	Org.Verb.2-74/5
–	$(CH_3)_3SnCH_2CH_2CHCH_2$	Sn:	Org.Verb.2-75/6
–	$(C_2H_5)_2Sn(CH_3)(CHCH_2)$	Sn:	Org.Verb.3-86/7
$C_7H_{17}NOS_4Sn$	$Sn(SCH_2CH_2S)_2 \cdot (CH_3)_3NO$	Sn:	MVol.C6-82/3
$C_7H_{17}NOSn$	$(CH_3)_3SnCH_2CON(CH_3)_2$	Sn:	Org.Verb.2-20/1
$C_7H_{17}NO_2S$	$[(CH_3)_4N][O_2CSC_2H_5]$	C:	MVol.D4-237
–	$[(CH_3)_4N][SOCOC_2H_5]$	C:	MVol.D4-237
$C_7H_{17}NO_3S$	$C_4H_9NHSO_2OC_3H_7$	S:	S-N-Verb.1-97
–	$(i\text{-}C_4H_9)NHSO_2OC_3H_7$	S:	S-N-Verb.1-97
$C_7H_{17}NS_3$	$[(C_2H_5)_3NH](HCS_3)$	C:	MVol.D4-162, 217
$C_7H_{18}I_2NiOP_2$	$CONi(P(CH_3)_3)_2I_2$	Ni:	Org.Verb.1-121
$C_7H_{18}I_2Sn_2$	$[(CH_3)_3Sn]_2Cl_2$	Sn:	Org.Verb.5-80
$C_7H_{18}MnN_5O_5$	$[MnOH(CH(COCH_3)_2)(C_2H_7N_5)H_2O]OH$	Mn:	MVol.D1-106/7

$C_7H_{18}N_2O_3SSi$	$O(C_2H_4)_2NSO_2NHSi(CH_3)_3$	S:	S-N-Verb.1-172
$C_7H_{18}N_2SiSn$	$(CH_3)_3SnC(N_2)Si(CH_3)_3$	Sn:	Org.Verb.2-17/8
$C_7H_{18}N_5P$	$[(CH_3)_2N]_3PNCN$	F:	PerFHalOrg.5-23
$C_7H_{18}OSn$	$(CH_3)_3SnCH(CH_3)CH_2CH_2OH$	Sn:	Org.Verb.2-50/1
–	$(CH_3)_3SnCH_2CH_2CH(CH_3)OH$	Sn:	Org.Verb.2-50/1
–	$(CH_3)_3Sn(CH_2)_4OH$	Sn:	Org.Verb.2-50
$C_7H_{18}OTi$	$(CH_3)_3Ti(OC_4H_9\text{-t})$	Ti:	Org.Verb.1-85
–	$(CD_3)_3Ti(OC_4H_9\text{-t})$	Ti:	Org.Verb.1-85
$C_7H_{18}O_2Sn$	$(C_2H_5)_2(CH_3OCH_2)SnOCH_3$	Sn:	Org.Verb.5-204
$C_7H_{18}O_3Sn$	$C_4H_9Sn(OCH_3)_3$	Sn:	Org.Verb.6-246
$C_7H_{18}O_3Ti$	$CH_3Ti(OC_2H_5)_3$	Ti:	Org.Verb.1-44, 48
$C_7H_{18}O_4Ti$	$CH_3OTi(OC_2H_5)_3$	Ti:	Org.Verb.1-48
$C_7H_{18}O_5STi$	$CH_3SO_2Ti(OC_2H_5)_3$	Ti:	Org.Verb.1-48
$C_7H_{18}Sn$	$(CH_3)_2Sn(C_2H_5)(C_3H_7)$	Sn:	Org.Verb.3-76
–	$(CH_3)_3SnC(CH_3)_3$	Sn:	Org.Verb.2-49
–	$(CH_3)_3SnCH(CH_3)CH_2CH_3$	Sn:	Org.Verb.2-49
–	$(CH_3)_3SnCH_2CH(CH_3)_2$	Sn:	Org.Verb.2-48
–	$(CH_3)_3SnC_4H_9$	Sn:	Org.Verb.2-47/8
–	$(C_2H_5)_3SnCH_3$	Sn:	Org.Verb.2-157
$C_7H_{19}ISiSn$	$(CH_3)_3SnCHISi(CH_3)_3$	Sn:	Org.Verb.2-14, 16
$C_7H_{19}ISn_2$	$[(CH_3)_3Sn]_2CHI$	Sn:	Org.Verb.5-81
$C_7H_{19}N_9Ni_2$	$(NC)Ni(NH_3)_3C_4Ni(NH_3)_3(CN) \cdot HCN$	Ni:	Org.Verb.2-269
$C_7H_{19}PSn$	$(CH_3)_3SnCHP(CH_3)_3$	Sn:	Org.Verb.2-16, 18
$C_7H_{20}INSn$	$[(CH_3)_3SnCH_2N(CH_3)_3]I$	Sn:	Org.Verb.2-16
$C_7H_{20}N_2O_2SSi$	$(C_2H_5)_2NSO_2NHSi(CH_3)_3$	S:	S-N-Verb.1-172
$C_7H_{20}O_2Sn$	$(CH_3)_3SnCH_2(OCH_3)_2CH_3$	Sn:	Org.Verb.2-15
$C_7H_{20}SiSn$	$(CH_3)_3SnCH_2Si(CH_3)_3$	Sn:	Org.Verb.2-13
$C_7H_{20}Sn_2$	$[(CH_3)_3Sn]_2CH_2$	Sn:	Org.Verb.5-80
$C_7H_{21}INiP_2$	$CH_3Ni(P(CH_3)_3)_2I$	Ni:	Org.Verb.1-12, 19
$C_7H_{21}IrN_6O^{2+}$	$[Ir(NH_3)_5(NHCOC_6H_5)]^{2+}$	Ir:	SVol.2-65
$C_7H_{21}NO_3SSi_2$	$(CH_3)_3SiN(CH_3)SO_2OSi(CH_3)_3$	S:	S-N-Verb.1-98
–	$[(CH_3)_3Si]_2NSO_2OCH_3$	S:	S-N-Verb.1-98
$C_7H_{21}NTi$	$(CH_3)_4Ti \cdot N(CH_3)_3$	Ti:	Org.Verb.1-113, 117
$C_7H_{21}N_3SSi^+$	$(CH_3)_3SiNS(N(CH_3)_2)_2{}^+$	S:	S-N-Verb.1-210
$C_7H_{21}N_3Ti$	$CH_3Ti(N(CH_3)_2)_3$	Ti:	Org.Verb.1-52, 57
–	$CD_3Ti(N(CH_3)_2)_3$	Ti:	Org.Verb.1-52, 57
$C_7H_{21}N_4SSi_2{}^+$	$[(CH_3)_3SiN]_2SN_2CH_3{}^+$	S:	S-N-Verb.1-210
$C_7H_{21}PTi$	$(CH_3)_4Ti \cdot P(CH_3)_3$	Ti:	Org.Verb.1-113
$C_7H_{22}N_3SSi_3{}^+$	$S[NSi(CH_3)_3]_2NSiHCH_3{}^+$	S:	S-N-Verb.1-207
$C_7H_{24}N_6NiS_3$	$Ni(C_2H_8N_2)_3CS_3$	C:	MVol.D4-225
$C_7H_{24}N_6S_3Zn$	$Zn(C_2H_8N_2)_3CS_3$	C:	MVol.D4-224
$C_{7.5}Cl_2H_{7.5}N_{1.5}OSn$	$SnOCl_2 \cdot 1.5\ C_5H_5N$	Sn:	MVol.C5-197
$C_{7.5}Cl_4H_{7.5}N_{1.5}U$	$UCl_4 \cdot 1.5\ C_5H_5N = 2\ UCl_4 \cdot 3\ C_5H_5N$	U:	SVol.E1-28, 33
$C_{7.5}Cl_4H_{17.5}N_{2.5}O_{2.5}Te$	$TeCl_4 \cdot 2.5\ HCON(CH_3)_2$	Te:	SVol.B3-174
$C_{7.5}Cl_4H_{17.5}N_{2.5}O_{2.5}U$	$UCl_4 \cdot 2.5\ HCON(CH_3)_2$	U:	SVol.E1-99/100
$C_{7.5}FeH_8N_5$	$Fe(N_2C_3H_3)_2 \cdot 0.5\ N_2C_3H_4$	Fe:	Org.Verb.B3-193
$C_{7.5}H_3K_4MnN_6O_{0.5}S_2Se_4$	$K_4Mn(NCS)_2(NCSe)_4 \cdot 0.5\ (CH_3)_2CO$	Mn:	MVol.D1-56
$C_{7.6}H_{26.6}I_4N_{3.8}Sn$	$SnI_4 \cdot 3.8\ (CH_3)_2NH$	Sn:	MVol.C5-170